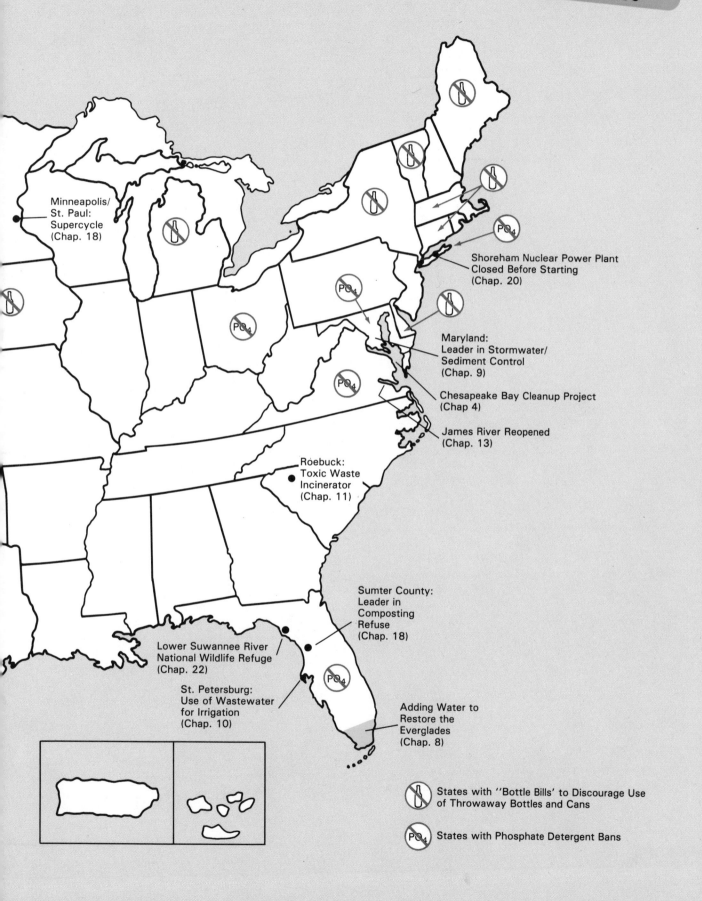

Minneapolis/
St. Paul:
Supercycle
(Chap. 18)

Shoreham Nuclear Power Plant
Closed Before Starting
(Chap. 20)

Maryland:
Leader in Stormwater/
Sediment Control
(Chap. 9)

Chesapeake Bay Cleanup Project
(Chap 4)

James River Reopened
(Chap. 13)

Roebuck:
Toxic Waste
Incinerator
(Chap. 11)

Sumter County:
Leader in
Composting
Refuse
(Chap. 18)

Lower Suwannee River
National Wildlife Refuge
(Chap. 22)

St. Petersburg:
Use of Wastewater
for Irrigation
(Chap. 10)

Adding Water to
Restore the
Everglades
(Chap. 8)

States with ''Bottle Bills'' to Discourage Use
of Throwaway Bottles and Cans

States with Phosphate Detergent Bans

THIRD EDITION

ENVIRONMENTAL SCIENCE

THIRD EDITION

ENVIRONMENTAL SCIENCE
The Way the World Works

Bernard J. Nebel

Department of Biology, Catonsville Community College

PRENTICE HALL, *Englewood Cliffs, NJ 07632*

Library of Congress Cataloging-in-Publication Data

Nebel, Bernard J.
  Environmental science : the way the world works / Bernard J.
Nebel.—3rd ed.
    p.   cm.
  Bibliography: p.
  Includes index.
  ISBN 0-13-282203-2
  1. Ecology. 2. Human ecology. 3. Pollution—Environmental
aspects. I. Title.
QH541.N39   1990
574.5—dc20   89-8482
                                                  CIP

Editorial/production supervision: *John Morgan and Nancy Bauer*
Interior design and page layout: *Maureen Eide*
Cover photo: *Edmund Nagele/FPG*
Manufacturing buyer: *Paula Massenaro*
Illustrations by Vantage Art
Photo editor: *Lori Morris-Nantz*
Photo research: *Tobi Zausner*

Photo credits: **1:** *Tom Hollyman/Photo Researchers;* **11:** *Len Rue, Jr./Photo Researchers;* **13:** *Pat and Tom Leeson/Photo Researchers;* **43:** *Frank J. Miller/ Photo Researchers;* **68:** *Leonard Lee Rue III/Photo Researchers;* **89:** *Peter B. Kaplan/Photo Researchers;* **111:** *United Nations Photo;* **113:** *Earth Satellite Corporation, Science Photo Library/Photo Researchers;* **153:** *Bjorn Bolstad, Peter Arnold, Inc.;* **155:** *Townsend P. Dickinson/Photo Researchers;* **190:** *Townsend P. Dickinson/Photo Researchers;* **221:** *Shirley Richards/Photo Researchers;* **225:** *Jerome Wexler/Photo Researchers;* **248:** *Lowell Georgia/Photo Researchers;* **271:** *B. Krueger/Photo Researchers;* **294:** *Eunice Harris/Photo Researchers;* **319:** *Tom McHugh/Photo Researchers;* **346:** *N. Tully/ Sygma;* **355:** *Jack Fields/Photo Researchers;* **370:** *Ken Brate/Photo Researchers;* **393:** *Stephen J. Krasemann/Photo Researchers;* **395:** *George Holton/Photo Researchers;* **425:** *Roger A. Clark, Jr./Photo Researchers;* **441:** *Sam C. Pierson, Jr./Rapho, Photo Researchers;* **464:** *Bertrand/ Explorer/Photo Researchers;* **487:** *Courtesy Pacific Gas and Electric;* and **517:** *Alese and Mort Pechter/The Stock Market.*

© 1990, 1987, 1981 by Prentice-Hall, Inc.
A Division of Simon & Schuster
Englewood Cliffs, New Jersey 07632

Printed in the United States of America

10 9 8 7 6 5 4 3 2

ISBN: 0-13-282203-2

Prentice-Hall International (UK) Limited, *London*
Prentice-Hall of Australia Pty. Limited, *Sydney*
Prentice-Hall Canada Inc., *Toronto*
Prentice-Hall Hispanoamericana, S.A., *Mexico*
Prentice-Hall of India Private Limited, *New Delhi*
Prentice-Hall of Japan, Inc., *Tokyo*
Simon & Schuster Asia Pte. Ltd., *Singapore*
Editora Prentice-Hall do Brasil, Ltda., *Rio de Janeiro*

To the environmental movement
of the 1990s

Brief Contents

Contents   ix

Case Studies   xv

Preface   xvii

Introduction: Using Science to Understand and Solve Environmental Problems   1

Part One   WHAT ECOSYSTEMS ARE AND HOW THEY WORK   11
1   Ecosystems: What They Are   13
2   Ecosystems: How They Work   43
3   Ecosystems: What Keeps Them the Same? What Makes Them Change?   68
4   Ecosystems: Adaptation and Change or Extinction   89

Part Two   POPULATION   111
5   The Population Problem: Its Dimensions and Causes   113
6   Addressing the Population Problem   135

Part Three   SOIL, WATER, AND AGRICULTURE   153
7   Soil and the Soil Ecosystem   155
8   Water, the Water Cycle, and Water Management   190

Part Four   POLLUTION   221
9   Sediments, Nutrients, and Eutrophication   225
10   Water Pollution Due to Sewage   248
11   Toxic Chemicals and Groundwater Pollution   271
12   Air Pollution and Its Control   294
13   Acid Precipitation, the Greenhouse Effect, and Depletion of the Ozone Shield   319
14   Risks and Economics of Pollution   346

Part Five   *PESTS AND PEST CONTROL*   353
15   The Pesticide Treadmill   355
16   Natural Pest Control Methods and Integrated Pest Management   370

Part Six   RESOURCES: BIOTA, REFUSE, ENERGY, AND LAND   393

17   Biota: Biological Resources   395

18   Converting Refuse to Resources   425

19   Energy Resources and the Energy Problem   441

20   Nuclear Power, Coal, and Synthetic Fuels   464

21   Solar and Other "Renewable" Energy Sources   487

22   Lifestyle, Land Use, and Environmental Impact   517

Epilogue   543

Bibliography   546

Appendices   561

Glossary   571

Index   591

Contents

Case Studies   *xv*

Preface   *xvii*

Introduction: Using Science to
Understand and Solve
Environmental Problems   *1*

Introduction   3
Evaluating Information   5
   What Is Science?   5
   The Role of Instruments in Science   9
   Science and Technology   9
   Unresolved Questions and Controversies   9
Science and Value Judgments   10
Environmental Science and Its Application   10

Part One

WHAT ECOSYSTEMS ARE AND
HOW THEY WORK   *11*

1 Ecosystems: What They Are   *13*

Definition and Examples of Ecosystems   15
   Box 1   Temperate Forest Biome   18
   Box 2   Grassland Biome   19
   Box 3   Desert Biome   20
   Box 4   Coniferous Forest Biome   21
   Box 5   Tundra Biome   22
   Box 6   Savanna Biome   23
   Box 7   Tropical Rain Forest Biome   24
Structures of Ecosystems   25
   Biotic Structure   25
   Abiotic Factors   35
Why Do Different Regions Support Different
Ecosystems?   37
   Climate   37
   Other Abiotic Factors and
    Microclimate   39
   Biotic Factors   39
   Physical Barriers   41
   Biotic and Abiotic Interactions   41
Implications for Humans   41

2 Ecosystems: How They Work   *43*

Elements, Life, Organization, and
Energy   45
   Organization of Elements in Living and
    Nonliving Systems   45
   Considerations of Energy   48
   Matter and Energy Changes in Organisms   54
Principles of Ecosystem Function   57
   Nutrient Cycling   57
   Energy Flow   61
   Flow of Energy and Decreasing Biomass
    at Higher Trophic Levels   62
Implications for Humans   64
Case Study: Biosphere II   67

3 Ecosystems: What Keeps
Them the Same? What Makes
Them Change?   *68*

The Key Is Balance   70
   Ecosystem Balance Is Population
    Balance   70
   Mechanisms of Population Balance   72
Case Study: Demise of the American
Chestnut   75
   Ecosystem Change: Succession   82
Implications for Humans   87

4 Ecosystems: Adaptation
and Change or Extinction   *89*

How Species Change and Adapt—or Don't   91
   Some Basic Genetics   91
   Gene Pools and Their Change   94
   Development of Ecosystems   100
   Evolutionary Succession   102
   Limits of Change: Making It or Not
    Making It   104
Implications for Humans   106
Case Study: Seeds for Change   109

Part Two

POPULATION   *111*

5 The Population Problem: Its Dimensions and Causes   *113*

Dimensions of the Population Problem 115
  The Exploding Human Population 115
  Rich Nations and Poor Nations 115
  Population and Poverty 118
  Population, Poverty, and the Environment 123

Case Study: The Impact of Population on Tropical Forests 124

The Population Explosion: Its Cause and Potential Solution 124
  Birth Rates, Death Rates, and the Population Equation 125
  Cause of the Population Explosion 127
  The Solution: Lower Fertility Rates 129

6 Addressing the Population Problem   *135*

Improving Lives of People 137
  Increasing Food Production: Successes and Limits 137
  The Hungry: A Problem of Food-Buying Capacity 139
  Food Aid and the Utterly Dismal Theorem 140
  Economic Development 142

Reducing Fertility 144
  Leaders of Less-Developed Nations Support Reducing Fertility 145
  People of Less-Developed Countries Want Fewer Children 145
  Effectiveness of Family Planning, Health Care, and Education 146

Case Study: A Lesson Learned from Bangladesh 148

  Additional Economic Incentives 149
  Promoting Family Planning and the Abortion Controversy 150
  Contraceptive Technology 151
  Costs of Family Planning 151

Actions You Can Take 151

Part Three

SOIL, WATER, AND AGRICULTURE   153

7 Soil and the Soil Ecosystem   155

Plants and Soil 157
  Critical Factors of the Soil Environment 157
  Growing Plants Without Soil 162
  The Soil Ecosystem 162
  Mutual Interdependence of Plants and Soil 169

Losing Ground 170
  Bare Soil, Erosion, and Desertification 170
  Causes of Losing Ground 173
  Dimensions of the Problem 178
  Preventing Erosion and Desertification 182

Implementing Solutions 186

Case Study: Haiti: In the Land Where Hope Never Grows 188

8 Water, the Water Cycle, and Water Management   *190*

Water 192
  Physical States of Water 192
  Evaporation, Condensation and Purification 193

The Water Cycle 194
  Water into the Atmosphere 194
  Precipitation 195
  Water over and through the Ground 197
  Summary of the Water Cycle 198

Human Dependence and Impacts on the Water Cycle 200
  Sources and Uses of Fresh Water 200
  Consequences of Overdraft of Water Resources 202

Case Study: Upsetting the Everglades 206
  Obtaining More versus Using Less 209
  Potential for Conservation and Reuse of Water 209
  Effects Caused by Changing Land Use 210
  Stormwater Management 215

Implementing Solutions 219
  Cheap Water and Waste 219
  What You Can Do 219

Part Four

POLLUTION   221

9 Sediments, Nutrients, and Eutrophication   225

Eutrophication 228
    Two Kinds of Aquatic Plants 228
    *Upsetting the Balance by Nutrient
      Enrichment* 229
    *Natural versus Cultural
      Eutrophication* 232
    Combatting Eutrophication 232
Case Study: Eutrophication of Lake Erie 232
Sources of Sediments and Nutrients 235
    Sources of Sediments 235
    *Impacts of Sediments on Streams and
      Rivers* 236
    Sources of Nutrients 240
    *Loss of Wetlands and Bulkheading of
      Shorelines* 241
Controlling Nutrients and Sediments, and
  What You Can Do 242
    *Best Management Practices on Farms,
      Lawns, and Gardens* 242
    *Sediment Control on Construction and
      Mining Sites* 244
    Preservation of Wetlands 246
    *Banning the Use of Phosphate
      Detergents* 246
    Advanced Sewage Treatment 247

10 Water Pollution Due to Sewage   248

Hazards of Untreated Sewage 250
    Disease Hazard 250
    Depletion of Dissolved Oxygen 250
    Eutrophication 250
Sewage Handling and Treatment 251
    Background 251
    Conventional Sewage Treatment 252
    Alternative Systems 262
Case Study: The Overland Flow Wastewater
  Treatment System 265
Taking Stock of Where We Are and What You
  Can Do 267
    Progress and Lack of Progress 267
    Impediments to Progress 268
    What You Can Do 269
    *Monitoring for Sewage
      Contamination* 269

11 Toxic Chemicals and Groundwater Pollution

Sources of Groundwater Pollution 274
Toxic Chemicals: Their Threat 274
    What Are the Toxic Chemicals? 274
    The Problem of Bioaccumulation 275
    Synergistic Effects 277
Environmental Contamination with Toxic
  Chemicals 277
    *Major Sources of Toxic Chemical
      Wastes* 278
    *Background of the Toxic Waste
      Problem* 278
    *Methods of Land
      Disposal* 278
    Problems in Managing Land Disposal 280
    Scope of the Problem 282
Cleanup and Management of Toxic
  Wastes 284
    *Assuring Safe Drinking Water
      Supplies* 284
    *Cleaning Up Existing Toxic Waste
      Sites* 284
    Groundwater Remediation 285
    *Management of Wastes Currently
      Produced* 285
    *Future Management of Hazardous
      Wastes* 287
Case Study: Canada's White Whales Are
  Dying 288
    *Reducing Occupational and Accidental
      Exposures* 289
    What You Can Do 292

12 Air Pollution and Its Control   294

Background of the Air Pollution Problem 296
Major Air Pollutants and Their Effects 300
    Major Pollutants 300
    *Adverse Effects of Air Pollution on
      Humans, Plants, and Materials* 300
Case Study: How Air Pollution Is Affecting
  Trees 306
Sources of Pollutants and Control
  Strategies 307
    *Pollutants: Products of
      Combustion* 307
    Setting Standards 310
    *Major Sources and Control
      Strategies* 311
    Taking Stock—What You Can Do 314
Indoor Air Pollution 317

13 Acid Precipitation, the Greenhouse Effect, and Depletion of the Ozone Shield   319

Acid Precipitation   321
  Understanding Acidity   321
  Extent and Potency of Acid Precipitation   324
  The Source of Acid Precipitation   324
  How Acid Precipitation Affects Ecosystems   325
  Depletion of Buffering Capacity and Anticipation of Future Effects   329
  Effects on Humans and Their Artifacts   332
  Strategies for Coping with Acid Precipitation   332

The Greenhouse Effect   336
  The Heat-Trapping Effect of Carbon Dioxide   336
  Sources of Carbon Dioxide Additions   336
  Other Greenhouse Gases   337
  Degree of Warming and Its Probable Effects   338
  Is the Greenhouse Warming Effect Here?   339
  Strategies for Coping with the Greenhouse Effect—What You Can Do   339

Case Study: The Greenhouse Effect   340

Depletion of the Ozone Shield   339
  The Nature and Importance of the Ozone Shield   339
  The Formation and Breakdown of the Ozone Shield   341
  The Source of Chlorine Atoms Entering the Stratosphere   342
  The Ozone "Hole"   343
  Coming to Grips with Ozone Depletion   344
  What You Can Do   345

14 Risks and Economics of Pollution   346

The Cost-Benefit Analysis   347

Problems in Performing Cost-Benefit Analysis   349
  Estimating Costs   349
  Estimating Benefits and Performing Risk Analysis   349
  Problems in Comparing Costs and Benefits   350

The Need for Regulation and Enforcement   351
Conclusion   351

Part Five
PESTS AND PEST CONTROL   353

15 The Pesticide Treadmill   355

Promises and Problems of Chemical Pesticides   357
  Development of Chemical Pesticides and their Apparent Virtues   357
  Problems Stemming from Chemical Pesticide Use   357
  The Pesticide Treadmill   363

Attempting to Protect Human and Environmental Health   365
  The Federal Insecticide, Fungicide, and Rodenticide Act (FIFRA)   365
  Shortcomings of FIFRA   365
  Nonpersistent Pesticides: Are They the Answer?   367
  What You Can Do   369

16 Natural Pest Control Methods and Integrated Pest Management   370

The Insect Life Cycle and Its Vulnerable Points   372

Methods of Natural or Biological Control   372
  Control by Natural Enemies   373

Case Study: Experience in Finding New Natural Enemies   377
  Genetic Control   378
  The Sterile Male Technique   378
  Cultural Controls   382
  Natural Chemical Control   384
  Economic Control versus Total Eradication   386
  Conclusion   386

Integrated Pest Management   386
  Socioeconomic Factors Supporting Overuse of Pesticides   386
  Addressing Socioeconomic Issues   388
  Integration of Natural Controls   389
  Prudent Use of Pesticides   390
  What You Can Do to Promote Ecological Pest Management   390

Part Six

RESOURCES: BIOTA, REFUSE, ENERGY, AND LAND 393

17 Biota: Biological Resources 395

Natural Biota: To Preserve or Destroy? 397
Values of Natural Biota 397
Assaults Against Natural Biota 404
Case Study: Last Stand for Africa's Elephants 408
Conservation of Natural Biota and Ecosystems 412
General Principles 412
Where We Stand with Regard to Protecting Biota 416
Is Saving Species Enough—The Need to Focus on Ecosystems 421
Saving Tropical Forests 422
What You Can Do 424

18 Converting Refuse to Resources 425

The Solid Waste Crisis 427
Background of Solid Waste Disposal 427
Problems of Landfills 429
Improving Landfills: Trying to Fix a Wrong Answer 430
Escalating Costs of Landfilling 430
Solutions 430
The Recycling Solution 430
Case Study: The Pot at the End of the Trash Rainbow 435
Composting 436
Waste-to-Energy Conversion 438
Reducing Waste Volume 438
Integrated Waste Management 440
Changing Public Attitudes and Lifestyle—What You Can Do 440

19 Energy Resources and the Energy Problem 441

Energy Sources and Uses in the United States 444
Background 444
Electrical Power Production 446
The Current Situation 449
The Energy Dilemma (or Crisis): Declining Reserves of Crude Oil 451

Formation of Fossil Fuels 451
Exploration, Reserves, and Production 452
Declining U.S. Reserves and Increasing Importation of Crude Oil 453
The Crisis 453
Response to the Oil Crisis of the 1970s 454
Victims of Success: Setting the Stage for Another Crisis 457
Preparing for Future Oil Shortages 458
Promoting a Sound Energy Policy: What You Can Do 462
Case Study: The Valdez Oil Spill 459

20 Nuclear Power, Coal, and Synthetic Fuels 464

Nuclear Power: A Dream or Delusion? 466
Nuclear Power 468
Radioactive Materials and their Hazard 474
Economic Problems with Nuclear Power 478
More Advanced Reactors 479
Coal and Synthetic Fuels 483
Coal 483
Synthetic Fuels 483

21 Solar and Other "Renewable" Energy Sources 487

Solar Energy 489
Use of Direct Solar Energy 489
Case Study: The Davis Experience 496
Use of Indirect Solar Energy 503
Geothermal, Tidal, and Wave Power 510
Geothermal Energy 510
Tidal Power 512
Wave Power 512
Conclusions 512
Need for Conservation 514
Energy Policy 514
What You Can Do 516

22 Lifestyle, Land Use, and Environmental Impact 517

Urban Sprawl: Its Origins and Consequences 519
The Origin of Urban Sprawl 519
Environmental Consequences of Urban Sprawl 522

*Social Consequences of Urban
   Sprawl*   526
Conclusion   532
Making Cities Sustainable   532
Cities Can Be Beautiful   532
Cities Can Be Sustainable   533
*Slowing Urban Sprawl and Aiding
   Urban Redevelopment*   536
What You Can Do   540
Case Study: An Integral Urban House   541

Epilogue   *543*

Bibliography   *546*

Appendices   *561*

A   Environmental Organizations   561
B   Units of Measure   563
C   Some Basic Chemical Concepts   564

Glossary   *571*

Index   *591*

Case Studies

2   Biosphere II   67

3   Demise of the American Chestnut   75

4   Seeds for Change   109

5   The Impact of Population on Tropical Forests   124
     Werner Fornos

6   A Lesson Learned from Bangladesh   148
     Charlene B. Dale

7   Haiti: In the Land Where Hope Never Grows   188
     Cristina Garcia—Time

8   Upsetting the Everglades   206
     Paul C. Parks

9   Eutrophication of Lake Erie   232
     Edward J. Kormondy

10   The Overland Flow Wastewater Treatment System   265

11   Canada's White Whales Are Dying   288
     Peter Benesh

12   How Air Pollution Is Affecting Trees   306
     James J. MacKenzie

13   The Greenhouse Effect   340
     David Rind

16   Experience in Finding New Natural Enemies   377
     David Pimentel

17   Last Stand for Africa's Elephants   408
     Eugene Linden, Carter Coleman, Roger Browning—Time

18   Value Added: The Pot at the End of the Trash Rainbow   435
     Neil N. Seldman

19   The Valdez Oil Spill   459

21   The Davis Experience   496
     Jonathan Hammond

22   The Integral Urban House   451
     Helga Olkowski

Preface

As people became aware of the seriousness of environmental issues for the first time in the 1960s a common question was: How long do we have? How long will it be before we reap the tragic consequences of our environmental disregard? The answer was 30 to 35 years. Now, as we approach 30 years later this answer haunts me. As we are beginning to experience a hotter climate, holes in the protective ozone layer over the poles, toxic chemicals in groundwater throughout the nation, famines in Africa, food contaminated with pesticide residues, and extinction of untold numbers of species as forests recede before growing populations, I think the prediction was not far off. We simply do not have another 30 or even 10 years to debate and study the issues. It is necessary to start taking the bold steps necessary to save the biosphere. Over the next 30 years, we must either create a sustainable society, or witness the decline of civilization on Earth.

That is the bad news. The good news is that the issues have been studied and technologies have been developed, at least to point of adequate understanding and workability, so that we can achieve a sustainable society. We simply need to get on with the task. *Environmental Science: The Way the World Works*, 3rd ed. is a text about how to do it.

Environmental Science: The Way the World Works, 3rd ed. has been totally rewritten and updated in all areas. The easy-to-read style and the step-by-step organization systematically leading from simple facts to comprehension of complex issues, which has made this text popular with typical freshman-level nonmajors, has been retained and strengthened. Careful editing has reduced the overall length of the text while allowing the inclusion of updated material. A large number of illustrations have been redone, new ones added, and the use of color has been extended to further enhance the book's appeal.

Case Studies have been added to provide additional interest and diversity. Most of these have been written by guests describing their personal experience. This provides more than a detailed example of an issue under discussion. It also shows what individuals can do and how they can contribute. Case studies are in tinted boxes so that they can be found easily and used to draw readers into the book.

The theme, which is emphasized throughout the text, is *sustainability*. Environmental issues are put in the context of making progress toward this goal. In each Part, students will learn how certain current trends, such as growing population, soil erosion, depletion of resources, and deforestation, are not sustainable, and how they must be modified to enable a sustainable civilization. In addition to discussing progress that is being made, each chapter now concludes with a list of specific things the student, or anyone else, can do to promote the adoption of sustainable practices described.

SYNOPSIS

Part One (Chapters 1 to 4) develops the concepts of what natural ecosystems are and how they function. It has been enhanced by highlighting specific ecological principles that make ecosystems sustainable. At the end of each chapter, I point out how all our environmental problems are basically the consequence of ignoring these basic principles and that sustainability will be gained by recognizing and adhering to them. The focus in Chapter 4 has been shifted from evolution *per se* to how genetic diversity underlies the ability to change and adapt to new conditions. It highlights the principle that maintaining genetic diversity is necessary for survival.

Part Two (Chapters 5 and 6) focuses on the fact that population balance is a key factor in sustainability and how the exploding human population is not just a concern for the Third World. If populations anywhere destroy their ecosystems, as they are currently doing, the entire biosphere of Spaceship Earth is in jeopardy. Gaining an understanding of the origins of the population explosion, students will see the need to intensify efforts to reduce birth rates, and the means for doing it are emphasized.

Part Three (Chapters 7 and 8) focuses on soil and water which are the most fundamental resources for sustainability because they underlie food production. Students will learn how natural systems work to continually renew these resources and that this renewability depends on adhering to certain principles.

They will understand how soil and water are being degraded and depleted by agricultural practices that ignore or disregard these principles. Methods for sustainable agriculture are thus made clear.

Part Four (Chapters 9 to 14) focuses the major pollution issues (eutrophication, sewage, hazardous wastes, air pollution, acid rain, the greenhouse effect, and ozone depletion). Using a separate chapter for each major kind of pollution aids students in gaining a thorough understanding of each issue, its consequences, and what needs to be done to solve it. However, the length of these chapters has been reduced, without sacrificing content. Sediments and eutrophication (formerly Chapters 9 and 11) have been combined and integrated into a single chapter (9).

Part Five (Chapters 15 and 16) shows the student the nonsustainability of traditional chemical pest control methods and describes the various methods of natural or biological control, which are sustainable.

Part Six (Chapters 17 to 22) is devoted to critical resources beyond soil and water. In Chapter 17, students will learn why conserving natural ecosystems and species is not just a matter of aesthetic principle; it is essential for the sustainability of our human society. Chapter 18 describes the need and the progress that is being made in recycling refuse as an integral part of sustainability. In Chapters 19 to 21, students will gain a clear understanding that our dependence on fossil fuels and nuclear power is not sustainable because of depletion of reserves and/or pollutants or waste products that are produced. They will also learn that there is a sustainable alternative in various solar technologies that are on the threshold of wide-scale use, if they are suitably supported.

The last chapter, Lifestyle, Land Use, and Environmental Impact, places new emphasis on how our urban-sprawl, energy-intensive, commuting lifestyle is at the root of much of our environmental degradation, resources depletion, and pollution. Students will learn that, ultimately, sustainability will demand certain changes in lifestyle, but that such changes may markedly enhance the overall quality of life.

PEDAGOGICAL AIDS

Organization. As in the second edition, each chapter of *Environmental Science: The Way the World Works* is rigorously organized and outlined according to a hierarchy of learning objectives, which facilitates students' gaining a clear, comprehensive understanding of each topic.

Concept Frameworks. Each chapter is preceded by a *Concept Framework* which consists of the outline of the chapter and parallel study questions page-keyed to the text. The concept frameworks may be used as: (1) a convenient preview of the chapter; (2) a learning guide; (3) a reference to quickly locate particular sections; (4) as a summary and review.

Vocabulary Words. New terms appear in boldface types where they are first introduced and defined, and later in italics to emphasize the term's use in context. Then, all such words are entered in the Glossary.

Illustrations and Photographs. Nearly 500 photographs and line drawings, many in four-color, enhance the attractiveness of the text. Moreover, all illustrations have been prepared and photographs selected especially to help convey particular concepts and increase understanding. They are all keyed to specific points in the text.

Appendixes. A list of environmental organizations; the metric system and equivalent English units, energy units and equivalents; and basic chemical concepts are included.

SUPPLEMENTARY MATERIALS

The following supplementary items are available to accompany *Environmental Science: The Way the World Works*, Third Edition:

Instructor's Edition Written by John P. Harley, the instructor's edition is a special printing of a textbook created for instructors, and is not available for students. It consists of material which precedes the student text. This material is designed to assist instructors and graduate assistants in the teaching of the course. It contains chapter outlines, summaries, lecture suggestions, discussion questions, and it highlights sections of the text which might require special attention.

Test Item File The complimentary test item file includes over 1,300 test items that are designed specifically to accompany this text. It is available in both printed form and on an IBM or Apple disk.

Study Guide Written by Clark E. Adams and Bernard J. Nebel, this study guide includes chapter overviews and outlines of the material to be covered in a chapter-by-chapter format. The guide also includes completion and matching exercises, and self-tests, so that students can monitor their own progress as they proceed through the course.

Laboratory Manual Written by John P. Harley and Bernard J. Nebel, this manual contains exercises and activities which will expand the students' appreciation of the environment. It is appropriate to be used in a laboratory course on environmental science or, if there are no separate lab meetings, it can be used as a supplement to the text to give the course a more practical flavor.

Transparency Pack This transparency pack includes 64 full-color transparencies which reproduce the illustrations as they appear in the text. An additional 30 transparency masters supplement the acetates. Both the acetate transparencies and the additional masters are free to adopters. Please ask your Prentice Hall representative for a list of the illustrations in the transparency pack.

Acknowledgements

My very special thanks and appreciation go to the many people who contributed special talents, hard work, and support to the completion of this third edition. Especially I wish to commend and thank:

Janet Hanna who did the "rough" (really very finished) drawings for all the new and revised illustrations that appear in this edition, and much support in other ways.

Mary Kintner, Beth Beck, and Angie Biederman for untold hours of typing and other office work, and their real caring and support for the project.

Jean Byars for editing the first draft, Holly LaMon and Gary Kaiser for proofreading/editing the final manuscript.

All those who contributed Case Studies: Dr. Jim MacKenzie, World Resources Institute; Ms. Charlene Dale, International Child Health Foundation; Dr. David Rind, NASA, Goddard Institute for Space Studies; Dr. Jonathan Hammond, University of Illinois; Ms. Helga Olkowski, Integral House Project; Dr. Edward J. Kormondy, University of Hawaii-Hilo; Dr. David Pimentel, Cornell University; Dr. Werner Fornos, The Population Institute; Dr. Paul C. Parks, Florida Wildlife Federation; and Mr. Neil Seldman, Institute of Self-Reliance.

All those who reviewed the manuscript: Peter S. Dixon, University of California, Irvine; David Pimentel, Cornell University; Fred Racle, Michigan State University; Ronald Ward, Grand Valley State Colleges, Allendale, Michigan; and Richard Andred, Montgomery County Community College, Blue Bell, Pennsylvania.

The people at Prentice Hall: my editor, Bob Sickles, whose enthusiasm for this project kept it almost on schedule despite an often tardy author. My production editor, John Morgan, for his careful attention to all the details of getting a manuscript into a finished book. Doug Gower, for copy editing the manuscript.

My colleagues at Catonsville Community College, particularly Carol Daihl, Steven Simon, Gary Kaiser, David Jeffrey, Barbara Carr, and David Hargrove, for their interest and for providing support in many ways over the years.

Additionally, I wish to acknowledge again all those who contributed to the earlier editions, and to whom I remain extremely grateful for their help: Nancy Minkoff, Edward Kormondy, Jean Nebel, Emily (McNamara) Bookholtz, David Hunley, Mary Beyer, Terri Leonnig, Lynn Carr, Kathy (Yaw) Zegwitz, Joan Truby, Byron Daudelin, Kai A. Nebel, Jack Anderson, Joseph Newcomer, Bruce Welch, Tamra Nebel, Christopher Nebel, Olive Blumenfeld, and Udo Essien.

Countless people who have been most helpful in providing information, reports, and photographs. I apologize for not listing them individually.

All my students for continually providing the incentive and the testing ground for this work.

Bernard J. Nebel

Introduction: Using Science to Understand and Solve Environmental Problems

CONCEPT FRAMEWORK

■ Outline

■ Study Questions

I. INTRODUCTION   3

II. EVALUATING INFORMATION   5

  A. What Is Science?   5

    1. The Scientific Method   5

    2. Observations and Facts   5

    3. Hypotheses and Their Testing   6

    4. Controlled Experiments   7

    5. Formulation of Theories   8

    6. Principles and Natural Laws   8

  B. The Role of Instruments in Science   9

  C. Science and Technology   9

  D. Unresolved Questions and Controversies   9

III. SCIENCE AND VALUE JUDGMENTS   10

IV. ENVIRONMENTAL SCIENCE AND ITS APPLICATION   10

1. Define: sustainable, environmental impacts, trade offs. Are current human interactions with the environment sustainable? Give evidence and explain.

2. Is it possible to change? Describe the environmental movement of the 1960s and 70s and progress made. What factors for change are now in place? Define: borrowed time, sustainable development.

3. Define: ecology, environmental science. Describe how following ecological principles will enable sustainable development. What ingredient is necessary to promote rapid change toward sustainable development?

4. What is the purpose, here, of studying the scientific method?

5. Define science. All scientific information is based on what process? State a key tenet of science.

6. The scientific method begins with what and ends with what?

7. What are "facts"? What is required of observations before they are accepted as facts?

8. What is done when a question cannot be answered by direct observation? What is the procedure? Define hypothesis.

9. What are the essential features of a controlled experiment? What are the fallacies of an experiment that is not controlled, has more than one variable, or is based on only one or two individuals? Do experimental results need to be verified? Does one always need to do an experiment in a laboratory?

10. What is a theory?

11. What are principles and natural laws? What is their value?

12. What roles do instruments play in science?

13. Distinguish between pure science and technology. How are they mutually reinforcing?

14. Where and why does legitimate controversy exist in science? How is such controversy gradually resolved? What continues to promote controversy after all reasonable doubt is removed?

15. What are value judgments? Give examples. Can science provide answers to value questions? Can it help? How?

16. Explain how the study of environmental science relates to achieving sustainable development.

INTRODUCTION

What kind of environment do you want to live in?

If you were asked to describe the kind of environment you would like to have, you would probably mention such things as clean water, clean air, productive soil and agriculture, freedom from hazardous wastes, and spacious natural areas with abundant wildlife. Most of all, you probably want our relationship to the environment to be **sustainable**. This is to say that we would like to be able to look toward the future with a feeling of confidence that these essential qualities are *not* being depleted or degraded but are being maintained and renewed so that they will be available in the future in at least the same abundance as they are now.

But what is your perception of the actual situation?

Through the news media you are constantly bombarded by warnings of impending if not present environmental calamities. Air and water are being polluted. Food is contaminated with pesticides. Acid rain is threatening lakes and forests. Soil is being degraded by erosion. The earth's protective ozone layer is being destroyed. The climate of the earth is warming due to carbon dioxide emitted from the burning of fuels—the greenhouse effect. Our water supplies may contain or be threatened with toxic wastes. Wildlife is dwindling: thousands of species are being pushed toward the brink of extinction and hundreds have already gone over the edge. Numerous resources are being depleted by overuse. Overpopulation threatens to lead the world into an era of unremitting famines as are already occurring in Ethiopia. In short, our current relationship with the environment is anything but sustainable. Our actions are leading to depletion and degradation which, if continued, could ultimately result in an environment that is no longer capable of supporting life.

Thus, we see that *what we have* is conspicuously different from *what we want*. There is an old Chinese proverb that is profound in its simplicity.

If you are not getting what you want, change what you are doing.

Is it possible to make such a change?

I am convinced that the answer is: Yes, we can! If I did not believe this, I would not be spending my time writing this book and teaching my own courses in environmental science. I am further convinced that each one of you who studies environmental science can and will make a difference. Changes come about from the way we perceive and understand problems and the values we put on different aspects of life, and these things do change. Our values will change as we gain more understanding of the problems, and acting on different values will produce change. Indeed, with regard to the environment, shifts in values and resulting changes are well underway.

Historically, humans pursued goals, such as obtaining a resource, manufacturing a product, or building a highway, with a focus on achievement of that goal alone. We simply did not concern ourselves with **environmental impact**, the side effects on the environment. Such a focus is not inherently bad; it has served to produce all the human-made good things which we enjoy. And, when the human population and production levels were small relative to the size of the earth, the environmental impacts were taken as acceptable **tradeoffs**. That is, natural areas were so abundant that it felt as if we were trading up (getting more value than we were giving) as we sacrificed them and certain amounts of environmental quality such as clean air and water to achieve the benefits of our goal.

But this process is obviously not sustainable in a finite world. As population and production levels have grown, environmental impacts became more severe and widespread, and the natural environment has become more limited. In the 1960s society began to recognize and feel threatened by the global nature of pollution problems. We felt that the sacrifices in environmental quality were no longer an acceptable tradeoff. Values shifted and action ensued.

Numerous laws aimed at reducing air and water pollution and alleviating other environmental concerns were passed. The Federal Environmental Protection Agency (EPA), which is charged by Congress to protect the nation's land, air, and water systems, was established in 1972, and parallel agencies were formed at state and local levels. Hundreds of environmental organizations were formed by citizens. Hundreds of scientists turned their talents toward investigating environmental problems. Private enterprise formed new businesses and developed new products in pollution control, waste disposal, and many other fields. This remarkable rise in public awareness and the actions which followed during the 1960s and 1970s have earned this era its place in history as the **environmental movement**.

Tremendous progress during the 1960s and 1970s notwithstanding, it is conspicuous that all the problems were not solved. The public, however, had temporarily exhausted their ability to focus on the environment. The early and mid-1980s proved to be a time of declining public interest in environmental issues. However, scientific research, technology, and the development of environmental organizations continued. We now have a much better understanding of the problems, what needs to be done, and technologies to solve them. A framework of laws, regulatory agencies, and citizen organizations staffed by thousands of knowledgeable, dedicated professionals is in place. Textbooks have been written and courses are being taught. The 1980s can be seen as a period of consolidation and growing professionalism.

Now, as we enter the 1990s, there is a new urgency. The world population has nearly doubled since 1965. What were in the 1960s and 1970s only vague predictions concerning acid rain, the destruction of the ozone shield, and climatic warming due to carbon dioxide emitted from burning fuels are now all-too-conspicuous realities; soil, water, and numerous other critical resources have been pushed closer to the limits. An "environmental movement of the 1990s" can, however, promote swift and profound changes because the groundwork has been done. Over the next 30 to 50 years (your lifetime) it can take us from our present course of *nonsustainable development* to a course of *sustainable development*.

Let me make these two courses very clear by analogy. A man may leap from the top of a high cliff with the conviction that he is going to fly. He may even experience some euphoria of success and declare as he passes the halfway point: "so far so good." But his euphoria, we recognize with certain gallows humor, is based on **borrowed time**. His "flight" is not sustainable.

Obviously, ignoring or defying gravity is not the way to sustainable flight. But, accepting and understanding gravity and additional principles of aerodynamics, and working with them, is. Thus we do have sustainable flight through the use of various aircraft which are consistent with these principles.

Much of the same may be said of our interactions with the environment. We are in many ways carrying on human development in ignorance or defiance of basic principles governing the perpetuation of living systems. We are in many ways living on *borrowed time*, and the many environmental problems confronting us are saying that this course is not sustainable. Just as we did not give up our attempts to fly when confronted by gravity, however, we need not abandon human development and return to primitive living. But, we do need to understand the principles involved and pursue our development in a manner that is consistent with these principles. Very simply, the concept of **sustainable development** is to meet the needs and aspirations of the present without compromising those of the future.

Fortunately, the necessary information is available. In the mid-1800s, scientists began to study and discover the principles of how plants and animals interact with each other and their environment. This area of biology became known as **ecology**. In the

FIGURE 0–1
Preserving the natural environment is the only pathway that is sustainable. Sustainable development means using soil, water, air, and biological and mineral resources in ways such that they are not degraded or depleted but that quality and amounts are preserved for future generations.

1960s it became widely recognized that ecological principles and theories did not apply only to wild plants and animals in natural environments. They apply also to humans, and on a global scale. This offshoot of ecology, the study of ecological principles and their application to the human situation, the keys for sustainable development, has become known as **environmental science**. A 1987 report from the National Academy of Science argues that "It is not the lack of ecological information that leads to poor environmental planning, but simply the lack of its proper application." In short we have the information necessary to make human development sustainable.

If we have the knowledge, why have we not already made the necessary changes? First, the knowledge and understanding is relatively recent. There has always been some lag time between the acquisition of information and the implementation of policy based on that information. Then, there has been considerable progress, and changes are to a degree underway. But finally, scientific understanding is one thing; decisions of what to do and when to do it are something else. Clearly we all want a sustainable environment. However, people also have many other values regarding such things as the use of materials and energy and the disposal of wastes. Many of these *values* are conflicting in ways that are not fully recognized, and conflicting values impede progress.

The environmental movement of the 1990s, the massive movement toward sustainable development, will depend on widespread public understanding and awareness, adoption of new values, and demands for and participation in change. Without public understanding and support, change is virtually impossible; with public demand, change cannot be resisted.

Each one of you contributes to the nebulous thing we refer to as "public opinion" or "societal values." Therefore,

Each one of you will be a participant in the movement toward sustainable development.

EVALUATING INFORMATION

As you probably already know, confronting environmental issues means also confronting conflicting information. How do you know whom to believe? Such conflicting information often paralyzes and prevents any action at all. It is extremely important, then, that we have a means of evaluating information and deciding for ourselves whom to believe. Therefore, I am devoting the rest of this chapter to a discussion of science and the scientific method. The purpose of this approach is to provide you with an understanding both of how scientific information is derived and of the tools you can use to evaluate that information.

What Is Science?

Science refers to both a particular way of gaining knowledge and the knowledge that is gained by this method. As students in a classroom we generally gain all kinds of information in essentially the same way. Teachers and textbooks provide the information and we study and learn it as well as we can. But where and how was that information originally obtained? Contrasting science subjects such as physics, chemistry, biology, and earth science with the humanities such as religion, philosophy, art, music, and literature, we find the roots come mainly from two distinct directions.

In the case of the humanities we find the origin in some person, who through some gift of inspiration and/or exceptional talent, gave us the teachings or composition which so affected us emotionally or spiritually that we have been treasuring and passing on those gifts ever since. While we may try to emulate the famous teacher, artist, writer, or composer, how they actually conceived their work remains a mystery.

In the case of science, our search also leads back to particular people, the scientists who made the discoveries. But here we can understand how they made their discoveries. They were based on their making certain *observations* and logical deductions based on those observations. Any of us could make the same observations and, through logical deduction, reach the same conclusions. Thus, in science, what becomes even more significant than the scientist (though we still respect his or her contributions) is the recognition that *all the information we call science is based on observations and logical deductions from those observations.* Hence, it is all open to being checked out and verified. It is absolutely counter to science to take anything on faith. The key tenet in science is:

Check it out!

Over the years the art of compiling scientific information has been refined into a series of steps that is called the **scientific method**.

The Scientific Method

The scientific method begins with observations of things and events, and progresses through the formulation and testing of theories. We shall look at each step in turn.

Observations and Facts

As noted, all scientific information is based on *observation* and is subject to objective verification ("checking it out"). In science, **observations** are restricted to include only impressions that are gained through one or more of the five basic senses in their normal state: seeing, hearing, touching, smelling, and tasting. This restriction is made because only such observations lend themselves to measurement, testing, and verification, and this is a most critical aspect. Even with the best of intent our senses may be deceived. Magic shows capitalize on such illusions or deceptions.

Therefore, before an observation is accepted as factual, it must be subject to confirmation or **verification**. That is, other investigators must be able to repeat the observation, perhaps using different techniques and tests, and confirm, independently, that the original observer was accurate in his or her observation. Observations that do not stand this test of verification are not accepted as scientific fact. For example, some claim to have seen spaceships with alien creatures, but this information is not generally accepted as factual because such observations have not been verifiable. On the other hand, observations that do stand the test of verification become accepted as **scientific fact**, but even here they remain open to further checking at any time (Fig. 0–2). It is this restriction to verifiable observations that gives scientific information its reputation of being "exact," "factual," or "objective."

It is interesting to note that science is not factual or objective because of any particular power of science or scientists; it is because science restricts itself to considering only things and events which can be observed in an objective way and eliminating any information that cannot be verified. Emotional feelings such as love, beauty, and spirituality are just as real and important to us, but since such things do not lend themselves to objective observation and measurement, they remain in the subjective realm outside of science.

The rule concerning verification applies not only here but to all steps of the scientific method. It is the most important tool you or anyone else can use to discriminate between accurate and inaccurate information. A good demand to make before accepting any information is:

Show me the evidence!

Hypotheses and Their Testing

A great deal of scientific information can be accumulated through observations alone without resorting to experiments. An almost automatic response, however, is to desire explanations for our observations. In some cases an answer may be found by simply making another observation. For example: What is causing the noise inside the box? Don't sit there like a dummy; open it and look! But, often the cause is not evident from direct observation. For example: the patient is sick; what is causing the illness?

The next step of the scientific method is to make *educated guesses* regarding the cause and then proceed to eliminate the wrong guesses through suitable tests or experiments. By weeding out wrong answers you are sooner or later left with the right answer(s). Each educated "guess" is called a **hypothesis**.

A specific thought process that is used in testing the validity of a hypothesis is the *"if . . . then . . ."* phrase. One reasons that *if* the hypothesis is true, *then* such and such should logically follow. If the "then" is not found, the hypothesis is discarded and the procedure is repeated on the next hypothesis. For example, a doctor makes an educated guess—a hypothesis—that the patient's illness is due to a virus. Experiments are then conducted to test for the presence of a virus. If a virus is found, it lends support to the virus hypothesis. If it is not found, the hypothesis is discarded and the doctor moves on to hypothesizing and testing for other causes. Hypotheses that are inconsistent with further observations are discarded and hypotheses that remain logically consistent with further observations are gradually accepted as true. It is a little like constructing a multiple choice question, the choices being all the possible hypotheses the scientist can think of, then gradually testing and eliminating the wrong ones. Perhaps they

FIGURE 0–2
All science is based on observations. Facts are observations that are confirmed by others and may be reconfirmed by anyone.

All Scientific Information Is Based on Careful, Thorough Observations Using the Basic Five Senses.

Further, Observations Must Be Verifiable by Others before They Are Accepted as Factual.

all prove invalid. This would only mean that scientists had not yet come up with the right guess. Many fields of scientific endeavor are blocked for years in this way. Great credit and admiration go to the scientist who conceives of the hypothesis that ultimately tests out as valid.

Controlled Experiments

An experiment to test a hypothesis must be carefully designed to stand up to the question: How can you be sure the observed results are due to the factor you hypothesize rather than to other unrecognized factors? The key is that experiments to test hypotheses must be *controlled*. By **controlled** we mean that the experiment must consist of two groups, a *test* group and a *control* group. The experiment is designed so that these two groups are exactly the same in every respect except for the single factor being tested. Results, positive or negative, can then be attributed to the single factor being tested (Fig. 0–3). Without a control group or with several differences between groups, you will be unable to interpret the results because you will have no way of knowing which fac-

tors or combination of factors is responsible for the result.

Two additional points concerning controlled experiments deserve emphasis. First, note that the experiment involves *groups* as opposed to *individuals*. There is always some variability (difference) among individuals. Therefore, if the "group" is only one, two, or a few individuals you are in danger of confusing natural individual differences with the effects of the treatment. The larger the group the better. Second, the results from a single experiment, especially when they are based on relatively small groups, should not be taken as conclusive evidence. They must stand the test of verification through repetition and confirmation (Fig. 0–4).

In some cases, performing controlled experiments would be prohibitively time consuming, costly, or impossible. Humans, for example, don't readily lend themselves to controlled experiments with regard to the effects of harmful or potentially harmful substances. However, there are so many humans living under such an array of different conditions and/

FIGURE 0–3
The uncontrolled experiment cannot be interpreted because any number of factors—singly or in combination—may be responsible for the lack of growth observed. The controlled experiment demonstrates that lack of nitrogen is responsible for the lack of growth, because this is the only factor that differs between the two groups.

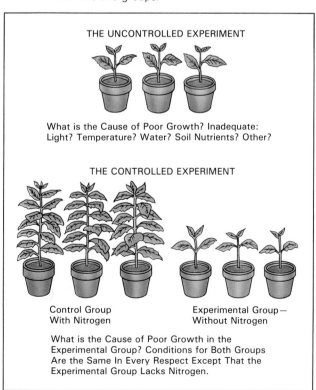

THE UNCONTROLLED EXPERIMENT

What is the Cause of Poor Growth? Inadequate: Light? Temperature? Water? Soil Nutrients? Other?

THE CONTROLLED EXPERIMENT

Control Group With Nitrogen   Experimental Group— Without Nitrogen

What is the Cause of Poor Growth in the Experimental Group? Conditions for Both Groups Are the Same In Every Respect Except That the Experimental Group Lacks Nitrogen.

FIGURE 0–4
Until it is subjected to testing and verification, an idea or hypothesis cannot be taken as fact. Controlled experiments have not shown that any kind of music or singing affects plants. What are other factors involved here?

or engaging in different habits, that investigators can generally identify persons for suitable experimental and control groups already in the normal population. For example, one can find enough smokers and non-smokers to test hypotheses regarding the effects of smoking on various aspects of health without resorting to experiments in the laboratory. Investigations of this nature are called **epidemiological** studies.

We can use these ideas to discriminate between accurate and inaccurate information. The key questions that should be asked are: Was there a control for the experiment or epidemiological study or is the information just an untested hypothesis or based on an uncontrolled experiment? If the information is based on a controlled experiment, what was the size of the groups involved and has the finding been confirmed? For example, a few years ago a compound—laetrile—gained the widespread reputation of being a cure for cancer. The Food and Drug Administration came under bitter attack for not allowing its use as a cancer cure. The truth of the matter was that controlled studies showed that people's cancer progressed just as rapidly under laetrile treatment as with no treatment at all.

Formulation of Theories

Individual observations and experiments only answer extremely specific questions. From small pieces of information, however, a larger picture, a *theory*, emerges. A **theory**, then, is a concept that provides a logical explanation for a certain body of facts. Constructing a theory is a process similar to a detective finding clues (observations) and then fitting them together into a picture (a theory) of "who dunnit." A theory is not a fact itself because it does not lend itself to direct observation. Nevertheless, theories may be tested and, pending the results, confirmed or rejected.

Theories are tested in much the same way that hypotheses are tested. The theory will suggest further aspects that should be observable directly or through experimentation—the "*if . . . then . . .*" reasoning. If observations are made that are contrary to the theory, the theory is either modified to incorporate the new findings or it is rejected in favor of an alternative theory that provides a rational explanation. In either case, a theory is developed that is consistent with all observations and experiments and which can be used to reliably predict outcomes in the sense of "*if . . . then . . . .*" When theories reach this state, we have every reason to believe that they represent a correct interpretation of reality. For example, we have never seen atoms as such, but innumerable observations and experiments are coherently explainable by the concept that all gases, liquids, and solids consist of various combinations of only slightly more than a hundred kinds of atoms. Hence, we fully accept the *atomic theory of matter*.

Principles and Natural Laws

In the course of observing the results of experiments, certain precise trends or patterns may be seen in the data. For example, in dropping objects and measuring their speed at various points in their fall, it is observed *without exception* that all objects accelerate at exactly the same rate (exclusive of air resistance). Thus, we observe a basic principle underlying the free fall of objects. When we discover such a principle underlying the behavior of things, we refer to it as a **natural law**. In this case, of course, we are referring to the *law of gravity*. Another example comes from the observation that in chemical reactions atoms are only rearranged; they are not created, destroyed, or changed. This principle is referred to as the *law of conservation of matter*. *Laws of thermodynamics* refer to principles discovered regarding energy changes.

Since natural laws are the expression of basic principles, which have no known exceptions, their predictive value is tremendous. Any attempt which violates or ignores a natural law can be predicted with virtual certainty to end in failure. Failure may not be instantaneous; there is generally a certain amount of "borrowed time" between the violation and the ultimate result as we noted in the case of the man jumping off the cliff.

When trying to judge whether a certain course of action or process will work or not, look for consistency with natural laws. This is so obvious in the case of the man jumping off the cliff that it is laughable. However, even people who should know better persist. Thus, the space shuttle *Challenger* was sent off in defiance of the known effect of temperature on O-ring seals. The borrowed time was 90 seconds. In the mid-1980s, a person was promoting his "backyard" invention of a new motor that produced more power than it consumed, a violation of the second law of thermodynamics. By playing the role of a cantankerous genius at odds with an envious "establishment," he gained considerable support in the popular press and raised several million dollars for development of his device while at the same time avoiding any critical testing of it. Finally, even his backers insisted on a demonstration, at which point his device proved to be a hoax. He has since dropped out of sight—along with the several million dollars. The moral of the story is: Be knowledgeable about the fundamental natural laws, and when you see actions or proposals that are in contradiction, be highly skeptical.

**Stupidity, ignorance, fraud, and deceit
are much more likely
than exceptions to natural laws.**

The Role of Instruments in Science

The use of complex instruments often gives science an aura of mystery. Yet there is nothing mysterious about scientific instruments. They have been developed to serve one of three purposes. First, they may *extend* our powers of observation. For example, microscopes and telescopes enable us to see smaller or more distant objects than is possible with the naked eye, and various detectors enable us to see (or hear) the evidence of X-rays and radio waves. All instruments used in science are themselves subjected to tests and verification until we are confident that they are giving us a real representation of natural phenomena as opposed to creating their own images or illusions.

Second, instruments as simple as a meter stick or thermometer or as complex as a radio telescope enable us to *quantify* our observations; that is, they enable us to *measure* exact quantities. For example, we may feel cold but a thermometer enables us to measure and quantify exactly how cold it is. We may feel an electric shock, but an appropriate meter enables us to say exactly what the voltage is. Comparisons and verification of different observations and events would be impossible if it were not for this measurement and quantification.

Finally, instruments help us achieve conditions and perform manipulations required for controlled experiments, such as maintaining constant temperature or humidity.

Science and Technology

Basic or **pure science** is the gaining of knowledge and understanding for its own sake without consideration as to how it will be used or even if it will be used. It is motivated only by the desire to gain knowledge. **Technology**, on the other hand, is the application of scientific knowledge to gain a specific objective. It is motivated by the desire to solve a particular problem or achieve a specific result. Technology may muddle forward by a trial-and-error approach and gradually build on the experience of what works and doesn't work. Some highly successful inventions have been developed in this way. There are, however, even more failures. The trial-and-error approach is generally exceedingly wasteful in terms of time, money, and occasionally human lives.

On the other hand, once principles, natural laws, and theories are understood, they can be efficiently applied by technologists to achieve desired re-

sults. For instance, it is conceivable that we could have sent men to the moon using the trial-and-error process, but at what cost? By adhering strictly to the theories and principles involved, we were able to do it on the first attempt.

Therefore, technological development is aided by understanding gained through basic science. But basic science also benefits from technology in two primary ways. First, technology serves to develop new instruments that may extend our ability to observe and test hypotheses. Second, technology is the "proving ground" for new theories and principles. The fact that a technological innovation works demonstrates that the theories and principles on which it is based are correct. For example, the fact that we did send men to the moon and back underscored that our theories concerning planetary motions and gravitational forces were correct. Thus, science and technology are mutually supporting.

In conclusion, the scientific method enables us to gain an understanding of how the world works. In turn, by using technology and working within the framework of this understanding, we can achieve desired objectives. We do not, for example, "wish" people to the moon and back. But, by working within the framework of the theories and natural laws involved, we are able to achieve this result.

Unresolved Questions and Controversies

Obviously, science is an ongoing process. Each new experiment or instrument for observation yields new information regardless of whether it supports or refutes the hypothesis in question. In turn each new piece of information raises more questions and so on. Much study and/or experimentation is required before all alternative hypotheses are adequately tested and firm conclusions are reached. Progress is particularly slow when the questions involve events such as climatic changes or dieoffs of forests, which do not lend themselves readily to controlled experiments. Honest controversy may be widespread in an area as long as there are two or more plausible hypotheses which have not been adequately tested.

When is a hypothesis or theory adequately tested and accepted as correct? Recall that finding the right hypothesis or theory occurs largely through the process of elimination. Incorrect hypotheses or theories are eliminated through experimentation which shows that they are inconsistent with certain observations, whereas hypotheses or theories that remain consistent with further observations gain credibility. But who is to say that we have thought of all the possible alternatives or performed all the critical tests? Perhaps there is another hypothesis/theory that will explain the observations even more clearly and it will finally be determined that our currently held theory is really inconsistent with observations and therefore wrong.

This is a real handicap of science. It is incapable of providing absolute proof of any theory. What occurs is that increasing amounts of evidence are accumulated which support one theory and tend to refute others. Somewhere along this line most people will decide (a value judgment, not science) that the evidence is adequate and conclude that the given theory is correct and act accordingly. However, there are others, particularly those with some vested interest, who simply capitalize on the inability of science to provide absolute proof while at the same time implying that it can. The Tobacco Institute, a lobbying association for the tobacco industry, provides a prime example. It continually makes the point that the connection between smoking and ill health effects is not *proven* and that more studies are necessary. By continually harping on the lack of absolute proof and simply ignoring the overwhelming amount of evidence supporting the connection between smoking and ill health effects, it has been highly successful in delaying restrictions on smoking.

You may also hear a theory attacked with the argument that since such-and-such a theory is not a proven fact, an alternate theory is just as valid. Don't allow yourself to be misled by such arguments. Although it is true that a theory falls short of being a proven fact, one theory may have overwhelming evidence supporting it while another may have little if any support or may even be contrary to evidence. How do you know which side of a controversy to believe, then?

Look at the actual evidence, weigh it for yourself. Beware of the voice that may be more concerned with its vested interest than with the truth.

SCIENCE AND VALUE JUDGMENTS

Scientific endeavors yield increasing understanding concerning the objective world and how it functions, and we may use this knowledge to achieve certain objectives. But science does not tell us what objectives to strive for. Decisions such as what to do, when to do it, and which thing or course of action is better are known as *value judgments*. They come from the moral, religious, ethical, and emotional sides of our being. For example, through science we have gained

an understanding of atoms; however, whether or not we use this understanding to build nuclear weapons is a decision based on values and judgments quite apart from science. Yet, understanding gained through science may aid us in making value decisions by helping us to see the consequences of certain actions. The "if . . . then . . ." reasoning applies again. For example, *if* we have a full-scale nuclear conflict, *then* virtually all life on earth will be destroyed.

Further, science itself does not demand that we adhere to principles or theories. We can, for instance, willfully or carelessly ignore certain principles and theories of construction. Science can only predict that the outcome of such behavior will be the collapse of the building.

ENVIRONMENTAL SCIENCE AND ITS APPLICATION

How does what we have learned here relate to environmental science? Through the study of interactions of plants and animals in their natural environments, scientists have uncovered basic *ecological principles* or *laws* underlying the sustainability of life on earth. It should be clear that if our human system, involving agriculture, industry, and so on, is to be sustainable, it must be operated in accordance with ecological laws and principles. Yet, we are, in large part, operating our human system according to notions and whims that bear little relationship to ecological natural laws. This is like building a skyscraper disregarding construction principles. Indeed, the increasing environmental and resource problems and crises we face are the consequence and the evidence that this course of action is not working. Indeed it may be on the verge of total collapse. Collapse may be avoided, however, and a course of sustainable development may be achieved by recognizing the basic ecological principles and adapting our human system accordingly.

Part I of this text is devoted to developing an understanding of ecological theory and principles and recognizing how they apply to the human situation. The parts following address specific environmental and natural-resource problems, and how they may be resolved by recognizing and adhering to ecological principles.

Then, it will be up to all of us to value the goal of sustainable development enough to make the efforts and sacrifices necessary to bring it about.

Part One
WHAT ECOSYSTEMS
ARE AND HOW THEY WORK

In 1968 astronauts returned with photographs of the earth taken from the moon. These photographs made it clear as never before that the earth is just a sphere suspended in the void of space. It is like a self-contained spaceship on an everlasting journey. There is no home base to which to return for repairs, more provisions, or disposal of wastes; there is just the continuous radiation from the sun. Indeed, the term "Spaceship Earth" was coined by futurist Buckminster Fuller as a result of this new perspective on our planet.

Who is at the controls of Spaceship Earth? Unfortunately, no one! But Spaceship Earth is equipped with an amazing array of self-providing mechanisms. Enormously diverse plant and animal species interact in ways such that each obtains its needs from and provides for the support of others. Air and water are

constantly repurified and recycled. Then there are self-regulating mechanisms as well, which tend to keep all the systems in balance with each other.

But now problems are arising. In particular, the human species is multiplying out of all proportion to others. This is placing greater and greater demands on all systems and, at the same time, it is undercutting their productivity through pollution and overexploitation. The natural regulatory mechanisms are being upset. It is clear that such behavior aboard Spaceship Earth cannot be sustained without catastrophic consequences. Nor can we afford the happy-go-lucky luxury of trial-and-error learning when the fate of the whole world is at stake. We must gain an understanding of how Spaceship Earth works and then we must learn to conduct our activities within this context.

Here in Part I our objective is to provide a general framework of understanding concerning the way our spaceship works. This understanding is gained through a study of natural ecosystems: what they are, how they function, how they are regulated, and how they develop and change. In keeping with the scientific method, we shall approach each area by describing the basic observations that have been made and showing how these observations have led to the formation of operating theories and principles. Finally, the understanding of these theories and principles will enable us to see more clearly where current trends are headed and how certain human activities must be modified if modern society is to be sustained.

1
Ecosystems:
What They Are

CONCEPT FRAMEWORK

▪ Outline

I. DEFINITION AND EXAMPLES OF ECOSYSTEMS
    15

II. THE STRUCTURE OF ECOSYSTEMS   25

  A. Biotic Structure   25

    1. Categories of Organisms   25
       a. Producers   25

       b. Consumers   27

       c. Detritus Feeders and Decomposers   28

    2. Feeding Relationships: Food Chains, Food Webs, and Trophic Levels   31

    3. Non-Feeding Relationships   33

       a. Mutually Supportive Relationships   33
       b. Competitive Relationships   35

  B. Abiotic Factors   35

    1. Optimum, Zones of Stress, Limits of Tolerance   35

    2. Law of Limiting Factors   36

III. WHY DO DIFFERENT REGIONS SUPPORT DIFFERENT ECOSYSTEMS?   37

▪ Study Questions

1. Define the terms *biosphere, species, ecosystem, plant community, biome, ecology, ecologists.*

2. Name, describe, and give the general location of major biomes of North America.

3. What is the human ecosystem? How does it fit into the other biomes?

4. Define the following: *structure, biota, biotic structure of an ecosystem, and abiotic.*

5. Name the three major categories of organisms that make up an ecosystem.

6. What are *producers?* What is their role? Name and describe the key process they carry on. Distinguish between organic and inorganic.

7. What are *consumers?* Give examples that show their diversity. Name and define various subcategories of consumers.

8. Define *detritus.* How do detritus feeders and decomposers differ from other consumers? How do decomposers differ from other detritus feeders? What two major categories of organisms are decomposers?

9. Define and distinguish between: a *food chain,* a *food web, trophic levels,* and *biomass.* Place various categories of organisms in their respective trophic levels. Contrast the relative biomass which is found at each trophic level. Why does this occur?

10. Name and describe various nonfeeding relationships.

11. Define and name various *abiotic* factors.

12. With respect to any abiotic factor, define *optimum, zone of stress, limits of tolerance.* Do these differ for different species?

13. Define *limiting factor.* What observations lead to the law of limiting factors? State this law.

14. Considering any species, define *population density.* How will it change with changing abiotic factors?

A. Climate   37

B. Other Abiotic Factors and Microclimate   39
C. Biotic Factors   39
D. Physical Barriers   41
E. Biotic and Abiotic Interactions   41

IV. IMPLICATIONS FOR HUMANS   41

15. What are the major abiotic factors of climate? How do these differ in different parts of the world? Contrast this with optimums and limits to tolerance of different species. Why do different biomes occupy different regions? Define: *microclimate.* Cite how other abiotic factors may influence distribution of species. How may distribution of species be limited by biotic factors; by physical barriers? Describe how biotic and abiotic factors may interact in limiting the distribution of a species.

16. What is likely to happen to an ecosystem if any abiotic or biotic factor is altered?

17. Contrast the human ecosystem with other ecosystems. What are the similarities? What are the differences, particularly in terms of limiting factors? Discuss how humans have overcome the usual factors that limit other species. What is the result in terms of balance between the human ecosystem and other ecosystems? Is this sustainable? Explain why not. Is there an alternative? What is meant by sustainable development? Is it possible for humans to impose their own limits? Cite such efforts that are already underway.

Spaceship Earth is unique among the planets of the solar system. In a thin layer where air, water, and earth come together and interact, there are strange and wonderful things—living things, and some of them are us. We refer to this layer of living things interacting with air (atmosphere), water (hydrosphere), and earth (lithosphere) as the **biosphere**. All living things, including us, depend on maintaining the integrity of the biosphere. Alter any aspect of the biosphere too much and it could collapse. Atmosphere, hydrosphere, and lithosphere would continue to be, but their interactions would no longer include living things. How are we humans endangering the integrity of the biosphere? What do we need to change to prevent this catastrophe?

Fully comprehending these questions and their answers demands a basic understanding of living and nonliving parts of the biosphere and how they function together to support the whole. Much as cells are the functional units of all living organisms, *ecosystems* are the functional units of the biosphere. By observing and studying ecosystems, we can gain an understanding and appreciation of the biosphere. Our objective in this chapter is to understand what ecosystems are and how they may be determined, controlled, or changed by different conditions.

DEFINITION AND EXAMPLES OF ECOSYSTEMS

Throughout this and further chapters, the word *species* will be used frequently. The word **species** is both singular and plural. It refers to specific kind(s) of plants, animals, and/or microbes (microscopic organisms). What constitutes a "kind" is best defined as the ability to interbreed. Members of the same species will breed to reproduce their kind; members from different species cannot. Thus horses and dogs are different species, but all breeds of dogs belong to one species because they can interbreed. Such subdivisions of species are referred to as different races (animals) or different varieties (plants).

An **ecosystem** may be defined as a grouping of various species of plants, animals, and microbes interacting with each other and their environment. Furthermore, the interrelationships are such that the entire grouping may perpetuate itself, perhaps indefinitely. This definition is a very condensed description of what is observed in nature and can be best understood by considering some examples with which you are probably already familiar.

A quick tour across the United States shows deciduous (leaf-shedding) forests in the East, which

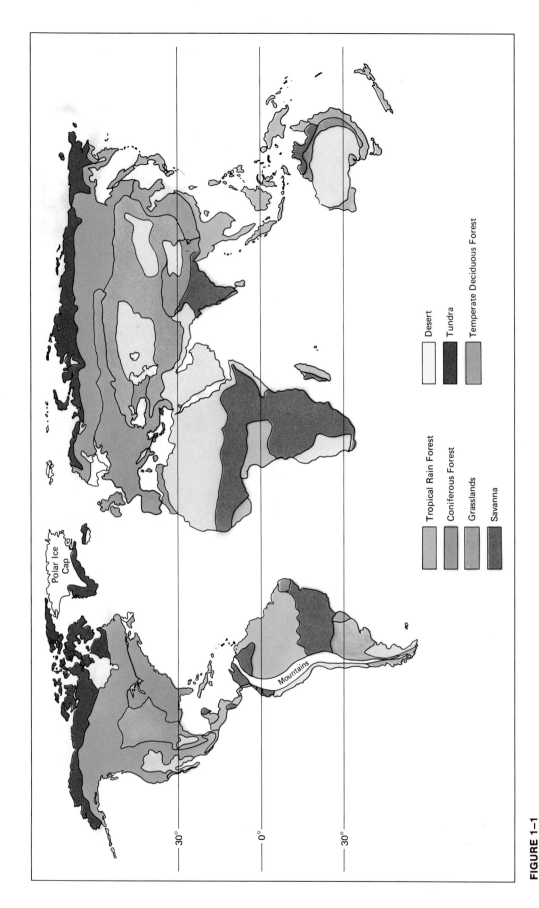

FIGURE 1–1
World distribution of major biomes. Frequently these major biomes are divided into a number of subtypes. (Reproduced by permission from D. K. Northington and J. R. Goodin, *The Botanical World*, St. Louis: Times Mirror/Mosby College Publishing, 1984.)

turn brilliant colors in the fall before the leaves drop; prairies or grasslands in the Central States; deserts with distinctive cacti in the Southwest; and evergreen, coniferous forests in the northern and western Mountain States. Going a little farther, across northern Canada and on the tops of mountains there are treeless expanses called tundra, and in equatorial regions we find tropical rainforests.

Looking at these examples more closely, we note that each is characterized by a distinctive **plant community** or grouping of particular plants. For example, various species of deciduous trees and associated plants make up the deciduous forest, whereas various kinds of grasses and associated plants dominate the prairies. Each plant community supports a particular array of animals. We tend to focus on large animals, deer in eastern deciduous forests, bison on the prairies (before they were killed off by early settlers), moose in conifer forests, and caribou on the arctic tundra, for example. However, smaller animals, such as mice, birds, various species of insects, earthworms, and other such organisms, are far more abundant in terms of both numbers and total combined weight or **biomass**. Finally, but even less conspicuous, an array of **microbes** (microscopic organisms, mainly bacteria and fungi) will be found in each system feeding on dead plant and animal material.

Each of these examples represents a distinctive grouping of plants, animals, and microbes interacting with one another and their environment. Further we know that these groupings existed long before humans came on the scene and, if they were not disturbed by humans, would continue to exist, perhaps indefinitely. Thus, each grouping along with its environment is an *ecosystem*.

Very large terrestrial ecosystems such as the ones mentioned above are referred to as **biomes**. Some further description of these biomes is given in Boxes 1–7, and Figure 1–1 shows their locations in North America and the world. Each major ecosystem or biome generally contains a number of smaller but related ecosystems within it. Thus, an ecosystem may be very extensive, covering millions of square miles ($1 \text{ mi}^2 = 2.5 \text{ km}^2$), or it may be as small as a patch of woods. The significant point is that it can be defined, at least for study purposes, as a more or less specific grouping of plants and animals interacting with one another and their environment. There are many kinds of aquatic ecosystems as well—streams, rivers, lakes, ponds, marshes, and swamps—each with its distinctive array of plants and animals and microbes. Similarly, oceans may be divided into separate ecosystems such as coral reefs, continental shelves, and deep oceans.

The division of the landscape into different ecosystems is rather arbitrary. They seldom have distinct boundaries. Rather, one ecosystem or biome blends into the next through a transitional region that contains many of the species and characteristics of the two adjacent systems (Fig. 1–2). Indeed, the blending of adjacent systems in a transitional region may create unique environments that support distinctive plants

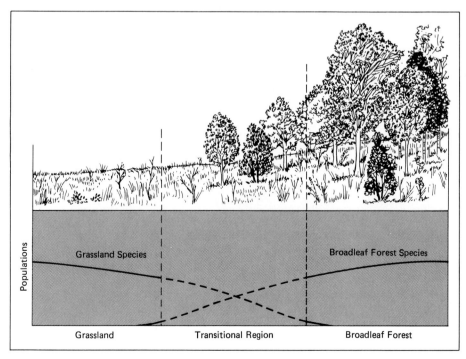

FIGURE 1–2
Ecosystems are not isolated from one another. One ecosystem blends into the next through a transitional region that contains many species common to the two adjacent systems.

Populations

Grassland Species

Broadleaf Forest Species

Grassland

Transitional Region

Broadleaf Forest

Box 1

TEMPERATE FOREST BIOME

Region: Western Europe, East Asia, Eastern United States.

Climate: Seasonal, with temperatures below freezing during winter (not below − 12 C though).

Rainfall: Annual range 30–80 inches.

Major Vegetation: Broad-leaved deciduous trees (120–150 feet tall), oak, hickories, maples, and other trees that shed their leaves in the fall and regrow them in the spring; shrubby undergrowth, lichens and mosses.

Animals: Abundant microbiota in the soil; mammals such as white-tailed deer, porcupines, raccoons, opossums, rabbits, squirrels, shrews; birds—e.g., warblers, woodpeckers, thrushes, owls, hawks; snakes, frogs, and salamanders; fish—trout, perch, bass, catfish.

Other: Biota well adapted to seasonality—hibernation, migration, or inactivity during coldest months; long growing season of 4–6 months; natural predators— wolves, bobcats, gray foxes, and cougars are largely eliminated due to hunting and habitat destruction.

Mixed oak-hickory stand in West Virginia. (Photo by William Bierley)

Box 2
GRASSLAND BIOME

Region: Central North America, Russia, parts of Africa, Australia, and Southeastern South America.

Climate: Seasonal.

Temperature: Moderate to hot summers, cold winters with subfreezing temperatures.

Rainfall: 10–30 inches annually.

Major Vegetation: Grass species—tall grasses, 6 feet or greater in some prairies of North America; short grasses, not exceeding 1.5 feet—e.g., the steppes in Russia; sparse bushes and occasional trees in moist areas.

Animals: Large grazing mammals—bison, antelopes (N. America), wild horses (Eurasia), kangaroos (Australia), giraffes, zebras, white rhinoceroses, antelopes (Africa), and predators—e.g., coyotes, lions, leopards, cheetahs, hyenas; variety of birds and small burrowing mammals such as rabbit, prairie dog, and aardvarks.

Other: Most of the grassland in North America has been converted into vast fields of crops—corn, wheat, soybeans; and wild grazing animals replaced with domesticated cattle, sheep, and goats.

Western Nebraska. (Earth Scenes/Lynn M. Stone)

Box 3

DESERT BIOME

Region: Parts of Africa—the Sahara; parts of the Mideast and Asia; the Great Basin and Southwest United States; Northern Mexico; any region with less than 10 inches annual rainfall; usually around latitudes of 30 N and 30 S.

Climate: Very dry.

Temperature: Hot days and cold nights; varies with latitude.

Rainfall: Less than 10 inches a year.

Major Vegetation: Widely scattered thorny bushes and shrubs, occasional cacti, and small flowers that quickly carpet the desert floor after brief rains; extensive shallow root systems as well as deep taproots (as long as 100 feet) provide means of access to scarce rainfall and groundwater.

Animals: Number of rodents (e.g., kangaroo rat), lizards, toads, snakes, and other reptiles (e.g., gila monsters), owls, eagles, vultures, many small birds, and numerous insects.

Other: Deserts cover over ⅓ of the earth's land surface, and are growing each year—due to human activity, as well as climate changes.

Saguaro cactus and creosote bush in the Sonoran desert in Organ Pipe National Monument in southern Arizona. (Tamra Nebel)

Box 4
CONIFEROUS FOREST BIOME

Region: Northern portions of North America, Europe, and Asia.

Climate: Extremely long and cold winters, much of the precipitation falls as snow.

Major Vegetation: Coniferous (evergreen) forests, especially of spruce, fir, and pine.

Animals: Large herbivores—e.g., moose, mule deer, caribou (which migrate down from the tundra during fall), elk; smaller herbivores—e.g., snowshoe hare, red squirrels, and variety of rodents; timber wolf, lynx, red fox, grizzly bear and black bear, wolverine, mink, other; many biting flies and mosquitoes during short summer.

Other: The taiga has many lakes, ponds, and bogs; needles and tree trunks litter the forest floor, since the cold greatly slows decomposition.

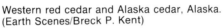

Western red cedar and Alaska cedar, Alaska.
(Earth Scenes/Breck P. Kent)

Box 5

TUNDRA BIOME

Region: North of the taiga in the Northern Hemisphere.

Climate: Bitter cold, limited sunlight.

Temperature: Average annual is −5 C (23°F); few weeks of summer thaws ground to depths of only 3 feet.

Rainfall: Low precipitation, less than 10 inches annually.

Major Vegetation: Low-growing lichens, mosses, grasses, sedges, and dwarf shrubs.

Animals: Year-round are small burrowing mammals—i.e., lemmings that cycle from scarcity to superabundance and back; predators (camouflaged white during the long winters) such as white foxes, lynxes, ermine, snowy owls, grizzly bears; large herbivores, such as caribou, reindeer, and musk ox, which slowly migrate south in the fall.

Other: During the short summer a large number of migrating birds, especially waterfowl, nest in the tundra to feed on the swarms of insects and freshwater invertebrates.

Alaska. (Animals Animals/David C. Fritts)

Box 6
SAVANNA BIOME

Region: Subequatorial Africa and South America, much of southern India.

Climate: Hot and dry much of the year, abundant rain during wet season.

Temperature: High average temperatures.

Rainfall: 30–65 inches per year during the wet season.

Major Vegetation: Grasslands with scattered deciduous trees.

Animals: Large grazing mammals such as antelope, zebra, rhinoceros, giraffes, wildebeests; predators such as lions, leopards, and cheetahs.

This photo is Masai Mara, Kenya, Africa. (Animals Animals/Zig Leszczynski)

Box 7

TROPICAL RAIN FOREST BIOME

Region: Northern South America, Central America, Western and Central Equatorial Africa, Southeast Asia, Northwest Coastal Australia; various islands in the Indian and Pacific Oceans.

Climate: Nonseasonal, due to equatorial location.

Temperature: Annual average greater than 17 C (50 F), mean 28 C (82 F).

Rainfall: Almost daily—heavy annual average greater than 94.5 inches (240 centimeters).

Major Vegetation: Hundreds of species of towering trees, the tallest being 60 meters or so; epiphytes—plants high in the trees whose roots do not reach the soil; and lianas—or woody vines that are rooted in the soil and climb to the treetops, forming a dense canopy.

Animals: More species than all other biomes combined; exotic, colorful insects; amphibians, reptiles, and birds are especially abundant—e.g., lizards, parrots, snakes, frog, macaws; monkeys and small mammals; colorful fish in waterways.

Other: The soil is generally thin and poor, with most of the nutrients embodied in the shallow-rooted vegetation; some areas systematically flood (e.g., Amazon Basin); rainforest is currently the most exploited and endangered biome, as human activity persists rampantly.

Tropical rain forest, Caribbean National Forest, Puerto Rico. (Photo Researchers/Maurice and Sallye Landre)

and animals as well as those that are common to both ... "es" in every ecosystem, the organisms on one adjoining ecosystems (Fig. 1–3). Thus, transition regions may be studied as ecosystems in their own right.

Ecosystems are not isolated. Many kinds of plants and animals may be found in two or more different ecosystems, and some species, such as migratory birds, may inhabit different ecosystems at different times of the year. What happens in one ecosystem will definitely affect another. For example, water draining from the land may carry soil sediments and nutrients that upset aquatic ecosystems. Thus, all ecosystems are interconnected and interdependent.

We humans, with our agricultural plants and animals, pets, and so on are also a grouping of plants and animals interacting with one another and the environment. This is an ecosystem too, the **human ecosystem**, and we interact with all the other ecosystems on earth. Therefore, all the ecosystems on earth including the human ecosystem are interconnected and make up a single entity, the *biosphere*.

The study of ecosystems and the interactions that occur among organisms and between organisms and the environment is the science of **ecology**; scientists who conduct these studies are **ecologists**.

THE STRUCTURE OF ECOSYSTEMS

As we are interested in preserving the integrity of the biosphere, we wish to first discover how it works. Our study begins with an overview of the *structure* of ecosystems. Structure refers to parts and the way they fit together to make the whole. There are two

hand and the environmental factors on the other. All the organisms—plants, animals, and microorganisms—in the ecosystem are referred to as the **biota** (*bio* = life). The way the categories of organisms fit together is referred to as the **biotic structure**. In contrast, the nonliving, chemical and physical factors of the environment are referred to as **abiotic**.

Biotic Structure

Categories of Organisms

Despite the diversity of ecosystems, from rainforests to deserts, ecologists find that they all have a similar biotic structure. That is, they all have the same basic categories of organisms that interact together in the same ways. The major categories of organisms are *producers, consumers, detritus feeders* and *decomposers*.

Producers. Producers are mainly green plants which carry on *photosynthesis*. **Photosynthesis** is the process in which green plants use *light energy* to convert carbon dioxide (which they absorb from air or water) and water into sugar and release oxygen as a byproduct. Plants are able to manufacture all the additional complex molecules that make up their bodies from the sugar and a few additional *mineral nutrients* that they absorb from the soil or water (Fig. 1–4). The molecule that plants use to capture light energy for photosynthesis is **chlorophyll**, which is a green pigment. Hence plants that carry on photosynthesis are easily identified by their green color. They range in diversity from microscopic single-celled algae through medium-sized plants such as grass and cacti to giant trees. Thus, every major ecosystem, both aquatic and

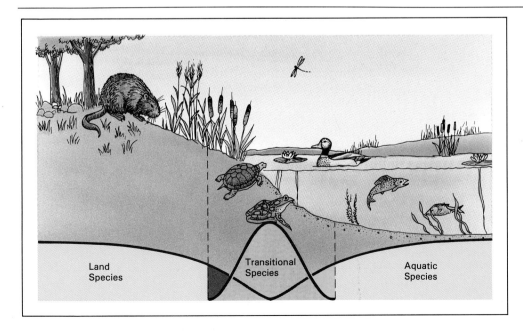

FIGURE 1–3
The transitional region may create a unique habitat that harbors specialized species. Thus, the transitional region itself may form a unique ecosystem.

Land Species

Transitional Species

Aquatic Species

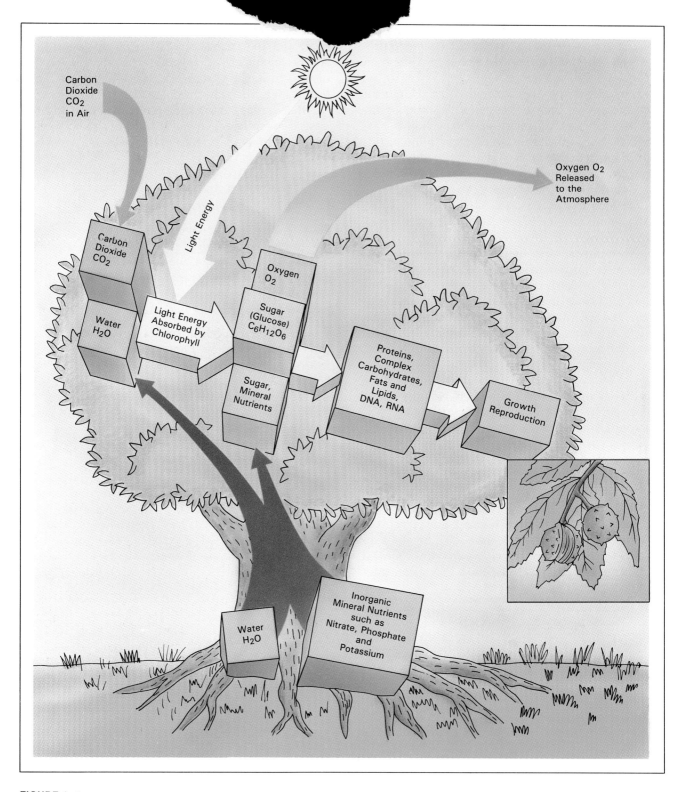

FIGURE 1–4

Producers. Green plants, which contain chlorophyll, can absorb light energy and use it to produce glucose from carbon dioxide and water, releasing oxygen as a byproduct. The glucose, along with a few additional mineral nutrients from the soil, is used in the production of all plant tissues leading to growth.

terrestrial, has its particular green plants, which carry on photosynthesis. In some cases the green may be overshadowed by additional red or brown pigments. Thus, red algae and brown algae also carry on photosynthesis.

The simple chemicals that make up air, water, and the minerals of rock and soil are referred to as **inorganic**. In contrast, the complex chemicals such as protein, fats, and carbohydrates, which make up tissues of plants and animals, are referred to as **organic** (Fig. 1–5). Thus, plants that carry on photosynthesis use light energy to *produce* all the complex *organic* chemicals for their bodies from the simple *inorganic* chemicals (carbon dioxide, water, mineral nutrients) present in the environment. As this occurs *energy* from light is incorporated into the organic compounds as well as the chemical elements from the environment.

All animals and other organisms that are not producers feed on complex organic material as their source of both energy and nutrients. Thus, plants that carry on photosynthesis are significant because they produce the organic matter or "food" which supports all the rest of the organisms in the ecosystem. Hence, plants that carry on photosynthesis are called *producers*. Not all plants are producers, however; all fungi (mushrooms, molds, and other such organisms) and a few higher plants, such as Indian pipe (Fig. 1–6), do not have chlorophyll and do not carry on photosynthesis. Like animals, they must feed on organic matter.

Indeed all organisms in the biosphere can be divided into two categories on this basis. Those organisms, mainly photosynthetic plants, that can produce their own organic material from inorganic raw material using an environmental energy source are producers. All others, which must consume organic material as a source of energy and nutrients, are *consumers*.

Consumers. **Consumers** embrace a tremendous variety of organisms ranging in size from microscopic bacteria to blue whales and include such diverse groups as protozoans, worms, fish and shellfish, insects and related organisms, reptiles, birds, and mammals including humans. For the purpose of understanding ecosystem structure, consumers are divided in various subgroups according to their food source.

FIGURE 1–5
Organic and *inorganic*. Simple states of matter found in air, water, and rock and soil minerals are said to be *inorganic*. The complex states found in plant and animal tissues are said to be *organic*. The change from inorganic to organic is enabled by producers using light energy in the process of photosynthesis. Organic materials will break down to inorganic materials again through burning or digestion, releasing energy. Chemically, organic compounds involve carbon-carbon and carbon-hydrogen bonds not found in inorganic materials.

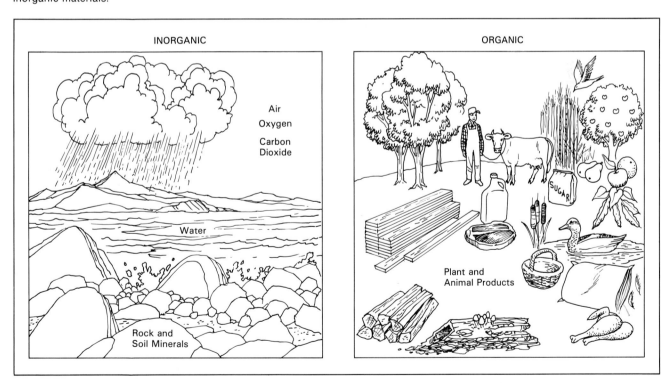

INORGANIC

Air
Oxygen
Carbon Dioxide

Water

Rock and Soil Minerals

ORGANIC

Plant and Animal Products

FIGURE 1–6
Indian pipe, a flowering plant that is not a producer. It does not carry on photosynthesis but derives its energy from other organic matter as do animals. (Photo by author.)

Animals that feed directly on producers, whether as large as elephants or as small as mites, are called **primary consumers**. Animals that feed on primary consumers are called **secondary consumers**. For example, a rabbit that feeds upon carrots is a primary consumer; a fox that feeds upon rabbits is a secondary consumer. There may also be third, fourth, or even higher levels of consumers. Certain animals may occupy more than one position on the consumer scale. For instance, humans are primary consumers when they eat vegetables, secondary consumers when they eat beef, and third-level consumers when they eat fish that feed on other organisms that in turn feed on algae.

Primary consumers, those animals that eat only plant material, are also called **herbivores**. Secondary and higher orders of consumers are called **carnivores**. Those that feed on both plants and animals are called **omnivores**.

In a relationship in which one animal attacks, kills, and feeds on another, the animal that attacks and kills is called the **predator**; the animal that is killed is called the **prey**. Together, the two animals are said to have a **predator-prey** relationship.

Parasites are another important category of consumers. **Parasites** are organisms that become intimately associated with their "prey" and feed on it over an extended period of time, typically without killing it (at least not immediately) but usually doing it harm. The plant or animal that is fed upon is called the **host**; thus we speak of a **host-parasite** relationship. A tremendous variety of organisms may be parasitic. Various worms are the most common examples, but bacteria and other microorganisms that cause disease of plants or animals are really highly specialized parasites. Many serious plant diseases and some animal diseases (such as athlete's foot) are caused by parasitic fungi. Even some plants such as dodder (Fig. 1–7a) are parasitic on other plants. (Ecologically speaking, parasitic plants must actually be considered consumers.) Virtually every major group of living things including mammals (vampire bats) has at least some members that are parasitic. Parasites may attach and live inside or outside their host as the few examples shown in Figures 1–7b and 1–7c illustrate.

These categories of consumers are summarized in Figure 1–8.

Detritus Feeders and Decomposers. Dead plant and animal material from sources such as natural leaf fall in forests, the dieback of grasses in unfavorable seasons, the fecal wastes of animals, and so on is called **detritus**. There are many organisms that are specialized to feed on detritus, and we refer to such consumers as **detritus feeders**. Examples include vultures, earthworms, millipedes, crayfish, termites, ants, wood beetles, and so forth. As with regular consumers, one can identify *primary* detritus feeders (those that feed directly on detritus), *secondary* detritus feeders, and so on.

Finally, much of the detritus in an ecosystem, particularly dead leaves and wood, does not appear to be eaten as such, but rots, decays, or decomposes. Actually, rotting or decomposition is the result of feeding activity of *fungi* and *bacteria*. Most species of bacteria do not cause disease but are harmless detritus feeders. **Fungi** include such organisms as molds, mushrooms, shelf fungi, coral fungi, and puffballs (Fig. 1–9). The part we recognize as the mushroom, shelf, or puffball is just the fruiting body or reproductive structure and is only a small portion of the whole organism. Attached is an extensive network of microscopic rootlike filaments, called **mycelia**, that penetrate through the dead leaves, wood, or other detritus. The mycelia secrete digestive enzymes that break down the detritus into simpler molecules which are then absorbed for nourishment (Fig. 1–10). While mushrooms occasionally appear to be growing on inorganic soil, their mycelia are actually feeding upon

(a)

FIGURE 1–7
Diversity of parasites. Nearly every major biological group of organisms has at least some members that are parasitic on others. Shown here are (a) dodder, a plant parasite (Photo by Steve Simon); (b) intestinal tapeworm, an endoparasite (USDA photo); (c) lamprey, an ectoparasite (U.S. Fish and Wildlife Service photo).

(b)

(c)

FIGURE 1–8
The most common feeding relationships among organisms.

FIGURE 1–9
Decomposers. Thousands of species of fungi, a few of the major types of which are shown here, are *decomposers*. They are plantlike organisms, but they feed on dead organic matter, much like animals. The result of their feeding is observed as the rotting, decay, or decomposition of the dead organic material, such as wood or dead leaves. Many species of bacteria are also decomposers. (Photos by author.)

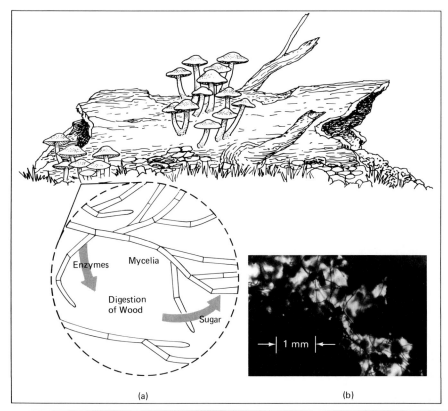

FIGURE 1–10
Rotting or decomposition of organic matter is generally the result of the feeding action of fungi and or bacteria. (a) Diagram showing the feeding process of fungal mycelia. (b) Photograph of mycelia among particles of dead wood (100 × magnification). (Photo by author.)

organic material in the soil. Bacteria, which are microscopic, single-celled organisms, receive their nourishment in the same way.

Fungi and some bacteria produce **spores**, reproductive cells, in tremendous abundance, and their microscopic size allows them to be carried easily by air currents. Hence, they are present virtually everywhere in the environment. Therefore, their growth and hence the rotting or decay of organic matter occurs wherever suitable conditions of temperature and moisture prevail unless specific measures are taken to prevent it. Consider, for example, what happens to food that is not preserved.

Since fungi and bacteria are so distinctive, they are generally placed in a subcategory apart from other detritus feeders; they are called **decomposers**. However, in every ecosystem all detritus feeders and decomposers perform the same role. They feed on and, in the process, break down dead organic matter. Bacteria and fungi are primary detritus feeders. In turn they are fed upon by such organisms as protozoans, mites, insects, and worms living in the soil or water (Fig. 1–11). (Many of the fungi, such as button mushrooms, are also considered a great delicacy by people.) When a fungus or other decomposer dies, its body becomes part of the detritus and the source of energy and nutrients for yet other detritus feeders or decomposers.

In summary, despite the apparent diversity of ecosystems, they all have *structural similarity*. They can all be described in terms of photosynthetic plant producers, various categories of consumers, and detritus feeders and decomposers. Some organisms do not fit neatly into a single category but act in different roles at different times. Insect-eating plants are an interesting example. Nevertheless, the biotic structure can still be defined in these terms.

Feeding Relationships: Food Chains, Food Webs, and Trophic Levels

In describing the biotic structure of ecosystems it becomes evident that major interrelationships among organisms involve feeding relationships. We can identify innumerable pathways where one organism is eaten by a second which is eaten by a third and so on. Each such pathway is called a **food chain**. A few simple examples are illustrated in Figure 1–12.

While it is interesting to trace such pathways, it is important to recognize that food chains seldom exist as isolated entities. More commonly an herbivore population feeds on several different kinds of plants and, in turn, may be preyed upon by several different carnivores. Consequently, virtually all food chains are actually interconnected, forming a complex *web* of feeding relationships. Indeed, the term **food web** is used to denote this actual complex pattern of interconnected food chains.

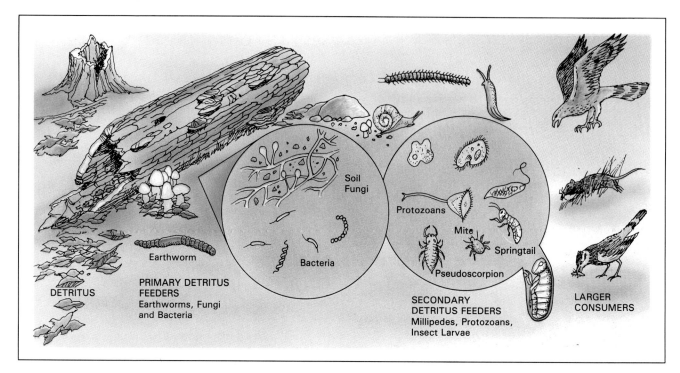

FIGURE 1–11
Detritus-based food web. Decomposers are not the end of the
line. Fungi and bacteria, which feed on detritus, support many
other organisms living in the soil and these, in turn, may be
fed upon by larger consumers.

FIGURE 1–12
Simple food chains. Food (nutrients and energy) is transferred
from one organism to another along pathways known as *food
chains*. However, food chains seldom exist as isolated entities
in nature. Instead, nearly all food chains are interconnected to
form a complex food web.

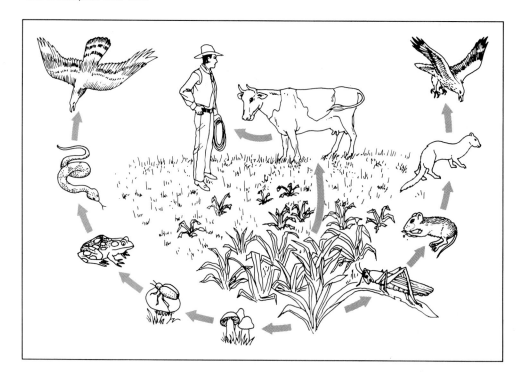

Despite the number of theoretical food chains and the complexity of food webs, there is a simple overall pattern. All food chains basically lead from producers to primary consumers (or primary detritus feeders) to secondary consumers or detritus feeders and so on.

These basic levels, producers, and various levels of consumers or detritus feeders are called **trophic levels**. *Trophic* literally means "feeding"; hence, trophic levels are feeding levels. All producers belong to the *first* trophic level. All primary consumers, whether feeding on living or dead producers, belong to the *second* trophic level, organisms feeding on these belong to the *third*, and so on. We can visualize all feeding relationships as a flow of nutrients and energy through a series of trophic levels. A diagrammatic comparison of a food chain, a food web, and trophic levels is shown in Figure 1–13.

How many trophic levels are there? There are usually not more than three or four discernable trophic levels in any ecosystem. This answer comes from straightforward observation. The total combined weight or *biomass* of organisms at each trophic level can be measured by collecting (or trapping) and weighing suitable samples. It is found that the biomass is in the order of 90–99 percent less at each higher tropic level. For example, if the biomass of

producers in a grassland is 10 tons (1 ton = 2000 lbs) per acre, the biomass of herbivores will be no more than 200 pounds and that of carnivores no more than 20 pounds. As you can see, you can't go through very many trophic levels before the biomass approaches zero. Depicting this graphically gives rise to what is commonly called a **biomass pyramid** (Fig. 1–14).

The reason for the declining biomass at higher trophic levels is largely because much of the food consumed is broken down for energy at each trophic level. Relatively little is converted into the body of the consumer. This and other factors involved will be discussed further in Chapter 2.

All food chains must start with producers. Without producers to constantly replenish the supply of organic matter, an ecosystem would soon eat itself into nonexistence.

Non-Feeding Relationships

Mutually Supportive Relationships. The overall structure of ecosystems is dominated by feeding relationships as we have seen above. In feeding relationships we generally think of one species benefiting and the other being harmed to a greater or lesser extent. However, there are many cases in which there is a mutual benefit to both species. This phenomenon is called **mutualism**. A classic example is seen in the group of plants known as lichens (Fig. 1–15). (Mosses should not be confused with lichens. Mosses have distinct stems with many minute, bright green leaves.

FIGURE 1–13
Three ways of representing the transfer of nutrients and energy. Each single pathway from the bottom to the top is a food chain. All interconnected food chains are the food web. The basic feeding levels, are trophic levels.

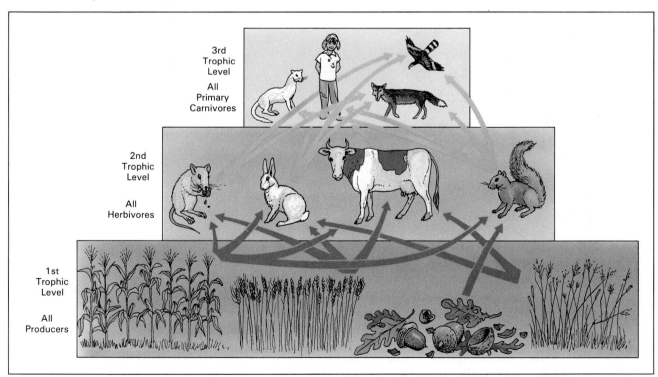

3rd
Trophic
Level

All
Primary
Carnivores

2nd
Trophic
Level

All
Herbivores

1st
Trophic
Level

All
Producers

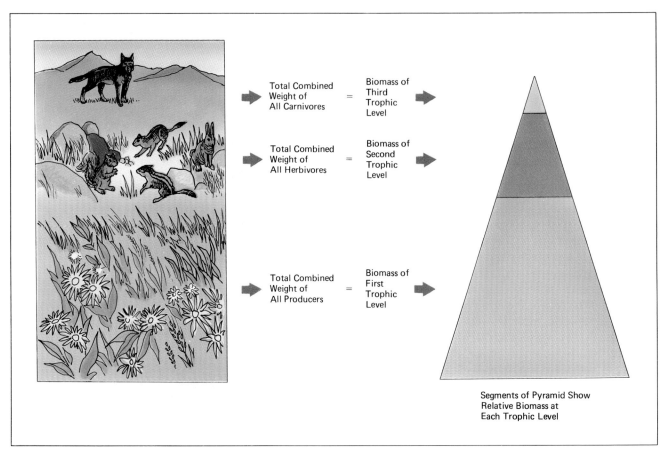

FIGURE 1–14
Biomass pyramid. A graphical representation of the biomass
at successive trophic levels has the form of a pyramid.

FIGURE 1–15
Lichens. The crusty-appearing
"plants" commonly seen
growing on rocks or the bark
of trees are actually composed
of a fungus and an alga
growing in a mutualistic
relationship. (USDA-Soil
Conservation Services.)

Lichens are generally grey-green in color and are usually scaly or crusty in appearance.) Lichens are actually comprised of two organisms, a fungus and an alga. The fungus provides protection for the alga, enabling it to survive in dry habitats where it could not live by itself; the alga, which is a producer, provides food for the fungus.

Another example is the relationship between flowers and insects. The insects benefit by obtaining nectar from the flowers; the plants benefit by being pollinated in the process. A third example is observed in coral reefs. Certain fish are immune to the predatory nature of the coral and hence are able to feed on detritus in and around the coral. The fish thus benefit by having access to a food source; the coral benefits by being cleaned.

Such relationships are also called *symbiotic* relationships. However, **symbiosis** (*sym* = together, and *bio* = living) refers to any intimate relationship between two organisms. Hence, symbiosis includes parasitism as well as mutualism.

Mutually supportive relationships, however, go far beyond the very close kinds of relationships illustrated above. For example, plant detritus provides most of the food for decomposers and soil-dwelling detritus feeders such as earthworms. Thus, these organisms benefit from plants, but the plants also benefit because the acitivity of these organisms is instrumental in releasing nutrients from the detritus and in maintaining soil quality, a subject that will be discussed further in Chapter 7.

In another example, many birds benefit from vegetation by finding nesting materials and places among trees. But the plant community also benefits because the birds feed on and reduce the populations of many herbivorous insects.

Even in predator-prey relationships some mutual advantage exists. The killing of individual prey, which are usually weak or diseased, may benefit the population at large by keeping it healthy and preventing it from becoming so abundant that it overgrazes the environment.

Competitive Relationships. Plants growing close together must compete for water, nutrients, light, and space itself. This competition between different plant types may have a significant effect in determining the character of an ecosystem, as will be explained further in the next section. However, animals in natural ecosystems are seldom, if ever, in direct competition except in the case of introduced species. Competition is minimized by the fact that animal species are adapted to feed on different things, in different locations, and/or at different times.

In concluding this discussion it should be clear that no organism lives as an entity unto itself. Every organism lives and can only live in relationships with other organisms in the context of an ecosystem. *Sustaining ecosystems is central to sustaining life itself.*

Abiotic Factors

As noted, organisms are one side of the ecosystem; the environment is the other. The chemical and physical factors of the environment are referred to as **abiotic** (*a* = not; *bio* = life) factors. They include light, temperature, water, wind, chemical nutrients, pH (acidity), salinity (saltiness), and fire. Of course, organisms are affected by all these factors simultaneously. The degree to which each is present or absent profoundly affects the ability of organisms to survive, but different species are affected differently. This has profound effects on the entire ecosystem. We shall come back to give examples of how this works shortly. First, we need to understand two fundamental principles regarding how organisms respond to abiotic factors.

Optimum, Zones of Stress, Limits of Tolerance

Different species thrive under different conditions. For example, some plants like it very wet; others like it relatively dry. Some like it very warm; others do best in cooler situations, and so on. Laboratory experiments bear this out even more clearly.

Plants are grown in a series of chambers in which all the abiotic factors can be controlled; thus, a single factor such as temperature can be varied in a systematic way while all other factors are kept constant. Note that this kind of controlled experiment is necessary to distinguish the effect of temperature from the effect of other factors. The results show that as temperature is raised from a low point, which fails to support growth, the plants grow increasingly well until they reach some maximum. Then, if temperature is raised still further, the plants become increasingly stressed; they do less well, suffer injury, and die. This is pictured graphically in Figure 1–16.

The point which supports the maximum growth is called the **optimum**. Actually, this may be a range of several degrees; thus there is an **optimal range**. The entire range from the minimum to the maximum which supports any growth is called the **range of tolerance**. Points at the high and low ends of the range of tolerance are called the **limits of tolerance**. Between the optimal range and the limits of tolerance there is increasing stress or **zones of stress**, until at the limits of tolerance survival is precluded.

Similar experimentation has been done to test other factors, and the results invariably follow the same general pattern. Of course, not every species has been tested for all factors; however, the consis-

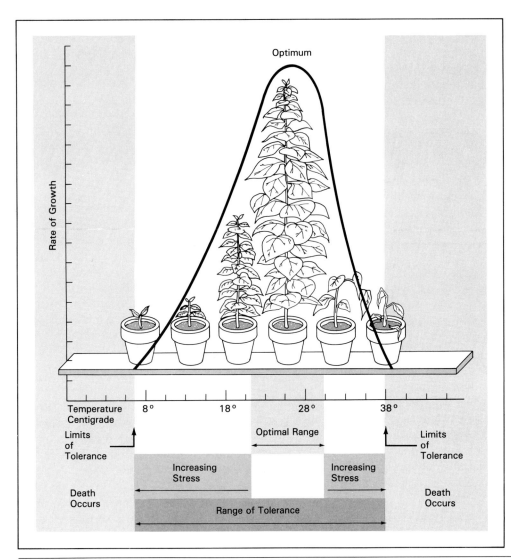

FIGURE 1-16
For every factor influencing growth, reproduction, and survival, there is an optimum, or best level. Above and below the optimum there is increasing stress until survival is precluded at the limits of tolerance. The total range between the high and low limits is a range of tolerance. Optimum limits of tolerance differ for different species.

tency of such observations leads us to conclude that this is a fundamental biological principle. *Each species (both plant and animal) has an optimum, zones of stress, and limits of tolerance with respect to every environmental factor.*

Additionally, this line of experimentation demonstrates that species may differ markedly with respect to the points or values at which the optimum and limits of tolerance occur. For instance, what may be the optimal amount of water for one species may stress a second and result in the death of a third. Some plants cannot tolerate any freezing temperatures (i.e., any exposure to 0°C [32°F] or less is fatal). Others can tolerate slight but not intense freezing, and some actually require several weeks of freezing temperatures in order to complete their life cycles. The same can be said for all other factors. While optimums and limits of tolerance may differ for different species, there may be considerable overlap in their ranges of tolerance.

Law of Limiting Factors

Another principle can be seen in the same observations and experimentation as described above. Note that we described testing *just one* factor at a time while all others were maintained within the optimal range. Thus, we observe what is now known as the **Law of Limiting Factors:** *Only one factor being outside the optimal range will cause stress and limit the organism.*

Such a factor is termed the **limiting factor.** Keep in mind that the limiting factor may be any one of the many different factors affecting growth, and it may be a problem of too much or too high as well as too little or too low. For example, plants may be stressed or killed by overwatering or overfertilizing as well as by underwatering or underfertilizing, a common pitfall for amateur gardeners.

The Law of Limiting Factors was first elucidated and expressed by Justus von Liebig in 1840 in connection with his observations regarding the effects of chemical nutrients on plant growth. He observed that restricting any one of the many different nutrients involved had the same effect. It limited growth. Thus, this Law is also called **Liebig's Law of Minimums**.

Of course further observations since Liebig's time have shown that the concept includes all factors which affect organisms. This includes biotic as well as abiotic factors. For example, the limiting factor may be competition from another species, or the presence of a predator or parasite. Indeed in agricultural production, such "pests" often are the limiting factor; consequently billions are spent on their control. Also, the whole concept applies to animals as well as to plants.

WHY DO DIFFERENT REGIONS SUPPORT DIFFERENT ECOSYSTEMS?

Armed with the understanding of limiting factors, we may now address questions such as: Why do different regions support different ecosystems? And what prevents one ecosystem from taking over another?

The basic reason is that abiotic conditions differ in different regions. Assuming a species has access to a region, its **population density** (number of individuals per unit area) will be greatest where all conditions are optimal. Its population density will taper off, but it will still be present where one or more factors are in the species' zone of stress. Finally, the species will not be present in any area where even one factor is outside its limit of tolerance (Fig. 1–17). We shall explore this in more detail.

Climate

Two abiotic factors, temperature and precipitation (rainfall and/or snowfall) are responsible for delimiting all the major terrestrial biomes cited earlier in this chapter. The long-term temperature and rainfall conditions for an area is what we call the **climate**. Climates in different parts of the world vary widely. Annual precipitation may vary from virtually zero to over 100 inches per year, and it may be evenly distributed (about the same number of inches each month) or it may all occur in certain months, a wet season, leaving the remainder of the year without rain, the dry season. Average temperature may range from subzero to close to 90°F (38°C) and may be fairly uniform throughout the year (regions near the equator), or there may be a seasonal fluctuation. Further, these temperature and rainfall conditions may be put together in almost any combination.

In turn, different climates support different biomes. The major climatic characteristics supporting the biomes discussed above are given in Boxes 1–7 (pp. 18–24). Water is the main factor responsible for

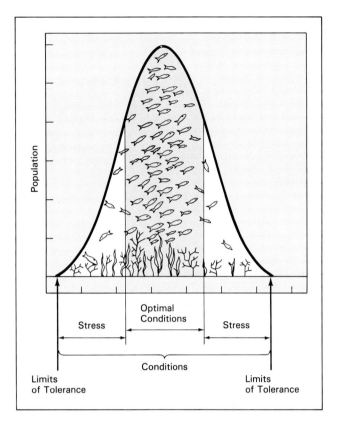

FIGURE 1–17
Individuals will be most abundant where conditions are optimal and less numerous as conditions are less favorable. Individuals will not be found beyond the limits of tolerance.

the separation of biomes into forests, grasslands, and deserts. This occurs as follows: Below about 40 in. (100 cm) of rain per year, many trees become stressed and most reach their limit of tolerance at about 30 in. (75 cm). Grasses, however, have a much lower limit of tolerance, about 10 in. (25 cm) and many species of cacti and other specialized desert plants do well with as little as 2 to 4 in. (5 to 10 cm) per year. Consequently, regions with more than 30 in. of rainfall per year typically support forests. Regions with 10 to 30 in. of rainfall typically support grasslands, and regions with less than 10 in. are only sparsely vegetated with species such as cacti, sagebrush, and tumbleweed, which are renowned for their drought tolerance. We recognize such areas as deserts. In intermediate ranges of rainfall, as one might expect, forests grade into grasslands and grasslands grade into deserts as the ranges of tolerance for the species involved overlap.

Temperature also plays a major role in demarcating major ecosystems. However, except for extreme cold, which gives rise to tundra or polar ice caps, the effect of temperature is largely superimposed on that of rainfall. That is, 30 inches or more of rainfall will support a forest, but temperature will determine the kind of forest. Spruces and firs cope best with the severe cold, heavy snows, and short growing seasons found in northern regions (Canada) and/or at high elevations. Deciduous trees, which also cope well with a freezing winter but require a longer summer growing season, predominate in temperate latitudes (central United States). Broadleaf evergreen tree species, which are extremely vigorous and fast growing but which cannot tolerate freezing temperatures, predominate in the tropics (equatorial latitudes). Likewise, any region receiving less than about 10 inches of rain per year will be a desert, but hot deserts have different species from those found in cold deserts.

Temperature also exerts some influence by its effect on the evaporation of water. Water evaporates faster at high temperatures. Consequently, the transitions from deserts to grasslands and from grasslands to forests are found at higher precipitation levels in hot regions and at lower precipitation levels in cold regions.

Finally, in the northern region of North America and Asia, the winters are so severe and summers are so short that only the snow and the top few inches of soil thaw. Below this the ground remains permanently frozen, a condition known as **permafrost**. Permafrost limits the northward extension of spruce-fir conifer forests because it precludes deep roots, but the small hardy plants which characterize the tundra still thrive. Of course, at still colder temperatures the tundra gives way to polar ice caps. These interactions between temperature and rainfall are diagrammatically illustrated in Figure 1–18.

FIGURE 1–18
Abiotic factors and major biomes. Moisture is generally the overriding factor determining the type of biome that may be supported in a region. Given adequate moisture, an area will generally support a forest. Temperature, however, determines the kind of forest. The situation is similar for grasslands and deserts. At cooler temperatures there is a shift toward less precipitation because lower temperatures reduce evaporative water loss. Temperature becomes the overriding factor only when it is severe enough to sustain permafrost.

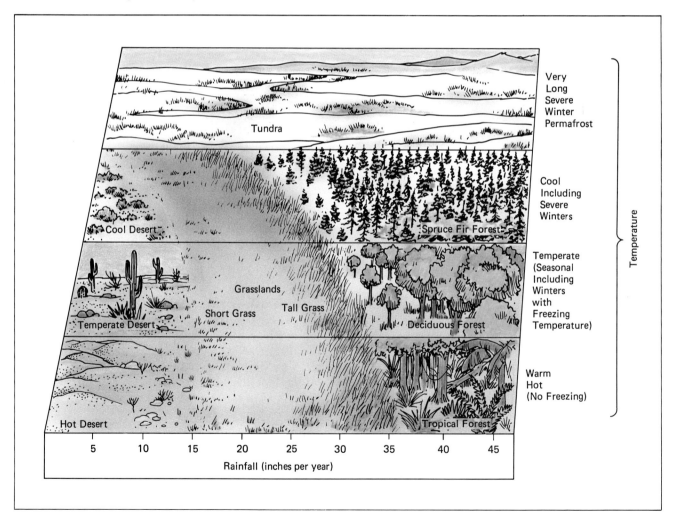

Other Abiotic Factors and Microclimate

A number of abiotic factors such as terrain, wind, and soil type act in large part by indirectly influencing temperature and/or moisture as is illustrated in Figure 1–19. Thus, a localized area may have temperature and moisture conditions that are significantly different from the overall climate for the area, which is necessarily an average. The conditions in a specific localized area are referred to as the **microclimate**. Different microclimates create diversity within the biome of the region. For example, within deciduous forests of eastern United States, the tree species on warmer, dryer, south-facing slopes differ from those on cooler, moister, north-facing slopes.

Soil type may also cause diversity through affecting the availability of moisture. In eastern United States, oaks and hickories generally predominate on rocky, well-drained soils, whereas beeches and maples are found on the richer soils which hold more moisture. In the transitional region between desert and grassland (10–20 inches of rainfall per year), a soil with good water-holding capacity will support grass, whereas a sandy soil with little ability to hold water will support only desert species.

In certain cases, an abiotic factor other than rainfall or temperature may be the primary limiting factor. For example, the strip of land adjacent to the coast frequently receives a salty spray from the ocean, a factor that relatively few plants can tolerate. Consequently this strip is frequently occupied by a unique community of salt-tolerant plants (Fig. 1–20). Relative acidity or alkalinity (pH) may also have a profound overriding effect. This is particularly significant in view of acid rain.

Biotic Factors

Limiting factors may also be biotic, that is, influence from other species. Grasses can obviously thrive under rainfall of more than 30 inches. However, when the rainfall is great enough to support trees, the grass gets shaded out. Thus, the factor that limits grasses from taking over high rainfall regions is biotic, overwhelming competition from taller species. Distribution of plants may also be limited by the presence of certain herbivores, particularly insect species and parasitic fungi.

The concept of limiting factors also applies to the distribution of animals. Again, the limiting factor

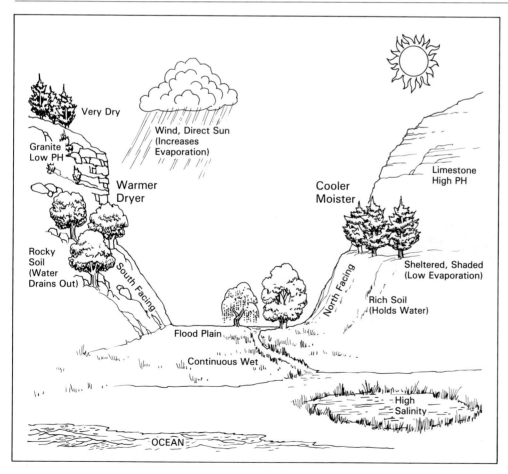

FIGURE 1–19
Abiotic factors such as terrain, wind, and soil type create different environments or microclimates by influencing temperature and moisture. Certain other factors such as pH and salinity exert a direct influence.

FIGURE 1–20
Specific abiotic factors may determine the type of vegetation.
Along the Pacific coast in central California, usual vegetation
is precluded by salt spray, but certain succulent species which
are salt-tolerant thrive. (Photos by author.)

may be biotic—lack of food, predators, parasites; or abiotic—lack of open water for drinking, locations for nesting, too cold, etc.

Physical Barriers

A final factor which may limit the spread of a species into another ecosystem is the existence of a physical barrier that the species is unable to cross such as an ocean, desert, or mountain range. Thus species making up the ecosystems on separate continents or remote islands are quite different despite similar abiotic factors. When such barriers are overcome, for example by humans transporting a species from one continent to another, the introduced species may make a successful "invasion." But a successful invasion by a foreign species usually means an ecological disaster because the "success" of the invader necessarily implies upsetting if not totally destroying the existing food web. Examples of this will be explored in Chapter 3.

Biotic and Abiotic Interactions

While we may be able to single out a particular factor which is limiting in a given situation, it should become clear that no factors act alone. The final result is always the result of numerous interactions between all the different abiotic and biotic factors. Sometimes the pathways between the initial cause and the final limit is long and devious indeed. For example, a prolonged hot, dry spell may cause birds to die, which allows insects, which are eaten by the birds, to flourish, which kills plants, which affects other consumers, and so on. Indeed, any factor which affects one species will in turn affect other species through the various interrelationships that exist in the ecosystem. Most significantly, factors which affect producers will invariably affect all other organisms through the food web.

You are probably beginning to feel overwhelmed by all the interactions that go on in ecosystems. If so, you have achieved a significant point: recognition that the complexity is infinite. The conclusion that should be clear is that *all ecosystems are maintained by a delicate interplay of limiting factors affecting all the species.* It should also be clear that it is impossible to alter any factor without, to some degree, affecting these limits. Consequently: *Altering any factor, abiotic or biotic, will invariably affect limits and set into motion a chain reaction with far-reaching consequences.*

IMPLICATIONS FOR HUMANS

Primitive humans survived in small tribes as **hunter-gatherers**, which means that they "lived off the land" catching wildlife and gathering seeds, nuts, roots, berries, and other plant foods as they could find. Settlements were never large and were of relatively short duration because as one area was "picked over" the tribe was forced to move on. As hunter-gatherers, humans were much like other consumers (omnivores) in natural ecosystems.

About 10,000 years ago, however, a very significant change occurred: humans began to develop agriculture. Agriculture is essentially a process of taking particular species out of the wild, clearing space, and providing other conditions to grow them preferentially. Plants are protected from competition (weeds) and other would-be consumers and additional nutrients (fertilizer) and water may be provided. Similarly, animals are protected from predators and given food for optimal growth. Over the years many agricultural plants and animals have been greatly modified through selective breeding so that now they are quite different from their wild ancestors, but the basic process remains unchanged.

With the development of agriculture, humans began to create their own distinctive *human ecosystem* apart from natural ecosystems. This was a major turning point in human history. For the first time it enabled human food supply to be stabilized and more or less assured. It also allowed specialization of labor; some could be farmers, while others were freed to pursue other endeavors. With the assurance of food supply and specialization of labor, permanent living places (villages and later cities) could be formed. People could indulge in the production of additional material goods and other pursuits that gradually have led to the complex society we have today.

In terms of vigor (ability to grow, multiply, and spread), our human ecosystem has been exceptional. It has enabled our population to grow at least 10,000-fold from a few hundred thousand to over 5 billion, and move into all parts of the globe. It is not that the principle of limiting factors, which limit natural ecosystems, does not apply to our human ecosystem. It is that our capacity to reason and to manipulate the physical world has enabled us to overcome the usual limiting factors, at least for the time being. Consider that the most prevalent limiting factors affecting wild organisms are food, water, predators/parasites, suitable habitat, and competition from other species. We overcome these factors by:

1. Producing abundant food, although distribution may still be problematic.

2. Creating reservoirs and distributing water both for ourselves and our crops (irrigation).

3. Overcoming our predators and most disease-causing organisms.

4. Constructing our own habitats and heating/cooling them as desired.

5. Overcoming competition of other species.

Thus, without the imposition of natural limits our human ecosystem has grown and spread and the process is still going on as we continue to cut, burn, plow up, and bulldoze natural ecosystems to make more and more room for our own. Further impacts on natural ecosystems are occurring as a result of pollution and of exploiting them for commercially valuable items such as furs and hardwoods. No ecosystem on earth has escaped some degree of human intrusion, and many smaller ecosystems have been totally destroyed. Even some major biomes, grasslands for example, have been all but obliterated, and tropical forests are rapidly suffering the same fate (an area of roughly 200 mi by 200 mi (40,000 mi^2 or 10,000 km^2) is currently being destroyed each year).

I am not suggesting that a certain amount of exploitation of the natural environment is not necessary. To be sure, we owe virtually all the creations and creativity of civilization to supporting ourselves with the human ecosystem; without it we would still be primitive humans living in caves subject to the whims of nature. At the same time, however, we need to look at where the process is leading and determine if that is where we want to go, or if we want to change directions at this juncture. Recall the principles that only *one factor being outside the optimal range will cause stress and limit an organism and that altering any factor, biotic or abiotic, will set into motion a chain reaction with far-reaching consequences.* Our current trends are toward changing conditions on a global scale through such factors as acid rain, depletion of the ozone shield, and warming of the climate due to carbon dioxide emissions from the burning of fuels. We are changing conditions even more dramatically in large areas through various more intense forms of pollution and the diversion of water. Total destruction of ecosystems for human development is self-evident. Additional damage to natural ecosystems is being caused by direct exploitation of various plant and animal species for short-term commercial gain, and the accidental or intentional introduction of species from one ecosystem into another.

These trends are rapidly closing in on (and in some cases have already arrived at) severe alterations, if not total annihilation, of natural ecosystems with the extinction of their species. The survival of many thousands, perhaps millions, of plant and animal species is threatened by current trends. Do we want to look forward in the next 50 years to a world in which most rare plants, animals, and ecosystems are reduced to museum displays and photographs of extinct species, or do we want to change directions?

It is not just the aesthetics of natural species and ecosystems that we need to be concerned about. How far can we push changes before we find ourselves irrevocably trapped in the consequences? Our ability to control our personal environment enables us to protect ourselves against the direct effects of climatic changes, for example, but this is not true for our agricultural crops. Air pollution, climatic change, destruction of ozone, acid rain, and altering certain biotic factors may well unhinge our ability to produce food.

The human ecosystem is in a state of rapid prolific growth, yet we have not changed nor can we avoid the basic principle of limiting factors. To a large extent we are supporting our human ecosystem through the exploitation of water, soil, and energy resources. As limits are reached social conflicts arise, such as wars and famine, until eventually the whole structure may collapse into a desolate world. Is this where we want to go, headlong to the limits and then let the pieces fall where they may, or do we want to change directions?

Fortunately there is an alternative path: it is toward sustainable development. A major feature of sustainable development must be to accept limits of our own choice short of the natural limits which may be caused by resource depletion or ecosystem upset. The limits we must accept are not limits of technology, creativity, or compassion. They are limits of numbers and our impact on natural ecosystems. Efforts in this direction are already underway. This is what family planning, preservation of natural areas, land-use restrictions, conservation, and pollution control are all about. These activities, and how you can support and participate in them, will be described in more detail in later parts of this book.

Other important aspects of sustainable development involve the use of materials and energy. We turn our attention to these in Chapter Two.

2
Ecosystems:
How They Work

CONCEPT FRAMEWORK

■ Outline

I. ELEMENTS, LIFE, ORGANIZATION, AND ENERGY
45

A. Organization of Elements in Living and
Nonliving Systems   45

B. Considerations of Energy   48
   1. Matter and Energy   50

   2. Energy and Organic Matter   53

C. Matter and Energy Changes in Organisms   54
   1. Producers   54

   2. Consumers   55
      a. Energy Role of Food   55

      b. Nutritive Role of Food   56

      c. Material Consumed but not Digested   56

   3. Detritus Feeders and Decomposers   56

II. PRINCIPLES OF ECOSYSTEM FUNCTION   57
   A. Nutrient Cycling   57

   1. The Carbon Cycle   58

   2. The Phosphorous Cycle   58

■ Study Questions

1. Define: *atom, molecule.* List six elements that are especially significant to living things and point out where they are also found in the environment. Contrast molecules making up air, water, and minerals with molecules of living systems. Give the chemical distinction between organic and inorganic molecules. Contrast growth and decay (or burning) in terms of what is happening to organic molecules; to individual atoms. State the *Law of Conservation of Matter.*

2. Define and distinguish between *matter* and *energy*; between *kinetic* and *potential energy.* Cite how forms of energy may be interconverted. Define a unit for measuring energy. As energy is converted, what happens to the total amount? The amount of useful energy? Explain. Relate this to the *Laws of Thermodynamics.* What must occur to move a system toward higher potential energy?

3. Relate changes between organic and inorganic states to inputs and outputs of kinetic and potential energy.

4. Cite the movement and changes of matter and energy that occur in photosynthesis and growth of green plants.

5. How does food provide energy. Name the process involved. What are the wastes released?

6. Describe the nutritive role of food. What compounds must be present for food to have nutritive value? Define: *"junk foods," malnutrition, starvation.*

7. Define and relate: fiber, cellulose, fecal wastes, material that is consumed but not digested.

8. Describe the movement and changes of energy and matter that occur as decomposers feed on detritus.

9. State the *first principle of ecosystem functioning.* Describe how it is in harmony with the *Law of Conservation of Matter.*

10. Trace possible pathways of carbon atoms through various rounds of the *carbon cycle.*

11. Do the same for *phosphorus* atoms.

3. The Nitrogen Cycle		61
B. Energy Flow		61
C. Flow of Energy and Decreasing Biomass at Higher Trophic Levels		62
III. IMPLICATIONS FOR HUMANS		64
CASE STUDY: BIOSPHERE II		67

12. Do the same for *nitrogen* atoms. What must occur before nitrogen in the air can be used by plants? Contrast natural and human means of fixing nitrogen.

13. Contrast the carbon, phosphorus, and nitrogen cycles in terms of *gas* and *mineral phases*. What are the implications in terms of recycling mineral nutrients such as phosphate?

14. State the *second basic principle of ecosystem function*. State three attributes of *solar energy*.

15. State three reasons why the biomass at higher trophic levels must be less than that at lower trophic levels. State the *third principle of ecosystem function*.

16. Describe how trends in human civilization are counter to each of the ecological principles noted above. What are the actual or potential consequences of this? Policy decisions should focus on what?

$\boxed{\text{I}}$n Chapter 1, we learned how ecosystems are groupings of plants, animals, and microbes interacting with each other and their environment. In this chapter, we will focus more on the principles of how natural ecosystems are able to perpetuate themselves indefinitely without suffering from resource depletion or pollution from their own wastes. By the same token we will see how many of our environmental problems stem from our attempt to operate our human ecosystem at variance with these principles.

ELEMENTS, LIFE, ORGANIZATION, AND ENERGY

Understanding how ecosystems perpetuate themselves demands comprehension of some basic chemical principles regarding atoms, elements, and energy and how they interact to form more complex arrangements. A discussion of atoms of different elements and how they bond to form molecules and compounds of various gases, liquids, and solids is given in Appendix C. You may find it helpful to study this appendix first to gain background necessary for understanding this material.

Organization of Elements in Living and Nonliving Systems

Many early biologists and chemists studied living things expecting to find some particular substance or "vital essence" that was responsible for imparting "life" to organisms. No such substance and no evidence that such a substance or element exists has ever been found. Scientists find the same **elements** or kinds of atoms in living things as are present in air, water, and rock and soil minerals. Furthermore, of the 96 elements (kinds of atoms) that occur in nature only about 20 are found in living organisms. These elements and where they occur in the environment are given in Table 2–1. The most significant of these elements and their chemical symbols are: *carbon (C), hydrogen (H), oxygen (O), nitrogen (N), phosphorus (P),* and *sulfur (S)*. You can remember them by the acronym N. CHOPS.

There is, however, a chemical feature that does distinguish between living and nonliving things. It is in the complexity of *molecules*. A **molecule** is defined as a chemical unit of two or more atoms bonded together, and it is the smallest unit of any given compound which still has all the qualities of that compound. In air, water, and minerals, molecules are relatively simple. Clean dry **air** is a simple *mixture* of molecules of three important gases: oxygen (O_2), nitrogen (N_2), and carbon dioxide (CO_2) as shown in Figure 2–1. A few other gases, which have no biological importance, are also present. **Water** (H_2O) is composed of molecules each formed by two hydrogen atoms bonding to an oxygen atom as shown in Figure 2–2. Rock and soil minerals are made up of dense clusters of two or more kinds of atoms bonded together by the attraction between positive and negative charges on the atoms (Fig. 2–3).

There are significant interactions between air,

Table 2–1

Elements Found in Living Organisms and Their Biologically Important Locations in the Environment

ELEMENT (KIND OF ATOM)	SYMBOL	BIOLOGICALLY IMPORTANT MOLECULE OR ION IN WHICH THE ELEMENT OCCURS[a]		LOCATION IN THE ENVIRONMENT		
		NAME	FORMULA	AIR	DISSOLVED IN WATER	SOME ROCK AND SOIL MINERALS
Carbon	C	Carbon Dioxide	CO_2	X	X	
Hydrogen	H	Water	H_2O		(Water Itself)	
Oxygen (required in respiration)	O	Oxygen Gas	O_2	X	X	
Oxygen (released in photosynthesis)	O_2	Water	H_2O		(Water Itself)	
Nitrogen	N	Nitrogen Gas	N_2	X	X	
		Ammonium Ion	NH_4^+		X	X
		Nitrate Ion	NO_3^-		X	X
Sulfur	S	Sulfate Ion	SO_4^{2-}		X	X
Phosphorus	P	Phosphate Ion	PO_4^{3-}		X	X
Potassium	K	Potassium Ion	K^+		X	X
Calcium	Ca	Calcium Ion	Ca^{2-}		X	X
Magnesium	Mg	Magnesium Ion	Mg^{2+}		X	X
Trace Elements[b]						
Iron	Fe	Iron Ion	Fe^{2+}, Fe^{3+}		X	X
Manganese	Mn	Manganese Ion	Mn^{2+}		X	X
Boron	Bo	Boron Ion	Bo^{2+}		X	X
Zinc	Zn	Zinc Ion	Zn^{2+}		X	X
Copper	Cu	Copper Ion	Cu^{2+}		X	X
Molybdenum	Mo	Molybdenum Ion	Mo^{2+}		X	X
Chlorine	Cl	Chloride Ion	Cl		X	X

NOTE: These elements are found in *all* living organisms: plants, animals, and microbes. Some organisms require certain elements in addition to these. For example humans additionally require sodium and iodine.
[a] A molecule is a chemical unit of two or more atoms bonded together. An ion is a single atom or group of bonded atoms that has acquired a positive or negative charge as indicated.
[b] Elements of which only small or trace amounts are required.

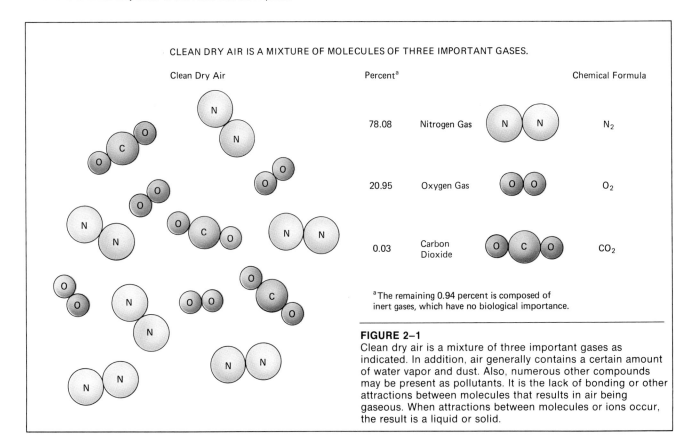

CLEAN DRY AIR IS A MIXTURE OF MOLECULES OF THREE IMPORTANT GASES.

Clean Dry Air

Percent[a]		Chemical Formula
78.08	Nitrogen Gas	N_2
20.95	Oxygen Gas	O_2
0.03	Carbon Dioxide	CO_2

[a] The remaining 0.94 percent is composed of inert gases, which have no biological importance.

FIGURE 2–1
Clean dry air is a mixture of three important gases as indicated. In addition, air generally contains a certain amount of water vapor and dust. Also, numerous other compounds may be present as pollutants. It is the lack of bonding or other attractions between molecules that results in air being gaseous. When attractions between molecules or ions occur, the result is a liquid or solid.

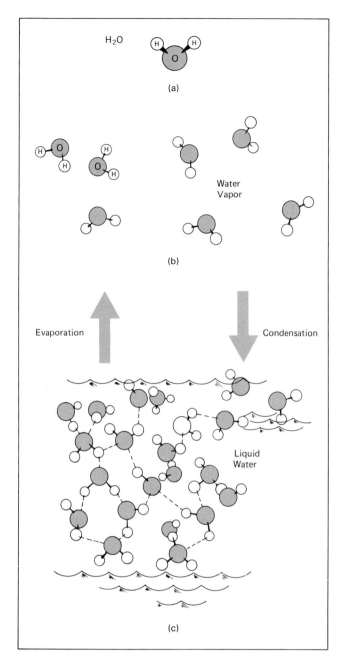

FIGURE 2-2
(a) Water consists of molecules, each of which is formed by two hydrogen atoms bonded to an oxygen atom, H_2O. (b) In water vapor molecules are separate and independent. (c) In liquid water a weak attraction between water molecules (dotted lines) gives the liquid property.

water, and minerals: molecules from the air may dissolve in water, water molecules may enter the air as water vapor, and various minerals may dissolve into and recrystallize from water solution (Fig. 2–4). Nonetheless, the molecular combinations remain relatively simple.

In contrast, the chemical structure of living things is based on complex molecules such as proteins, carbohydrates, fats, lipids, sugars, and nucleic acids (such as DNA, the genetic material). These and other molecules that comprise the tissues of living things are constructed in large part from carbon atoms bonded together into chains with hydrogen atoms attached. Certain other elements, namely oxygen, nitrogen, phosphorus, and sulfur may be involved also, but the key common denominator is carbon-carbon and/or carbon-hydrogen bonds. The complexity of such molecules is fantastic—some may contain millions of individual atoms—and their potential diversity is infinite. Indeed, the diversity of living things is a function of the diversity of such molecules.

These carbon-based molecules, which comprise the tissues of living organisms, are called **organic molecules**. (Don't miss the similarity between the words *organic* and *organism*.) Since the common denominator of all these molecules is carbon-carbon and/or carbon-hydrogen bonds, the term **organic molecule** (or **compound**) has come to mean *any* molecule with carbon-carbon and/or carbon-hydrogen bonds. Conversely, all compounds without such bonds are termed **inorganic**. Causing some confusion is the fact that all plastics, and countless other human-

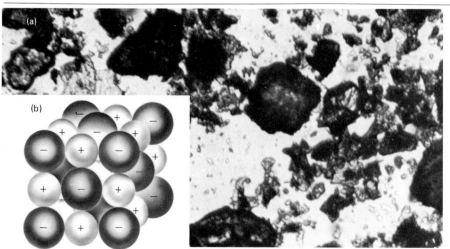

FIGURE 2-3
Rock and soil minerals. (a) Mineral particles of soil enlarged × 100. Minerals, as seen in rock or inorganic soil particles, are composed of dense clusters of atoms of two or more elements. The atoms (or small groups of atoms) have acquired a positive or negative charge and are bonded together by the attraction between opposite charges as shown in the insert. (b) Atoms or small groups of atoms with a charge are called *ions*. Various minerals are formed by different combinations of elements, some of which form positive (+) ions, others of which form negative (−) ions. (Photo by author.)

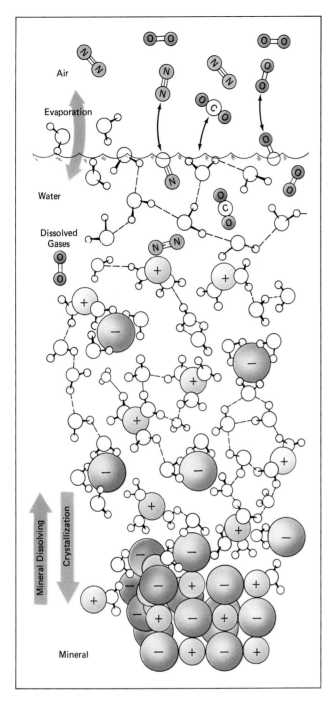

FIGURE 2–4
Interrelationships between air, water, and minerals. Minerals and gases dissolve in water, forming solutions. Water evaporates into air, causing humidity. These processes are all reversible. As water evaporates, minerals in solution recrystallize. Water vapor in the air condenses to re-form water.

made compounds are based on carbon-carbon bonding and are chemically speaking, organic compounds—although they have nothing to do with living systems. We resolve this confusion in places where there is doubt by referring to the compounds

of living organisms as **natural organic compounds** whereas the human-made ones are **synthetic organic compounds**. In any case, *organic* compounds do not exist in air, water, or minerals to any appreciable extent except where they have been introduced as a result of the activities of living organisms; this includes human pollution as well as natural processes.

In conclusion, what we see in contrasting living and nonliving things is not different kinds of atoms—the kinds of atoms (N, C, H, O, P, S) are the same—but a profound difference in the way in which the atoms are organized or bonded into various molecules or compounds. In air, water, and minerals, atoms are found in relatively simple *inorganic* compounds. In living organisms, atoms are organized into complex *organic* compounds. In turn these organic compounds make up the various parts of cells which make up the tissues and organs of the body (Fig. 2–5).

On a chemical level, then, the fundamental life processes of growth and reproduction of producers may be seen as their taking carbon, hydrogen, oxygen, and other elements from simple inorganic compounds present in the environment and rebonding them into the complex arrangements of organic molecules which make up the tissues of their bodies. Consumers do the same sort of chemical rearranging starting with molecules present in food. Conversely, the processes of decay or burning involve a breakdown of the complex organic molecules and the rearrangement of the atoms into the simple inorganic molecules. For example, as sugar is burned its carbon, hydrogen, and oxygen atoms re-form into carbon dioxide and water (Fig. 2–6). Since additional oxygen generally enters into the process it is also referred to as **oxidation**.

Did you note that we have described both growth and decay or burning in terms of simply rearranging atoms, not altering them in any other way? Indeed, this is what is observed in all chemical reactions. It is one of those observations that is so invariable that it is interpreted as a natural law, the **Law of Conservation of Matter**. Stated specifically it says that: *In chemical reactions atoms are never created, destroyed, or changed one into another; they are only rearranged to form different molecules and compounds.* (In high-energy nuclear reactions, atoms may be changed and this forms the basis of nuclear power discussed in Chapter 20. However, in chemical and biological reactions such reactions do not occur, and so with this proviso, the Law of Conservation of Matter holds.)

Considerations of Energy

More is involved than just the rearrangement of atoms, however. Chemical reactions also involve the

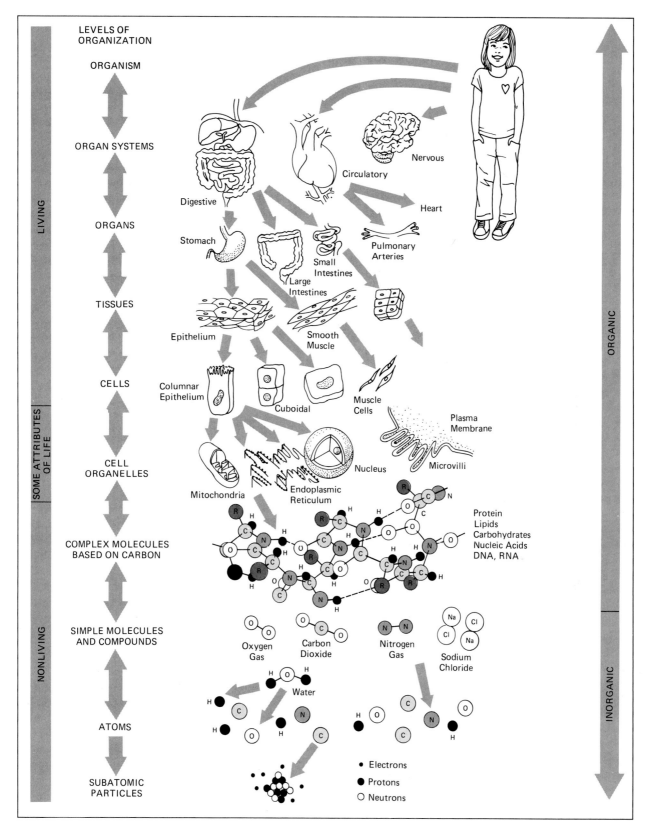

FIGURE 2–5
Life can be seen as a hierarchy of organization of matter. In the inorganic sphere, elements are in very simple arrangements of molecules of the air, water, and minerals. In living organisms, they are arranged in very complex organic molecules which, in turn, make up cells which comprise the whole organism.

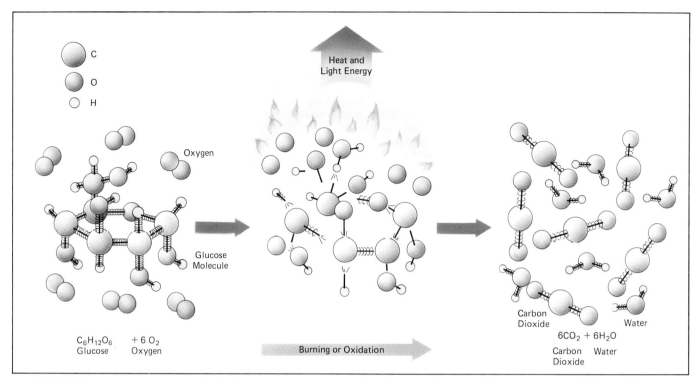

C

O

H

Heat and
Light Energy

Oxygen

Glucose
Molecule

$C_6H_{12}O_6$ + 6 O_2
Glucose   Oxygen

Burning or Oxidation

Carbon
Dioxide   Water

$6CO_2 + 6H_2O$
Carbon   Water
Dioxide

FIGURE 2–6

Burning or oxidation of glucose. Burning or oxidation of complex organic molecules involves their breakdown and the rearrangement of their atoms into simple molecules. Generally oxygen is used in the process, and carbon dioxide and water are the end products. Note that no atoms are created or destroyed; they are only rearranged. Also extremely important in this reaction is the release of energy, potential energy contained within the glucose molecule. All heterotrophs or consumers, including humans, derive their body energy from the oxidation of organic molecules, such as glucose, through a process called *cell respiration*.

absorption or release of *energy*. To grasp this concept we must first be clear concerning the distinction between matter and energy.

Matter and Energy

Matter is defined as anything that occupies space and has mass, that is, can be weighed when gravity is present. This includes all solids, liquids, and gases. The particles that make up atoms, the subatomic particles known as protons, neutrons, and electrons, and thus the atoms themselves, are the basic units of matter. In contrast, common forms of **energy** include *light* and other forms of *radiation, heat, movement*, and *electrical power*. Note that these do not occupy space, nor do they have mass. If you wish to argue this point, just put some light in a container, take it into a dark

room, and weigh it for me. But light, heat, and other forms of energy are real. They *do* things; they *affect* matter causing changes in its *position* or *state*. A release of energy in an explosion causes things to go flying, a change in position. Heating water causes it to boil and change to steam, a change in state. Heating an object makes its molecules vibrate more rapidly, increasing molecular *movement*. Thus, energy is defined in terms of what it does. Physicists define the "doing" as work. Thus, *energy* is defined as the *ability to do work*.

Energy is commonly divided into two categories: *kinetic* and *potential* (Fig. 2–7). **Kinetic energy** is energy in action or motion. Light, heat, motion and electrical current are all forms of kinetic energy. On the other hand, **potential energy** is energy in storage. A substance or system with potential energy has the capacity or *potential* to release one or more forms of kinetic energy. A stretched rubber band, for example, has potential energy; it can send a paper clip flying. Numerous chemicals such as gasoline and other fuels release kinetic energy—heat, light, and movement— when ignited. The potential energy contained in such chemicals or fuels is called **chemical energy**.

There are innumerable ways in which energy may be changed from one form to another. How many examples can you think of in addition to those

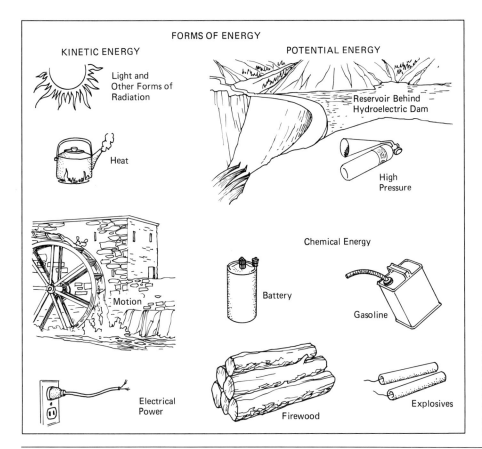

FORMS OF ENERGY

KINETIC ENERGY

Light and Other Forms of Radiation

Heat

Motion

Electrical Power

POTENTIAL ENERGY

Reservoir Behind – Hydroelectric Dam

High Pressure

Chemical Energy

Battery

Gasoline

Firewood

Explosives

FIGURE 2–7
Energy. Energy is distinct from matter in that it neither has mass nor occupies space, but it has the ability to act on matter, changing its position and/or its state. Kinetic energy is energy in one of its active forms. Potential energy refers to systems or materials which may release kinetic energy.

listed in Figure 2–8? Especially important is the fact that kinetic energy may be converted to potential energy, as in charging a battery, as well as the other way around.

Since energy does not have mass or occupy space it cannot be measured in units of weight or volume, but it can be measured in other kinds of units. One of the most commonly used units is the **calorie**, which is an amount of heat, namely the *amount of heat required to raise the temperature of 1 gram (approximately 0.03 oz) of water 1 degree Celsius (1.8 degrees Fahrenheit)*. But, since this is a very small unit, it is frequently more convenient to speak in terms of kilocalories (1 kilocalorie = 1000 calories), the amount of heat required to raise one liter (1.057 qt) of water 1 degree Celsius. Food calories (often capitalized as Calories to distinguish them from "small" calories) are actually kilocalories. Any form of energy can be measured in calories by suitably converting it to heat and measuring that heat in terms of a rise in temperature of water.

Countless experiments involving the measurement of energy as it is converted from one form to another yield an invariable result. In changing energy from one form to another the total amount remains the same. This observation is now interpreted as another natural law, the **First Law of Thermodynamics**. Simply stated: *Energy is neither created nor destroyed, but it may be converted from one form to another*. (Again high-energy nuclear reactions which involve mass-energy conversions need not be considered here because such reactions do not occur in the course of usual chemical, biological, or physical processes.)

Because energy is not destroyed, and because it may be converted from one form to another, it might seem that we should be able to design a system that would run continuously on recycled energy and not need additional energy inputs. Many attempts at this have been made with one result common to all—they don't work. They fail because of two additional observations. First, in every energy conversion at least part of the energy is converted to heat, and second, heat always flows toward cooler surroundings (Fig. 2–9). Hot food, for example, always cools off losing its heat to the cooler room, never the other way around. Thus, whenever energy is converted from one form to another some of it is always lost—the portion that is converted to heat. The result is now accepted as another natural law, the **Second Law of Thermodynamics:** *In any energy conversion, you will end*

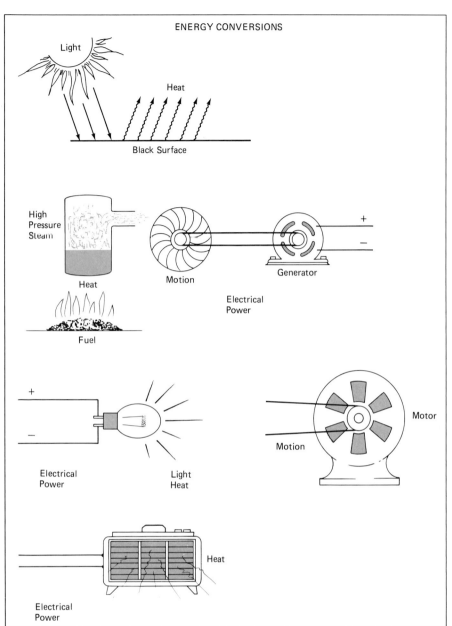

ENERGY CONVERSIONS

FIGURE 2–8
The first law of thermodynamics. Energy is neither created nor destroyed but any form can be converted into any other.

Why Doesn't It Work?

FIGURE 2–9
The second law of thermodynamics. Energy cannot be recycled because every conversion involves a loss of heat which cannot be recovered because heat only flows toward cooler surroundings. Consequently, the power output of a generator, for example, will always be less than the power input because of the loss of energy as heat.

up with less useable energy than you start with because of the inevitable loss of heat that occurs. Often the loss exceeds 50 percent. Said another way, most energy conversions are less than 50 percent efficient. Many are in the range of only 1 to 10 percent efficient, the other 90 to 99 percent being lost as waste heat in the process. For example, when burning coal to produce electricity, only 30–40 calories of electricity are produced for every 100 calories of coal burned.

Therefore, to keep any system running requires a continuous input of new energy. Perpetual motion machines, systems that will run without inputs of energy, exist only in the imagination. Any operating machine or system left without inputs of energy will gradually run out of energy and come to a stop.

The Laws of Thermodynamics do not prevent a certain component from gaining potential energy, a battery being charged for example. This simply represents a conversion of kinetic energy, electrical power in this case, to potential chemical energy. The conclusion which is demanded, however, is that the amount of kinetic energy *used* in the process will be *greater* than the amount of potential energy stored because of the loss which will occur. One will only get out of the battery about 90 percent of what is expended in charging it.

Energy and Organic Matter

Now we are going to relate these concepts of matter and energy to organic molecules, organisms, and ecosystems. All organic molecules, which make up the tissues of living organisms (wood, fat, sugars, starches, and so on), contain more than the atoms of carbon, hydrogen, and other elements that comprise them. They also contain *potential energy.* This is evident by the simple fact that they burn. The heat and light of the flame is their potential energy being released as kinetic energy. On the other hand, try as you might, you will not be able to get energy out of inorganic molecules such as carbon dioxide, water, and minerals. Indeed these materials are used as fire extinguishers. This is evidence that such materials are essentially void of potential energy.

Thus, the production of organic molecules from inorganic raw materials represents a *gain* in potential energy. This will not occur without suitable inputs of kinetic energy. Conversely, whenever organic molecules are broken down, their potential energy will be released as some form of kinetic energy, which may be used for other functions (Fig. 2–10). However, because of the inevitable loss of heat that occurs with each conversion, the amount of useable energy available at the end will always be less than the amount at the beginning. Let us relate this specifically to the categories of organisms in an ecosystem.

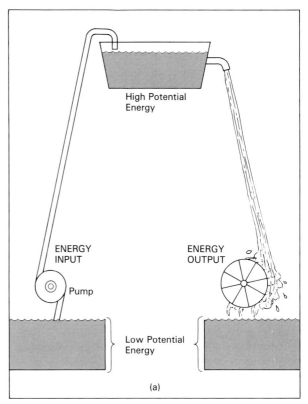

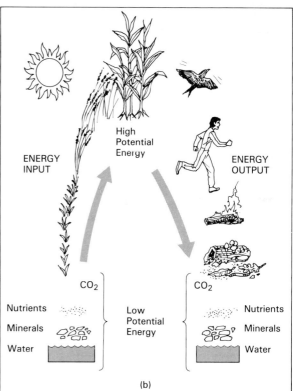

FIGURE 2–10
(a) With energy from the pump, water has become a source of high potential energy; it can be used as a source of energy to turn a mill and produce work. (b) Likewise, nutrients, minerals, and water have low potential energy. Plants use the sun's energy by way of photosynthesis to convert these into the organic molecules of their bodies, which have high potential energy. As the plant is consumed, burns, or decays, this energy is released, leaving only the low potential energy components—nutrients, minerals, and water.

Matter and Energy Changes in Organisms

Producers

As we noted in Chapter 1, the producers in every major ecosystem are plants that carry on photosynthesis. In photosynthesis, plants take the carbon and some oxygen atoms from carbon dioxide molecules, which they absorb from the air (or water solution in the case of aquatic plants), and take hydrogen atoms from water molecules. They use these carbon, oxygen, and hydrogen atoms to construct molecules of the simple sugar called glucose. As you can see in Figure 2–11, a molecule of glucose is constructed from 6 carbon atoms, 12 hydrogen atoms, and 6 oxygen

atoms, hence its formula, $C_6H_{12}O_6$. The construction of one molecule of glucose requires 6 molecules of carbon dioxide (CO_2) and 6 molecules of water to provide the 6 carbon atoms and the 12 hydrogen atoms respectively. Among these molecules of carbon dioxide and water, there are 18 oxygen atoms, whereas only 6 are needed. The extra oxygen atoms are given off as molecules of oxygen gas (O_2), 6 for every molecule of glucose formed. Note our accounting for each atom. What natural law demands that we do this? (See page 51).

Now the energy. Glucose is an organic molecule with high potential energy, whereas the potential energy in carbon dioxide and water is nil. Can we get energy from nowhere? Not according to what natural law? Do you see now, the requirement for light in photosynthesis? Light is kinetic energy. Photosynthesis involves its capture, conversion, and storage as potential energy in glucose molecules. We have noted that the light energy is initially captured by being absorbed by the green pigment chlorophyll.

Producers do not violate the second law of thermodynamics either. In their production of high-po-

FIGURE 2–11
Producers. Producers are remarkable chemical factories. Using light energy, they make glucose from carbon dioxide and water, releasing oxygen as a byproduct. Oxidizing some of the glucose to provide additional chemical energy, they convert the remaining glucose with certain additional nutrients from the soil into other complex organic molecules used in growth.

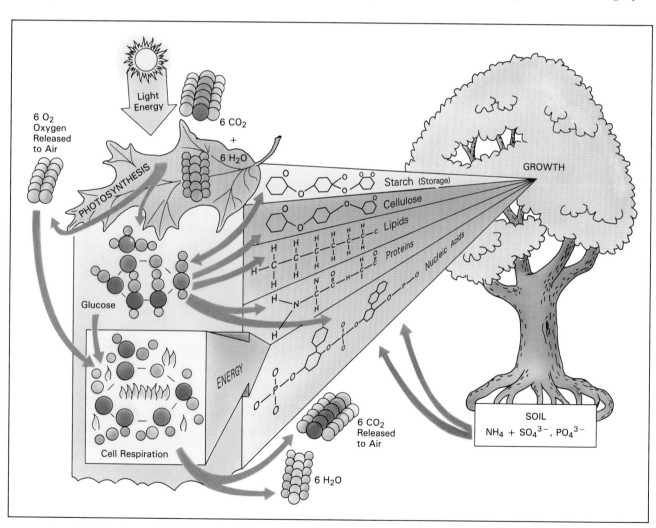

tential-energy organic compounds one finds that only 1 to 5 calories worth of organic chemicals are produced for every 100 calories worth of light energy falling on the plant. Thus, they are really only 1 to 5 percent efficient in converting light energy to stored chemical energy.

The glucose produced in photosynthesis serves two major functions in the plant. First, it serves as a raw material for growth. Glucose provides the carbon-hydrogen portion of all the organic molecules that make up root, stem, leaf, and other tissues of the plant. Additional elements such as nitrogen, phosphorus, potassium, and calcium are incorporated in the processes. Assuming they are present—plants won't grow if they are not—these other elements are absorbed from the soil (or water solution) as they are required.

Second, glucose is used to provide energy for all plant functions. Just as it takes energy, in the form of labor, to build a house, the plant requires energy in order to grow (build plant tissue). Plants also use energy to absorb nutrients (nitrogen, phosphorus, etc.), and for other plant functions. These functions occur 24 hours a day. The plant obtains energy for these functions from the potential energy stored in the glucose molecules. In other words, only a portion of the glucose that plants manufacture during photosynthesis is used as the raw material for growth; another portion is broken down again to release energy required for plant functions. The breakdown occurs through the process of *cell respiration*, which will be described shortly. Finally, most plants may store glucose by converting it to starch, as in potatoes, or oils, as are present in seeds.

Since sunlight is the most all-pervasive and abundant energy source in the environment, it is logical that the producers in all major ecosystems are photosynthetic plants which use this energy source. However, a few high-potential-energy inorganic chemicals also exist; hydrogen sulfide (H_2S) is the most notable example. Some bacteria are able to break down such chemicals and use the released energy for the synthesis of organic molecules from carbon dioxide and water in much the same way as plants use light energy. This process is known as **chemosynthesis**, indicating that the energy source is from chemicals as opposed to light (*photo*). In deep ocean rifts, where light does not penetrate but where hydrogen sulfide is abundant, unique ecosystems have been found which are based on chemosynthetic bacteria as the primary producers.

Consumers

Energy Role of Food. Consumers, which are animals, exhibit a considerable output of kinetic energy. We ourselves are a good example as we run our errands, play sports, give off body heat, and so on. Even a "couch potato" exhibits kinetic energy output in such things as heart pumping, chest movement for breathing, body heat, and other functions. That energy comes from the potential energy contained in organic molecules that are consumed or eaten as food. Thus, a considerable portion of what we or any other consumer eats is broken down to release its potential energy. It is this energy that powers all our activities and body functions and is finally given off (lost) as body heat.

The process through which organic molecules are broken down to release their potential energy is called **cell respiration**. Overall, it is the *reverse of photosynthesis*. Glucose is broken down in the presence of oxygen gas, and the carbon, hydrogen, and oxygen atoms are reformed into carbon dioxide and water as the formula below indicates.

$$C_6H_{12}O_6 + 6O_2 \longrightarrow$$
Glucose   Oxygen

$$6CO_2 + 6H_2O + Energy$$
Carbon Dioxide   Water

The process is called *cell* respiration because it takes place in each individual cell of the body. Food that is eaten is **digested**, which means that it is effectively "chopped" into smaller molecules within the stomach and intestine. Glucose comes from the digestion of starch and other carbohydrates and sugars. The smaller organic molecules which result from digestion are absorbed into the circulatory system and delivered to the individual cells of the body. Oxygen gas, which is absorbed by lungs or gills, is delivered to cells by the circulatory system. Thus, the energy released through cell respiration enables each particular muscle cell, nerve cell, kidney cell, liver cell, and so on to perform its particular functions. The waste, carbon dioxide, moves into the circulatory system and is eliminated through the lungs (or gills). Of course, all the cells of the body function together to make the whole organism.

Can you see why increasing activity makes you breathe harder and results in your "working up an appetite?" The increased activity demands increased cell respiration to provide the energy, and the other things occur as a result.

The overall reaction for cell respiration, you may note also, is the same as that for simply burning glucose (see page 48). Thus it is appropriate to speak of "burning" our food for energy. The distinction between burning and cell respiration is that cell respiration actually takes place in about 20 small steps that release the energy in small "packets" that are suitable for the power needs of each cell. If a cell were to release all the energy from glucose molecules in single "bangs" as occurs in burning, it would be like heating and lighting a room with large firecrackers. Energy yes, but not the most useful.

Proteins, fats, and oils (vegetable oils) also may be broken down readily to yield their potential energy. These molecules are chemically "chopped up" in digestion, further altered in the cell, and fed into the same cell respiration pathway. Fats and oils are especially high in potential energy; of all organic molecules, they have the highest calorie content per unit of weight. In addition to the energy yield, carbon dioxide and water again result as wastes from the carbon-hydrogen portions of the molecules. When other elements such as nitrogen or phosphorus are present, as is the case of proteins and nucleic acids, these are also left as wastes. Waste nitrogen, phosphorus, and so on are removed in the urine (or analogous waste in other kinds of animals). Because these wastes are mineral in nature, they will not pass off as gas through the lungs; but they are soluble in water and consequently are removed in water solution.

Nutritive Role of Food.   Another portion of food that is eaten, digested, and absorbed, of course, provides the raw material for growth, maintenance, and repair of body tissues. To serve in this capacity certain amino acids (present in proteins), a host of special organic molecules we call vitamins, and a number of minerals including such elements as calcium, iron, potassium, and phosphorus must be present. If any one or more of these specific nutrients is not present in the diet, various "diseases" of **malnutrition** will develop regardless of how many calories worth of food are consumed.

Sufficient amounts of these nutrients will be present in a balanced diet including whole grains, meats, vegetables, and dairy products. However, in most highly processed "foods" these nutrients are entirely or almost entirely absent.

Here you may see the problem of "junk foods" such as potato chips, sodas, candies, baked goods such as cake and donuts, and alcohol. Rich in fat and/or sugar, they are very high in calories (potential energy), but they contain essentially none of the necessary nutrients for body growth, maintenance, and repair. Consequently, a diet high in junk foods may meet your energy needs, but lead you into a condition of severe malnutrition. The malnutrition may lead you to "feel a lack of energy" even though calorie consumption is fully adequate just as an engine that is out of tune won't deliver power even though it has plenty of gasoline.

Indeed a diet high in junk foods is likely to oversupply calories. Your need for calories depends largely on how active you *choose* to be. Foodstuffs with a calorie content in excess of what your body is using are converted to and stored as body fat leading to weight gain. Thus, a diet high in junk foods is very likely to lead to *both* being fat and malnourished. On the other hand, consuming fewer calories than the body is using forces the body to obtain the energy from breaking down its own tissues, mainly body fat but also protein. **Starvation** is this process carried to an extreme.

The point of a balanced diet is that it supplies both calories and nutrients in adequate but not excessive amounts. Further, it is important to maintain a balanced diet because most nutrients are not stored in the body. Excessive amounts taken at any one time are simply excreted or broken down for their calorie content. In a balanced diet, 80 to 90 percent of what is digested and absorbed is used for its calorie content. But, even though the protein-vitamin-mineral nutrient portion is the smaller portion (10–20 percent), it is of vital importance.

Material Consumed but not Digested.   Finally, a portion of what is eaten is not digested but simply passes through the digestive system and out as **fecal** wastes. For consumers that eat plants, such waste is largely the material of plant cell walls, **cellulose**. We refer to it as *fiber, bulk,* or *roughage.* Some is necessary for the intestines to have something to push through to keep them clean and open. However, most consumers cannot get nutrient or energy value from cellulose because they are unable to digest it (the exception being cows and certain other grazing animals). This is why humans can't use wood or other coarse plant material as food; it is almost entirely cell walls or cellulose. (These aspects of food are summarized in Fig. 2–12.) This brings us to detritus feeders and decomposers.

Detritus Feeders and Decomposers

There is little difference in principle between detritus feeders and decomposers and the consumers discussed above. Detritus, you will recall, is dead plant and animal material, dead leaves, wood, dead animals, fecal wastes, and so on. But, it is still organic! As such, it is high in potential energy and nutrients and is a good food source for those organisms that can digest it. Digesting it is the key. Detritus is largely dead plant material, which consists mostly of cellulose and, therefore, can't be digested by most consumers as noted above. Various species of fungi, bacteria, and certain other microorganisms (decomposers) are unique in being able to digest cellulose into its constituent glucose molecules. Many large, grazing animals such as cattle actually maintain such bacteria in their digestive systems. This is how they digest detritus. You can see how this is a mutualistic relationship in which both cattle and bacteria benefit.

Once detritus is digested, detritus feeders and decomposers use it as a source of energy and nutrients. As described above the major portion is broken down through cell respiration to release energy for their life functions while another portion is used as the raw materials for the actual growth of tissues of the organism. The waste products, resulting from cell respiration, are again carbon dioxide, water, and inorganic forms of nitrogen, phosphorus, and other elements that may be present. In the energy conversions there is also the inevitable loss as heat. Indeed, the heat may be considerable. (You may observe a rotting manure pile "steaming" on a cool day.)

Some bacteria and yeasts (members of the fungi group) are capable of a modified form of cell respiration which is significant. In the absence of oxygen they are capable of deriving sufficient energy to sustain themselves from the partial breakdown of organic molecules. (Oxygen is necessary for a complete breakdown to carbon dioxide and water.) The resulting wastes in this case are compounds such as alcohol (C_2H_6O), methane gas (CH_4), and acetic acid (vinegar). Indeed, these natural processes are exploited to produce these compounds commercially and are generally referred to as fermentation.

PRINCIPLES OF ECOSYSTEM FUNCTION

Nutrient Cycling

In looking at the various inputs and outputs of producers, consumers, detritus feeders, and decomposers, have you noted how they fit together? Organic material and oxygen, which are produced by photosynthetic plants, are exactly what is needed by consumers: food to eat and oxygen to breath. And, the wastes from consumers, carbon dioxide and minerals excreted in the urine, are exactly the nutrients needed by plant producers. Indeed, urine is a good fertilizer if properly diluted (Fig. 2–13). Herein we observe the

FIGURE 2–12
Consumers. Only a portion of the food ingested by a consumer is assimilated into body growth, maintenance, and repair. A larger amount is oxidized through cell respiration to provide energy for assimilation, movements, and other functions. Waste products of oxidation are carbon dioxide, water, and various mineral nutrients. A third portion is not digested but passes through, becoming fecal wastes.

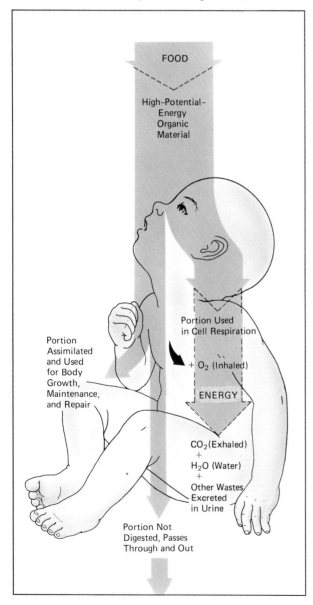

FIGURE 2–13
Animal wastes are plant fertilizer. The oxidation of food for energy leaves the inorganic nutrients needed by plants. The photograph shows a dog urine spot on a lawn. The "donut" of dark green grass is where urine has been diluted to optimal concentration; the grass in the center has been killed by overfertilization by concentrated urine. (Photo by author.)

first basic principle of ecosystem function: *Resources are supplied and wastes are disposed of by recycling all elements.*

Do you see how this is in harmony with the Law of Conservation of Matter? Since atoms are neither created nor destroyed, nor converted one into another, they can be reused indefinitely because they never "wear out." This is exactly what natural ecosystems do; they recycle the same atoms over and over again. We can see this even more clearly by focusing on the pathways of three key elements: carbon, phosphorus, and nitrogen. Since these pathways do lead in a circle they are respectively known as the *carbon cycle*, the *phosphorus cycle*, and the *nitrogen cycle*.

The Carbon Cycle

For descriptive purposes, it is convenient to start the carbon cycle (Fig. 2–14) with the "reservoir" of carbon dioxide molecules present in the air and dissolved in water. Through photosynthesis, carbon atoms from carbon dioxide molecules are incorporated into glucose and then into the other organic molecules that make up all plant tissues. Through food chains they then move into and become the tissues of all the other organisms in the ecosystem. However, it is unlikely that a particular carbon atom will be passed through many organisms on any one cycle because at each step there is a considerable chance that the consumer will use the organic molecule in its cell respiration to

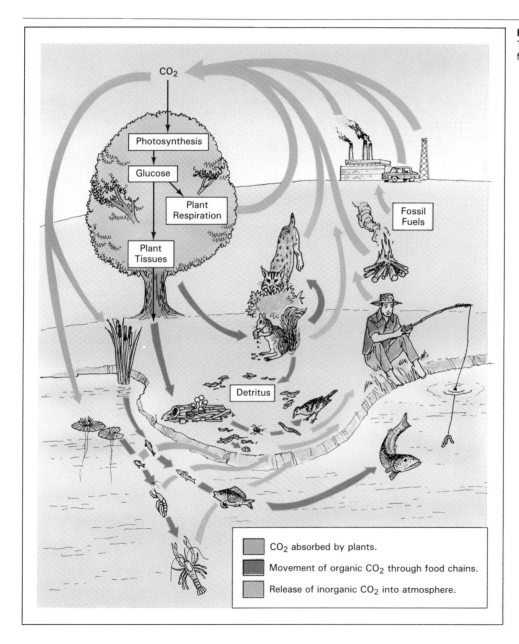

FIGURE 2–14
The carbon cycle. See text for explanation.

gain its energy content. When this occurs the carbon atoms are released back to the environment in molecules of carbon dioxide, thus completing one cycle, but of course ready to start another. Likewise, burning organic material returns the carbon atoms to the air in carbon dioxide molecules.

Another interesting and important aspect of the carbon cycle is that in ancient geological times (hundreds of millions of years ago) much of the organic matter produced in photosynthesis was neither consumed nor decomposed; it accumulated and was gradually buried under sediments. As a result of millions of years under heat and pressure in the earth,

this detritus has been converted to crude oil, natural gas, or coal, the particular product depending on the plant material involved and the conditions of heat, pressure, and length of time in the earth. We are now mining or pumping these **fossil fuels** in huge quantities to run our industrialized society. In burning these fuels, we are, in one sense, completing the carbon cycle. But we are also increasing the concentration of carbon dioxide in the air since the release far exceeds reabsorption. This has serious climatic implications (see greenhouse effect, Chapter 13).

The Phosphorous Cycle

The phosphorous cycle is diagramatically illustrated in Figure 2–15. The element phosphorus is crucial in

FIGURE 2–15
The phosphorus cycle. See text for further description.

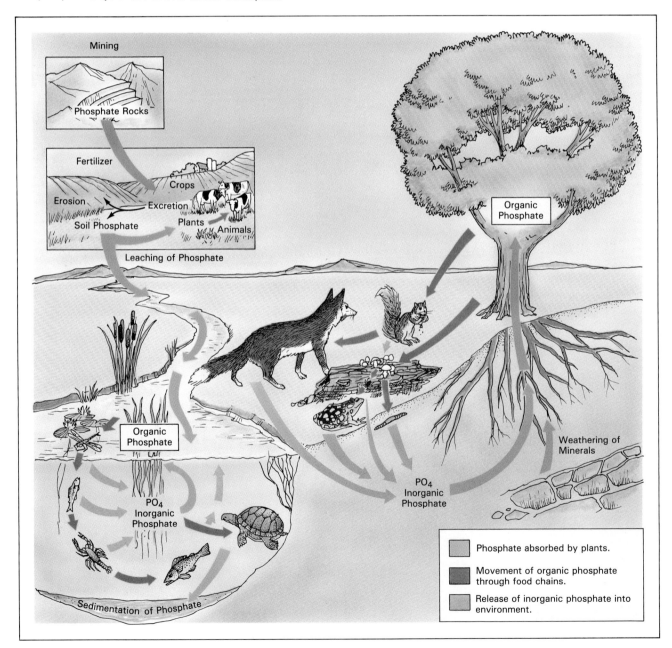

the structure of genes and in molecules that transfer energy within cells. Phosphorus exists in various rock and soil minerals as the inorganic ion phosphate (PO_4^{-3}). Phosphate dissolves in water but does not enter air. Plants absorb PO_4^{-3} from the soil or water solution and incorporate it into various organic compounds. Phosphate bonded into organic compounds is frequently referred to as **organic phosphate**. Through food chains, organic phosphate is passed from plants to all the other organisms in the ecosystem. But again, at every step there is a high likelihood that the compounds containing the phosphate will be oxidized for energy in cellular respiration and, as this occurs, the phosphate is released in urine or similar waste back to the environment where it may be reabsorbed by plants to start another cycle.

Do you note a major difference between the carbon cycle and the phosphorus cycle? The carbon cycle has a gas phase, carbon dioxide in the air; therefore, wherever carbon dioxide is released it will mix into and move through the air where it can be reabsorbed by plants and repeat the cycle. Such "free return" does not occur with phosphate, which has no gas phase. As phosphate gets into waterways, it enriches, indeed overenriches, aquatic ecosystems. But note there is virtually no return. Birds feeding on fish may return some phosphate to land through their droppings, but this will be a small portion of the total and it will also be close to the shore. Also, ocean sediments may be uplifted by geological processes, but this only occurs over millions of years.

Consequently, phosphate and other such mineral nutrients from the soil are, for the most part, only recycled in ecosystems insofar as the wastes which contain them are deposited on the soil from which they come. In a natural ecosystem, this is basically what happens. However, we have constructed our human ecosystem such that crops, which contain the nutrients withdrawn from the soil, are shipped long distances to human consumers. Then human wastes carrying the nutrients are discharged into waterways. We shall discuss the long-term implications later.

FIGURE 2–16
The nitrogen cycle. See text for details.

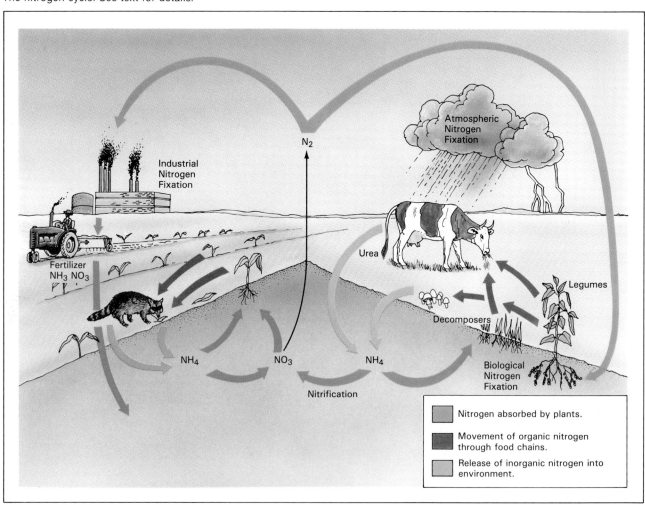

The Nitrogen Cycle

The nitrogen cycle (Fig. 2–16) is somewhat more complex because it has both a gas phase and a mineral phase. The main reservoir of the element nitrogen is the air, which is about 78 percent nitrogen gas (N_2). Curiously, however, plants cannot utilize nitrogen gas directly from the air; the nitrogen must be in a mineral form such as ammonium (NH_4^+) or nitrate (NO_3^-). Fortunately, a number of bacteria and also certain blue-green algae (cyanobacteria) can convert nitrogen gas to the ammonium form, a process called **nitrogen fixation**. Most important among these nitrogen-fixing organisms is a bacterium called *Rhizobium* that lives within nodules on roots of legumes, members of the pea-bean family of plants (Fig. 2–17). This is another good example of mutualism. The legume provides the bacteria with a place to live and food (sugar), and gains a source of necessary nitrogen in return. From the legumes, fixed organic nitrogen may be passed to other organisms in the ecosystem through food chains.

As proteins and other organic compounds containing nitrogen are broken down for energy in cell respiration, nitrogen is excreted, generally in the ammonium (NH_4^+) form. Additional bacteria may convert it to the nitrate form (NO_3^-), but either form may be reabsorbed by *any* plant. Thus it is transferred to nonlegume plants and recycled further as a mineral nutrient. However, it does not remain in the "mineral" phase indefinitely; another kind of bacterium in the soil gradually changes the nitrate form back to nitrogen gas. Some nitrogen gas is also fixed in the air by discharges of lightning and comes down with rainfall, but this is estimated to be only about 10 percent of what is fixed by the biological process.

All natural ecosystems, then, depend on nitrogen-fixing organisms; legumes with their symbiotic bacteria are, by far, the most important. The legume family includes a huge diversity of plants ranging from clovers (common in grasslands) through desert shrubs to many trees. Thus every major terrestrial ecosystem, from tropical rainforest to desert and tundra, has its representative legume species. It is also interesting to note that legume species are generally the first plants to recolonize a burned-over area. Without them all production would be sharply impaired because of lack of available nitrogen. The nitrogen cycle in aquatic ecosystems is similar, but here blue-green algae are the most significant nitrogen fixers.

Only humans have been able to bypass the need for legumes in growing nonlegume crops such as corn, wheat, and other grains. We do this by fixing nitrogen in chemical plants. Synthetically produced ammonium and nitrate are major constituents of fertilizer. However, the high cost of artificially fixed nitrogen is causing many farmers to readopt the natural process of alternating legumes with nonlegume crops.

Energy Flow

The flow of energy through ecosystems is in full harmony with the laws of thermodynamics. Ecosystems may be seen basically as systems of converting energy from one form to another. Namely, energy is converted from light to stored chemical energy by photosynthetic plants, and this chemical energy is reconverted to various forms through food chains. At each step, a portion of the chemical energy (food) is broken down to release its potential energy and, as this energy is used to perform the organism's functions or "work," it is gradually converted to and lost as heat. Thus, we observe a *flow of energy* through ecosystems entering as light, performing work, and exiting as heat.

FIGURE 2–17
Root nodules. Nitrogen fixation, the conversion of nitrogen gas in the atmosphere to forms that can be used by plants, is carried out by bacteria that live in the nodules of roots of plants of the legume family. (USDA)

Indeed, the whole earth involves a massive flow of energy entering as sunlight, **solar energy**, sustaining life and then exiting as heat. Most of the solar energy falling on the earth is converted directly to heat as it is absorbed by water or bare ground which in turn heats the air. This heating serves as the driving force for the water cycle, wind, and ocean currents—our weather!—but gradually the heat radiates into outer space where it is lost (Fig. 2–18).

The key point of ecosystems is that they are "plugged into" this natural flow of solar energy and, vast and complex as they are, they only utilize a small portion of it. This brings us to the **second basic principle of ecosystem function**: *Ecosystems run on solar energy, which is exceedingly abundant, nonpolluting, relatively constant, and everlasting.*

Let's consider each of these attributes of solar energy in more detail:

Exceedingly Abundant.   Plants utilize only about 0.5 percent of the total solar energy falling on earth. If humans were to get all their energy from the sun, it would be equivalent to an even smaller percentage. This is to say that the amount of solar energy far exceeds any foreseeable demand. Since all energy is eventually converted to heat, using more of the solar energy for functional purposes would not change the dynamics of the biosphere.

Nonpolluting.   The pollutants from the nuclear reactions that produce solar energy are conveniently left behind on the sun some 93 million miles away in space. Thus, solar energy is pure energy; with no

mass it does not entail polluting byproducts as is the case in getting energy from fossil or nuclear fuels here on earth.

Constant.   Solar energy will always be available in the same amounts (vast) at the same price (free). It is not subject to political embargos or increasing prices as supplies become more limited. It is diffuse, however, and needs to be concentrated to serve our energy needs, but this drawback is being overcome.

Everlasting.   Astronomers tell us that the sun will burn out in another few billion years, but this is hardly meaningful in our time frame. Consider that humans, from the earliest prehistoric traces, have been on earth about 3 million years. This is 0.3 percent of one billion. By this comparison we have 99.7 percent of our time still to go; each 100 years will be just another 0.01 percent.

Flow of Energy and Decreasing Biomass at Higher Trophic Levels

Our discussion regarding the flow of energy also enables us to see more clearly why biomass decreases with each successive trophic level. As each consumer, detritus feeder, or decomposer breaks down a fraction of its food to release its energy, the total biomass is reduced by that fraction. However, two additional factors also contribute. First, any population can be looked at as a **standing biomass** (total mass of existing individuals), to which there is a yearly addition by

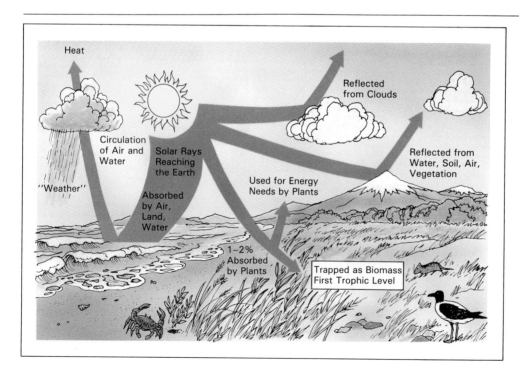

FIGURE 2–18
Only a small portion of the solar energy hitting the earth is absorbed and trapped by photosynthesis, but this is then the total energy available to the rest of the ecosystem.

growth and reproduction and a subtraction by death and/or consumption. If the standing biomass is to remain constant, as it does in a stable ecosystem, it follows that herbivores, for example, eat no more than the yearly production of the producers. To take more would be to deplete the stock of producers (overgraze). Second, a significant portion of what is consumed is not digested but simply passes through and out again as fecal waste. When these two factors are considered along with the fact that the larger portion of what is consumed and digested is broken down for energy, then you can readily see why the biomass of herbivores will be but a tiny fraction of the biomass of producers. The same can be said in terms of suc-

cessive trophic levels (Fig. 2–19). Herein we observe the **third basic principle of ecological function**: *Large biomasses cannot be supported at the end of long food chains. Increasing population means moving closer on the food chain to the source of production.*

It is interesting that science fiction writers commonly ignore this principle and portray huge monsters, carnivores at that, living in barren deserts. They also portray food chains that are circular and supposedly self-supporting in the absence of producers. It should be clear that the second law of thermodynamics makes such things impossible. Such systems, even if they were initiated, would quickly starve into extinction. These three basic principles of ecosystem

FIGURE 2–19
Decreasing biomass at higher trophic levels. The decrease results from the fact that much of the preceding trophic level is "standing crop" that is not available for consumption and much of what is consumed is broken down for energy.

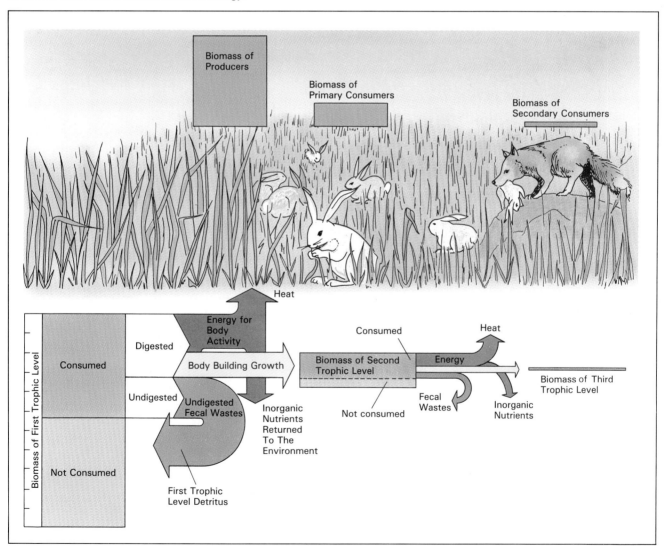

function, nutrient cycling, energy flow from the sun, and decreasing biomass are summarized in Figure 2–20.

IMPLICATIONS FOR HUMANS

As described in Chapter 1, the turning point from humans being primitive hunter-gatherers to building civilizations hinged on the development of agriculture some 10,000 years ago. In more recent times we have been outstandingly successful at building and expanding food chains which have enabled human populations to flourish over the short run. However, we have been much less cognizant of other principles which sustain ecosystems over the *long run*.

A large share of the environmental problems we currently face are the result of not giving adequate attention to the principles described in this chapter. Let's look at our lack of conformity with each.

First Principle. *Natural ecosystems gain resources and dispose of wastes by recycling all elements.* The Law of Conservation of Matter has been known for over 200 years, yet we operate our ecosystem as if we believed that elements will continually be created at one location and disappear "out of sight, out of mind" at another. We mine phosphate rock, for example, from where there are particularly rich but inevitably limited

FIGURE 2–20
Nutrient cycling within, and energy flow through, ecosystems. Arranging organisms according to feeding relationships and depicting energy and nutrient inputs and outputs of each shows that there is a continuous recycling of nutrients within the ecosystem and a continuous flow of energy through the system.

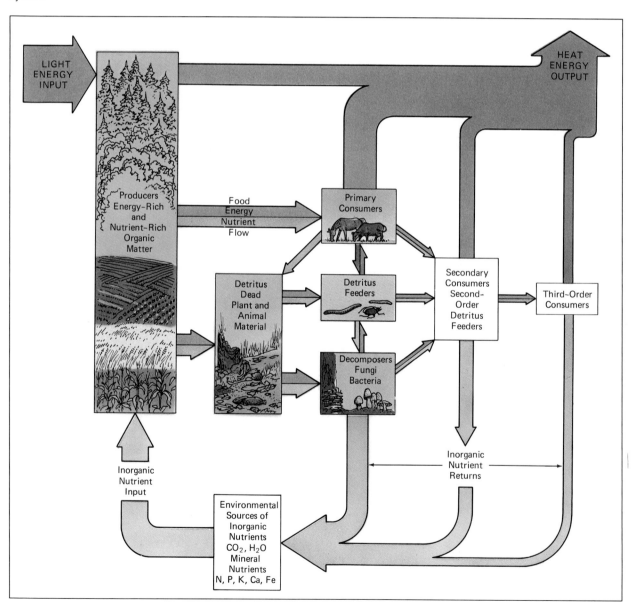

deposits. We apply it to agricultural crops to enhance their growth; we eat the crops and then flush our wastes down drains that eventually dump the phosphate into natural waterways from which there is no natural return (Fig. 2–21). It should be obvious that this is an unsustainable policy. Sooner or later the deposits will be exhausted. Even more significant at present is that aquatic ecosystems in lakes, bays, and estuaries around the world are being upset and destroyed by the overenrichment with phosphate and nitrogen (See Chapter 9.).

The same is true for other elements that are the "nutrients" for maintaining industry, mercury for example. In the absence of recycling, the inevitable result is resource depletion on one hand, and pollution on the other. With many industrial elements the pollution problem is extreme toxicity as opposed to overenrichment.

Second Principle.   *Natural ecosystems sustain themselves by running on solar energy which is exceedingly*

abundant, nonpolluting, constant, and everlasting. Until the Industrial Revolution, humans supplemented their own labor with animals, wood, wind, and water power. These are all indirect sources of solar energy. It has only been in the last 250 years that we have used fossil fuels. Yet, in this brief time we have brought ourselves to the threshold of depletion. Even more, burning them is responsible for numerous pollution problems including what may be catastrophic changes in the climate due to the carbon dioxide emitted. Nuclear power, which we have developed in the last 40 years, promises little improvement. Clearly, our nonsolar alternatives are nonsustainable.

Third Principle.   *Large biomasses cannot be supported at the end of long food chains.* The human population has been growing phenomenally over the last 100 years or so and continues to grow at close to 90 million people per year. Yet many people, especially in developed countries, feed largely on the third trophic level, eating meat. Since 10 to 20 pounds of edible grains are consumed in producing one pound of meat, this places an additional burden on agriculture (Fig. 2–22). Something in the order of tenfold more land must be cultivated to provide people with a high

FIGURE 2–21
In contrast to applying the ecological principle of nutrient recycling, human society has developed a pattern of one-directional nutrient flow. There are increasing problems at both ends.

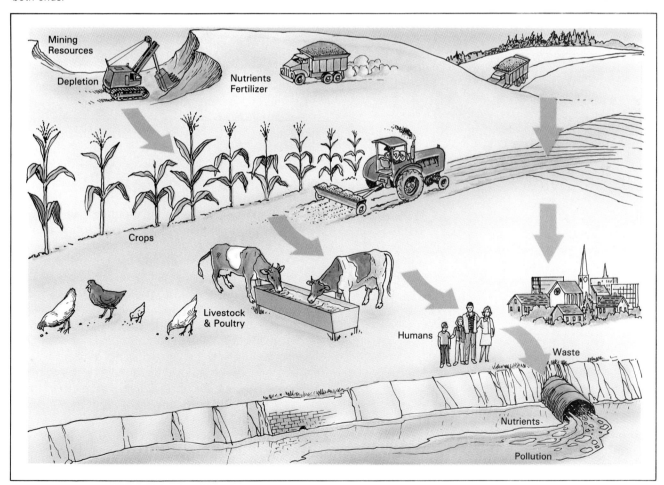

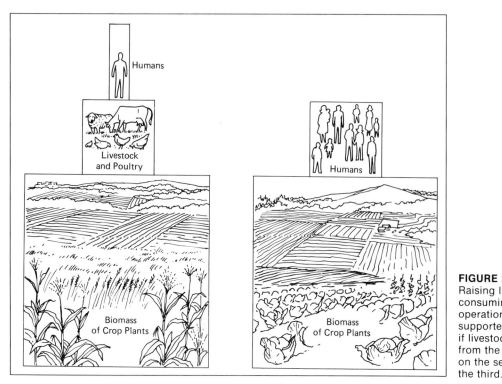

FIGURE 2–22a
Raising livestock and poultry is a food-consuming, not food-producing, operation. Many more humans can be supported on the same agricultural base if livestock and poultry are eliminated from the food chain and humans feed on the second trophic level rather than the third.

FIGURE 2–22b
Pounds of grain and soy fed to get one pound of meat, poultry, or eggs (USDA).

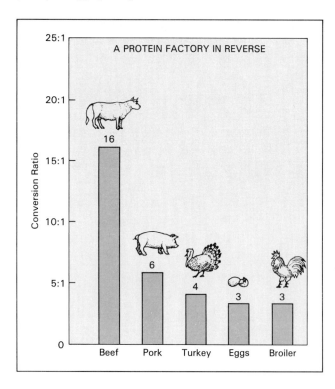

We noted in the Introduction that developing a working system based on certain theories is the ultimate test and proof of those theories. Constructing a small aquatic biosphere has become a common environmental science student project. Aquatic plants, a few small fish which will feed on the plants, and a few snails (detritus feeders) can be sustained indefinitely in a *sealed* aquarium providing just light for photosynthesis. In Arizona a project, *Biosphere II*, is underway which includes humans in the sealed environment (See Case Study 2).

The conclusion is self-evident. Sustainable development can only be attained on a foundation of recycling and using nonpolluting everlasting solar energy sources. We have learned these principles from studying natural ecosystems, and proven them with tests and there is some progress toward implementing them into our society. Solar energy technologies are on the threshold of widescale commercialization and growing numbers of communities are starting to recycle treated sewage effluents and refuse. Still, because of ignorance and inertia, the majority still clings to the nonsustainable ways. Clearly there are roles to be played in changing this. Policy decisions must focus on promoting recycling, developing solar energy alternatives, and maintaining the productivity of the land. These issues will be discussed in more depth and specific pointers will be given in later chapters.

meat diet. The consequence is that the added pressure on production is causing land to be destroyed by erosion and overgrazing. Increasing population and decreasing productivity of land is an obvious collision course.

Sometime in the early 1990s, at a location 30 miles north of Tucson, Arizona, four men and four women will enter an air-tight, glass and steel structure covering an area of little more than two football fields. The door behind them will be closed and sealed and this enclosure will be their world for the next two years. No, this is not a prison; the men and women are explorer-scientists who worked hard for this assignment. They are the select crew of *Biosphere II*, which is a prototype space station undergoing its testing here on earth. Space Biospheres Ventures, a private company, developed Biosphere II in order to gain information and test systems that may be used for setting up permanent, self-sustaining human habitats on the moon and other planets. It will also provide valuable information for managing "Biosphere I," our own Spaceship Earth.

What is this new space station, *Biosphere II*, like? Conspicuously absent are the traditional tanks of oxygen, bottles of water, crates of foodstuffs, and barrels for wastes, because however large, these traditional supports have obvious limits. All the systems of *Biosphere II* have been designed according to the principles learned in studying the ecosystems of our own earth. In addition to the human "Biospherians," *Biosphere II* will contain some four thousand species of natural plants, small mammals, birds, reptiles, insects, and soil microbes. Studied and selected for their compatibility and ability to provide sustainable food webs and recycling of wastes, these species will be arranged into ecosystems creating samples of tropical rainforest, savannah, scrub forest, desert, fresh and saltwater marshes, and a mini-ocean complete with a living coral reef (Fig. 2–23). Water will be recycled

and purified by sun louvers that will drive convection currents of warm air across the ocean causing evaporation. The moisture picked up here will be condensed creating high rainfall over the tropical forest. The water will then trickle back toward the marshes and ocean through soil filters providing a continuous supply of purified fresh water for both humans and ecosystems. Carbon dioxide from respiration will be reabsorbed and oxygen will be replenished through photosynthesis. Thus, these "natural" ecosystems will provide a basic ecological stability for the atmosphere and hydrosphere of *Biosphere II*.

The Biospherians, however, aside from monitoring the health of the natural ecosystems, will be occupied mainly in another section where they will live in modern apartments and carry on intensive agriculture to produce vegetables for both themselves and for feeding a few goats and chickens, which will supply milk, eggs, and meat. Aquaculture (fish farming) will supply additional protein. They will have well-equipped laboratories for monitoring the status and progress of *Biosphere II* and conducting research. They will be able to send and receive electronically any information, including movies, but this will be their only connection with the outside world.

With everything sealed in the relatively small space of *Biosphere II*, it is obvious that there is no place for the dumping of wastes without destroying the system. Therefore, all human and animal excrements and other wastes will be treated and decomposed such that the nutrients will be recycled back to support the growth of plants, which in turn feed the fish, animals and humans. Nor can the use of toxic chemicals such as pesticides be permitted. Pest control will be achieved by biological means. Neither are pollution-causing energy sources such as burning fuels tolerable. All power for cooking, lighting, and running equipment will be provided by solar cells, which convert light energy directly into electrical power without generating any waste products.

These lessons of living within a space station will be, perhaps, the most important lessons to come from *Biosphere II* because, in fact, we already do live within a sealed space station. Spaceship Earth is only slightly larger, but no less vulnerable in the long run to our careless ways.

Excerpted from information provided courtesy of Space Biospheres Ventures

Architectural model of Biosphere II. (Peter Menzel)

3

Ecosystems:
What Keeps Them the Same?
What Makes Them Change?

CONCEPT FRAMEWORK

■ Outline

I. THE KEY IS BALANCE   70

  A. Ecosystem Balance Is Population Balance   70

    1. Biotic Potential versus Environment Resistance   70

    2. Density Dependence and Critical Numbers   72

  B. Mechanisms of Population Balance   72
    1. Predator-Prey and Host-Parasite Balances   72

CASE STUDY: DEMISE OF THE AMERICAN CHESTNUT   75

    2. Specialization to Habits and Niches   74
      a. Animals   74
      b. Plants   76

    3. Competition among Plants and Plant-Herbivore Balances   77

    4. Fire   78

    5. Territoriality   81

  C. Ecosystem Change: Succession   82

    1. Primary Succession   83
    2. Secondary Succession   83
    3. The Climax Ecosystem   85
    4. Degree of Imbalance and Rate of Change: Succession, Upset, or Collapse   85

II. IMPLICATIONS FOR HUMANS   87

■ Study Questions

1. What is the key for ecosystems remaining the same? changing?

2. Each species in an ecosystem is represented by what? How must birth rate and death rate for each population in a stable ecosystem compare?

3. Define and contrast *biotic potential* and *environmental resistance.* What are the components of each? State the principle regarding stable populations. What occurs if the balance is upset?

4. What is meant by a factor being *density dependent?* How does this serve to balance populations? Why are human impacts more likely to cause extinctions? Define what is meant by a *critical number* of a certain species.

5. Describe how a *predator-prey* or *host-parasite relationship* may balance the population of each. Define *natural enemies.* Why are balances involving several natural enemies more stable? State the principle. Are balances "automatic?" Give examples illustrating lack of balance.

6. Define and distinguish between *habitat* and *niche.*

7. Give examples illustrating how plants and animals being specialized to particular niches and/or habitats minimizes competition among species. How does this increase ecosystem diversity and help maintain ecosystem stability?

8. Define *herbivory.* Describe how herbivory leads to diversity and stability.

9. Give examples of how certain ecosystems are maintained by periodic burning.

10. Define *territoriality.* How does territoriality limit population density of certain populations? How does this promote ecosystem stability?

11. Give examples of how natural ecosystems may undergo gradual change. Define and give examples of *primary* and *secondary succession;* a *climax ecosystem.*

12. How do human-induced changes differ from natural succession? Contrast succession, upset, collapse.

13. Is the human ecosystem diverse? stable? Explain. Is sustainable human development possible without maintaining biological diversity in the natural world? Explain. What can you do to promote maintaining biological diversity?

We have learned that each ecosystem is a dynamic structure of hundreds or even thousands of species of producers, consumers, detritus feeders, and decomposers interacting with one another through feeding and other relationships. The system sustains itself by cycling nutrients and drawing on the continuous flow of energy from the sun. But, what prevents herbivores from devouring all the producers? What prevents carnivores from eliminating their prey? What prevents one species from eliminating others in competition? In short, how is the stability of an ecosystem maintained?

In Chapter 1 we learned the general principle that various *limiting factors* keep species from spreading beyond areas to which they are best adapted. We will expand upon this concept to gain a clearer picture of how ecosystems sustain their composition in spite of the apparent conflicts and competitions among the many species involved. There are major lessons to be learned concerning what we need to do to make our own human ecosystem sustainable.

THE KEY IS BALANCE

No forces or rigid structures exist that prevent ecosystems from changing. In fact, they may and do change, even drastically, as conditions are altered. The one thing that enables ecosystems to sustain a given composition of species over long periods of time is that all the dynamic relationships are in *balance*.

Ecosystem Balance Is Population Balance

Each species (plant, animal, or microbe) found in an ecosystem actually exists as a **population**—that is to say an interbreeding, reproducing group. Otherwise, the species would disappear from the ecosystem as its members died. Further, an ecosystem remaining stable over a long period of time implies that each population remains more or less constant. Any continuing increase or decrease in any population would be observed as a change in the ecosystem. A population remaining constant means that its reproductive rate is equalled by its death rate. Thus, the problem of ecosystem balance boils down to a problem of how death rate is balanced with reproductive rate for each species in the ecosystem.

Biotic Potential versus Environmental Resistance

Maintaining or increasing a population depends on more than **reproductive rate** (number of live births, eggs laid, or seeds or spores set in plants) by itself.

Equally important is **recruitment,** making it through the growth stages and becoming part of the breeding, reproducing population. A very high reproductive rate may achieve little population growth if recruitment is very low. Fish and shellfish, for example, typically lay thousands or even hundreds of thousands of eggs, but only a tiny percentage survive to reach adulthood. The same may be said for plants, which typically set huge numbers of seeds. The same objective of maintaining or increasing the population may be achieved by a drastically lower reproductive rate combined with a higher recruitment rate. Humans, for example, have relatively low reproductive rates but high recruitment rates through parental care.

Other important factors that lend to increasing a population include the ability of animals to migrate or of seeds to disperse to similar habitats in other locations, the ability to adapt to and invade other habitats, defense mechanisms, and resistance to adverse conditions and disease. All the factors that contribute to a species increasing its numbers are referred to as its **biotic potential.**

Although different species have different strategies regarding biotic potential, one phenomenon is invariably the case. Every species has sufficient biotic potential to rapidly increase its population *if all environmental factors are favorable.* In fact, the population growth in such situations may be so rapid that it is commonly referred to as a **population explosion.**

Populations in natural ecosystems generally do not "explode" because all conditions are seldom favorable for any extended period of time. One or more abiotic factors such as unfavorable temperature, pH, salinity, amount of available water, and/or one or more biotic factors such as predators, parasites, disease organisms, or lack of sufficient food become *limiting*. The combination of all these abiotic and biotic limiting factors is termed **environmental resistance** (Fig. 3–1).

The impacts of environmental resistance hit hardest on the very young, since they are the most vulnerable to predation, disease, lack of food (or nutrients) or water, and adverse conditions. Reproduction may not be affected, but recruitment is. In more severe cases dieoff of reproducing adults also occurs. In conclusion, whether a population increases, decreases, or remains constant is the result of a balance between biotic potential adding individuals and environmental resistance removing individuals (Fig. 3–2). Indeed, this is the **principle of population change:** *A change in population of a species is the result of a dynamic balance between its biotic potential and the environmental resistance it faces.*

The balance between biotic potential and environmental resistance is *dynamic* or continuously read-

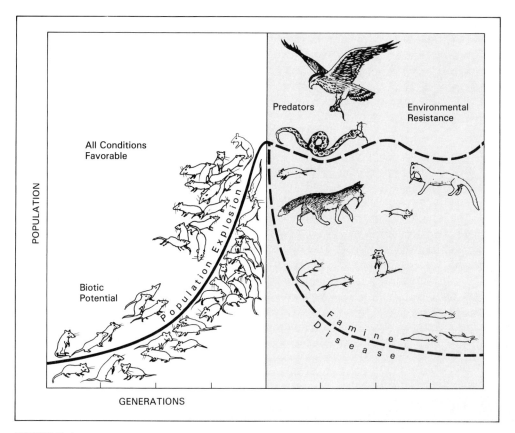

FIGURE 3–1
All organisms have a biotic potential (reproductive capacity) which will result in a rapid population increase when all conditions are optimal. Graphically, the growth curve for such a period has a J-shape and is termed a population explosion. However, conditions never remain optimal indefinitely. Factors of environmental resistance lead to increasing dieoffs such that the growth curve levels off giving an overall S-shape. Or, if the population explosion leads to depletion of certain essential resources, the population explosion will be followed by a population crash giving a spike-shaped curve.

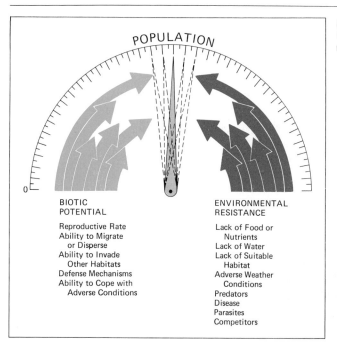

FIGURE 3–2
Population is a balance between factors that increase numbers and factors that decrease numbers.

justing as factors of environmental resistance are seldom constant for very long. For example, a drought may cause a population to die back one year, but in following normal years it will recover, and this cycle may be repeated indefinitely. Balance is a relative phenomenon. Some balances fluctuate very little, others may fluctuate widely, but as long as decreased populations restore their numbers, the system may be said to be balanced. But, the question remains: What maintains the balance within a certain range on the "scale?" What prevents a population from going into an explosion or into extinction? Indeed neither possibility is ruled out. Both extinctions and explosions are seen in nature, more so with human impacts. Population explosions are temporary. However, as the World Wildlife Fund emphasizes in their slogan, *extinction is forever*.

Density Dependence and Critical Numbers

In natural ecosystems balance is generally maintained within a certain range because factors of environmental resistance are **density dependent.** That is, as **population density** (number of individuals per unit area) increases, environmental resistance becomes more intense, increasing the rate or percentage of dieoff and causing the population to cease growing —if not decline. Conversely, as population density decreases, environmental resistance is generally mitigated allowing the population to recover.

Human impacts often result in extinction because they are not density dependent. Impacts such as ecosystem destruction, habitat alteration, pollution, and exploitation can be just as intense at low populations as at high. Further, the biotic potential of many species depends on a certain minimum population base, a herd of deer, a pack of wolves, a flock of birds, or a school of fish for example. If a population is pushed below a certain **critical number** necessary to support the group, biotic potential fails and extinction is virtually assured.

Note that extinction may be imminent even when large numbers of organisms in a species are alive and well—as individuals. For example, thousands of pet owners may gloat over their pet boa constrictor or parrot. Yet isolated from their natural ecosystem and each other, they are but caged victims awaiting extinction.

Mechanisms of Population Balance

Let us turn our attention to some specific mechanisms which provide population balance in nature. For the sake of study, it is necessary to focus on one mechanism at a time, but keep in mind that in natural eco-

systems all of the mechanisms are working in concert to create the overall balance. Knowledge of these mechanisms will make us aware of how ecosystems may be upset and the consequences which may result.

Predator-Prey and Host-Parasite Balances

Balance may be provided through predator-prey relationships. A classic example is the predation on rabbits by the lynx, a member of the cat family. When the rabbit population is low, each rabbit can find abundant food and plenty of places to hide and raise offspring. In other words, the rabbits' environmental resistance is relatively low, and their population increases despite the presence of the lynx predator. However, as the rabbit population increases, there is relatively less food and fewer hiding places per rabbit. More rabbits provide easier hunting for the lynx so that, with plenty of rabbits to feed their young, the lynx population begins to increase. In short, the higher rabbit population runs into *increasing environmental resistance* in the form of limited food and shelter and increased predation. As a result, it begins to fall. As the rabbit population falls, the amount of food and shelter available to each rabbit again increases. Also, surviving rabbits are those that are most healthy and best able to escape from the lynx. Hunting becomes harder for the lynx; many of them starve and their population begins to fall. These factors sum up to *lower environmental resistance* for the rabbits, and their population increases again, repeating the cycle. These events explain the fluctuating but continuing balance found between the rabbit and lynx populations (Fig. 3–3).

While large predators such as the lynx draw attention, there are relatively few situations where they are the primary controlling factor. Much more abundant and ecologically important in population control is a huge diversity of parasitic organisms ranging in size from tapeworms, which are many feet long (Fig. 1–7) to microscopic disease-causing protozoans, fungi, bacteria, and viruses. All species of plants, animals, and even microbes themselves may be infected by parasites.

In terms of population balance, parasitic organisms act in the same way as large predators. As the population density of the host organism increases, parasites and their *vectors* such as disease-carrying insects, have little trouble finding new hosts and infection rates increase causing dieback. Conversely, when population density of the host is low, transfer of infection is impeded and there is a great reduction in levels of infection, which allows the population to recover again.

Parasites may not kill their host, but they gen-

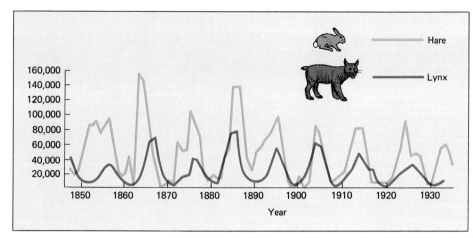

FIGURE 3–3
Predator-prey relationship creates balance between predator and prey populations. Data based on pelts received by Hudson's Bay Company. (From D. A. MacLulich, University of Toronto Studies, Biological Series No. 43, 1937.)

erally weaken it and make it more vulnerable to adverse conditions and/or attack by larger predators. It is commonly observed that the animals killed by large predators are infected with parasites whereas animals killed by hunters are generally healthy.

In a food web any given organism is generally affected by numerous predators and/or parasites. Consequently, the balance can be thought of more broadly as a balance between an organism and its **natural enemies.** Balances between an organism and its several natural enemies in a complex food web are much more stable and less prone to wide fluctuation, because different natural enemies come into play at different population densities. Also when the preferred prey is at a low density the population of the natural enemy may be supported by its feeding on something else. Thus, the lag time between increase of the prey population and that of the natural enemy is diminished. These factors have a great damping effect on the rise and fall of the prey population, as illustrated in Figure 3–4. The wide swings in populations as noted in the rabbit-lynx case are generally typical of very simple ecosystems involving relatively few species. Here is seen the **principle of ecosystem stability:** *Species diversity provides ecosystem stability.*

Predator-prey or host-parasite balances are not "automatic." They have developed over many thousands, even millions, of years. The species involved have adapted to *each other,* as well as to their environment, in such a way that a given natural enemy is incapable of completely eliminating its prey. Note that the lynx couldn't catch all the rabbits. Yet, the natural enemy provides limits at higher population densities. The lynx did keep the rabbits from going beyond a certain population density. Similarly species have some resistance to parasites and disease organisms which are naturally present so that they are not entirely eliminated by these organisms, yet they are limited by them. Thus, parasites will tend to regulate

the population of their host at a certain population density.

We can see the importance of this mutual adaptation through observing situations where it is absent. The organisms in ecosystems that have developed in isolation from each other, as on different

FIGURE 3–4
a) A population balance between an organism and a single natural enemy is prone to wide fluctuations because there is a lag between the increase of the prey population and the buildup of the predator population. b) When two or more natural enemies are involved and they come into play at different population densities, the fluctuations tend to be dampened out.

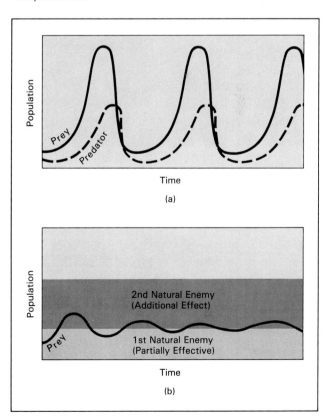

continents or on remote islands, have evolved the same kinds of balances, but the balances for one system generally do not match the balances of another. Thus, an introduced species may not encounter natural enemies or any other factors of environmental resistance that are sufficient to control it. The result is that its population explodes and this, in turn, may cause tremendous damage including the extinction of many species in the receiving system. Classic examples include the introduction of rabbits to Australia (below) and introduction of the chestnut blight, a parasitic fungus, to North America (see Case Study 3).

In 1859 rabbits were introduced from England to Australia for sport shooting. The Australian environment proved favorable and the Australian carnivore, the dingo, was poorly adapted to catching rabbits—it could not run fast enough. The result was that the rabbit population exploded and devastated vast areas of rangeland by overgrazing (Fig. 3–5).

Many of the most severe insect pests of agricultural crops and forests, the Japanese beetle and the gypsy moth, for example, are species from other ecosystems that do not have effective natural enemies in the ecosystem to which they were introduced.

Predator-prey or host-parasite balances have also been upset by introducing predators that are *too* effective. Domestic cats introduced into island ecosystems have often proved to be overly effective in comparison to native animals. Many unique species of island wildlife have been completely exterminated by cats.

Introducing a natural enemy to control an introduced "pest" has been used successfully in some cases, but it is not as simple as it seems on the surface. There is no guarantee that the introduced natural enemy will focus its attention on the target pest. For example, foxes were introduced in Australia to control the rabbits, but the foxes also found they could catch other Australian wildlife more easily than rabbits. Hence, they went their own way and seldom bothered the rabbits. Later a rabbit parasite was introduced which did prove successful. This topic of controlling pests with natural enemies is discussed further in Chapter 17.

Specialization to Habitats and Niches

Habitat is where an organism lives, that is, forest, grassland, swamp, or inside another organism (parasites). **Niche** refers to its "occupation," that is, where, when, and on what it feeds, where it builds its nests, and so on.

Animals. Many animals within an ecosystem appear to be competing as they strive for food and shelter. Actually, competition is much less than apparent because they occupy different *niches*. That is to say, they are adapted and specialized and thus limited to feeding on different things or in different locations and/or at different times of day. For example, robins, which eat worms, are not competing with woodpeckers, which eat insects in dead trees, and neither is competing with sparrows, which eat seeds. Flycatchers (birds) and bats both feed on flying insects, but they are not competing because flycatchers feed during the day and bats feed at night. The giraffe is able to graze on treetops; therefore, it does not compete with other grazing animals around it. Each animal having its characteristic niche minimizes competition between species. Thus there is little tendency for one species to cause the extinction of another.

Animals being adapted and specialized to different niches brings into play the concept of limiting factors (Chapter 1). As an animal attempts to feed or carry on other activities outside its particular niche, it faces increasing stress and limits, that is, environmental resistance. Thus, it is a relatively strong com-

FIGURE 3–5
Results of rabbits overgrazing the Australian ecosystem. On one side of a rabbit-proof fence there is lush pasture; on the other side, it is barren. Rabbit-proof fences were built over thousands of miles but proved unsuccessful in stopping movement. (Australian Information Service photo.)

Prior to the 1900s, the American chestnut was the dominant tree in deciduous forests of the eastern United States. As many as one-fourth of the trees were chestnuts. Mature trees were commonly 30 in. (75 cm) in diameter and 100 ft (30 m) tall. The chestnut was highly valued for its timber, which was strong, light, straight-grained, and resistant to rotting; its clean burning firewood; its aesthetic qualities as a shade tree; and its prolific production of chestnuts, which were a major food for forest wildlife as well as a prized food for many people (Fig. 3–4). It was said that many mountain people lived on the chestnut trees alone since they provided food, shelter, and fuel.

But in 1904, forester W. H. Merkel noticed that some of the trees in the New York Zoological Society's park in the Bronx were dying, and the bark of the trunks had ruptures or cankers exuding orange-colored spores. The problem was quickly identified as a

American chestnut. An important species of tree in eastern U.S. deciduous forests was essentially eliminated by an introduced disease, the chestnut blight. (National Archives photo No. 95-6-250527.) Insert shows fruit of chestnut tree. (U.S. Forest Service photo.)

parasitic fungus *Endothia parasitica*, which was common on Chinese and Japanese chestnuts but until then unknown in North America. Apparently it had been unwittingly introduced along with some imported Chinese chestnuts recently planted in the park. The spores of this parasitic fungus, or chestnut blight as it is commonly called, enter cracks in the tree and the mycellia grow between the wood and the bark, destroying the growing cells and cutting off sap flow. The spore-forming bodies of the fungus break out through the bark causing the cankers and the cycle is repeated.

Chinese chestnuts have enough natural resistance so that, while they are frequently infected, they are seldom severely damaged. Unfortunately, the American chestnut trees, which had never been exposed to the blight, had no such resistance. The fungus completely girdled and killed them and rapidly spread from tree to tree. All efforts to control it proved fruitless. From 1904 until about 1950 the disease spread out over the chestnut's entire range from Maine to Alabama leaving only dead trunks behind.

(continued)

Such is the hazard of upsetting the delicate balances which sustain all ecosystems.

Spaces left by the chestnuts were generally filled in by oaks and other species, so that most people today don't recognize the loss. However, had the American chestnut remained with us, it undoubtedly would be supporting a multi-billion dollar timber industry, and wildlife in forests would be much more abundant. Its loss has been considered as the greatest botanical disaster in history and efforts are being made to restore it.

The roots of many of the killed trees continued to live and send up saplings, some of which survived enough years to produce a few nuts before they too were killed back by the blight. For many years, various programs encouraged people to collect and plant these nuts in hopes that sooner or later one would yield a resistant tree. None did, and gradually the root stocks, exhausted after 50 years of trying to regrow, have died too. Today, of an estimated 3.5 billion trees originally, fewer than 100 are still producing nuts.

Other avenues, however, may yield more success. It was hoped that cross breeding the American chestnut with the Chinese chestnut, which is a far smaller tree with no timber value, would yield a resistant tree with the stature of the American parent. Unfortunately the initial result was an inferior tree with little resistance. However, by repeatedly crossing and selecting and recrossing the progeny of these trees, some trees are now coming along that appear to have most of the desirable traits of the American chestnut and sufficient resistance to withstand the blight. In another approach, a strain of the fungus blight has been found which is far less harmful. Apparently this "hypovirulent" strain is infected with a virus that reduces the vigor of the fungus. If blighted trees are inoculated with the hypovirulent strain, the fungus blight becomes infected and the trees recover. The problem is to get the hypovirulent strain to spread to provide natural protection. Researchers in both of these areas are now in a race to achieve success before the last remnants of the once great American chestnut fade into extinction. It shows that balances once upset are difficult, and may be impossible, to restore despite all our knowledge and technology.

petitor in its particular niche, but it offers weak if any competition outside its niche.

Again, adaptation and specialization of organisms to individual niches is something that has evolved over millions of years and is specific to each individual ecosystem. Introduced species may push native species into extinction by competing for their niches. A few examples are:

☐ Starlings and house sparrows, introduced to North America from Europe, have displaced native species such as bluebirds from many areas because of their more aggressive nature.

☐ Wild burros are overgrazing desert ecosystems in the West and are displacing bighorn sheep and small mammals.

☐ Native species of fish have been displaced from many rivers and lakes by the introduction of foreign species, such as carp.

☐ Goats introduced into many islands have caused the extinction of many species by being more effective grazers.

Unfortunately, the list of such examples goes on and on. Of all the introductions that have been made, very few have hit the balance of being able to survive and reproduce within the new system without upsetting it.

Plants. This concept also applies to plants. Abiotic conditions vary from location to location, and plants have corresponding adaptations, some liking a moister situation, others requiring it to be dryer; some requiring shade and cool, others thriving in hot sun, and so on. A plant may be a strong competitor in its particular environment, but as it spreads and moves out of its specialized environment, it faces increasing stress. It is a strong competitor where conditions are optimal for it, but it is much weaker where conditions are less favorable.

The following example may be seen throughout the central portion of Eastern deciduous forests: sycamore trees are found along river banks and flood plains (areas that are occasionally flooded), and oaks are found on the valley sides. The sycamore is best adapted to very wet soils and can withstand frequent flooding. Oaks, on the other hand, are best adapted to well-drained soils and cannot withstand frequent or prolonged flooding. Sycamore seeds spread up the valley sides and, in the absence of oaks and other competing species, they will grow there. However, when oaks are present, they win the competition in this location. Similarly, acorns fall on the flood plain, but oak seedlings and saplings cannot tolerate excess water and they are killed back by flooding. The result is that sycamores win the competition on the flood plain. Thus, each species is able to win on its own "home ground," and hence the competitive balance may be maintained indefinitely (Fig. 3–6).

The maintenance of such balances assumes, of course, that the underlying conditions remain unchanged. What would you expect to happen, for example, if the river were dammed and the flood plain became a drier habitat?

Additionally, one portion of the plant community may create additional habitats within it. Deciduous forests, for example, create a cool, shady environment on the ground which the trees cannot utilize. Numerous mosses, ferns, and other plants grow on the ground, occupying this space. The diversity of plants, in turn, creates more niches and thus supports a corresponding diversity of consumers. All the consumers in their various niches are the same consumers as those involved in predator-prey and host-parasite balances which provide stability to the overall system.

Competition among Plants and Plant-Herbivore Balance

We have noted that subtle differences in microclimate and other abiotic factors account for maintaining different plant species in an ecosystem. However, this cannot be the full answer because diversity of plant species is also seen even where the environment is exceptionally uniform. Tropical forests are phenomenal in this respect; literally hundreds of different tree species may be found intermixed and scattered throughout apparently uniform areas. These trees are obviously competing for light, nutrients, and water.

How is the competitive balance maintained in situations like this?

A factor that plays an important role in maintaining this diversity in a plant community is **herbivory.** This refers to plants being fed upon by any of numerous kinds of herbivores, which include thousands of species of insects and parasites, even more than large animals.

Insects, parasites, and other small herbivores are generally restricted to feeding on one or a few plant species and follow the same kind of density-dependent pattern as was described for a natural enemy and its prey. When a plant population is at a high density, a specific insect pest, for example, has a "bonanza." With such an abundant food supply the pest population may "break away" from its natural enemies and explode. The pest population simply multiplies faster than its natural enemies can keep up, and an event commonly called an **outbreak** occurs. The result is that the plant population is virtually wiped out. Then, the pest population crashes due to lack of food and its natural enemies regain control. With the plant population at a low density (surviving individuals widely scattered), the pest has trouble finding its next meal, its population does not explode, natural enemies keep it in check, and herbivory is relatively light. But, the episode may be repeated if the plant population regains high density. Thus, systems which lack diversity are prone to population "booms and busts."

The low density of the first plant, however, implies vacant space that may be occupied by other kinds of plants. If each has the same relationship with a particular herbivore as the first, they will all end up coexisting in the same area, despite competition

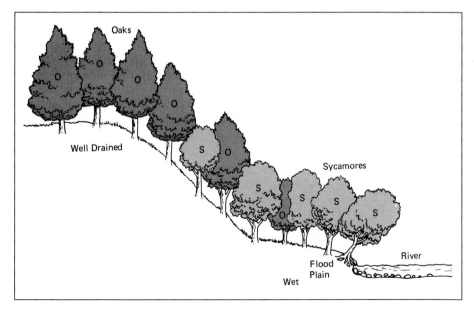

FIGURE 3–6
Competing species coexist in the same area by occupying different habitats. Oaks are better adapted to well-drained soils. Sycamores are better adapted to wet soils and can withstand flooding.

among them, because if any one of them begins to gain dominance it also gains higher density and is cut back by an outbreak of its particular pest (Fig. 3–7). The most stable and sustainable situation, then, is a diverse system which includes numerous species all at relatively low density. Again we see the evidence of the principle that *species diversity provides ecosystem stability.*

As with predator-prey balances, balances among competing plant species are not automatic but are the product of many thousands of years of evolutionary development. A plant introduced from another ecosystem may not fit into the balance. There are numerous examples where introduced plants have "gone wild," meaning that they have proven too vigorous to be held in check by competition of native plants or natural enemies in the receiving system. The result is that they overgrow and eliminate native plants in the ecosystem. This has tragic consequences for the animals also, as the entire food web is disrupted. A couple of examples:

In 1884, the water hyacinth, originally from South and Central America, was introduced into Florida as an ornamental flower. It soon escaped into waterways, where it had little competition and few natural enemies. It has proliferated to the extent of making navigation difficult or impossible (Fig. 3–8a). Millions of dollars have been spent attempting to get rid of this weed, but with little success. Kudzu, a vigorous vine introduced into the Southeast to control erosion, is winning the competition against everything else and taking over whole forests (Fig. 3–8b).

Fire

Fire is an abiotic factor that needs particular mention. A raging forest fire consumes virtually every living thing and is obviously destructive (Fig. 3–9). Unfortunately, about 70 years ago forest and range managers interpreted this potential destructiveness of fire to mean that all fire is bad and embarked on fire prevention programs that eliminated fires from many areas. Unexpectedly, however, the elimination of fire did not preserve all ecosystems in their existing state. Relatively dry pine forests of the western United States that were originally clear and open became cluttered with trunks and branches of trees that died in the normal aging process. These became the breeding ground for wood-boring insects that proceeded to attack live trees. In pine forests of the southeastern United States, economically worthless scrub oaks and other broadleaf species grew up under and proceeded to displace the more valuable pines. Semiarid grasslands were gradually taken over by scrubby, woody species that hindered grazing.

It is now recognized that fire, since it may be started by lightning, is a *natural* abiotic factor. Different plants have varying degrees of adaptation or resistance to it. In particular, grasses and pines have their growing buds located deep in the center of leaves or needles where they are protected from fire

---

FIGURE 3–7
Species diversity provides ecosystem stability. A high density of a single species is prone to being killed back by a pest outbreak leaving room for invasion by other species (*top*). As this process is repeated a diverse ecosystem results.

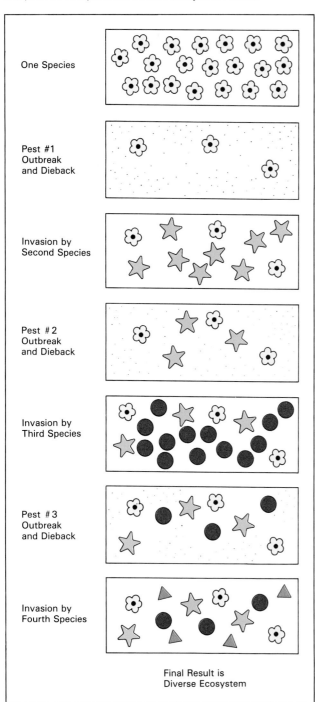

One Species

Pest #1 Outbreak and Dieback

Invasion by Second Species

Pest #2 Outbreak and Dieback

Invasion by Third Species

Pest #3 Outbreak and Dieback

Invasion by Fourth Species

Final Result is Diverse Ecosystem

FIGURE 3–8
Introduced species don't have balanced relationships with their new partners. (a) Water hyacinth overgrowing waterways. (U.S. Department of the Army Corps of Engineers photos.) (b) Kudzu overgrowing forests. (USDA-Soil Conservation Service photo.)

whereas broadleaf species such as oaks have their buds in exposed locations where they are sensitive to fire damage (Fig. 3–10). Consequently, where these species exist in competition, periodic burning may be instrumental in maintaining a balance in favor of pines and/or grass by cutting back other species. In the absence of fire, the broadleaf species gradually win the competition and the balance tips in their direction.

In a relatively dry ecosystem, where natural decomposition is slow, fire may also play a valuable role in releasing nutrients from dead organic matter. Some

FIGURE 3–9
(a) Controlled ground fires are virtually impossible after a significant amount of deadwood has accumulated. The ground fire shown here is becoming a crown fire. (USDA-Soil Conservation Service photo.) (b) Fires that burn from the ground to the treetops are highly destructive as shown here in the aftermath of a forest fire. (U.S. Forest Service photo.)

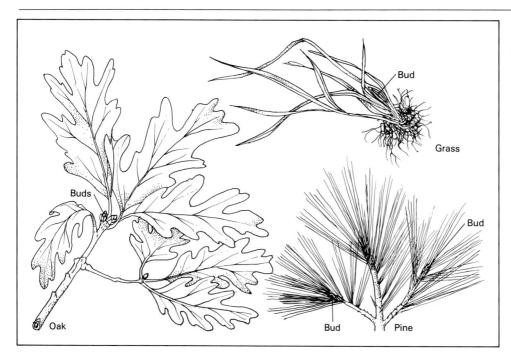

FIGURE 3–10
Position of growing bud gives different degrees of protection against fire. Species such as pines and grasses with protected buds are fire resistant; oaks do not have protected buds and are not fire resistant.

species actually depend on fire. The cones of lodgepole pine, for example, will not release their seeds until they have been scorched by fire.

Fire is now being used again as a tool in the management of rangelands (Fig. 3–11) and conifer forests. Particularly, ground fires, which simply burn along the ground, remove the dead litter, eliminate the breeding ground of wood-boring insects, and kill broadleaf seedlings but do not harm mature trees or wildlife (Fig. 3–12). Furthermore, if ground fires occur every few years, there will be relatively little accumulation of dead wood, and little danger that a ground fire will ignite a crown fire. In those forests, however, where all fire has been prevented for more

FIGURE 3–11
Fire is an important factor in maintaining grasslands because the grass, with buds protected in the ground, survives and regrows while woody species and many weeds are killed. (USDA-Soil Conservation Service photo.)

FIGURE 3–12
Ground fire. Far from being harmful, periodic ground fires as shown here are necessary to preserve the balance of pine forests. They remove excessive fuel and kill competing species. Note the absence of deadwood and competing species. (USDA-Soil Conservation Service photo.)

than 60 years, so much deadwood has accumulated that, if it does catch fire, it will almost certainly set off a crown fire. This was a major factor in the fires occurring in the summer of 1988, particularly in Yellowstone National Park. However, even in these cases, the fires serve to clear the dead wood and sickly trees that provide a breeding ground for pests, release nutrients, and provide for a fresh ecological start. After the fire the forest will recover to be more healthy and diverse. It will be unfortunate if this experience causes us to go back to the no-burn policy that created the highly flammable situation and many forest pest problems as well.

It is important to emphasize that fire is effective in maintaining only certain kinds of ecosystems. In deciduous forests and tropical forests that have achieved balance without fire, even ground fires can serve only to destroy the humus-rich topsoil and permit harmful erosion and loss of nutrients.

Territoriality

Many species of animals have behavioral adaptations that tend to keep their population within resource limits. A commonly observed trait is **territoriality.** For example, in many species of songbirds, the males "stake out" a territory at the time of nesting. Their song actually has the function of warning other males to keep away. The territory defended is such that it assures the male and his mate of being able to gather enough food to successfully rear a brood. One can readily see the advantages of territoriality. However, while some members are successful, other members are unable to find and/or defend a territory and are prevented from breeding. Therefore, the reproduction of the population is balanced with the environment required to support it.

Additional forms of territoriality can be observed in other animals. For instance, wolves defend a territory as a pack rather than as individuals. When the pack gets beyond a certain size it splits or individuals are forced out; if these splinter groups are not able to form a pack with a territory of their own, they will be forced into nonbreeding, if not death.

Ecosystem Change: Succession

When a factor of the ecosystem is changed, some species are disfavored and their populations dwindle and may be eliminated altogether. Other species, however, are favored by the change and their populations increase. Still other species may invade the area for the first time. Overall, a change leads to one or more species being displaced by other species, a process called **succession.** Succession is a natural phenomenon although it is frequently the result of human-induced changes. Natural succession occurs when the growth of one community causes changes that make the environment more favorable to a second community. As the second community develops, conditions become less favorable to the first. This results in a natural, gradual transition from the first to the

second system. There may be additional stages before a lasting balance among species is reached.

Primary Succession

If the area has not previously been occupied, the process of initial invasion and then progression from one ecosystem to another is referred to as **primary succession.** A classic example is the gradual invasion of a bare rock surface by what eventually becomes a forest ecosystem. Bare rock is an inhospitable environment. There are few places for seeds to lodge and germinate, and if they do, the seedlings are killed by lack of water or exposure to wind and sun on the rock surface. However, moss is uniquely adapted to this environment (Fig. 3–13). Its tiny spores, specialized cells that function reproductively, can lodge and germinate in minute cracks, and it can withstand severe drying simply by becoming dormant. With each bit of moisture it grows and gradually forms a mat that acts as a sieve, catching and holding soil particles as they are broken from the rock, or as they blow or wash by. Thus, a layer of soil, held in place by the moss, gradually accumulates (Fig. 3–14a). The mat of moss and soil provides a suitable place for seeds of larger plants to lodge, and the greater amount of water held by the mat permits their germination and growth (Fig. 3–14b). The larger plants in turn serve to collect and build additional soil, break up the rock via rooting, and eventually there is enough soil to support shrubs and trees. In the process, the fallen leaves and other litter from the larger plants smother and eliminate the moss and most of the smaller plants that initiated the process. Thus, there is a gradual succession from moss through small plants and finally to trees. Erosion, earthquakes, landslides, and volcanic eruptions

FIGURE 3–13
Moss growing on bare rock. Moss can invade bare rock because its spores can lodge on the surface. It survives intense heat and dryness by simply becoming dormant. With each bit of moisture, it resumes growth. (Photo by author.)

(a)

(b)

FIGURE 3–14
Primary succession on bare rock. (a) Moss holds soil. (b) Seed plants invade the moss-soil mat. (Photos by author.)

expose new rock surfaces so that primary succession is always occurring somewhere.

Another example of primary succession is the gradual filling and invasion of lakes by forest ecosystems. As streams or rivers enter large bodies of water, their velocity slows, agitation decreases, and soil particles carried by the water settle to the bottom. When sediments build up to within a few feet of the water surface, various aquatic plants, such as water-lilies and cattails, root in the bottom and send their leaves to the surface. The roots of these plants stabilize the sediments, and the dense strands of stems and leaves act as a filter that traps increasing amounts of sediments. Also, as the plants grow and die, their dead remains accumulate and add further to the sediments. (Decomposition in such environments is very slow; therefore, accumulation of the dead organic matter does occur.) Floodwaters and plant production continue to add sediments, finally bringing the level of the sediments higher than the average water level and providing an environment that is dry enough for grasses, then shrubs, and finally trees. Thus, the land ecosystem slowly invades a body of water and aquatic organisms are gradually eliminated. In time a lake may become completely filled and the area can become indistinguishable from the surrounding ecosystem (Fig. 3–15).

In the course of geological history, new lakes were formed by glaciation, crustal movements, or the blockage of river courses by landslides. In recent times, humans have formed many new lakes by building dams. All these lakes are destined to be gradually filled by the process of succession.

Secondary Succession

When an area is cleared, as for agriculture, and then abandoned, the dominant ecosystem of the area will generally return through a series of well-defined stages. Since this is the reestablishment of an ecosystem that was originally present, the process is termed **secondary succession**. A classic example is the progression from abandoned agricultural fields back to broadleaf trees that occurs in deciduous forest regions of the eastern United States (Figs. 3–16a, 3–16b).

On an abandoned agricultural field, crabgrass is predominant among the initial invaders. Crabgrass is particularly well adapted to invading bare soil. Its seeds germinate in the spring, and it grows and spreads rapidly by means of runners; moreover, it is exceptionally resistant to drought. However, in spite of its vigor on bare soil, crabgrass is easily shaded out by taller plants. Consequently, taller weeds and grasses, which take a year or more to develop, eventually take over from the crabgrass. Next, young pine

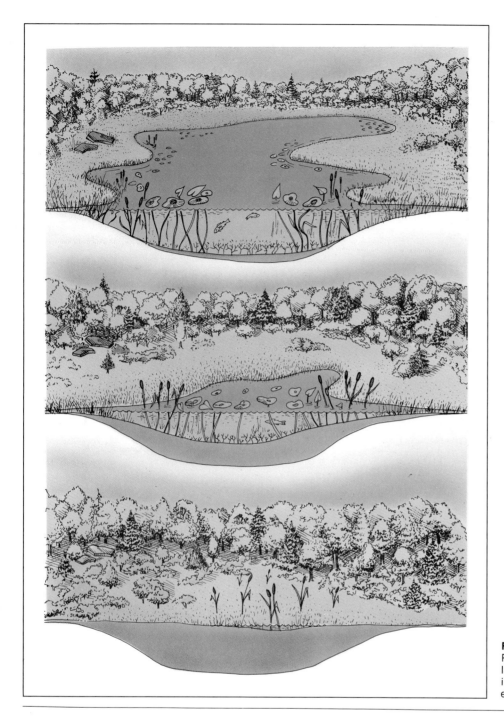

FIGURE 3–15
Primary succession. Ponds and lakes are gradually filled and invaded by the surrounding land ecosystem.

trees, which are well adapted to thrive in the direct sunlight and heat of open fields, gradually develop and shade out the smaller, sun-loving weeds and grasses, eventually forming a pine forest. But pine trees also shade out their own seedlings. The seedlings of deciduous trees, not pines, develop in the cool shade beneath the pine forest (Fig. 3–16b). Consequently, as the pines die off (their life span is 40 to 100 years), they are replaced by oaks, hickories, beeches, maples, and others that characterize Eastern deciduous forests. The seedlings of the latter continue

to flourish beneath the cover of their parents, providing a stable balance and completing the process of succession.

Both primary and secondary succession imply that spores and seeds of the various plants, and breeding populations of the various animals, are present to invade the area as conditions become suitable for them. If not, succession will be blocked or altered accordingly. Also, it is assumed in secondary succession that a fertile soil base is present at the beginning. If this soil base has been destroyed by ero-

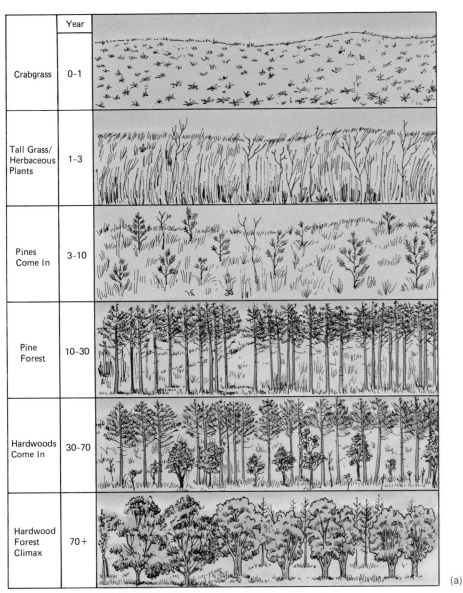

	Year	
Crabgrass	0-1	
Tall Grass/ Herbaceous Plants	1-3	
Pines Come In	3-10	
Pine Forest	10-30	
Hardwoods Come In	30-70	
Hardwood Forest Climax	70+	

(a)

(b)

FIGURE 3-16
Secondary succession. Reinvasion of an agricultural field by a forest ecosystem occurs through a series of stages.
b) Hardwoods (oak species) growing up under and displacing pines in eastern Maryland. (Photo by author.)

sion or some other means, it may be necessary for the system to start over in a manner similar to primary succession (Fig. 3–17).

The Climax Ecosystem

Succession finally reaches a point at which all the species present continue to reproduce in proportion to one another and no further change occurs. This balanced state is called the **climax,** and the system is called a **climax ecosystem.** The nature of the climax ecosystem differs according to the prevailing abiotic factors of the region. The climax in hot dry areas is a desert ecosystem; in hot wet areas it is a tropical rainforest. The major biomes described in Chapter 1 are the climax ecosystem typical of each region.

It is important to emphasize, however, that even a climax system is in no way static or rigid; it is simply a system in which all the species involved have reached a balance with one another and their environment. In this dynamic balance there is continual adjustment and readjustment as conditions and populations fluctuate from year to year. Then, over thousands of years, conditions not only fluctuate around an average, there are trends of gradual change in climate and other abiotic or biotic factors. Such changes

introduce a certain degree of imbalance, and consequently certain populations will increase and others will decrease accordingly in the process of establishing a new balance. Thus, even climax ecosystems are probably always in a state of at least very gradual change or succession. For instance, some 10,000 years ago, during the last ice age, tundra existed in the eastern United States where there are now deciduous forests.

Degree of Imbalance and Rate of Change: Succession, Upset, or Collapse

The rate of change, whether it occurs slowly over the course of many thousands of years or rapidly over the course of just a few years, is proportional to the degree of imbalance. *Succession* implies a slow, gradual change such that the degree of imbalance at any given time is not great. Thus, there is a gradual but orderly displacement of some species by others, but a diverse ecosystem of many interacting species is maintained throughout.

On the other hand, there may be very pronounced sudden changes that lead to a population explosion of one species at the expense of most other species in the system. This is referred to as an **ecological upset** as opposed to succession. Introduction of foreign species leading to problems such as those discussed earlier are examples of upsets. Another example is the discharge of nutrient-rich wastes into waterways, leading to an explosion of undesirable algal growth.

FIGURE 3–17
Copper Hill, Tenn. The forest in this area was originally killed by fumes from a copper-smelting operation. Although the fumes are no longer present, reinvasion and succession has been blocked by severe soil erosion. (U.S. Fish and Wildlife Service photo.)

Finally, changes may be so drastic or of such a nature that almost nothing survives. Such an event is referred to as *collapse of the ecosystem*. After the dieoff or collapse there may be an invasion of the area by species that are able to tolerate the new condition, but the invasion of species is not the cause of the dieoff. In fact, it initiates a new round of succession. Collapse of an ecosystem may be caused by such factors as pollution from one or more toxic substances or the diversion of a waterway causing wetlands to dry up.

It is significant to note that with few exceptions such as earthquakes and volcanic eruptions, natural changes are generally gradual, leading to succession. On the other hand, human-induced changes are frequently sudden and/or drastic, leading to upsets or collapse.

IMPLICATIONS FOR HUMANS

We have observed the principle that *populations are the result of a dynamic balance between biotic potential and environmental resistance*. How does this relate to us and the human ecosystem?

As hunter-gatherers, the human population was held in check by factors of environmental resistance as were other animal populations. With the advent of agriculture, a major factor of environmental resistance for consumers, food shortages, was greatly lessened. Still, the human population grew very slowly and suffered occasional setbacks because of diseases, especially ones such as small pox and diphtheria which killed numerous infants and children. During the mid-1800s, infectious diseases began to be controlled through vaccinations and improved sanitation. With this additional major factor of environmental resistance reduced, the human population entered a phase of explosive growth which is still continuing. The human population explosion is the result of increased recruitment; far fewer infants and children are dying of disease. Birth rate is actually down somewhat due to contraceptive practices, but not enough to counter the increase in recruitment (see Chapter 5).

In a fundamental sense, then, the human population has not behaved differently from other populations. Predictably, it has exploded, as would any population when environmental resistance is reduced. What differs is that we reduced environmental resistance through our own manipulations. Further, in affecting the balancing factors regarding our own population, we have managed to upset virtually all other natural balances as well.

We have observed that balances between species in natural ecosystems are density dependent. As populations decline, pressures of environmental resistance tend to moderate. Where humans are involved, however, this feedback control does not work because humans, supported by agriculture, can and do continue exploitation down to the total extinction of species and even entire ecosystems. In this sense, we are like an introduced species without natural enemies.

Ways in which we are vastly reducing and causing the extinction of innumerable species include:

- ☐ Wholesale destruction of natural ecosystems for human development, for example, clearing of forests for agriculture and housing.
- ☐ Diverting and damming waterways, flooding some areas and draining others.
- ☐ Discharging pollutants into air and water. For example, acid rain, a pollutant from burning coal, is causing the collapse of both aquatic and terrestrial ecosystems.
- ☐ Intentional and unintentional introduction of species from one ecosystem into another.
- ☐ Overgrazing with domestic livestock, which upsets balances (Fig. 3–18).
- ☐ Deforestation for firewood and lumber.
- ☐ Discharging nutrient-rich wastes into waterways.
- ☐ Intentional elimination of predators.

FIGURE 3–18
The range on the left side has been lightly grazed; the range on the right side has been heavily grazed. Grazing tips the balance between grasses and other species. Species resistant to grazing gain dominance. (USDA-Soil Conservation Service photo.)

☐ Use of pesticides, which cause unintentional side effects.

☐ Overhunting, overfishing, overcutting, overtrapping, overcollecting of particular plant and animal species for profit or sport.

All these activities cause the reduction of populations and extinction of species, and hence reduce the diversity of the biosphere. Further, our human ecosystem itself is an extremely *simple system*. Over 80 percent of all food produced is based on just five plant species: wheat, corn, rice, soybeans, and sugar cane. Each of these major crops, as well as most minor ones, are grown in **monoculture**. That is, a single species (even a single variety of a species) is grown over a large, highly uniform acreage. This has certain economic advantages in terms of planting and harvesting, and production is immense. However, recall the principle, *Species diversity provides ecosystem stability*.

The world's agricultural system, involving very few producer species, is extremely unstable. Crop varieties must continually be crossed with wild varieties to maintain vigor (see Chapter 17). Even so, we use in the order of a million tons of toxic chemical herbicides (weed killers) and other pesticides per year to maintain crops against the attack of pests. This approach is not sustainable. There is a crying need to develop a more diversified agriculture and use natural balances for control of pests. Many biologists, agriculturalists, and other scientists are working on this (see Chapter 16), but it is a long-term effort, and it cannot be achieved if we allow the biological diversity of natural ecosystems to be sacrificed in the meantime.

What can you do?

☐ First and foremost, recognize that continuing the trends of an exploding human population with consequent destruction, exploitation, and loss of natural ecosystems and loss of species diversity is not sustainable. It is creating an increasingly unstable biosphere more prone to upset and collapse. The need to bring our own population into balance through birth control is most urgent and will be addressed in more detail in Chapters 5 and 6. However, at the same time, much can be done to protect the biological diversity of natural ecosystems.

☐ Recognize that it is consumers who are willing to pay high prices for exotic "pets" including fish, reptiles, mammals, and various plants or their parts such as furs, tusks, and wood, who provide the incentive to overexploit natural populations and push them toward extinction. Do not buy or even condone the ownership of such "pets" or products.

☐ Join and support organizations such as the World Wildlife Fund/Conservation Foundation, Nature Conservancy, Environmental Defense Fund, and the Audubon Society (see Appendix A) that have conservation and species diversity as a major focus. Numerous local organizations working toward saving particular areas in their regions also exist and can use your support. (see Chapter 17 and 22).

☐ Support legislation focused on this issue. The first bill, the National Biological Diversity Conservation and Environmental Research Act, which makes a start toward maintaining and restoring biological diversity in the United States, was passed in 1988. But this bill is only a beginning. If such programs are going to expand and succeed, members of Congress must receive letters from their constituents giving support.

The biological basis for species diversity and its importance is explored further in Chapter 4.

4

Ecosystems: Adaptation and Change or Extinction

CONCEPT FRAMEWORK

◾ Outline

I. HOW SPECIES CHANGE AND ADAPT
—OR DON'T 91
  A. Some Basic Genetics 91
    1. Traits and Genes 91

    2. Genetic Variation 91
      a. Sexual Reproduction 91

      b. Mutations 91

  B. Gene Pools and Their Change 94

    1. Change through Selective Breeding 94

    2. Change through Natural Selection 96

      a. Specialization to Niches and Habitats 98

      b. Speciation 98

  C. Development of Ecosystems 100

  D. Evolutionary Succession 100

  E. Limits of Change: Making It or Not Making It 104

II. IMPLICATIONS FOR HUMANS 106

CASE STUDY: SEEDS FOR CHANGE 109

◾ Study Questions

1. Define: *species, traits, genes, genetic traits, genetic makeup, heredity.* How are genetic traits acquired?

2. The genetic makeup of an organism depends on how many sets of genes? Are they identical? Describe how the sets of genes behave in sexual reproduction. What is the result in terms of genetic variation among offspring?

3. Define: *mutations.* Are mutations specific or random? Explain how mutations lead to genetic variation.

4. Define: the *gene pool* of a species.

5. Contrast the number of individuals conceived versus the number that survive and reproduce in turn. How may the gene pool, and thus genetic traits, be altered through selective breeding? In how many ways may it be altered?

6. Does a form of selective breeding occur in nature? What is the basis for selection and propagation of some genetic traits over others? Define: *natural selection, survival of the fittest.*

7. Define: *niche, habitat.* Explain how natural selection generally leads to adaptation and specialization to specific niches and habitats.

8. How may two or more species be derived from one? Define: *speciation.*

9. Explain how all organisms in an ecosystem are undergoing natural selection simultaneously. How do ecosystems develop balanced relationships? How do species originally present affect the development of an ecosystem? State the pertinent principle.

10. Through natural selection, species become adapted to: present conditions? future conditions? Explain. Name and describe the three possibilities for a species when conditions change. Explain what is meant by evolutionary succession. Cite the evidence for it.

11. Contrast the features of a species that increase the chances for survival against those that decrease the chances. How does the degree and rate of change influence the situation?

12. How are humans affecting the degree and rate of environmental changes? What species are most likely to survive? to perish?

13. Is the human organism adaptable to change? is agriculture? Explain. How will changing conditions and loss of wildlife most affect humans? Explain how humans are currently at a major evolutionary turning point. Speculate regarding future human evolution on Earth.

In Chapter 3, we observed that sustainable balances among species depend on species being *adapted* to their environment and to one another. In order to sustain the biosphere, it is necessary to preserve these adaptive balances and, so far as possible, build them into our own human ecosystem. In order to appreciate the importance of this problem, we need to understand how species become adapted and their capabilities and limitations in this regard.

You may note that the adaptation of species is encompassed in the theory of evolution as set forth by Charles Darwin in 1859. However, we are going to approach the subject from the point of view of modern genetics. This will both give us a more concrete understanding of what is involved and will reveal certain additional principles.

HOW SPECIES CHANGE AND ADAPT —OR DON'T

Some Basic Genetics

Traits and Genes

A key observable feature of all species is that there is variation among the individuals. This is obvious in humans. We come in a wide range of different sizes, shapes, and colors, and we have widely differing abilities quite apart from any training or learning. Similar variation exists among the individuals of all other species: rabbits, elephants, oak trees, houseflies, and so on. The only reason we tend to think of all houseflies, for example, as identical, is simply because we do not examine them closely. When we do, it is not hard to find slight differences in size, color, and any other *trait* we choose to focus on.

The term **trait** refers to any characteristic of physical appearance including both outward appearance and inward characteristics of internal anatomy. It also refers to characteristics such as tolerance to heat or cold and disease resistance, and abilities such as running, swimming, and so on. Further, traits refer to instinctive behavioral characteristics such as a spider spinning a web, a bird flying south for the winter, or gentle versus aggressive behavior. To be sure, many traits may be modified or developed more fully by learning or training. Nevertheless, the underlying basis of traits is hereditary or genetic, which is to say that individuals are conceived with certain traits. Again, the key observation that can be made about any species is that whatever *trait you choose to consider, you will find variation among the individuals of that species.*

The last 40 years of biological research have revealed that the hereditary component of all traits, re-

gardless of species, is encoded in molecules of DNA, which reside in each cell of the body. These DNA molecules are the organism's genes (Fig. 4–1). We speak of all of them together as the **genetic makeup** of the organism.

Genetic Variation

Sexual Reproduction. The *genetic makeup* of virtually all organisms, including ourselves, consists of not one but *two* complete sets of genes. It is similar to the shoes in your closet, which can be said to consist of two sets: a set of "right" shoes and a set of "left" shoes. With genes there is no right and left as such, but the gene-pairs may be different, or they may sometimes be the same. For example, one of the pair may be a gene for blood type A while the other is for blood type O, or both may be the same. Then, another individual may have blood type genes A and B. Note that no individual will have more than two genes for a particular characteristic, but in the population at large there may be many "versions," or **alleles,** of a particular gene. The same is true for all the many thousands of genes which comprise the genetic makeup of an organism. Thus, the traits that an organism shows are the reflection of all its genes interacting together.

As cells divide in the process of body growth, each new cell receives an exact copy of both sets of genes. However, a different cell division process occurs in the formation of ova (eggs) and sperm. Egg and sperm cells end up with just *one* of the two sets of genes. However, it is a random assortment of "rights" and "lefts" so that each egg and each sperm will differ in terms of the particular members of the various gene pairs.

As a sperm enters an egg, the process called **fertilization,** their respective sets of genes combine so that the individual conceived has two complete sets. Since each egg and sperm carried a random assortment of "rights" and "lefts," each fertilization will result in a different combination of genes and this combination will be different from that of either of the parents (Fig. 4–2). Consequently, each offspring and, therefore, each member of the population (with the exception of identical twins) will have a different genetic makeup. This genetic variation is the basis for inborn variations in traits. Inborn variations are in contrast to differences that may result from an injury, learning, or training. Learning or training may develop a certain inborn genetic trait, but it does not change the genetic trait itself.

Mutations. Additional genetic variation may come into the picture through *mutations.* **Mutations** are random "accidental" changes in the DNA, and they may

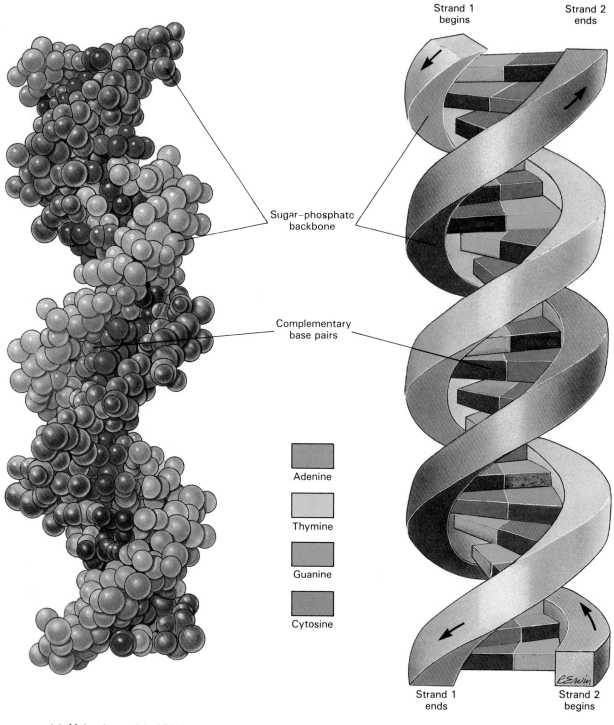

Strand 1 begins   Strand 2 ends

Sugar-phosphate backbone

Complementary base pairs

Adenine

Thymine

Guanine

Cytosine

Strand 1 ends   Strand 2 begins

(a) Molecular model of DNA

(b) Ribbon model of DNA

FIGURE 4–1

The structure of DNA (*deoxyribonucleic acid*), the genetic material for all organisms. (a) A model showing the actual arrangement of atoms. (b) A simplified model showing the key aspects of the structure. The sequence of base pairs, which form the "steps" down the spiral, provide a code that, in the organism, is translated into the production of specific proteins, which determine the organism's structure and functions. A single gene involves several hundred base pairs. In cell division the DNA molecule "unzips," coming apart between the base pairs, and each side serves as a template for the synthesis of the "other" side. The end result is two identical molecules. (From F. Martini, *Fundamentals of Anatomy and Physiology,* Englewood Cliffs, N.J.: Prentice Hall, 1989.)

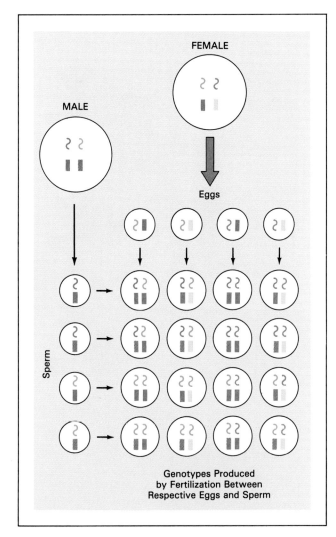

FIGURE 4–2
Genetic variation results from sexual reproduction. Adult individuals carry two sets of genes represented in this illustration by the 2-shape and the 1-shape. Different colors represent different variations or alleles of the particular genes. Eggs and sperm are produced in a way such that they receive just one set of genes but it is a random assortment of "rights" and "lefts." Then, in fertilization any egg may be fertilized by any sperm, again resulting in two sets of genes. Note the number of different genetic combinations that may result from just two genes, each involving three alleles. How many additional combinations could be obtained in the next generation just with these alleles? (Answer: total combinations = 81.) Now consider that most organisms have in the order of a million genes. Is there any limit to genetic variation?

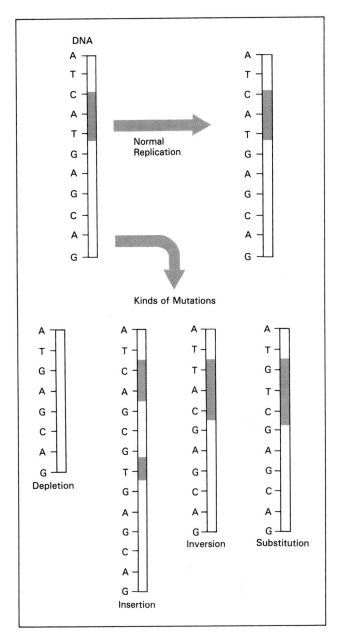

FIGURE 4–3
Mutations. Generally the DNA molecule is replicated exactly so that all cells resulting from divisions carry exactly the same genes. Occasionally, however, changes occur or are caused by factors such as radiation. All changes in DNA are called mutations. The four basic kinds of changes illustrated may involve any number of base pairs. Such mutations create new variations or alleles of the gene in which they occur. Of course, most mutations result in a less-functional gene and are therefore considered harmful.

involve any gene or genes (Fig. 4–3). Mutations may occur spontaneously (with no apparent cause). Though such mutations are relatively rare, this is how different alleles of genes (blue eyes vs. brown eyes for example) apparently originated and continue to originate. But, we now know that mutations may also be caused at much higher frequencies by high-energy radiation such as that from X-rays or radioactive materials, by various drugs, and by a wide variety of chemicals used in industry or contained in industrial wastes. Unfortunately, these induced mutations are frequently of a sort that results in cells dividing in an uncontrolled way resulting in cancer, or insofar as they occur in eggs or sperm, they may result in severe birth defects in the next generation as they interfere with normal development.

In summary, through the occurrence of mutations and through the segregation and recombination of genes that occur in the course of sexual reproduction, genetic variation among individuals is inevitable. **Cloning** is the process of creating new individuals without going through the sexual process. Propagation of plants from cuttings is an example. In this case, all the new individuals are produced through the regular cell division process in which all the DNA is copied exactly. Thus all members of the clone are genetically identical, barring mutations.

Gene Pools and Their Change

We have noted that each individual has two complete sets of genes and hence may have two different alleles of any particular gene. However, other individuals may have still other alleles of the gene. Thus, in the population at large, there may be a considerable number of alleles for each of the many thousands of genes that are involved in the genetic makeup of an individual.

Suppose you could take a sample of the genes from every individual in a population and put all these samples together. What you are visualizing here is the *gene pool* for the population. The **gene pool** of a species is the total of all the different alleles of each gene that exist in the entire population of the species. Some alleles in the pool are very common, and almost every individual has a copy of these. Other alleles, however, may be carried by only a few individuals. As an individual reproduces, copies of its alleles are passed on to its offspring. Conversely, if an individual does not reproduce, the alleles it carries are eliminated when it dies. Consequently, the gene pool for a species is not static; it may change from one generation to the next. Suppose, for example, that a gene pool for a species starts with an even number of "big-A" and "little-a" alleles, two different versions of some particular gene. If individuals carrying two "big-A" alleles reproduce more than those carrying two "little-a" alleles, the "big-A" allele will become more abundant in the population while the "little-a" allele becomes less common. Similarly, if a new gene arises by mutation and the individual and subsequent offspring reproduce more than average, that new gene will become more and more prevalent in the population. On the other hand, any new allele that inhibits reproduction of the individual carrying it will be weeded out of the population when the individual dies (Fig. 4–4).

Change through Selective Breeding

Gene pools may be changed in desired ways through *selective breeding*. Agricultural and domestic breeds of plants and animals have been derived from wild

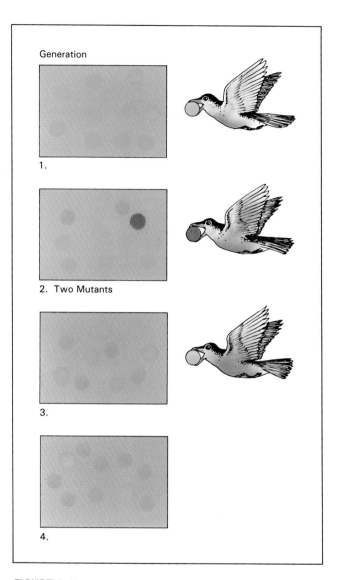

FIGURE 4–4
New alleles, which arise by mutation, may enhance or hinder survival. A new allele that makes the individual more able to survive (less conspicuous to predators in this case) and reproduce will become increasingly common in the population. Conversely, a new allele that makes the individual less apt to survive (more conspicuous to predators) will be weeded out of the population.

ancestors by this process. Breeders first envision the characteristics that they would like to achieve in a given species; a dog with short, squat legs that is able to wiggle down gopher holes to aid hunters for example. Then, examining the existing population of dogs, they select individuals that show the traits (short squat legs) a little more than others, and use them as the parents for the next generation. The offspring tend to be like the parents, but some express the particular trait more than the parents. Those individuals that show the trait the most are selected as the parents for another generation. This process, called **selective breeding,** is repeated over and over again until gradually the desired traits are developed; dachsunds are the result of this process, illustrated in Figure 4–5.

Breeders are doing their selective breeding on the basis of traits (leg length in this case). However, since traits are the manifestation of a certain genetic makeup, what is actually occurring is the selection and propagation of certain alleles in the gene pool which lead to short, squat leg structure while weeding out alleles that give taller structure. That the DNA in the gene pool has actually changed in the process can be verified by techniques which enable one to compare the DNA of the starting population with that of the developed breed.

It is also significant to note that the same starting gene pool can be manipulated through selective breeding in any number of different ways. It simply depends on selecting and breeding for different traits. Thus all the different breeds of dogs from great danes to Chihuahuas are derived from the same wild dog gene pool, and the same is true for multiple breeds of other species (Fig. 4–6).

Further modification of the gene pool may be accomplished through crossbreeding with a different but closely related species. (More distantly related species do not interbreed.) This technique is called **hybridization** and the resulting offspring are referred to as a **hybrid.** Any hybrid is, however, a random

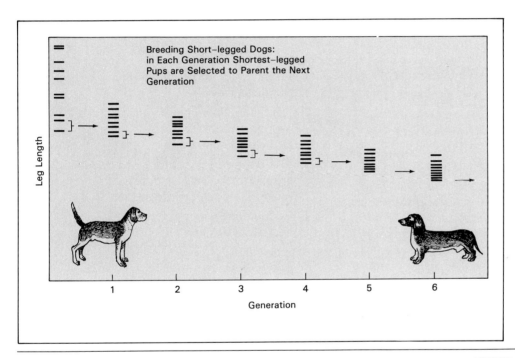

FIGURE 4–5
Breeding for short-legged dogs. When the shortest-legged dogs are selected to be parents of the next generation, over a number of generations the result is dogs with shorter and shorter legs.

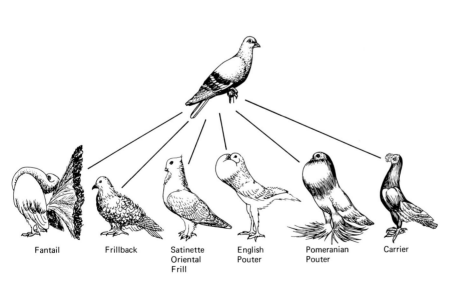

FIGURE 4–6
Breeds of pigeons developed by selective breeding. The wild rock pigeon of Europe, top, is thought to be the ancestor of all the domesticated breeds shown here. (Redrawn from W. W. Levi, *The Pigeon,* Sumter, S.C.: Levi Publishing Co., 1957.)

assortment of genes, half from each of the two parents. To achieve a desirable assortment of genes/traits, the breeder must again proceed with selective breeding. For example a breeder may cross a domestic variety of wheat with a wild relative in order to introduce enhanced disease resistance. However, undesired characteristics such as diminished productivity are introduced as well and these must be sorted out through further selective breeding.

In just the last few years, technology has been developed that enables specific genes, segments of DNA, to be isolated from one species and introduced into another directly without crossbreeding. The potential is tremendous since it both shortcuts the time involved in selective breeding and overcomes the hurdle of not being able to cross anything but closely related species. This technology, referred to as **genetic engineering,** is the subject of considerable controversy, however, because the ultimate results of modifying gene pools to this extent are not entirely predictable.

Change through Natural Selection

Wild populations in nature are also continually undergoing a selection process. Recall from Chapter 3, that every population is a balance between its biotic potential adding individuals to the population and factors of environmental resistance causing their demise. A critical feature is that every species has a reproductive rate (babies borne, eggs laid, or seeds set) that appears "excessive." What keeps populations from exploding is high dieoffs during the recruitment phase. Most individuals conceived don't live to become reproducing adults but succumb along the way to predation, disease, lack of food or nutrients, lack of sufficient water, or other factors of environmental resistance.

It is interesting to look at this in terms of actual numbers. Consider a population of 100 individuals (50 male/female pairs). If this population is to remain constant, these 100 individuals must ultimately be replaced by 100 of their offspring, no more and no less. If the population is 50 pairs of fish, each producing 20,000 fertile eggs, there will be a total production of 1 million eggs ($50 \times 20,000 = 1,000,000$). Yet, only 100 live to become reproducing adults, a survival of just 1 in 10,000. Even animals with the lowest of reproductive rates are capable of producing 10 to 20 offspring over their lifetimes, implying an average survival of 1 in 5 to 1 in 10 to maintain a stable population. The only time that the percentage of survival may be higher is during periods of population explosions, and such episodes never go for very many generations—environmental resistance reimposes the low-percentage survival situation.

In nature, then, essentially every generation of every species is subjected to an intense selection for *survival and reproduction.* Very simply, individuals that survive and reproduce pass their alleles on to the next generation, while the alleles of those that don't survive are eliminated from the pool. Thus the gene pool for every species is constantly undergoing a process of **natural selection,** in which new alleles that provide or enhance a trait which aids survival and reproduction will become increasingly widespread in the population. Simultaneously, genes that are less effective in promoting survival and reproduction are gradually weeded out of the population as individuals carrying them fail to survive or reproduce.

It is hardly surprising then, that virtually all traits of all organisms serve in one way or another to support the survival and reproduction of that species. They can be grouped accordingly:

1. Adaptations for coping with climatic and other abiotic factors.
2. Adaptations for obtaining food and water in the case of animals, or nutrients, energy, and water in the case of plants.
3. Adaptations for escaping from or protection against predation, and resistance to disease-causing or parasitic organisms.
4. Adaptations for finding or attracting mates in animals, or pollination and setting seed in plants.
5. Adaptations for migration in animals or seed dispersal in plants.

The "bottom line" for any trait is: Does it work to support survival and reproduction of the organism? If it does, it will be developed and enhanced through the natural selection process regardless of what it is. Consequently, among various organisms we find numerous different traits which accomplish the same function. For example, the ability to run fast, fly, or burrow, and features such as quills, thorns, and obnoxious smell or taste all support the function of reducing predation and are seen in various organisms. The same may be said regarding other functions (Fig. 4–7). Indeed, it is instructive to select any species (plant, animal, fungus, or bacterium) and list its specific traits and note how each one benefits survival by supporting one or more of the above functions.

You will note that "survival and reproduction" are lumped together. Of course an organism won't reproduce if it doesn't survive. However, it is possible for an organism to live to "ripe old age" without ever reproducing. When this occurs the genes of that individual, beneficial to survival as they may be, are eliminated from the gene pool of the species as surely

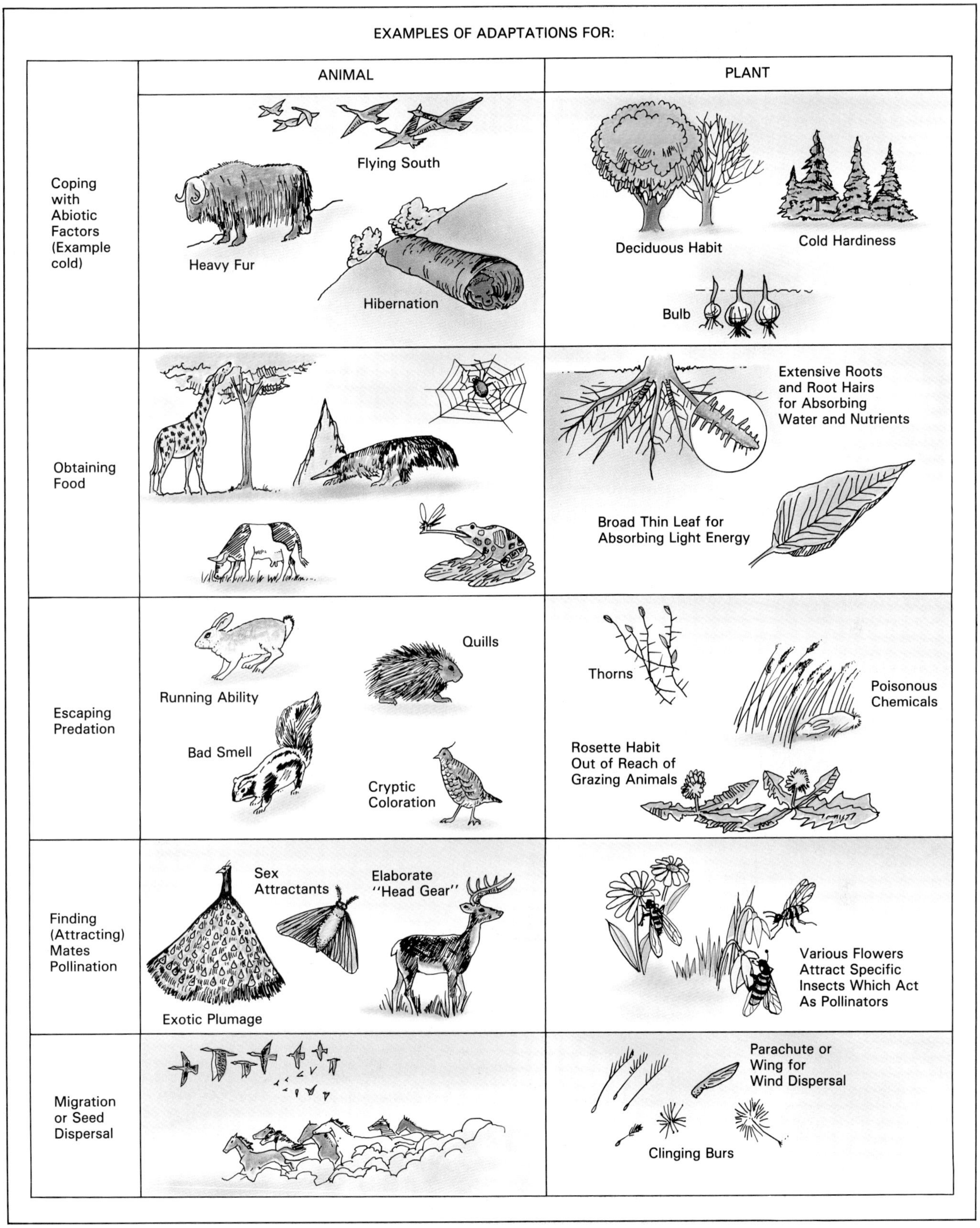

EXAMPLES OF ADAPTATIONS FOR:

	ANIMAL	PLANT
Coping with Abiotic Factors (Example cold)	Flying South, Heavy Fur, Hibernation	Deciduous Habit, Cold Hardiness, Bulb
Obtaining Food		Extensive Roots and Root Hairs for Absorbing Water and Nutrients, Broad Thin Leaf for Absorbing Light Energy
Escaping Predation	Running Ability, Bad Smell, Quills, Cryptic Coloration	Thorns, Poisonous Chemicals, Rosette Habit Out of Reach of Grazing Animals
Finding (Attracting) Mates Pollination	Sex Attractants, Elaborate "Head Gear", Exotic Plumage	Various Flowers Attract Specific Insects Which Act As Pollinators
Migration or Seed Dispersal		Parachute or Wing for Wind Dispersal, Clinging Burs

FIGURE 4–7

Examples of adaptations. Every species can be viewed as a complex of adaptations enabling the individuals to: (1) cope with abiotic factors; (2) obtain food and water, in the case of animals, or nutrients, energy, and water, in the case of plants; (3) escape from or gain protection against predation; (4) find and attract mates and reproduce in the case of animals or pollinate and produce seeds or spores in the case of plants; and (5) migrate or disperse seeds or spores. A multitude of different features will accomplish these functions.

as if the individual had died in infancy. For genes to be passed on and have any part in future generations, reproduction must occur; survival by itself will not suffice. Further, the genetic effect on the next generation is directly proportional to the number of offspring that are produced and that survive and reproduce in turn.

Specialization to Niches and Habitats. Through natural selection, traits which aid survival will be gradually developed until there is optimal adaptation to the *conditions present*. Thus, natural selection results in species gradually becoming increasingly modified and specialized to the particular habitats and niches in which they exist. Thus, anteaters become increasingly well adapted to feeding on ants, giraffes become increasingly well adapted to feeding on treetops, and so on.

As organisms become better adapted to their habitat and niche, the rate of change may be expected to diminish. For example, when natural selection has led to the point where all skunks emit an odor so noxious that would-be predators can't tolerate it, then there is no longer selection pressure for increasing the offensiveness of the odor. When ancestors of giraffes started grazing on trees, there would have been strong selection for the ability to graze higher and higher until all the population could graze on most of the trees available. After that, there is no longer selection for increasing neck length. Indeed, some counterselection comes into play because the very tall stature of giraffes presents severe problems in equalizing blood pressure from the head to the body when the animal bends over to drink water. The same may be said of any number of other traits. In a species that has become well-adapted to its habitat and niche, natural selection acts to preserve the status quo because any further increase or decrease in any trait may have a negative effect on survival (Fig. 4–8).

Speciation. However, conditions do not remain constant indefinitely. Biotic and/or abiotic factors may change or a population may migrate or disperse to a location where conditions are somewhat different. If any factor in the new situation is beyond the population's limit of tolerance, the population dies out. However, if at least some individuals survive, albeit under stress, there is a strong selection of those alleles that provide better adaptation (survival) in the face of the new conditions (Fig. 4–9). Then, as new alleles providing better adaptation to the new conditions arise and are reproduced, there is increasing modification until changes in the population may be so pronounced that we recognize it as a different species from the original. This process is called **speciation**. A

FIGURE 4–8
Selection pressure. As long as a situation persists in which a taller, longer-necked animal gets more food and hence enhances its survival advantage, there is selection pressure and hence evolution toward the taller animal. When a longer neck is of no additional benefit, selection pressure and further modification stop. The same can be said for any other trait of plants or animals.

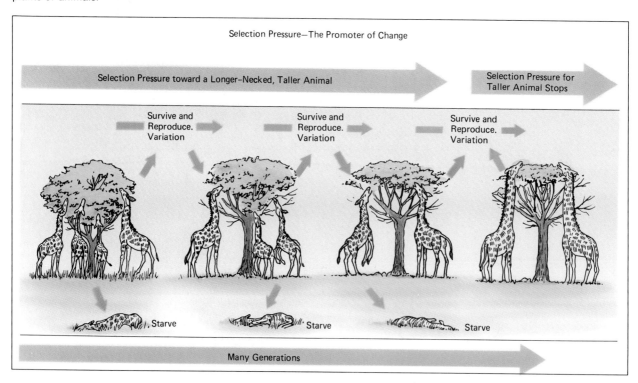

Selection Pressure—The Promoter of Change

Selection Pressure toward a Longer-Necked, Taller Animal

Selection Pressure for Taller Animal Stops

Survive and Reproduce. Variation

Survive and Reproduce. Variation

Survive and Reproduce. Variation

Starve

Starve

Starve

Many Generations

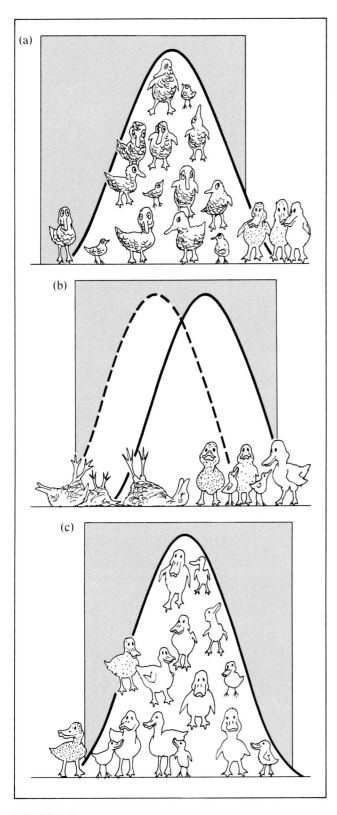

FIGURE 4–9
(a) Changing conditions (background) may cause most of the population, which was adapted to the previous conditions, to die out. A few having variations that adapt them to the new conditions survive (b) and replenish the population. The traits in the surviving population (c) thus differ from those of the original population.

classic example, development of the horse, is shown in Figure 4–10.

There are innumerable examples where we can all but see this process of modification and speciation occurring in nature. For example, in alpine regions (high elevations above the treeline) virtually all plants have a low, dense, mosslike vegetative structure. The flowers on these plants, however, are very similar if not identical to those of tall plants at lower elevations (Fig. 4–11). One can visualize that as the plant dispersed up the mountain, heavy snow and abrasive ice crystals driven by strong wind sheered off and killed plants with alleles for tallness. However, plants with alleles for a shorter, stouter form survived and reproduced. Continuing this process over many generations resulted in a low mosslike body form. In some cases populations showing all the intermediate stages continue to coexist (Fig. 4–12). In other cases, populations showing the intermediate stages have died out leaving the two related but distinct species. For example, there is not one maple or oak species but many related species.

The same situation may be seen among animals. Take foxes, for example; in the arctic, selection pressures favor individuals that have variations such as heavier fur and shorter tail, legs, ears, and nose, all of which help conserve body heat. In southern regions, selection pressures regarding adaptation to temperature are the reverse; the above variations would actually be harmful because in warmer climates the fox needs to dissipate excessive body heat. If fox populations in the two regions interbreed, the resulting genetic mixing will preserve them as one species. However, if they are separated in such a way that interbreeding does not occur, natural selection may eventually produce differences great enough that the populations of the two areas become different species. Observe the arctic fox and gray fox shown in Figure 4–13.

In nature we observe that each species generally occurs not singly but in a group of related species. This fact is explained by the process of speciation.

In summary, through reproduction and limited survival of offspring, we observe gradual modification of the gene pool. As this process progresses over many generations the changes may be such that we define the modified population as a new species.

What we have described is exactly the concept presented by Charles Darwin in 1859, the evolution of species through "natural selection" and "survival of the fittest," the "fittest" being those individuals with variations of traits that best enable their survival and reproduction. Darwin deserves great credit for constructing a picture of what was occurring purely from empirical evidence without any knowledge of genes or genetics, information that wasn't discovered until some years later.

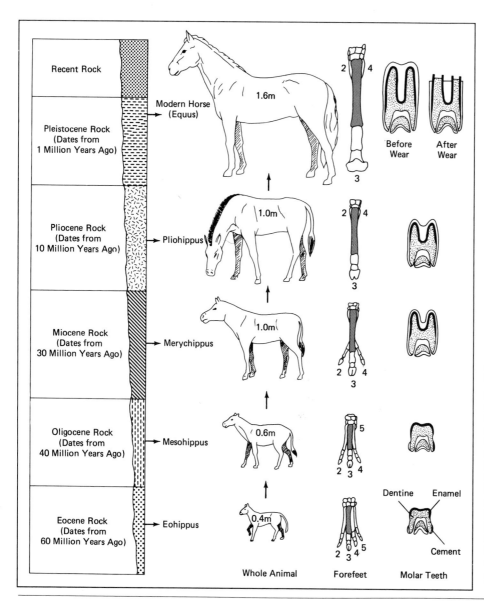

FIGURE 4–10
The evolution of the horse. The ancestor of the horse, *Eohippus*, existed in moist forests browsing on tender buds and shoots. As the climate became dryer forests gave way to grasslands, an ecosystem that favors a larger animal with a stronger leg structure that can outrun predators and travel longer distances between water holes and grazing areas. It also favors larger, stronger teeth that can grind the coarser grasses. Fossil evidence shows these adaptations occurred gradually, leading to the modern horse from *Eohippus*. (Redrawn from G. deBeer, *Atlas of Evolution*, p. 49, New York: Thomas Nelson, Inc., 1964.)

Development of Ecosystems

In our discussion to this point, it has been necessary to focus on one species, at a time. However, it is important to remember that every species exists, and can only exist, in the context of an ecosystem which involves many species (producers, consumers, detritus feeders, and decomposers) interacting together. Therefore, a major portion of the adaptation of every species involves adaptation to other species in the ecosystem as well as to the abiotic factors of the environment. For example, in a predator-prey relationship, natural selection is both increasing alleles in the prey that give it greater protection against the predator, and increasing alleles in the predator that enhance its ability to capture the prey. The same is true in all biotic relationships.

How are balances created and maintained in the course of this process? You can answer this question for yourself by seeing the answer to another more obvious question: If there is a lasting imbalance between two species, what will inevitably happen to the species on the "downside?" Obviously it will be forced into extinction. Whether the species on the "upside" is adversely affected by this depends on its ability to survive without the other species. Consequently, there is a kind of natural selection against any relationship that doesn't balance. It is simply eliminated from the ecosystem by the downside becoming extinct. The only relationships that will survive over long periods of time are balanced relationships, as we studied in Chapter 3.

Another feature profoundly affects the way in which ecosystems develop. Note that the process of

(a)

(b)

(c)

(d)

FIGURE 4–11
Adaptation to different climates. The white daisy (a) and wild mustard (b) are common tall weeds
of open fields in moist temperate climates. In alpine tundra (c) closely related species with ground-hugging
moss-like habit capable of surviving severe wind and ice are found (d). We can
presume that the latter developed from the former through the gradual process of natural selection
and survival of the fittest as they were exposed to harsh alpine conditions. (Photos by author.)

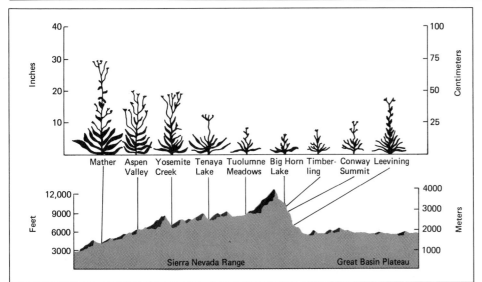

FIGURE 4–12
Speciation in progress. The plants
shown in the illustration are still
one species as demonstrated by
the fact that they are all capable
of interbreeding. But
environmental factors clearly
favor different variations in
different areas. Given time (1000 +
generations), the high-altitude and
low-altitude plants may diverge
enough to become separate
species. (Redrawn from J.
Clausen, D. Keck, and W. Hiesey,
Carnegie Institute of Washington
Publication No. 581, Washington,
D.C., 1948.)

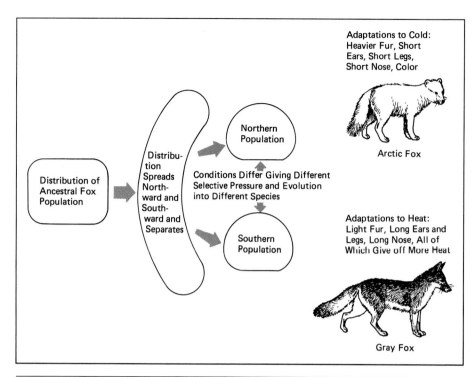

FIGURE 4–13
A population spread over a broad area faces different selective pressures in different regions. Eventually this may result in development of different species. Drawing shows a hypothetical distribution of the ancestral fox population that speciated into the arctic fox and the gray fox.

Within the figure:

Distribution of Ancestral Fox Population

Distribution Spreads Northward and Southward and Separates

Conditions Differ Giving Different Selective Pressure and Evolution into Different Species

Northern Population

Southern Population

Adaptations to Cold: Heavier Fur, Short Ears, Short Legs, Short Nose, Color

Arctic Fox

Adaptations to Heat: Light Fur, Long Ears and Legs, Long Nose, All of Which Give off More Heat

Gray Fox

---

selection is a process of modifying gene pools of existing species. Neither nature nor humans have any way of creating an entirely new gene pool or species from scratch. *What exists can be modified. What does not exist cannot be created.*

Consequently, how an ecosystem develops depends on what species are present at the beginning of the process. This explains why we find very different species making up the ecosystems of different continents and remote islands even though abiotic factors may be similar. For example, when the Galapagos Islands volcanically rose from the sea about 10,000 years ago, mammals were not present but tortoises were. Consequently, natural selection, acting on variations in tortoises, resulted in tortoises being modified and adapted into the niche of being the prime herbivore (Fig. 4–14a). Similarly, as what is now Australia split away from other land masses about 60 million years ago, mammals were not present but marsupials (mammal-like animals that carry their young in a pouch) were. Thus, marsupials, namely kangaroos, evolved into the various niches occupied by mammals on other continents (Fig. 4–14b). The Hawaiian Islands also have countless unique species which are a reflection of modification of the particular species that were the first to inhabit the islands.

This should reinforce your understanding of the point we made in Chapter 3. Namely that while each

species has developed balanced relationships in its own ecosystem, those balances may not apply when it is introduced into another ecosystem. Species in relatively young ecosystems (tens of thousands versus tens of millions of years old), such as those that exist on islands of recent geological origin, are particularly vulnerable because their competitive ability is relatively undeveloped. They stand to be totally eliminated by more highly developed species brought from the continents by humans (Fig. 4–15).

Evolutionary Succession

Through natural selection a species may become increasingly well adapted to coping with predators, parasites, climatic conditions, and other abiotic and biotic factors present. However, species other than humans have absolutely no way of anticipating future changes, much less preparing for them.

Consequently, as any biotic or abiotic factor is changed (e.g., a species from another ecosystem is introduced or the climate changes), each species that is ill-adapted to the new situation faces three possibilities (Fig. 4–16).

1. **Migration.** Part of the population may be able to migrate and find an environment where conditions are still optimal, and thus continue in the new location.

(a)

FIGURE 4–14
Evolution of ecosystems. Natural selection can only act on and modify species that are present. Therefore, species that fill particular niches differ depending on what species were present as evolution of the ecosystem began. (a) On the Galapagos Islands where mammals were not present but tortoises were, tortoises evolved into the niche of being the prime grazing animal. (Photo by author.) (b) Similarly, in Australia, with the absence of mammals, the marsupials present (kangaroos) evolved as the prime grazing animal. (Australian information Center photo.)

(b)

2. **Adaptation.** The gene pool of the species may contain sufficient variation that some individuals will tolerate the new condition, survive, and reproduce. Natural selection over succeeding generations will lead to a population that is increasingly well adapted to the new condition.

3. **Extinction.** If none of the individuals in the population can escape the new condition by migration and the new conditions are outside the limits of tolerance for all individuals, then extinction of the population and the gene pool it represents is the inevitable result. Indeed, the fossil record is replete with now extinct species. Dinosaurs are only the most well known.

As some species become extinct and surviving members of others reproduce, adapt, and speciate through natural selection, we can visualize a picture of **evolutionary succession,** that is, the earth being occupied by different species at different times

(a)

(b)

(c)

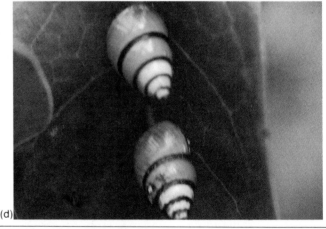

(d)

FIGURE 4–15
Because of isolation from other ecosystems the ecosystem on the Hawaiian islands has developed thousands of species which are found nowhere else. Over 500 of these are in danger of extinction due to the introduction of more vigorous competing species. (a) Silversword. (b) Hawaiian Goose. (c) Hawaiian Stilt. (d) Oahu Tree Snail. (a, Photo Researchers/ Stan Goldblatt © 1980; b, Eugene Kridler, Hawaii Fish and Wildlife Service; c, Hawaii Fish and Wildlife Service; d, Hawaii Fish and Wildlife Service.)

through the ages. Of course, this is exactly the evidence we find in the fossil record (Fig. 4–17).

Limits of Change: Making It or Not Making It

As conditions change, we see that some species may adapt and speciate while others die off and become extinct. What are the criteria for making it or not making it?

For adaptation, first, some individuals must be able to survive and reproduce despite the initial change. This depends on two factors: (1) the amount of genetic diversity in the gene pool of the species, and (2) the degree of change. A species with a high degree of genetic diversity (many different alleles) may be expected to have some members that can tolerate a fair degree of change in almost any aspect of their environment. Conversely, if a species has very little variation in its gene pool, almost any change

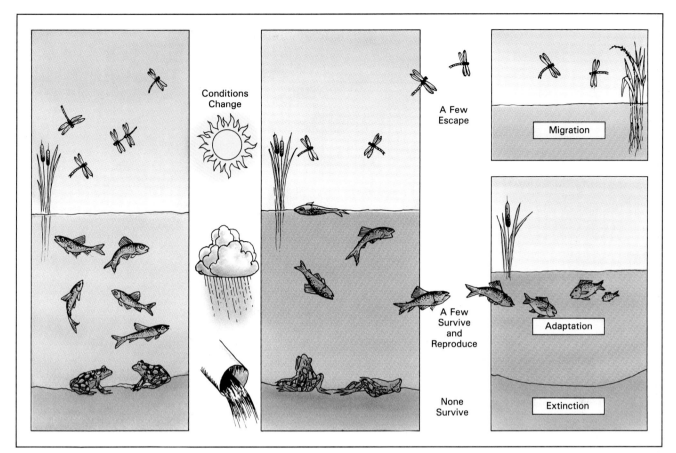

FIGURE 4–16
Species may be well adapted to their environment, but what occurs if one or more abiotic or biotic factors change?

may result in its extinction because alleles allowing it to tolerate the new conditions simply don't exist in the population. But, it is conspicuous that the degree of change is equally significant. If changes are very slight and/or occur very slowly, most species may survive and adapt. As the degree of change increases, greater genetic diversity is required for survival (Fig. 4–18). Of course, a change may be so cataclysmic, nuclear war for example, that no species survive. Another basic principle emerges: *Survival of a species depends on maintaining its genetic diversity and on minimizing changes.*

An additional factor enters into genetic diversity and change. It is geographic distribution. A population with a wide geographic distribution is likely to contain a considerable degree of genetic diversity and vice versa. Likewise some areas of a wide geographic range are likely to be more or less isolated from the change so that populations of the species may survive in some regions while being eliminated in others.

Second, given that some individuals have the traits (alleles) that enable them to survive the initial

stage, replenishment of the population and further adaptation to the new conditions depend on reproductive capacity, because *change only takes place with the selection that occurs in each generation.* For example, a pair of insects commonly produce several hundred offspring and complete a life cycle from egg to reproducing adult in a few weeks. Thus insects have a reproductive capacity which is at least 1000-fold greater than birds, which may raise two to six fledglings per year. Thus insects accomplish the same degree of adaptation to new conditions in one year as birds might accomplish in 1000 years. Is it surprising then, that insect populations rapidly adapt and become resistant to pesticides used against them while other wildlife remains threatened? Again, the changed conditions, pesticides in this case, do not cause adaptive mutations. Mutations are random changes. The adaptive traits, if they exist, stem from genetic variations already present in the gene pool of the species. If they are not present, extinction occurs.

Third, size of the organism is important. Smaller organisms, such as flies, can sustain a population in a garbage pail, while larger organisms often require large areas of habitat to survive. Panda bears, for example, need many square miles of bamboo forest.

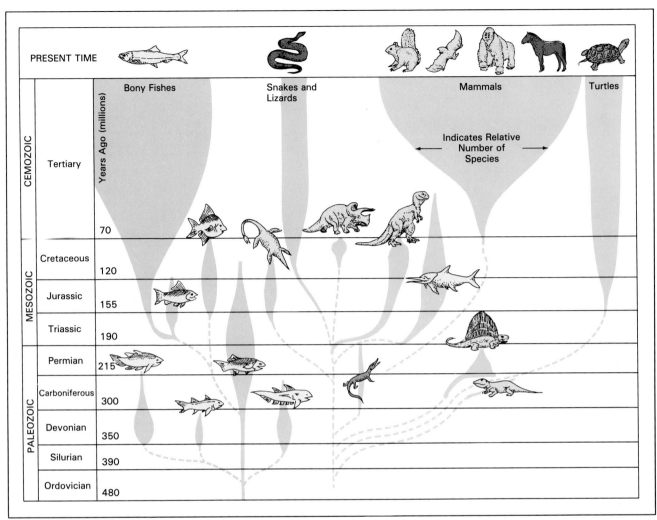

FIGURE 4–17
Survival of no species is assured. The fossil record is replete with examples of new species emerging, spreading, and diverging into many more species, many or all of which may fade into extinction.

IMPLICATIONS FOR HUMANS

Study and dating of the fossil record shows that the origin of humans on earth is a very recent phenomenon. Just how recent can be better appreciated by the following: Imagine the entire time period of organic evolution (some 4 billion years) condensed into a single year where each day represents about 11 million years (Fig. 4–19). In this time frame, most of the year, up to mid-September, is taken up by the evolution of primitive bacteria-like organisms. However, on about September 1—1.4 billion years ago—the first complex cells typical of present-day plants and animals were formed and then the pace quickened. All the major invertebrate phyla of marine organisms de-

veloped through September and October, and the first vertebrates (forerunners of fish) developed during the first part of November (some 450 million years ago). The Age of Fish and the Age of Amphibians, each lasting roughly 100 million years, occupy most of November. By the first of December, amphibians are giving way to the giant reptiles (dinosaurs) that dominate the earth during the first half of December. Then toward the end of the third week of December (60 million years ago), dinosaurs fade into extinction and mammals and birds become the dominant animals. Finally, the first humans appear about 4:00 P.M. on December 31 (3 million years ago), but agriculture is only established in the last two minutes of the year, (10,000 years ago) and the explosion of technology and knowledge of the last 200 years is represented by the last *2 seconds.*

Looking at the fossil record, we see that evolutionary succession has been punctuated by major events that have changed the following course of evolution. The development of vertebrates giving rise

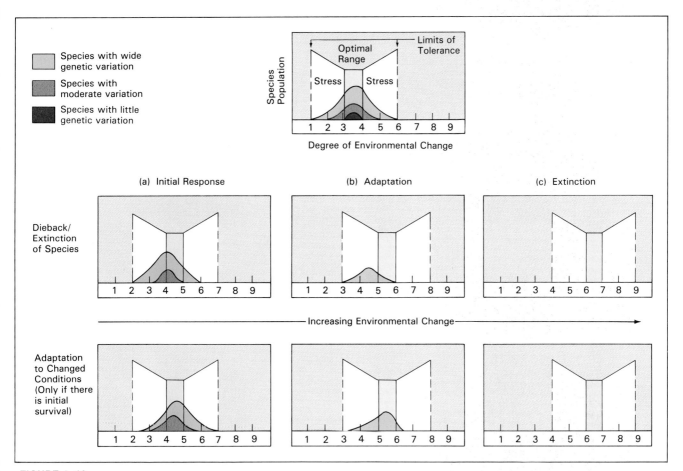

Legend:
- Species with wide genetic variation
- Species with moderate variation
- Species with little genetic variation

Species Population

Optimal Range — Limits of Tolerance

Stress | Stress

1 2 3 4 5 6 7 8 9

Degree of Environmental Change

(a) Initial Response | (b) Adaptation | (c) Extinction

Dieback/Extinction of Species

————————— Increasing Environmental Change —————————→

Adaptation to Changed Conditions (Only if there is initial survival)

FIGURE 4–18

Whether or not a species survives changing conditions depends on both the degree of change and the amount of genetic variation within the population. Small populations with little genetic variation (*red*) have very little ability to adapt and may be pushed into extinction with modest changes (*a*), while species with more variation (*green* and *blue*) may adapt to greater changes (a and b). Very marked changes, however, will cause extinction of all species (c).

to the Age of Fish and developments giving rise to land animals, reptiles, and then mammals are such events. The advent of humans is the most recent "major event," and we *Homo sapiens* have certainly changed and continue to change the face of the earth.

However, is there anything in theory indicating that humans are the culmination or the endpoint of the evolutionary process? The answer is definitely no! To the contrary, we see that all the processes of genetic change leading to new species or extinction are ongoing. *The one factor that determines longevity of a species on earth is its reaching and maintaining a stable balance with other species in an ecosystem that has efficient recycling of nutrients and sustainable energy flows.* Recognizing this, we see that modern humans could be a mere flash on the evolutionary scene—a few seconds compared to two or three weeks for dinosaurs—be-

cause we have developed a system based not on balance but on extensive and continuing exploitation of soil, energy, nutrient, water, and other limited resources.

Up to this time the trend of modern civilization has been basically animalistic in that we have exploited our environment to our maximum capability with relatively little thought of future consequences. Specifically, our wholesale destruction and alteration of natural ecosystems is undercutting the species diversity needed to maintain a stable biosphere. Even short of extinction, it is diminishing the gene pools of alleles which are necessary to enable species to adapt to change. This includes our own agricultural plants and animals. Recall that the drought of 1988 resulted in a 38 percent loss of the U.S. corn crop. Adapting crop plants to environmental changes requires introducing alleles from wild populations. If these are destroyed, what then?

Undercutting the ability of species to adapt to change is bad enough in itself because changes in climate and other factors do occur gradually even without the influence of humans. However, we humans additionally are embarked on precipitating environmental changes on a global scale through

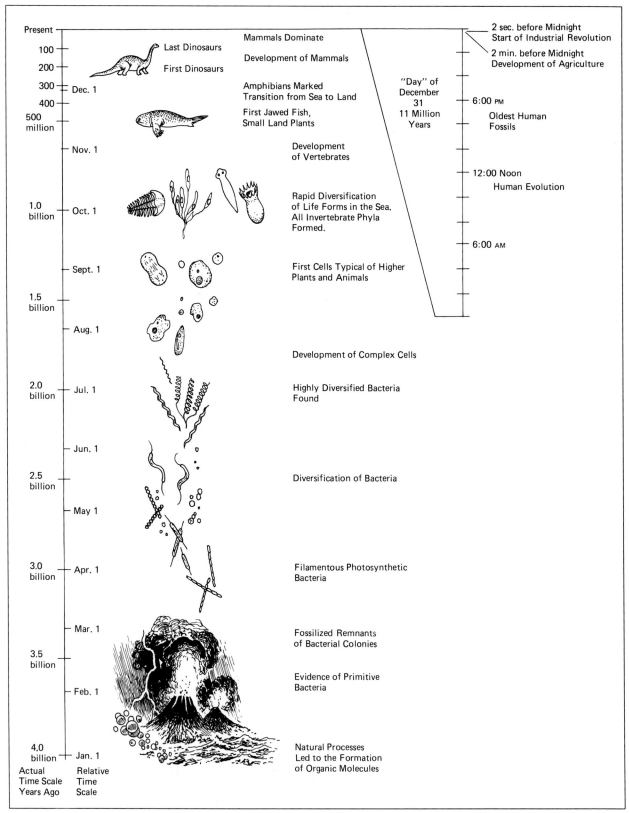

Present
100
200
300 — Dec. 1
400
500 million

Nov. 1

1.0 billion — Oct. 1

Sept. 1

1.5 billion

Aug. 1

2.0 billion — Jul. 1

Jun. 1

2.5 billion

May 1

3.0 billion — Apr. 1

Mar. 1

3.5 billion

Feb. 1

4.0 billion — Jan. 1

Actual Time Scale Years Ago

Relative Time Scale

Last Dinosaurs
First Dinosaurs

Mammals Dominate
Development of Mammals

Amphibians Marked Transition from Sea to Land
First Jawed Fish, Small Land Plants

Development of Vertebrates

Rapid Diversification of Life Forms in the Sea. All Invertebrate Phyla Formed.

First Cells Typical of Higher Plants and Animals

Development of Complex Cells

Highly Diversified Bacteria Found

Diversification of Bacteria

Filamentous Photosynthetic Bacteria

Fossilized Remnants of Bacterial Colonies

Evidence of Primitive Bacteria

Natural Processes Led to the Formation of Organic Molecules

"Day" of December 31 11 Million Years

2 sec. before Midnight Start of Industrial Revolution
2 min. before Midnight Development of Agriculture

6:00 PM
Oldest Human Fossils

12:00 Noon
Human Evolution

6:00 AM

FIGURE 4–19
Contrasting the geological time scale with a single year gives one a better appreciation for the relative amount of time taken for various evolutionary stages. Note that two-thirds of the time is taken in the development of cells; then the pace quickens. The origin and development of humans occur only in the last 8 hours of the last day, and progress since the Industrial Revolution occupies only the last 2 seconds of the "year."

such factors as warming the climate with additions of carbon dioxide and other pollutants to the atmosphere (the greenhouse effect), destruction of the ozone shield, and acid rain. To put these changes in perspective, climatologists estimate that the greenhouse effect will change the climate more in the next 50 years than would occur naturally over the next 2000 years, an increase in the rate of change by forty fold.

Insects and weeds probably have the genetic capacity to adapt to such accelerated rates of change, but many larger plants and animals probably do not. Even modern agriculture with its very narrow genetic base is extremely vulnerable. A more diverse agriculture such as existed in premodern times would be much more resilient, but, except for the efforts of some dedicated scientists, this too is at risk. (See Case Study 4.)

The point should be clear. Diminishing genetic diversity on the one hand and precipitating drastic environmental changes on the other is hardly the direction toward a sustainable biosphere. Hence, we find ourselves at the threshold of another "major evolutionary event" which will determine the future

course of life on earth. Will it be collapse of the biosphere marked by extinction of most species and the beginning of a new "Age of Insects and Weeds," or will it be that we humans, at this juncture, learn to hold back our exploitive powers and develop a sustainable human ecosystem in balance with other species on earth and progress toward exploring the full extent of our imagination and spirit? Only by developing a sustainable, balanced human ecosystem can science, technology, culture, indeed, civilization itself, continue to be maintained and advanced. Major events in evolution only occur in the order of once every 100 million years. To understand that we are alive at this time and participating in such an event is a profound concept. Yet no less is the case. In the next 50 years or so it is our choice to either develop a sustainable human ecosystem or witness collapse.

There seems little choice as to the preferable course. Now armed with the basic principles for ecosystem sustainability and balance, we move on to *apply* these principles to aspects of the human ecosystem—and to build on foundations that have already been laid.

Case Study 4

SEEDS FOR CHANGE

Botanist Dr. Gary Paul Nabhan, Assistant Director of the Desert Botanical Garden in Phoenix, Arizona, and cofounder of Native Seeds/SEARCH spends a considerable portion of his time combing through herbaria, botanical archives, and the writings of early explorers and missionaries. When he turns up a report of a food plant that he recognizes as being no longer commonly cultivated, he sets out on an expedition to see if it still exists. He often finds a subsistence farmer in a remote area who nods his head and says something like: "Yes, we used to grow that but no longer. We buried the last seeds with my mother when she died many years ago." Sadly, another agricultural species has become extinct. Occasionally, however, the search is successful. He finds a family that is still growing the seeds passed down through the generations. From such finds, Dr. Nabhan brings some seeds back and distributes them to members of his organization, Native Seeds/SEARCH, who grow them and keep these species alive. What is the point of this effort?

You can select and grow plants suited to the conditions or you can change the conditions to suit

the plants. Having limited ability to control the conditions, premodern cultures selected and cultivated plants that had exceptional ability and tenacity to produce even in unfavorable conditions without the benefits of irrigation, pesticides, or fertilizer.

Along the Missouri River in North Dakota, Indian families grew numerous species of corn and squashes. On the dry, wind-shifting sands of the Colorado Plateau, Hopi farmers tended sunflowers. In the dry lands of southwestern North America, Indians grew hundreds of species of beans. Sending their roots as much as six feet into the earth to find moisture, these beans have a high yield despite scorching 115° F heat and almost no rain. In Peru, each valley grew its particular variety of potato or other tuber. Thus, ancient people survived for millennia by growing these remarkably adapted crops. These crops are often referred to as "heirloom vegetables" by horticulturists because they were grown only in one particular locality and their seeds were passed down generation after generation.

Modern agriculture, however, has taken the approach of changing the conditions to fit the plant. It

Navajo cornfield in Canyon de Chelly in northern Arizona shows how native seeds thrive without irrigation. (Bill Steen, Elgin, Arizona.)

has focused on using relatively few high yielding varieties and obtaining maximum yields by intensive use of pesticides, fertilizer, and irrigation. In the process heirloom species have been abandoned and many lost forever. Of all the food plants that were grown by native Americans at the time of Columbus, roughly 75% no longer exist and many more are at risk, says Dr. Nabhan.

Such prestigious scientific organizations as the U.S. National Academy of Sciences and the United Nations Food and Agriculture Organization now recognize that the course of modern agriculture is exceedingly tenuous. Because of its very narrow genetic base, modern agriculture holds little potential for tolerating harsh conditions or adapting to changes in climate. Therefore, with the climate likely to become hotter and dryer because of the greenhouse effect and depletion of water resources modern agriculture is exceedingly vulnerable. Recall the 38 percent loss of the corn crop that occurred as a result of the drought of 1988. Thus, we may come again to depend on the diversity and resilience of heirloom species and their tenacious ability to produce under stressful conditions.

Then, humankind may look with gratitude to people like Dr. Nabhan and organizations such as Native Seeds/SEARCH, whose efforts kept such species from extinction.

From articles by and personal communication with Gary Paul Nabhan, Assistant Director of the Desert Botanical Garden in Phoenix, Arizona, and a cofounder to Native Seed/SEARCH.

Part Two
POPULATION

An exploding human population is rapidly overwhelming and destroying natural ecosystems, systems that are the heart of biological diversity and central to maintaining the stability and sustainability of our biosphere. Therefore, central and crucial to sustainable human development is curtailing the growth in human *numbers* and lessening the negative impact of those numbers on the natural environment. Only by gaining a balanced population can we continue to develop human culture, technology, and, indeed, civilization itself. In the words of Philip Handler, past president of the United States National Academy of Sciences:

> I cannot believe that the principal objective of humanity is to establish experimentally how many human beings the planet can just barely sustain. But I can imagine a remarkable world in which a

limited population can live in abundance, free to explore the full extent of man's imagination and spirit.

The relationship between human beings and impact on the environment involves more than just population numbers; it also involves *lifestyle*. An affluent lifestyle utilizing many material goods involves the exploitation of both resources to produce those goods and, in many cases such as cars, additional energy resources to run them. Production of material and energy resources, burning fuels, and ultimately scrapping the products when they are worn out, results in wastes which cause pollution. Both the exploitation of resources and the resulting pollution extract heavy tolls on the environment. Obviously a simpler lifestyle, riding a bicycle rather than driving a car for example, will have less negative impact. But, the connection between lifestyle and negative environmental impact is not a simple "one to one" relationship either.

The negative impact of lifestyle may be either greatly lessened or made worse by the degree of **ecological regard** which is exercised. For example, practicing ecological regard means the negative impact of an affluent lifestyle can be greatly decreased by such measures as substituting resources that involve less impact, conservation, recycling products, and by suitable pollution control measures. On the other side, driving ATVs (all terrain vehicles) through wilderness areas, which is highly destructive of natural ecosystems, is an example of minimal ecological regard. It greatly exacerbates the negative environmental impact of an affluent lifestyle.

Without ecological regard, even a simple "down-to-earth" lifestyle is not automatically low impact. Consider clearing a section of forest to grow your own food and obtain your own firewood. This not only destroys habitat and hence the wildlife on the cleared land itself but, unless well managed, your land may erode making your soil nonproductive and filling an adjacent stream with sediment and destroying its aquatic life. Additional negative impacts may result from other activities you carry on. Thus, a "pioneer" lifestyle without environmental regard may have a more severe impact on the natural environment than "high-tech" living in a city. Contrary to popular belief, city living is often more resource and energy efficient than suburban or rural lifestyles. Rowhousing and apartment complexes can be heated, cooled, and served with utilities and transportation much more efficiently than individual homes. The relationship among these three factors, population, lifestyle, and ecological regard may be summarized by the following formula.

$$\frac{\text{ENVIRONMENTAL}}{\text{IMPACT}} = \frac{\text{POPULATION} \times \text{LIFESTYLE}}{\text{ECOLOGICAL REGARD}}$$

This formula states that the environmental impact caused by humans is a function of population (number of people) exacerbated by lifestyle but moderated by ecological regard. The important thing to note here is that the human impact on the environment may be influenced as much by lifestyle and ecological regard as by actual numbers. Discussions regarding population and how many people the earth can or cannot support are really meaningless without considering lifestyle and ecological regard.

When lifestyle and environmental regard are considered, we find that the world really has two different and distinct "population problems." One is found in the industrialized countries (the nations of Western Europe, United States, Canada, Australia, Japan, and Russia) where the problem is not numbers so much as pollutants, wastes, and other impacts deriving from affluent lifestyles. The "other" problem is found in less-developed nations, which include the majority of the world's population. Here impoverished people, in their pursuit of immediate day-to-day survival, are cutting forests, overgrazing rangeland, and overcultivating croplands. This allows erosion which, in addition to eventually destroying the productivity of the land, leads to siltation and destruction of the aquatic ecosystems of streams, rivers, and lakes. Obviously it is a collision course that cannot be sustained. In essence, for survival in the short-term, the less-developed countries are committing ecological suicide in the long-term. The recurring famines in Ethiopia are an example of what will become increasingly widespread if this trend continues.

To the crisis of Third World ecological destruction, some would say, in effect: Let's just worry about our own problems. If "they" destroy their environment and starve to death as a result, that is "their" problem. Others object, however, that such an attitude is like rich people standing in the bow of a ship and jeering at the poor people in the stern: Hey, your end of the boat is sinking!

Actually, this analogy is not so far-fetched. We are all on one ship, Spaceship Earth, and the same biosphere interconnects all of us. If the poor nations of the earth cut all their tropical forests, this will so diminish biological diversity and otherwise upset the stability of the biosphere that it is more than likely that we will all go down together.

Consequently, moving toward sustainable human development demands that we address the population-poverty issue of the less-developed countries. This is the topic of Chapters 5 and 6. Later chapters will focus on the various resource-pollution issues which are more pertinent to highly developed countries. What is clear is that even at the current world population level, human life cannot be sustained at anything approaching a decent lifestyle of adequate food, clothing, and shelter, without considerably more ecological regard leading to better management and protection of the natural environment.

5

The Population Problem:
Its Dimensions
and Causes

CONCEPT FRAMEWORK

■ Outline

I. DIMENSIONS OF THE POPULATION
   PROBLEM   115
   A. The Exploding Human Population   115

   B. Rich Nations and Poor Nations   115

   C. Population and Poverty   118

      1. Fertility and Population Profiles   118

      2. Growth Momentum   119

      3. Population Growth Undercuts Economic
         Gain and Worsens Debt Crisis   122

   D. Population, Poverty, and the Environment 123

CASE STUDY: THE IMPACT OF POPULATION ON
TROPICAL FORESTS   124
II. THE POPULATION EXPLOSION: ITS CAUSE
    AND POTENTIAL SOLUTION   124
    A. Birth Rates, Death Rates, and the
       Population Equation   125

    B. Cause of the Population Explosion   127

       1. Reduction in Infant and Childhood
          Mortality   127

       2. Change from Prereproductive to
          Postreproductive Death   127

       3. The Minor Impact of Postreproductive
          Longevity   129

       4. The Minor Impact of Accidents and
          Natural Disasters   129

■ Study Questions

1. Draw a graph illustrating how the human population has grown from early times to present and how it is continuing to grow. Can such growth be sustained?

2. On a map, point out which countries are highly, more-, and less- developed. Describe the distribution of wealth and population among these countries.

3. In what countries are populations growing most rapidly? Where are they reaching stability?

4. Define: *fertility rate, replacement fertility.* Contrast replacement fertility and actual fertility for more- and less-developed countries. Define, describe, and compare *population profiles* for more- and less-developed countries. Predict future population growth for more- and less-developed countries based on current population profiles, fertility rates, and death rates.

5. Describe how populations of less-developed countries will respond over the next 60–80 years if their fertility is cut to replacement immediately. Define: *growth momentum.* How is the proportion of rich and poor in the world likely to change over the next 50 years?

6. Describe how population growth undercuts economic gains.

7. Explain why it is inadvisable for us in more-developed countries to simply ignore the population problem of less-developed countries.

8. Define: *crude birth rate, crude death rate.* Given these statistics for a population, calculate its percentage rate of increase and its doubling time. Cite the actual statistics for the more- and less-developed countries.

9. Give the range of CBR and CDR for the human population prior to the 1800s.

10. What significant events occurred between the mid-1800s and mid-1900s? Describe how reduction in infant and childhood mortality affected survivorship, CDR, CBR, and population growth.

11. Explain why the change from pre- to postreproductive death is particularly significant in terms of population.

12. Define: *postreproductive longevity.* Describe why this has relatively little impact on population growth.

13. Describe why accidents and natural disasters have little impact on population growth.

C. The Solution: Lower Fertility Rates   129

    1. The Demographic Transition   129

    2. Reasons for Disparity in Fertility Rates between Developed and Developing Countries   131

14. What is the most acceptable solution to the population problem?

15. Describe the demographic transition. Has it been completed in more-developed countries?—less-developed countries?

16. Cite and describe factors that cause people in agrarian societies to favor having more children while people in industrialized societies favor having fewer. Compare the history of current more-developed and less-developed countries and, on this basis, explain why the disparity in fertility rates exists. Cite additional factors that now compound the problem.

The objective of this chapter is to provide an understanding of the dimensions of the human population explosion and its causes. Only through such understanding can we address the problem effectively.

DIMENSIONS OF THE POPULATION PROBLEM

The Exploding Human Population

Over the last 150 years or so the human population on Spaceship Earth has grown and continues to grow at a phenomenal, explosive rate. The numbers speak for themselves.

From the dawn of human history until the beginning of the last century the world population hovered around a few hundred million, growing slowly then suffering setbacks due to epidemics of disease and intermittent famines. It was roughly 1830 before world population reached the 1 billion mark. During the 1700s and 1800s, however, a remarkable turnabout occurred. Human population changed from a condition of slow growth punctuated by setbacks to one of explosive growth. By 1930, just 100 years after reaching the first billion, the population had increased to over 2 billion. Barely 30 years later (1960) it reached 3 billion. And in only 15 more years (1975) it had climbed to 4 billion. Then, 12 years later (1987) it crossed the 5 billion mark, and growth is continuing, adding about 90 million people (births minus deaths) per year. This is equivalent to the populations of all the major cities of the United States combined (Fig. 5–1a). During the last couple of decades, however, the percentage *rate of growth* began to slow. Still, with the huge population base, even the lower rate of growth continues to add absolute numbers faster. Thus, even with the current trend of slowing growth

rate, the 6 billion mark will be crossed in 1999 and, unless dramatic changes occur, this growth pattern will go on well into the twenty-first century before leveling off at 10 billion by the end of the next century (Fig. 5–1b).

These projections into the future, however, are derived from simply extending lines on graph paper based on current directions. They do not consider any of the looming ecological questions of whether the biosphere can sustain these numbers. Given that the biosphere is on the verge of dramatic climatic shifts and its resources are being rapidly degraded and destroyed under the pressures of 5 billion, it seems dubious that it can support the growth to, much less sustain, 10 billion, unless there are dramatic changes in both *lifestyles* and *ecological regard*. Sustainable human development will require slowing the rate of population growth even more and also exercising more ecological regard.

Where are all the additional millions, indeed billions, of people being put and how are they being fed, clothed, housed, educated, and otherwise cared for? To answer this question it is necessary to recognize that we live in a world with tremendous economic disparity between nations.

Rich Nations and Poor Nations

The world is commonly divided into three main economic categories.

1. *Highly developed, industrialized* or *high-income* countries: namely, the United States, Canada, Japan, Australia, and the countries of Western Europe and Scandinavia.

2. *Moderately developed* or *middle-income* countries: mainly countries of Latin America (Mexico, Central and South America), North and West Africa, and East Asia.

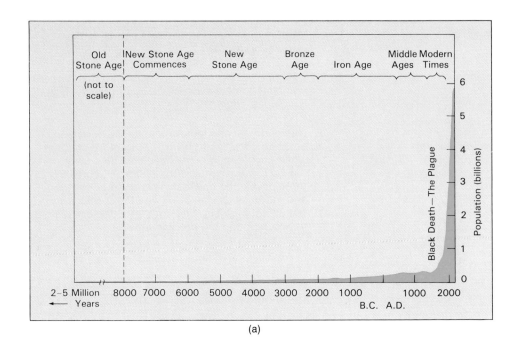

(a)

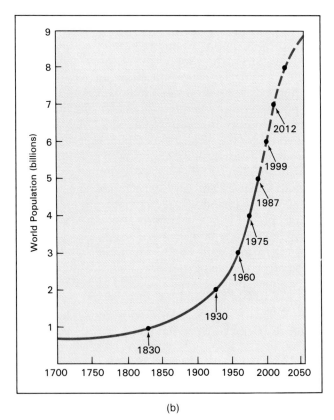

(b)

FIGURE 5–1
The population explosion. (a) For most of human history the human population grew very slowly, but in modern times it has suddenly "exploded" as shown. Data prior to 1800 are estimates. (Reprinted by permission from E. J. Kormondy, *Concepts of Ecology*, 3rd ed., Englewood Cliffs, NJ: Prentice-Hall, Inc., 1984.) (b) The growth of the human population is shown in more detail from 1700 to 1989 and estimated to 2050.

3. *Low-income* countries: mainly the countries of East and Central Africa and India. The People's Republic of China is still placed in this category, but it may soon move into Category 2. These low-income nations are also known as the "Third World."

Socialist countries, with the exception of China, are generally considered in a separate category, as are a few countries like Saudi Arabia—where most people are poor but the national income is high due to exporting oil. The world map shown in Figure 5–2 shows these groups of countries with their populations and data regarding average incomes.

The high-income, industrialized nations are also commonly referred to as **developed countries** while middle- and low-income countries are often grouped and referred to as **developing countries**. Also, industrialized countries are referred to as **more** or *highly developed countries* (HDCs), while middle- and low-income countries are grouped as *less-developed countries* (LDCs).

The disparity in distribution of wealth among these nations and their people is mind-boggling. Highly developed nations hold just 25 percent of the world population; yet they control about 80 percent of the world's wealth. Thus, developing countries, which have 75 percent of the world's population, have only about 20 percent of the world's wealth. This disparity is illustrated in the following analogy. Imagine the world's economic wealth as a plate of 20 cookies on a table. Twenty persons, each representing about 5 percent of the world's population, surround the table. It would seem fair that each person receive

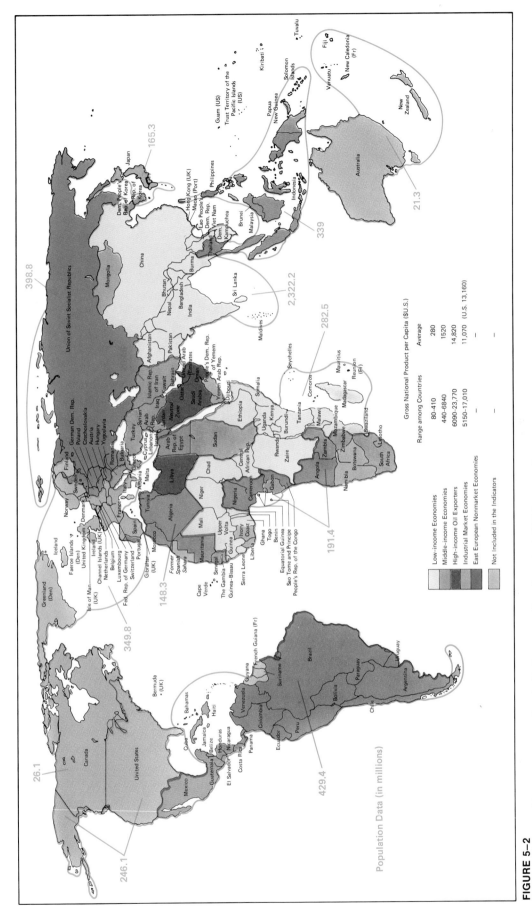

FIGURE 5–2

Nations of the world grouped according to their gross national product (GNP) per capita. GNP per capita is a general indicator of standard of living. Population of various regions is also shown. (From the *World Development Report 1984*. Copyright © 1984 by the International Bank for Reconstruction and Development/The World Bank. Reprinted by permission of Oxford University Press, Inc., New York. Population data from the Population Reference Bureau.)

Gross National Product per Capita ($U.S.)

	Range among Countries	Average
Low-income Economies	80–410	280
Middle-income Economies	440–6840	1520
High-income Oil Exporters	6090–23,770	14,820
Industrial Market Economies	5150–17,010	11,070 (U.S. 13,160)
East European Nonmarket Economies	—	—
Not Included in the Indicators	—	—

Population Data (in millions)

one cookie, but this is not the situation. Five of the twenty, representing the populations of the highly developed nations, take 16 of the cookies to divide among themselves (one person being the U.S. population, which is now close to 250 million). This leaves just 4 cookies for the other fifteen people who represent the poorer 75 percent of the world's population. But these fifteen (the developing nations) do not divide the 4 remaining cookies equally. Five people—representing the populations of moderately developed countries—take 3 cookies, leaving just one cookie for the remaining ten people who represent the 2.5 billion people living in Third World countries.

Of course, the distribution of wealth within each country is also disproportionate. Between 10 and 15 percent of the people in highly developed countries are recognized as **poor** (unable to afford adequate food, shelter, and/or clothing), and about 10 percent of those in less-developed countries are wealthy. Nevertheless, the economic concerns of the average person in a highly developed country center on owning a comfortable home, car, several television sets, and other amenities. Food is mostly a problem of avoiding eating too much (Fig. 5–3a). On the other hand, large portions of the population in Third World countries and considerable numbers in moderately developed nations are desperately poor. They have *inadequate* food and/or shelter and no amenities. Their primary economic concern is simply day-to-day survival (Fig. 5–3b). Worldwide, at least a billion people, one out of every five, fall into this category. Particularly in the poor nations of Africa and Asia malnutrition and starvation-related deaths are common, especially among infants and children.

During the 1960s and 1970s development efforts resulted in some mitigation of this economic disparity, but in the last decade those efforts were slackened and the trend was reversed. In the 1980s there was a considerable widening of the gap as the rich grew richer and the poor grew poorer, and it is this trend that continues currently.

Population and Poverty

Yet it is within the less-developed countries that the population explosion is most intense; they are growing at rates that, if sustained, will double their present populations in 25 to 35 years. In contrast, barring immigration, populations in highly developed nations are approaching stability (no growth).

Fertility and Population Profiles

The key factor that underlies the disparity in population growth rates is **total fertility rate,** the average

(a)

(b)

FIGURE 5–3
(a) In highly developed countries food is thought of in terms of what is wanted to eat. (b) UN giving food to malnourished children in Bangladesh. (a, Bob Gambarelli; b, United Nations)

number of children being born to each woman in her lifetime based on current statistics. With modern methods of disease control, most children who are born survive to adulthood and have children in turn. Assuming for a moment that all children survive, a fertility rate of 2.0 leads to a stable population because the two children will replace the mother and father when they die. Fertility rates of less than two will lead to a declining population because the parental generation is not fully replaced, and fertility rates of more than two will lead to growing populations because the number of parents increases with each generation. The fertility rate that just replaces the parents and gives a stable population is known as the **replacement fertility.** Considering the number of children who die before reaching adulthood, the actual replacement fertility for highly developed countries is 2.03, and for less-developed countries it is 2.20 because infant and childhood mortality is somewhat greater.

Comparing these figures with the actual situation: the fertility rate for highly developed nations is now 1.9, somewhat below the replacement level. Populations are still growing, however, because higher fertility rates in the past created a large generation which, despite low fertility, is now producing numbers of children that exceed the numbers of elderly who are dying. However, population stability and decline may be anticipated as the present large population of parents ages, dies, and is not fully replaced by their children. The fertility rate for less-developed countries, however, is 4.8 excluding China, 4.1 including China. This is more than twice their replacement fertility and is leading to a doubling of these populations each generation.

Over time, differences in fertility rates lead to strikingly different *population profiles*. The **population profile** refers to the age makeup of the population. It is typically presented as a bar graph showing the number of people in each age group, usually by 5-year intervals, males on one side and females on the other (Fig. 5–4). Note in Figure 5–4 that the population profile for highly developed countries (Fig. 5–4b) is column-shaped showing that populations of children, young adults, mature adults, and senior adults are about equal. This comes from maintaining fertility at or near the replacement level so that each rising group just replaces the one before it.

In contrast, the population profile for developing countries has a pyramid shape which stems from each succeeding group of parents producing a number of children that is nearly double their own number (Fig. 5–4a). This creates a population that is composed predominantly of young people—a relatively small portion is mature or senior adults. It should not escape attention that in less-developed countries nearly 40 percent of the total population is under 15 years of age and another large portion is young adults lacking significant training and/or experience. The task of educating, training, and providing productive jobs for these people is monumental.

Can you see how future populations may be predicted from the current population profile and fertility/mortality statistics? Simply visualize that every 5 years the bars move up one position. Upper bars are reduced and the top bar is removed as elderly people die, and a new bar representing the number of births for the period is added at the bottom. If fertility rates remain constant, the population profile for less-developed countries will become a larger and larger pyramid as each succeeding generation of parents is larger and produces a still larger generation of children (Fig. 5–4c). The population profile for highly developed countries, however, remains a similarly-sized column (Fig. 5–4d).

Growth Momentum

Even if fertility rates of developing countries dropped to the replacement level immediately (an extremely unlikely event), their populations would continue to grow for some time before stabilizing. This would occur because even with the "ideal" fertility rate of 2.0, the current population of children under age 15 will still produce their own number of offspring, which will greatly exceed the dieoffs among the relatively small number of people in older age groups. This situation will persist for 60 to 70 years, until the present young people reach old age and begin dying off. At that time, the profile will have roughly the shape of a straight column, the population of old people roughly equaling the population of young people. With this population structure, the deaths among the old will balance the reproduction of the young. But at this point, the population will be two to three times larger than it is now (Fig. 5–4c).

Thus, populations of less-developed countries have a **growth momentum,** due to their current large populations of young people, that will effectively double their populations even if those young people exercise low fertility. The distinction, of course, is that with low fertility their populations will reach stability over the next 60 years or so, while if high fertility is maintained the population will double and redouble again until environmental limits such as famine, disease, and social upheavals take their toll.

Actually, family planning efforts throughout the world have succeeded in reducing fertility rates significantly over the past few decades. Assuming this trend continues, replacement fertility may be reached

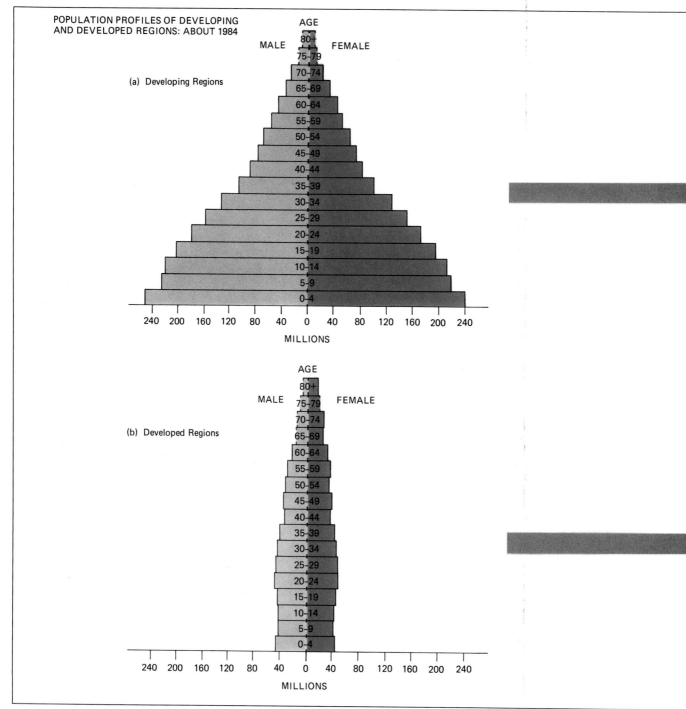

FIGURE 5–4

Population profiles for developing and developed regions, 1984 and estimates for 2024. (a) All developing countries combined. (b) All developed countries combined. Note that (a) and (b) are drawn on the same scale; each bar indicates actual numbers of people. (From Leon Bonvier, "Planet Earth, 1984–2034: A Demographic Vision," *Population Bulletin*, 39, No. 1 [Population Reference Bureau, Inc.] Washington, D.C. 1984.) (c) The population of less-developed regions in 40 years (2024). Total length of bars is assuming current birth rates continue. Lighter portion only is assuming fertility rates decline to 2.0 by 2004, a highly optimistic assumption. (d) Population of developed regions in 2024 assuming fertility continues at current rate 2.0. Note that even with favorable assumptions the population of less-developed regions will increase enormously while that of developed regions will increase little, placing a much greater burden on the developed world.

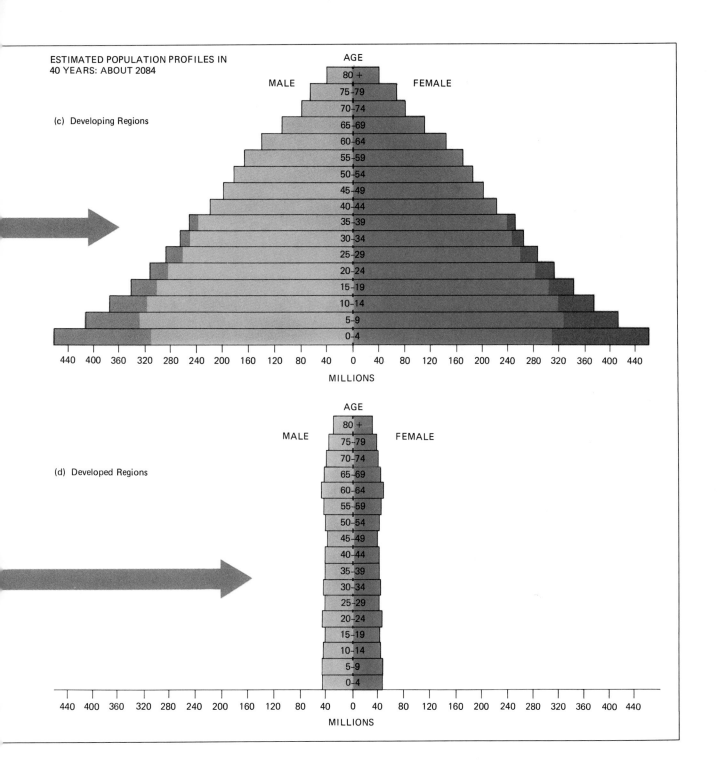

ESTIMATED POPULATION PROFILES IN
40 YEARS: ABOUT 2084

(c) Developing Regions

AGE

MALE   FEMALE

| 80 + |
| 75-79 |
| 70-74 |
| 65-69 |
| 60-64 |
| 55-59 |
| 50-54 |
| 45-49 |
| 40-44 |
| 35-39 |
| 30-34 |
| 25-29 |
| 20-24 |
| 15-19 |
| 10-14 |
| 5-9 |
| 0-4 |

440 400 360 320 280 240 200 160 120 80 40 0 40 80 120 160 200 240 280 320 360 400 440

MILLIONS

(d) Developed Regions

AGE

MALE   FEMALE

440 400 360 320 280 240 200 160 120 80 40 0 40 80 120 160 200 240 280 320 360 400 440

MILLIONS

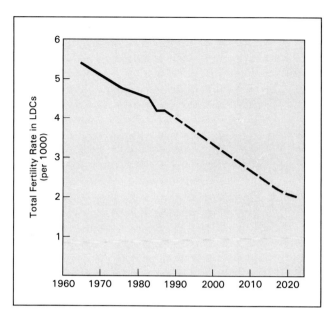

FIGURE 5–5
Fertility rates in less-developed nations have declined in recent years. Assuming the trend continues, the world may reach replacement fertility around 2020.

in the developing world by 2025 (Fig. 5–5). However, given the growth momentum of less-developed countries, their populations will continue to grow markedly until around 2080.

Consequently, the fact we confront over the next 50–70 years is rapid growth of populations in less-developed countries while populations in highly developed countries grow little if at all. (Fig. 5–6). Thus, developed countries are coming to represent a smaller and smaller percentage of the world's population. In another 50 years, developed countries will probably represent only 10 percent of the world's population, as opposed to the current 25 percent. Further, the rich may be even richer while the poor are poorer, due to pressures described in the following.

Population Growth Undercuts Economic Gain and Worsens Debt Crisis

One measure of average standard of living is the **gross national product per capita,** which is given by dividing a country's gross national product (total value of goods and services exchanged) by its population. You can see that economic gains in developed countries,

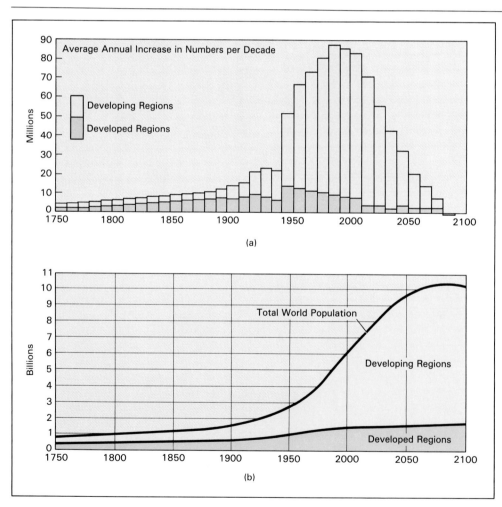

FIGURE 5–6
(a) The higher fertility rate in developing nations is causing their population to grow much faster than developed nations. (b) This is creating a world in which 90% of the population will live in developing nations, mostly in poverty. Thomas W. Merrick, "World Population in Transition," *Population Bulletin,* Vol. 41, No. 2 (Washington, D.C.: Population Reference Bureau, Inc., Jan. 1988 reprint), p. 4.

where populations are not growing, translate directly into increasing standards of living, while economic gains in less-developed countries must be divided among more and more people such that standards of living may actually decrease.

Said another way, the *real* economic growth of a nation in terms of providing for its people is the difference between economic growth and population growth.

$$\begin{matrix} \text{ECONOMIC} \\ \text{GROWTH} \end{matrix} - \begin{matrix} \text{POPULATION} \\ \text{GROWTH} \end{matrix} = \begin{matrix} \text{REAL ECONOMIC} \\ \text{GROWTH} \end{matrix}$$

For example, if the economies of both the United States and Kenya, a poor African nation, are growing at 2.5 percent per year, the United States, which has a population growth rate of 0.7 percent, is getting ahead by 1.8 percent per year (2.5 − 0.7 = 1.8) while Kenya, which has a population growth rate of 4.0 percent, is falling behind by a similar percentage per year (2.5 − 4.0 = −1.5).

The economic advancement of less-developed countries is now further handicapped by the **debt crisis.** During the 1960s and 1970s less-developed countries financed economic expansion largely with loans from industrialized countries and considerable progress was made. But today, their total indebtedness exceeds 1 trillion dollars ($1,000,000,000), and they have left the "buy now" phase and entered the inevitable "pay later" phase. Many less-developed nations are now paying more than half of all their earnings in interest, leaving little for further devel-

opment. Moreover, lending agencies are reluctant to extend credit because it is unlikely that they will be able to repay current loans, much less additional loans. Strapped for funds, many less-developed nations are in economic crises with high unemployment, inflation rates topping 100 percent per year, and housing, schools, hospitals, and public services deteriorating.

Between high fertility rates and paying most of their earnings in interest to developed nations, you can readily see that, far from closing the gap between rich and poor nations, the rich are getting richer while the poor are getting poorer.

Addressing the question posed at the start: Where are all the additional millions—indeed billions—of people being put?—the answer is mostly into poor, less-developed countries. And to the question: How are they being fed, clothed, housed, educated, and otherwise cared for?—the answer all too often is—inadequately (Fig. 5–7).

Population, Poverty, and the Environment

Lest we feel that we (in the developed nations) can simply ignore this situation as "their problem," it is these growing populations of the less-developed nations who, for the sake of their day-to-day survival, are overgrazing, overcultivating, and overcutting forests for firewood and committing other acts of eco-

FIGURE 5–7
Scene from a village in Benin, West Africa. Cou Provence Rural Development project. (Photo by Joseph Hadar/World Bank.)

THE IMPACT OF POPULATION ON TROPICAL FORESTS

Perhaps in no other country is the interrelationship between resources, the environment, and population more evident than in Brazil, now the world's sixth largest nation. With a population today of 145 million, its population is expected to soar to 180 million by the year 2000 and, with current growth rates, is expected to double in just 33 years.

The situation in Brazil is a little better than in the 90 nations that are now expected to double their populations in less than 30 years. But with the world itself expected to double its population in the next 40 years alone, Brazil becomes a classic case of pressures upon the resource base, the environment, and a nation's population-carrying capacity.

In an effort to pay the interest on the prohibitive debt it has incurred, now the world's largest, Brazil is selling off its precious natural resources, the hard woods contained in the tropical rainforests of the Amazon. Most people don't realize it but 50 percent of our rainforests are located in just three countries—Indonesia, Brazil, and Zaire.

In a 3-month period of 1988, the Amazon forest saw 80,000 square miles cut down. That is equal in size to all of Denmark, Belgium, Switzerland, and Austria.

This rape of the earth's oxygen-cleansing forests was even further exacerbated by the reliance upon slash-and-burn methods of ground clearing. Such clearing techniques resulted in 620 million tons of carbon gases and particulate matter being thrown into the atmosphere, adding to the world's greenhouse problem.

Brazil is now on the threshold of recognizing the interrelationship between excessive population growth, environmental deterioration, and appropriate management of its resource base. Brazil's newly passed constitution guarantees family planning as a basic human right.

The Brazilian government abandoned its traditional policy encouraging population growth in the late 1970s after the adverse effects of overpopulation on the nation's economy and environment became apparent. In 1980, then President Joao Baptista Figueiredo acknowledged in his inaugural address, "The success of social development programs depends largely on family planning."

The government, once hesitant to promote family planning because of the opposition of the Catholic Church, the military, and certain nationalists, has recently begun to support such efforts indirectly. Today, most services are provided through an effective network of private family-planning organizations, and affordable oral contraceptives are widely available without prescription at Brazilian pharmacies. (Ninety percent of oral contraceptives and condom users in Brazil purchase their supplies at privately operated pharmacies.)

Consequently, fertility has fallen in Brazil. Brazilian women now have on average just over three children, down from an average of six children in the 1960s. Sixty-five percent of Brazilian couples now use birth control, and progress continues to be made in extending contraceptive availability to the country's rural areas. In Brazil's largely agrarian northeastern

logical folly. Thus, once-productive lands are deteriorating into worthless desert (see Chapter 7), which threatens not just the less-developed nations' survival but the entire biosphere (see Case Study 5). Sustainable development, if we are to achieve it, will demand more than standing by and watching to see what happens. It will demand concerted actions toward both reducing fertility and protecting the environment.

Before we get into those actions, however, we need to understand why the population explosion is centered on poor nations, or conversely, how developed nations have all but achieved population stability.

THE POPULATION EXPLOSION: ITS CAUSE AND POTENTIAL SOLUTION

Dozens of factors may affect births and deaths—disease, war, family and social traditions, economics, religious traditions, moral beliefs, and so on. Still, aside from immigration and emigration, any change in size of a population boils down to subtracting the number of deaths from the number of births. Any change in population (growth or decline) is given by the difference between these two numbers. We wish to look at the population explosion in terms of why

Cornfield and owner's hut in an area cleared by burning, Palenque, Mexico. (Tom McHugh/Photo Researchers)

sector, for example, contraceptive use among married women now stands at 53 percent, up from 37 percent in 1980. Despite such encouraging trends, however, nearly half of all Brazilian women who had been pregnant in the past five years reported in 1986 that their last pregnancy had been unwanted or unintended.

Interestingly, the opposition to family planning in Brazil has come not only from the Church but also from the political left, apparently in part because some leftist radicals view continued overpopulation, with its incident poverty and destabilizing effects, as an important component of any future revolution.

If Brazil is to achieve replacement level fertility by 2000, a goal the government has endorsed, use of family planning must spread to 74 percent of Brazilian couples. Given the widespread acceptance of family planning in Brazil, and the government's recent efforts to help, this is not an unrealistic goal.

WERNER FORNOS, *The Population Institute*

and how birth rates and death rates have changed over time.

Birth Rates, Death Rates, and the Population Equation

All countries record births and deaths so that such information is available. In order to compare rates of population growth in different countries, it is customary to divide the populations into groups of 1000 and present the data as the average number of births and deaths *per 1000 per year*. These figures are known as the **crude birth rate (CBR)** and **crude death rate** (CDR), respectively. The term "crude" is used because there is no consideration given to what proportion of the population is old or young, male or female, which will have obvious impact as noted above. Subtracting the CDR from CBR gives the **natural increase** (or decrease if the number of deaths is greater than the number of births) to distinguish it from changes resulting from immigration and emigration.

CBR	−	CDR	= Natural Increase (or Decrease)
Number of *Births* per 1000 per Year		Number of *Deaths* per 1000 per Year	in Population per 1000 per Year

The rate of growth (or decline) can be converted to a percentage by dividing the result by 10.

$$CBR - CDR = \text{change}/1000.$$
$$\text{Divided by } 10 = \text{change per 100 or percent.}$$

Looking at actual statistics (Table 5–1) we find that highly developed nations, averaged as a group, have a CBR of 15 and a CDR of 9, giving a growth rate of 0.6 percent per year.

$$15 - 9 = 6 \text{ per } 1000.$$
$$\text{Divided by } 10 = 0.6 \text{ percent.}$$

Based on their current population of 1.2 billion, this rate of increase results in 7 million additional people each year in the developed world. On the other hand, less-developed nations as a group have a CBR of 31 and a CDR of 10, giving a growth rate of 2.1 percent per year. This higher growth rate and a higher population base, 3.9 billion, results in an increase of nearly 83 million people per year. Statistics for a number of individual countries are given in Table 5.1. Observe that population growth rates in developed countries are very low or even negative, while those in low and moderate income nations are relatively high.

Table 5–1

Population Data for Selected Regions or Countries in 1988

REGIONS	TOTAL FERTILITY RATE	CRUDE BIRTH RATE*	CRUDE DEATH RATE*	GROSS NATIONAL PRODUCT (PER CAPITA, U.S. $)
World	3.6	28	10	$3,010
More Developed	1.9	15	9	10,700
Less Developed	4.1	31	10	640
Less Developed excluding China	4.8	35	12	780
Africa	6.3	44	15	620
Asia (excluding China)	4.3	32	11	1,480
Latin America	3.7	29	8	1,720
North America	1.8	15	9	17,170
Europe	1.8	13	10	8,170
Highly Developed Countries				
United States	1.8	16	9	17,500
West Germany	1.4	10	11	12,080
France	1.8	14	10	10,740
England	1.8	13	10	8,920
Japan	1.7	11	6	12,850
Australia	1.9	15	7	11,910
Moderately Developed Countries				
Mexico	4.0	30	6	1,850
Brazil	3.4	28	8	1,810
Argentina	3.3	24	9	2,350
South Korea	2.1	19	6	2,370
Costa Rica	3.6	34	5	1,420
Less Developed Countries				
China	2.4	21	7	300
India	4.3	33	13	270
Pakistan	6.6	43	15	350
Ethiopia	7.0	46	15	120
Kenya	8.0	54	13	300
Zambia	7.0	50	13	300
Rwanda	8.5	53	16	290

* Births/deaths per 1000 per year.

It is also common to express population growth in terms of **doubling time,** that is, how long it will take the population to *double* at its current rate of growth. The significance of doubling the population is that all housing, agricultural output, industry to provide jobs, power production, water supplies, sewage collection and treatment systems, schools, health care facilities, and so on must be likewise doubled in the same time just to maintain the r⁻ˡ⁻ᵗⁱᵛe standard of living. And all faciliti⁻⁻ ⁻⁻ ⁻ ⁻ ⁻ d in a *shorter* time if th⁻⁻ ⁻⁻ ⁻⁻ is ⁻⁻ ⁻⁻ ⁻e.

T⁻ ⁻ o⁻⁻ ⁻ne ⁻⁻⁻⁻⁻ ng the percent rat⁻ ⁻⁻⁻⁻⁻⁻ ⁻⁻ulation of highly ⁻⁻⁻⁻⁻⁻ 117 years (70 divi⁻⁻ ⁻⁻⁻⁻⁻ ⁻wth rate of .6 pe⁻ ⁻⁻⁻⁻⁻ ⁻g countries, on ⁻⁻⁻⁻⁻ 33 years if current ⁻⁻⁻⁻⁻ ues (70 divided by ⁻⁻⁻⁻⁻ ng time is much le⁻⁻ ⁻⁻⁻⁻⁻ Kenya are 53 and ⁻⁻⁻⁻⁻ ⁻h rate of 4 percent ⁻⁻⁻⁻⁻ ⁻, will result in dou⁻⁻ ⁻⁻⁻⁻⁻ars.

53 − 1⁻

Try imagining ⁻⁻⁻⁻⁻ ⁻rt life in a develo⁻⁻ ⁻⁻⁻⁻⁻ ⁻u can readily see ⁻⁻⁻⁻⁻ ⁻n offset by increas⁻⁻ ⁻⁻⁻⁻⁻ f living in such a ⁻⁻⁻⁻⁻

Cause of the P⁻⁻⁻⁻⁻

Recall from Chapt⁻ ⁻⁻⁻⁻⁻ ductive potential t⁻ ⁻⁻⁻⁻⁻ plosion if a high p⁻ ⁻⁻⁻⁻⁻ adulthood and repr⁻ ⁻⁻⁻⁻⁻ ulations in check is ⁻⁻ ⁻⁻⁻⁻⁻ which cause the die⁻ ⁻⁻⁻⁻⁻ young before they re⁻ ⁻⁻⁻⁻⁻ last century or so th⁻ ⁻⁻⁻⁻⁻ human species as well⁻ ⁻⁻⁻⁻⁻

Prior to the late 18⁻ ⁻⁻⁻⁻⁻ parents to have seven t⁻ ⁻⁻⁻⁻⁻ one to three made it to ⁻ ⁻⁻⁻⁻⁻ eases such as smallpox⁻ ⁻⁻⁻⁻⁻ diphtheria, scarlet fever⁻ ⁻⁻⁻⁻⁻ cough took heavy tolls. A⁻ ⁻⁻⁻⁻⁻ ilies were sometimes left⁻ ⁻⁻⁻⁻⁻ other words, birth rates⁻ ⁻⁻⁻⁻⁻ order of 40–50. But, becau⁻ ⁻⁻⁻⁻⁻ among infants and young⁻ ⁻⁻⁻⁻⁻ rates were almost equally as⁻ ⁻⁻⁻⁻⁻ human population grew slo⁻⁻

Reduction in Infant and Childhood Mortality

In the mid-1800s, however, the work of Louis Pasteur and others revealed that epidemic diseases were caused by microorganisms, discoveries which led to effective means of disease prevention, control, and cure. Particularly significant innovations were vaccinations against most childhood diseases and improvements in sanitation which keep sewage-borne pathogens from contaminating water and/or food supplies. Then, in the 1900s, the discovery of antibiotics provided what at the time seemed like a miraculous means of cure. The most profound result of these innovations was a dramatic reduction in infant and childhood mortality. In developed countries today, it is unusual for parents to lose a child to disease, whereas 100 years ago it was uncommon not to.

This decrease in infant and childhood mortality can also be seen as an increase in **survivorship.** Survivorship is an expression of how long members of a group of 1000 newborns survive. A survivorship curve is constructed from statistics of the age of people when they die. Curve A in Figure 5–8 is typical of the situation before the advent of modern medicine. It shows a high dieoff during the years of infancy and childhood; only about half of those born reach 20 years of age. Then, the decline is gradual to the limit of longevity. Curve B in Figure 5–8 expresses the modern situation. Note that it shows that over 95 percent of those born survive to age 40 or beyond; a rapid dieoff does not begin until after age 50.

The prevention, control, and cure of childhood diseases, then, resulted in a dramatic reduction of death rates. Crude death rates dropped from 40–50 per 1000 per year down to about 10, the portion of older people in the population who die. Crude birth rates, however, remained at 40–50. Here is the disparity that initiated the population explosion. It is highly significant that it results from a drop in death rate alone. Fertility and birth rates have not increased; in fact they have come down—as we will discuss later—but not as much.

Change from Prereproductive to Postreproductive Death

People obviously still die. Why don't deaths from cancer and heart disease, for example, play the same role as did childhood diseases? The answer lies in whether a person dies *before* having children (*pre*reproductive death) or *after* having children (*post*reproductive death).

To illustrate, consider two cases in which parents traditionally have four children. In the first case, suppose that two of the four children in each generation die before they reach reproductive maturity

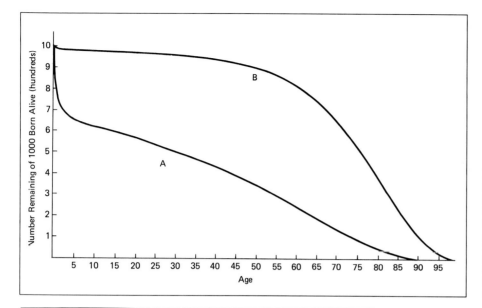

FIGURE 5–8
Survivorship. Starting with a population of 1000, survivorship curves indicate the number that will reach each age. (a) Survivorship of American nonwhite males (1902), typical of human population prior to modern health care. (b) Survivorship of U.S. population (1976). (Data from U.S. Department of Health, Education and Welfare, Public Health Service.)

FIGURE 5–9
Effect of prereproductive and postreproductive death on population growth. (a) Prereproductive death holds growth in check. (b) Postreproductive death allows population to grow.

(Fig. 5–9a). The original parents then have just two children who reach maturity. These children produce four grandchildren, but only two of these reach maturity. As the older generation dies, there is a replacement of two for two and a stable (nongrowing) population results.

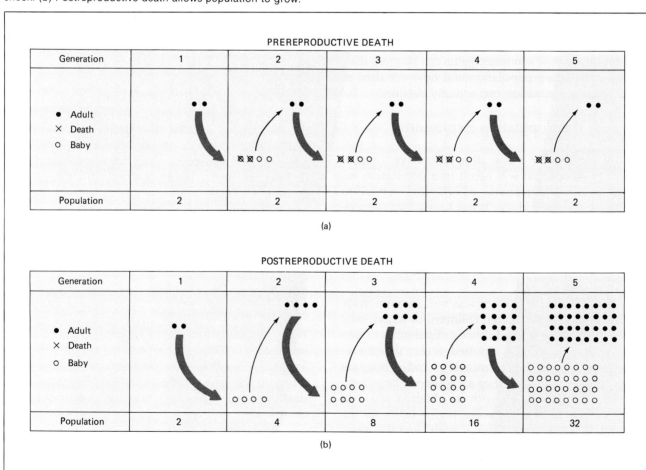

(a)

(b)

On the other hand, if all four of the children live to sexual maturity and reproduce, the four children produce eight grandchildren and these produce 16 great-grandchildren, and so on (Fig. 5–9b). There is a doubling of the population each generation, the death of the older generations notwithstanding. *Post-reproductive deaths can never be a controlling factor in population growth.*

The Minor Impact of Postreproductive Longevity

Indeed, the postreproductive **longevity** of people has minimal impact on population growth and even that impact is temporary. Figure 5–10 illustrates the point. To simplify, it is assumed in the figure that there are four children per family and that the generation time is 20 years. Life expectancies of 60, 80, and 40 years are compared. In each case, the population starts with a total of 14 people: 2 grandparents, 4 parents, and 8 children. With a life expectancy of 60 years, the oldest generation is removed from the graph with the addition of each new generation, keeping three generations in the picture (Fig. 5–10a). Note that the population doubles each generation: from 14 to 28 to 56, and so on. In Figure 5–10b, the life expectancy is increased to 80 years; thus, four generations are kept in the picture. Old people remaining in the population for another generation initially increase the total population by about 7 percent, but after this, the population still only doubles each generation: 30 to 60 to 120, and so on. Finally, in Figure 5–10c, the life expectancy is decreased to 40 years, thus removing a generation of older people from the population. This initially results in a population about 15 percent less than when the life expectancy is 60 years (24 compared to 28); but after this reduction, the population still proceeds to double each generation.

Consequently, whether diseases of maturity such as cancer and heart disease are cured so that people live another 20 to 30 years, or whether they worsen, causing more people to die in their forties and fifties, will have little long-term impact on population growth. It will, however, have a social and economic impact.

The Minor Impact of Accidents and Natural Disasters

Likewise, accidents and natural disasters will not control population as is sometimes suggested. Again, such factors are not directed at prereproductive death and, despite the social and economic impact of such losses, they have a relatively small effect on the overall population increase. For example, our yearly loss in U.S. car accidents, about 50,000, is replaced within 10 days. Several years ago a tidal wave swept a coastal rice-growing region of India and about half a million persons were killed, a tremendous loss in terms of a natural disaster. However, India's natural rate of increase replaced this loss in about 30 days. Even the wars since World War II have not had a significant lasting impact on population numbers. In the Vietnam War, about 45,000 American lives were lost. Given the natural increase for the U.S. population of about 150,000 per month, such a loss is made up in about 10 days, or three weeks if men alone are considered. Even the current loss of 3 to 6 million people per year to famine and malnutrition is small when compared to the world's natural increase of about 90 million per year. The loss again represents about two weeks of natural increases.

In summary, the historic human condition was one of high birth rates and high death rates (childhood mortality). High growth rates were initiated by a drop in childhood mortality while birth rates remained high.

The Solution: Lower Fertility Rates

Of course population stability might be achieved by death rates rising again, but the thrust of all our human endeavors is to prevent this from happening. The only humane way to control population growth is to reduce birth rates to a level that matches the lower death rate. Indeed, this has already occurred in the more-developed nations. Fertility rates have come down so that birth rates are very near the replacement level (Fig. 5–11). Fertility rates have also come down somewhat in developing nations but not as much; hence rapid growth continues.

The Demographic Transition

Examining birth rates and death rates in developed countries over the last 200 years, then, we see a striking transition from a "primitive" stability given by high birth rates and high death rates to a "modern" stability resulting from low death rates and low birth rates. This phenomenon is known as the **demographic transition.** It has four distinct phases as shown in Figure 5–12.

Phase I is the condition before disease prevention in which the birth rate is high but infant and childhood mortality is correspondingly high, so that the population grows slowly, if at all.

Phase II is the period in which society learns to control the diseases that are primarily responsible for high infant and childhood mortality. In this phase, childhood mortality rates drop dramatically but the birth rate remains high, giving rise to rapid population growth.

Phase III is the period in which social and/or economic changes cause fertility rates to come down. At

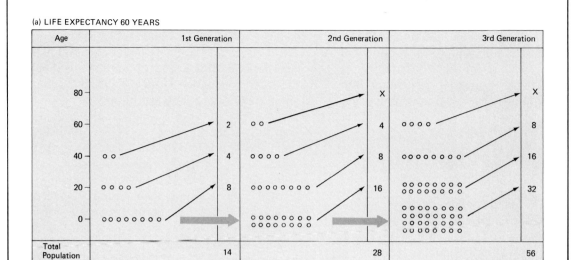

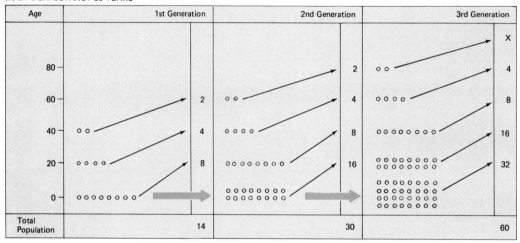

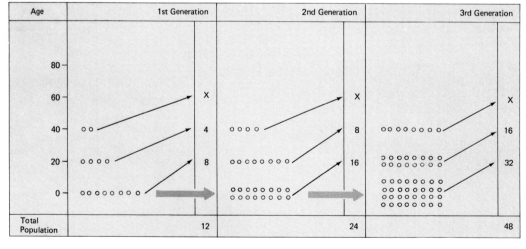

FIGURE 5–10

Effect of longevity on population growth. Four-child families and a 20-year generation time are assumed in each case. (a) Assumes a 60-year life span; note that population doubles each generation. (b) Assumes an 80-year life span; note that population is only slightly larger and just doubles every generation. (c) Assumes a 40-year life span; population is 15 percent smaller but still doubles every generation.

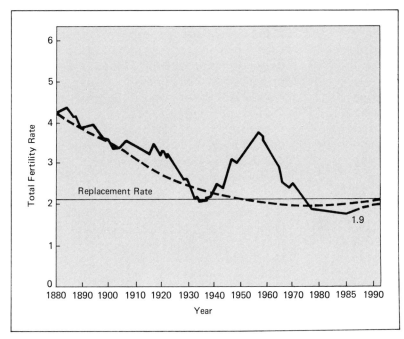

FIGURE 5–11
Fertility rate in the United States (solid line). The trend (dotted line) has been downward since 1880 to a replacement level. The peak in the 1945–1965 period is the anomalous post-WW II "baby boom" and is not part of the overall trend. (From Population Reference Bureau, Inc., Washington, D.C.)

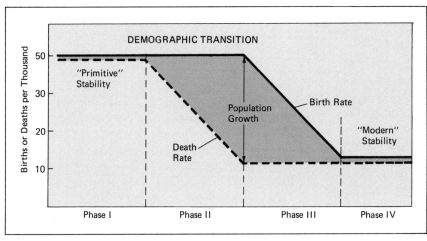

FIGURE 5–12
Demographic transition. The four phases of the demographic transition are shown in this idealized graph. Phase I is characterized by a stable population with a high birth rate and high death rate. Phase II is characterized by a dropping death rate but continued high birth rate, leading to an increasing rate of population growth. Phase III is characterized by a falling birth rate, which restores population stability. Phase IV is characterized by a new stability given with both low birth rate and low death rate. Highly developed countries are nearing the end of Phase III. Developing countries are still in the early stages of Phase III.

the end of this phase, population stability is again achieved because low infant and childhood mortality is balanced by low birth rates.

Phase IV is the new stability given by low birth rates and low death rates.

Again, industrialized countries have essentially completed Phase III of the demographic transition; fertility rates have dropped to the point such that their populations will reach a new stability with low birth rates matching low death rates. However, developing nations are only in the early states of Phase III. Death rates have come down, but birth rates, although coming down, are still significantly above death rates. What are the social and economic factors that have led people in industrialized countries to lower their fertility? Understanding these factors will give us insight into fostering lower fertility in less-developed nations, a goal that must be reached if we are to achieve sustainable human development.

Reasons for Disparity in Fertility Rates between Developed and Developing Countries

Fertility, the number of children a couple has, is dependent on two basic factors:

1. The number of children the couple desires to have (assuming the number desired is less than reproductive capacity).
2. The availability of effective contraceptive techniques.

A large number of social and economic factors,

in addition to love and fulfillment, come to bear on a couple's decision regarding the number of children they desire. Many of these socioeconomic factors hinge on differences between an agrarian society and an industrial society. For a number of reasons, the agrarian society tends to favor large families, while the industrial society tends to favor small families:

1. **Children: An Economic Asset or Liability.** On a family farm, children can be a great economic asset. A child as young as 4 can provide significant help with lighter chores such as feeding the chickens and collecting eggs, and 12-year-olds often do the same work as an adult. Children are also a tremendous help in taking the produce to market and selling it. In an urban setting, however, opportunities for children to contribute to the economic welfare of the family are extremely limited but their costs for food, clothing, and education are conspicuous. Stated simply, agrarian families tend to desire more children because they are an economic asset and urban families desire fewer children because they are an economic liability.

2. **Old-Age Security.** Industrialized societies have come to offer and even require that people belong to various health care and pension plans that will provide for their care in times of sickness and old age. In agrarian societies people are self-employed; no such plans exist. It is traditional, and expected, that children will provide whatever care and support their parents may need. Consequently, people in agrarian societies desire and look toward children as their old-age security, and it follows that more children mean more security.

3. **Educational and Career Opportunities.** Industrial societies and urban settings generate a wide variety of educational and career opportunities that people can take advantage of to get ahead. Thus, people in industrial societies tend to delay marriage and/or put off having children in order to take advantage of these opportunities. The result is that they forgo the most fertile years of their lives (the late teens and early twenties) and ultimately have fewer children than might otherwise be the case. Conversely, an agrarian society offers relatively few opportunities beyond carrying on the traditional farming that people learn as they are growing up. Thus they tend to marry and start having children earlier, leading to larger numbers overall.

4. **Status of Women.** In agrarian societies it is generally felt that the first and foremost role of a woman is to bear and rear children. Women are frequently ostracized and even legally barred from careers in business and industry. The only fully acceptable role for women in the eyes of men is to rear a large family. Women, therefore, have little choice but to focus their abilities on doing so. Alternatively, the women's rights movement in industrialized societies has opened most careers to women as well as men so that many women are making the choice to pursue careers and limit child bearing accordingly.

5. **Religious Beliefs.** Some religions promote large families. People in urban settings, perhaps as a result of being more exposed to different cultural values, tend to set aside such dogma, if not the entire religion. On the other hand, people in agrarian settings tend to maintain such traditions.

6. **Availability of Contraceptives.** Finally, desires for fewer children are unlikely to be met unless safe, acceptable, and effective contraceptives are readily available. In industrialized countries such contraceptives are readily available, whereas in less-developed countries, especially in rural areas, they are not.

Prior to the Industrial Revolution, all societies were basically agrarian. They had to be; without mechanization the majority of the population had to be farmers to produce enough food for the society. During the last half of the 1800s through the first half of the 1900s, three things occurred more or less simultaneously in what are now the highly developed, industrialized countries. As we have already noted, the progress of disease prevention and cure was bringing down death rates (childhood mortality). In the same period, however, farms were being mechanized and cities were growing industrially so that labor was moving from the farms to take jobs in the cities. As more people were adopting the urban lifestyle, fertility rates were coming down more or less in parallel with death rates. The result is that a severe population explosion never developed in what are now the industrialized countries (Fig. 5–13a). Finally, the opening up and expansion of the Western World, particularly the United States and Canada, took the population pressure off European countries.

Today's less-developed countries, however, are the former subjects of eighteenth- and nineteenth-century colonialism. While the countries of Europe and America were industrializing, what are now less-developed nations were dominated and kept agrarian

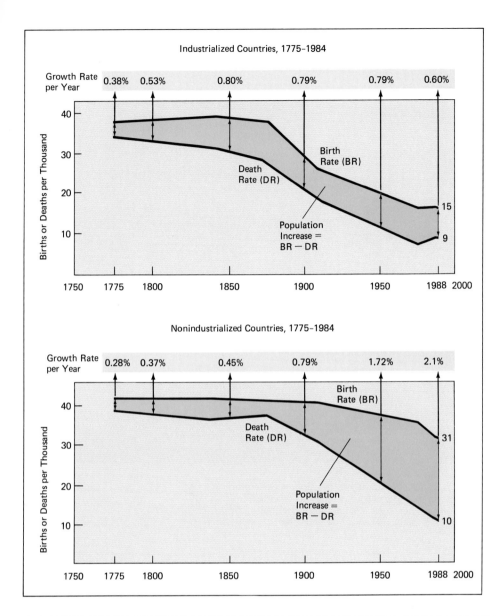

FIGURE 5–13
The increase in world population has been due to a drop in death rate, not to an increase in birth rate. In fact, birth rates have also declined somewhat. The increasing rate of population growth is seen in the increasing gap between birth rate and death rate in nonindustrialized countries. (Redrawn with permission of Population Reference Bureau, Inc., Washington, D.C.)

to produce cash crops (coffee, sugar, tea, tobacco, cotton, bananas, etc.) for the "mother" countries. Most of these colonialized countries only gained independence in the mid-1900s. Consequently, the condition of high birth rates and high death rates persisted, in large part, until World War II. Following the war, there was a massive effort to improve world health. The result was a sudden and precipitous drop in childhood mortality as children received vaccinations and antibiotics while all the conditions favoring high fertility remained constant. Thus, the majority of the countries of Africa and Asia began their independence with a legacy of a population explosion underway and virtually no industrialization to provide the socioeconomic factors to reduce fertility.

Additional factors now compound the problem.

Insufficient land to accommodate the growing population is leading to overgrazing, overcultivation, and soil erosion, which is undercutting the productivity of the land. Natural resources, particularly forests, are being sacrificed for firewood and meager foreign exchange, an unsustainable endeavor. Large numbers of people (mostly young), unable to make a living on the land, are moving to cities. However, with minimal industrialization there are no jobs for them; nor is there adequate housing, clean water, or sewage facilities. The result is mushrooming Third World cities with millions of people, unemployed and malnourished, living in squalid, inadequate housing or even on the streets (Fig. 5–14).

It is to these problems that we turn our attention in Chapter 6.

FIGURE 5–14
Slum area of Bombay, India. Coupled with growing population, modernization of agriculture in nonindustrialized countries is causing a massive migration from rural areas to cities. But cities can provide neither enough jobs nor housing. The result is slums, the extent and squalor of which are difficult for Americans to imagine. Homeless thousands sleep in the streets. (United Nations)

6
Addressing the
Population Problem

CONCEPT FRAMEWORK

▪ Outline

I. IMPROVING LIVES OF PEOPLE   137
    A. Increasing Food Production: Successes and Limits   137

    B. The Hungry: A Problem of Food-Buying Capacity   139

    C. Food Aid and the Utterly Dismal Theorem   140

    D. Economic Development   142
       1. Large-Scale Centralized Projects   142
       2. Decentralized Projects—Appropriate Technology   143
       3. Appropriate Technology and Agriculture   144

II. REDUCING FERTILITY   144

    A. Leaders of Less-Developed Nations Support Reducing Fertility   145
    B. People of Less-Developed Countries Want Fewer Children   145
    C. Effectiveness of Family Planning, Health Care, and Education   146

CASE STUDY: A LESSON LEARNED FROM BANGLADESH   148

    D. Additional Economic Incentives   149

    E. Promoting Family Planning and the Abortion Controversy   150

    F. Contraceptive Technology   151

    G. Costs of Family Planning   151

III. ACTIONS YOU CAN TAKE   151

▪ Study Questions

1. List and describe major ways in which food production has been increased over the last 40 years. Describe how each of these methods is at or near its limits.

2. Give evidence that ''the hungry'' are a problem of economic distribution rather than food production capacity.

3. Describe how food aid undercuts food production in less-developed countries. Name the phenomenon. Discuss why aid must focus on economic development.

4. List two basic kinds of development aid. Describe the features of each. What is the most important area in which to focus development?

5. List and describe the factors that are supportive of reducing fertility. What are the three areas of assistance that lead to lower fertility? Are leaders and/or people in less-developed countries supportive of reducing fertility?

6. Describe: (a) the need for family planning, health care, and education in less-developed countries; (b) what these consist of; and (c) how each lends to lowering fertility. Can fertility be reduced without resorting to coercive measures?

7. What kinds of additional incentives have been used in China to reduce fertility?

8. Describe three situations in which the life of the mother may be endangered by continuing a pregnancy. What may be the consequences of denying a legal abortion? How may other dependent children also be affected?

9. Name and describe new contraceptives being developed.

10. Contrast military armaments and stabilizing population growth as methods toward assuring a sustainable future. Compare the monies going into supplying armaments versus providing family planning services. What may be a cost-effective method toward attaining sustainable development?

11. List and discuss actions you can take toward promoting a stable world population.

Impoverished people will impoverish the environment in their desperation to survive, and an impoverished environment can only support impoverished people. To break this vicious cycle we need to strive toward: (1) improving the lives of people, (2) reducing fertility, and (3) protecting the environment.

IMPROVING LIVES OF PEOPLE

Increasing Food Production: Successes and Limits

The first and foremost prerequisite toward improving the human condition is to make sure that everyone has adequate nutrition. Without adequate nutrition all other factors are moot. Nearly 200 years ago (1798), when the world population was still less than 1 billion, British economist Thomas Malthus pointed out that population has the capacity to increase geometrically, as was shown in Figure 5–1b, while agriculture, which is dependent on a finite amount of arable land, is limited. Malthus predicted that, in the absence of checks, the world was headed toward catastrophic famines as human populations outpaced agricultural capacity. As the population has grown, numerous others have echoed this cry.

Yet, despite a six- to sevenfold increase in population, the percentage of people affected by malnutrition or starvation (10 to 20 percent) is probably no higher now than it was 200 years ago. Remarkably, expansion of agricultural production has kept pace with and even exceeded the growth of the population. Particularly following World War II, as the population explosion was recognized, there was an all-out effort to increase food production. As can be seen in Figure 6–1, efforts were successful; food production since World War II outpaced population in both developed and developing nations, with few exceptions. Between 1950 and 1984 world grain production (the most basic foodstuff for both direct human consumption and feeding livestock) increased about 2.6-fold from 624 to 1645 million tons per year, moving ahead of population growth by 40 percent. In developed countries, especially the United States, the increase in grain production over population was dramatic, leading to huge surpluses that the United States gave away or sold at low prices to those in need.

But the world population is now growing at a rate of about a million people every four days (90 million per year)—faster than ever before in history. Whether agricultural production can continue its rapid pace of expansion to supply these growing numbers bears examination.

The increases in agricultural production of the last 40 years have been accomplished mainly by the following:

☐ Bringing additional land into cultivation

☐ Increasing the use of fertilizer

☐ Increasing the use of irrigation

☐ Increasing the use of chemical pesticides

☐ Substituting new genetic varieties of grains that, with adequate fertilizer and water, doubled or tripled yields per acre.

According to many agricultural experts, expanding production by these methods has reached, or even exceeded, sustainable limits.

Land Area Devoted to Grain Production. From 1950 to 1981, grain land was increased by 24 percent, but much of this land was highly susceptible to erosion or arid and could not sustain agriculture. Considerable portions are now being abandoned because erosion or depletion of water resources have rendered

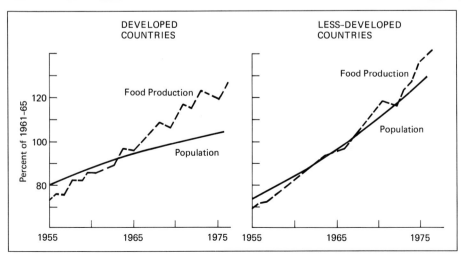

FIGURE 6–1
Food production and population in developed and less-developed countries. (Redrawn with permission from S. Wartman and R. W. Commings, Jr., *To Feed This World*, Baltimore: Johns Hopkins University Press, 1978. Based on data from USDA, Economic Research Service.)

them no longer productive or they are being retired to save them from this fate. Additional acreages are being taken from agriculture to build houses, industries, and highways. As a consequence of these trends and the lack of new arable lands that can be brought into production, the total area in grain production fell about 7 percent from 1981 to 1988, and the downward trend seems certain to continue.

An even greater proportion of increased production, however, has come from increasing yields per acre. Yields per acre have been increased by use of irrigation, fertilizer, pesticides, and substituting higher yielding varieties. But these factors, too, are reaching limits.

Irrigation. From 1950 to 1980, irrigated acreage increased about 2.6 times. It is still expanding, but at a much slower pace due to lack of water resources (Fig. 6–2). More ominous, much of the present irrigation is not sustainable because groundwater resources are being depleted. In addition, production is being adversely affected on as much as one third of the world's irrigated land, and farmers are being forced to abandon major tracts because of waterlogging and accumulation of salts in the soil, consequences of irrigating where there is poor drainage (see Chapter 8).

Fertilizer. Nutrients are frequently the limiting factor in plant growth. Consequently, providing additional nitrogen, phosphorus, potassium, and other nutrients (fertilizer) increases yields, but only up to the optimum. From 1950 to 1984, fertilizer use grew about ninefold. In the beginning 15 to 20 additional tons of grain were gained from each additional ton of fertilizer used. Now, however, farmers are providing near optimal levels and there is little to be gained from adding more. In practice, it is not cost-effective to add fertilizer to the point of maximum production; it is only economical to add it up to that point at which returns from increased yield more than pay for the fertilizer used. Additionally, high levels of fertilizer make plants more vulnerable to attack by insects and other pests, and the washing away of fertilizer leads to water pollution. Thus, increasing yields through increasing fertilizer also seems to be at or near its limits.

Pesticides. Chemical pesticides developed after World War II provided better pest control and increased yields accordingly. However, the level of control has not improved since that time, and there has been some backsliding in control because pests have become resistant to each new pesticide developed. Also, there are efforts to reduce the use of pesticides because of side effects to human and environmental health. Progress is being made toward developing natural means of control that will be environmentally *safe*, but these will be unlikely to increase yields above those achieved by chemical methods.

High-yielding varieties. In the 1960s, plant geneticists developed new varieties of wheat and rice which, with optimal fertilizer and water, would give yields of double to triple those of traditional varieties. As these varieties were introduced throughout the world, production soared (Fig. 6–3). Indeed, it was

---

FIGURE 6–2
A farmer attempting to irrigate the desert finds that there is not enough water, and soil erosion is severe. This land should have been left in range. (USDA-Soil Conservation Service photo)

(a)

(b)

FIGURE 6-3
Comparison of an old variety of wheat (a) (photo by Phi Farnas/Photo Researchers) with a new short-stemmed variety (b) (photo by J. Latta/Photo Researchers). The first of the more productive short-stemmed varieties was developed under the direction of Norman Borlaug. This advance was heralded as the beginning of the "Green Revolution."

hailed as the **green revolution.** These high-yielding varieties remain the backbone of production. However, now most of their potential has been realized and plant geneticists confess that they do not have any more "super-high-yielding" varieties in the wings to repeat the performance. Even breakthroughs in genetic engineering such as introducing nitrogen-fixing genes into wheat and corn may enable farmers to reduce their costs for fertilizer, but they will not increase yield per acre.

Climate. The climate from World War II to 1980 was exceptionally stable and ideal for agriculture in most parts of the world. Then three severe droughts occurred drastically affecting harvests in the years 1980, 1983, and 1988. Corn harvest in the United States was cut 38 percent by the 1988 drought, an unprecedented loss. Fortunately, the world and particularly the United States had ample stocks from previous surpluses. Drawing these down averted actual shortages but left stocks at a 41-day supply, which is considered a dangerously low level. To be safe, there should be about a 100-day supply, one season's harvest. If another year similar to 1988 occurs before stocks are replenished, which will take at least three years, real hardship will be imposed. Based on their computer models, some respected climatologists are predicting that such droughts will become increasingly commonplace as the climate warms due to the greenhouse effect (see Chapter 13). At best, climate cannot do more than return to the "ideal" of the 1950–1980 period. Expecting climatic change to increase production is out of the question.

On balance, grain production per capita in Africa has been on a downward course since 1975, more than reversing the gains since 1950 (Fig. 6–4a). On a worldwide basis, production per capita declined from 1984 to 1988 with absolute declines registered for 1987 and 1988, the last statistics available.

Fish harvests from the oceans have not increased significantly since 1970 indicating limits here as well (Fig. 6–4b). With harvests only holding steady or actually headed downward while population is growing by 90 million per year, Malthusian alarms are being raised again.

The Hungry: A Problem of Food-Buying Capacity

Assuring that people have adequate nutrition involves more than average production per capita. The problem of starvation and malnutrition persisted in the world despite increases in per capita production through the 1960s and 1970s and huge surpluses produced in the United States. Food production is only half the ticket; people must also be able to buy it. The plight of the hungry is caused by their inability to buy

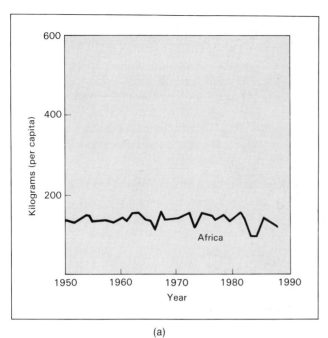

(a)

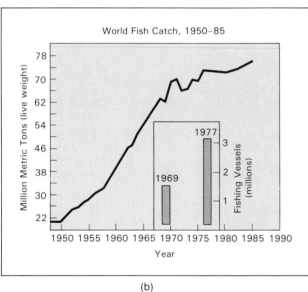

(b)

FIGURE 6–4
(a) Due to exploding population, grain production per capita achieved only modest increase in the 1950s and 1960s. Since 1975 the trend has been downward. (Source: *State of the World 1985* and *State of the World 1987*, copyright Worldwatch Institute.) (b) The sea is not an unlimited source of food waiting to be harvested. World catches have not increased significantly since 1970, despite increasing fishing effort, indicating that the seas are already being overfished, at least in terms of familiar species. (Data from National Marine Fisheries Service, National Oceanic and Atmospheric Administration, U.S. Department of Commerce)

the food produced, not because the food does not exist.

Frances Moore Lappe, in her book *Food First*, points out that there is a great deal of slack in the agricultural system. Huge amounts of the best agricultural land in less-developed countries are con-

trolled by rich plantation owners who grow cash crops such as sugar, coffee, tea, rubber, cotton, bananas, and tobacco. In Brazil, for example, 2 percent of the landowners hold 60 percent of the arable land in cash crops, or keep it idle, while 70 percent of the *rural* households have little or none, according to a 1985 report. This is a continuing legacy of nineteenth-century colonialism. If land reforms were mandated so that this land was put back in the hands of the peasant farmers, it would be producing food and benefiting the incomes, as well as the diets, of the rural poor.

Additionally, there is the factor of "eating high on the food chain." Much of the grain produced, which could be used for direct human consumption, is used to feed livestock. There is a loss of about 90 percent of food value in the conversion to meat (see Chapter 2). It is estimated that if people in developed countries minimized or eliminated meat from their diets and ate grains instead, as people in Third World countries already do, there would be enough grain to feed another 2–3 billion people.

In conclusion, even in the face of limited or even decreasing grain production, there is basically enough agricultural capacity to feed the world's growing population for some time. The problem is that the purchasing power of the highly developed world effectively diverts food from the hungry. Farmers will logically grow and sell products to the market that net them the most profit. Because people will pay $6.00 a pound for steak, meat producers are able to pay more for grain to feed their animals than the hungry can afford to pay to feed themselves. Similarly, since the affluent are willing to pay high prices for cash crops and rich plantation owners control production, the land is kept in cash crops rather than converting to food production. Effectively, then, the buying power of the developed world results in a disproportionate amount of agricultural production bypassing the poor and flowing toward the affluent (Fig. 6–5).

Shortfalls in grain production, if they occur in the years ahead, will cause an escalation of food prices and force many more people into the ranks of the hungry. Still at this stage in history, it will be more a problem of economic distribution than inability to produce adequate food.

Food Aid and the Utterly Dismal Theorem

Numerous humanitarian efforts to end world hunger have been mounted and sustained since World War II. The United States has been a world leader in giving surpluses or selling them at low prices. A number of serious famines have been moderated or averted by

FIGURE 6–5
Due to the greater purchasing power, the bulk of agricultural production including cash crops, grain-fed beef, and many luxury items go to developed nations while poor people in less-developed nations suffer from malnutrition and starvation.

these efforts. As virtuous as such efforts seem on the surface, however, routinely supplying food aid in an attempt to alleviate chronic hunger in developing countries has been self-defeating. As authorities S. Wortman and R. Cummings write in their book, *To Feed This World:*

> The food generosity of industrial countries, whether in their own self-interests (disposing of food surpluses) or under the mantle of alleged distributive justice, has probably done more to sap the vitality of agricultural development in the developing world than any other single factor.

This situation has occurred because people will not pay more than they have to for food. Therefore, provision of foreign food for free, or at a lower price than that of domestically produced food, undercuts the local market. In effect, local farmers must compete economically with free or low-cost imported food. When they cannot earn a profit, they stop producing and eventually enter the ranks of the poor. The cycle continues as people who sell goods to the farmer also suffer when the farmer loses buying power. In the long run, the entire local economy deteriorates. Hence, the availability of free food, while well-intended, often aggravates the very conditions of poverty and hunger that it is meant to alleviate. Meanwhile, population pressures continue to build and the magnitude of the problem increases. Economist Kenneth Boulding observed and described this phenomenon in 1957 and called it **the utterly dismal theorem.**

Even if the United States does not give food away, its surpluses aggravate the situation. The U.S. price support system for commodities, which effectively subsidizes the production of surpluses, climbed from about $6 billion in 1955 to over $25 billion in 1987. These surpluses are put on the world market at a cost which is actually below the cost of production in many developing countries. Thus, developing countries may buy food from foreign countries more cheaply than their farmers can produce it, putting their own farmers out of business and driving them into the ranks of the desperately poor.

The governments in some developing nations have led themselves into the trap of the utterly dismal theorem with their internal policies. The problem is particularly evident in newly independent countries of Africa where leaders have put ceilings on food prices in an effort to maintain the favor and support of large urban populations. The policy has made producing food unprofitable and forced farmers out of business. As a result, these countries are increasingly short of food. This phenomenon also contributes to the migration of rural farm families to urban centers as described in Chapter 5.

It also contributes to environmental and ecological deterioration. Poor farmers struggling for survival are pressed to overgraze, overcultivate, cut forests, and farm steep slopes, all practices which lead to severe soil erosion. This is the most tragic result because as erosion eats away the soil it becomes less and less productive until it is essentially worthless desert and must be abandoned (see Chapter 7). As noted above, this is a major factor in the current trend of losing agricultural land.

In conclusion, it is necessary to focus more on economic development to alleviate poverty and on reducing fertility rates. The hunger problem will be solved when a limited number of people can afford adequate food at a price which enables farmers to maintain their land, that is, **sustainable agriculture.**

Economic Development

It has long been recognized that promoting economic development of developing countries provides several advantages:

☐ Improving the lives of the poor is a humanitarian goal that may be pursued for its own sake and which produces its own rewards.

☐ Economic development is a means toward reducing fertility rates because as industrialization/urbanization occurs people desire fewer children.

☐ Greater prosperity in developing countries provides greater markets and hence additional prosperity and opportunity for people in industrialized countries as well.

Through providing monetary loans and technical assistance, industrialized nations have promoted economic development in less-developed nations since World War II, again with varying degrees of success. The experience we have gained may be summarized as follows. There are basically two kinds of development projects: large-scale centralized projects and decentralized projects.

Large-Scale Centralized Projects

Examples of large-scale centralized projects include construction of hydroelectric dams, nuclear power plants, modern industrial plants, modern hospitals, and high-speed highways. The advantages of such projects are: they are relatively easy to administer; lenders or donors need only interact with a few people in the recipient nation; progress on these projects is easy to measure and monitor; and they result in very conspicuous end products that both grantors and recipients can point to with pride.

These projects may also substantially increase the gross national product (GNP) of the recipient nations so that, on the books, it looks as if the country grew wealthier. Unfortunately, the appearance may be deceiving. Too often there is little change in the general condition of poverty. For example, a new highway is an obvious advantage to those with cars but creates additional hazards for the majority of the population who use it on foot or with oxcarts, and the poor cannot afford the services at the new hospital or buy electrical appliances or power from the new power plant.

In many cases, these centralized projects actually aggravate poverty. A modern textile mill, for instance, may provide some well-paying factory jobs, but it may also put hundreds of individual weavers out of business. This problem has been especially acute in connection with agriculture. A common thought in industrialized nations is that poor countries would prosper if they modernized agriculture. Thus, industrialized nations have promoted the use of modern agricultural machinery. However, the multitude of peasant farmers in these countries have small plots of land and cannot make use of large tractors and other machinery. Such machinery goes to already wealthy farmers with large landholdings enabling them to dismiss laborers, cut costs, and acquire even more land, thus pushing increasing numbers of people into unemployment and ever deeper poverty.

At the same time, such mechanization generally has not increased food production in developing countries. The highest production per acre is achieved on small plots where land is carefully tended and harvested by hand. Further, mechanization undercuts both diversification and varied timing of food supply. When numerous independent farmers tend small plots, diversity of crops and variation in times of planting and harvest occur, especially in the tropics where few seasonal restrictions exist. Individual farmers then sell and trade fresh produce in large, open markets (Fig. 6–6). With this direct farmer-to-consumer exchange, there is little need for middlemen and storage facilities. As farming becomes mechanized, however, it becomes impractical to raise a diverse variety of crops and harvesting takes place at one time. Thus, both the food and economic advantages of the local market are undercut, putting further strain on the poor.

Most seriously, too many centralized projects have promoted nonsustainable agriculture. For example, irrigation projects which boosted production in the short run are now forcing abandonment of lands due to waterlogging, salting of soils, or deple-

FIGURE 6–6
Open-air market in Equador. Numerous small, independent farmers bring their produce and crafts to these huge market places for sale and trade. The system encourages individuality and creates tremendous diversity. (Photo by author.)

tion of water supplies (see Chapter 7). In Africa, drilling deep wells to provide more watering holes for cattle allowed people to put more cattle on the land, but this led in turn to destruction of the grassland due to overgrazing.

Finally, these large-scale projects have huge price tags which were paid for by loans also from the developed world. This has created the debt crisis described in Chapter 5. Major funding organizations such as the World Bank and the International Monetary Fund are now beginning to look at loan requests more carefully in terms of *sustainable development*.

Decentralized Projects—Appropriate Technology

By the early 1960s, the problems inherent in large centralized projects were apparent, and it was recognized that to effectively alleviate poverty, projects must be on a scale such that the poor can immediately benefit and the benefit must be sustainable. The most common term for this new approach is **appropriate technology.**

A listing of attributes and a few examples best explains the concept of appropriate technology.

- ☐ It should be directly relevant to the economic and social needs of the recipients, but it should not upset the existing social structure.
- ☐ It should not require a high degree of training or skill but should involve the kinds of techniques that one person can teach others.
- ☐ It should utilize local resources that are plentiful and inexpensive.
- ☐ It should not require expensive, centralized workplaces; existing homes, farms, and small shops should suffice.
- ☐ It should allow many individuals to use their own talents and minds, express their creativity, and develop self-reliance.

Here are a few examples.

Handloom versus Textile Mill. We have seen that building a modern textile mill was inappropriate because it put many local people out of work and provided only a few jobs. In contrast, introduction of an improved handloom has allowed women to produce more cloth in their homes while caring for children.

The social structure was not upset and these women increased their earning power.

Handmade Brick Dwellings versus a Large Apartment Complex. Inappropriate technology was almost employed in a situation where the government planned to clear a poor slum area and build a huge complex of apartments. This project would have drastically altered the existing social structure. The appropriate technology involved teaching people to make bricks from the local mud. The technique spread quickly, and soon nearly every family in the area was making bricks and building a simple but adequate dwelling.

More Efficient Wood Stoves. Many people in less-developed countries are dependent on firewood for cooking and women spend much of their day gathering wood, denuding the landscape in the process. Introduction of a more efficient wood stove that consumes only half the wood is an appropriate technology which alleviates the situation.

Appropriate Technology and Agriculture

Nowhere can appropriate technology be applied with more advantage than to developing a sustainable agriculture. Techniques for utilizing and recycling wastes back into the soil to improve productivity and protect the soil from erosion abound. A few examples:

☐ Use of wastes as mulches to maintain soil moisture, prevent erosion, and improve soil quality.

☐ Introduction of polycropping techniques—the growing of several things simultaneously in the same plot so that harvesting of one thing and planting of another is going on continuously without leaving the soil bare.

☐ Anaerobic digesters to dispose of agriculture waste and produce methane gas for cooking, a process common in China.

☐ Use of small ponds and agriculture wastes for high-density fish culture, also common in China.

☐ Reforestation with tree species that will protect the soil from erosion as well as being good producers of both edible fruit and firewood.

As such techniques improve the productivity and incomes of numerous farmers, the entire local economy is boosted, because farmers, in turn, buy more products made by townspeople, promoting a cycle of economic improvement.

The concept of appropriate technology, however, is not entirely without drawbacks. Most seriously, it may be viewed by people of developing nations as an effort to prevent them from entering the modern industrialized world. In fact, some mix of centralized projects and appropriate technology is probably in order. One example that is commonly cited is the need for power plants and electrification. These are essential for development because without them people are handicapped simply by the lack of light when it comes to education, reading, studying, and learning new techniques and skills. Whether projects are small or large, however, it is now recognized that conservation of the environment and protection of basic soil and water resources, that is, *sustainable development*, must be the primary consideration.

FIGURE 6–7
Appropriate technology utilizes local labor, skills, and materials to improve conditions and incomes of local people. Here, workers in western Ethiopia are washing seed to counteract plant disease. (United Nations photo)

REDUCING FERTILITY

Regardless of how economic development is pursued, it is evident that such efforts are undercut by continuing population growth, since any economic gain must be consistently divided among more people. Economic development alone cannot be counted on to bring fertility rates down when they are so high that they virtually preclude gains and development. There must be direct efforts toward reducing fertility rates.

Much of the evidence and experience of the last twenty years shows that fertility rates may be reduced substantially through suitable efforts in *family planning, health care,* and *education,* the economic situation notwithstanding. Further, the social climate has changed significantly in recent years. As the crisis of overpopulation has become more evident, both political leaders and citizens of less-developed nations have become more supportive of and receptive to family planning programs.

Leaders of Less-Developed Nations Support Reducing Fertility

More than 100 nations representing over 95 percent of the world's population now have family planning programs, and political leaders are giving increasing attention to such programs. In 1987, a "Statement on Population Stabilization" was signed by 45 heads of state of developing countries and presented to the U.S. Congress. The statement reads in part:

> Degradation of the world environment, income inequality, and the potential for conflict exist today because of over-consumption and over-population. If this unprecedented population growth continues, future generations of children will not have adequate food, housing, medical care, education, earth resources, and employment opportunities . . .
>
> We believe the time has come now to recognize the worldwide necessity to stop population growth within the near future and for each country to adopt the necessary policies and programs to do so. . . . We call upon donor nations and institutions to be more generous in their support of population programs in those developing nations requesting such assistance.

In their call for donors to be "more generous in their support," developing nations are not looking for a free ride. Already 75 percent of the funding for family planning programs is provided by the developing countries themselves. Fifty percent of the population of these countries is under 20 years of age, largely poor, and lacking in education. Reaching this population of young people rapidly rising into their reproductive years is imperative and will require help from developed countries.

People of Less-Developed Countries Want Fewer Children

As a legacy of agrarian societies, people in less-developed countries have wanted more children, but faced with poverty and lack of opportunity, *they want fewer children than they are having.* They also appreciate that having a few healthy children will meet their desires for old-age security and economic help better than several sickly children. They can readily see the virtue of focusing their scant resources on raising two or three healthy children as opposed to having many children who, for lack of adequate resources, are ill-fed and ill-cared for. This is the general conclusion of an extensive survey conducted in 61 less-developed

nations between 1972 and 1984. Specifically, 50 percent of those women interviewed wanted no further children, and an additional 25 percent wanted to delay their next pregnancy for at least two years.

However, 60 percent of these women who wanted to delay or not have future pregnancies did not know of *any* family planning source, and the 40 percent who did know often experienced inadequate service. It was frequently too far away, too busy to see them, out of supplies, or the supplies were too costly. Thus contraceptive use in less-developed countries, while increasing in some, is still low (Fig. 6–8).

People in developed countries often take basic knowledge regarding family planning and the availability of contraceptives for granted. It is urgent to appreciate the fact that huge populations of young people in developing countries are uneducated, lacking even rudimentary reading and writing skills. They do not have even the most basic understanding of the reproductive process, much less knowledge of birth control or of the availability of contraceptives. Again it is desperately important to reach this pop-

FIGURE 6–8
Trends in contraceptive prevalence, 1970–83, in selected countries. Note that while contraceptive use is increasing, it is still very low in most developing countries. Thus, there is a tremendous opportunity to decrease fertility through promoting family planning and supplying contraceptives. (From the *World Development Report 1984.* Copyright © 1984 by the International Bank for Reconstruction and Development/The World Bank. Reprinted by permission of Oxford University Press, Inc.)

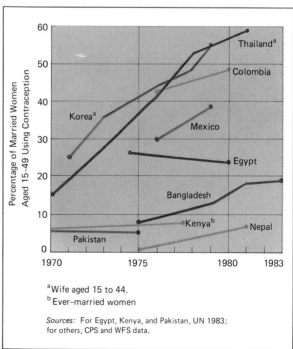

[a]Wife aged 15 to 44.
[b] Ever-married women

Sources: For Egypt, Kenya, and Pakistan, UN 1983; for others, CPS and WFS data.

ulation with information and contraceptives as they move into their reproductive years.

Effectiveness of Family Planning, Health Care, And Education

In actual practice, family planning involves:

☐ Counseling of singles or couples regarding the reproductive process and the pros and cons of various contraceptive techniques to avoid unwanted pregnancies.

☐ Supplying contraceptives that individuals or couples choose.

☐ Counseling regarding achieving the best possible pre- and postnatal health of mother and child. Emphasis is on good nutrition, sanitation, and hygiene.

☐ Counseling regarding the health advantages of spacing children. Emphasis is on breastfeeding which provides both ideal nutrition and acts as a natural contraceptive since nursing mothers generally do not ovulate. Nursing, and the contraceptive protection it provides, may be extended til the child is 2½–3 years old. Such spacing may cut total fertility in half.

☐ Since all contraceptive techniques have certain drawbacks, some unwanted pregnancies occur despite the best of intentions. Also women already carrying an unwanted fetus may come to family planning seeking assistance. Abortions are *not* advocated or promoted but they may be made available as an option of last resort—with the purpose of preventing illegal abortions to which the woman might otherwise turn, endangering her life or health, and the health of existing children who will not be cared for if the mother gets sick or dies. Maternal injuries and deaths from legal abortions are nil.

Health care is an integral part of or is closely allied with family planning. In industrialized countries we tend to take excellent health care of mothers and infants for granted; we think of improved health care primarily in terms of curing diseases and extending the lives of people in later life. But in developing nations, the situation is very different. Poor health and mortality are common among mothers and children and frequently involve poor nutrition and infectious diseases resulting from poor sanitation and hygiene. As these conditions are addressed from a purely medical standpoint, the effectiveness of delaying childbearing and spacing children is stressed.

FIGURE 6–9
Total fertility rate by years of education of the woman in selected developing countries. Even a few years of elementary education has a significant effect on reducing total fertility. The apparent increase between 0 and 1–3 years seen in some cases probably indicates that the group of women with no education tends to include more women that are unable to bear children for other physical and/or mental reasons. (From the *World Development Report 1984*. Copyright © 1984 by the International Bank for Reconstruction and Development/The World Bank. Reprinted by permission of Oxford University Press, Inc.)

Thus health care, especially of mothers and children, ties in with and supplements family planning efforts. As couples begin to rear healthy, vigorous children, the desire for more than two or three begins to drop. This is expanded upon in Case Study 6.

The same principal holds for education. As their cultures are undergoing rapid change, parents can see that an educated child will be more likely to achieve success and provide old-age security than an uneducated child. Sending a child to school demands suitable clothing, pencils, paper, and other items. Thus, poor parents are immediately in the position of having to allocate scant resources. They can see that their desire for educated children conflicts with their desire for many children and they begin to want fewer.

We are not talking about higher education. Rates of illiteracy in some less-developed countries are as high as 80 percent. Simply teaching people, especially women, to read and do simple calculations is tremendously important in order that they can follow printed directions regarding health care, nutrition, and contraception provided by family planning and health care services. One of the biggest handicaps that such services face is the low levels of literacy.

In addition, the availability of education to women opens up opportunities and allows them to pursue goals other than bearing children. Figure 6–9 shows that decreasing fertility correlates with increasing years of education of the wife. Note that only seven years of schooling (through middle school) is involved in this study (Fig. 6–10). Fertility rates dropped significantly in many developing countries in the decade from 1972 to 1982 even without significant improvement in economic development. These drops in fertility are attributed to family planning efforts.

Thailand, an Asian country of 55 million people, provides a particularly good example of the effectiveness of family planning. The total fertility rate in Thailand has dropped from over 7 in 1960 to about

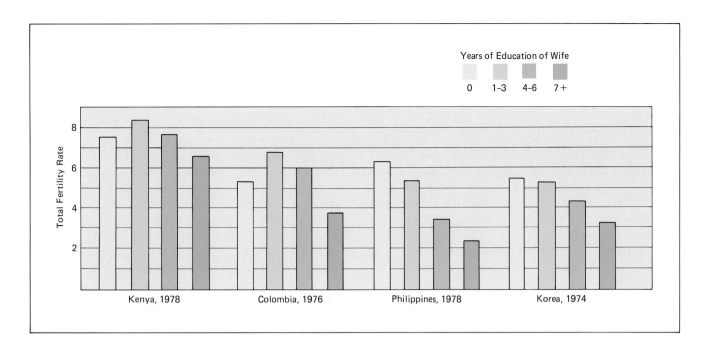

FIGURE 6–10
Fertility in relation to income in developing countries, 1972 and 1982. There is a correlation between rising income and decreasing fertility rates, but it is a weak correlation. Additional significant factors involved are improvements in education, health care, and promotion of family planning services. (From the *World Development Report 1984.* Copyright © 1984 by the International Bank for Reconstruction and Development/The World Bank. Reprinted by permission of Oxford University Press, Inc.)

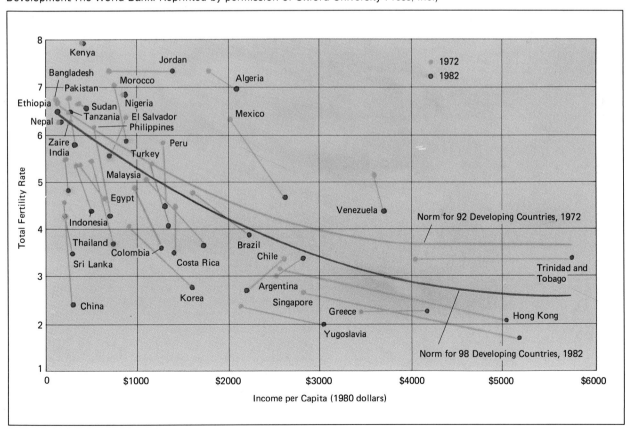

A LESSON LEARNED FROM BANGLADESH
Better Health and Nutrition Saves Lives and Helps Slow Population Growth

My own lessons in international health began in 1982 when I visited Bangladesh for the first time. Bangladesh, a warm and beautiful delta at the mouth of two of the largest rivers in the world, the Ganges and the Bramaputra, is one of the poorest countries in the world and a country with one of the highest populations per square mile. The country is approximately the size of Wisconsin and 110 million people live there. Imagine half the people in the United States living in Wisconsin! Over half of these people live on less than $100 per year.

I worked for an international research center that does much of the background work for the UNICEF and World Health Organization health programs, and the International Centre for Diarrhoeal Disease Research, Bangladesh (ICDDR,B). This center has been working in Bangladesh for over 25 years to find the causes and possible cures for the major killer of children: diarrheal disease. In 1987, these preventable diseases and related causes accounted for nearly 15 million child deaths. The center was established in the early 1960s by the United States government so that scientific studies could be done right in the area where these diseases are common, rather than in laboratories far removed from the problems.

The work I did at the ICDDR,B, included helping establish a national program for fighting rampant diarrheal epidemics. That meant considerable travel around the country to the rural field areas where cholera and other diarrheal diseases still kill thousands of children every year. In Bangladesh, about 300 children die every day from dehydration and related causes. Worldwide, over 10,000 children die daily from preventable causes and most of these deaths are from diarrhea—a condition in America that is considered trivial but, in reality, is one of the top five causes of hospitalization.

At one large research site, the Matlab Bazaar, the ICDDR,B studies how certain health treatments and preventions affect the survival of children. The ICDDR,B has worked in this area for over twenty-five years, trying different methods to save lives, reduce population growth, and improve child health. The Matlab area's population of about 180,000 has been studied through carefully controlled scientific demographic methods. The area studied has been divided (on paper!) into two main groups which both receive free medical care. One group of 90,000 people also receives education and certain health interventions which include certain immunizations, education about birth control methods, breast feeding and oral rehydration therapy, a simple and low-cost solution that replaces essential body fluids lost during diarrheal illness—the major killer here.

One of the questions I am often asked as Executive Director of the International Child Health Foundation, an organization dedicated to saving the greatest number of child lives at the lowest cost, is "what are we going to do with all the children saved?" People often think that health programs that save lives mean more children living in already difficult situations. "Why help? Why," they say, "when it means people in poor areas or countries will just keep having more children, more than they can feed or clothe? Why help when that means the situation will be even more desperate in years to come?" In Matlab, we found the answer.

The lesson learned at Matlab Bazaar in rural Bangladesh—a lesson that was dramatic to me—was

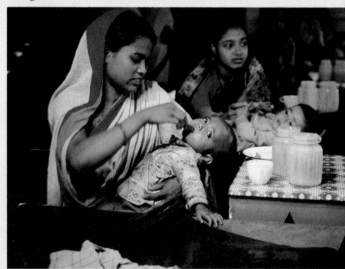

that when mothers learn they can prevent the deaths of their children, when mothers find they are able to save their children's lives, they are willing to have fewer children. Those children will be better nourished and healthier and, therefore, less susceptible to disease.

In its 1987 Annual Report, UNICEF acknowledged that it is parents who are the most important child health workers—the "front line" in health. The UNICEF report was based, in part, from facts learned in a small and remote village in Bangladesh.

For further information, do write us or one of the following organizations: UNICEF, 3 UN Plaza, New York, New York 10017; The Population Council, One Dag Hammarskjold Plaza, New York, New York 10017; Save the Children Federation, 54 Wilton Road, Westport, Connecticut 06880; or the U.S. Agency for International Development, Office of Health, U.S. Department of State, Washington, D.C. 20523; International Child Health Foundation, P.O. Box 1205, Columbia, Maryland 21044.

CHARLENE B. DALE
International Child Health Foundation

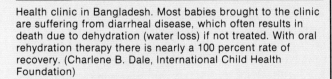

Health clinic in Bangladesh. Most babies brought to the clinic are suffering from diarrheal disease, which often results in death due to dehydration (water loss) if not treated. With oral rehydration therapy there is nearly a 100 percent rate of recovery. (Charlene B. Dale, International Child Health Foundation)

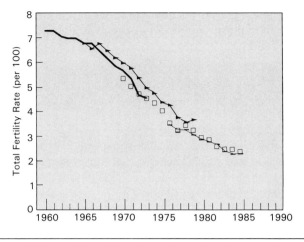

FIGURE 6–11
The total fertility rate in Thailand is falling. This remarkable decline over the last 30 years is attributed mainly to vigorous promotion and free distribution of contraceptives. Different lines in the graph represent separate surveys, but all show close agreement. (Source: Population Reference Bureau, Inc. and Thailand Demographic and Health Survey 1987.)

2.4 now, among the lowest in the less-developed countries (Fig. 6–11). Much of this remarkable success has been achieved simply through the vigorous promotion and free distribution of contraceptives. The percentage of Thai people in reproductive age groups using modern, effective methods of contraception is among the highest in the world, about equal to that in the United States and many European countries.

In conclusion, experience shows that provision of family planning services, health care for mothers and children, and education result in significant reduction in fertility at very modest cost. Simply making these services more widely available would be an extremely cost-effective method of reducing fertility.

Additional Economic Incentives

The survey noted above and Thailand's experience indicates that fertility in most less-developed nations can be reduced substantially (from the current 4.8 to 3 children per family) just by filling the unmet needs for family planning. However, a country may choose to provide various economic incentives and/or deterrents aimed at reducing fertility still further. China, the largest developing country, with over 1 billion people, is the prime example. Some years ago, China's leaders recognized that unless population growth was controlled, the country would be unable to live within its limits. Their goal was a one-child family, and to achieve this the authorities instituted an elaborate array of incentives and deterrents. The prime ones are as follows:

- Paid leave to women who have fertility-related operations, namely sterilization or abortion procedures.
- A monthly subsidy to one-child families.
- Job priority for single children.
- Additional food rations for single children.
- Housing preferences for single-child families.
- Preferential medical care to parents whose only child is a girl. (There is a strong preference for sons in China and parents generally wish to have children until at least one son is born.)

Penalties for an excessive number of children in China include:

- Repayment of bonuses for the first child if a second is born.
- Payment of a "tax" for a second child.
- Payment of higher prices for food for a second child.
- Denial of maternity leave and paid medical expenses after the first child.

Along with improving economic opportunities, these incentives and deterrents have helped China achieve a precipitous drop in its total fertility rate from about 4.5 in the mid-1970s to 2.6 in 1982, and 2.4 currently. Because of this unique achievement, and its tremendous size, China is frequently excluded from the demographic categories of developing countries.

It should be evident however, that such measures can only be applied by the particular nation itself. One nation cannot impose such sanctions on another, but it can condone and support such actions. Even the government in power must give consideration to what its people will accept and tolerate; otherwise the government is asking for trouble. For example, one government of India was voted out of power largely because of its policy of promoting sterilization.

Promoting Family Planning and the Abortion Controversy

The abortion issue and, as a result, family planning have become the center of extreme controversy between anti-abortionists or "right-to-life" groups and "pro-choice" groups. Pro-choice advocates are not for abortions; they are for women having the right to choose whether or not to have a child, and allowing this choice to include the option of a safe abortion— if she should become pregnant against her wishes.

Right-to-life advocates wish to exclude the option of abortions regardless of circumstances.

However, it is often not a simple choice between a legal abortion and a healthy child being born some months later. A woman with an unwanted pregnancy is often motivated to seek an abortion. If legal abortions are not available, women frequently resort to illegal abortions, which are often clumsily performed by untrained persons under unsterile conditions, or women may attempt to induce an abortion themselves. Worldwide an estimated 155 to 204 thousand women die each year from infections and injuries received in illegal or self-induced abortions. In some poor African and Asian countries, the chance of death during pregnancy is as high as 3000 for every 100,000 live births compared to between 2 and 10 per 100,000 in developed countries. At least half of these deaths in less-developed countries are caused by illegal or self-induced abortions. The other half of maternal deaths occur mostly in one of the following *high-risk* situations: when the mother is under age 16 or over age 40; when a pregnancy occurs less than one year after a previous birth; when there have already been 4 or 5 previous pregnancies. For every woman who dies, many suffer serious, long-term, often permanent health problems. Further, when a mother with dependent children dies or is debilitated, there is almost always additional suffering, malnutrition, and not infrequently deaths among these children. This human suffering and death must be seen as a backdrop behind uncontrolled fertility.

The focus of family planning is on providing information, techniques, and means to avoid unwanted and/or risky pregnancies. However, when a woman does become pregnant against her desires or in a high-risk situation, the option of an abortion may be made available and a woman is allowed to choose this alternative. Since the chance of injury or death from legal abortions, which are performed by skilled practitioners under sterile conditions, is nil, the availability of a legal abortion serves to protect the woman from the hazards of attempting to carry a risky pregnancy to term or, even more, from the risks of illegal or self-induced abortion to which she might otherwise resort.

Nevertheless, because of the connection with abortions, right-to-life advocates are vigorously attacking family planning itself. Their success in curtailing family planning programs is having a result which is exactly the opposite to their stated purpose. By cutting back family planning services, more, not fewer, women become pregnant against their desire and seek abortions. Further, where legal abortions are not available, many women resort to illegal or self-induced abortions, or attempt to carry risky pregnancies to term. The tragic result is an increase in injury, suffering, and death of women and children. This is a reality which right-to-life advocates are loath to but should acknowledge.

Contraceptive Technology

A perfect contraceptive—one which is 100 percent effective, does not require remembering, is fully and immediately reversible, has no painful or harmful side effects, and is freely available to all women—would make the abortion issue largely moot. Abortion is not a desirable substitute for contraception. It only comes into the family planning issue because "perfect" contraceptives do not exist and even existing contraceptives are not consistently available to the poor.

Encouragingly, contraceptive technology is advancing. Perhaps most promising is Norplant, a small capsule that, when implanted in a woman's arm, releases hormonal chemicals that prevent pregnancy. One implant may last up to five years but may be reversed any time by removing the implant. Norplant is already approved by the United Nations World Health Organization and is being introduced into Asia, Africa, and Latin America. It is extremely cost-effective since once a village in a developing country has been serviced by a medical team, there is no need to return at frequent intervals. Other similar implants and contraceptive injections are being developed.

A French pharmaceutical company has developed an "after-the-fact" drug known as RU486 which prevents a pregnancy from developing if taken within six weeks of the first missed period. The drug has been approved for use in France and China and is undergoing tests in other countries. However, in 1988, the company received vigorous protests and threats of boycotts from U.S. right-to-life groups, because the drug effectively causes a spontaneous abortion. This led the company to announce that it would halt production, but the French government intervened, saying that the company had an obligation to produce the drug to protect the welfare of women who might otherwise seek unsafe methods of abortion. However, it was withdrawn from the American market.

Costs of Family Planning

There are over 400 million people in the developing world who want to limit the size of their families but do not have the knowledge or means to do so, and this number is growing. Over the next 20 years 3 billion people will be coming into their reproductive years (recall population profile, Fig. 5–4). Any chance of reducing fertility depends on reaching these people with family planning services. The World Population Institute estimates that this could be accomplished with total expenditures of $2.4 billion per year starting in 1989, and rising to $4.1 billion in 1998; the U.S. share would be $600 million per year at the start.

These sums are tiny when compared to the more than trillion (1000 billion) dollars per year the world spends on armaments. They are also small when compared to the costs of environmental degradation and human misery which will result from not doing anything. For example, the U.S. government spent $409 million on emergency aid to Africa between 1985 and 1987, costs which might have been avoided if smaller sums had been spent on family planning in the preceding years. Further money spent on emergency relief does not repair the ecological damage and deterioration which underlie the famines. Indeed, long-term security depends just as much on stabilizing population and protecting the environment as on protection from outside enemies. Thus, family planning may be seen as an extremely cost-effective way of avoiding huge tolls in the future.

In the period from World War II to 1984, the United States provided world leadership and financial support for family planning programs. Then in 1984, at the international conference on population in Mexico City, just as the rest of the world leaders were recognizing the extreme urgency of population control and asking for family planning assistance, the United States, siding with right-to-life advocates, announced an about-face. Less than 1 percent of U.S. support for family planning was being used to perform abortions. However, more than just denying funding for abortions, the United States charted a new policy of denying funds to any organization that even mentioned to women, regardless of circumstances, that abortion might be one of their options. The consequent loss of U.S. support has severely undercut the world's most prominent family planning organizations—The United Nations Fund for Population Activities, International Planned Parenthood Federation, and Family Planning International Assistance, as well as others. More than 340 million couples in 65 countries are adversely affected by this policy, estimates The Worldwatch Institute.

Of course, as environmental degradation occurs as a result of overpopulation, everyone will be adversely affected.

ACTIONS YOU CAN TAKE

There are a number of direct actions which you can take toward promotion of a stable world population:

☐ Most importantly, write the President and your Senators and Congressperson, asking that the United States return to its former policy of supporting family planning activities, in particular that funding to the above mentioned organizations be restored.

☐ Additionally, request that the government provide more support for the development and testing of new, more effective contraceptives.

☐ You can join The Planned Parenthood Federation or other family planning organizations and have your membership dollars provide support.

☐ There are many temporary internships and career opportunities with family planning organizations, which you may consider.

Bear in mind that population is only one factor in the equation relating the impact of people on the environment; the impact may be moderated greatly by environmental regard or exacerbated by lack of it. Population momentum will cause populations to grow considerably even with the best of fertility control. Therefore, in addition to efforts toward stabilizing population, it will be vitally important to promote conservation of soil, water, and other natural resources, control pollution, which adversely affects these resources as well as human health, and protect natural ecosystems. It is to these issues that we turn our attention in subsequent chapters.

Looking at all the problems and all that needs to be done, it is easy to become overwhelmed and wonder: What can I do? Let me just leave you with this thought: *Waiting for someone to tell you what you should do prepares you to be a menial laborer. Looking around and seeing what you can do and doing it prepares you to be a leader and manager.*

Part Three

SOIL, WATER, AND AGRICULTURE

Fertile soil and adequate water are humankind's most important resources because they form the basis for virtually all food production. Additionally, they are the basis for all timber and wood products, natural fibers, beverages, and countless other commodities. Abundant soil and water resources underlie a prosperous civilization; impoverished soil supports only impoverished people—or none at all.

Both soil and water are "renewable" resources, meaning that natural processes will maintain them indefinitely, but only given proper care and conservation. Humans tend to ignore the need for proper care and conservation and "mine" these resources until they are exhausted, and then suffer the consequences. Archaeologists have determined that the downfall of many prosperous, ancient civilizations was caused not by enemies from the outside but by

slow ecological suicide from the inside—failure to maintain soil and water resources. For example North Africa, which was the grain-growing region that supported the Roman Empire, is now largely desert. This deterioration probably played a significant part in sapping the vitality of the Empire. Similarly, loss of soil fertility due to erosion seems to have been a key factor in the decline of the prosperous Mayan civilization of Central America.

Yet, far from learning from these lessons of the past, we seem bent on repeating them on a global scale. Under the pressures of rapidly growing population, measures for long-term conservation have been largely ignored, and erosion has reached unprecedented levels. The Worldwatch Institute estimated that worldwide soil loss from erosion in 1984 was 26 billion tons and growing. This is equivalent to all the topsoil on some 23 million acres, enough land to support 150 million people. Unlike earthquakes and volcanic eruptions, this disaster is unfolding gradually, but per capita grain production in Africa has been on the decline since 1967 and there are already several million "ecological refugees," starving, penniless people migrating from their homelands because the soil no longer supports their existence. Massive soil erosion is threatening the economic future of India, China, the Soviet Union, and a number of other countries, developed and developing alike. The United States is one of the few countries that still produces quantities of food significantly beyond the needs of its own population. But ominously, soil degradation is occurring here as about 3 billion tons per year are lost to erosion. Likewise, water resources are overdrawn in many areas and exhaustion of water supplies for irrigation is likely to impose further constraints on agriculture.

Obviously, growing population and deteriorating soil and water resources do not add up to a sustainable future. However, we do have the knowledge and the ability to maintain soil and conserve water. Our objective in Part III is to examine fundamental aspects of soil and water so that we can both appreciate the nature of the problems and understand the kinds of management procedures that need to be adopted to achieve a sustainable society.

7
Soil and the Soil Ecosystem

CONCEPT FRAMEWORK

■ Outline

■ Study Questions

I. PLANTS AND SOIL — 157
A. Critical Factors of the Soil Environment — 157

   1. Mineral Nutrients and Nutrient-Holding Capacity — 157

   2. Water and Water-Holding Capacity — 158

   3. Oxygen and Aeration — 160

   4. Relative Acidity (pH) — 160

   5. Salt and Osmotic Pressure — 160

B. Growing Plants without Soil — 162

C. The Soil Ecosystem — 162
   1. Soil Texture: Size of Mineral Particles — 162

   2. Detritus, Soil Organisms, Humus, and Topsoil — 165

   3. Other Soil Biota — 167

   4. Organic versus Inorganic Fertilizer — 167

D. Mutual Interdependence of Plants and Soil — 169
II. LOSING GROUND — 170
A. Bare Soil, Erosion, and Desertification — 170

B. Causes of Losing Ground — 173
   1. Overcultivation — 173
   2. Overgrazing — 173
   3. Deforestation — 174

1. List and describe aspects of the soil that are critical for maintaining plant growth.

2. List four nutrients that plants need from the soil. How are they added to and taken from the soil? Define: *weathering, leaching.* Describe the significance of nutrient-holding capacity.

3. Explain why plants need continuous access to water to maintain growth. Define: *infiltration* and *water holding capacity* and tell why each is important.

4. Define *soil aeration* and explain why it is important. Describe two factors that may prevent aeration.

5. What is *pH?* What is the range that will support life?

6. Describe how salty water will prevent plant growth.

7. Name and describe a system for growing plants without soil. What do such systems prove regarding what plants need from the soil?

8. Define *soil texture.* Name and describe the three categories which comprise soil texture. What is *loam?* Describe how different soil textures affect water, nutrients, aeration, and workability. What are the best soil textures? Are they not ideal by themselves?

9. Define *humus.* Where does it come from? Describe how topsoil is derived from subsoil, detritus, and soil organisms. Contrast the characteristics of topsoil and subsoil. Name and describe what occurs if insufficient detritus is added to the soil. How is plant growth affected by loss of topsoil? Why?

10. Name certain other soil organisms and tell how they may affect plant growth.

11. Distinguish between and give pros and cons of *organic* and *inorganic fertilizer.*

12. Describe the interdependencies that exist between plants and the soil ecosystem.

13. Name and describe various kinds of *erosion.* How are soil characteristics changed by erosion? What happens to productivity? Why is "desertification" an apt name for the phenomenon?

14. List and define three major causes of erosion. What are the economic/social pressures behind each?

4. Irrigation, Salinization, and Desertification 177

C. Dimensions of the Problem 178

D. Preventing Erosion and Desertification 182
   1. Preventing Erosion: Traditional Techniques 182

   2. Preventing Erosion: New Techniques 182
      a. No-Till Agriculture 182
      b. Perennial Crops 184
   3. Limit Grazing 184
   4. Reforestation and Rehabilitation of Desertified Lands 185
   5. Countering Salinization 186

III. IMPLEMENTING SOLUTIONS 186

CASE STUDY: HAITI: IN THE LAND WHERE HOPE NEVER GROWS 188

15. Describe how irrigation may lead to loss of productivity.

16. Discuss and give examples illustrating the dimensions of the desertification problem. Is it preventable?

17. Name and describe traditional techniques for preventing erosion.

18. Name and describe new techniques for preventing erosion.

19. How may salinization be avoided?

20. Discuss what you might do to promote solutions. Describe what is meant by *debt-for-nature swaps*.

$\boxed{\text{M}}$aintaining productive soil for agriculture, range-lands, and forests is the most fundamental aspect for sustainable human development. Yet every year we are allowing the equivalent of 23 million acres (an area the size of a square 190 miles on each side) to slip into barren nonproductivity because of various abuses or shortsighted mismanagement.

We will address in this chapter the abuses and mismanagement practices that cause this to happen, what can prevent them, and how we can restore barren lands to productivity. We will begin with a description of the critical aspects of soil that are necessary to support productive plant growth. We can then understand how productivity is lost and how it may be restored.

PLANTS AND SOIL

Critical Factors of the Soil Environment

In order to sustain plants, the soil environment must supply the plants' need for mineral nutrients, water, and oxygen. The pH (relative acidity) and salinity (salt concentration) of the soil environment are also critically important.

Mineral Nutrients and Nutrient-Holding Capacity

Plants need mineral nutrients such as nitrate (NO_3^-), phosphate (PO_4^{3-}), potassium (K^+) and calcium (Ca_2^+) as was described in Chapter 2. Except for nitrogen compounds, which are derived from the at-mosphere through the nitrogen cycle, all mineral nutrients are originally present as part of the chemical composition of rocks along with nonnutrient chemicals such as silica, aluminum, and oxygen (see Fig. 2–3). However, plants cannot absorb these nutrients as long as they are bonded in the rock structure. The rock must be broken down so that the nutrient ions are released into a water solution or a loosely bound state (see Fig. 2–4).

Rock, referred to as **parent material,** is naturally broken down by *weathering.* **Weathering** includes all the physical forces such as freezing and thawing, heating and cooling, the abrasive action of sand particles carried by wind or water, biological factors such as pressure exerted by roots growing into small cracks, and various chemical reactions.

As nutrient ions are released, they may be absorbed by roots. They may also be washed away by water percolating through the soil, a process called **leaching.** Leaching not only causes a loss of soil fertility, it also contributes to pollution. Consequently the capacity of soil to bind and hold the nutrient ions, so that they will not leach but may be absorbed by roots, is just as important as the initial supply. This is the soil's **nutrient-holding** or **ion-exchange capacity.**

While weathering is the original source of nutrients, it is much too slow to support anything approaching normal plant growth. In natural ecosystems, the major supply of nutrients is from the decay of detritus and animal wastes, that is, the **recycling of nutrients** as described in Chapter 2 and illustrated for phosphorus in Figure 2–15. If nutrient-holding capacity is lost, nutrients leach away and fertility drops.

In agriculture systems, there is an unavoidable removal of nutrients through harvesting, since nutrients are a portion of the plant material. Consequently, nutrients are routinely replaced or supplemented by the addition of **fertilizer**. **Inorganic** or **chemical fertilizers** are mixtures of necessary mineral nutrients. **Organic fertilizers** are plant and animal wastes—such as manure—which contain nutrients that are released as wastes decay. Pros and cons of these fertilizers will be discussed later.

Even with fertilizer, the nutrient-holding capacity of the soil remains vitally important. First, leaching of fertilizer is an obvious economic loss. Even more serious, leached nutrients go into bodies of water causing serious pollution problems, as will be de-

scribed in Chapter 9. Figure 7–1 provides a summary of these ideas concerning nutrients.

Water and Water-Holding Capacity

Plants have tiny pores in the leaves, which permit the absorption of carbon dioxide (CO_2) and the release of oxygen (O_2) in photosynthesis. These pores, however, also allow the escape of water vapor from the moist cells inside the leaf. This loss of water vapor from leaves, called **transpiration,** accounts for at least 99 percent of a plant's need for water; less than 1 percent of the water absorbed is used in photosynthesis. If there is inadequate water to replace transpiration loss, plants wilt. The wilted condition con-

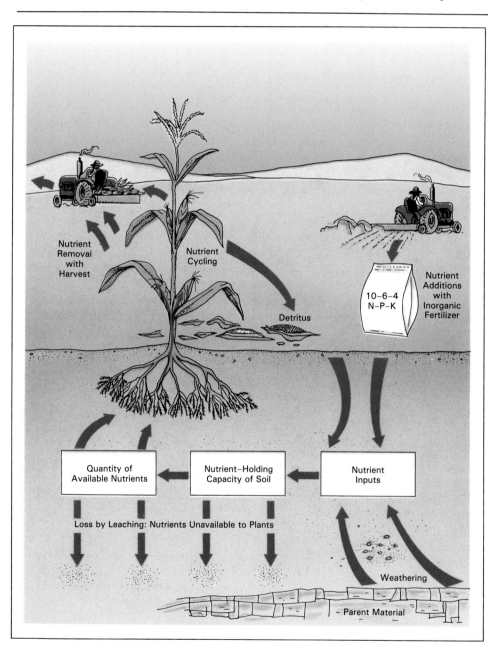

FIGURE 7–1

Plant-soil-nutrient relationships. Soil nutrients may come from decay of detritus, additions of fertilizer, and very slowly from weathering of parent material. Whether these nutrients remain in the soil until they are absorbed by plants or are lost from the system through leaching depends on the nutrient-holding capacity of the soil.

serves water and may hold off total dehydration and death for some time, but it also shuts off growth. Consequently, to keep most plants flourishing requires a considerable amount of water. A field of corn, for example, transpires an equivalent of a layer of water 17 in. (43 cm) deep in a single growing season.

First, it is obvious that if rainfall runs off the surface rather than soaking in, it won't be useful. Therefore, **infiltration,** the capacity of soil to allow water to soak into the surface, is highly significant. Second, since the roots of most plants do not go very deep into soil, water that trickles down through the soil more than a few feet (much less in the case of small plants) becomes unavailable. Therefore, between rains plants depend on a "reservoir" of water that is held in the surface soil, as water is held in a sponge. The size of this reservoir is determined by the **water-holding capacity** of the soil. Given the same intermittent precipitation, soils with good water-holding capacity can hold sufficient water to sustain plants over a considerable dry spell. Plants on soil with poor water-holding capacity, however, may suffer during even a short dry spell because the small reservoir is quickly exhausted.

Finally, the reservoir of water held in the soil may be depleted by its *evaporation* from the soil surface as well as by the plants using it. A mulch or vegetative covering which retards evaporation is advantageous.

To summarize, the ideal situation is a soil with good infiltration, good water-holding capacity, and a cover which minimizes evaporative water loss. These aspects of soil water are summarized in Figure 7–2.

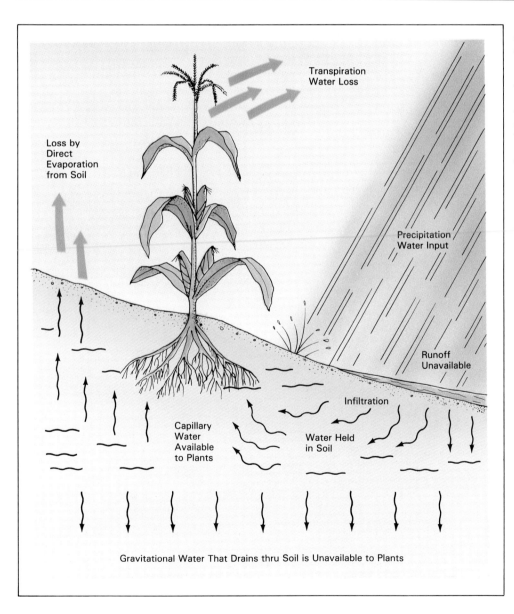

FIGURE 7–2
Plant-soil-water relationships. Water lost from the plant in transpiration must be replaced from a reservoir of capillary water held in soil. The size of this reservoir, beyond amount and frequency of precipitation, depends on the soil's ability to allow water to infiltrate, hold water, and minimize direct evaporation.

Transpiration Water Loss

Loss by Direct Evaporation from Soil

Precipitation Water Input

Runoff Unavailable

Infiltration

Capillary Water Available to Plants

Water Held in Soil

Gravitational Water That Drains thru Soil is Unavailable to Plants

Oxygen and Aeration

Roots require energy for growth and absorption of nutrients, and this energy is derived from the oxidation of glucose in the process of *cell respiration.* Cell respiration requires oxygen and produces carbon dioxide as a waste. Roots must be able to absorb oxygen from and expel carbon dioxide into the soil environment because, with few exceptions, plants have no means of transporting these gases. Consequently, the ability of soil to allow the diffusion (passive movement) of oxygen from the atmosphere into the soil and the reverse movement of carbon dioxide is another important aspect of the soil environment. This property is referred to as soil **aeration** (Fig. 7–3).

Two ways in which aeration is commonly blocked, and plants retarded or killed as a result, are *compaction* and *waterlogging.* **Compaction** refers to the packing down of soil until air spaces become too limited to allow diffusion (Fig. 7–4). **Waterlogging** is overwatering so that all the spaces in the soil are saturated with water. This is also commonly referred to as "drowning" plants.

A few plants such as bald cypress, mangrove, Spartina (marsh grass), sedges, and reeds, which can diffuse oxygen through the stem to the roots, thrive in the waterlogged soils of wetlands.

Relative Acidity (pH)

pH refers to the units and scale used to measure relative acidity and alkalinity (basicity), neutral or pH 7 being in the middle of the scale. This will be discussed more fully in Chapter 13 as will the consequences of changing pH through acid rain. For now it suffices to say that most plants as well as animals require a pH near neutral, and most natural environments provide this.

Salt and Osmotic Pressure

To function properly, all cells of living organisms must maintain a certain amount of water, a feature referred to as **water balance.** However, cells have no way of pumping water itself in or out. Instead, water balance is determined by the relative salt concentration inside and outside the cell membrane. Water molecules are attracted to salt ions. The cell membrane inhibits the passage of salt, but water moves rapidly through the membrane toward the higher salt concentration, a phenomenon called **osmosis** (Fig. 7–5).

Cells control their water balance by adjusting their internal salt concentration, and water follows passively by osmosis. If the salt concentration outside the cells is too high, the cell cannot absorb water;

FIGURE 7–3
Soil aeration. Plant roots are dependent upon an exchange of oxygen and carbon dioxide in respiration. In turn, these gases must be able to diffuse through the soil between soil particles.

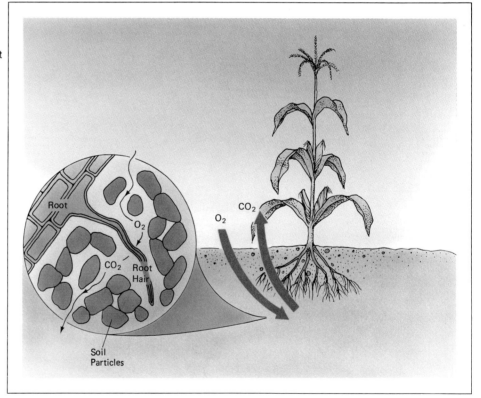

FIGURE 7–4
Poor soil aeration reduces plant growth. The cut through the soil shows that the upper layer has been severely compacted. For the soybean plant on the left, the compacted layer was broken and loosened by tillage; for the plant on the right it was not. The small size of the plant on the right shows that compacted soil severely retards growth, an effect largely due to poor soil aeration. (USDA—National Tillage Machinery Laboratory photo)

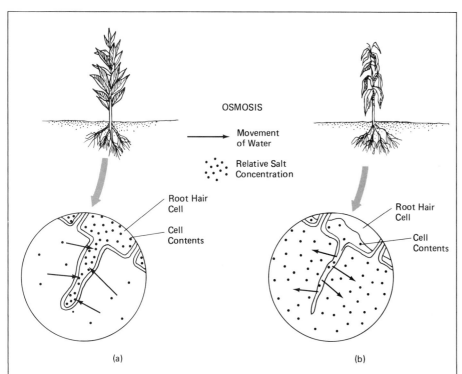

FIGURE 7–5
Soil salt and movement of water by osmosis. (a) Cells absorb potassium salts but exclude sodium salts. When soil water is fresh, roots absorb potassium such that salt concentration inside cells is greater than outside. The result is movement of water into cells, keeping the plant turgid (nonwilted). Cell walls keep cells from bursting. (b) If there is a high concentration of salt outside, especially sodium salts, which cells exclude, water is drawn out of the cells, causing wilting and death of cells.

indeed, water may be drawn out of the cells by osmosis resulting in dehydration and death of the plant. This is why freshwater species cannot tolerate seawater, which has a salt concentration of about 3.5 percent. Marine species have various adaptations that enable them to maintain water balance in seawater; they are killed by the same process in reverse when put in freshwater. Most land plants require fresh water. As soil becomes salty, its ability to support plants decreases. Highly salted soils are virtual deserts supporting no life at all.

Growing Plants without Soil

Plants do not require any "secret ingredients" from the soil except the aspects described above. This is demonstrated by the practice of hydroponics, the culturing of plants without soil. In most hydroponic systems, plants are rooted in a bed of pea-sized gravel or coarse sand that is irrigated more or less continually with a circulating nutrient solution (Fig. 7–6). Thus, the roots have continuous access to water, mineral nutrients, and good aeration. Plants cultured hydroponically do exceedingly well, and tests indicate that they are just as nutritious as those grown on good soil.

FIGURE 7–6
Hydroponics, growing plants without soil. One method of hydroponic culture is to grow plants in a bed of gravel. A solution containing all the necessary mineral nutrients is pumped up into the gravel every eight hours and allowed to drain back to the storage tank. The gravel periodically wetted with nutrient solution provides an environment of water nutrients and air that is ideal for roots. Costs, however, preclude hydroponics being used for general food production.

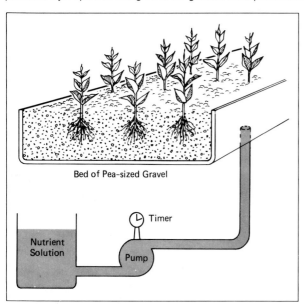

Bed of Pea-sized Gravel

Timer

Nutrient Solution

Pump

The concern for soil then is one of cost. The labor and material used in hydroponic systems are so expensive that food produced must command a very high price. For example, to support growing wheat hydroponically would require bread to sell for about $10 a loaf. The only way that we can have food and other agricultural products at a price we can afford is to preserve soils under natural conditions in such a way that most of the requirements are provided free of charge.

The Soil Ecosystem

On examining a productive topsoil you will find a complex combination of the following:

☐ Mineral particles.

☐ Detritus, or dead organic matter from vegetation and animal wastes, in various stages of decay.

☐ A host of living organisms, ranging from the decomposers (fungi and bacteria) to larger detritus feeders (earthworms, snails, and insects), forming a complex food web supported by the detritus.

A soil that supports the productive growth of plants, then, is much more than "dirt." It is a complex ecosystem (Fig. 7–7). Maintenance of this soil ecosystem is essential to maintaining the productivity of the soil. We shall look at each of these components in more detail and see how they work together to make a productive soil.

Soil Texture: Size of Mineral Particles

Rock exposed to the chemical and physical factors of the environment is gradually broken down or *weathered*. The resulting particles are classified as sand (the most coarse), silt, and clay (the finest, see Table 7–1). These particles make up the mineral component of soil. The size of the particles is defined as **soil texture.** Some soils are virtually pure sand, silt, or clay, but commonly there is a mixture which can be determined by shaking a small amount of soil with water in a large test tube and then allowing it to settle. Since particles settle according to weight, sand particles settle first, silt second, and clay last. The proportion of each can then be measured.

The *soil texture triangle* shown in Figure 7–8 allows one to plot any combination of sand, silt, and clay and determine the texture classification for any soil. A classification of considerable importance is **loam,** which has a mineral composition of roughly 40 percent sand, 40 percent silt, and 20 percent clay.

Soil texture has a profound effect on infiltration,

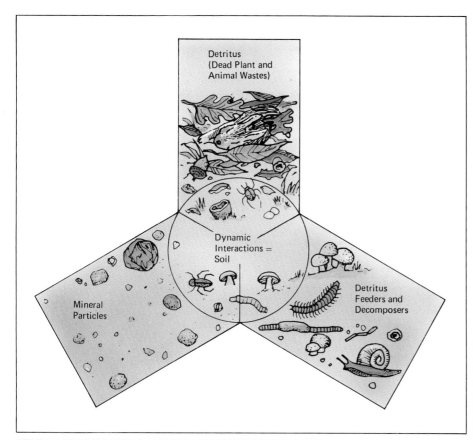

FIGURE 7–7
A productive soil is more than just "dirt." Maintaining soil involves a dynamic interaction between mineral particles, detritus, and detritus feeders and decomposers. Altering any of the three main factors may have drastic effects on the soil.

water-holding capacity, nutrient-holding capacity, and aeration. Larger particle size also means larger spaces between particles. Consequently, infiltration and aeration are very good with large particle size and worsen as particle size decreases. Hence, infiltration and aeration are excellent in sandy soils and very poor in clayey soils. Silty soils are intermediate.

However, water-holding and nutrient-holding capacity have the opposite relationship to particle size. Water- and nutrient-holding capacity improves as particle size decreases because water molecules

and nutrient ions stick to surfaces and surface area increases as particle size decreases. Imagine a rock being broken in half again and again. Every time it is broken, two new surfaces are exposed, one on each side of the break, but the overall weight or volume of material is not changed. Thus, in a given volume of soil, the smaller the particles, the greater the overall surface area and the greater the nutrient- and water-holding capacities (Fig. 7–9).

Soil texture also affects **workability,** the ease with which a soil can be cultivated. This has an important bearing on agriculture. Clayey soils are very difficult to work, because with even modest changes in moisture they go from being too sticky and muddy to being too hard to break. Sandy soils are very easy to work because they do not become muddy when wet, nor hard and bricklike when dry.

These relationships between soil texture and various properties are summarized in Table 7–2. What is the most productive kind of soil? Recall the principle of limiting factors. The poorest attribute is the limiting factor; the very poor water-holding capacity of sandy soils, for example, may preclude agriculture altogether. The best texture proves to be silt or loam in which there is a compromise between properties of sand and clay. This "best," however, is

Table 7–1
USDA Classification of Soil Particles

NAME OF PARTICLE	DIAMETER (mm)[a]
Very Coarse Sand	2.00–1.00
Coarse Sand	1.00–0.50
Medium Sand	0.50–0.25
Fine Sand	0.25–0.10
Very Fine Sand	0.10–0.05
Silt	0.05–0.002
Clay	Below 0.002

[a] mm = millimeter. 1 mm = 1/25 in., about the thickness of a dime.

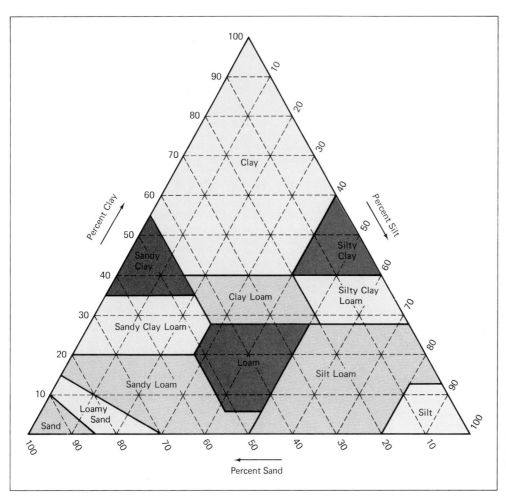

FIGURE 7–8
Soil texture classes. Soils are classified according to texture on the basis of the percentage of sand, silt, and clay that they contain. (USDA.)

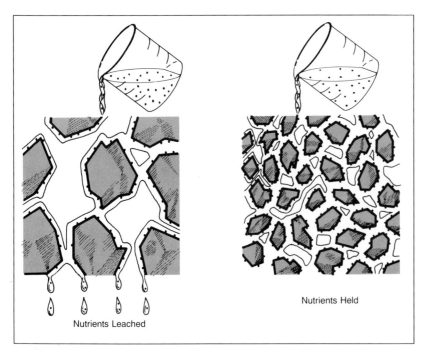

FIGURE 7–9
Soil water- and nutrient-holding capacity increase as particle size decreases. Both water and nutrient ions (represented by dots) tend to cling to surfaces, and smaller particles have relatively more surface area.

Table 7-2
Relationship between Soil Texture and Physical and Chemical Properties

SOIL TYPE	WATER INFILTRATION	WATER-HOLDING CAPACITY	NUTRIENT HOLDING CAPACITY	AERATION	WORKABILITY
Sand	Good	Poor	Poor	Good	Good
Silt	Medium	Medium	Medium	Medium	Medium
Clay	Poor	Good	Good	Poor	Poor
Loam	Medium	Medium	Medium	Medium	Medium

really only "medium." We will now consider how to maximize all factors.

Detritus, Soil Organisms, Humus, and Topsoil

A truly remarkable improvement in all the aspects of soil occurs as *humus* is integrated with the mineral portion. As each organism feeds it leaves a certain amount of undigested waste. **Humus** is the residue of undigested organic matter which remains after the bulk of detritus has been consumed. A familiar example is the black or dark brown, spongy material remaining in a dead log after the center has rotted out (Fig. 7-10).

FIGURE 7-10
Humus is organic material that is more resistant to decomposition and therefore remains after the major portion of decomposition has occurred, as in this tree stump. Gradually humus also decomposes to inorganic materials. (Photo by author.)

The natural process of integrating humus into the mineral portion of soil begins with the accumulation of dead leaves, roots, and other detritus on and in the soil. Detritus supports a complex food web including numerous species of bacteria, fungi, protozoans, mites, insects, millipedes, spiders, centipedes, earthworms, snails, slugs, moles, and other burrowing animals (Fig. 7-11). As these organisms feed, they not only reduce the detritus to humus; their activity also mixes and integrates it with the mineral particles and develops what is called *soil structure*. For example, as earthworms feed on detritus they ingest inorganic soil particles as well. As much as 15 tons of soil per acre (37 tons/hectare) may pass through earthworms each year. As the mineral particles go through the gut, they become thoroughly mixed and "glued" together with the nondigestible humus compounds. Thus, the sand, silt, and clay particles are bound together with humus into larger clumps and aggregates. The burrowing activity of such animals as mice keeps the clumps loose. This loose, clumpy characteristic is referred to as **soil structure** (Fig. 7-12).

Humus formation and development of soil structure occurs mainly in the upper 8 to 12 inches (2 to 3 decimeters) of soil, the zone in which the soil organisms are active. Thus, a layer of dark-colored soil with a clumpy, aggregate structure develops on top of the lighter-colored, humus-poor, compacted soil. This layer of humus-rich soil is called **topsoil;** the soil below is **subsoil.** A careful cut through a natural, undisturbed soil reveals this layering, referred to as the **soil profile** (Fig. 7-13).

Humus has phenomenal nutrient- and water-holding capacity, as much as 100-fold greater than clay on the basis of weight, and the clumpy aggregate structure greatly enhances infiltration, aeration, and workability. Regardless of soil texture, then, attributes are enhanced with humus and the soil structure it imparts. Sandy soils may be given significant water-holding capacity, clayey soils may be given sufficient aeration and infiltration, and loamy and silty soils may be enhanced in all regards (Table 7-3).

The difference between the ability of topsoil and subsoil to support plant growth has been tested by growing plants on adjacent plots, one of which has

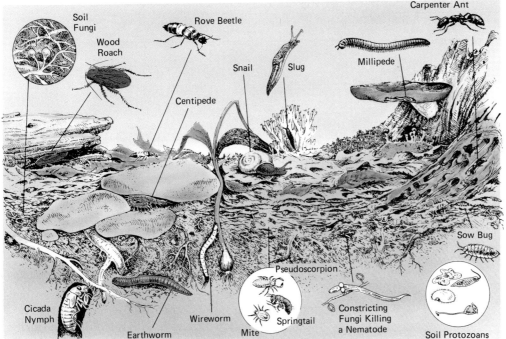

FIGURE 7–11
Soil organisms. The major groups of soil organisms are depicted in the illustration. It is the action of these organisms that reduces detritus to humus, intimately mixes the humus with soil, and in the process develops soil structure. (From Robert Leo Smith, *Ecology and Field Biology*, 2nd ed. [Figure 17–5, p. 540]. Copyright 1966, 1974 by Robert Leo Smith. Reprinted by permission of Harper & Row, Publishers, Inc.)

(a)

FIGURE 7–12
Humus and the development of soil structure. (a) On the left is a humus-poor sample of loam. Note that it is a relatively uniform, dense "clod." On the right is a sample of the same loam but rich in humus. Note that it has a very loose structure composed of numerous aggregates of various sizes. (Photo by author.) (b) A diagrammatic illustration of the difference.

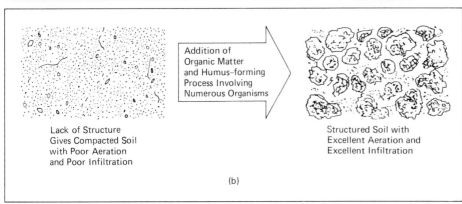

Addition of Organic Matter and Humus-forming Process Involving Numerous Organisms

Lack of Structure Gives Compacted Soil with Poor Aeration and Poor Infiltration

Structured Soil with Excellent Aeration and Excellent Infiltration

(b)

FIGURE 7–13
Soil profile. A cut through soil generally reveals a layer of loose, dark topsoil overlying light-colored, compacted subsoil. The topsoil layer results from the addition of organic matter and the activity of soil organisms. (USDA photo.)

decomposition at a rate of about 20 to 50 percent of its volume per year depending on conditions. Consequently, without periodic additions of ample detritus, humus will gradually decompose. As this occurs, there is a collapse of the soil structure and other attributes which it imparts. This loss of humus and the consequent collapse of topsoil is called **mineralization,** because what is left is just the gritty mineral content—sand, silt, and clay—devoid of humus (Fig. 7–14). House plants, for instance, need to be repotted periodically because the humus-rich "potting soil" mineralizes. Much more serious, however, are the huge areas of tropical forests cleared for agriculture that have become nonproductive due to the mineralization of the soil.

Other Soil Biota

A number of other interactions between plants and soil biota exist. One is a mutually beneficial symbiotic relationship between the roots of some plants and certain soil fungi. Masses of fungal filaments called **myocorrhizae** surround the roots. Some of the filaments penetrate and draw some nourishment from the plant, but they also spread through the soil and help the plant absorb nutrients (Fig. 7–15). Another important relationship is the role of certain soil bacteria in the nitrogen cycle discussed in Chapter 2.

Not all soil organisms are beneficial, however. Nematodes, small worms that feed on living roots, are highly destructive to a number of agricultural crops. In a flourishing soil ecosystem, however, nematode populations may be controlled by other soil organisms, such as a fungus, which forms little snares to catch and feed on nematodes (Fig. 7–16).

Organic versus Inorganic Fertilizer

Among gardeners and farmers, there is considerable dispute concerning the use of organic versus inorganic fertilizers. Understanding the process and importance of topsoil formation enables us to clarify the issues of this debate. First, hydroponic culture of plants provides a clear demonstration that plants prosper with only inorganic nutrients in water solution. There is no evidence that organic compounds from the soil are required for the plant's nutrition.

had the topsoil removed. Results are striking. Removal of topsoil resulted in an 85 percent decline in crop yield. In other words, if top soil is lost production may be only 15 percent. Conversely, restoring a good topsoil from the subsoil base may enhance the productivity by as much as sixfold.

This is an extremely important concept because humus is not permanent; it is natural organic material and, despite its resistance to digestion, is subject to

Table 7–3
Comparison of Humus-Poor and Humus-Rich Loam

LOAM TYPE	WATER INFILTRATION	WATER-HOLDING CAPACITY	ION-EXCHANGE CAPACITY	AERATION	WORKABILITY
Humus-poor	Medium	Medium	Medium	Medium	Medium
Humus-rich	Excellent	Excellent	Excellent	Excellent	Excellent

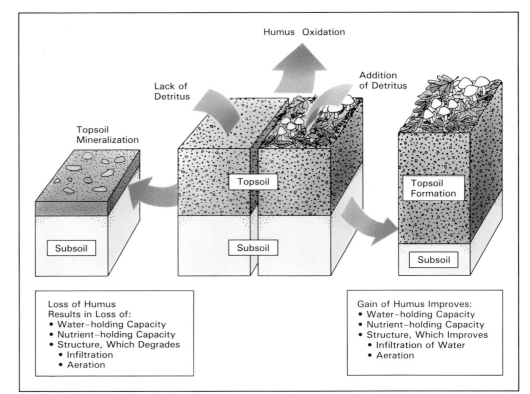

FIGURE 7-14
Topsoil formation or mineralization. Topsoil is the result of a dynamic balance between organic additions leading to humus formation and loss of humus through decomposition and oxidation referred to as mineralization. Soil characteristics important in supporting plants will vary accordingly.

Humus Oxidation

Lack of Detritus

Addition of Detritus

Topsoil Mineralization

Topsoil

Topsoil Formation

Subsoil

Subsoil

Subsoil

Loss of Humus
Results in Loss of:
• Water–holding Capacity
• Nutrient–holding Capacity
• Structure, Which Degrades
  • Infiltration
  • Aeration

Gain of Humus Improves:
• Water–holding Capacity
• Nutrient–holding Capacity
• Structure, Which Improves
  • Infiltration of Water
  • Aeration

FIGURE 7-15
Mycorrhizae. In many plants, soil fungi, or mycorrhizae, aid in the absorption of nutrients. Drawing shows cross-section of rootlet with mycorrhizae forming a connection between the root cells and soil particles.

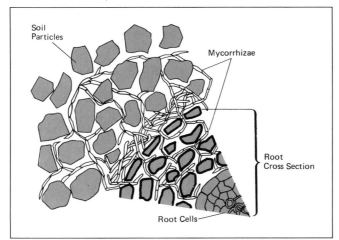

Soil Particles

Mycorrhizae

Root Cross Section

Root Cells

FIGURE 7-16
Soil nematode (roundworm), a root parasite, captured by the constricting rings of the predatory fungus *Arthrobotrys anchonia*. (Courtesy of Nancy Allin and O. L. Barron, University of Guelph.)

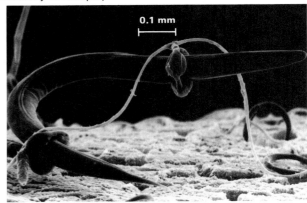

0.1 mm

Since optimal amounts of the inorganic chemical nutrients can be provided very efficiently and economically by using chemical fertilizers, they have an obvious advantage. The disadvantage of chemical fertilizer comes in allowing it to substitute for additions of detritus. Without the addition of sufficient detritus, soil organisms starve, humus content declines, and all the desirable attributes decline accordingly as topsoil mineralizes and soil structure collapses.

On the other hand, exclusive use of organic material may not provide sufficient amounts of certain required nutrients—especially organic mulches such as bark or wood chips, which provide protection from erosion and conserve soil moisture but lack nutrients. Soil microorganisms may actually absorb nutrients from the soil in the course of feeding on such material, thus "robbing" the soil of nutrients in the short run.

The Agricultural Extension Service, which has an office in every county in the United States, provides a service that tests soil samples and makes recommendations regarding needed soil nutrients. Following such recommendations, judicious use of some inorganic fertilizer along with organic material to protect and maintain the topsoil can provide the best of both worlds. Finally, growing a legume crop, which fixes nitrogen (see Chapter 2) in alternation with other crops, a practice known as **crop rotation,** may serve to add both nitrogen and organic matter to the soil.

Mutual Interdependence of Plants and Soil

You should now be able to visualize the mutual interdependence which exists between soil and the plants growing on it. Plants provide the food for a soil ecosystem that serves to maintain the topsoil in a condition that best supports the growth of the plants (Fig. 7–17).

Plants protect the soil in other ways, too. Most prominently, a vegetative cover protects the soil from erosion, a topic that we shall address in detail in the next section, and a cover of organic material greatly

reduces evaporative water loss while still allowing infiltration. This and the high water-holding capacity of humus-rich topsoil maximizes the percentage of precipitation available for plant growth.

Like all ecological relationships, the interrelationship between plants and soil is a *dynamic balance,* not a fixed situation. Soil deterioration sets in motion a *vicious cycle* that becomes ever more difficult to halt or reverse. Degraded soil slows plant production, which means less detritus for humus production and protection from erosion and evaporative water loss. This exposure leads to further soil deterioration and even less plant production until the land becomes totally barren (Fig. 7–18). The cycle can be reversed with suitable mulching and replanting of a vegetative cover, but it is expensive and painfully slow. It can require several decades or hundreds of years. Recognizing this underscores how imperative it is that we do what is necessary to preserve soil. The following sections explore what we must do to ensure such soil preservation.

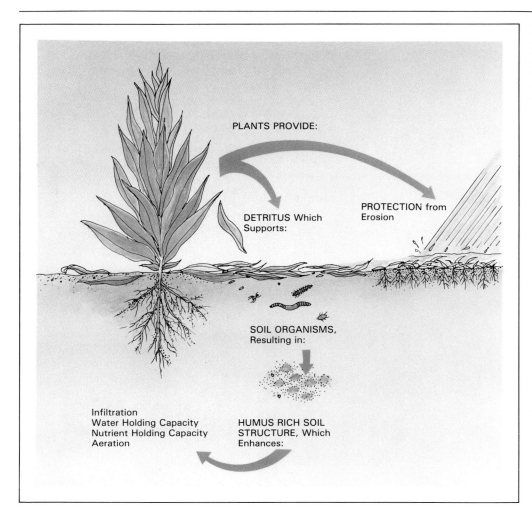

FIGURE 7–17
A mutual relationship exists between plants and the soil environment. Attributes of soil are enhanced by inputs of detritus derived mainly from dead plant wastes.

PLANTS PROVIDE:

DETRITUS Which Supports:

PROTECTION from Erosion

SOIL ORGANISMS, Resulting in:

Infiltration
Water Holding Capacity
Nutrient Holding Capacity
Aeration

HUMUS RICH SOIL STRUCTURE, Which Enhances:

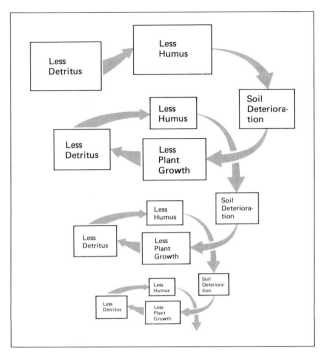

FIGURE 7–18
Soil degradation may become a vicious cycle. Degraded soil supports fewer plants, permitting further soil degradation.

LOSING GROUND

If we wish to have sustainable human development, we must recognize negative impacts on soil and adopt practices to control them.

Bare Soil, Erosion, and Desertification

By far the most destructive force on soil is **erosion,** the process of soil particles being picked up and carried away by water or wind. The removal may be slow and subtle, as soil is gradually blown away by wind erosion, or it may be dramatic, as gullies are completely washed out by water erosion in a single storm (Fig. 7–19).

Normally, a cover of vegetation or natural litter (dead leaves and other detritus) provides protection against all forms of erosion. The energy of falling raindrops is dissipated against such material enabling the water to trickle gently into the soil without disturbing its structure. Where infiltration is preserved, runoff is reduced. Runoff that does occur is slowed by its movement through the vegetative or litter mat so that it does not have sufficient energy to pick up soil particles. Grass is particularly good for erosion control because when volume and velocity of runoff increase, grass, well-anchored in the soil, simply lies down, making a smooth mat over which the water can flow without disturbing the soil underneath. Similarly, vegetation slows the velocity of wind and holds soil particles (Fig. 7–20).

Unfortunately, plowing and cultivating, overgrazing, and deforestation, all of which are occurring on massive scales in the name of supporting the world's population, render the soil bare and are causing unprecedented erosion that can only undercut future productivity. First, we will focus on the process of erosion itself, and the magnitude of the problem, so that we can better understand how it undercuts

FIGURE 7–19
Agricultural fields are often subject to severe erosion because the crop does not provide a complete protective cover. (USDA)

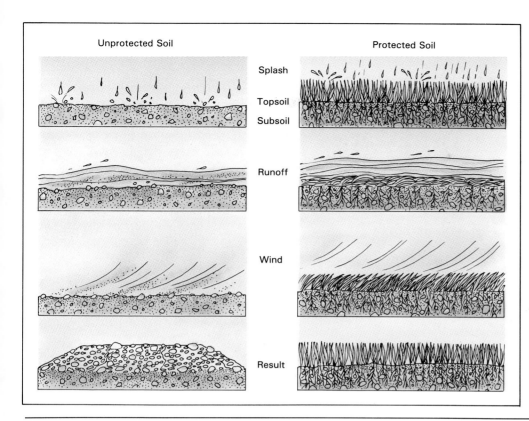

Unprotected Soil   Protected Soil

Splash

Topsoil

Subsoil

Runoff

Wind

Result

FIGURE 7–20
Erosion. Bare soil is extremely vulnerable to erosion. The splash of falling raindrops breaks up soil aggregates into individual particles. The finer particles of humus, clay, and silt, are readily carried away by runoff or wind, leaving only a layer of coarse sand, stones, and rocks, which are resistant to erosion on the surface. A vegetative cover protects soil from all forms of erosion.

productivity. Then we will address alternative practices and policies which can reduce erosion.

Water erosion starts with **splash erosion,** the effect of raindrops splashing against bare soil (Fig. 7–21). Recall the clumpy, aggregate structure of topsoil. When a raindrop hits bare topsoil, it blasts the clumps and aggregates apart, and the separated soil particles wash into and clog pores and spaces between other aggregates, thereby decreasing both aeration and infiltration. The effect is called **puddling** because the decreased infiltration may lead to water standing in puddles and drowning plants. More seriously, the decreased infiltration results in more water running off the surface causing further stages of erosion.

As water flows over the surface, it picks up and carries soil particles. A uniform loss of soil from the surface is called **sheet erosion.** Frequently, small stones protect the soil under them from the combined effects of splash and sheet erosion and are left sitting on pedestals giving mute testimony to the amount of soil lost (Fig. 7–22).

As further runoff occurs, the water converges into rivulets and streams, which have greater volume, velocity, and energy, and hence greater capacity to pick up and carry soil particles. Thus, channels of various sizes are gouged out. When the result is many small channels, it is called **rivulet erosion;** when the

result is a few large channels it is called **gully erosion** (Fig. 7–23).

A very important feature of erosion is that it *differentially* picks up and removes particles according to weight, the finer, lighter humus and clay particles being the first to go, while coarse sand, stones, and rocks remain. (People commonly believe that clay will not erode easily because they visualize clay as hard, bricklike clods. Try pouring water over a clod of dry clay and you will observe that the water running off is very muddy in appearance. This attests to how easily individual clay particles are separated from the clod and brought into water suspension.)

The more severe the force of wind or water, the heavier the particles that may be removed, but always there is the *differential* removal. Consequently, as erosion removes finer materials, the soil becomes progressively more coarse (sandy, stony, and rocky). Such coarse soils are frequently a reflection of past or ongoing erosion (Fig. 7–24). Did you ever wonder why deserts are full of sand? The sand is what remains; the finer, lighter clay and silt particles have been blown away.

Recall that clay and humus are the most important components for water- and nutrient-holding capacity. As these components are removed by erosion, topsoil is destroyed and water- and nutrient-holding capacity are greatly reduced. In regions which receive

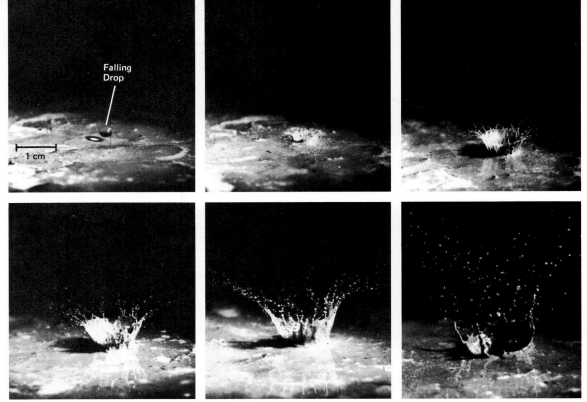

FIGURE 7–21
Splash erosion. Raindrops hitting bare soil have a destructive effect. They break up soil structure and seal soil pores. In turn, aeration and infiltration are decreased and surface runoff is increased. (USDA—Soil Conservation Service photos.)

FIGURE 7–22
Sheet erosion. Stones protect the soil under them against splash and sheet erosion. As soil is eroded away, stones are left on pedestals, the height of which shows how much soil has been removed. (USDA—Soil Conservation Service photo.)

FIGURE 7–23
Rivulet and gully erosion. Water flowing over a surface tends to collect in rivulets of increasing volume, velocity, and energy. Soil is accordingly eroded in a pattern of converging rivulets and gullies of increasing size. (USDA—Soil Conservation Service photo)

FIGURE 7–24
Erosion removes small soil particles more easily than large ones; therefore, soil subjected to erosion becomes progressively more sandy and stony. Coarse, stony soil remaining after wind erosion; overgrazing removed vegetation allowing erosion. (Russ Kinne/Photo Researchers)

Causes of Losing Ground

The major causes behind turning the soil bare, allowing erosion, and desertification are: (1) overcultivation, (2) overgrazing, and (3) deforestation. Salting of the soil through irrigation is another cause of desertification.

Overcultivation

Traditionally, the first step in growing crops has been, and to a large extent still is, to plow. The reason for plowing is to control weeds, wild plants which are better competitors for nutrients and water than the crop plants. Without weed control, a grower gets a field of weeds with little if any yield from crop plants. By turning the top layer upside down, weeds are buried and smothered. The drawback is that soil is exposed to wind and water erosion. Further, it may remain bare for a considerable part of the year before the crop forms a complete cover; after harvest, the soil again may be left largely exposed to erosion. On slopes, runoff and erosion are particularly severe. In regions of minimal rainfall, where the soil is often dry, wind erosion may extract a heavy toll regardless of topography.

It is ironic that plowing and cultivation are frequently deemed necessary to "loosen" the soil to improve aeration and infiltration. All too commonly the effect is the reverse. Splash erosion destroys the soil's aggregate structure and seals the surface so that aeration and infiltration is decreased and runoff is increased. The weight of tractors used in plowing may add to the compaction. Plowed earth also loses a greater amount of water by evaporation, aggravating the problems of reduced infiltration and water-holding capacity.

Overgrazing

Grasslands that have too little rainfall to support cultivated crops have traditionally been used for grazing livestock. Unfortunately, such lands are frequently subjected to **overgrazing.** That is, the grass is grazed faster than it can regenerate resulting in its death and exposure of the underlying topsoil to erosion. Wind erosion and consequent desertification of these regions is particularly severe (Fig. 7–25).

Desertification of rangeland is not a linear phenomenon. Ranchers may think their herds are within the **carrying capacity,** the number of animals that the land can support without a deterioration of the ecosystem. However, with intensive grazing there is less detritus to generate soil humus. Thus, there may be gradual mineralization of soil over many years that is not readily recognized. Then, because of gradual deterioration of soil, and perhaps an abnormally dry

only 10 to 30 inches (25–75 cm) of rainfall per year, the loss of water-holding capacity resulting from erosion is exceedingly serious. Such regions originally supported productive grasslands. With loss of water-holding capacity, they degenerate until they are able to support only drought-resistant, desert species. In other words, as a result of erosion, the land may lose productivity until it is effectively like a desert. The term **desertification** is now used to denote this process, and lands that have lost productive capacity as a result are called **desertified.**

In summary, soil that is left unprotected by a vegetative cover suffers erosion from wind and water. As topsoil is destroyed and removed, productivity drops drastically; recall experiments that showed an 85 percent reduction in productivity resulting from loss of topsoil. Finally, the end point may be nothing more (or less) than a barren "desert" landscape that supports virtually no growth at all.

FIGURE 7–25
Desertification. Desertification is a two-stage process. First, overgrazing removes the protective cover of grass. Then erosion reduces the water holding and nutrient content of the soil by selectively removing humus and clay. Thus the potential for recovery and or reclamation is drastically reduced as the eroded soil can only support a few desert species. (USDA—Soil Conservation Service photos)

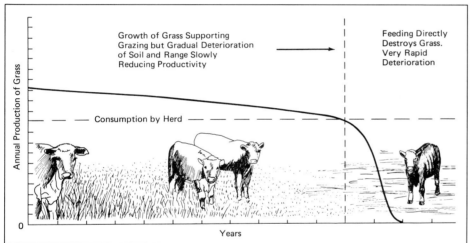

Growth of Grass Supporting Grazing but Gradual Deterioration of Soil and Range Slowly Reducing Productivity

Feeding Directly Destroys Grass. Very Rapid Deterioration

Consumption by Herd

Annual Production of Grass

Years

(a)

FIGURE 7–26
(a) Overgrazing begins with a slow deterioration of rangeland, gradually reducing the production of grass. The end may come with extreme suddenness as production of grass fails to keep up with what is being eaten and cattle graze the land into barrenness. (b) Desertification in the Sahel, Africa. These cattle were slaughtered before they died of hunger. (Photo Researchers/© 1975 Victor Englebert)

(b)

year, grass production drops below the grazing level. As the land is grazed to barrenness, erosion sets in and the vicious cycle of increasing soil degradation and dropping plant productivity accelerates. Thus, apparently productive range lands go into desertlike barrenness with surprising and often catastrophic suddenness (Fig. 7–26). Huge areas of agricultural and range lands are already partially desertified (Table 7–4). The suddenness with which the final stage of desertification can occur does not speak well for the future.

Deforestation

A forest cover is particularly efficient in preventing erosion and holding water since it breaks the fall of raindrops and allows the water to infiltrate into a litter-covered, loose topsoil. Investigators at Hubbard

Table 7–4
Productive Drylands Desertified[a]

LAND AREA	AREA (MILLION ACRES)	PERCENT DESERTIFIED[b]
Sudano-Sahelian Africa	1175	88
Southern Africa	755	80
Mediterranean Africa	250	83
Western Asia	335	82
Southern Asia	750	70
USSR in Asia	725	55
China and Mongolia	762.5	69
Australia	1222.5	23
Mediterranean Europe	175	39
South America and Mexico	702.5	71
North America	962.5	40
Total	**7815**	**61**

[a] Includes rangelands and croplands with minimum rainfall.
[b] Total of lands moderately (loss of productivity less than 25 percent) to very severely (loss of productivity greater than 50 percent) desertified.
SOURCE: *World Resources 1986*. Washington, D.C.: World Resources Institute.

Brook Forest in New Hampshire have found that run-off from a forested slope is as much as 50 percent less than that from a comparable grass-covered slope. Likewise, forests are particularly efficient at reabsorbing and recycling the nutrients released from decaying detritus. The same investigators found that leaching of nitrogen, for example, increased as much as 45-fold after clearing forests. Thus, clearing forests not only renders the soil subject to erosion, but problems are compounded by nutrient loss. Also, soil and water running from the slopes causes flooding and siltation of agricultural and aquatic ecosystems in the lowlands.

Forests are cleared for three main reasons: (1) to permit agriculture, (2) to obtain structural wood for building and for other wood and paper products, and (3) to obtain firewood for cooking and heating. Industrialized countries went through vast deforestation before and during the Industrial Revolution and suffered the consequences (Fig. 7–27). Much land now has low productivity as a result. In the northwestern United States and some other areas, clearcutting of forests and the resulting consequences are still occurring (Fig. 7–28).

In general, however, industrialized countries have recognized such problems and are currently reforesting areas at rates equal to or exceeding rates of deforestation. This is made possible in large part by the following:

☐ Low population growth rate and increasing agricultural production per acre has essentially eliminated the need for more agricultural land.

FIGURE 7–27
Erosion is extremely severe on forested slopes that have been clearcut. Deforestation led to the erosion seen here. (Photo by Victor Englebert Photo Researchers.)

FIGURE 7–28
Erosion is extremely severe on forested slopes that have been cut. Olympic National Forest, Washington. (Photo Researchers/ © 1987 Calvin Larsen)

wasteful and inefficient. For agriculture, forests are cleared by burning with no attempt to make use of the wood. Wood stoves are inefficient as are the processes used for producing charcoal.

☐ Forests are being exploited for commercially valuable hardwoods with a viewpoint of maximizing short-term income with little or no view toward long-term replanting and management.

☐ Finally, for about two-thirds of the people in developing countries (some 2.5 billion total), the modern energy era has not arrived; they still depend on firewood for cooking and heating (Fig. 7–29). Even in cities many use charcoal, which is produced by people in rural areas by cutting and charring the wood. For about 60 percent of these people (1.5 billion), forests are being cut at rates faster than they can grow back. In some countries cutting exceeds grow-back by a factor of 5 to 1. Already many have reached such desperate straits that they are burning animal dung

FIGURE 7–29
Some 2.5 billion people, mostly in less-developed countries, still depend on firewood for cooking. Women, to whom the chore falls, may have to walk many miles and spend most of their day seeking and carrying wood.

☐ Utilization of waste wood (trimmings, chips, and sawdust) to make pressed-board products have reduced demand.

☐ More forests are being managed to increase production.

☐ Fossil fuels and electricity have made use of wood for fuel optional.

In developing countries, however, essentially all these factors are reversed.

☐ High population growth rates and lack of other employment leads increasing numbers to clear forests on erosion-prone slopes and take up **subsistence farming**—farming just for their own survival.

☐ Wood use in developing countries is often

and natural litter which they collect off the ground. Consequently, the land is not only deforested; all other protective and humus-nutrient–providing detritus is also being removed. Expanding areas of desertified earth, which produce neither food nor fuel, surround villages and urban centers of many less-developed countries. The potential for increased human suffering can only be surmised.

In too many cases over-exploitation of forests is promoted by shortsighted government policies. For example, Brazil is actually subsidizing the conversion of rainforest to cattle ranches.

Irrigation, Salinization, and Desertification

Irrigation, supplying water to growing areas by artificial means, has served to dramatically increase crop production in regions which typically receive inadequate rainfall. Irrigated lands in both developed and developing countries total about 330 million acres (130 million hectares). Traditionally, water has been diverted from rivers through canals (Fig. 7–30) and flooded through furrows in fields (Fig. 7–31). In recent years **center pivot irrigation,** in which water is pumped from a central well through a gigantic sprinkler that slowly pivots itself around a well, has become much more popular (see Fig. 8–17).

FIGURE 7–30
Irrigation channel. Such channels are constructed to carry water from where it is plentiful to where it is needed for irrigation. (Gary M. Handsher/Photo Researchers.)

FIGURE 7–31
Flood irrigation. A traditional method of irrigation is to periodically flood fields with water. (USDA photo.)

FIGURE 7–32
Many acres of irrigated land are now worthless because of the accumulation of salts left behind as the water evaporates, a phenomenon known as *salinization.* (Photo courtesy of Texas Agr. Exp. Sta. El Paso.)

In either case, irrigation may lead to **salinization,** an intolerable increase in salinity (saltiness), because even the best irrigation water contains at least 200–500 parts per million salt, dissolved from the earth. Additional salts may be leached from the minerals in the soil itself. As irrigation water leaves by evaporation or transpiration, salts in solution remain behind and may accumulate in and on the soil to the point of precluding plant growth (Fig. 7–32). Salinization is considered another form of desertification.

As shown in Table 7–5, 30 percent of all irrigated land already has been salinized, negating the value of both the irrigation project and the land, and an additional 2.5–3.7 million acres are salinized each year. In the United States, the problem is especially acute in the lower Colorado River Basin area and in the San Joaquin Valley of California where some 400,000 acres (160,000 hectares) have been rendered nonproductive, representing an economic loss of more than $30 million per year. Adding to the problems, water supplies are being depleted by withdrawals for irrigation. More will be said about this in Chapter 8.

Dimensions of the Problem

Erosion and its consequences are not a new phenomenon. The food that sustained the Roman Empire was largely grown on rich, fertile lands in North Africa, lands which are now mostly desert. Many historians conclude that erosion and consequent desertification of these lands were no small factors in the collapse of the Roman Empire. In the United States, about 20

Table 7–5
Irrigated Land Desertified by Salinization

LAND AREA	AREA (MILLION ACRES)	PERCENT DESERTIFIED
Sudano-Sahelian Africa	7.5	30
Southern Africa	5.0	30
Mediterranean Africa	2.5	40
Western Asia	20.0	40
Southern Asia	147.5	35
USSR in Asia	20.0	25
China and Mongolia	25.0	30
Australia	5.0	19
Mediterranean Europe	15.0	25
South America and Mexico	30.0	33
North America	50.0	20
Total	**327.5**	**30%**

SOURCE: *World Resources 1986.* Washington, D.C.: World Resources Institute.

FIGURE 7–33
Erosion causes loss of soil productivity. The rocky, stony soil to the right of the fence has such poor water-holding capacity that it will no longer support the growth of grass. The condition results from initial destruction of the protective grass cover by overgrazing; then erosion has removed fine material leaving only coarse material. Note the drop of 10–15 cm (4–6 in.) at the fence line. Since the eroded soil now only supports desert species, the process is frequently referred to as *desertification*. (USDA photo.)

percent of our original agricultural land has been rendered nonproductive by erosion (Fig. 7–33). China has lost about one-third of its land to erosion, and the story is similar or even worse for many other countries.

Yet, we have not learned the lesson. Indeed, erosion is gaining momentum as population and economic pressures push people to deforest and cultivate hillsides and marginally dry lands, and to adopt other practices of intensive cropping that may increase yields in the short run but allow extra erosion. The processes of weathering and soil formation vary greatly with climate and composition of the parent material. On an average, however, new soil will be formed at a rate of about 5 tons per acre (12.5 tons per hectare) per year, which is equivalent to a layer of soil about $1/64$ inch (0.4 cm) thick. Hence, soils can sustain an erosion rate of up to 5 tons per acre per year and still remain in balance. The hard fact is that much of the globe's crop and forest land is not within this balance.

Where enough measurements have been made to allow realistic estimates, they often show erosion at 2 to 10 times the tolerable rate and, in some cases, they affect entire countries (Table 7–6). Nor is erosion just a problem of developing countries. In the early 1980s, nearly 40 percent of U.S. croplands were incurring unacceptable rates of erosion. In effect, U.S. farmers were sacrificing about 6 tons of topsoil for every ton of grains produced (Fig. 7–34). Worldwide, the loss of soil from croplands has reached some 23 billion tons per year, according to a 1984 estimate by the Worldwatch Institute. This is equivalent to all the topsoil on some 23 million acres (9.2 million hectares), an area about the size of the state of Indiana.

Erosion is a particularly insidious phenomenon since the first 20 to 30 percent of the topsoil may be lost with only marginal declines in productivity,

Table 7–6
Cropland Soil Erosion in Selected Countries

COUNTRY	AREA	PERCENTAGE OF AREA AFFECTED	RATE OF EROSION (METRIC TONS PER ACRE PER YEAR)
United States	Cropland (All)	19	3.84
China	Loess Plateau Region	6.4	4.4–100.4
India	Seriously Affected Cropland	27	30
Peru	Entire Country	100	6
Ethiopia	Central Highland Plateau	43	13.6
Madagascar	Cropland	79	10–100
Nepal	Entire Country	100	14–28

SOURCE: *World Resources 1986*. Washington, D.C.: World Resources Institute.

FIGURE 7–34
The poor cover given by corn allows very significant erosion.
(J. P. Jackson/Photo Researchers.)

ductive drylands (regions that receive minimum rainfall to support agriculture) on various continents have been at least moderately desertified, meaning that some loss of productivity already is being observed (Fig. 7–35 and Fig. 7–36).

In Ethiopia, the sequence has already run its course. In 1978, a report from the U.S. embassy there observed that an ecological tragedy was in the making because more than 1 billion tons of topsoil per year were flowing from Ethiopia's productive highlands, largely due to cutting forests for firewood. Now most of the region is thoroughly desertified and will not support even subsistence farming, much less produce extra for market. The recurrent famines in Ethiopia (1985 and 1988) are the result. A number of other countries in Africa and Asia seem destined to follow this course in the near future unless drastic steps toward soil conservation are taken.

The loss of topsoil and the decline in productivity creates additional problems. The soil that erodes from the land washes into and clogs the channels of rivers downstream, destroying productive aquatic ecosystems and causing flooding of lowland agricultural areas. Additionally, water running off rather than soaking in leads to a depletion of groundwater and other water resources as will be described more fully in Chapter 8. Then, eroding soils are often too compacted and/or unstable to allow the seedlings to take hold. Their roots cannot penetrate and they are washed away or killed as the surface layer dries. Thus, desertified areas do not regenerate by themselves, but remain bare—continuing to erode indefinitely.

which may be compensated for by additional fertilizer and favorable distribution of rains that make up for the losses in nutrient and water-holding capacities of the soil. However, as loss of topsoil continues, the loss of productivity and vulnerability to drought becomes increasingly pronounced. This is particularly disturbing because already 40 to 90 percent of pro-

FIGURE 7–35
Hillside overgrazed by sheep. You can see the evidence of erosion and loss of productivity referred to as desertification. (Yva Momatuk/Photo Researchers.)

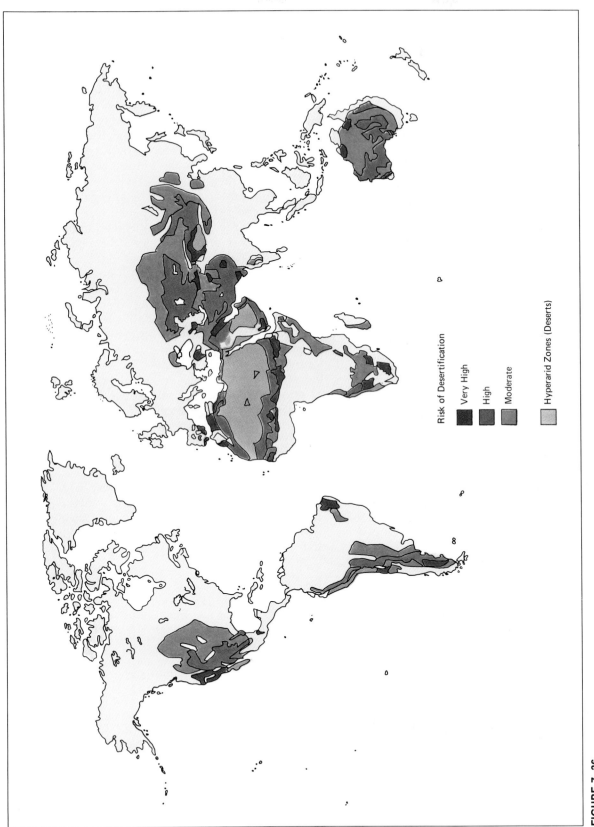

FIGURE 7–36
Deserts and areas subject to desertification. Throughout the world, overgrazing and/or deforestation are causing vast low-rainfall areas to degenerate into deserts. Reprinted with permission from "Desertification: Its Causes and Consequences," United Nations Conference on Desertification, Nairobi, 1977. Copyright 1977, Pergamon Books Ltd.

Risk of Desertification

- Very High
- High
- Moderate
- Hyperarid Zones (Deserts)

Preventing Erosion and Desertification

Erosion and desertification are the result of an unconscionable lack of environmental regard. Means are available to control and prevent them.

Preventing Erosion: Traditional Techniques

Traditional methods of controlling soil erosion include *contour farming, strip cropping, shelter belts* or *windrows,* and *terracing.*

Contour farming is the plowing and cultivating along the contour at right angles to the slope. Thus, water is caught between furrows giving it more time to soak in, checking runoff and erosion (Fig. 7–37). **Strip cropping** is cultivating alternate strips, generally leaving strips of grass or hay between strips of a cultivated crop like corn. Erosion from the cultivated strip is caught and held in the grass strip, thus preventing overall erosion. Strip cropping is especially effective in reducing wind erosion (Fig. 7–38). Strip cropping and contour farming are often used together. **Shelter belts** or **windrows,** rows of trees planted around fields, also slow wind and reduce wind erosion. **Terracing** is the grading of slopes into a series of steps (Fig. 7–30) so that water does not run down the slope. Very steep slopes were elaborately terraced by ancient civilizations, but such terracing is not compatible with modern machinery (Fig. 7–39).

Preventing Erosion: New Techniques

No-Till Agriculture. Recall that the purpose of plowing and cultivation is weed control. Chemical herbicides (weed killers), which were first developed in the mid-1960s, provided an alternative. This **no-till agriculture** is now a routine practice over much of the eastern United States and other areas. The field is first sprayed with an herbicide; then the planting apparatus pulled behind a tractor accomplishes several operations at once. A steel disc cuts a furrow through the mulch of dead weeds, drops seed and fertilizer into the furrow, and then closes it (Fig. 7–40). The new seedlings come up through the mulch of dead weeds and do just as well or better than those on a cultivated seedbed, presumably because of the better water conservation under the mulch (Fig. 7–41). At harvest, the process is repeated and the previous crop becomes the detritus and mulch cover for the next. Thus, the soil is never left exposed, erosion is minimized, and there is always detritus that maintains and regenerates topsoil.

No-till agriculture has several additional advantages to the farmer. First, there is a tremendous savings in time (labor expense). Second, there is a savings in energy (tractor fuel), since the farmer accomplishes in one operation what formerly required at least three—plowing, disking, and planting. Third, the single pass with a tractor minimizes soil compaction. Fourth, a farmer is able to get his crop in at the optimal time in the spring, whereas formerly he was often delayed by having to wait for the soil to become dry enough to plow. The early planting and the ability to plant a second crop immediately on harvesting the first allows double cropping, the growing of two successive crops in the same year. These factors generally give no-till methods an economic advantage in addition to controlling erosion.

However, there are also disadvantages. Most conspicuous is the use of chemicals which may have undesirable side effects. A number of serious pests including mice, slugs, insects, and plant pathogens that were formerly controlled by plowing may winter

FIGURE 7–37
Contour farming. Cultivation up and down a slope promotes water running down furrows and may lead to severe erosion. The problem is reduced by plowing and cultivating along the contours at a right angle to the slope. In this case strip cropping is also being utilized. (USDA photo.)

(a)

(b)

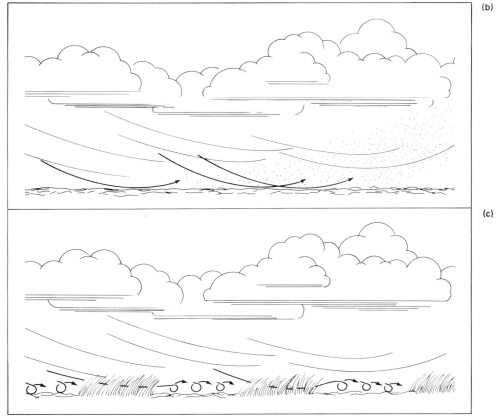

(c)

FIGURE 7–38
(a) Soil deposited at fenceline attests to wind erosion occurring on the open cropland to the right. (Dean Krakel/Photo Researchers.) (b) Wind erosion is most severe when wind is unimpeded. (c) Stripcropping (leaving alternate strips in grass) causes wind eddies near the surface and greatly reduces erosion.

FIGURE 7-39
Terraced rice fields in the mountains of the Philippines. These terraces were constructed by ancient civilizations and are only compatible with hand agriculture. (Philippine Ministry of Tourism.)

over in the undisturbed soil and old crop residues left on the surface. In addition to herbicides, other chemicals may be required to control these pests. In all, no-till procedures require two to four times more pesticides than conventional farming techniques. Also, weeds gradually become resistant to the herbicides, requiring the use of increased dosages. Finally, the use of herbicidal chemicals, which require special handling and equipment, are neither practical nor safe in small-scale operations.

The principles of no-till agriculture, however, may be applied to home gardens or subsistence farming in developing countries without the use of herbicidal chemicals. A generous covering with an organic mulch of leaves, grass clippings, or other waste organic matter serves to

- ☐ protect the soil from erosion
- ☐ suppress the growth of weeds (weeds that come through are easily pulled and added to the mulch)

- ☐ conserve soil moisture
- ☐ provide organic matter and nutrients for maintaining and building a rich loose topsoil.

The mulch can be separated just enough to insert seeds or seedlings, and they thrive without doing anything else to the soil.

Perennial Crops. All the major crops (corn, wheat, beans) that support humans are **annuals**—plants that grow, set seed, and then die in a single year. Each year's crop demands preparing the soil and restarting from seed. Our dependence upon annuals is more a happenstance of history than a basic necessity. It may be possible to develop **perennials**—plants that grow from the same root stock year after year—that will produce as abundantly. Perennial species would provide enormous protection from soil erosion, without the use of chemicals, to say nothing of additional savings of the labor and energy now spent in replanting seed each year.

Limit Grazing

The solution to desertification caused by overgrazing is self-evident. Grazing must be limited to what the rangeland can sustain, or it will be limited by the land

(a)

(b)

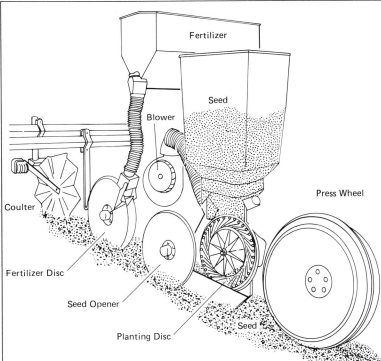

FIGURE 7–40
No-till planting. (a) The no-till planting machine is pulled behind a small tractor and performs several operations at once. (USDA—Soil Conservation Service photo.) (b) Schematic diagram of the apparatus. The direction of movement of the machine in this drawing is to the left. The first wheel is a fluted coulter that opens a narrow band in the untilled soil, usually cutting also through the residue of the preceding crop, which is left on the ground as a mulch. The coulter is followed by a disc that applies fertilizer. Here the fertilizer is dry, but in some machines it is applied as a liquid. (Herbicide can also be applied at about the time of planting to control weeds.) Next comes a disc, offset from the fertilizer disc by about 2 inches, that opens the furrow in which the seed is to be planted. It is followed by the planting unit, which receives the seed from the container above it. The seed is forced into the slots in the planting disk by air from a blower; the air also holds each seed in place until the slot approaches the ground and the seed drops into the furrow. The last wheel presses the soil down over the seed. Several units of this kind are normally ganged together in one machine. (From Glover B. Triplett, Jr., and David M. Van Doren, *Agriculture Without Tillage.* Copyright © Jan. 1977 by Scientific American, Inc. All rights reserved.)

turning into desert. The grazing of large animals is not necessary for meat production in developing countries. A variety of small animals that can be raised on other agricultural wastes lend themselves much better to the small-scale subsistence farming that exists.

Reforestation and Rehabilitation of Desertified Lands

The need for forest conservation, better management to increase forest productivity, and reforestation requires no elaboration. Trees and shrubs used as shel-

FIGURE 7–41
Crop plants planted by no-till methods grow as well as or better than on fields cultivated in the traditional way. (USDA)

ter belts or strips on slopes are particularly effective in reducing erosion on adjoining areas and supply considerable fuel wood. Reforesting desertified areas does take sustained effort, however, since soil must be mulched to hold moisture and reduce further erosion and seedlings must be dug in and nurtured over several years to become well established. As vegetation becomes reestablished, natural processes will work to regenerate topsoil. Many individuals have made it a labor of love to buy an old farm with badly eroded soil and gradually bring it back to a productive state. Israel has achieved spectacular success in bringing the desert, which in that region is really a product of desertification, back to agriculture. Successful projects have been accomplished in a number of developing countries also, but a severe handicap in many others is the tenuousness of landownership and protection. People and communities engaging in reforestation must be assured that they will gain the benefits of their efforts.

Countering Salinization

Salinization can be avoided, or even reversed, if sufficient water is applied to leach the salts down through the soil. However, unless there is suitable drainage, the soil will only become a waterlogged quagmire in addition to being salinized. At great expense, artificial drainage may be installed as shown in Figure 7–42, but then attention must be paid to the draining of the salt-laden water. The wildlife in the Kesterson Wildlife Preserve in California has been all but destroyed by pollution from irrigation drainage.

The installation of dip irrigation systems, now common in the southwestern United States, will at least mitigate the progression of salinization and water shortages (Fig. 8–21). In another approach, plant breeders are attempting to develop varieties of crop plants that are more salt tolerant. The potential of this approach is demonstrated by the fact that countless marine and seashore species are adapted to such conditions. If similar adaptations can be bred into commercial species, then perhaps salinized areas can again become somewhat productive. However, many adaptations to high salinity involve the plant accumulating so much salt that it becomes inedible.

IMPLEMENTING SOLUTIONS

In the early 1930s in the United States, lack of soil conservation combined with drought conditions to create the infamous Dust Bowl, which extended from

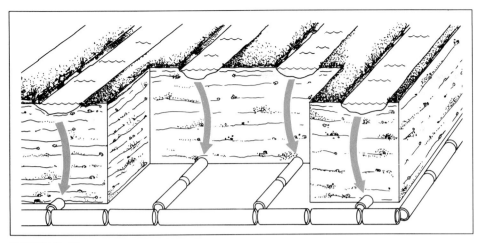

FIGURE 7–42
Drainage of irrigated land. To prevent salinization, excess water with salts must be drained away. This may necessitate a system of tiles as shown here and additional environmental problems may be caused by the discharge of such high-mineral-content drainage into natural waterways.

Texas to Illinois. In the words of John Steinbeck in *The Grapes of Wrath*, "The dust lifted out of the fields and drove great plumes into the air like sluggish smoke. . . . Dawn came but no day . . . and the wind cried and whimpered over the fallen corn." Farmers went bankrupt, tenant farmers migrated in mass to California, and the dusty earth settled on policymakers as far east as Washington, D.C., providing the impetus to pass the Soil Conservation Act of 1935, which created the Soil Conservation Service (SCS). The SCS assisted farmers in establishing what are now the traditional soil conservation techniques discussed above. This was a major step toward stabilizing the nation's topsoil and restoring the economy of the Great Plains. The invaluable service continues through an SCS extension office in every county of the United States. However, the SCS can only provide assistance; it has no "teeth" to enforce conservation.

During the 1970s, pressures to produce more food for the growing world population, introduction of huge tractors that increase the efficiency of farming on vast, uninterrupted acreages, and dimmed memories of the Dust Bowl era led many farmers to abandon soil conservation. Some 30 million acres of windrows, fallow strips, and slopes were sacrificed and put into cultivation. As well as increasing short-term profits and production, however, erosion increased by about 50 percent to rates equaling if not exceeding the Dust Bowl period.

Again, there has been an effective response to this latest episode of drastic soil erosion. The problem was recognized and publicized by environmentally

focused organizations, notably the American Farmland Trust, Rodale Press, and the Worldwatch Institute. Leaders supported by the members of environmental organizations including the Sierra Club, National Audubon Society, Conservation Foundation, and others lobbied through Congress to win passage of the Food Security Act of 1985. The two major provisions of this act are:

1. To convert 40 million acres of highly erodible cropland into a "Conservation Reserve" of forest and/or grass by 1990. Farmers are paid about 50 dollars per acre per year for land placed in the reserve.
2. Farmers with erodible land must develop and implement a soil conservation program to remain eligible for price supports and other benefits provided by the government.

By 1987, about half the requisite land had been placed in the reserve. This achieved about 30 percent of the target goal for erosion reduction, indicating that in the long-term the program may have outstanding success.

In this historical sketch, it should not escape your attention that individuals like you can have an impact on creating and promoting effective environmental policy through membership in environmental organizations. Of course, wherever you are involved with gardening or farming, or come in contact with those who are, you can use and promote the use of soil conservation techniques. But what can we do to promote soil conservation and sustainable agriculture in less-developed countries?

Many developing countries, where erosion is most severe, do not have any equivalent of the SCS, much less economic incentives for farmers to practice

HAITI: IN THE LAND WHERE HOPE NEVER GROWS
Why a Dirt-Poor Nation Is Destined to Remain So

At first light Mercius Pierre rises and leaves the stuffy interior of his mud hut. He opens the door and window that have been shut tight against strangers and the loups-garous, the werewolves that stalk the nights of the superstitious. Pierre's movement stirs up the rats nesting in the thatched roof, as well as his wife Annaise and three young daughters. Annaise lights a fire with a few scarce twigs, then boils coffee with the last drops of water from a gourd and sweetens it with a piece of sugarcane. Her daughter Melina, 6, places the gourd on her head and begins a morning-long walk to a well. Mercius, meanwhile, picks up his wooden hoe, balances it on his shoulder and scuttles down the mountainside to till a field of millet for a *gros neg*, a landowning peasant. If he is lucky, he will earn 60¢ for his day's work.

So begins another day in the "other republic," as rural Haiti is known. Governments come and go in Port-au-Prince, but daily life in the western hemisphere's poorest country remains a tedious grind, with little chance for Mercius and the hundreds of thousands of other landless peasants to improve their lot. Hope flared briefly in 1986, when Haitians rebelled and forced "President-for-Life" Jean-Claude Duvalier into exile. Since then, the government has changed hands three times, most recently last month, when a coup installed the regime of Lieut. General Prosper Avril. No matter how good Avril's intentions are, however, Haiti is so dirt poor, literally, that it may never flourish again.

A woman searching for firewood on a northwestern plain, once a rain forest. (Maggie Steber/JB Pictures Ltd.)

soil conservation. They are where the United States was in the Dust Bowl days. Thus, in addition to family planning to reduce population growth rates, U.S. foreign policy and aid should be directed at providing an SCS-like structure for offering assistance with soil conservation, reforestation, and sustainable agriculture. The U.S. Peace Corps and numerous private organizations are becoming involved in reforestation and sustainable agricultural projects. Thus, there are numerous opportunities for you to become directly involved, and you can certainly write your local congressperson to support and promote policies toward these ends.

Financial incentives for conservation in less-developed countries can be provided by **debt-for-nature swaps.** Recall that most developing countries are

RURAL POVERTY

As desperate as life is in Port-au-Prince's slums, a truer picture of Haiti's plight emerges in the countryside, where some 75% of the country's 6.3 million people live. Land is both the hope of these peasants and the yoke that dooms them to poverty. Over the years, land parcels have shrunk to handkerchief size through repeated division among descendants and illegal seizures by landowners. Even the practice of voodoo has had an effect: some peasants have been forced to sell their land to pay for elaborate religious rituals for dead relatives.

Decades of misuse have left the earth spent and barren; today only 2% of Haiti is forested. The rape of trees began in colonial times with the export of hardwoods, used for the production of everything from dyes to ships. These days trees are the peasants' only real cash crop. A muddy brown ring surrounds Haiti's coast as the topsoil erodes and dissolves into the turquoise Caribbean, leaving behind what amounts to tropical desert. Reforestation efforts are outpaced by the country's demand for charcoal, a critical fuel in the urban areas.

Victims of hurricanes, drought, debts, superstition and disease, peasants are constantly preyed upon. Those with a bit of land are hesitant to improve it for fear of attracting the attention of covetous *gros negs*, who often hire corrupt lawyers to steal the land on one pretext or another. The rural police, notaries and Tonton Macoutes also seize property with a flourish of phony documents and a bag of city tricks. Even those who try to help the peasants often end up hurting them. When African swine fever hit the pig population of Haiti several years ago, Haitian authorities, under U.S. insistence, slaughtered all the peasants' hardy black Creole pigs. Unable to afford the new, imported white pigs or provide for their finicky tastes, most peasants suffered a severe decline in their standard of living.

FLEEING THE COUNTRY

The lure of escape is beginning to replace the dream of land ownership for many rural Haitians. On the beach outside the southern town of Petite-Rivière-de-Nippes, peasants are building three large boats, each capable of carrying 100 illegal immigrants to Florida. A youngster eyes the boats wistfully. "There is nothing for us here," he says. Some peasant families sell their land or chip in money to send their smartest relative to be a "boat person." They are dispatched to America in the hope they will find jobs and send money home. When the boat people return, they are often shocked at how life has deteriorated. "I cried and cried to see how poor my family was," said a young man who works as a busboy in Orlando. "I gave them all I had and left penniless after only three days."

Back in the Central Plateau, Annaise is preparing her family's single daily meal. She straightens her back, picks up a 20-lb. pestle and begins the rhythmic pounding of two handfuls of *petit mil*, or sorghum, in a wooden mortar. She cooks the meal in an ancient black iron bowl, scraping the remains from the bottom. None of her children are able to attend school this year, she says, because she cannot afford the registration fee of less than $3. "I don't sing anymore," she adds quietly. "I'm sad."

By Cristina Garcia
Reported by Bernard Diederich/Central Plateau
Time, November 7, 1988

---

struggling under debts to the United States and other developed countries. (See the *debt crisis*, Chapter 5.) Since it is unlikely that most developing countries will be able to repay these debts in any case, the idea is being promoted to trade (forgive) debt in exchange for mandatory conservation of forests and other natural ecosystems and suitable policies and support of conservation projects. You can write your congress-person asking that they support the concept of debt-for-nature swaps.

We spend billions on military aid and additional millions on famine relief and aiding refugees of ecological collapse. Yet without developing a sustainable agriculture and forests there is nothing any kind of aid can do other than drift away with the soil into the tragedy of desertification (see Case Study 7).

8

Water, the Water Cycle, and Water Management

CONCEPT FRAMEWORK

■ Outline

I. WATER   192
   A. Physical States of Water   192

   B. Evaporation, Condensation, and
     Purification   193

II. THE WATER CYCLE   194
   A. Water into the Atmosphere   194

   B. Precipitation   195

   C. Water Over and Through the Ground   197

   D. Summary of the Water Cycle   198

III. HUMAN DEPENDENCE AND IMPACTS ON
    THE WATER CYCLE   200
   A. Sources and Uses of Fresh Water   200

   B. Consequences of Overdraft of Water
     Resources   202

     1. Overdraft of Surface Waters   202
       a. Inevitable shortages   202
       b. Ecological effects   203
CASE STUDY: UPSETTING THE EVERGLADES   206
     2. Overdraft of Groundwater   204
       a. Falling water tables and depletion   204
       b. Diminishing surface water   207
       c. Land subsidence   207
       d. Saltwater intrusion   208
   C. Obtaining More versus Using Less   209

   D. Potential for Conservation and Reuse
     of Water   209

     1. Irrigation   209

■ Study Questions

1. Name the two forces that affect the physical state of water and describe how they act to produce *ice, liquid,* and *vapor* states.

2. Describe *evaporation* and *condensation* and how they act to purify water.

3. What are the main sources of water vapor entering the atmosphere? Define *relative humidity.*

4. What causes condensation of water vapor and precipitation? What air movement gives rise to high precipitation? Where does this typically occur? Define a *rain shadow.*

5. Define: *runoff, infiltration, infiltration-runoff ratio, surface water, capillary water, gravitational water, percolation, groundwater, water table, aquifer, recharge area, spring, seep.* Describe the movement of water into and through the ground using the above terms. What water do plants generally draw upon? What water is obtained from wells?

6. Trace three different "loops" of the water cycle. How are they similar? How do they differ? Contrast the quality of surface water with that of groundwater. Explain the reason for the difference. Define *leaching.* Why are bodies of water with no outlets salty?

7. From where and how is water obtained? For cities and industries, most of it is used for what? What are the limiting factors and effects? Can these be avoided? How? Explain why industrial and municipal use of water is called *nonconsumptive* while irrigation is called *consumptive.*

8. Give examples that show we are overdrawing surface waters. Describe the consequences.

9. Give examples that show we are overdrawing groundwater. Name, describe, and give examples illustrating the consequences.

10. What will be the consequences of continuing the trend of obtaining more water? Is it sustainable?

11. Describe and give examples illustrating how we could use much less water for irrigation, in homes and industries.

2. Municipal Systems	210
E. Effects Caused by Changing Land Use	210
1. Effects of Increasing Runoff	212
a. Stream bank erosion	212
b. Flooding	212
c. Increased pollution	212
2. Effects of Decreased Infiltration	215
F. Stormwater Management	215
IV. IMPLEMENTING SOLUTIONS	219
A. Cheap Water and Waste	219
B. What You Can Do	219

12. Explain how development affects the infiltration/runoff ratio. List and describe the effects of increasing runoff. Of decreasing infiltration.

13. Describe the traditional way of handling stormwater. What are the consequences? What is the modern concept of stormwater management? Describe means of retaining stormwater and reestablishing infiltration.

14. What can be done and what can you do to promote conservation of water and mitigate water-related problems? Why is the price of water a prime issue? Could we pay more?

[A]s scientists probe other planets and ask if life exists elsewhere in the solar system, a prime factor they look for is *water*. Without water life cannot exist. There is plenty of water on Spaceship Earth; about 70 percent of the earth's surface is covered by oceans and seas, but this water is salty. All major terrestrial ecosystems including the human ecosystem depend on fresh water, water that has a salt content of less than 0.01 percent (100 parts per million). Fresh water is much more limited—less than 1 percent of the global water supply is fresh—and growing human populations are overtaxing and polluting this invaluable resource.

Our objective in this chapter is to examine our freshwater resources in order to understand how we are undercutting them and what we can do to conserve and manage them in a sustainable way.

WATER

All the water on earth is constantly in the process of being repurified and recycled. To understand this, we must first address some of the chemical-physical aspects of water.

Physical States of Water

You are familiar with the three physical states of water: solid (ice), liquid (water itself), or gaseous (water vapor). These three states are the result of different degrees of interaction between water molecules. Two forces are involved: one is a weak attraction between water molecules (H_2O) called **hydrogen bonding** which results from the hydrogen atom of one molecule being attracted to the oxygen atom of the next. Hydrogen bonding tends to hold water molecules together. The second force is **kinetic energy,** the energy of vibrational movement that is inherent to all atoms and molecules. Kinetic energy tends to separate molecules from one another. The degree to which molecules are held together, and hence the physical state of water, depends on the relative balance between hydrogen bonding and the kinetic energy of the molecules (Fig. 8–1). Hydrogen bonding is a constant attraction; kinetic energy, however, varies with temperature: the warmer the temperature, the greater the kinetic energy.

Below freezing, 0° C (32° F), the kinetic energy of water molecules is small compared to hydrogen bonding. Thus, the hydrogen bonding holds the molecules in a rigid position with respect to one another, resulting in ice. As temperature increases, the increasing kinetic energy of the molecules literally shakes the structure apart and the result is melting. However, as a hydrogen bond breaks at one point, it reforms at another. Thus, molecules "slide" about one another but basically remain held together. This results in the liquid state of water.

Finally, at the boiling point, 100° C (212° F), molecules gain sufficient kinetic energy to break the hydrogen bonds altogether, and they go off into the air as free (unattached) water molecules, a process we observe as **evaporation.** We speak of water molecules in the air as **water vapor,** and the amount of water vapor in the air is commonly measured as **humidity.**

All these processes are reversible. **Condensation** is basically a reversal of evaporation. As temperature and corresponding kinetic energy decrease, water molecules in the vapor state are held together by hydrogen bonding as they come in contact. If temperature is low enough, water vapor may go directly

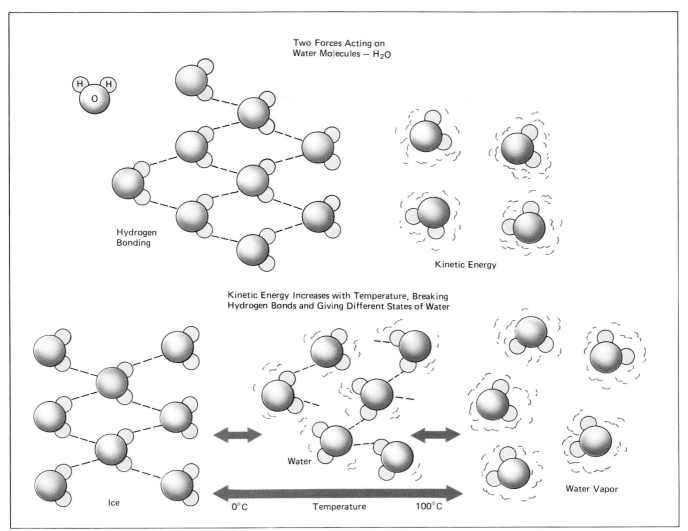

FIGURE 8–1
Physical states of water. Water molecules are affected by two forces: hydrogen bonding, which tends to hold them together, and kinetic energy, which tends to break them apart. As kinetic energy, which increases with temperature, overcomes hydrogen bonding, different states of water result.

from the vapor state to the solid or crystalline state as in the formation of frost in nature or in a freezer (Fig. 8–2).

These changes in the physical state of water are the basis of the water cycle of the earth. Water vapor is constantly entering the atmosphere by evaporation and then returning to earth through condensation and precipitation.

Evaporation, Condensation, and Purification

Evaporation and recondensation of water also entail an important process of purification called **distillation.** When water evaporates, only water molecules leave the surface; salts and other materials in solution remain behind (except in the case of dissolved gases or volatile liquids such as alcohol). When the water vapor is recondensed it is pure water. Indeed, the most chemically pure water for use in laboratories is obtained by this method (Fig. 8–3). Thus, the earth and its atmosphere act like a gigantic still. Water from salty seas and other sources is constantly evaporated, and hence purified, recondensed in the atmosphere, and returned to earth as fresh water. All the natural fresh water on earth is from this source.

Humans can purify water by distillation or other means. The problem is one of economics. The energy costs of distillation, for example, are such that producing large volumes of distilled water would be prohibitively expensive.

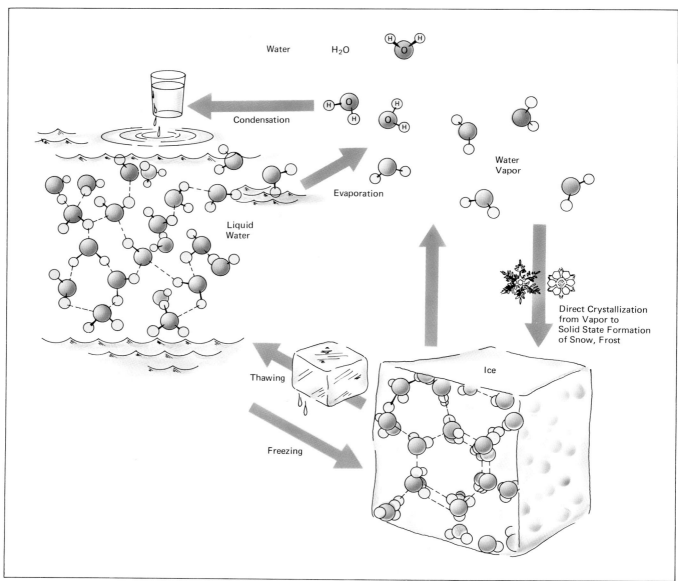

FIGURE 8–2
Physical states of water. At normal temperatures, kinetic energy is such that hydrogen bonds (dotted lines) break and re-form, producing the liquid state. As water is cooled, molecules lose energy and attractions become rigid, forming ice. As water is heated, molecules gain sufficient energy to break the weak attraction between them and they become water vapor. As water vapor is cooled, the attractive force again holds molecules together, resulting in the re-formation of water or condensation. In all these changes in state, the water molecules themselves remain the same, H_2O.

THE WATER CYCLE

The **water cycle,** also called the **hydrological cycle,** of the earth is represented in Figure 8–4. The cycle basically consists of water entering the atmosphere through evaporation and returning through condensation and precipitation. However, there are additional aspects that bear more consideration.

Water Into the Atmosphere

Since oceans cover about 70 percent of the earth's surface, it is not surprising that the largest amount of water vapor enters the atmosphere by evaporation from the ocean surfaces. Additional water evaporates from lakes, rivers, moist soil, and other wet surfaces; over vegetated land, large amounts of water enter the atmosphere by **transpiration** from plants. The combination of both evaporation and transpiration is called **evapotranspiration.**

The amount of water vapor in the air is com-

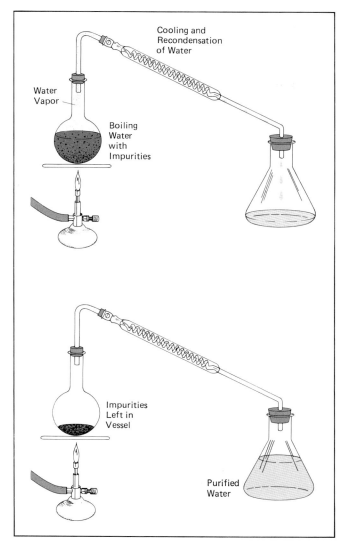

FIGURE 8–3
Purification of water by distillation. When water evaporates, only water molecules leave the surface; impurities remain behind. Hence, water may be purified by distillation, a process of boiling the water and recondensing the vapor.

monly spoken of (and felt) as *humidity.* The total amount of water vapor that air can hold varies with temperature. Therefore humidity is generally measured as **relative humidity,** the amount of water vapor in the air compared to the total amount that the air could hold at that temperature expressed in percent. For example, a relative humidity of 60 percent means that the air contains 60 percent of the amount of water vapor that it could hold at that temperature.

The warmer the air, the more readily it will pick up moisture and the more water vapor it will hold. This is why you use hot air to dry your hair. Conversely, the cooler the air the less moisture it will

hold. When air with a maximum amount of water vapor is cooled, the water molecules *condense;* they come together forming droplets of water. The droplets tend to form on existing surfaces, thus the 'sweating' of a cold glass or the formation of dew on grass. In the atmosphere, water molecules condensing on dust particles result in the formation of mist, fog, or clouds. As these droplets, or ice crystals when the temperature is below freezing, get larger, rain or snow occurs (Fig. 8–5).

Precipitation

We noted in Chapter 1 that the amount of precipitation is a primary factor in determining the type of ecosystem an area can support. The distribution of precipitation over the earth, which ranges from near zero in some areas to more than 3 meters (120 in.) per year in other areas, is basically dependent upon patterns of heating and cooling of the earth's atmosphere.

As air rises, it cools. Thus, high precipitation typically occurs under rising air currents. Conversely, as air descends or remains stable, it warms. Thus, there is minimal precipitation where air currents are descending or when an air mass is stable. As air masses rise and fall with various storms, precipitation is intermittent and variable. However, there are two situations that cause rising air (hence high precipitation) and descending air (hence low precipitation) more or less continuously.

First, solar heating of the earth is most intense in equatorial regions where sunlight hits the earth most perpendicularly. The air is heated, in turn, by the warm earth and the warmed air expands and rises. The rising air cools because of the further expansion allowed by the lower pressure at higher altitudes and the loss of heat into outer space. Finally, the cooling air cannot contain its water vapor, and rainfall results. Thus, equatorial regions have consistent high rainfall. But, the air which rises over equatorial regions must come down again. It descends over subequatorial regions (25 to 35 degrees north and south of the equator) resulting in deserts (Fig. 8–6). The Sahara of Africa is the prime example.

The second situation occurs where trade winds (winds that blow almost continuously from the same direction) hit mountain ranges. As moisture-laden air encounters a mountain range, it is deflected upward causing cooling and high precipitation on the windward side. As the wind crosses the range and descends on the other side, however, it becomes warmer and increases its capacity to pick up moisture. Hence, deserts occur on the leeward side of mountain ranges. The dry region downwind of a mountain is referred to as a **rain shadow** (Fig. 8–7). The most se-

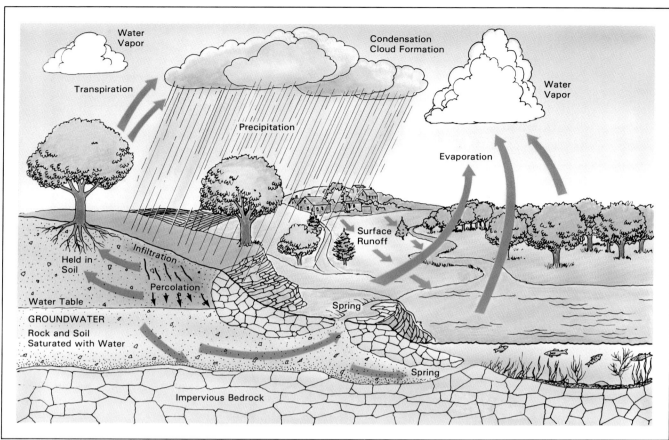

FIGURE 8–4
The water cycle. The water on earth is continuously recycled through evaporation or transpiration, condensation, and precipitation.

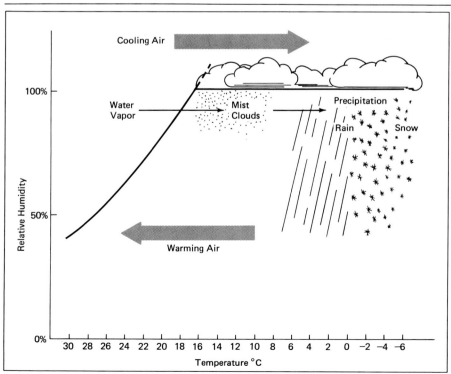

FIGURE 8–5
Given a certain amount of water in the air, relative humidity increases as temperature drops. When air cools below the point of saturation (100%), condensation and precipitation result. When air warms, there is a drop in relative humidity.

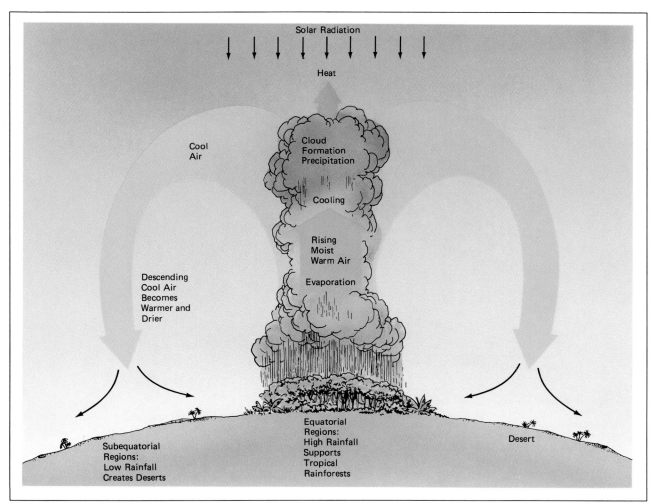

FIGURE 8–6
Equatorial tropical rainforests and subequatorial deserts.
Solar radiation causes maximum heating in equatorial regions
and produces rising currents of moist air. As the moist air
cools, there is heavy precipitation over the equatorial regions
supporting tropical rainforests. The air then descends over
subequatorial regions. As it descends, it becomes warmer and
drier, resulting in subequatorial deserts.

vere deserts in the world are caused by the rain shadow effect. For example, the westerly winds, full of moisture from the Pacific Ocean, strike the Sierra Nevada mountain range in California. As the winds rise over the mountains, large amounts of water precipitate out, supporting the lush forests on the western slopes. Immediately east of the Sierra Nevada range lies Death Valley, a result of the rain shadow.

Water over and through the Ground

Water from precipitation landing on the ground may follow two alternative pathways. It may soak into the ground, **infiltration,** or it may run off the surface, **runoff.** We speak of the amount that soaks in com-

pared to the amount that runs off as the **infiltration/ runoff** ratio. Runoff flows over the surface into streams and rivers which make their way to the ocean, or other points of evaporation. All ponds, lakes, streams, rivers, and other waters on the surface are referred to as **surface waters.**

For water that infiltrates, there are also two alternatives. Water may be held in the soil, the amount depending on the water-holding capacity of the soil as was discussed in Chapter 7. This water, called **capillary water,** returns to the atmosphere by way of **evapotranspiration.**

Infiltrating water that is not held in the soil is called **gravitational water** because it is pulled by gravity and trickles or **percolates** down through pores or cracks in the earth. Sooner or later, however, gravitational water comes to an impervious layer of rock or dense clay. Free water accumulates, completely filling all the cracks, pores, and spaces above such an impervious layer. This accumulated water is called **groundwater,** and its upper surface is the **water table**

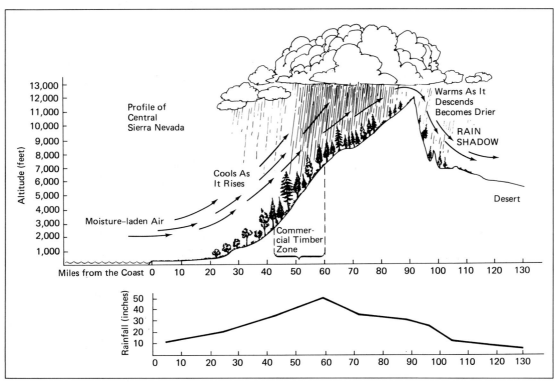

FIGURE 8–7
Rain shadow. Moisture-laden air cools as it rises over a mountain range, resulting in high precipitation on the windward slopes. Desert conditions arise on the leeward side as the descending air warms and tends to evaporate water from the soil. (Redrawn from *Trees: The Yearbook of Agriculture, 1949.* Washington, D.C.: USDA.)

(Fig. 8–8). *Gravitational* water becomes *groundwater* as it hits the water table in the same way rainwater becomes lake water as it hits the surface of the lake. Wells must be dug to below the water table; then groundwater, which is free to move, seeps into the well and fills it to the level of the water table.

Underground rock layers frequently slope, causing groundwater to move slowly like great underground rivers. The layers of porous material through which groundwater moves are called **aquifers.** The actual location of aquifers is complex. Layers of porous rock are often found between layers of impervious material and the entire formation may be folded or fractured in various ways. Thus groundwater may be found at various depths between layers of impervious rock. Also, the **recharge area,** the area where water actually enters an aquifer, may be many miles from where it is withdrawn. If the recharge area is at a higher elevation it may result in high pressure, which forces the movement of water through the aquifer. An aquifer with water under pressure is called an **artesian aquifer** (Fig. 8–9).

Drawn by gravity, groundwater may move through aquifers until it finds some opening to the surface. We observe such natural exits as *springs* or *seeps.* A **seep** is where water seeps out over a relatively wide area as opposed to a **spring** where it exits as a significant flow from a given point. In turn, springs and seeps feed streams, lakes, and rivers. Thus, groundwater rejoins and becomes part of surface water. A spring will only flow, however, if the water table is higher than the spring. Springs will go dry whenever the water table drops below the level of a spring.

Summary of the Water Cycle

In summary, the water cycle always consists of evaporation, condensation, and precipitation. But in completing the cycle there are three principal "loops": (1) the *surface runoff loop,* in which water runs off the surface and becomes part of the surface water system; (2) the *evaporation-transpiration loop,* in which water enters the soil and is held as capillary water and then returns to the atmosphere by way of evaporation from soil or through absorption by plants and transpiration; and (3) the *groundwater loop,* in which water enters and moves through the earth, finally exiting through springs, seeps, or wells, thus rejoining the surface water system.

The natural water cycle acts to continually purify and replenish freshwater systems. Purified by evaporation, precipitation is fresh. As precipitation hits the ground it may pick up impurities such as soil par-

FIGURE 8–8
Pathways of infiltrated water. Water that infiltrates into the soil and is held against further movement is called *capillary* water. When soil reaches its capacity of capillary water (field capacity), additional infiltrated water percolates downward under the pull of gravity and is called *gravitational* water. Water that accumulates above an impervious layer, saturating all the pore spaces in the soil, is called *groundwater*. The upper surface of the groundwater is called the *water table*. Groundwater is withdrawn by drilling wells to below the water table.

ticles, dissolved chemicals, and detritus and the microorganisms feeding on it. Therefore, runoff may be heavily polluted, but in natural ecosystems there is relatively little runoff. Most of the water infiltrates, and particles of dirt, detritus, and microorganisms are filtered out as it percolates through soil and porous rock.

Chemicals that dissolve, however, are not filtered out but are carried along or **leached.** Indeed, as water percolates through the earth, it may dissolve and leach a number of minerals. Underground caverns that exist in many parts of the country are a result of millennia of leaching limestone (calcium carbonate)

FIGURE 8–9
Artesian aquifer. Aquifers may lie between two impermeable layers and the recharge area may be at a higher elevation, so that water in the aquifer is under pressure from the higher head in the recharge area. Such an aquifer with water under pressure is an artesian aquifer.

(Fig. 8–10). In most natural situations, the minerals that leach into groundwater are not harmful. Indeed, calcium from limestone is considered beneficial to health. Thus, groundwater is generally high-quality fresh water that is safe for drinking. A few exceptions occur where there is leaching of minerals containing arsenic or other poisonous elements. Likewise, as groundwater exits through springs and seeps, it feeds and replenishes surface waters with high-quality fresh water.

We can visualize all the lands of the earth being continually bathed with a flow of fresh water coming down as precipitation and gradually moving over and through the surface. Salts and other soluble minerals in the earth are gradually carried toward points of evaporation where they are left as the water evaporates and returns for another cycle. The point of evaporation, the end of the line for the salts and other minerals in solution, is generally the oceans, but it may also be certain inland lakes, in arid regions, that have no outlets such as Great Salt Lake in Utah.

HUMAN DEPENDENCE AND IMPACTS ON THE WATER CYCLE

All the water we use must come out of the water cycle at one point or another. More striking, all the pollutants and wastes we produce may get into the water cycle. Further, urban development, agriculture, deforestation, and desertification may grossly increase runoff and decrease infiltration. How these human activities affect the water cycle and its ability to sustain modern civilization will be addressed in following sections and chapters. However, points of major

concern are depicted in Figure 8–16 so that you can begin to appreciate their number and diversity.

Sources and Uses of Fresh Water

Traditionally, humans have obtained most of their fresh water for homes, industries, and irrigation from surface waters. To assure more steady supplies, dams are built to create reservoirs, which hold water in times of excess flow and can be drawn down at times of inadequate flow. In addition, dams and reservoirs may provide for power generation, recreation, and flood control. Water is piped from the source to a treatment plant where its quality is improved (if necessary) and chlorine or another disinfectant is added to preclude the hazard of waterborne pathogens (disease-causing organisms) before distribution. Most of the water used in homes and industries is just "borrowed" in the sense that most of it is returned, albeit polluted (Fig. 8–11). The following scene occurs for innumerable cities and towns throughout the world; just change the names and places appropriately:

> The Patapsco River used to flow through a beautiful valley in central Maryland down to the Chesapeake Bay. Now, most of the Patapsco River is caught in Liberty Reservoir and it flows (after treatment) through our bathtubs, sinks, and toilets and then (after wastewater treatment) into the bay.

Such use of water leads to problems in three areas:

☐ The source has limitations in the amount of water it can provide.

FIGURE 8–10
Limestone caverns such as this are the result of groundwater leaching away limestone. Formations develop as limestone from seeping groundwater recrystallizes. (National Park Service photo by Fred E. Maug, Jr.)

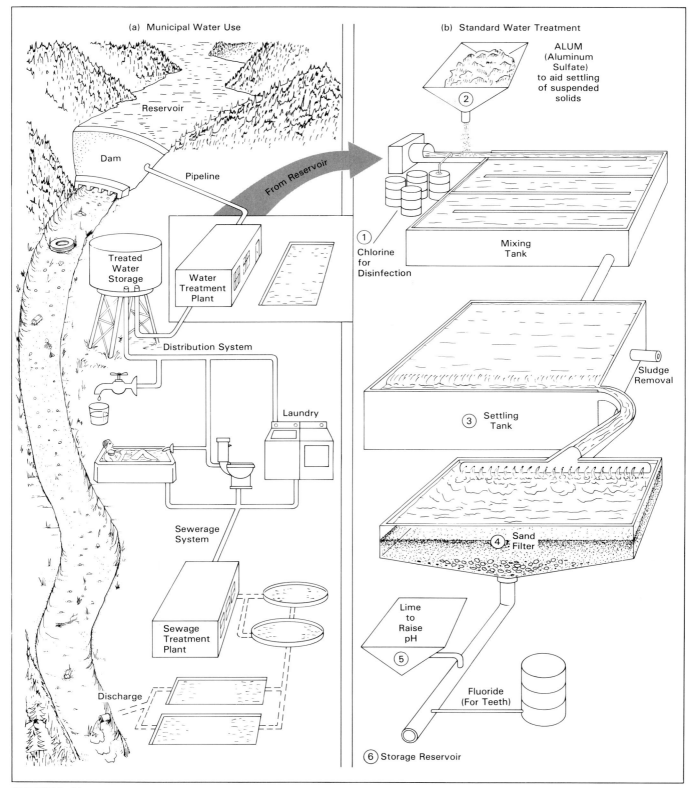

(a) Municipal Water Use

Reservoir

Dam

Pipeline

From Reservoir

Treated Water Storage

Water Treatment Plant

Distribution System

Laundry

Sewerage System

Sewage Treatment Plant

Discharge

(b) Standard Water Treatment

ALUM (Aluminum Sulfate) to aid settling of suspended solids

②

① Chlorine for Disinfection

Mixing Tank

Sludge Removal

③ Settling Tank

④ Sand Filter

Lime to Raise pH ⑤

Fluoride (For Teeth)

⑥ Storage Reservoir

FIGURE 8–11

a) Schematic diagram of municipal water use. Water is generally taken from a river, treated, used, then returned. b) Schematic diagram of a typical water treatment plant. Water is piped from a reservoir to the treatment plant. At the treatment plant, (1) chlorine is added to kill bacteria, (2) alum (aluminum sulfate) is added to coagulate organic particles, (3) the water is put into a settling basin for several hours to allow floc to settle, (4) it is then filtered through sand filters, (5) treated with lime to adjust the pH, and (6) put into a storage reservoir until delivery to your home.

□ Ecosystems downstream from that source may suffer as the water is diverted to other locations.

□ The water being returned is polluted presenting hazards for both human and environmental health.

Water used in the above way is not really consumed, however; with removal of pollutants, it may be reused any number of times. In fact, on major rivers such as the Mississippi, water is reused many times even without good pollution control. Each city in turn takes water from the river, treats it, uses it to flush away more wastes, and then returns it to the river, often with minimal treatment. Thus, each successive city has a higher load of pollutants to contend with and ecosystems at the end of the line may be severely affected by the pollution.

Industrial wastes add to the burden. Small industrial plants generally use water from the same municipal system as do homes, schools, and commercial establishments, and they flush wastewater into the same sewer system. Large industrial plants typically have been located on major rivers, lakes, and bays where they can withdraw unrestricted amounts of water, use it, and return it—polluted with their wastes.

However, the opportunity exists for good pollution control. With suitable treatment and removal of pollutants, the water may be recycled and reused any number of times and still not harm ecosystems on final discharge. Hence, such use of water is called **nonconsumptive** because the water is not really consumed. With suitable pollution control and recycling, the above problems can be avoided.

In contrast, irrigation is called a **consumptive** use because the water returns to the atmosphere by evaporation or transpiration and does not remain available for reuse. Most water for irrigation has come from diverting the flow of rivers through canals to agricultural fields (Figs. 7–30 and 7–31); no treatment is necessary. Again, there are the problems of limited supplies, and downstream ecosystems may suffer due to the diversion of water.

Growing population, expanding industrial output, and increasing agricultural production, especially since World War II, have created an ever-growing demand for fresh water. But, there are only so many practical sites for dams and reservoirs, and rivers can provide only so much. Also, building a dam and reservoir involves a tradeoff. The values of the natural river and the land flooded by the reservoir are sacrificed for the value of the water. Very few rivers remain undammed, and there is strong public pressure to keep these few in their natural state—as expressed by the passage of the Wild and Scenic Rivers

Act of 1968. In short, while demands have increased, the potential to develop more surface water resources has become increasingly limited and surface waters have become more polluted.

The answer to both augmenting supplies and obtaining higher-quality fresh water was to turn to tapping more groundwater. Since the end of World War II, hundreds of cities have drilled huge wells for municipal supplies, and millions of wells have been drilled for individual households in suburbs beyond municipal supply systems. Farmers have turned to center-pivot irrigation systems, a system in which water is withdrawn from a central well and applied to the field through a gigantic sprinkler that pivots itself in a circle around the well (Fig. 8–12 and 8–17). The use of such wells has increased tremendously in the last 20 years, creating an increase in agricultural production but also consuming huge amounts of groundwater. One system may use as much as 10,000 gallons (about 40,000 liters) per minute. Average per capita demands for water for various uses are given in Table 8–1.

Demands continue to increase, wells continue to be drilled, and dams and reservoirs continue to be built. However, in many regions, it is becoming increasingly evident that we are already using water at rates beyond sustainable limits, and beginning to reap increasing consequences.

Consequences of Overdraft of Water Resources

Overdraft of Surface Waters

Inevitable shortages. Occasional dry years in which river flows drop to abnormally low levels are an inevitable fact that must be taken into account in long-term planning. The rule of thumb is that more than 30 percent of a river's *average* flow cannot be taken without risking shortfalls on the average of once every twenty years. As more is taken, shortfalls will occur with increasing frequency and severity (Fig. 8–13). In large portions of the United States, withdrawal from surface waters already exceeds the 30 percent figure and additional regions are projected to exceed it in the near future (Fig. 8–14). The demand on some rivers exceeds 90 percent of the average flow, making chronic water shortages inevitable.

The Colorado River provides a dramatic example. Withdrawals for the city of Los Angeles and irrigation are essentially 100 percent of its average flow. The point at which the Colorado River once flowed into the Gulf of California is now seldom more than a dry bed.

FIGURE 8–12
Center pivot irrigation. Water is pumped from a central well. A self-powered boom rotates around the well, spraying water as it goes.

Table 8–1
U.S. Demands on Fresh Water (per capita)

USE	GALLONS USED PER PERSON PER DAY
Irrigation and Other Agricultural Use	700
Electrical Power Production	600
Industry	370
Residential Use	150

Ecological effects. As a river is diverted, ecological impacts may go far beyond the river itself. Wetlands along many rivers, no longer nourished by occasional overflows, have dried up, resulting in tremendous dieoffs of waterfowl and other wildlife that depended on these habitats. See Case Study 8 concerning the Everglades of southern Florida.

The problem also impacts **estuaries,** bays where fresh water from a river gradually mixes with seawater. Estuaries are among the most productive ecosystems on Earth; they are rich breeding grounds for many species of fish, shellfish, and water fowl. As river flow is reduced, there is less fresh water flushing the estuary. Its salt concentration increases, profoundly upsetting the ecology of the estuary. For example, the entire ecology in the upper end of the Gulf of California has been altered by the reduced flow of fresh water from the Colorado River.

A similar situation is occurring in Mono Lake, a 100-square-mile lake in east-central California. Mono Lake has no outlet, but a substantial inflow of fresh water from snowmelt off the Sierra Nevada Mountains immediately to the west once kept its water at modest salinity. It supported numerous species of wildlife, especially impressive, huge flocks of

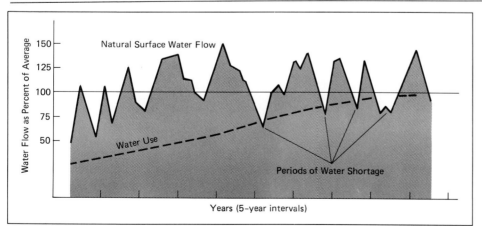

FIGURE 8–13
This hypothetical diagram shows natural variations in surface water flow and increasing human demands on water. Do water shortages occur because of droughts (periods of low flow) or because of excessive demands on the system?

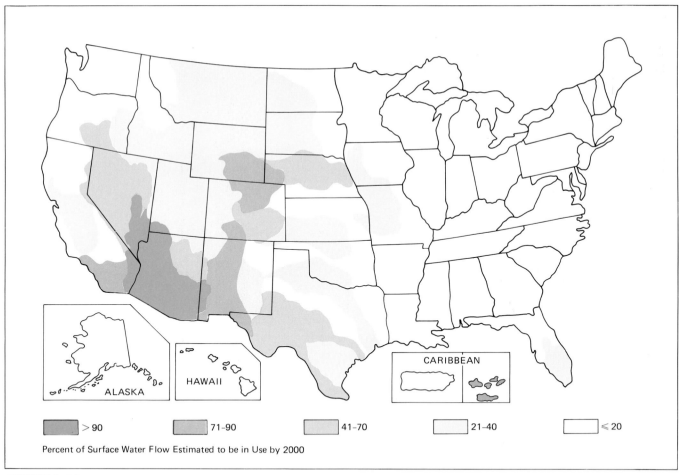

> 90   71-90   41-70   21-40   ≤ 20

Percent of Surface Water Flow Estimated to be in Use by 2000

FIGURE 8-14
No more than 30 percent of the average surface water flow
can be counted on to be available in 95 out of 100 years. By
the year 2000, large areas will be above the 30 percent level,
making severe water shortages inevitable. (U.S. Water
Resources Council.)

aquatic birds. But now, much of the freshwater inflow
is being diverted to support the growing water consumption of Los Angeles. As a result, Mono Lake is
losing more water in evaporation than it is receiving.
It is shrinking rapidly and becoming increasingly
salty (Fig. 8-15). If this situation is not altered, it will
become a small, dead lake, too salty to support life,
surrounded only by many square miles of barren salt
flats. The problem is not limited to the United States.
The Aral Sea, a 30,000-mi² inland sea in south central
Russia, has already dropped 40 feet and may dry up
altogether in another 20 years because of ill-advised
irrigation projects diverting freshwater inputs.

Overdraft of Groundwater

Groundwater is essentially a system of underground
reservoirs. The total volume of groundwater is estimated to exceed that of all surface water by 75-fold.

But groundwater reservoirs, like any other reservoir,
are depleted if rates of withdrawal exceed rates of
recharge, the rate at which water from precipitation
percolates down to the water table.

Falling Water Tables and Depletion.   Groundwater
withdrawal now exceeds recharge, and water tables
are dropping two to three feet or more per year over
large areas of the United States (Fig. 8-16). At the
very least, pumping costs are increased. Eventually,
wells will become dry holes.

The problem is especially acute in low rainfall
regions where rates of recharge are very low and the
demand on groundwater is high because of insufficient surface water. A prime example is the Great
Plains region (Texas, Oklahoma, Arizona, Colorado,
Kansas, and Nebraska) where the Ogallala aquifer is
extensively exploited for center-pivot irrigation of
crops such as corn and cotton (Fig. 8-17). It is predicted that 3.5 million acres (1.4 million hectares) in
this region will be abandoned or converted to dryland
farming (ranching and production of forage crops)
over the next 10 years because of depletion of water.
Many cities that draw municipal supplies from

FIGURE 8–15
With its freshwater input diverted to San Francisco, Mono Lake is drying up leaving grotesque salt forms from evaporation. (Americ Higushi/Stockpile)

FIGURE 8–16
Declining groundwater levels and related problems in the United States. Source: V. J. Pye et al., *Groundwater Contamination in the United States* (Philadelphia: University of Pennsylvania Press, 1983). Used with permission from The Academy of Natural Sciences, Philadelphia, PA.

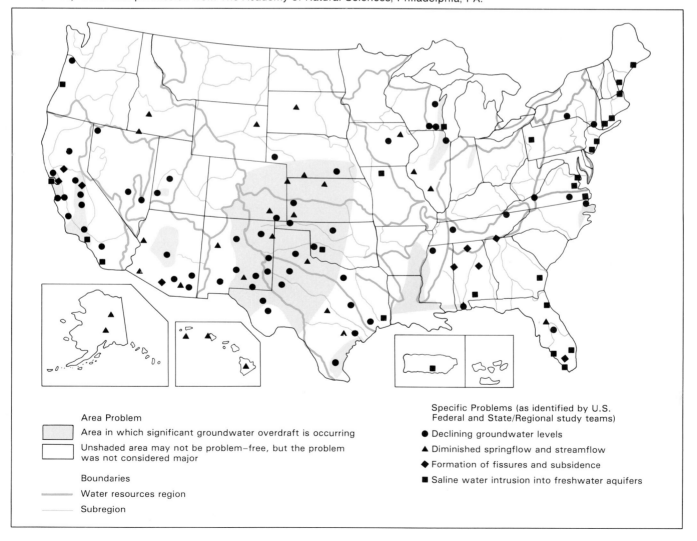

Area Problem

▨ Area in which significant groundwater overdraft is occurring

☐ Unshaded area may not be problem–free, but the problem was not considered major

Boundaries

—— Water resources region

—— Subregion

Specific Problems (as identified by U.S. Federal and State/Regional study teams)

● Declining groundwater levels

▲ Diminished springflow and streamflow

◆ Formation of fissures and subsidence

■ Saline water intrusion into freshwater aquifers

The Florida Everglades, often called the "River of Grass," is a flowing marsh of sawgrass and wet prairie species with embedded islands containing willow, myrtle and cypress. Tropical hardwood hummocks are found on islands of higher elevation. Larger than most people can imagine, this marsh is forty to sixty miles wide and one-hundred miles long. Incredibly flat land with a slope of only one vertical inch per mile has made this waterway broad; water cuts a narrow channel where there is significant relief in a drainage basin. Flatness creates a profound sense of openness from horizon to horizon, an image of limitless space seldom seen on land. In the monsoon season, towering white thunderheads rise into the blue sky with their shadows moving slowly across the marsh as the trade winds move them inland. Flocks of white egrets circling in the distance suspend time; one has a sense of creation unaffected by humans.

Famous for its subtle beauty, the "River of Grass" begins east and south of Lake Okeechobee . . . and flows south to the estuary of Florida Bay at the tip of the Florida peninsula. Spillover from the southeastern rim of Lake Okeechobee was the original Everglades headwaters. As part of a scheme to enable farming, canals were cut from the lake to the Gulf of Mexico and the Atlantic Ocean. This diverted flow from Lake Okeechobee and its 4000 square mile drainage basin and left the Everglades marsh to subsist on local rainfall. In addition, a 1000 square mile area of Everglades marsh was diked and equipped with huge drainage pumps. The rich, peat soils of this enclosed area are used mainly for sugar cane production, a crop made profitable by import quotas which limit U.S. sales from countries such as the Philippines and Dominican Republic where the standard of living often necessitates U.S. foreign aid and defense expenditures.

Roots and plant debris from about 5000 years of marsh production formed the peat soil. Peat accumulated because of the low rate of oxygen transfer through the overlying marsh water. Drained for farming, air enters the soil and the organic peat becomes food for aerobic microbes which oxidize it to carbon dioxide and water. The ash from this biochemical fire contains nitrogen and phosphorus bound when the soil was created. While the soil concentrations are small, about an inch per year of soil is consumed in this fashion and the mass of nutrients released is considerable. There are also contributions from crop fertilization. Rain percolating through soil into the drainage system extracts these nutrients. The pumps that drain the farming area discharge nutrient laden stormwater into the Everglades.

Native Everglades vegetation is a low nutrient ecosystem. The much higher concentrations of nitrogen and phosphorus in the agricultural discharges allow different plant species to dominate. The result is eutrophication of the marsh. A dense monoculture of cattails is displacing the native sawgrass and wet prairie vegetation. The periphyton community which is the base of the Everglades food web is entirely altered. The resulting marsh has very low dissolved oxygen values, very low biological diversity and is of little value to native species of invertebrates, fish, rep-

FIGURE 8–17
Aerial photograph shows the extent of center-pivot irrigation in Nebraska. As a result, groundwater is being depleted, which will bring an end to this kind of farming. (Earl Roberge/ Photo, Researchers.)

tiles and birds. As saturation occurs, eutrophication effects are spreading slowly downstream from the points of discharge like a green cancer. The analogy of wetlands as "kidneys" for the purification of stormwater is not necessarily appropriate for phosphorus. With no permanent storage mechanism to render phosphorus unavailable, the relative dominance of plant species will change and, in the extreme, hypereutrophication will result as in this example. A Kleenex is a better analogy. Once you blow your nose on it, it is forever changed.

Left unchecked, these nutrient effects will inevitably spread downstream through the remaining marsh into Everglades National Park. The Park is to be preserved as it was created for the enjoyment and study of future generations. The solution to this problem is economic and political rather than technical. Stormwater retention and treatment within the farming area would require land, other capital investment and diminish agribusiness profits.

PAUL C. PARKS, Ph.D.
Florida Wildlife Federation

The Everglades.
(John Rova/Photo Researchers)

groundwater will also face severe water shortages in the years ahead if current trends of overdraft continue.

Diminishing Surface Water. Surface waters are also affected by falling water tables. Recall that streams, rivers, and lakes are fed in large part by springs and seeps of outflowing groundwater. As the water table drops, the amount of water exiting from a spring will diminish and finally cease as the water table drops below the level of a spring. Consequently, dropping water tables result in diminishing flows from springs and, thus, diminishing of surface water flows as well. This exacerbates all the ecological problems discussed above.

Land Subsidence. Over the ages, groundwater has leached cavities in the earth, but, as these spaces are filled by water, water itself plays a role in supporting the structure of overlying rock and soil. As the water table drops, this support is lost and there may be a gradual settling of the land, a phenomenon known as **land subsidence.**

The rate of sinking may be on the order of 0.5 to 1 foot (15 to 30 cm) per year. In some areas of the San Jaoquin Valley, in California, land has settled as much as 30 feet (about 9 meters) because of groundwater removal. Land subsidence leads to the cracking of building foundations, roadways, and water and sewer lines (Fig. 8–18). In coastal areas, it causes flooding unless levees are built for protection. For ex-

FIGURE 8–18
Land subsidence. Removal of groundwater may allow ground to settle, resulting in severe property damage, such as the cracking foundations seen here. (USDA—Soil Conservation Service photo.)

ample, a 4000-square-mile (10,000 km²) area in the Houston-Galveston Bay region of Texas is gradually sinking because of removal of groundwater, and coastal properties are being abandoned as they are gradually inundated by the sea. Land subsidence is also a serious problem in New Orleans, in sections of Arizona, and in many other places throughout the world.

Another kind of land subsidence, the occurrence of a **sinkhole,** may be sudden and dramatic (Fig. 8–19). A sinkhole results when an underground cavern, drained of its supporting groundwater, suddenly collapses. Sinkholes may be 100 meters (300 ft) or more across and as much as 50 meters (150 ft) deep. Formation of sinkholes is particularly severe in the southeastern United States where groundwater has

eaten numerous passages through ancient beds of underlying limestone. An estimated 4000 sinkholes have occurred in Alabama alone, some of which have "consumed" buildings, homes, livestock, and sections of highways.

Saltwater Intrusion. Another problem resulting from groundwater removal is **saltwater intrusion,** also called **saltwater encroachment.** In coastal regions, springs of outflowing groundwater may lie under the ocean. As long as the water table on land is higher than the ocean level and the pressure in the aquifer is maintained, there is a net flow of fresh water into the ocean. Thus, wells near the ocean yield fresh water (Fig. 8–20a). However, a lower water table or rapid rates of groundwater removal may re-

FIGURE 8–19
Sinkhole. Removal of groundwater may drain an underground cavern until the roof, no longer supported by water pressure, collapses, resulting in a sinkhole such as this one in Alabama. (Department of the Interior, U.S. Geological Survey photo.)

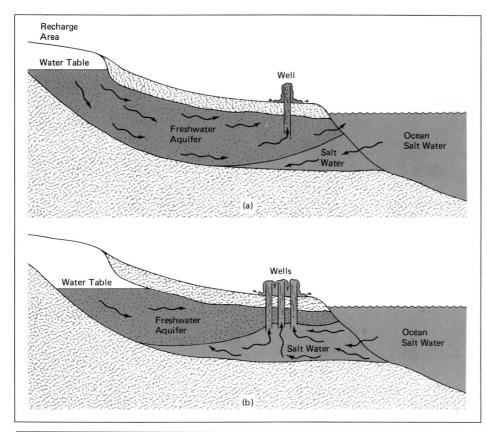

FIGURE 8–20
Saltwater encroachment. (a) Where aquifers open into the ocean, fresh water is maintained in the aquifer by the head of fresh water inland. (b) Excessive removal of water may reduce the pressure so that salt water moves into the aquifer.

duce the pressure in the aquifer, permitting salt water to push back into the aquifer and hence into wells (Fig. 8–20b). Saltwater intrusion is problematic at many locations along U.S. coasts (Fig. 8–16). Overdraft of groundwater and such ensuing consequences are by no means unique to the United States.

Obtaining More versus Using Less

Based on past trends, which show that per capita water consumption in the United States increased by about 50 percent between 1950 and 1975, economists project continuing increases. There are two contrasting points of view about meeting this problem: (1) We need more water, and (2) we can get along with less.

Consider the former. There are numerous plans on the drawing boards for ever more elaborate water diversion projects to bring water from where it is seen as not needed to where it is seen as required. One of the more grandiose, which has been seriously considered, is a huge dam on the Yukon River in Alaska, which would flood approximately 10 percent of the state, and a 2000-mile canal to bring the water to the mid- and southwest United States. Such plans consider only technological feasibility, not ecological tradeoffs, nor sustainability. Exploiting more and more water resources can only spread the problems

we have been discussing to additional areas and exacerbate them in regions already affected. Finally, they can borrow only a little more time before limits and shortages will be confronted again. In short, it is not a sustainable course of action. Fortunately, however, there is ample opportunity for balancing supply with demands through water conservation.

Potential for Conservation and Reuse of Water

Irrigation

In low rainfall regions of the country, the regions facing the most serious water shortages, about 85 percent of the water is used in irrigation. Hence, cutting the amount of water used in irrigation by only a few percent would make ample water available for the growing demands of cities. About 50 to 75 percent of the water applied to crops by flooding or sprinkling methods is wasted; it evaporates, percolates, or runs off rather than being absorbed by the plants. This waste is unnecessary. First, the crops most commonly grown under this wasteful irrigation are corn and cotton, crops that currently run huge surpluses that the government ends up buying to support the price. The irony should not escape notice that water resources are being depleted to grow crops that are not needed. The land could be returned to range, which was its natural ecological condition, or water-sparing crops could be grown. If irrigation is to be continued, water-saving *drip irrigation* systems are available and are coming into use in the southwestern United States. **Drip irrigation** systems are networks of plastic pipes with pinholes which literally drip water at the base of each plant (Fig. 8–21). Such systems have the added benefit of retarding salinization (Chapter 7).

With these alternatives available, why do farmers still engage in such wasteful water use for growing crops that are not needed? It is a paradox of ill-begotten government subsidies. In the interests of promoting agriculture, the government has built water projects (dams, reservoirs and canals to distribute water) and sells water to farmers for irrigation at nominal cost. Then, the government provides price supports that give the farmer a guaranteed price for the corn regardless of surpluses. As long as farmers get these subsidies it is not cost-effective to do other than waste water growing crops that are not needed.

Municipal Systems

People backpacking in water-short regions discover that they can get by on less than a gallon of water per day including needs for cooking and washing. In contrast, water consumption in modern homes averages around 150 gallons of water per person per day, for flushing toilets (3–5 gallons per flush), showers (2–3 gallons per minute), laundry (20–30 gallons per wash), and so on. If watering lawns is included, use goes up from the 150 gallon mark. It is conspic-

uous that there is considerable room for decreasing water without causing hardship.

Numerous cities are promoting water conservation through "public awareness" programs to meet shortages or accommodate more growth without developing new supplies. In addition to taking shorter showers and turning off the water while brushing teeth, water saving devices include toilets that use only 1.5 gallons per flush, dishwashers and laundry machines that use less water, and shower heads that deliver a finer spray. Beautiful gardens can be achieved with much less watering by landscaping with more drought-resistant species and by increased use of mulches to reduce evaporation.

Going further, *gray water recycling* systems are being adopted in some water-short areas. **Gray water,** the slightly dirtied water from sinks, showers, and bath tubs, is collected in a holding tank and used for such things as flushing toilets, watering lawns, and washing cars (Fig. 8–22). Also, some cities and industries are considering purifying and recycling waste water (see Chapter 10). Again, there is the question of cost-effectiveness, however.

Effects Caused by Changing Land Use

All the land area from which water drains into a particular stream or river is known as the **watershed** of that stream or river (Fig. 8–23). If the watershed is covered by natural forest or grassland with rich topsoil, most of the precipitation will infiltrate and serve to recharge the groundwater. Only exceptionally heavy or prolonged storms are sufficient to saturate

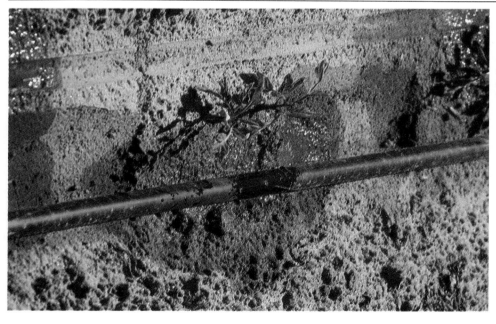

FIGURE 8–21
Drip irrigation. Irrigation is the most consumptive water use. Drip irrigation offers a conservative method of applying water, dripping it on each plant through a system of plastic pipes. (Lowell Georgia/Photo Researchers.)

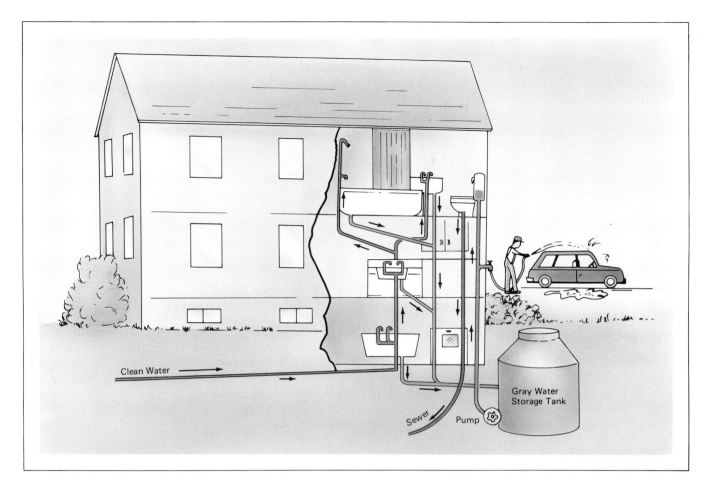

FIGURE 8–22
Gray water use. It is extravagant to use drinking-quality water for such uses as toilet flushing and lawn watering. Significant water conservation could be achieved by trapping waste water from tubs and sinks (gray water) and recycling it for such purposes.

FIGURE 8–23
A watershed is all the land area that drains into a particular stream or river. Dotted line depicts the watershed for this stream.

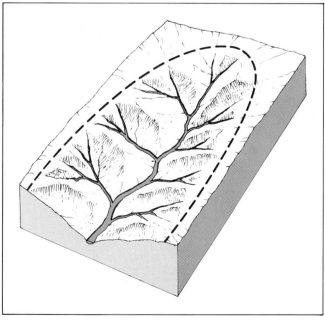

the ground and cause significant runoff. In turn, groundwater "leaks" out through springs and seeps. The water leaks out at about the same rate regardless of fluctuations in the amount of water in the reservoir. Thus, a stream draining a forested watershed is prevented from flooding during rains because water is being taken into the groundwater, and it continues to flow during dry periods because water continues to leak out. Such a stream is able to support not only a rich aquatic ecosystem; it also serves to support much of the surrounding terrestrial ecosystem as many species depend on the water directly for drinking or indirectly through various food chains.

Many human activities change the nature of the land surface such that the infiltration/runoff ratio is shifted to cause less infiltration and more runoff. Most conspicuously, urban and suburban development greatly increases runoff by creating innumerable hard, impervious surfaces such as roadways, parking lots, and rooftops. Even the soil of suburban lawns is generally compacted so that infiltration is decreased and runoff is significantly increased. Agricultural practices that cause the soil to become puddled and/or compacted, as well as clearcutting of forests and overgrazing by livestock, also lead to increased runoff. Whatever its cause, shifting the infiltration/runoff ratio in this way has numerous and far-reaching effects. On one side are the effects of increasing runoff; on the other are the effects of decreasing infiltration.

Effects of Increasing Runoff

Whereas infiltration fills groundwater reservoirs, runoff washes directly and immediately into streams. Storm drains funnel runoff from parking lots and city streets to the nearest stream bed (Fig. 8–24). Thus, with even a modest thunderstorm, a quiet stream may be changed into a surging torrent in a matter of minutes (Fig. 8–25).

Stream Bank Erosion. Natural stream channels are unable to handle the increased flow that rips and tears soil and rock from the stream bank. Trees, which normally stabilize the bank, are undercut and toppled into the stream, diverting water against and over the banks causing more erosion (Fig. 8–26). This process of **stream bank erosion** is a natural process but normally occurs very slowly so that regrowth of vegetation restores the bank. With the additional runoff, the restorative balance is upset and the erosion be-

comes an ongoing process. Additional loads of material may be contributed by flows from badly placed storm drain outlets, eroding gullies down valley sides (Fig. 8–27). Finer materials are washed away, but rocks, stones, and coarse sand are deposited in the bottom of the channel, raising the water level and causing it to erode more of the banks. The result is that the channel gets *wider* and *shallower* (Fig. 8–28). Indeed, a stream channel may become completely filled and blocked so that the water is diverted in a new direction across the valley floor (Fig. 8–29). This results in more erosion and the death of many trees as the soil is waterlogged. Gradually, a narrow tree-lined stream may be converted into a broad washway of fallen trees and drifts of sand and gravel sediment.

Flooding. As the runoff from small streams is funnelled together into larger streams and rivers, flooding occurs. Floods have always been a part of nature, however, with increased runoff, even a very modest storm may lead to a flood. Countless communities, many of them expensive new, suburban developments, have experienced flooding with increasing frequency and severity as expanding development has paved more of the upstream watershed; thus, flood damages have generally increased, despite flood control measures.

Increased pollution. The surge of increased runoff is also highly polluted. In contrast to the high-quality water from springs, which has been filtered through the earth, runoff carries all kinds of polluting mate-

FIGURE 8–24
Storm drains generally are direct pipes to the nearest convenient stream bed. Storm drains from residential area in background empty here into a stream bed. (Photo by author.)

(a)

(b)

FIGURE 8–25
Effect of development on stream flow. Before development, this stream maintained a generally modest flow of water throughout the year. Now, after development of the surrounding area with suburban homes, the flow fluctuates sharply between (a) high surges of runoff and (b) dryness. Photo (a) was taken during an average summer thunderstorm; photo (b), a few hours later. Note the bank erosion as well as the dryness. This situation is a typical result of development. (Photos by author.)

FIGURE 8–26
Stream bank erosion. The surges of increased runoff greatly accelerate erosion of stream banks. Note the many trees that have fallen because of undercutting. Fallen trees block the channel and cause still further erosion and undercutting. (Photo by author.)

FIGURE 8-27
Erosion from storm drain outlets. Frequently storm drains open onto valley sides, and water is allowed to find its own way down the slope. The gully seen here is the result of erosion by water coming from a storm drain outlet in the left background. (Photo by author.)

FIGURE 8-28
By increasing runoff and decreasing infiltration, development results in many undesirable changes in natural water flow as shown here.

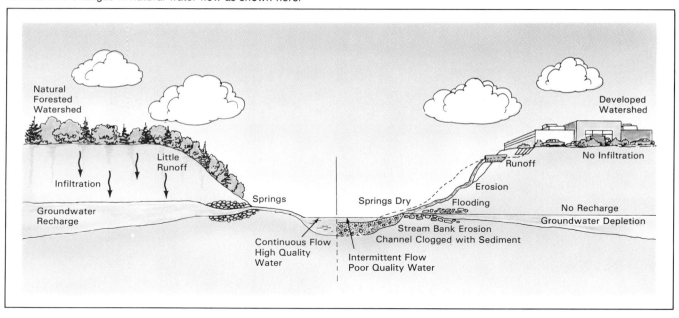

rials from the surface directly into streams and rivers. Such materials include:

- ☐ soil from erosion
- ☐ nutrients from fertilizer used on lawns, gardens, and agricultural fields
- ☐ insecticides and herbicides used on lawns, gardens, and crop lands
- ☐ fecal wastes and associated bacteria from pets and farm animals
- ☐ road salt and other chemicals from surface treatments or spills
- ☐ grime and toxic chemicals from settled vehicle exhaust and other air pollution
- ☐ oil and grease from roadways and parking lots, or disposal down storm drains
- ☐ trash and litter.

Indeed, surface runoff is now recognized as the major source of pollution for many rivers. This is es-

(a)

(b)

(c)

FIGURE 8–29
The channel of the stream shown here (a) has been so filled with sediment that the water is being forced to find new pathways over the valley floor, causing further erosion and ecological damage (b) and (c). (Photo by author.)

pecially true since pollutants in industrial and municipal sewage treatment plant discharges have been significantly reduced in recent years (see Chapters 10 and 11). A recent study conducted by Inform, a New York City-based research group, determined that runoff contributed nearly 100 tons of lead to the Hudson River per year in contrast to only a quarter of a ton coming from industrial and municipal discharges.

Effects of Decreased Infiltration

Water that runs off does not recharge the groundwater. In metropolitan areas, falling water tables may be caused as much or more by decreasing infiltration as by groundwater withdrawal. Thus, development, which increases runoff and decreases infiltration, exacerbates the problems of saltwater intrusion, land subsidence, and other problems related to falling water tables described above.

As water tables drop, flow from springs and seeps diminishes and streams may be reduced to stony beds in dry periods. Figure 8–30 graphically shows the flow of a stream as urbanization of its watershed occurred. Not only did peak flow during a rain increase; the flow after the rain also diminished. In other words, hard surfacing, which is so much a part of traditional development, changes streams from a condition of constant flow supporting a rich ecosystem to an ecologically destitute condition of rapid alternation between flood-producing surges and dryness. In effect, natural streams become little more than open storm drains; indeed, they have frequently been incorporated into the storm drain system by laying drain pipe in the stream bed and covering it over (Fig. 8–31). This is a particularly tragic loss in view of our desire to preserve and protect more of the natural ecology within an increasingly urbanized landscape.

Stormwater Management

The traditional approach in addressing stormwater has been to get rid of it as quickly as possible and, unfortunately, this attitude still prevails in most re-

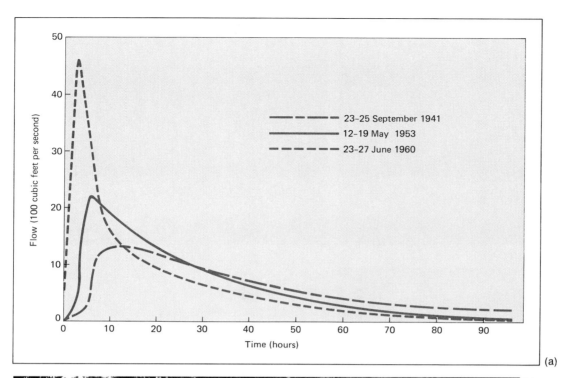
(a)

(b)

FIGURE 8–30

(a) Changes in stream flow occur with development. Curves are for similar storms on Brays Bayou in Houston, Texas, before, during, and after development. Note the increasing height of the surge occurring with the storm and also the decreasing volume of flow that occurs later in the cycle. (b) Because of increasing runoff from new development in the watershed, many established areas experience flooding where none occurred before. Peachtree Creek, Atlanta, Georgia. (a, from D. Van Sickle, in *Effects of Watershed Changes on Stream Flow.* W. Moore and C. Morgan (Eds.) Austin, Texas: University of Texas Press, 1969; b, Photo Researchers/ © 1982 Melissa Hayes English)

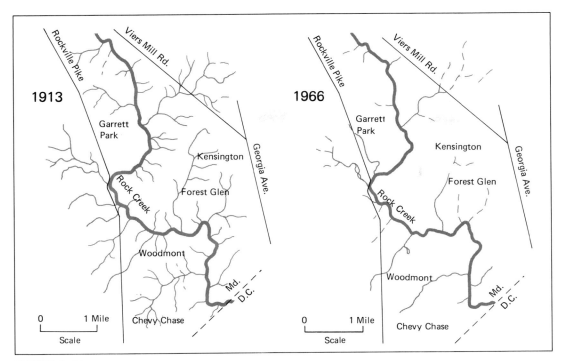

FIGURE 8–31
With urbanization, streams become storm drains. This 26 square mile section of the Rock Creek watershed in Maryland, now a heavily populated suburb of Washington, was rural in 1913, with many small tributaries fed by springs and seeps. Ensuing development, carried out in ignorance of natural processes, covered most of the old acquifer recharge areas with pavements and rooftops, so that more precipitation ran rapidly off the land instead of soaking in and flowing out gradually into streams. Flooding during storms and loss of flow at other times caused most of the tributaries to be covered over as storm sewers; of 64 miles of natural-flowing stream channels that existed in 1913 in this section, only 27 miles could be found above ground in 1966. (National Park Service.)

gions. Parking lots, roadways, lawns, and so on are graded to quickly shed water into storm drains, which funnel the water into the nearest convenient stream. If streambank erosion and flooding become too problematic, streams are **channelized,** that is, the channel is dredged out, straightened or made into gradual sweeping curves, and lined with rock or concrete to prevent bank erosion (Fig. 8–32). The channelized stream carries both water and sediment more efficiently, so, the hazard of flooding in the area is greatly reduced.

Ecologically, however, a channelized stream has little if any semblance to the natural stream it replaces. Water flow continues to alternate between surges and dryness, and, in addition, all the crevices and pools that might have supported aquatic and other wildlife during dry periods are eliminated. The channel may even be a hazard to wildlife since animals that fall in may not be able to get out. Aesthetically and func-

tionally, a channelized stream is effectively an open storm drain. Nevertheless, thousands of miles of streams have been channelized and the process continues. Furthermore, water removed from one area is put somewhere else. Consequently channelization often amounts to moving a flood from one place to another.

In contrast, the modern ecological concept of **stormwater management** is to keep the stormwater near where it falls and preserve the natural infiltration/runoff ratio. Runoff after development should not exceed the ratio before development. A number of techniques that have been developed to achieve this goal include:

☐ Dry wells and trenches—broad, shallow, rock-filled wells or trenches that receive runoff and then let it percolate into the ground.

☐ Swales with barriers and depressions—grading a lawn area so that would-be runoff accumulates in broad shallow depressions from where it infiltrates into the ground.

☐ Parking lots with porous surfaces.

☐ Retention ponds—building into the development a pond that will collect the runoff and drain it away slowly. Stormwater retention ponds may additionally support pockets of wildlife (Fig. 8–33), or they may provide a practical source of water for nondrinking purposes such as car washing or lawn watering, thus lessening

FIGURE 8-32
Channelized stream. To decrease flooding and erosion, stream channels may be dredged, straightened, and lined with concrete, a process that may simply transport the problem downstream; it also destroys all semblance of the natural stream ecosystem. (USDA—Soil Conservation Service photo.)

FIGURE 8-33
Stormwater reservoirs. Rather than letting the excessive runoff from developed areas cause environmental damage, it may be funnelled into such reservoirs from where it may drain away slowly or infiltrate. The design of the standpipe allows this reservoir to hold and slowly drain away excess water while retaining some to create a useful pond.

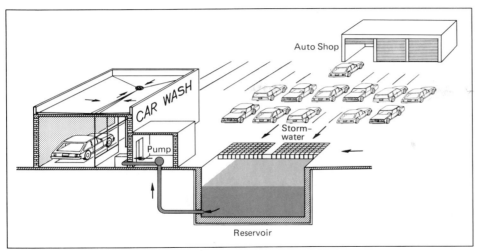

FIGURE 8-34
Stormwater may be a cheap source of water for many uses; here, runoff from a car dealership is being stored and used to wash cars.

the demand on the municipal system (Fig. 8–34). Large stormwater retention reservoirs can double as recreational facilities (Fig. 8–35). During a storm the recreation area around the reservoir can store floodwater. Recreational equipment can be made so that it will not be damaged by the high water, and there is little inconvenience to people since most would leave the area at the time of the storm, whether it was flooded or not.

☐ Rooftops of buildings and parking lots may be built so that they "pond" the water and let it trickle away slowly.

IMPLEMENTING SOLUTIONS

Cheap Water and Waste

Current trends in water use are clearly not sustainable. Increasingly severe costs entailed in loss of wildlife, land subsidence, and saltwater encroachment are being paid in addition to facing running out. Yet such vast opportunities exist for conserving and recycling water that exploiting resources to the point of depletion and/or ecological collapse is unconscionable.

Water conservation has been promoted for years, and many communities have found it necessary to impose restrictions on watering lawns and washing cars in time of drought. Los Angeles has enacted a regulation that all new toilets must be of a water-saving type. But talk alone has a short-term and limited effect, and regulations get into more and more bureaucracy that is increasingly unwieldy to manage and enforce. In short, few people can be talked into saving water when the costs or inconveniences of saving it outweigh the money saved.

Yet, most cities continue to provide water to consumers for only a tiny fraction of a penny per gallon so that, even with extravagant water use, the cost is still nominal. Indeed, the cost is basically the cost of treatment and piping; water itself is considered free. Therefore, there is little economic incentive to save water regardless of the virtues. A much more direct solution is: *Raise the price of water. An "ounce" of economic incentive has more effect than many "pounds" of talk.*

Hardship on the poor can be avoided by having a sliding scale so that cost per gallon increases with the number of gallons used and increasing costs may be phased in. For example, Arizona has passed a law (the 1980 Groundwater Management Act) that will gradually raise prices high enough to prevent the water table from dropping further. Already consumption has been reduced 21 percent from 205 to 161 gallons per day per capita, and water still only costs about one cent for 7 gallons. Additional value placed on water will also encourage the retention and use of stormwater and the reuse of gray water. Similarly, water subsidies to farmers should be phased out to the extent of balancing withdrawals with recharge and preventing further degradation of ecosystems. Nothing less is sustainable.

What You Can Do

In the 1970s, environmentalists in Maryland held numerous meetings and other events which served to educate, support, and prod the leadership to attend to the problems of water resources. The legislature was informed that stormwater retention and maintenance of the water table would be the least expensive way to avoid future costs of flooding, saltwater intrusion, and land subsidence. Gradually, regulations have been put into effect which require that the

FIGURE 8–35
Lake Needwood flood control project. (a) This lake and surrounding area is a favorite recreational area in Montgomery County, Maryland, a suburban area of Washington, D.C. In addition, the area is a flood control reservoir. (Air Photographics, Inc., Silver Spring, Maryland.) (b) In the event of a storm, the dam can hold up to an additional 40 feet of water and allow it to drain down gradually, preventing flooding and erosion downstream. (c) Recreational equipment is designed to withstand flooding. (Photos by David Hunley.)

infiltration/runoff ratio after development does not exceed that before development. The techniques used for stormwater management are those discussed above. Maryland stormwater control legislation and regulations may now serve as a model for the rest of the country.

Investigate the status of water resources, including stormwater management, in your region and determine what problems are present. Identify organizations that are involved with these issues. Join and lend your knowledge and support in promoting state or local legislation that will address these issues.

Clean Water Action Project, Inc. (Appendix A) has many affiliates nationwide in this area.

Monitor the use of water in your own home (find the pertinent information on your water bill). What is the average water use per person per day in your household? Can you cut this to 60 gallons per person per day, considered to be a conservation goal in many regions? Investigate where stormwater from your own home and yard goes. Does it mostly run off or infiltrate? Install a system, perhaps just a rain barrel at a downspout, to capture and use stormwater for lawn/garden watering and car washing.

Part Four
POLLUTION

Pollution has become a household word that calls up images of despoiled water, air, or land. Yet pollution is actually much more complex. It eludes simple definition because it may involve hundreds of factors that stem from numerous sources. The Environmental Pollution Panel of the President's Science Advisory Committee in its 1965 report, *Restoring the Quality of Our Environment*, defined environmental pollution as "the unfavorable alteration of our surroundings, wholly or largely as a by-product of man's action."

Some "alterations," such as contamination of air or drinking water, may directly affect human health and well-being. Other alterations may affect humans in a more indirect way, such as carbon dioxide emissions affecting the climate, which in turn may affect food production; or changes in the nutrient levels causing some populations to die out and others to explode. Also, alterations such as litter on beaches may simply decrease opportunities for recreation and enjoyment of nature. The key point is simply that an "unfavorable alteration of our surroundings" occurs.

Some of the more significant categories of pollution include:

☐ Sediments eroding from soil that adversely affect aquatic ecosystems.

☐ Oversupply of nutrients from sewage and fertilizer runoff that causes overgrowth of undesirable algae and upsets existing ecosystems.

☐ Disease-causing organisms from sewage wastes that contaminate drinking or recreational water supplies; organic wastes from sewage or other sources that cause depletion of dissolved oxygen and suffocation of aquatic life.

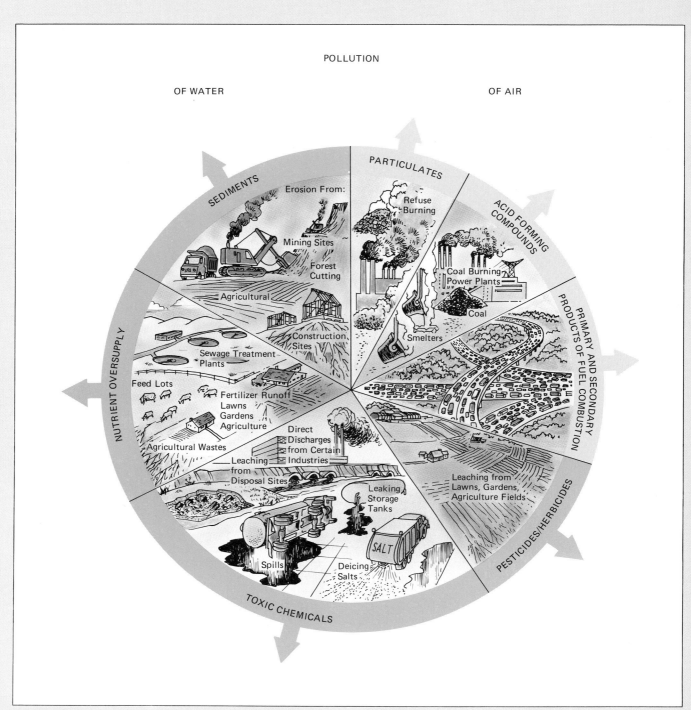

FIGURE IV–1
Pollution may be defined as any chemical or material out of place. Thus, chemicals and materials that are useful in one place cause pollution as they are discarded in or migrate to places where they are not wanted and where they may cause damage to environmental and human health. The figure shows the major categories of pollution and illustrates their most important sources.

☐ Toxic chemicals from wastes and pesticides that contaminate water supplies.

☐ Direct and indirect products from the combustion of fuels, affecting air quality and causing acid precipitation, which affects soil and water.

☐ Radioactive wastes and materials from the production of nuclear power and weapons which may potentially contaminate air, water, and/or soil.

☐ Emissions of carbon dioxide and ozone depleting chemicals that may alter the climate and affect all species on earth.

As this list illustrates, the maxim "don't pollute," which is taught to elementary school children, is a gross oversimplification of the pollution problem. As ecologist Edward Kormondy points out:

[Pollutants are] normal by-products of people as purely biological organisms and as creative social beings. They are the organic and inorganic wastes of metabolic and digestive processes and of creativity in protecting and augmenting the production of crops, of warming houses, clothing the body and harnessing the atom. . . . Solutions do not and cannot lie solely in removing the cause because as long as humanity exists, it will have by-products. Rather, answers lie in intelligent management of that production and through regulating the unfavorable alteration of our surroundings.*

Indeed, every organism in a natural ecosystem

* Kormondy, Edward J., *Concepts of Ecology*, 3rd Ed., Prentice Hall, 1984, p. 247.

produces potentially polluting waste products. What makes natural ecosystems sustainable is that the wastes from one kind of organism become the food and/or raw materials of another. In balanced ecosystems, wastes do not accumulate to produce "unfavorable alterations"; they are broken down and recycled.

Through much of their history humans have relied on the same natural processes to dispose of their wastes. But the situation has become extremely unbalanced. Exploding human population coupled with increasing use of materials and energy have led to enormous volumes of wastes and other materials being discharged into the environment. Even when materials are biodegradable, that is, of a kind that can be assimilated and recycled by organisms, sheer volumes overwhelm the capacity of natural systems to cope. Aggravating the problem is the production of increasing amounts and kinds of *nonbiodegradable* materials, which are *not* readily broken down and assimilated by natural processes.

Clearly, pollution involves so many different factors from so many different sources that there is no single or simple remedy (Fig. IV–1). In each situation the pollutant(s) causing the problem must be identified, sources determined, and then appropriate control strategies developed and implemented. This is a complex and difficult task. It should hardly be surprising that despite efforts and significant progress in some areas, huge problems remain. In Part IV we shall address pollution issues that are of greatest and most widespread concern and show what is being done and what remains to be done to solve the problems. Additional problems such as those involving pesticides, refuse, and radioactive wastes will be considered in later chapters.

9
Sediments, Nutrients, and Eutrophication

■ Outline

I. EUTROPHICATION 228

   A. Two Kinds of Aquatic Plants 228

   B. Upsetting the Balance by Nutrient
     Enrichment 229
     1. The Oligotrophic or Nutrient-Poor
       Condition 229
     2. Nutrient Enrichment, Eutrophication 230

   C. Natural versus Cultural Eutrophication 232

CASE STUDY: EUTROPHICATION OF LAKE ERIE 232

   D. Combatting Eutrophication 232
     1. Attacking the Symptoms 233
       a. Chemical treatments 233
       b. Aeration 233
       c. Harvesting algae 234
     2. Controlling Inputs 235
II. SOURCES OF SEDIMENTS AND NUTRIENTS 235
   A. Sources of Sediments 235
   B. Impacts of Sediments on Streams
     and Rivers 236
     1. Damage to Aquatic Ecosystems
       of Streams and Rivers 237

     2. Filling of Channels and Reservoirs 238

   C. Sources of Nutrients 240

   D. Loss of Wetlands and Bulkheading of
     Shorelines 241

III. CONTROLLING NUTRIENTS AND
    SEDIMENTS AND WHAT YOU CAN DO 242

   A. Best Management Practices on Farms,
     Lawns, and Gardens 242

   B. Sediment Control on Construction and
     Mining Sites 244

   C. Preservation of Wetlands 246

   D. Banning the Use of Phosphate Detergents 246

   E. Advanced Sewage Treatment 247

■ Study Questions

1. What was the cause of the dieoff of seagrasses in Chesapeake Bay? Name the phenomenon.

2. Name and describe two major categories of aquatic vegetation. Define and describe: *benthic plants, phytoplankton.* Where does each get its nutrients?

3. Explain how nutrient enrichment shifts the balance. Name the process. Discuss the question: Is a eutrophic lake really dead?

4. Distinguish between *natural* and *cultural eutrophication.*

5. List and describe ways of combatting *eutrophication.* Which methods address only symptoms? Which get at the cause?

6. List and describe sources of sediments.

7. Describe the impact of sediments on streams and rivers and their ecosystems.

8. Describe the impact of sediment on channels and reservoirs.

9. List and describe sources of nutrients.

10. Explain how loss of wetlands and bulkheading of shorelines affects nutrients and sediments.

11. Distinguish between *point* and *nonpoint* sources of nutrients.

12. Give examples of best management practices for controlling nutrients and sediments.

13. Describe means of controlling sediments on construction and mining sites.

14. Discuss the value of preserving wetlands.

15. Discuss the value of banning phosphate detergents.

16. Discuss the need to remove nutrients from sewage waste water.

The Chesapeake Bay (Fig. 9–1) is North America's largest estuary and prior to the 1970s its most productive, yielding many millions of pounds of fish and shellfish and supporting vast flocks of varied waterfowl. Most of the food chains supporting this rich bounty had their origin in the seagrasses, half a million acres of underwater "grass" growing on the bottom a few feet beneath the surface. The beds of seagrass provided not only food but also habitat for spawning and shelter for the young fish and shellfish, and dissolved oxygen for them to breath.

But in the early 1970s, the seagrasses in all the major rivers and subestuaries leading into the bay started dying. By 1975, the dieback was dramatic. By 1980, they were gone, except in the main stem of the lower bay. Fish, shellfish, and waterfowl, which had depended on the grasses, declined accordingly. Even more devastating, the bottom waters became depleted of dissolved oxygen causing huge numbers of fish, lobsters, oysters, and others to be suffocated. What caused the dieoff of seagrasses and the depletion of dissolved oxygen in the Chesapeake?

A team of scientists from the University of Maryland and the Virginia Institute of Marine Science supported by grants from the Environmental Protection Agency investigated the problem. Toxic chemicals from industry were ruled out because, while problematic in certain locations, they could not be responsible for a dieback throughout the bay. Herbicides used on farmlands were suspected, but tests showed that they did not reach damaging levels except in small ditches and streams receiving drainage directly from farm fields. Then, investigations turned to light, and this proved to be the key. The waters of the Chesapeake had become increasingly murky or cloudy, and the cloudiness was persisting over extended periods of time. The reduced light was cutting back photosynthesis, and the seagrasses were dying as a result. What was causing the murkiness? It was *sediments*, clay particles in suspension, and *phytoplankton*, species of algae that grow as single cells floating and dividing in the water.

With loss of the seagrasses, dissolved oxygen was no longer being supplied by their photosynthesis. But even more, bacterial decomposers, feeding on the abundance of dead seagrass, phytoplankton, and other organic matter, were consuming dissolved oxygen making it unavailable to fish and shellfish.

The Chesapeake Bay has fallen prey to a process called *eutrophication* (pronounced, *yoo-tro-fuh-kay-shun*). This is not a unique situation. In the last 40 years many thousands of ponds, small lakes, and even some large lakes have suffered this fate and the

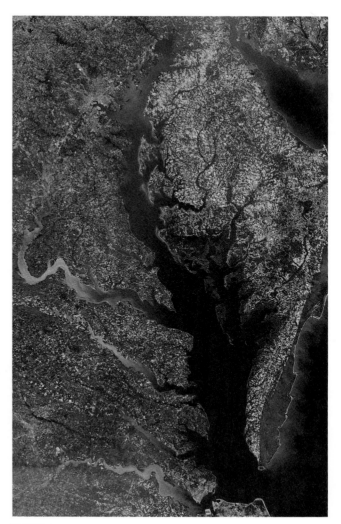

FIGURE 9–1
Photograph of Chesapeake Bay area taken by the Earth Resources Satellite at an altitude of 540 miles. Gray areas are cities, Washington, D.C., just below Baltimore in the upper left. Light tan or reddish rectangles are agricultural fields. Dark green is forest. The dark reddish green on the lower east side of the bay is remaining tidal wetlands. Note the extent of urbanization and agriculture in the watershed. Both are sources of nutrients and sediments. The tan color of two rivers in the lower left are a result of sediments eroding after a rain. The white haze in the lower part of the photograph is clouds. (Satellite imagery provided by EOSOT, Landham Maryland. © Chesapeake Bay Foundation, 1987.)

problem is spreading. Unless checked, numerous estuaries and entire sections of coastal ocean waters may follow the course of the Chesapeake resulting in drastic reductions of fish and shellfish harvests around the world. The objective of this chapter is to understand the problem of eutrophication in more detail, to see how it can be checked and how bodies of water such as the Chesapeake may be restored.

EUTROPHICATION

Two Kinds of Aquatic Plants

To understand eutrophication more thoroughly, we need to consider two distinct life forms of aquatic plants: *benthic plants* and *phytoplankton*.

Benthic plants (from *benthos*, "deep") are aquatic plants that grow attached to or rooted in the bottom. All common aquarium plants and seagrasses are examples (Fig. 9–2a). As a group they are also referred to as **submerged aquatic vegetation** or **SAV** for short. They will thrive in nutrient-poor water because they get their nutrients such as phosphorus and nitrogen from the bottom sediments, but they depend

on sufficient light penetrating through the water for their photosynthesis.

The depth from the surface to the limit of adequate light to support photosynthesis is known as the **euphotic zone.** In very clear water this may be nearly 100 feet (about 30 m). However, as water becomes more **turbid** or cloudy the euphotic zone is reduced; in extreme situations it may be reduced to a matter of a few inches (1 in. = 2.5 cm).

Phytoplankton (from *phyto* = plant, and *plankton* = floating) consists of numerous species of algae that grow as microscopic, single cells or small groups or "threads" of cells that maintain themselves near or even on the surface (Fig. 9–2b). Since they are near or on the surface, turbid water is of little consequence to phytoplankton. Indeed, high population of phy-

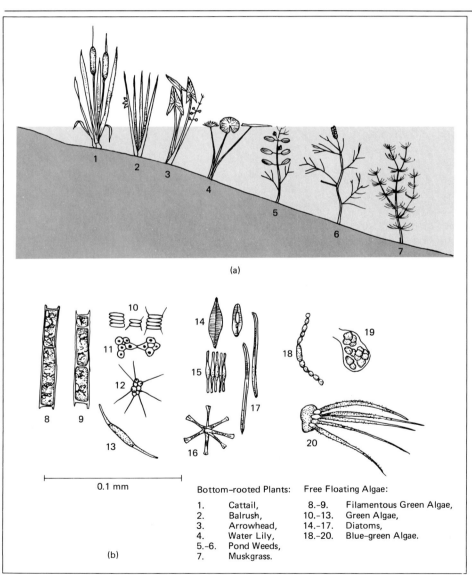

(a)

(b)

0.1 mm

Bottom-rooted Plants:
1. Cattail,
2. Balrush,
3. Arrowhead,
4. Water Lily,
5.–6. Pond Weeds,
7. Muskgrass.

Free Floating Algae:
8.–9. Filamentous Green Algae,
10.–13. Green Algae,
14.–17. Diatoms,
18.–20. Blue-green Algae.

FIGURE 9–2
Two basic categories of aquatic plant life. (a) Benthic or bottom-rooted plants, also referred to as submerged aquatic vegetation. (b) Phytoplankton, various species of plants that exist as single cells or small groups or filaments of cells that float freely in the water. Drawings of phytoplankton are enlarged 500 to 1000-fold. Benthic plants withdraw nutrients from the sediments and hence do well in nutrient-poor water, while phytoplankton depends on nutrients dissolved in the water. (From: *Fundamentals of Ecology*, 3rd ed., by Eugene P. Odum. Copyright © 1971 by W. B. Saunders Co. Copyright 1953 and 1959, W. B. Saunders Co. Reprinted by permission of CBS College Publishing.)

toplankton is a major cause of turbidity. In extreme situations, water may become literally pea-soup green (or tea-colored depending on species involved) and a scum of phytoplankton may float on the surface absorbing essentially all the light. (Fig. 9–3). But, since they are not connected to the bottom, populations of phytoplankton must get their nutrients from the water. Lack of nutrients in the water limits their growth accordingly.

Can you see how a balance between benthic plants and phytoplankton may be shifted with nutrients in the water? In the following, we shall see that this is what humans have done.

Upsetting the Balance by Nutrient Enrichment

The Oligotrophic or Nutrient-Poor Condition

Natural terrestrial ecosystems are very efficient at reabsorbing and recycling nutrients and at preventing runoff and erosion. Thus, in natural conditions rela-

tively small quantities of nutrients and sediments are leached or eroded from the land into waterways. In turn, fed by clear rivers and streams, the natural condition of lakes and estuaries is generally **oligotrophic,** meaning the water is *nutrient-poor.* This limits the growth of phytoplankton, but benthic plants may thrive to a depth of 30 feet (10 m) or so, the usual extent of the euphotic zone.

Oxygen from the atmosphere is very slow to dissolve in and mix through the water. Therefore, in addition to providing food and habitat, benthic plants are also crucial in maintaining a high dissolved oxygen level in deeper water because the oxygen from their photosynthesis is dissolved directly into the water. Thus, a nutrient-poor body of water may maintain a rich, diverse ecosystem of fish and shellfish supported by benthic plants (Fig. 9–5a).

Nutrient Enrichment, Eutrophication

As erosion and leaching occur, a body of water is gradually filled with sediments and enriched with nutrients. The nutrient enrichment promotes the growth of phytoplankton which, in turn, makes the water more turbid and shades out the benthic plants, a problem compounded by sediments. The oxygen produced by photosynthesis of the phytoplankton supersaturates the upper water and escapes to the sur-

FIGURE 9–3
In nutrient-rich water, phytoplankton may grow so thickly that it forms a dense scum over the surface. However, even without forming a scum, phytoplankton may grow so thickly that benthic plants are severely shaded. (Stephen Collins/ Photo Researchers)

face. On a calm, sunny day, you can often observe bubbles of oxygen entrapped by filamentous algae being released at the surface. Thus, the photosynthesis of phytoplankton does not replenish the dissolved oxygen of deeper water.

Still more is involved. Phytoplankton has a high **turnover rate,** which means its continuing rapid growth and reproduction becomes balanced by dieoff, leading to an unusually high accumulation of detritus. In deep water, where the bottom is below the euphotic zone, the "rain" of dead phytoplankton deposits detritus where there was virtually none before. As decomposers, mainly bacteria, feed on the detritus, they also consume oxygen in their respiration, depleting the dissolved oxygen in the water. When dissolved oxygen is no longer available, bacteria shift and continue to thrive by fermentation. Thus bacteria can deplete and hold the dissolved oxygen at absolute zero as long as there is detritus to support their growth. Consequently, dissolved oxygen may be extra high at the surface from photosynthesis of phytoplankton but be depleted at lower levels by the respiration of decomposers (Fig. 9–4).

Nutrients absorbed by the phytoplankton are carried to the bottom in the "phytoplankton rain," but they are released again as the detritus decomposes. With convection currents caused by cool, denser water sinking and warmer, less-dense water rising, nutrients are brought back up to repeat the process.

In summary, **eutrophication** refers to nutrient enrichment that promotes the growth of phytoplankton. This increases the turbidity and causes the dieoff of benthic plants, depletion of dissolved oxygen, and consequent suffocation of fish and shellfish that inhabit deeper waters. The increasing turbidity is compounded by sediments (Fig. 9–5).

Eutrophic lakes have been called "dead," but, biologically, this is a misnomer because the total production of biomass by phytoplankton may greatly outweigh that of the previous benthic plants. In turn, there may be large populations of certain fish that feed on the phytoplankton and avoid the oxygen-depleted deep water. For example, in Chesapeake Bay, populations of bay anchovy and menhaden—plankton feeders—are at all-time highs, but since they are small and oily these species are not good for eating or sport fishing. It is more correct, then, to view eutrophication as a change of ecosystems from a diverse system based on benthic vegetation to a simple one based on phytoplankton. However, since the phytoplankton-based system does not support the aesthetic pleasures of swimming, boating, and sport fishing, it may be considered "dead" in terms of these values. Also, if the lake is a source of drinking water, its value may be greatly impaired because algal cells rapidly clog water purification filters and may cause a foul taste.

FIGURE 9–4
Dissolved oxygen levels typical of oligotrophic and eutrophic conditions are shown. (Redrawn by permission: Robert Leo Smith, *The Ecology of Man: An Ecosystem Approach*, 2nd ed., p. 281. Harper and Row, Pub., Inc., 1976.)

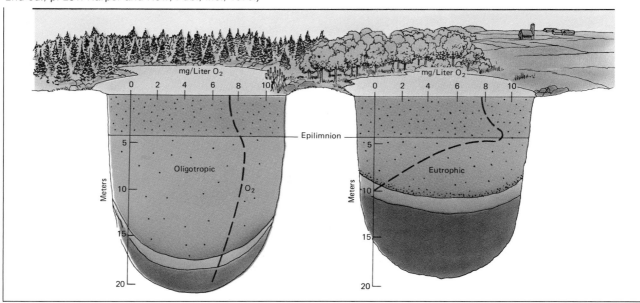

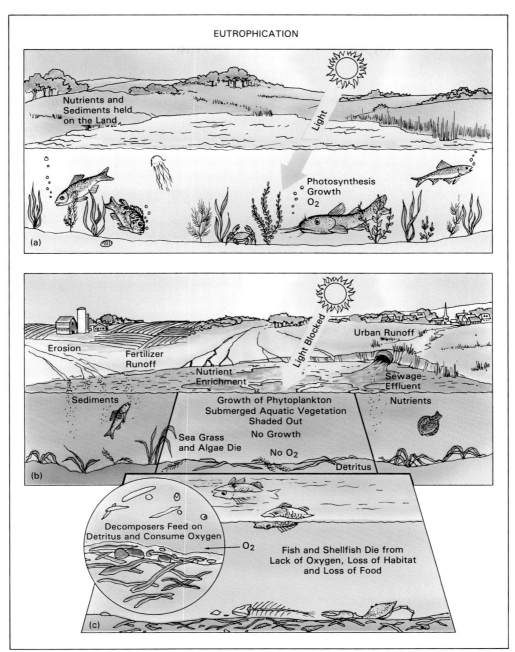

EUTROPHICATION

(a)

Nutrients and Sediments held on the Land

Light

Photosynthesis Growth O₂

(b)

Erosion

Fertilizer Runoff

Nutrient Enrichment

Light Blocked

Urban Runoff

Sewage Effluent

Nutrients

Sediments

Sea Grass and Algae Die

Growth of Phytoplankton Submerged Aquatic Vegetation Shaded Out

No Growth

No O₂

Detritus

(c)

Decomposers Feed on Detritus and Consume Oxygen

O₂

Fish and Shellfish Die from Lack of Oxygen, Loss of Habitat and Loss of Food

FIGURE 9–5

(a) The process of eutrophication. Natural bodies of water are generally maintained in a nutrient-poor condition because nutrients are held on the land by natural ecosystems. Submerged aquatic vegetation getting nutrients from bottom sediments provides habitat, food, and dissolved oxygen for fish and shellfish. (b) Nutrients from sewage effluents and , fertilizer runoff, stimulate the growth of phytoplankton which shade out the submerged aquatic vegetation depriving fish and shellfish of habitat, food, and dissolved oxygen. The shading is compounded by sediments from erosion. (c) Decomposition of dead algae consumes additional oxygen causing fish and shellfish to suffocate.

EUTROPHICATION OF LAKE ERIE

Lake Erie, the shallowest and smallest of the Great Lakes, extends 240 miles from Toledo, Ohio, to Buffalo, New York. The Detroit River, which services the industrial city of Detroit, Michigan, is the conduit for its major source of water, Lake Huron. Other rivers serving such industrial cities as Toledo, Cleveland, Erie and Buffalo and draining agricultural land also feed into Lake Erie.

As with any ecosystem, Lake Erie was and is subject to change due to the natural aging process known as succession. In aquatic ecosystems, succession is often also called eutrophication. Freshwater ecosystems are oligotrophic, that is, nutrient-deficient, unproductive and detritus-free. They undergo succession into eutrophic ecosystems which are nutrient-rich, highly productive and detritus-rich.

Earliest records indicate that Lake Erie was an oligotrophic lake with low primary productivity and oxygen abundant in deep waters. Wetlands, marshes and bays were of considerable extent and provided habitats for ample fish spawning. But all that was to change, not by natural eutrophication, but by cultural eutrophication. Cultural eutrophication is succession accelerated by human activity, and it began its great acceleration in Lake Erie about 1910.

It was at this time that a large population growth began in the drainage area affecting the Lake. With people came human and industrial wastes and corresponding accelerating increases in nutrients, most notably phosphates. This nutrient enrichment resulted in burgeoning populations of phytoplankton, especially nuisance algae like *Cladophora* which accumulated in mats up to 50 feet in width and 2 feet in depth. These algal blooms increased the amount of detritus which in turn led to significant depletion of oxygen in the deeper waters. Substantial increases in Coliform bacteria resulted in the closure of most of the excellent swimming beaches. Tubificid worms, oligochaetes tolerant of low oxygen conditions, replaced burrowing mayflies as the dominant bottom fauna. Lake herring, sauger, blue pike, whitefish and walleye—the base of commercial fishing in the Lake—declined dramatically and were replaced by less table-favored varieties such as yellow perch, carp, channel catfish and suckers.

The natural eutrophication that might have taken thousands of years was culturally accelerated in the space of less than fifty years. Can it be reversed? Not really, but it can be and has been slowed. Beginning about 1970, government restrictions on nutrient loading from municipal sewage and detergents (the primary source of phosphates) were imposed and improved shoreline and fisheries management and other steps were undertaken. Very few beaches are now closed to recreation, and the fishing industry is recovering. However, given the rate of water flow, it

Natural versus Cultural Eutrophication

Given geological time periods, bodies of water are gradually enriched with nutrients and filled with sediments leaching and eroding from the land. Thus, eutrophication of bodies of water is part of the natural aging process as was described under "Succession" in Chapter 3. Humans have vastly accelerated this process. We have brought changes in a few decades that normally would occur only over several thousands of years. Thus, we can speak of *cultural* eutrophication, caused and accelerated by humans, as opposed to *natural* eutrophication. Lake Erie provides another prime example of cultural eutrophication (see Case Study 9).

It is an almost automatic reflex reaction to blame pollution problems on industries dumping toxic compounds. Significantly, eutrophication is due to what are generally considered benign substances, soil particles, and fertilizer nutrients. It emphasizes the point made in Chapter 3: *Changing any natural factor can grossly upset the balance of an ecosystem.*

Combatting Eutrophication

There are two general approaches to combatting the problem of eutrophication. One is to *attack the symptoms,* the growth of algae and/or the lack of dissolved oxygen; the other is to *get at the root cause,* excessive inputs of nutrients and sediments. Attacking the symptoms continues to be pursued in some cases, but it is of dubious merit and becomes cost-prohibitive.

Recreation on Lake Erie, largely precluded when the Lake became eutrophic, has returned since programs to reduce nutrient inputs have led to recovery. (Mark Reinstein/ TSW-Click/Chicago, Ltd.)

may take twenty years to remove 90 percent of the wastes, and hundreds of years from the larger Great Lakes, Superior and Michigan.

Lake Erie is a classic case study in cultural eutrophication, but it is not unique. The story has been repeated many times in many places around the world. It is a sad commentary on unthinking human activity. But, intelligent use of our natural resources for ourselves can mean a precious heritage and legacy for future generations.

EDWARD J. KORMONDY
University of Hawaii-Hilo

We should be aware of its shortcomings so that we are not tempted to repeat mistakes.

Attacking the Symptoms

Chemical Treatments. Herbicides have been used very successfully in agriculture to eradicate unwanted plants. Therefore, an obvious extrapolation was to believe that unwanted algae also could be eliminated by chemical treatment (Fig. 9–6), and thousands of tons of chemicals were spread on ponds and lakes in the 1960s and 1970s. The general finding was that chemicals do not work. The planktonic algae, especially blue-green species which are the most obnoxious, prove to be among the most resistant of all organisms. If enough chemical is added to kill them, it also kills everything else in the water. When herbicide concentrations dissipate, the algae are among the first species to reappear because the root cause, high nutrient level, has not been changed. To date, no chemical has been found that will selectively kill algae and not harm other aquatic plants and animals.

Nevertheless, copper sulfate is currently being used to control algae growth in some water-supply reservoirs where algae otherwise imparts a bad taste to the water and causes excessive clogging of purification filters. But, above trace amounts, copper is known to be highly toxic to all organisms. Therefore, you should remain skeptical of the long-term effects of this practice. Is it sustainable?

Aeration. Installing a mechanical aeration system in a eutrophic lake or pond will keep the dissolved oxygen high and at least prevent the fishkills due to suf-

FIGURE 9–6
Spraying water with herbicides to inhibit undesired plants. What will be the side effects? (FAO photo by S. Baron.)

focation (Fig. 9–7). In addition, aeration may also reduce the level of phosphate by causing it to change to insoluble forms, and it does not cause any undesirable side effects. However, the energy costs to operate an aeration system on a large body of water are prohibitive.

Harvesting Algae. Residents may remove algae by hand from community lakes; harvested algae makes a good organic fertilizer. A raft made by lashing a few planks across two canoes, as shown in Figure 9–8, is remarkably effective for such an effort. A lake's appearance is greatly improved by removing unsightly mats of algae and other litter that collect in various places, but the bulk of phytoplankton, microscopic cells in suspension, does not lend itself to harvesting. Filtering does not work either because the algae clogs filters quickly. Finally, on a large body of water such as Lake Erie or Chesapeake Bay, the logistics, much less the costs, of such measures quickly puts them out of the realm of possibility.

FIGURE 9–7
Lakes and reservoirs may be aerated to avoid the consequences of eutrophication. Aeration also aids in stabilizing phosphates in sediments. (USDA—Soil Conservation Service photo.)

FIGURE 9–8
To fight the use of chemical herbicides to control algae, these residents in Columbia, Maryland, resorted to harvesting algae by hand. (Photo by Bruce Fink.)

Controlling Inputs

It should be clear that real control of eutrophication requires decreasing nutrient and sediment inputs. Sources of sediments and nutrients must be identified and evaluated, and then suitable cost-effective methods of control must be implemented.

SOURCES OF SEDIMENTS AND NUTRIENTS

We have seen that eutrophication is caused by inputs of nutrients and it is compounded by sediments. Experience shows that if nutrient and sediment inputs are curtailed, existing quantities will gradually stabilize or be flushed from the system and bodies of water may eventually recover to the oligotrophic condition.

Sources of Sediments

The source of all sediments is soil erosion. Wherever erosion occurs, it adds to the load of sediments being carried by streams and rivers to lakes, bays, estuaries and, finally, the ocean. The major points of erosion are the following.

Croplands. Even with relatively good erosion control, each acre (0.4 hectare) loses 3 to 5 tons of soil per year and thus contributes this amount of sedi-

ment to waterways. Without good soil conservation each acre may contribute upwards of 20 tons per year.

Overgrazed Rangelands. Pastures and rangelands that have been grazed to the point of creating bare spots, and cattle pathways, cause erosion gullies, contributing similar amounts of sediment.

Deforested Areas. Steep deforested slopes may lose and contribute hundreds of tons of sediment per acre to waterways (Fig. 7–27).

Construction Sites. Approximately 1.6 million acres (0.6 million hectares) of land in the United States are affected each year by development of housing tracts, highways, and other construction. The subsoil exposed by construction activity has little capacity for infiltration, and slopes created by such activities are frequently steep and unstable. The runoff and resulting erosion is often severe. Water-gullied embankments along highways under construction are probably familiar to everyone (Fig. 9–9). Losses of 1000 tons per acre are not uncommon, and may be as high as 10,000 tons per acre. In other words, the soil lost from a construction site during one year may exceed what would be lost over 20,000 to 40,000 years under natural conditions.

Surface Mining Sites. In the United States some 3 million acres have been disturbed by surface mining. Worse, mining activities increase each year as we turn more to coal to meet our energy needs. Erosion from

FIGURE 9-9
The gullies in the embankments of this highway under construction attest to severe erosion. Such construction activities may be the most significant source of sediments entering waterways. (Photo by author.)

mining sites is similar to that from construction sites, but mines are generally larger and remain open for longer periods of time (Fig. 9–10). A federal law requiring reclamation of strip-mined areas was passed in 1977, but it has been unevenly enforced and attempts to weaken it persist. In the meantime, some 2 million acres of old mined soils still remain exposed to ongoing erosion.

Gully and Streambank Erosion. Gully erosion from poorly placed storm drain outlets has increased runoff from development; furthermore, accelerated streambank erosion caused by flooding is another source of sediment (discussed in Chapter 8).

Miscellaneous Sources. Notice bare patches around schools, homes, and shopping centers, and along highways where the earth has been denuded or where reseeding after construction did not take. Such areas often do not revegetate by themselves but remain open and erode indefinitely (Fig. 9–11). These patches may not be very large individually, but taken together they contribute significantly to the sediment problem.

Which of these sources of sediment is most prevalent will depend upon the activities in the given region.

Impacts of Sediments on Streams and Rivers

Sediment starts as soil eroding from the land, but the particles of sand, silt, clay, and humus are quickly separated by the agitation of flowing water, and different particles are carried at different rates. Coarse

FIGURE 9-10
Erosion from mining sites. Strip mining leaves massive amounts of earth exposed to the forces of erosion. As a result, sedimentation of waterways draining such areas is often severe. Further pollution of waterways results from the leaching of various chemicals from the disturbed rock formations. (Francis Current/Photo Researchers.)

FIGURE 9-11
Following construction there are frequently numerous patches, such as this one on a highway embankment, that fail to be adequately stabilized. The continuing erosion from such patches adds significantly to the sediment burden of waterways. (Photo by author.)

material, if not left on the land, settles out and tends to clog the channel where stream velocity slows. Silt is carried along more readily but tends to deposit at the mouth where velocity slows more as the river enters the larger body of water. The very fine clay and humus particles, however, may remain in suspension for days even in calm water, and hence are readily carried downstream and throughout large lakes or estuaries, leading to the problem of cutting off light and compounding eutrophication. However, the negative impacts of sediments begin immediately in streams and rivers.

Damage to Aquatic Ecosystems of Streams and Rivers

When erosion is slight, streams and rivers draining in the area run clear. They support producers, algae and other aquatic plants, that attach to rocks or root in the bottom. These producers, and miscellaneous detritus from fallen leaves and so on, support a complex food web including bacteria, protozoans, worms, insect larvae, snails, fish, and crayfish. These organisms keep themselves from being carried away by attaching to rocks or, as in the case of fish, seeking shelter behind or under rocks. Even fish that maintain their position by active swimming occasionally need such shelter to rest.

The sediment load has compound effects upon this ecosystem. Clay and organic particles in suspension not only make the water look muddy; they reduce light penetration and the rate of photosynthesis. As the sediment settles, it coats everything and continues to block photosynthesis (Fig. 9–12). It also kills the animal organisms by clogging their gills and feeding structures. Eggs of fish and other aquatic organisms are particularly vulnerable to smothering by sediment.

Equally destructive is the **bedload** of sand and silt, which is not readily carried in suspension but is gradually washed along the bottom. As particles roll and tumble along, they scour organisms from the rocks. They also bury and smother the bottom life and fill in the hiding and resting places for fish and crayfish. Aquatic plants and other organisms are prevented from reestablishing themselves because the bottom is a continually shifting bed of sand (Fig. 9–13). Further streambank erosion, aggravated by sediments filling the channel, cause the stream or river to become wider but shallower. Many additional pollutants enter from runoff as was discussed in Chapter 8. During times of low water, the river may be little more than a wide expanse of sediment deposits with a minor rivulet meandering through (Fig. 9–13). Little of the natural ecosystem can survive in a stream subjected to high sediment loads.

Sediments do not receive the attention the news media give to hazardous wastes and certain other pollution problems; land disturbance is so widespread, however, that few streams and rivers escape the harsh impact of excessive sediment loads. Consequently, sediment is considered to be the foremost pollution problem of streams and rivers. Further, as erosion throughout the world increases due to overcultivation, overgrazing, and deforestation (Chapter 7), the destruction of aquatic ecosystems by sediments is worsening also.

Filling of Channels and Reservoirs

Sediment also causes serious economic problems. In Chapter 8, we discussed how sediments fill and clog stream channels and aggravate problems of streambank erosion and flooding. Additionally, water supply reservoirs are filled, shipping channels are made impassable, and irrigation canals are clogged. The task is unending because dredged areas soon fill in with new sediments, and present dredging efforts do not address all the problem areas. Many millions of cubic meters of water storage capacity in reservoirs are lost each year because of sedimentation (Figs. 9–14a and b). This loss will intensify future water shortages because there are few new locations that are suitable for reservoirs. Therefore, the need to dredge reservoirs may be unavoidable.

Along with the costs of dredging, there is the problem of disposing of the dredged material. It can't be trucked back to the land because sediment is not the same as the soil that originally eroded from the land. Sand, silt, and clay settle out at different rates and settle in different locations. The magnitude of the problem is illustrated by Baltimore Harbor in Maryland. Baltimore Harbor needed to be dredged, but the

FIGURE 9–12
The "S" on the rock is where I have brushed away the sediment with my finger. This shows the amount of sediment deposited from an upstream construction site. Such sediment can severely damage aquatic organisms attached to rocks.

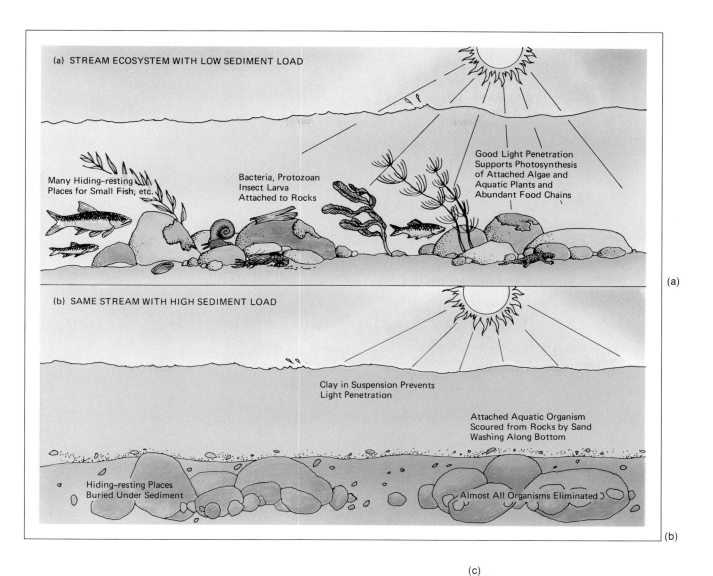

(a)

(b)

(a) STREAM ECOSYSTEM WITH LOW SEDIMENT LOAD

Many Hiding-resting Places for Small Fish, etc.

Bacteria, Protozoan Insect Larva Attached to Rocks

Good Light Penetration Supports Photosynthesis of Attached Algae and Aquatic Plants and Abundant Food Chains

(b) SAME STREAM WITH HIGH SEDIMENT LOAD

Clay in Suspension Prevents Light Penetration

Attached Aquatic Organism Scoured from Rocks by Sand Washing Along Bottom

Hiding-resting Places Buried Under Sediment

Almost All Organisms Eliminated

(c)

FIGURE 9–13
Negative impacts of sediments on the aquatic ecosystems of streams and rivers. (a) The ecosystem of a stream that is not subjected to a large sediment load. (b) The changes that occur when there are large sediment inputs. (c) A river channel choked with sediment from upstream erosion. The sand bars seen here, which shift and move with high water, constitute the bedload. What will this do to aquatic life? Platte River at Lexington, Nebraska. (Charles R. Belluky/Photo Researchers)

(a)

(b)

FIGURE 9–14
Lake Como, Minnesota, was entirely filled with sediments in the ten-year period from 1926 (a) to 1936 (b). The sediments came from the erosion resulting from timber clearing and plowing up and down slopes. (USDA—Soil Conservation Service photos.)

project was held up for 15 years because of controversy over what to do with the dredged material. In this case, the material is a "mucky ooze" of fine clay with 90 percent water and a generous mixture of sewage and industrial chemical wastes that had accumulated in the harbor over the years. The solution finally accepted was to create an offshore island of the dredged material.

The Soil Conservation Service estimates that the impacts of sediments cost the United States over $6 billion each year, but much of the damage goes uncorrected. However, costs are more than just financial. The process of dredging inevitably stirs up and redistributes the fine sediment, thus initiating another round of sediment pollution.

Sources of Nutrients

Nutrients—ions such as nitrate, phosphate, and potassium—cling to particles of clay and humus; recall the nutrient-holding capacity of the soil. Therefore,

nutrients invariably come along with sediments, and any source of sediments is also a source of nutrients. In addition, there are large quantities of nutrients from:

☐ Leaching of fertilizer applied to agricultural fields.

☐ Leaching of fertilizer applied to lawns and gardens.

☐ Leaching and runoff of animal wastes from feedlots, dairy barns, horse stables, and other locations where animals are kept at high density. Recall (Chapter 2) that the nutrients absorbed by plants are excreted in the urine as animals metabolize food for energy. Therefore, any types of animal waste, including human, are a rich source of nutrients.

☐ Runoff of pet wastes from urban/suburban areas. The density of pets in urban neighborhoods is at least 100 times higher than populations of wild animals of similar size in a natural ecosystem. Pet excrements, along with fertilizer leaching from lawns and gardens, make runoff from urban/suburban areas a major source of nutrients.

☐ Human excrements. Finally, the density of humans living in an urban area is at least 1000 times higher than populations of wild animals of similar size in a natural ecosystem. Consequently, the load of nutrients from human excrements is tremendous. Most human sewage in developed countries is collected and treated. However, sewage treatment, as it presently exists in most towns and cities, does not remove nutrients. They are discharged into waterways, with the waste water making sewage treatment plants a major source of nutrients (see Chapter 10).

☐ Detergents containing phosphate. Phosphate is frequently the limiting nutrient in aquatic ecosystems; therefore its presence is especially critical. It has been estimated that only about 30 percent of the phosphate in waste water comes from human excrements. Sixty percent comes from detergents. Traditional soaps combine with calcium in water to form an insoluble precipitate or "curd"—the ring around the bathtub is an example. When modern washing machines, which use the spin-dry principle, were introduced, the curd remained on the clothing. Therefore, soaps were replaced by detergents. Detergent molecules consist of a long hydrocarbon chain, which is fat-soluble, at one end, and a highly water-soluble group, phosphate, at the other (Fig. 11–10). This molecule lifts fatty or oily substances and brings them into water solution and is not precipitated by hard water. Phosphate is the favored group for the water soluble end since it is inexpensive and nontoxic, but substitutes are now available.

☐ Acid precipitation. In 1988, the Environmental Defense Fund reported research showing that as much as one-fourth of the nitrogen entering Chesapeake Bay is from acid precipitation. Nitric acid (H^+NO^{3-}) is one of its major constituents and, apart from the acidity that is caused by the H^+, the NO^{3-}, or nitrate portion, is a nitrogen nutrient.

Which of these sources is most significant will obviously differ depending on the number of farms, farming activities, and population centers in the region. For the Chesapeake Bay watershed, for example, it is estimated that contribution of nutrients from farm runoff, urban runoff, and sewage treatment plants is split more or less equally, about one-third from each.

If the contribution of nitrogen from acid precipitation is as significant as the Environmental Defense Fund believes, it becomes a four-way split. Acid precipitation is discussed in detail in Chapter 13.

Loss of Wetlands and Bulkheading of Shorelines

Not only do human activities lead to much higher inputs of sediments and nutrients, we have also largely destroyed the mechanism that provides for their natural control—*wetlands*. **Wetlands** are land areas that are naturally covered by shallow water at certain times and more or less drained at others. Depending on the depth and permanence of water they are divided into marshes, swamps, and bogs, and they may be either fresh or salt water.

At times of flooding, wetlands along and especially at the mouths of streams and rivers receive the overflow water, which is rich in nutrients and sediments. Stilled by slow movement through the wetlands, sediments with bound nutrients settle out and purified water percolates into the groundwater. Thus, wetlands play a vital role in filtering nutrients and sediments out of water before it enters lakes, bays, and estuaries, and they are also important in groundwater recharge. Additionally, wetlands, enriched with the nutrients, are tremendously productive ecosystems supporting flocks of waterfowl and other wildlife.

Tidal wetlands are broad expanses of marshy grassland in coastal regions, which are shallowly covered at high tide but drained at low tide. These are tremendously important in removing sediments and nutrients from estuaries. Sediments and nutrients are carried by rising tides onto the wetlands. With falling tide, water drains through the meshwork sieve of grass leaves, roots, and soil, leaving most of the sediments and nutrients. Again the wetlands utilize the nutrients to become the most productive of all ecosystems.

Moderation of wave action is another function of shoreline wetlands which, in turn, serves to stabilize sediments and nutrients. Storm or boat waves are gradually dampened and stilled as they pass through the stems and leaves of grass on wetlands. Again, outgoing water is filtered and sediments are removed and stabilized (Fig. 9–15).

But a common human attitude toward wetlands is that they are too wet to plow or build on, and too dry to go boating on. Consequently, they have been viewed as "waste land" good only for changing to be used in "better" ways. Over 50 percent of the wetlands in the United States have been destroyed by draining or filling. Along Chesapeake Bay for example, over 50 percent of the shoreline has been

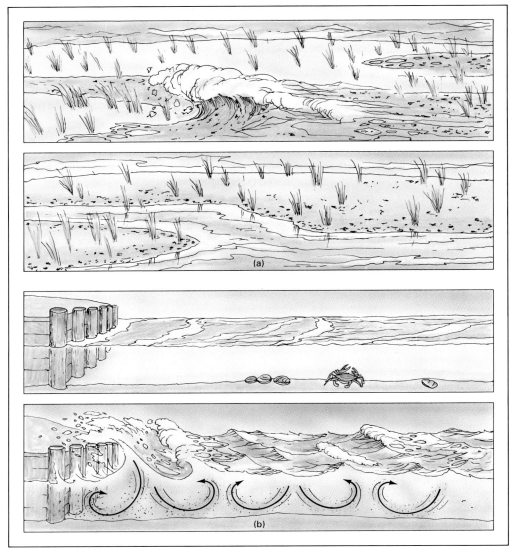

FIGURE 9–15
(a) Wave and tidal action bring sediment-laden water into wetlands where sediments are filtered out as water recedes. (b) When shoreline is bulkheaded, wave action hitting bulkhead stirs up the bottom sediments so that water remains muddy. Thus, bulkheading shorelines greatly aggravates the problem of sedimentation.

dredged, filled, and bulkheaded—a process that entails dredging to a depth of 3 to 4 feet, using the dredged material to build up the other portion, and stabilizing the edge with a wall or bulkhead.

The advantage in creating more usable bayfront property is obvious. But the cleansing capacity of the wetlands is lost, and waves smacking against the bulkheads creates turbulence that stirs up sediments and keeps them in suspension, cutting off light, photosynthesis, and growth of the submerged aquatic vegetation indefinitely.

CONTROLLING NUTRIENTS AND SEDIMENTS AND WHAT YOU CAN DO

Best Management Practices on Farms, Lawns, and Gardens

The runoff and leaching from rural and urban areas is referred to as **nonpoint** pollution since it comes from the whole area in general as opposed to a specific point such as the effluent pipe of a sewage treatment plant. The latter are called **point sources.** The cornerstone of federal water pollution control legislation is the *Clean Water Act of 1972.* But until recently, efforts under this act have focused on point sources both because they were easy to identify and measure, and the extremely large contributions from nonpoint sources were not fully appreciated. When the Clean Water Act was reauthorized in 1987, however, a new section (section 319) was added which requires states to develop *management programs* to address nonpoint sources of pollution.

Controlling nonpoint pollution basically demands that individual farmers and homeowners recognize the contribution of soil erosion and/or nutrient runoff on their property to the overall problem of eutrophication, and adopt "best management practices." **Best management practice** is a catch-all term that implies examining all methods of soil conservation discussed in Chapter 7 and adopting the most effective for each situation. Thus, keeping the ground covered with vegetation or mulch to prevent erosion, keeping a high humus content in the soil, and using organic fertilizers to help retain nutrients would be at the top of the list.

Where fields are directly adjacent to streams, rivers, or lakes, owners should plant buffer strips of trees between their fields and the waterways in order to catch and reabsorb the nutrient-rich leachate (Fig. 9–16). Where animal wastes from feedlots, dairy barns, or horse stables drain directly into waterways,

ponds should be constructed to intercept the nutrient-rich runoff. Such water can then be recycled as irrigation water, returning the nutrients to the soil (Fig. 9–17). Rather than letting animals wade into streams to drink and also excrete wastes, farmers should provide water troughs so that animals will drink and excrete their wastes on the land. The need for replanting deforested or overgrazed areas is self-evident.

Best management practices around homes are equally important since urban runoff is a major source of nutrients. Revegetate or mulch all spots of bare ground. Be sure that lawn fertilizer is not overused, and substitute organic fertilizer for chemical fertilizer. Pet wastes should be composted, or pets should be taken to urinate and defecate on a compost pile that is eventually applied to a garden. Planting trees wherever possible will enhance infiltration of water and reabsorption of nutrients.

State management programs are likely to include these measures, perhaps with various incentives to get people to comply. However, it will finally come down to largely voluntary compliance, and this will depend on people's understanding and appreciation of the problem. Spreading this understanding is perhaps the most important role you can play.

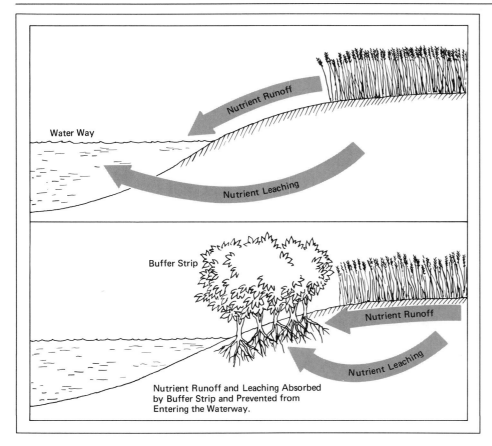

FIGURE 9–16
Mitigation of eutrophication by buffer strips of trees. Maintaining a buffer strip of trees along waterways helps to absorb and hold nutrients that would otherwise leach from agricultural fields into the waterway.

Water Way

Nutrient Runoff

Nutrient Leaching

Buffer Strip

Nutrient Runoff

Nutrient Leaching

Nutrient Runoff and Leaching Absorbed by Buffer Strip and Prevented from Entering the Waterway.

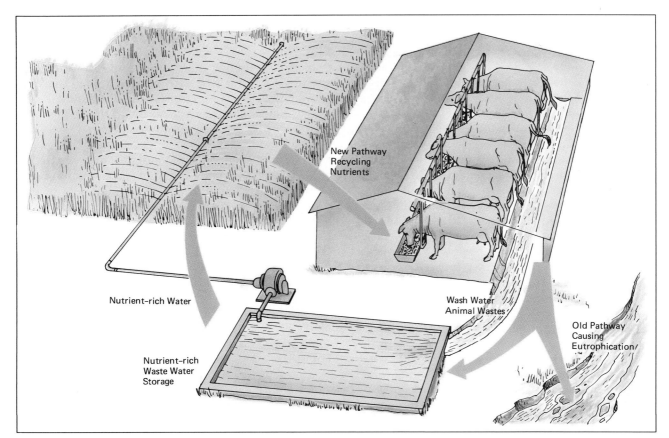

New Pathway
Recycling
Nutrients

Wash Water
Animal Wastes

Old Pathway
Causing
Eutrophication

Nutrient-rich Water

Nutrient-rich
Waste Water
Storage

FIGURE 9–17
A collection pond for dairy barn washings. If washings from
animal facilities such as dairy barns are flushed into natural
waterways, they may contribute significantly to eutrophication.
This may be avoided by collecting the flushings in ponds from
which both the water and nutrients may be recycled. Many
farmers use such flushings for irrigation.

Sediment Control on Construction and Mining Sites

Many techniques may be used to reduce the loss of
soil from construction sites. A practice that aggravates
the problem is that the entire site is often left bare
and open to erosion for the entire period of construction. Alternatively much of the site can be brought to
final grade and restabilized with grass immediately.
This technique is particularly appropriate in highway
construction (Fig. 9–18).

FIGURE 9–18
Stabilizing earth with vegetation. In this photograph note that
little if any erosion occurred where the bank was protected by
grass, but severe erosion occurred where it was left bare.
Leaving the sowing of grass until the very end of the
construction process allows the maximum amount of erosion.
Alternatively, erosion can be largely controlled by completing
the final grading and sowing grass as soon as possible.
(USDA—Soil Conservation Service photo.)

To control sediments that will inevitably erode, a **sediment trap,** which is essentially a pond into which runoff from the site is channeled, can be constructed at the lower edge of the site. As water with sediment enters, its velocity is reduced, sediment settles, and sediment-free water flows out over a rock dam or through a standpipe (Fig. 9–19). After construction is complete the sediment trap may be converted into a stormwater retention reservoir. On small sites, the lower perimeter of the site may be diked with bales of straw, which filter the runoff and remove sediment. Or, a plastic fence (Fig. 9–20) may be used to catch water and direct it to a sediment trap.

These techniques trap and hold at least the heavier sand and silt sediments—the clays tend to wash on through. The trapped sediment may then be redistributed over the site.

To many developers, sediment control is just an added expense they want to reduce as much as possible. Erosion control on construction sites must therefore be promoted through legislative action and legal enforcement. Under Maryland State Law, sediment control regulations are added to other building codes and construction requirements so that developers must submit sediment-control plans along with other plans for site development. These plans are re-

(a)

FIGURE 9–19
Sediment control on construction sites. (a) A typical sediment trap. Runoff carrying sediment flows into a "pond" constructed at the lower edge of the site. Sediment settles (foreground) while water overflows the rock dam (where the person is standing). Such sediment traps are temporary structures removed after construction is completed. (b and c) Construction of a larger pond with a standpipe overflow will function both as a sediment trap during construction and as a permanent stormwater retention reservoir after construction. (Photos by the author.)

(b)

(c)

FIGURE 9-20
Plastic fencing used to control sediment runoff from small areas. (Photo by the author)

viewed by the government and modified if necessary before a building permit is issued. Then, the site is inspected periodically to be sure sediment traps or other devices are installed and maintained properly. Storm water control legislation works similarly and is important also to reduce gully and stream bank erosion. Also, nutrient runoff is reduced insofar as storm water is checked and allowed to filter back into the ground.

Inspection and enforcement are frequently the weak link. It is not uncommon to find sediment traps that are nonfunctional for lack of proper installation or maintenance. This is an issue where local citizen action can be effective. Save Our Streams and Trout Unlimited are two citizen organizations whose members take it upon themselves to act as "inspectors" overseeing construction projects in their areas. When they find uncontrolled erosion threatening waterways, they report it to the authorities. If the authorities fail to take action, they report the situation to the local media. Media publicity frequently brings action when all else fails.

Does your region have similar sediment and stormwater control legislation? If not, you can join a conservation organization that is lobbying for such legislation and lend your support. If so, you can make yourself one of the "inspectors" noted above.

Regarding the numerous small patches of eroding earth noted before, the best, and in many cases only, solution is for a few people to get together and do some raking, mulching, and reseeding, because government bureaucracies tend to be too large and

cumbersome to deal with such small areas effectively. The Soil Conservation Service, which maintains an office in every county of the United States, does not have the work force to do the work itself but is generally enthusiastic about providing advice and technical assistance and sometimes seed or seedlings.

Preservation of Wetlands

There are some state and federal regulations that restrict development of wetlands. Yet, they remain inadequate and pressures from developers are such that wetlands continue to be sacrificed. Many other alternatives exist for development. What is promoted as a "need" to develop wetlands is often simply an attempt to support an outmoded policy of profligate use of resources and energy, for the short-term enhancement of a developer's profit at the long-term expense of environmental quality. Is this sustainable? Wherever wetlands are being developed, there are now environmental organizations working for their preservation. Some people are also trying to rehabilitate wetlands by replanting vegetation where it has been destroyed. Which side will you support?

Banning the Use of Phosphate Detergents

When the connection between eutrophication and phosphate detergents was recognized in the 1960s, the detergent industry developed a number of substitutes. Nevertheless, for economic reasons, they have continued to market phosphate detergents preferentially. It has proven necessary to legislatively ban the sale of phosphate detergents in order for the changeover to occur. Such "phosphate bans" are now in effect in the Great Lakes region, on Long Island, Maryland, Virginia and some other areas. The ban in the Great Lakes area was implemented in the mid-1970s when Lake Erie was becoming eutrophic and, since the ban, the lake has improved considerably. Maryland enacted its phosphate ban in 1985 after eutrophication was shown to be a major factor in the decline of Chesapeake Bay. This has reduced the phosphate in discharges from sewage treatment plants by 35–40 percent. A similar ban is being promoted in Pennsylvania, which contributes about two-thirds of the flow to the Chesapeake Bay through the Susquehanna River, and in many other areas. Unfortunately, such legislation is vigorously opposed and stalled by the detergent industry despite the availability of no-phosphate substitutes. They claim

that substitutes are not as effective and will not be accepted by the public. Yet, I have not met anyone who noticed any change in packaging, pricing, or cleaning efficiency when the ban went into effect in Maryland.

You can join the organization that is promoting a phosphate detergent ban in your region, and lend your support to the lobbying effort. In the meantime, read the fine print on detergent packages and select a brand that is phosphate-free.

Advanced Sewage Treatment

There has been much upgrading of sewage treatment plants over the last two decades. Still, most plants do not remove nutrients from the waste water. The need for advance sewage treatment, which includes nutrient removal, speaks for itself. The process of sewage treatment and how this may be achieved is the subject of Chapter 10.

10
Water Pollution Due to Sewage

CONCEPT FRAMEWORK

■ Outline

I. HAZARDS OF UNTREATED SEWAGE   250

  A. Disease Hazard   250

  B. Depletion of Dissolved Oxygen   250

  C. Eutrophication   250

II. SEWAGE HANDLING AND TREATMENT   251
  A. Background   251

  B. Conventional Sewage Treatment   252

    1. Raw Sewage   252

    2. Treatment Steps   253
      a. Preliminary treatment   253
      b. Primary treatment   253
      c. Secondary treatment   254
      d. Advanced treatment   255
      e. Disinfection   259

    3. Sludge Treatment   259
      a. Anaerobic digestion   259
      b. Composting   260

  C. Alternative Systems   262
    1. Using Nutrient-Rich Water for Irrigation   262
    2. Individual Septic Systems   266

CASE STUDY: THE OVERLAND FLOW WASTEWATER
TREATMENT SYSTEM   265

III. TAKING STOCK OF WHERE WE ARE
AND WHAT YOU CAN DO   267
  A. Progress and Lack of Progress   267

  B. Impediments to Progress   268
    1. Contamination with Industrial Wastes   268
    2. Public Apathy   268

  C. What You Can Do   269

  D. Monitoring for Sewage Contamination   269

■ Study Questions

1. List hazards of discharging raw sewage into waterways.

2. Describe why raw sewage presents a disease hazard.

3. Explain why raw sewage may deplete dissolved oxygen.

4. Explain why raw sewage may cause eutrophication.

5. Describe: how sewage was handled prior to 1800; problems that arose as a result; significant discoveries of the mid-1800s; remedies. Why are sewage and stormwater systems interconnected in old cities?

6. List and describe the steps of conventional sewage treatment.

7. Describe the contents of raw sewage.

8. Describe the facilities, how they work, and what is removed in each step.

9. What are the problems in using chlorine for disinfection? What are the alternatives?

10. What is *sludge?* Where does it come from? How may it be treated? Describe two methods and what useful products may be produced from each.

11. Describe alternative systems for handling sewage waste.

12. Describe what progress has been made in upgrading sewage treatment and in what areas is improvement still lacking.

13. Describe impediments to using sewage waste as a resource.

14. Describe what you can do to improve the handling of sewage in your community.

15. How may water be tested for the presence of raw sewage contamination?

Natural ecosystems are sustainable in that they recycle nutrients from plant detritus and animal excrements. In contrast we have created a one-way flow of nutrients from the land through crops to us, and then into lakes and estuaries as we discharge our treated wastes into waterways. We have seen (Chapter 9) that this is not sustainable because it is causing unacceptable eutrophication of lakes and estuaries. The obvious alternative is to take heed of the ecological principle and recycle our wastes. However, our one-way flow is not without reason. In recycling human wastes it is all too possible to "recycle" and spread numerous disease-causing bacteria, viruses, and other parasites as well. The objective of this chapter is to understand current methods of sewage treatment and how they may be amended to incorporate recycling of nutrients while not compromising public health.

HAZARDS OF UNTREATED SEWAGE

Disease Hazard

Untreated sewage presents a major public health hazard because humans and other animals may be infected with **pathogens** (disease-causing bacteria, viruses, or other *parasitic* organisms). Such infected individuals or animals may discharge large numbers of these pathogenic organisms or their eggs in their excrements. People may carry such low populations that they do not show disease symptoms, but they nevertheless act as carriers. If infected sewage wastes contaminate drinking water, food supplies, or water used for swimming, the parasites may gain access to and infect other individuals. In some cases, infection may be through food chains. Oysters, for example, may ingest and harbor the parasite, and then humans become infected by way of eating the oysters. Hence oyster beds contaminated with raw sewage are closed. In addition, it is recommended that certain foods always be cooked.

In general, pathogenic organisms survive only a few days outside of a host, and the number of organisms entering the body is an important factor in determining subsequent infection. Therefore, when populations are sparse, relatively little transfer of pathogenic organisms occurs because levels of contamination remain low and more time elapses between the elimination by one host and contact by the next. As populations become denser, however, the reverse is true. Humans tend to live and work in high-density urban situations, making themselves extremely vulnerable to the spread of pathogens.

Before the connection between disease and sewage-borne pathogens was discovered in the mid-1800s, disastrous epidemics were common in cities. For example, epidemics of typhoid fever that killed thousands of people were common in the United States until the turn of the century. Today, public health measures that prevent the disease cycle have been adopted throughout most countries. These measures involve (1) disinfection of public water supplies with chlorine or other agents, (2) personal hygiene and sanitation, especially in relation to preparation and handling of food, and (3) sanitary collection and treatment of sewage wastes. Many attribute good health to modern medicine, but it is much more due to these public health measures that we take too much for granted.

Depletion of Dissolved Oxygen

The discharge of raw sewage into waterways is not only a disease hazard, it may also deplete the **dissolved oxygen (DO)** causing collapse of the aquatic ecosystem. The reason is the same as described in Chapter 9. Organic matter present in sewage is readily consumed by decomposers and detritus feeders, which also consume oxygen in their respiration. When excessive amounts of detritus are present, these organisms consume dissolved oxygen faster than it enters the system, thus depleting the supply (Fig. 10–1). Indeed, the concentration of sewage wastes is often expressed in terms of its **biological oxygen demand** or **BOD,** the amount of oxygen that will be consumed in the course of its decomposition by decomposers.

The depletion of dissolved oxygen does not eliminate the bacterial decomposers, however, because they are capable of surviving by means of anaerobic respiration and fermentation. An *anaerobic* (without oxygen) body of water is not only incapable of supporting fish and shellfish, it also smells bad because many of the waste products of anaerobic metabolism have an unpleasant odor. Indeed, this is what gives sewage its characteristic smell.

In addition, depletion of dissolved oxygen may increase the hazard of microbial pollution. Many pathogenic organisms survive much longer in an anaerobic environment. They tend to die off quickly or are consumed in an oxygen-rich environment.

Eutrophication

Finally, even when pathogens and organic matter, BOD, have been removed, discharge of waste water with dissolved nutrients may still cause ecological upset through the process of eutrophication as was described in Chapter 9.

(a)

(b)

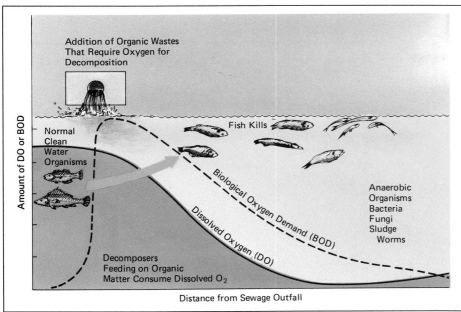

Addition of Organic Wastes
That Require Oxygen for
Decomposition

Fish Kills

Normal
Clean
Water
Organisms

Amount of DO or BOD

Biological Oxygen Demand (BOD)

Dissolved Oxygen (DO)

Anaerobic
Organisms
Bacteria
Fungi
Sludge
Worms

Decomposers
Feeding on Organic
Matter Consume Dissolved O_2

Distance from Sewage Outfall

FIGURE 10–1
(a) Raw sewage being discharged
into a stream. (b) Dissolved oxygen
(DO) and biological oxygen demand
(BOD). Fish and other aquatic
organisms depend on dissolved
oxygen for their respiration.
Additions of organic matter create an
additional biological oxygen demand
because bacteria (decomposers) that
feed on the organic matter consume
oxygen as well. Thus, additions of
organic wastes to waterways may
result in fish kills through depletion
of dissolved oxygen. After the
organic wastes are fully
decomposed, recovery of dissolved
oxygen will occur as atmospheric
oxygen slowly dissolves in the water.
(a, James Jackson/Photo
Researchers)

SEWAGE HANDLING
AND TREATMENT

Clearly, modern sewage treatment must address the
above three areas: pathogenic hazards, BOD, and nutrients. We begin with some background to understand how far we have come and how far we still need
to go.

Background

Prior to the late-1800s, the general means for disposing of human excrement was the outdoor privy (Fig.
10–2). Seepage from the privy not infrequently contaminated drinking water supplies and caused disease, especially in cities where privies and wells were
located near one another. The discoveries by Louis
Pasteur and others, in the mid-1800s, which showed

FIGURE 10–2
Outhouses were the widely used means of sewage disposal through the turn of the century and remain in use in many rural locations. This photograph also shows that dropping wastes into streams was accepted as a means of flushing them away. (USDA—Soil Conservation Service photo.)

that sewage-borne bacteria were responsible for many infectious diseases, led to intensive efforts to rid cities of wastes as expediently as possible. Cities already had drain systems for stormwater, but using these for human wastes had been prohibited. However, with the urgency of the situation, minds quickly changed. The flush toilet was introduced and sewers were tapped into storm drains. Thus, western societies initiated the system of flushing sewage wastes into natural waterways.

By the late 1870s, many waterways were so noxious with dead fish and foul odors resulting from depleted oxygen, and presented such a public health menace that it became clearly evident more had to be done, namely:

1. Develop facilities for treating the wastewater.
2. Have separate systems for stormwater and sewage because it is not practical to treat the total volume of combined stormwater and sewage.

Engineers designed facilities to remove pollutants from the combined sewage and stormwater and the first plants, in the United States, were built around the turn of this century at major sewage-stormwater outfalls. Gradually, regulations were passed requiring developers to install separate systems: *storm drains* for collecting and draining off rainwater into streams, and *sanitary sewers* that receive all the water from sinks, tubs, and toilets in homes and buildings and carry them to the sewage treatment plant. But progress has been extremely uneven and growing populations continually add to the burden. While tremendous progress has been made in some areas, cases of raw sewage overflowing with stormwater into waterways still abound. In other words, 100 years later we are still playing "catch-up." Before we address the remaining problems, however, let us turn our attention to the techniques that are currently used for sewage treatment.

Conventional Sewage Treatment

Raw Sewage

The sanitary sewer system brings all tub, sink, and toilet drains in our homes and buildings together like the twigs and branches of a tree connecting to the trunk. At the base of the "tree trunk," the total combination of all that goes into this collection system comes out—as **raw sewage** or **raw wastewater**. Because we use such large amounts of water to flush away small amounts of waste, or just run the water with no waste at all, raw sewage is about 1000 parts water for every part of waste, that is 99.9 percent water to 0.1 percent wastes. With the addition of stormwater, this is diluted still more. But the wastes or polluting materials in raw sewage are highly significant. They are divided into three categories:

☐ **Debris and grit.** *Debris* includes rags, plastic bags, and other objects flushed down toilets or washed into the system through storm drains where systems are still connected. *Grit* is coarse sand and gravel also entering mainly through storm drains.

- **Colloidal or organic material.** Includes both living organisms, the pathogens noted above as well as nonpathogenic detritus-feeding bacteria, and dead organic material from fecal matter, food wastes introduced through garbage disposal units, and cloth and paper fibers. The term *colloidal* expresses the fact that this material does not settle readily but tends to remain suspended in the water.

- **Dissolved materials.** Includes mainly nutrients such as nitrogen, phosphorus, and potassium compounds in solution. These come mainly from excretory wastes and are enriched with phosphate from detergents.

Treatment Steps

A sewage treatment plant must remove each of these categories if there is to be complete treatment. Debris and grit are removed by *preliminary treatment*. Colloidal material is removed by a combination of *primary* and *secondary treatment*. And dissolved nutrients are removed by *advanced treatment*.

Bear in mind that sewage treatment in any given case does not imply all four of these steps. Basically, these steps have been added on one after the other, as circumstances have demanded. Consequently there are some communities that still discharge raw sewage, others that go only as far as primary treat-ment and then discharge the water, and others that carry it through secondary treatment before discharge. Only a relatively few cities go all the way through advanced treatment.

Preliminary Treatment. Debris and grit would clog up and damage later treatment processes. Therefore, their removal is termed **preliminary treatment.** Debris is removed by letting raw sewage flow through a **bar screen,** a row of bars mounted about 1 inch (2.5 cm) apart (Fig. 10–3). Debris is mechanically raked from the screen and taken to an incinerator. Following the bar screen the water flows through a **grit chamber** or **grit settling tank,** a swimming pool-like tank where water velocity slows just enough to permit the grit to settle (Fig. 10–4). The settled grit is mechanically removed from these tanks and put in landfills.

Primary Treatment. Following preliminary treatment, the water moves on to primary treatment where it flows very slowly through large tanks called **primary clarifiers** (Fig. 10–5). While flowing slowly through these tanks, the water is nearly motionless for several hours. This permits the heavier particles of organic matter, about 30 to 50 percent of the total, to settle to the bottom where they can be removed. At the same time, fatty or oily material floats to the top and is skimmed off. All the material removed is known as **raw sludge;** its treatment and disposal will be considered shortly.

Note that primary treatment involves nothing more complicated than "putting polluted water in a

FIGURE 10–3
Bar screen. Channelling the waste water through a bar screen is the first step of preliminary treatment. The screen removes large pieces of debris. Such debris is removed from the screen by a mechanical rake. (Washington Suburban Sanitary Commission photo.)

FIGURE 10-4
Grit chamber, the second step of preliminary treatment. The velocity of flow through these chambers is slowed to 1 to 2 feet per second, which allows sand and other coarse grit to settle to the bottom while the water flows over the top edge (foreground). The grit is removed by mechanical plows scraping the bottom. (Photo by Tommy Noonan, Washington D.C.)

bucket, letting material settle, and pouring off the water." Nevertheless, it accomplishes significant removal of organic matter at minimal cost. The water leaving the primary clarifiers still contains 50 to 70 percent of the colloidal material that did not settle and nearly all the dissolved nutrients. Secondary treatment addresses removing the remaining colloidal or organic material, but not the dissolved nutrients.

Secondary Treatment. Secondary treatment is also called **biological treatment** because it employs organisms, natural decomposers and detritus feeders, that consume the organic matter and break it down through their cell respiration to carbon dioxide and water. Either of two types of systems may be used: *trickling filters* or *activated sludge systems.*

In **trickling filter** systems, the water is sprayed onto and allowed to trickle through a 6–8 foot (2–3

m) bed of fist-sized rocks (Fig. 10–6). As in a natural stream, this environment supports a complex ecosystem including bacteria, protozoans, rotifers, various small worms, and other detritus feeders attached to the rocks (Fig. 10–7). The organic matter in the water, including pathogenic organisms, is literally eaten up as it trickles by. Clumps of organisms that occasionally break free and wash from the trickling filters are removed when the water moves from trickling filters into secondary clarifiers, tanks similar to the primary clarifiers. Material that settles here is handled as raw sludge. Through primary treatment and the trickling filter system, 85 to 90 percent of the total organic matter is removed from the wastewater.

An increasingly common alternative method of secondary treatment is the **activated sludge system.** Here, water from primary treatment enters a tank that could hold several tractor trailer trucks parked end to end (Fig. 10–8). A mixture of detritus-feeding organisms, referred to as **activated sludge,** is added to the water as it enters the tank, and the water is vigorously aerated as it moves through the tank creating an oxygen-rich environment ideal for the growth of the organisms (Fig. 10–9). As the organisms feed, the biomass of organic matter, including pathogens, is reduced.

When the water leaves the aeration tank, it still contains the rich mixture of the feeding organisms. Therefore, from the aeration tank, the water goes into a secondary clarifier tank. Since the feeding organisms are usually clumped on bits of detritus, settling is relatively efficient. The settled organisms are the *activated sludge,* which is pumped back into the entrance of the aeration tank (Fig. 10–9b). Thus, the detritus-consuming organisms are continually recycled in the system while water, now with 90 to 95 percent of the organic matter removed, flows on. Excess activated sludge, which accumulates from the growth of the organisms, is generally combined and treated with the raw sludge from the primary clarifiers.

In constructing new wastewater treatment plants or in upgrading old ones, the activated sludge system is often selected simply because it is a more compact system. This is an important factor because available space is limited. Its one notable disadvantage, however, is the cost for large amounts of energy required for the aeration pumps. Trickling filter systems are generally gravity-fed systems, and their operation requires little if any additional energy.

Note that neither system of secondary treatment system removes dissolved nutrients, the major cause of eutrophication.

Prior to two decades ago little need was seen for treatment beyond secondary. Water from secondary treatment was disinfected with chlorine and dis-

(a)

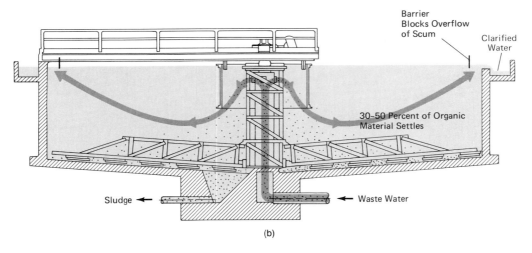

Barrier
Blocks Overflow
of Scum

Clarified
Water

30–50 Percent of Organic
Material Settles

Sludge ◄—   ◄— Waste Water

(b)

FIGURE 10–5
Primary clarifiers used for primary treatment. (a) The water enters these tanks at the center and exits over the wires at the edge. The slow velocity (1–2 inches per minute) of flow through the tanks permits 30 to 50 percent of the colloidal organic matter to settle while oil and grease rise. Settled organic material, raw sludge, is pumped from the bottom while oil and grease are simultaneously skimmed from the surface. Cross-section of clarifier. (Photo by Tommy Noonan, Washington, D.C.; b, Courtesy of Walker Process Division of C.B.I.)

charged into natural waterways, and this situation still predominates. As problems of eutrophication have become more severe, however, increasing numbers of cities are adding a further stage of treatment, *advanced treatment,* to remove nutrients.

Advanced Treatment. Water from secondary treatment enters advanced treatment where one or more

nutrients are removed. Numerous methods are available for this task. The water could be 100 percent purified by distillation or microfiltration; however, the problem is cost. The total flow is about about 150 gallons per person per day. Purifying this amount of water by such methods becomes prohibitively expensive. But cost-effective methods are also being developed and implemented. For example, phosphate may be removed by mixing lime (a calcium compound) into the water. Calcium combines chemically with phosphate to make an insoluble compound (calcium phosphate) which may be removed by filtration. Where phosphate is the limiting factor in eutrophication, phosphate removal is sufficient.

Since Washington, D.C., implemented advanced treatment a few years ago, the Potomac River has cleared up considerably. Aquatic vegetation and waterfowl are returning, a hopeful sign for Chesapeake Bay as a whole if other cities follow suit.

(a)

(b)

FIGURE 10–6
Trickling filters, secondary treatment. (a) The water from the primary clarifiers is sprinkled onto and trickles through a bed of 6 to 8 feet of rocks. (b) Various bacteria and other detritus feeders adhering to the rocks consume and digest the organic material in the water as it trickles by. The water is again collected at the bottom of the filters. (Photos by author.)

FIGURE 10–7
(a) Some of the organisms that are active in secondary treatment. (b) These organisms form a biomass pyramid of detritus feeders. Through this pyramid the biomass of organic matter entering the system is reduced by up to 90 percent.

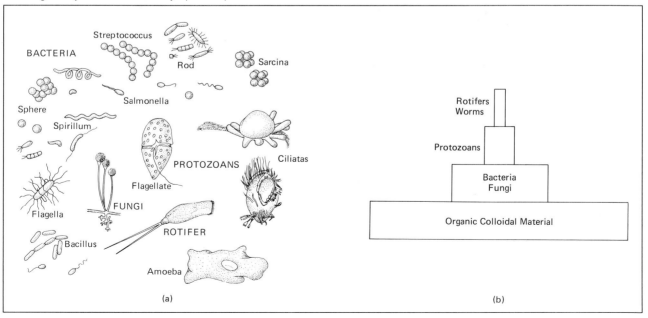

FIGURE 10–8
Aeration tank used in activated sludge treatment. As waste water from the primary clarifiers moves through the tank it is vigorously aerated by air forced up from the tubes along the bottom. (Photo by author.)

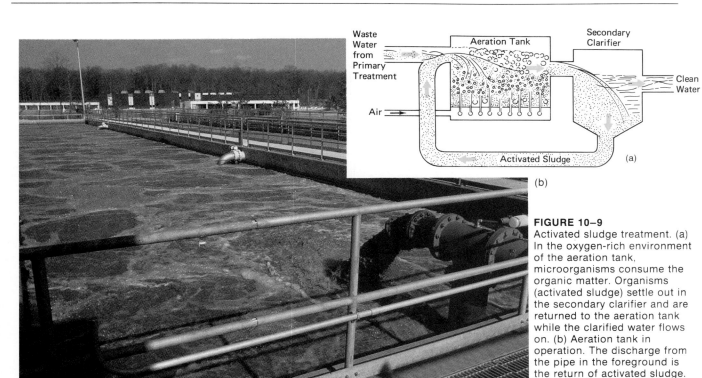

FIGURE 10–9
Activated sludge treatment. (a) In the oxygen-rich environment of the aeration tank, microorganisms consume the organic matter. Organisms (activated sludge) settle out in the secondary clarifier and are returned to the aeration tank while the clarified water flows on. (b) Aeration tank in operation. The discharge from the pipe in the foreground is the return of activated sludge. (Photo by author.)

With suitable advanced treatment, one can end up with a final effluent that is of drinking water quality. Whether such water will be recycled back into the municipal system will depend on the following question: Does the value of water justify the expense of pumping it back to the head of a system? As water shortages become more severe (see Chapter 8), the answer to this question is likely to become increasingly, "Yes."

Many people pale at the idea of recycling what they consider sewage, but we should remember that all water is recycled by nature in any case. Indeed, suitable advanced treatment may provide better-quality water than is obtained in drawing water from a river or lake into which cities have discharged raw sewage, as is currently the case in many situations. A summary of the sewage treatment steps is shown in Figure 10–10.

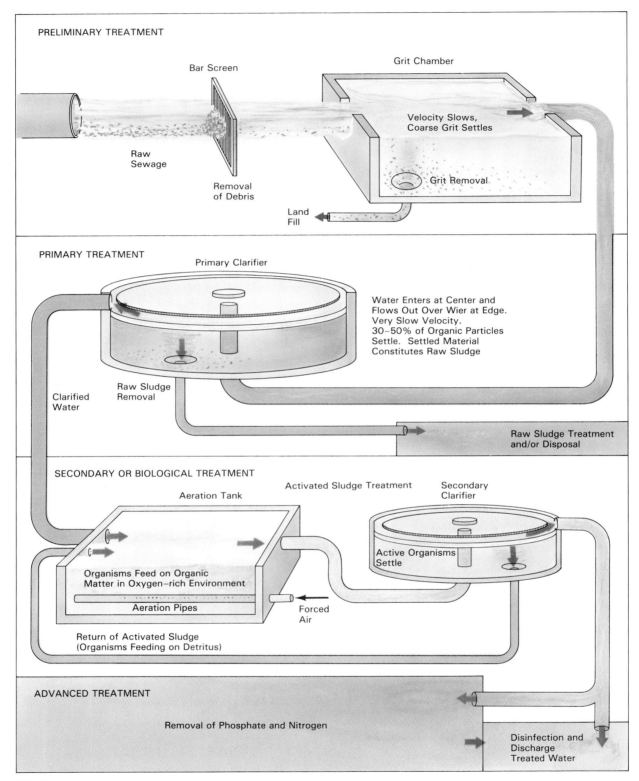

PRELIMINARY TREATMENT

Bar Screen

Grit Chamber

Raw Sewage

Removal of Debris

Velocity Slows, Coarse Grit Settles

Grit Removal

Land Fill

PRIMARY TREATMENT

Primary Clarifier

Water Enters at Center and Flows Out Over Wier at Edge. Very Slow Velocity. 30–50% of Organic Particles Settle. Settled Material Constitutes Raw Sludge

Clarified Water

Raw Sludge Removal

Raw Sludge Treatment and/or Disposal

SECONDARY OR BIOLOGICAL TREATMENT

Activated Sludge Treatment

Secondary Clarifier

Aeration Tank

Active Organisms Settle

Organisms Feed on Organic Matter in Oxygen-rich Environment

Aeration Pipes

Forced Air

Return of Activated Sludge (Organisms Feeding on Detritus)

ADVANCED TREATMENT

Removal of Phosphate and Nitrogen

Disinfection and Discharge Treated Water

FIGURE 10–10
Diagram is a summary of wastewater treatment through secondary treatment. Further purification processes, referred to as advanced treatment, may be added on after secondary treatment.

Disinfection. Whatever the degree of treatment wastewater receives, it is generally disinfected by adding chlorine gas, to kill any pathogenic organisms that may persist, as it is discharged into a natural waterway (Fig. 10–11). The use of chlorine gas (Cl_2) for disinfection presents environmental problems that need consideration. Chlorine is used because it is both effective and relatively inexpensive. However, it is extremely toxic and transportation is hazardous—posing a threat to humans. Second, it is even more toxic to some fish. Levels that are too low to measure have been found to inhibit the hatching of trout eggs and development of the embryos. Finally, it has been found that to some extent chlorine reacts spontaneously with organic compounds to form *chlorinated hydrocarbons*, which are organic molecules with chlorine atoms attached. Many of these compounds are toxic and nonbiodegradable, and some have been identified as compounds that cause cancer, abnormal development, and reproductive problems (see Chapter 11).

Disinfecting agents other than chlorine may be more satisfactory. One is ozone (O_3). Ozone is extremely effective in killing microorganisms, and in the process it breaks down to oxygen gas, which improves water quality. However, ozone is not only toxic, it is explosive. Thus, ozone must be manufactured at the point of use, a step that demands considerable capital investment and energy. With improved technology, however, costs might be comparable to the use of chlorine and safety hazards might be reduced. Another suggestion is to pass the water under an intense source of ultraviolet or other radiation that would kill microorganisms but would not otherwise affect the water. Also, after chlorine disinfection, other chemicals, such as sulfur dioxide may be added that react with chlorine to form inactive, harmless compounds.

Sludge Treatment

Recall that 30 to 50 percent of the organic matter present in raw sewage settled in the primary clarifiers (primary treatment). This settled material is called **raw sludge.** Pumped from the bottom of tanks, raw sludge is a black, foul-smelling, syrupy liquid of about 98 percent water and 2 percent organic matter, which includes many pathogenic organisms. Without treatment, the disease potential makes it definitely a hazardous material. However, with suitable treatment, it can be converted into a *nutrient-rich humus* that can be used as a soil conditioner or organic fertilizer. In nature, detritus is broken down and converted to humus by the activity of soil organisms. Likewise, sludge treatment involves letting bacteria and other detritus feeders work on it in either the absence of air, **anaerobic digestion,** or in the presence of air, **composting.** Excess sludges from secondary and advanced treatment may be added into these processes.

Anaerobic Digestion. In anaerobic digestion the raw sludge is put into large airtight tanks called **sludge digesters** (Fig. 10–12). Bacteria, which are naturally present, feed on the organic matter, but in the absence of air they carry on *anaerobic* respiration. An important waste product of anaerobic respiration is *biogas.* Although it contains carbon dioxide and compounds that give sewage its foul smell, **biogas** is about two-thirds **methane.** Natural gas, widely used for heating and cooking, is nearly pure methane. Because of its methane content, biogas is flammable and can be used for fuel as it is. In fact, it is commonly burned to heat the digesters themselves since the organisms do best when maintained at about 38° C (100° F).

After four to six weeks in the tank, digestion is more or less completed and what remains is called **treated sludge.** It is still a liquid, black with organic

FIGURE 10–11
Tanks of chlorine gas used in disinfecting waste water. Is this purification, or the addition of one more toxic substance? (Photo by author.)

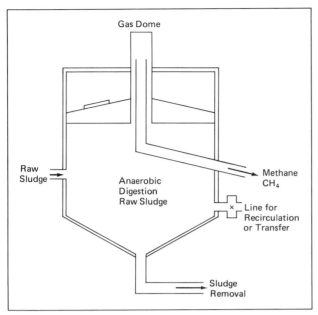

FIGURE 10–12
Treatment of raw sludge by anaerobic digestion. The raw sludge withdrawn from primary clarifiers (primary treatment) may be treated by placing it in airtight tanks for a period of 6 to 8 weeks. In the absence of oxygen, bacteria feed on the sludge (anaerobic digestion), producing biogas (about 60 percent methane) as a waste product and leaving a reduced amount of stable humus-like sludge (treated sludge) as a byproduct. (EPA.)

FIGURE 10–13
Use of treated sludge. The treated sludge remaining after anaerobic digestion is a humus- and nutrient-rich liquid that is an excellent soil conditioner. The vehicle shown is specially designed for applying sludge to soils. (Courtesy of Ag-Chem Equipment Co. Inc., Minneapolis, Minnesota.)

matter, but this organic matter is now relatively stable and odorless. In addition, pathogenic organisms have been greatly reduced if not eliminated so they no longer present any great hazard. Effectively, it is humus in nutrient-rich water solution.

Treated sludge may be applied directly to lawns and agricultural fields in the liquid state in which it comes from the digesters, thus providing benefit from both the humus and the nutrient-rich water (Fig. 10–13). Since 1979, a program operated by the Madison (Wisconsin) Metropolitan Sewerage District has recycled over 100,000 tons (dry weight) of sewage sludge to 27,000 acres of agricultural land for a fertilizer cost savings of over $1.2 million. Alternatively, the treated sludge may be filtered (Fig. 10–14a), leaving a semisolid humus **sludge cake** (Fig. 10–14b). Sludge cake is easy to stockpile, distribute, and spread on fields without requiring special equipment. However, since a large share of the nutrients are in water solution, much of the nutrient value goes with the filtered water, which is generally put back into the wastewater stream.

Composting. For composting, raw sludge is filtered, mixed with wood chips or other material to improve aeration, and piled in **windrows** (Fig. 10–15a). Additional air may be drawn through the piles to further increase aeration, as shown in Figure 10–15b, or piles may be frequently turned with machinery (Fig. 10–16). In the compost piles, bacteria and other decomposers and detritus feeders consume and break down the organic material leaving a rich humus-like material. Pathogenic organisms lose out in the competition, and heat produced through respiration is sufficient to kill them. As long as the piles are kept well

(b)

(a)

FIGURE 10-14
Filtering of treated sludge. (a) A vacuum inside the rotating drums sucks water from the sludge as it is applied to the belts. (Photo by Tommy Noonan, Washington, D.C.) (b) A semisolid sludge cake of humuslike material results. This material also has great value as a soil conditioner but unfortunately most is either dumped in landfills or burned. (Photo by author.)

FIGURE 10-15
Treatment of raw sludge by composting. (a) Raw sludge is mixed with wood chips, which absorb the excess water, and the mixture is placed in piles through which air is drawn. (b) Diagram of the piles shown in (a). In the aerated piles, bacteria reduce the sludge to a nutrient-rich, humuslike material that can be used as a soil conditioner. Heat produced by the aerobic respiration of bacteria is sufficient to kill pathogenic organisms. (USDA photo by Robert C. Bjork.)

(a)

(b)

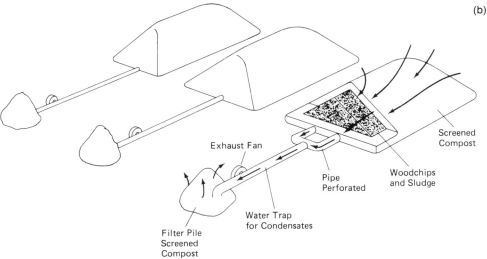

Exhaust Fan

Water Trap
for Condensates

Filter Pile
Screened
Compost

Pipe
Perforated

Woodchips
and Sludge

Screened
Compost

FIGURE 10–16
Composting, the conversion of sewage sludge and other organic wastes into humus-like material, may be aided by machinery. Machine shown straddles a windrow and the rotating drum with flails between the wheels turns, fluffs, and aerates the pile. (Courtesy of Eagle Crusher Company, Inc., Galion, Ohio.)

aerated, the waste products of respiration are just carbon dioxide and water. Obnoxious odors, which are compounds from *anaerobic* respiration, are negligible. After six to eight weeks of composting the resulting humus is screened out of the wood chips, which may be reused, and is ready for application to soil. Combined composting of sewage sludge and waste paper and yard wastes is gaining popularity (see Chapter 18).

"Humus" produced from sewage sludge is commonly available to the public free of charge. Numerous communities, especially in the Southwest, apply their sludge production on agricultural fields. Chicago has been using much of its sludge to reclaim soils ravaged by strip mining (Fig. 10–17). Milwaukee, which has a particularly rich sludge resulting from the brewing industry, pasteurizes, bags, and sells it as organic fertilizer under the trade name "Milorganite."

Alternative Systems

Note that overall process of sewage treatment described above still maintains the underlying concept of flushing wastes toward natural waterways. The thinking has been to add more and more steps to remove polluting materials but still discharge the water into a natural waterway. Increasingly, this underlying thinking is being challenged. After all, there is nothing inherently bad with nutrient-rich water; what is wrong is our failure to appreciate the ecosystem principle of recycling. With our one-way-flow system, we are putting the nutrients toward growing

or producing what we don't want—algae and eutrophication—instead of what we do want.

Using Nutrient-Rich Water for Irrigation

As an alternative to installing advanced treatment and throwing away the water, an increasing number of communities are turning to using the nutrient-rich wastewater for irrigation. For example, in St. Petersburg, Florida, nutrient-rich wastewater from secondary treatment, which once polluted Tampa Bay, now irrigates 4400 acres of urban open space from parks and residential lawns to a golf course. Revenues from the water sales help offset operating costs. Bakersfield, California, receives $300,000 a year income from a 5000-acre farm that is irrigated with its treated effluent. Claton County, Georgia, is irrigating 2700 acres of woodland with partially treated sewage. Hundreds of other similar projects are underway around the country. (See Case Study 10).

The nutrient-rich water does not have to go onto land to be useful. Highly productive *aquaculture* systems are also in use and being developed. **Aquaculture** involves constructing artificial "wetlands" and/ or ponds to entrap nutrient-rich water and the use of these for the intensive culture of one or more species. Ponds with water hyacinths are being used in Mississippi as a means of advanced wastewater treatment (Fig. 10–18). However, any number of useful aquatic plants can be grown. For example, water hyacinths and water lilies can be used for livestock feed. Cattails and other such "reeds" may be used in weaving "straw" mats, baskets, and so on. Or such plants may

FIGURE 10–17
Soil-conditioning value of treated sewage sludge. Plants on the right were grown on strip-mined soil treated with 10 percent sludge. Plants on the left were grown in untreated soil. (USDA photo.)

be anaerobically decomposed to produce methane, fermented to produce alcohol, or burned directly as a boiler fuel. Additional trophic levels including fish, shellfish, and waterfowl may be added (Fig. 10–19). In concept, the system is the same as a natural wetland. Plants are filtering out sediments, absorbing nutrients, and producing food and habitat for additional consumers. Water finally exiting the system is relatively pure. In the Philippines Islands and other parts of the world, sections of coastal wetlands have been diked off to create ponds. Nutrient-rich water flowing into the ponds supports a vigorous growth of algae and of fish that thrive on the algae (Fig. 10–20).

There are some obvious restrictions to using wastewater for irrigation or aquaculture. First there must be suitable areas nearby and preferably at lower elevation to receive the irrigation water. Otherwise prohibitive piping and pumping costs are incurred. Then, too, such systems are most workable in dry,

(*Text continued on p. 266*)

FIGURE 10–18
Natural wetlands may be converted into aquaculture using waste water as the primary source of nutrients. Photograph shows a water hyacinth farm in Lucedale, Mississippi. (Photo by N. D. Vietmeyer, National Space Technology Laboratory.)

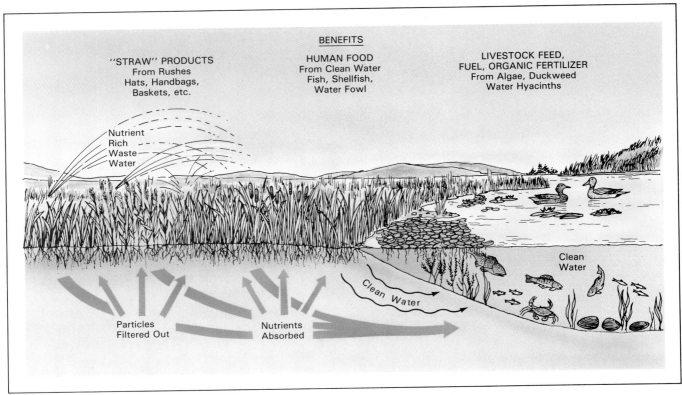

FIGURE 10–19
Advanced biological treatment. Nutrient-rich waste water may be spray-irrigated onto agricultural fields or natural wetlands. Nutrients are absorbed by plants and lead to enhancing their productivity while nutrient-free clean water seeps through into natural waterways. Plants grown under such irrigation may have various commercial uses as indicated.

FIGURE 10–20
Mariculture in the Philippines. Natural wetlands have been made into fishponds. (FAO photo by P. Boonserm.)

THE OVERLAND FLOW
WASTEWATER TREATMENT SYSTEM

We have seen that it is necessary to heed the ecological principle of recycling nutrients in wastes as opposed to the traditional practice of having them flow one way into bodies of water where they cause eutrophication. An alternative to the traditional wastewater treatment that accomplishes this objective is the overland flow system. A facility utilizing this principle has just been put into operation in the community of Emmitsburg, Maryland.

The raw wastewater, about 1 million gallons per day, from all the homes and commercial establishments of the community is first put through a pond where grit and heavier particles will settle. It is then irrigated onto the long narrow "fields" which you see in the photograph. These fields have about a foot of rich topsoil supporting a crop of reed canary grass, a forage grass that has a particularly voracious appetite for nitrogen and other nutrients. Below the topsoil, the subsoil is compacted clay, which is impermeable, and slopes gently downward away from the irrigation pipe. The waste water applied to the side of the field thus percolates through the topsoil across the field to a collecting gutter at the opposite side. In its passage natural organisms in the topsoil break down and utilize the organic wastes and maintain the richness and aggregate structure of the topsoil in the process, while the grass absorbs the nutrients. The water exiting into the collecting gutter is clear and nearly nutrient free. It is collected into another reservoir and spray irrigated onto forage crops so that none of the nutrients go to waste. The canary grass is periodically mowed and becomes feed for cattle. Thus, the nutrients make a complete cycle from wastewater to grass to beef to humans to waste water and again back to the soil.

You may ask how a small community in western Maryland happened to install a state-of-the-art ecological method of wastewater treatment. First, the town did need a new wastewater treatment facility to accommodate population growth. However, it was largely the ecological thinking of one farmer in the area who persuaded the town officials to go ecological. He volunteered his land, some two hundred acres, to be the final recipient of the the water and the nutrients. His forward thinking has created a situation in which everyone wins. He gets free water and nutrients for irrigation, Emmitsberg meets the standards for sewage treatment that will accommodate growth with a low-cost, low-maintenance system, and all benefit by not having the nutrients go into Chesapeake Bay where they would cause more eutrophication.

Overland flow wastewater treatment system. Wastewater is being irrigated onto fields. Note the lush growth of grass benefitting from the nutrients. See text for details. (Photo by the author.)

warm climates where soil is almost always able to receive the additional water and plants grow all year. However, suitable lagoons for storing the water over winter months and judicious choice of species may compensate. The biggest limitations come down to the imagination and the will to see the problem solved.

Individual Septic Systems

Countless homes in rural areas are not connected to a municipal sewer system. Instead, they have individual septic systems. A typical septic system consists of a septic tank and drain field (Fig. 10–21). The wastewater flows into the tank where the heavier particles settle to the bottom; it acts like a primary clarifier in standard sewage treatment. Water, still carrying much of the fine organic material and the dissolved nutrients, overflows into the drain tiles buried in the ground and gradually percolates into the soil. Bacteria gradually digest the organic material that settles in the tank, reducing it to a stable humus that must be pumped out every two to three years. Likewise, soil bacteria decompose the organic material that comes through the drain tiles. Some people establish successful vegetable gardens over septic drain fields, thus exercising the sound principle of recycling the nutrients.

A septic system from which the humus material is periodically removed from the tank may function indefinitely. However, organic material frequently enters the soil faster than it decomposes, gradually

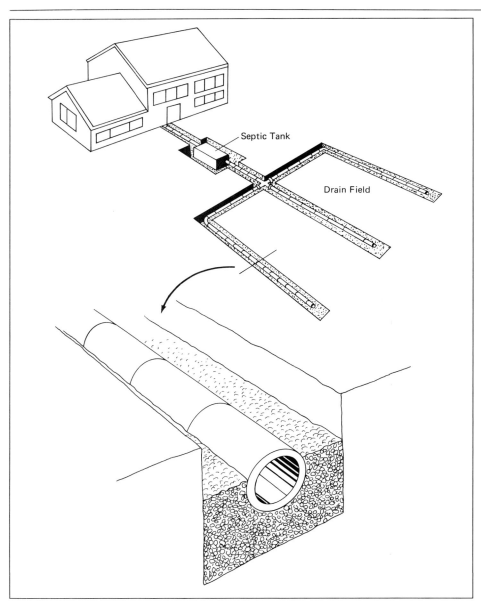

FIGURE 10–21
Individual sewage treatment. Septic tank and drain field. (USDA—Soil Conservation Service.)

Septic Tank

Drain Field

clogging the soil pores and forcing raw sewage to the surface where it causes objectionable odors, contaminates surface water, and is a general health hazard. If the lot is not large enough to relocate the drain field, little can be done about this problem except to try to get centralized sewage as soon as possible. In many areas of the country, expensive suburban developments using individual septic systems have become quite obnoxious due to this problem. If and how soon this problem may occur depends on the rate of loading. This is why one is advised not to have garbage disposal units in connection with septic systems or put disposable diapers or sanitary supplies into them.

An interesting alternative for the handling of personal wastes in individual homes is a **composting toilet.** An example is the Clivus Multrum developed in Sweden (Fig. 10–22). The Multrum receives only personal excrements and food wastes—no water other than urine. These wastes pass through a series of chambers as they decompose, and after two to four years they arrive at the final chamber as a stable, nu-

trient-rich humus that is suitable for application on lawns and gardens. There must be another means for disposing of bath, dish-washing, and other *gray* water, but since it is not contaminated by human excrements, this generally offers relatively little problem. It may be used for irrigation of lawns and gardens.

TAKING STOCK OF WHERE WE ARE AND WHAT YOU CAN DO

Progress and Lack of Progress

As noted earlier, the disease hazards of untreated sewage were well understood by the 1870s, and the technology for primary and secondary (trickling filters) treatments was developed in the late 1800s. Yet there was no great rush to solve and keep up with the expanding problem. Particularly in less-developed countries, modern sections notwithstanding, many live without treated water supplies or bathrooms. Excretory wastes just go into the gutters, which drain into streams and lakes where people

FIGURE 10–22
Clivus Multrum, a dry waste treatment system. Sewage and garbage wastes deposited in the Clivus Multrum decompose aerobically. A dry humus- and nutrient-rich compost is removed from the Clivus Multrum as a byproduct of treatment. (Courtesy of Clivus Multrum, W. Pa., Inc.)

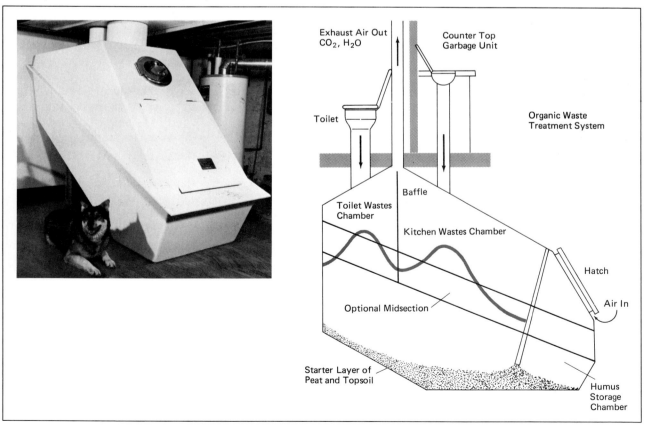

draw drinking water and wash clothes. Many suffer from sewage-born parasites that undercut their productivity, and diseases from "bad" water are a leading cause of infant and childhood mortality.

Even as late as the 1960s nearly a third of the United States' total raw sewage was still being discharged with stormwater into natural waterways—with no treatment whatsoever. Another third received only primary treatment before being discharged. Even in cities that did have good sewage treatment, many outfalls of raw sewage existed where pocket communities had never been connected into the central sewage collection system. Increasing population added to the problem by overburdening existing systems, causing leaks, backups, and overflows. Signs labeled "No Swimming! Polluted Water!" proliferated along formerly clean rivers and beaches.

Under pressure from public outcry and disgust, Congress passed the **Clean Water Act of 1972,** which provided a major thrust for moving ahead. The Clean Water Act provided federal money for new interceptor sewers to collect wastes, to separate storm and sewage systems, and to construct, expand, and upgrade sewage treatment plants. Nearly 50 billion dollars has been spent on these efforts since the act was passed. Now most sewage in the United States is processed at least through secondary treatment before discharge.

But the job is so enormous that there is still a long way to go. Some 150 communities are still discharging raw sewage into bays, lakes, and coastal waters. Notably San Diego and Boston still pipe sewage into the ocean after only primary treatment, but they are under EPA orders to stop. Due to remaining interconnections of stormwater and sewage collection systems, many plants are still overloaded and forced to discharge raw sewage in times of storms. Finally, despite the severe and worsening problem of eutrophication, relatively few cities have installed advanced treatment or alternative use of the nutrient-rich water.

Finally, many cities face tremendous problems in disposing the sludge removed in the course of primary and secondary treatment. Much ends up being landfilled or incinerated for reasons noted below, despite its potential value. New York City and surrounding areas still barge their raw sludge to sea and dump it. This practice is technically prohibited under the Clean Water Act, but they have been able to receive extensions to continue dumping until 1992, claiming they have no practical alternative. Also, much of the biogas produced in anaerobic digestion of sludge is simply torched (burned off) to get rid of it rather than utilizing it (Fig. 10–14).

Impediments to Progress

Contamination with Industrial Wastes

There is a serious impediment to using sewage sludges and nutrient-rich water for agricultural purposes, namely contamination. Industries frequently flushed wastes containing such toxic chemicals as lead, mercury, chromium, and nonbiodegradable organics into the sanitary sewer system. You can see that the conventional steps of sewage treatment will not affect such chemicals. Indeed, they poison the organisms used in secondary treatment and thus impede the process. Then, they bind to organic material and end up in the treated sludge making it unfit for agricultural use because these materials are poisonous to plants, as well. Likewise they may make the water unfit for irrigation. Much of the sludge in the United States is still unfit for agricultural uses for this reason.

However, also under the Clean Water Act, industries are now required to either **pretreat** their wastewater to remove toxic chemicals before discharging into municipal sewer systems, or to find alternative means of disposal. As standards for and enforcement of pretreatment become more rigorous, treated sewage sludge and wastewater will become increasingly available for fertilizer or irrigation.

Public Apathy

The biggest impediment in the way of upgrading sewage treatment and/or implementing the agricultural use of sewage-derived humus and nutrient-rich water is public apathy or even antagonism. Many people feel that what goes down the drain should be "out of sight, out of mind." There is a failure to make any connection between spotless, sanitary bathrooms and raw sewage in waterways. Combined water and sewer bills, which include the costs for sewage treatment, are only 5–10 dollars a month per household. For another 2–3 dollars per month, we could have "state of the art" sewage treatment. Yet, many people become adamant that sewer charges should not be increased, and municipalities find that referenda to raise money for sewage treatment facilities fail. Some people become so adamant against the use of sludge or wastewater for agricultural purposes that they will bring court action against individuals or municipalities to block their using it. It should not escape notice that while people are rejecting putting sewage-derived humus on lawns or agricultural fields, they are tolerating untreated sewage going into lakes and rivers that serve as drinking water supplies. Clearly, a major part of the problem is lack of public understanding regarding the problem and its solution.

What You Can Do

As a person who now understands the problems and solutions regarding sewage, you can be a significant force in spreading this understanding. Find out the situation in your own community. How much treatment does your sewage receive? Where is the water discharged? With what effects? Is advanced treatment in order? Are there possibilities for using the water for irrigation or aquaculture? What is done with sludge that is removed? If there is potential for ecological improvement in any of these areas, join environmental groups and political action committees and make these items part of their agenda if they are not already.

At the more personal level, if a suitable treated sludge is available, start using it on your own lawn and garden and encourage your friends and neighbors to do likewise. In housing built prior to the 1970s, downspouts and patio drains were commonly tapped into the sanitary sewers adding to the stormwater overflow problem. Check these drains for your own home and if you find that this is the case, disconnect and reroute them.

If we desire healthy rivers and lakes, we must become involved and find out what is being done and what is not being done to maintain or restore good water quality. We need to support and encourage public officials who are doing what they can and prod those who are not.

Monitoring for Sewage Contamination

In spite of general improvement in sewage collection and treatment, there are and always will be unrecognized leaks, breaks, or overflows in sewage systems. Therefore, the need to monitor for sewage contami-

FIGURE 10–23
The total coliform test. The number of fecal coliform bacteria in a sample of water is determined by: (a) inserting a filter disc in a filtering apparatus, and (b) drawing the water sample through the filter. Bacteria in the sample are entrapped on the filter disc. (c) The filter disc is then placed on a medium for growing bacteria and incubated at 38° C for 24 hours. In this period each *E. coli* bacterium entrapped on the disc multiplies to form a colony that is visible to the naked eye. (d) The number of bacteria that were present in the water sample is thus given by the number of colonies (seen in the photograph as spots) on the disc. Fecal coliform bacteria are identified by colonies that have a distinctive metallic-green sheen. (Photos by Bob Hudson.)

(a)

(b)

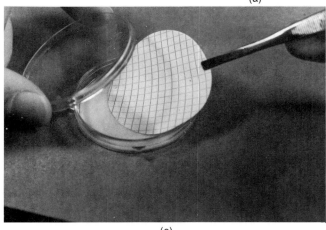

(c)

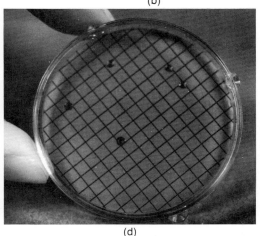

(d)

nation continues. It is worth understanding how this is done.

It would be exceedingly difficult, time consuming, and costly to test for each specific pathogen that might be present. Therefore an indirect method called the **fecal coliform test** has been developed. This test is based on the fact that huge populations of a bacterium called *E. coli* (*Escherichia coli*) normally inhabit the lower intestinal tract of humans and other animals, and large numbers are invariably excreted with fecal material. *E. coli* is not found in the environment except when it enters from animal excrement. It is not a pathogen; in fact our bowels would not function properly without it. However, since *E. coli* is invariably part of fecal wastes, its presence reveals a persistent source of raw sewage and, consequently, the potential presence of other sewage-borne pathogens. Conversely, the absence of *E. coli* organisms indicates that the water is probably free of sewage-borne, microbial pathogens.

The fecal coliform test, which is described in Figure 10–23, is a means of detecting and counting the actual number of coliform bacteria in a sample of water, usually 100 ml (about ½ cup). Thus, the test shows how much sewage pollution is present and the relative degree of hazard. For example, it is considered that for water to be safe for drinking the *E. coli* count should be zero, that is, absolutely no sewage

contamination. However, water with as many as 100 to 200 *E. coli* per 100 ml is still considered safe for swimming. Beyond this level a river may be posted as polluted, and swimming and other direct contact should be avoided. By comparison, raw sewage water itself (99.9 percent water/0.1 percent wastes) has counts in the millions.

When water is too polluted for a desired use, two possible approaches exist: disinfection and/or reduction of sewage sources. Drinking water supplies and swimming pools are generally disinfected with chlorine. However, there is obviously no way of disinfecting natural bodies of water without also killing everything else in the system. Therefore, there is no substitute for continuing and improving the quality of sewage treatment.

Unfortunately, however, we still tend to disregard the importance of sewage treatment. As long as something disappears down the drain, the general public believes it has been taken care of. Consequently, the question of whether adequate sewage treatment facilities are needed frequently becomes a matter of political debate about how to spend the taxpayers' money. It is important to recognize the connection between our sanitary bathroom and polluted beaches and waterways. Sewage treatment is not a political issue. It is a vital public health and environmental concern.

11

Toxic Chemicals and Groundwater Pollution

CONCEPT FRAMEWORK

■ Outline

I. SOURCES OF GROUNDWATER POLLUTION 274

II. TOXIC CHEMICALS: THEIR THREAT 274
   A. What Are the Toxic Chemicals? 274

      1. Heavy Metals 275

      2. Synthetic Organics 275

   B. The Problem of Bioaccumulation 275

   C. Synergistic Effects 277

III. ENVIRONMENTAL CONTAMINATION WITH TOXIC CHEMICALS 277

   A. Major Sources of Toxic Chemical Wastes 278

   B. Background of the Toxic Waste Problem 278

      1. Indiscriminate Discharge into Air and Water 278

      2. Shift to Land Disposal 278

   C. Methods of Land Disposal 278
      1. Deep Well Injection 278
      2. Surface Impoundments 279
      3. Landfills 280
   D. Problems in Managing Land Disposal 280
      1. Midnight Dumping 280
      2. Nonsecure Facilities 280

   E. Scope of the Problem 282

IV. CLEANUP AND MANAGEMENT OF TOXIC WASTES 284

   A. Assuring Safe Drinking Water Supplies 284

   B. Cleaning Up Existing Toxic Waste Sites 284

   C. Groundwater Remediation 285
   D. Management of Wastes Currently Produced 285

■ Study Questions

1. Cite the importance of groundwater. List five sources of its pollution.

2. Cite two major classes of toxic chemicals.

3. Give examples of *heavy metals* and their toxicity.

4. Define and give examples of *synthetic organic chemicals.* Name a class and subclass that is particularly hazardous. What atoms are included in a chlorinated hydrocarbon molecule?

5. Define the process of *bioaccumulation.* Why do heavy metals and halogenated hydrocarbons tend to bioaccumulate? What happens in a food chain? Describe and give an example of *biomagnification.*

6. Define and give an example of a *synergistic effect.*

7. Give the major sources of toxic chemical wastes.

8. Describe the traditional methods of disposing of wastes.

9. What brought a curtailment to indiscriminant discharges? What laws were (are) involved?

10. What alternatives for disposal were adopted? What results (good and bad) were gained?

11. Name and describe three methods of land disposal. Describe how each may be designed to protect groundwater but may fail to do so.

12. State two basic problems of land disposal. Describe the shortfalls that occurred in each area.

13. Discuss the scope of the toxic waste problem, giving volume of wastes and estimates of disposal sites that may be leaching wastes into groundwater.

14. List four major areas of the toxic waste problem which must be addressed.

15. What law protects you from toxic wastes in drinking water? What are the safeguards? the shortcomings?

16. What law provides for cleaning up toxic waste sites. Discuss progress and shortcomings.

17. Is *groundwater remediation* possible? How?
18. Cite two laws that apply to the current management of toxic wastes. Discuss the requirements and shortcomings of each.

  1. The Clean Water Act   285
  2. The Resources Conservation and
     Recovery Act   285
CASE STUDY: CANADA'S WHITE WHALES
ARE DYING   288

  E. Future Management of Hazardous Wastes 287
     1. Waste Reduction, Reclamation, and
        Recycling   287
     2. Incineration   289
     3. Biodegradation   289

  F. Reducing Occupational and Accidental
     Exposures   289
     1. Right-to-Know Legislation   290
     2. Accidents and Emergency Response   290

  G. What You Can Do   292

19. Describe the idea of *interim permits*. Why are they necessary? What is their shortcoming?

20. Discuss what can be done regarding future management of toxic wastes? What more environmentally sound methods of reducing or disposing of toxic wastes are available? Describe each.

21. Discuss laws that apply to reducing occupational exposures to toxic chemicals and injuries resulting from accidents involving hazardous materials.

22. Discuss what you can do to protect your own health and that of others from exposures to toxic wastes.

O ver the past several decades, groundwater has become an increasingly important resource. Between 1950 and 1980 groundwater withdrawals in the United States nearly tripled and currently provide about half of domestic needs and about 40 percent of irrigation needs. In addition, about 750,000 new wells are drilled each year. Traditionally, with few exceptions, groundwater has been of exceptional quality, meeting drinking water standards with no treatment.

Unfortunately, cases of high-quality groundwater becoming polluted with toxic chemicals are becoming increasingly common. As a result, some serious illnesses have occurred, wells have been closed, and people have been forced to bring water from other sources or to resort to costly purification processes (Fig. 11–1). Moreover, the threat of such pollution is becoming increasingly widespread. Indeed, groundwater pollution was recognized as one of the top environmental issues of the 1980s, and it undoubtedly will remain so through the 1990s and beyond. What are the toxic chemicals that are contaminating groundwater? Where are they coming from? What can be done to prevent further contamination? How can polluted groundwater be remediated? These are the questions to be addressed in this chapter.

FIGURE 11–1
Residents in McKeesport, Pa, getting army-purified water because the municipal supply was contaminated. As groundwater is contaminated with toxic materials, more and more people must seek other sources which may be inconvenient, costly, and limited in quantity. (Grapes Michaud/Photo Researchers.)

SOURCES OF GROUNDWATER POLLUTION

As water infiltrates and percolates through the soil, it will tend to dissolve and carry any soluble chemical into the groundwater. The soil will not filter out chemicals that are in solution. It is the old problem of *leaching* described in earlier chapters. Consequently, the general principle is: *Any chemical used on, disposed of, spilled, or leaked onto or into the ground can contaminate groundwater* (Fig. 11–2).

Major sources of groundwater pollution are currently recognized as the following:

☐ Inadequate landfills and other facilities where toxic chemicals have been dumped and from which they may leach into groundwater.

☐ Leaking underground storage tanks or pipelines. The leakage of gasoline from service station storage tanks is a particular problem.

☐ Pesticides and fertilizers used on croplands, lawns, and gardens.

☐ Deicing salt, used on roads.

☐ Waste oils used on dirt roads to keep dust down.

☐ Overapplication of sewage sludges or wastewater.

☐ Transportation spills.

Of all these, inadequate disposal and use of pesticides are considered to be the most widespread threat to groundwater.

TOXIC CHEMICALS: THEIR THREAT

The most insidious of all groundwater pollution problems involves certain *toxic chemicals* that may go undetected because of being at very low concentrations, but which may gradually accumulate in the body to cause many adverse health effects including cancer.

What Are the Toxic Chemicals?

Most toxic chemicals belong to one of two classes: *heavy metals* or *synthetic organic* chemicals.

FIGURE 11–2
Any chemical used, stored, spilled, or disposed of on or near the surface may leach into the groundwater. This illustration depicts some of the most significant sources of groundwater contamination.

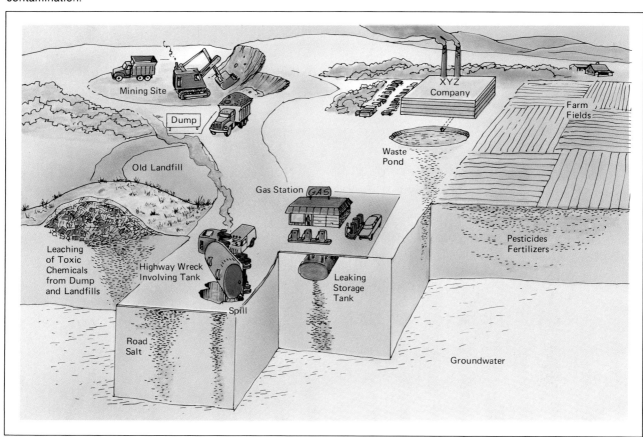

Heavy Metals

Heavy metals are metallic elements that in pure form are heavy. Lead, mercury, arsenic, cadmium, tin, chromium, zinc, and copper are examples. They are widely used throughout industry. Heavy metals are extremely toxic because, as ions or in certain compounds, they are soluble in water and may be ingested and absorbed into the body where they tend to combine with and inhibit the functioning of particular enzymes. Thus, very small amounts can have severe physiological or neurological consequences. The mental retardation caused by lead poisoning and the insanity and crippling birth defects caused by mercury are particularly well known.

Synthetic Organics

Recall that all the complex molecules that make up plants and animals are *natural* organic chemicals. In contrast, chemists have learned to make hundreds of thousands of organic (carbon-based) compounds that are the basis for all plastics, synthetic fibers, synthetic rubber, paintlike coatings, solvents, pesticides, wood preservatives, and innumerable other chemicals. Such human-made organic compounds are referred to as *synthetic* organic chemicals.

Many synthetic organic compounds are similar enough to natural organic compounds that they may be absorbed into the body and interact with particular enzymes or other systems, but here they cause problems. The body may not be able to break them down or metabolize them in any way; they are *nonbiodegradable*. The result is that they upset the system. With sufficient doses the effect may be acute poisoning and death. However, with low doses over extended periods, the effects are even more insidious, including **mutagenic** (mutation-causing), **carcinogenic** (cancer-causing), and **teratogenic** (birth-defect–causing) effects. In addition, they may cause serious liver and kidney dysfunction, sterility, and numerous other physiological and neurological problems.

A class of synthetic organics that is particularly troublesome is the **halogenated hydrocarbons,** organic compounds in which one or more of the hydrogen atoms have been replaced by an atom of chlorine, bromine, fluorine, or iodine. These four elements are classed as *halogens;* hence the name halogenated hydrocarbons.

Among all the halogenated hydrocarbons, those containing chlorine and referred to as **chlorinated hydrocarbons** are by far the most common. Such compounds are widely used in plastics (e.g., polyvinyl chloride), pesticides (e.g., DDT, kepone, and mirex), solvents (e.g., carbon tetrachlophenol), electrical insulation (e.g., PCBs or polychlorinated biphenyls), flame retardants (e.g., TRIS), and many other products. *PCBs* and *dioxin* are examples of chlorinated hydrocarbons that are notorious for their pollution hazard. Others are given in Table 11–1 and Fig. 11–3.

The Problem of Bioaccumulation

Both heavy metals and halogenated hydrocarbons are particularly insidious because they tend to *bioaccumulate*. **Bioaccumulation** means that small, seemingly harmless doses received over a long period of time may *accumulate* in the body, finally reaching toxic levels and causing harm. Bioaccumulation occurs because, first, these chemicals are *nonbiodegradable*. Heavy metals, being elements, cannot be broken down or destroyed by any chemical process. Chlorinated hydrocarbons may be broken down under certain conditions such as very high temperature but, in general, organisms lack enzymes for breaking them down. Second, these chemicals are readily absorbed into the body, but they are excreted very slowly if at all. The body's inability to excrete these substances occurs because heavy metals bind tenaciously with proteins, and halogenated hydrocarbons are very fat soluble and sparingly soluble in water. These characteristics prevent these chemicals from being flushed out of the body by urination. In essence, the body acts as a filter removing and accumulating these chemicals from all the food or liquids that pass through.

Bioaccumulation may be compounded in a food chain, as shown in Figure 11–4. Organisms at the bottom of the food chain absorb the chemical from the environment and accumulate it in their tissues. In feeding on these organisms, animals at the second trophic level receive a higher dose and accumulate still higher concentrations in their tissues, and so on up the food chain. Organisms at the top of the food chain may thus accumulate levels that are as much as 100,000 times higher than environmental concentrations. It is hardly surprising that such concentrations have lethal effects. This concentrating effect that occurs through a food chain is referred to as **biomagnification.**

One of the most distressing aspects of bioaccumulation and biomagnification is that you may have no indication it is occurring until the damaging level is reached, by which time it is too late to do much if anything about it.

The danger of bioaccumulation and biomagnification of chlorinated hydrocarbons became clear in the 1960s when it was discovered that diebacks in populations of many species of predatory birds, such as the bald eagle and osprey, were due to bioaccumulation of DDT, a chlorinated hydrocarbon pesticide. Numerous sport and commercial fishing areas have been closed because dangerous levels of PCBs

Table 11–1
Examples of Toxic, Synthetic Organic Compounds Frequently Found in Chemical Wastes

CHEMICAL	KNOWN HEALTH EFFECTS	MUTATIONS	CARCINOGENIC	BIRTH DEFECTS	STILL BIRTHS	NERVOUS DISORDERS	LIVER DISEASE	KIDNEY DISEASE	LUNG DISEASE
Benzene		X	X	X	X				
dichlorobenzenes		X			X	X	X		
hexachlorobenzene		X	X	X	X	X			
Chloroform			X	X	X		X		
Carbon tetrachloride			X		X	X	X	X	
Ethylene									
chloroethylene (vinyl chloride)		X	X			X	X		X
dichloroethylene		X	X		X	X	X	X	
tetrachloroethylene			X			X	X	X	
trichloroethylene		X	X			X	X		
Heptachlor		X	X		X	X	X		
Polychlorinated biphenyls (PCB's)		X	X	X	X	X	X		
Tetrachlorodibenzo dioxin		X	X	X	X	X	X		
Toluene		X			X	X			
chlorotoluenes		X	X						
Xylene				X	X	X			

Source: Adapted from "Hazardous Waste in America" by S. Epstein, L. Brown, and C. Pope. Copyright © 1982 by Samuel S. Epstein, M.D., Lester O. Brown and Carl Pope. Reprinted with permission of Sierra Club Books.

FIGURE 11–3
Chlorinated hydrocarbons are organic compounds in which chlorine (Cl) has been substituted for hydrogen (H).

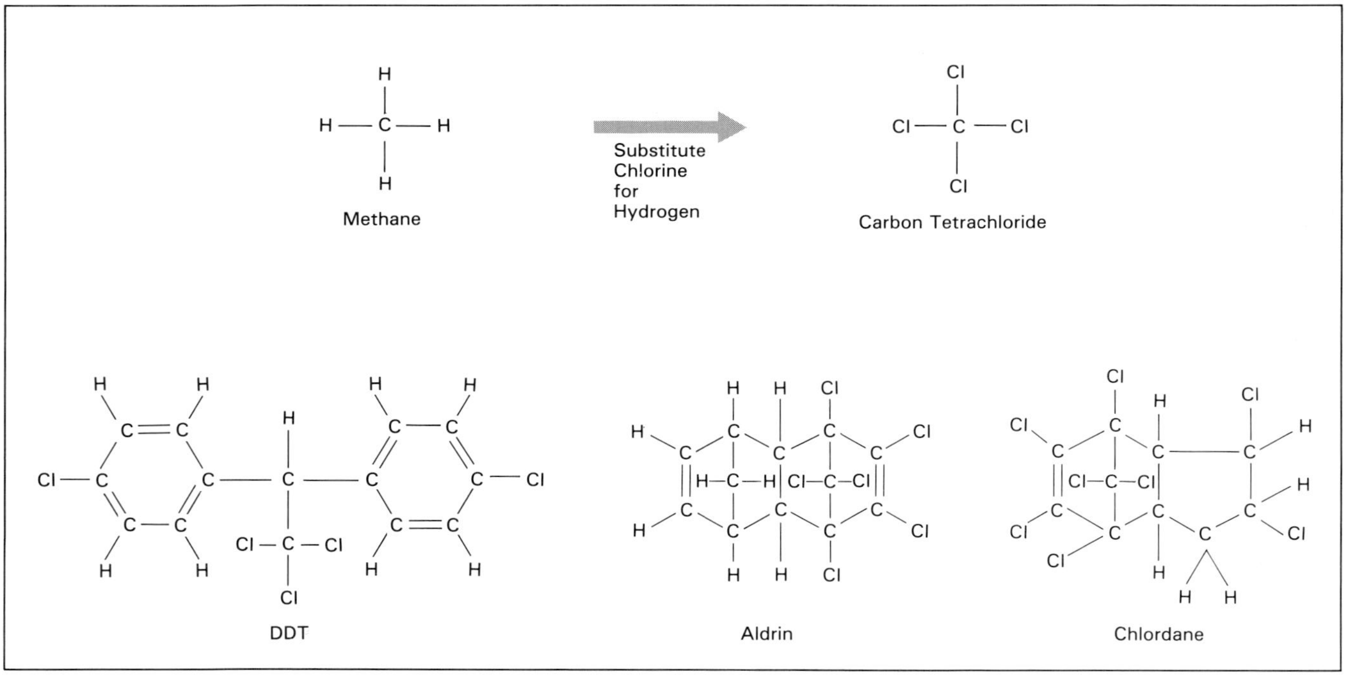

Methane → Substitute Chlorine for Hydrogen → Carbon Tetrachloride

DDT   Aldrin   Chlordane

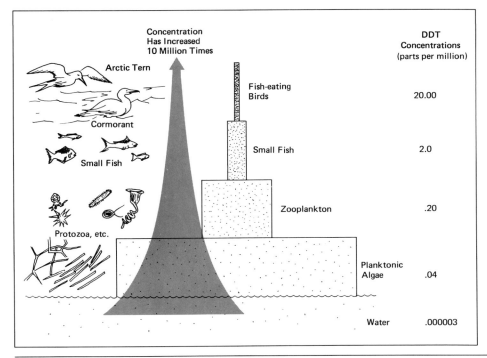

FIGURE 11-4
Biomagnification. When the phenomenon of bioaccumulation is put into the context of a food chain, each successive consumer receives a more contaminated food supply and, in turn, accumulates the contaminant to yet a higher level. For example, scientists have observed that the concentration of the pesticide DDT was magnified some 10 million-fold as it passed through the food chain shown.

and other chlorinated hydrocarbon compounds have accumulated in fish.

A tragic episode in the early 1970s known as the **Minamata disease** revealed the potential for biomagnification of mercury and other heavy metals. The disease is named for a small fishing village in Japan where the episode occurred. In the mid-1950s, cats in Minamata began to show spastic movements, followed by partial paralysis and, later, coma and death. At first this was thought to be a peculiar disease of cats, and little attention was paid to it. However, concern escalated quickly when the same symptoms began to occur in people; additional symptoms such as mental retardation, insanity, and birth defects were also observed. Scientists and health experts eventually diagnosed the cause as acute mercury poisoning.

A nearby chemical company was discharging wastes containing mercury into a river that flowed into the bay where the Minamata villagers fished. The mercury, which settled with detritus, was first absorbed by bacteria and then further concentrated as it passed up the food chain through fish to cats and humans. Cats had suffered first and most severely because they fed almost exclusively on the remains of fish. By the time the situation was brought under control, some 50 people had died and 150 had suffered serious bone and nerve damage. Even now, the tragedy lives on in the crippled bodies and retarded minds of Minamata descendants.

Synergistic Effects

Complicating the situation is the problem of synergistic effects. Toxic chemicals seldom occur singly. Often, as many as a dozen or more occur simultaneously, and frequently two or more chemicals act together to produce an effect that is greater than the sum of the effects caused by the two acting separately. This is known as a **synergistic effect.** An extremely serious synergistic effect has come to light recently. Certain halogenated hydrocarbons, and perhaps other toxic chemicals as well (one factor), weaken the immune system so that the organism may fall victim to various parasitic and disease organisms (the second factor). This is the suspected cause for a tragic dieoff of seal populations in the North Sea, which has occurred in recent years.

ENVIRONMENTAL CONTAMINATION WITH TOXIC CHEMICALS

The major source of environmental contamination with toxic chemicals is the disposal of chemical wastes. Use of pesticides is another prime source that will be considered further in Chapter 15.

Major Sources of Toxic Chemical Wastes

Wastes containing heavy metals come largely from metal refining, processing, and plating, and from the production of pigments for paints. Synthetic organic wastes come mainly from chemical and related industries that produce plastics, soap, synthetic rubber, fertilizers, synthetic fibers, medicines, detergents, cosmetics, paints, adhesives, pesticides, and explosives.

The waste chemicals are, or are contained in:

☐ Byproducts and "leftovers" of various chemical processes.

☐ Spent processing, cleaning, and lubricating materials.

☐ Wash water from cleaning finished products, equipment, and containers.

☐ Residues remaining in emptied nonreusable containers.

When it is not cost-effective to isolate, repurify and recycle chemicals from these sources, they are treated as wastes to be disposed of.

Background of the Toxic Waste Problem

Indiscriminate Discharge into Air and Water

Historically, chemical wastes have been disposed of as expediently as possible. It was once common practice to exhaust all combustion fumes up smoke stacks and to vent all evaporating materials into the air. All waste liquids and wash water contaminated with all kinds of materials were flushed into sewer systems, or directly into natural waterways. Many human health problems occurred, but they were either not recognized as being caused by the pollution or were simply accepted as the "price of progress." Indeed much of our understanding regarding human health effects of these materials is derived from these uncontrolled exposures. For example the expression "mad as a hatter" comes from the fact that people who made hats in the 1800s frequently became insane. The insanity, it was later found, was caused by the mercury used in the production process.

As production expanded, and as synthetic organics came into widespread use following World War II, many streams and rivers became essentially open chemical sewers. They were not only devoid of life; they were hazardous. For example, in the 1960s

the Cuyahoga River, which flows through Cleveland, Ohio, carried so much flammable material that it actually caught fire and destroyed seven bridges before it burned itself out.

Incidents such as this plus increasing recognition of adverse health effects and public outcry against these highly visible problems pushed Congress to pass the Clean Air Act of 1970 and the Clean Water Act of 1972. As a result of this legislation, industry has spent billions of dollars on pollution control equipment. Direct discharges of wastes into air and water were greatly reduced, water quality in many areas improved dramatically, and fish returned to many formerly polluted waterways. Progress notwithstanding, however, there is still a long way to go, as we shall discuss later.

Shift to Land Disposal

Curtailing the discharge of wastes into air and water, however, does not make them disappear; they must be put somewhere else. In the early 1970s, wastes taken out of air and water were largely redirected toward land disposal. Unfortunately, the potential shortcomings and dangers of this redirection were not adequately considered. Consequently, while air and surface-water quality improved during the 1970s, land disposal and the potential for groundwater pollution increased enormously. This will become clear as we discuss the methods of land disposal.

Methods of Land Disposal

There are three methods for land disposal of hazardous wastes: (1) *deep well injection*, (2) *surface impoundments*, and (3) *landfills*. Even with "state of the art" safeguards, some risk remains that these methods will lead to groundwater contamination. Worse, in the beginning safeguards were seldom applied, making groundwater contamination virtually inevitable.

Deep Well Injection

Some 57 percent of hazardous wastes are presently disposed of by deep well injection. This method involves drilling a "well" into dry, porous material below the groundwater (Fig. 11–5). In theory, hazardous waste liquids pumped into the well soak into the porous material and remain isolated from groundwater by impermeable layers. However, it is impossible to guarantee that fractures in the impermeable layer will not eventually permit wastes to escape and contaminate groundwater. Indeed, stresses produced by the introduction of wastes may even cause such fractures. Also, there are a number of other ways in which wastes can escape into groundwater as shown in Figure 11–5.

DISPOSAL OF HAZARDOUS WASTES BY DEEPWELL INJECTION

THEORY

A Well Is Drilled into a Dry Porous Layer and Wastes Are Pumped in. Contamination of Groundwater Is Prevented by Casing and Seal around the Portion of the Well That Penetrates Groundwater

Waste Storage

①

Injection Well

Casing

Seal

Shallow Groundwater

Impervious

Deep Groundwater

②

③

Impervious Layer

④

Dry Porous Strata

PRACTICE

1. Spills or Leaks of Wastes at Surface
2. Corrosion of Casing Allows Waste Escape
3. Inadequate Seal Permits Wastes to Back-flow
4. Fractures, Existing or Caused by Earthquakes or the Introduction of Fluids, Allow Wastes to Escape into Groundwater

FIGURE 11–5
Injection well, a technique that is used for disposal of large amounts of liquid wastes. The supposition is that toxic wastes may be drained into dry porous strata within the earth, where they may reside harmlessly "forever." Precautionary measures to make this method safe are listed on the left. Often the installation of these measures is inadequate and, even when it is adequate, potential for failure remains (right). (Adapted with permission from *Environmental Action* magazine, 1525 New Hampshire Ave. NW., Washington, D.C. 20036.)

Surface Impoundments

Another 38 percent of hazardous wastes are deposited in surface impoundments. This is the least expensive way to dispose of large amounts of water carrying relatively small amounts of hazardous materials (industrial wash water for example). The waste is discharged into a sealed pit or pond so that solid wastes settle and accumulate, while water evaporates (Fig. 11–6). If the bottom is well sealed and evaporation equals input, such impoundments may receive wastes indefinitely. However, inadequate seals may allow wastes to percolate into groundwater, exceptional storms may cause overflows, and volatile materials, organic solvents for example, can evaporate into the atmosphere adding to air pollution problems (see Chapter 12) and eventually coming down with rainfall to contaminate water in other locations.

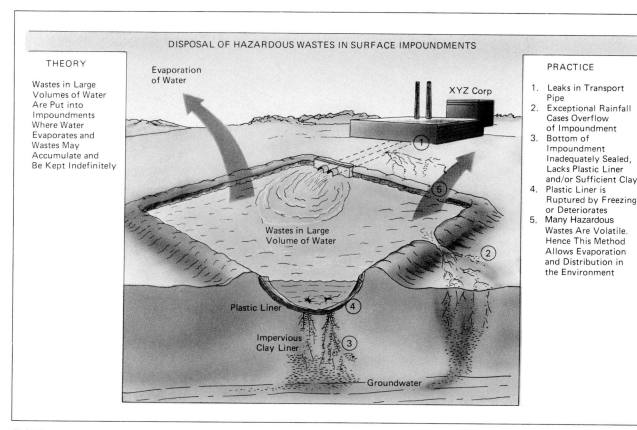

DISPOSAL OF HAZARDOUS WASTES IN SURFACE IMPOUNDMENTS

THEORY

Wastes in Large Volumes of Water Are Put into Impoundments Where Water Evaporates and Wastes May Accumulate and Be Kept Indefinitely

Evaporation of Water

XYZ Corp

Wastes in Large Volume of Water

Plastic Liner

Impervious Clay Liner

Groundwater

PRACTICE

1. Leaks in Transport Pipe
2. Exceptional Rainfall Cases Overflow of Impoundment
3. Bottom of Impoundment Inadequately Sealed, Lacks Plastic Liner and/or Sufficient Clay
4. Plastic Liner is Ruptured by Freezing or Deteriorates
5. Many Hazardous Wastes Are Volatile. Hence This Method Allows Evaporation and Distribution in the Environment

FIGURE 11–6
Surface impoundments, an inexpensive technique that is used for disposal of large amounts of lightly contaminated liquid wastes. The supposition is that only water leaves the impoundment, by evaporation, while wastes will remain and accumulate in the impoundment indefinitely. Precautionary measures required to make this method safe are given on the left. Frequently these measures are not installed or are inadequate, but even when they are adequate, potential for failure remains.

Landfills

When hazardous wastes are in a concentrated form, they are commonly put into drums and buried in **landfills.** If a landfill is properly lined, covered, and supplied with a means to remove **leachate,** material that leaks through, as shown in Figure 11–7, it is presumed safe; it is called a **secure landfill.** However, as noted in the figure, the various barriers are subject to damage or deterioration. Many experts feel it is only a question of time before the contents will leach from even the most secure landfills.

Problems in Managing Land Disposal

Two problems are inherent in land disposal. The first is ensuring that wastes get to disposal facilities. The second is ensuring that they stay there; that is, ensuring that disposal facilities are properly constructed, managed, and sealed. Problems in both areas became strikingly apparent during the 1970s.

Midnight Dumping

When the Clean Water Act restricted flushing chemical wastes into sewers or rivers, many companies were caught without alternative means of disposal. What were they to do with them? As stacks of drums of hazardous wastes "mysteriously" appeared in abandoned warehouses, vacant lots, or in municipal landfills, where their dumping is also illegal, it became clear that some highly disreputable individuals were taking advantage of the situation. For a fee, they were offering to dispose of the wastes. Then they were simply pocketing the money, leaving the toxic chemicals in any untended locations they could find, often under the cover of darkness, then disappearing without a trace. Thus, no one could be found who was legally responsible for the waste. This practice was dubbed **midnight dumping** (Fig. 11–8). Also, some companies simply stored wastes on their own properties with equally bad results (Fig. 11–9).

Nonsecure Facilities

Just as bad, much of the "legal" disposal was in landfills, impoundments, and wells that had few if any

FIGURE 11–7
(a) Secured hazardous waste landfill under construction in Alabama. (Nancy Shute.) (b) Precautionary measures to make this method safe are listed on the left. Frequently the execution of these measures is inadequate, but even when they are installed properly, many potentials for failure remain. (Adapted from an illustration by Rick Farrell, copyright ©. All rights reserved.)

(b)

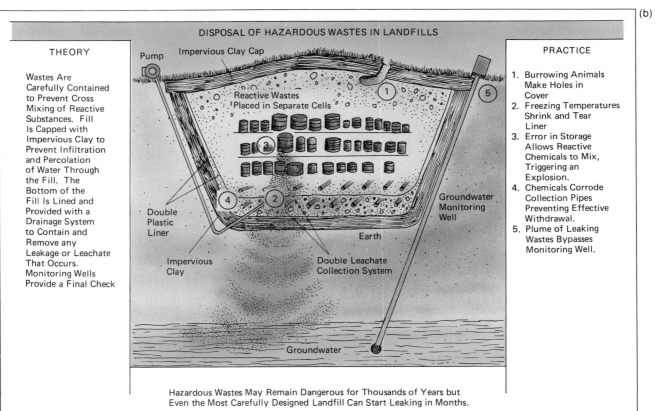

DISPOSAL OF HAZARDOUS WASTES IN LANDFILLS

THEORY

Wastes Are Carefully Contained to Prevent Cross Mixing of Reactive Substances. Fill Is Capped with Impervious Clay to Prevent Infiltration and Percolation of Water Through the Fill. The Bottom of the Fill Is Lined and Provided with a Drainage System to Contain and Remove any Leakage or Leachate That Occurs. Monitoring Wells Provide a Final Check

PRACTICE

1. Burrowing Animals Make Holes in Cover
2. Freezing Temperatures Shrink and Tear Liner
3. Error in Storage Allows Reactive Chemicals to Mix, Triggering an Explosion.
4. Chemicals Corrode Collection Pipes Preventing Effective Withdrawal.
5. Plume of Leaking Wastes Bypasses Monitoring Well.

Pump
Impervious Clay Cap
Reactive Wastes Placed in Separate Cells
Double Plastic Liner
Impervious Clay
Double Leachate Collection System
Groundwater Monitoring Well
Earth
Groundwater

Hazardous Wastes May Remain Dangerous for Thousands of Years but Even the Most Carefully Designed Landfill Can Start Leaking in Months.

of the precautions noted above to make them secure. The folly of this was dramatically brought to the public attention by "Love Canal." Love Canal was an abandoned canal bed near Niagara Falls, New York. In the 1930s and 1940s it served as a convenient burial site for thousands of drums of waste chemicals—over 20,000 tons in all. After the canal was filled and covered, the land was transferred to the city of Niagara and developers. Homes and a school were built on the edge of what had been the old canal, and the area of covered chemicals became a playground. Over the

years, parents observed that children who attended the school and came in contact with "black gooey stuff" oozing out of the ground had health problems ranging from chemical burns and skin rashes to severe physiological and nervous disorders. Even more alarming, residents began to note that an unusually high number of miscarriages and birth defects were occurring. In one neighborhood 9 out of 16 newborns had abnormalities. The average occurrence for such abnormalities is about 1 in 100. The situation climaxed in 1978 when health authorities finally came in and

FIGURE 11–8
Midnight dumping. In the past, large quantities of hazardous wastes were simply dumped and abandoned with no precautionary measures whatsoever. Frequently dumping was done under the cover of darkness to escape notice, hence the term *midnight dumping.* Thousands of such "midnight dumps" remain scattered about the country. (Mark Sherman/ Bruce Coleman, Inc.)

identified the black ooze as a potent mixture of numerous chlorinated hydrocarbons known to cause birth defects and other disorders in experimental animals.

Some $3 billion in health claims were filed against the City of Niagara, several hundred times the city's annual operating budget, and nearly 600 families demanded relocation at state expense. Eventually the state did purchase about 100 homes, but most people had only added frustration to their other problems.

Along with the episode at Love Canal, numerous lesser incidents of groundwater contamination from toxic wastes occurred. Many of these incidents were discovered only after people experienced "unexplainable" illnesses over prolonged periods. While they did not receive the media attention of Love Canal, they were nevertheless devastating to the people involved.

Scope of the Problem

The shocking fact is that prior to the mid-1970s no one was paying any real attention regarding how much and what kinds of wastes were being generated or where and how they were being disposed. That disposing of wastes in the ground might endanger public health through contaminating groundwater had not been considered. However, with the recognition of midnight dumping, the episode at Love Canal, and various other instances of groundwater contamination, toxic wastes quickly became a priority issue.

The Environmental Protection Agency now estimates that there are close to 150 million tons of hazardous wastes generated each year in the United States (close to two-thirds of a ton per person), and, in the late 1970s, as much as 90 percent was still being disposed of improperly, that is, in nonsecure facilities. For example, landfills did not have liners or leachate collection systems; some landfills were found where wastes were actually immersed in groundwater. Surface impoundments did not have sealed bottoms so percolation was going directly into groundwater. Injection wells were found where wastes were being deposited above or even into aquifers. Even in new sites designated for toxic wastes, EPA found that monitoring systems were inadequate to detect leakage should it occur.

The number of sites with inadequately contained chemical wastes is astounding. In 1987, the World Resources Institute estimated that in the United States alone, as many as 75,000 active industrial landfill sites along with 180,000 surface impoundments and 200 other special facilities may be possible sources of groundwater contamination. In most cases the contaminated area is relatively small, 200 acres or less. Nevertheless, as toxic compounds continue to leach and spread from these sites, significant portions of groundwater could become contaminated. Already, thousands of individual wells, some in every state, and some major municipal wells have been closed as a result. In Los Angeles seven municipal wells were closed because of toxic chemical contamination until purification devices could be installed. Furthermore, surveys show that drinking water supplies in many cities are tainted with synthetic organic chemicals. Very low levels are generally considered not dangerous; however, the level that constitutes danger when one is dealing with material that bioaccumulates is a matter of extreme controversy.

(a)

(b)

FIGURE 11-9
Nonsecure disposal of toxic wastes. In addition to midnight dumping, huge quantities of toxic wastes were just deposited on or in sites with few if any of the features required for security. Thousands of such nonsecured sites containing toxic wastes remain a serious threat to groundwater. (a) Hazardous waste stored in barrels near Chatfield Reservoir, Denver, Colorado, area. (Kent & Donna/Photo Researchers.) (b) Site clean up, Chemical Management, Elizabeth, N.J. (Courtesy of J. B. Moore, CH2M HILL.)

CLEANUP AND MANAGEMENT OF TOXIC WASTES

From the preceding discussion, it should be clear that there are four major aspects to the toxic waste problem.

- ☐ The need to assure safe drinking water and irrigation water supplies.
- ☐ The need to clean up thousands of existing sites from which toxic materials are already or may soon be leaching into the groundwater.
- ☐ Remediation of contaminated groundwater.
- ☐ The need to provide for proper management and disposal of hazardous wastes being generated now as well as those that will be generated in the future.

Considerable progress has been made in all four areas.

Assuring Safe Drinking-Water Supplies

To protect the public from the risk of toxic chemicals contaminating drinking water supplies, Congress passed the Safe Drinking Water Act of 1974. It mandates setting standards regarding allowable levels of pollutants and the monitoring of municipal water supplies. If specified toxic chemicals are found, supplies may be closed until adequate purification procedures or other alternatives are adopted. It also pro-

vides for proper location and construction of injection wells.

However, three major shortcomings remain. First, there is no provision for systematic monitoring of groundwater or private wells. Therefore, groundwater contamination is seldom recognized until people experience "unexplained" illness and report "funny" tasting or smelling water. Then, even after analyzing the water and verifying the presence of synthetic organic chemicals, action may not be taken because standards regarding what constitutes "unsafe" versus "safe" levels have not been set for many of the chemicals involved. Finally, closing wells is hardly a solution. It imposes the hardship of getting water from an alternative source or installing purification measures that may be prohibitively expensive (Fig. 11–10). Clearly, there is a need to have more extensive water testing and to identify and eliminate sources of pollution.

Cleaning Up Existing Toxic Waste Sites

A major federal program aimed at identifying and cleaning up existing waste sites was initiated by the Comprehensive Environmental Response, Compensation, and Liability Act of 1980, popularly known as **Superfund.** Through a tax on chemical raw materials, this legislation provided a fund of 1.6 billion over the period 1980–1985 to identify and clean up sites that posed a threat to groundwater. However, the Environmental Protection Agency's (EPA's) record in administering this program over the first five years was disgraceful.

Many feel that at least 10,000 to 15,000 sites are in immediate need of remediation to prevent serious groundwater contamination. Yet, getting a site on EPA's "Superfund list," an official list of sites eligible

FIGURE 11–10
When water supplies become seriously contaminated, there may be no alternative but to abandon the property, even though it may be financially ruinous to do so. (Photo by Thomas Busier.)

to be cleaned up under the Superfund program, has been a difficult and slow process, not without political favoritism involved. So far, fewer than 1 out of 10 sites in need of remediation have been officially listed. Some have made it to the list only to be deleted again. And even then, by 1987 only about a dozen sites had actually been cleaned up. EPA has been accused of being "an inch wide and a mile deep," meaning that certain sites have been very extensively studied and cleaned up to a super degree, while essentially no attention has been paid to other sites. Finally, EPA has been accused of making Superfund into a "super shell game," taking toxic wastes from one site and putting them into another, which may just perpetuate the problem at a later time.

Through the Superfund Amendments and Reauthorization Act of 1986 (SARA), Congress increased the fund to 8.6 billion and extended it for another five years. Still the questions remain about what sites should be tackled first, how clean is clean, and how to treat or dispose of wastes removed. Nevertheless, with the increased funding, we should expect movement in the right direction.

Groundwater Remediation

It has frequently been said that once groundwater is contaminated it is effectively lost forever, because there is no way to purify an aquifer and it may be hundreds of years before wastes are eventually flushed out. Fortunately, not everyone believed this. A new technology of **groundwater remediation** has developed in recent years and is expanding rapidly. In general, the techniques involve drilling wells, pumping the contaminated groundwater, passing it through chemical adsorption filters, and reinjecting it until the pollutants are flushed out (Fig. 11–11). In cases where biodegradable organics are involved, oxygen and organisms may be injected into contaminated zones. The organisms feed on and eliminate the pollutants as in secondary sewage treatment. Groundwater remediation is particularly applicable where contaminated areas are relatively small, such as leaks from gasoline storage tanks.

Management of Wastes Currently Produced

The Clean Water Act

The Clean Water Act remains the cornerstone legislation. Under it, standards are set as to allowable levels of various pollutants in wastewater discharges. However, this proves to be a difficult balancing act between the ideal, no pollution, and economic practicality. A company could meet a demand for immediate stopping of all pollution only by shutting down and putting all its employees out of work. To avoid this, the Clean Water Act allows the granting of **interim permits,** which effectively say: You can go on polluting for the time being, but you must reduce discharges by X amount by Y date. Thus a schedule is made to phase out the pollution over a certain period of time without causing undue economic hardship. The schedule for pollution reduction is arrived at by negotiation between the company and the EPA or the state equivalent. Fines may then be levied for noncompliance with the schedule, but also, the interim permit may be renegotiated and extensions granted.

You can probably see the difficulties. If the government is weak, which it was in this area under the Reagan administration, interim permits are both generous to industry and extended indefinitely. Second, there is lax monitoring so that many violations go unnoticed. Third, fines may be so minor that a company finds it considerably less expensive to pay the fine than do anything about its pollution. Finally, small firms, homes, and farms are exempt from regulation. Thus, progress notwithstanding, there is still considerable dumping of toxic wastes, and many waterways remain highly polluted with severe environmental consequences as a result (see Case Study 11). Seals in the North Sea are dying off apparently as a result of the synergistic interaction between toxic wastes and disease-carrying organisms. Fish in many areas have abnormalities and cancerous growths.

The Resources Conservation and Recovery Act

To address the problem of midnight dumping and wastes going into inadequate disposal facilities, Congress passed the Resource Conservation and Recovery Act of 1976 (RCRA). RCRA requires records to be kept on amounts and kinds of all hazardous wastes from their point of origin to their ultimate disposal, fondly referred to as "cradle-to-grave" tracking of hazardous wastes. Furthermore, all landfills or other facilities receiving hazardous wastes must be *permitted*, meaning that various standards for construction and operation must be maintained. Finally, the hazardous-waste generator remains liable indefinitely, meaning that should the wastes cause a problem at any time in the future, the company that produced them remains responsible for cleaning up the mess. This prevents the situation: Company A gave the wastes to hauler B and B is no longer in business, so no one is responsible.

Under RCRA regulations, numerous facilities that were receiving toxic wastes have been shut down, many reverting to Superfund sites in need of

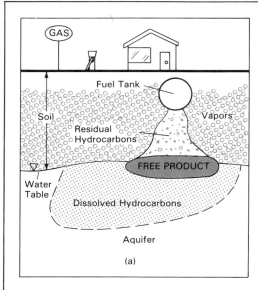
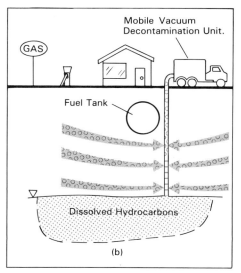
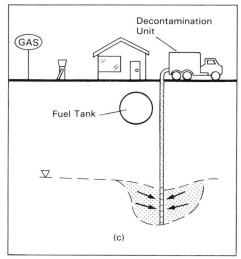

FIGURE 11-11
Groundwater remediation. (a) Typical subsurface contamination from a leak or spill of hydrocarbons such as gasoline. (b) After repairing the leak, vacuum extraction process causes free product and residual hydrocarbons in the soil and on the water table to evaporate and removes vapors preventing further contamination of the groundwater. (c) Contaminated groundwater is pumped and treated. Photograph of two vapor extraction wells, Vacuum Extractor Unit (silver box) and Vapor Water Separator (white tank). (Courtesy of Terra Vac., Inc., San Leandro, California)

remediation. And, very few new landfills meeting all the standards for liners, leachate collection systems, and monitoring systems have been opened. This has created a "crisis" in toxic waste disposal. With no better place to put the wastes, large amounts are still being allowed to go into inadequate facilities, especially those on the same property as the company producing them, under a system of interim permits as described above. A survey conducted by Congress in 1983 showed that only 24 of 8000 facilities met the full requirements.

Another issue is that significant quantities of hazardous and toxic wastes may escape regulation because operations that produce less than 100 kilograms (220 lb) per month are exempt from the law, as are homeowners and farmers. Surveys also show

that the quantity of pesticides and other toxic materials thrown away by homeowners, farmers, and small businesses is considerable. Sewage sludges are also not covered, although they may be heavily contaminated in cases where industries have been and may still be disposing of toxic wastes down sewers. Under the Clean Water Act such discharges were supposed to be phased out by now, but they continue in some cases under interim permits and illegally as noted above.

While federal law and its enforcement by EPA may be somewhat lax, a number of states are instituting much tougher standards. For example, California law now demands jail sentences rather than just fines for persons convicted of illegally disposing of toxic wastes.

Some companies are still working to circumvent the law rather than obey it, however. A new kind of midnight dumping has recently come to light. Shipments of toxic wastes, falsely labeled as products or raw materials, are going to less-developed countries where people, unaware and unsuspecting of their hazards, accept them being dumped without safeguards for token payments. Also numerous firms have moved their entire production facilities to less-developed countries where requirements for pollution control, safety precautions, worker protection, and so on are much weaker or nonexistent. Consequently, some Third World cities and rivers are gaining notoriety as being the world's most polluted. It was in this context that a leak at a Union Carbide pesticide manufacturing plant in Bhopal, India, in 1984 resulted in the release of an extremely poisonous gas, methyl isocyanate. Over 2000 people were killed immediately as the gas spread through the community (Fig. 11–12). Another 1300 have died since and over 20,000 have permanent injuries, largely blindness.

Also, under SARA (Community Right-to-Know legislation) companies are required for the first time to keep records of and report total emissions and discharges of toxic chemicals even though they are within the standards and guidelines of their permits under the Clean Air and Water Acts. Previously, it was only required that chemicals in discharges be below certain concentrations; the totals (concentration times volume of the emissions times the time period of discharge) were not calculated. Compiling and totaling this data for the first time in 1989, the Environmental Protection Agency came to the shocking recognition that some 13.5 billion pounds per year of toxic chemical wastes are still being *legally* discharged into the environment without control, 2.5 billion pounds into the air and the rest into the water and onto the land. This amounts to over 50 pounds per person per year. Clearly, this is unacceptable and demands further tightening of laws and standards.

There are constructive alternatives.

Future Management of Hazardous Wastes

It is widely recognized that, even at its best, *land disposal is not a sustainable solution*, because the lifetime of the toxic wastes will inevitably outlast the lifetime of protective barriers. And, there are alternatives.

Waste Reduction, Reclamation, and Recycling

The stringent requirements for new waste disposal facilities, the paperwork involved in the cradle-to-grave tracking, and the unending liability for the wastes, all add up to costs that are motivating companies to examine the potential for reducing waste outputs for economic if not humanitarian reasons.

Two general possibilities exist. One is to modify or change the production process so that smaller amounts of toxic waste products result. In many cases it has been possible to find nonhazardous substitutes. The second is to reclaim and recycle the toxic material from wastes. This is particularly applicable in the case of heavy metals, which can be removed from waste streams by a number of chemical processes, repurified, and reused.

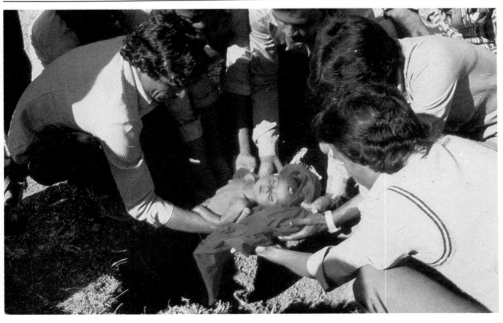

FIGURE 11–12
Many of the most disastrous "pollution" episodes are the result of accidents and/or carelessness. In 1984 a leak of methyl cyanate from a pesticide manufacturing plant in Bhopal, India, killed some 3300 persons and injured and blinded some 20,000 more. (Copyright Baldev-Sygma.)

CANADA'S WHITE WHALES ARE DYING

White beluga whales in Canada's St. Lawrence River—one of the world's most polluted waterways—are facing extinction, and scientists have been unable to pinpoint what is killing them. Last September, an international forum on the endangered belugas brought marine biologists, toxicologists, and environmentalists together in Tadoussac, Quebec. The conference ended without conclusion—and without optimism. One possibility, say researchers, is that a chemical soup of river pollutants weakens the whales' immune systems. The chemicals themselves could be killing the animals directly, too.

A century ago, the St. Lawrence was the home of 5,000 of the small white whales. Until 50 years ago, the Canadian government paid a bounty for each carcass, on the assumption that belugas were depleting fish stocks. During World War II, pilots in training used the whales for target practice. In the 1960s, large hydroelectric projects destroyed major areas of their natural habitat. As recently as 1968, they were hunted for oil, meat, and sport. Finally, in 1983, the Canadian government, thanks largely to a campaign by researcher Leone Pippard, known as "Our Lady of the White Whales," declared the belugas endangered.

Pippard, who lives on an island in the St. Lawrence, has studied and fought for the whales since 1975. "Because of the sewer [the St. Lawrence River] they live in, belugas are probably the most contaminated animals in the world," she says. "They've been in the area for 10,000 years, but we are on the verge of losing them."

Now that as few as 400 whales are left in the St. Lawrence, they have become big business for companies organizing whale-watching expeditions. "Seeing a whale changes your life. It turns you into an instant conservationist," one cruise organizer says. Last year, 50,000 visitors paid an average of $37 for a three-hour cruise on the St. Lawrence.

The origins of the river's poisons are hard to trace. Autopsies show that the pure white whales carry at least 24 different contaminants, including PCBs, achlorobenzene, hexachlorobenzene, heavy metals, mirex, and DDT. The animals have bladder cancer, pulmonary fibrosis, hepatitis, transitional cell carcinoma, and perforated ulcers. Many are so contaminated with PCBs that when they die, their corpses, under Canadian law, must be treated as toxic waste to keep the poison from returning to the environment.

The whales, sometimes called sea canaries because of their chirping song, can grow to 15 feet and weigh half a ton. Unlike other whales, belugas do not migrate from their St. Lawrence River habitat, east of Quebec City. Their home-sweet-home instinct makes

A beluga whale.
(A. W. Ambler/Photo Researchers)

it impossible for them to escape the pollutants in the river, which drains the Great Lakes. Nor can sick belugas be replaced by healthy ones from Alaska or the Canadian Arctic, because the latter would not survive the pollution. Canadian scientists say the toxins could originate in any U.S. state or in the two Canadian provinces along the Great Lakes and the St. Lawrence. Environmentalists say it is ironic that pressure from the U.S. has almost eliminated the commercial whale hunt, while at the same time American pollution may be contributing to the demise of beluga whales just a few miles north of the border.

Pierre Beland, who has performed 25 autopsies on belugas, says there may never be absolute proof that pollution kills them, since clinical studies are impossible with such large animals. In any case, he says, "there is a whole cocktail of chemical pollutants at work at once in the environment." But he believes that "the weight of public opinion will eventually negate the need for [absolute] proof," adding that popular pressure will compel governments to enact tough environmental protection laws. Many scientists, including Beland, believe that governments should not wait for scientific proof that pollution is destroying nature before stepping in.

Last summer, the Canadian and Quebec governments pledged more than $2.4 billion over the next 10 years to clean up the St. Lawrence and create a marine sanctuary for the belugas. However, even such major spending is likely to be too late. Scientists at another recent conference warned that toxic chemicals already accumulated in the Great Lakes will take years to flow out through the St. Lawrence. When they do, say these scientists, they will leave residues in the river bottom.

Meanwhile, researchers have recently discovered that the whales are suffering from a new form of contamination: a carcinogen called benzo-a-pyrene, or BaP, which is emitted during aluminum smelting. Tests conducted on beluga brain tissue at Tennessee's Oak Ridge Laboratories revealed massive amounts of BaP. Joseph Cummins, a geneticist at the University of Western Ontario, says, "There is no question that [BaP] is killing the belugas."

PETER BENESH
Observer News Service of London.
By permission of United Media

As a means to aid recycling and reuse, a number of states are now maintaining **waste inventories** or **exchanges,** open listings of each company's wastes. Through such listings, it may be found that the waste from one company is what another company can use. For example, ash in many cases is good for making bricks and ceramic products.

Incineration

Most synthetic organic compounds, like natural organic compounds, are flammable. Even certain flame-resistant chlorinated hydrocarbons can be broken down with oxygen to carbon dioxide, water, and harmless chlorine compounds. It is simply a matter of *retention time* and *temperatures* in the firebox.

Cement kilns have these conditions. Consequently most cement plants have recently adopted a second business—disposing of hazardous wastes. Organic wastes are mixed with the regular fuel oil and fed into the kiln, thus contributing fuel value in the process of being destroyed. Any resulting toxic ash is incorporated into the cement, which ultimately provides a safe "container."

Also, a number of companies have built or are in the process of building special incinerators specifically designed for destruction of chemical wastes. (Fig. 11–13). Unfortunately such efforts are frequently being blocked by people who see the facility as being a "toxic threat" to their community. Such people generally don't recognize that their actions are forcing the continuation of a much greater threat—disposal of toxic materials in inadequate landfills, impoundments, or injection wells.

Biodegradation

Synthetic organic compounds, as we have noted, are notorious for being nonbiodegradable. Gradually, however, bacteria are being discovered that will break down these compounds, albeit slowly. These organisms are being adapted through the techniques of "breeding" and selection similar in principle to those used in the development of agricultural plants and animals, so that we may soon have organisms that are fairly efficient in breaking down synthetic organic wastes. Then systems similar to those used in secondary sewage treatment (Chapter 10) may be used to dispose of toxic organic wastes.

Reducing Occupational and Accidental Exposures

Environmental contamination with toxic substances, since it involves lifelong exposures with unknown effects, poses a risk that must not be underestimated. However, the most severe exposures leading to injury and death continue to occur as a result of occupational

FIGURE 11–13
Hazardous waste incineration. Increasingly hazardous wastes are disposed of by incineration. When temperatures and retention times are sufficient, incineration will accomplish essentially 100 percent destruction of organic compounds. Photo is a thermal oxidation incinerator unit located in Roebuck, South Carolina. The process includes heat recovery, particulate removal and acid-gas scrubbing to allow the destruction of a wide variety of liquid waste materials. The entire process is a closed loop, computer controlled to insure that all waste is destroyed in an environmentally sound manner. (Courtesy of Thermal Oxidation Corporation, Roebuck, South Carolina)

exposures (working with and around toxic materials) and/or accidents. How are these areas being addressed?

Right-to-Know Legislation

Occupational Safety and Health Administration (OSHA) regulations require air quality standards and measures to protect workers. Still, a severe handicap exists because workers are often ill-informed regarding the nature and hazards of the materials they are working with. Nor, until recently, could they find out. Employers were not obligated to tell them and many withheld such information. This situation has been corrected by Title III of the Superfund Amendments and Reauthorization Act of 1986 (SARA, Title III), commonly referred to as "Right-to-Know." Under this legislation, employers are obligated to make such information available at least in the form of printed information. However, workers remain responsible for reading and understanding it.

Accidents and Emergency Response

As modern society uses increasing amounts and kinds of toxic materials, the stage is set for very simple mistakes or accidents to become wide-scale disasters. The above-mentioned disaster in Bhopal, India, is one example. Another occurred in 1973 when a few sacks of a fire retardant chemical got mixed up with an animal feed additive by a distributor in Michigan. The amounts actually consumed by livestock were not great. However, this chemical was PBB, a highly stable, bioaccumulating, halogenated hydrocarbon some five more times toxic than the notorious PCBs. As a result, numerous people, mostly farm families who consumed the eggs, milk, and meat from their animals, became sick, suffering varying degrees of nervous disorders. When the source of the problem was identified, about 500 farms had to be quarantined; 30,000 cattle, 1.5 million chickens, thousands of sheep and hogs, and tons of cheese, milk, and eggs had to be destroyed because of the contamination. The damage was estimated at about $100 million, not including compensation for individual human suffering. Moreover, the chemical remained in the Michigan ecosystem. Several years after the initial incident, reaccumulation from "unknown" sources still caused sporadic occurrence of PBB poisoning.

In still another incident, the entire town of Times Beach, Missouri, had to be evacuated and relocated and all existing structures destroyed after dirt roads were unwittingly sprayed with waste oil contaminated with dioxin that was later distributed throughout all the town's structures by a flood (Fig. 11–14).

Transport is an activity that is particularly prone to accidents. Over the course of a year, about 4 billion tons of hazardous materials, which include fuels, chemical raw materials, and products such as chlorine for water treatment, as well as toxic wastes, cross over roads, railroad tracks, and waterways in the United States. Numerous accidents involving tankers filled with hazardous chemicals have required the evacuation of residents. Many experts speculate that the fact that such accidents have not led to wide-scale injury and death is more a matter of good luck than good management.

But management and response is improving. The Department of Transportation has extensive regulations (DOT Regs) concerning the kinds of containers, and the packing of them, that can be used in the transport of various hazardous materials. This is to reduce the risk of spills, fires, poisonous fumes, and so on that may be generated in case of an accident.

On the other side is appropriate response to an accident. DOT Regs require that every individual container and the outside of a truck or railcar must carry a standard placard identifying the hazards (flammability, corrosiveness, potential for poisonous fumes, etc.) of the material inside. Increasingly, fire and/or police departments have special **emergency response** teams (Fig. 11–15). These are people who are specifically trained to know the potential dangers involved with each kind of hazardous material and how to respond to each to protect both themselves and the public. For example, can water, which may cause certain chemicals to explode, be used to contain a particular fire or is a chemical foam required? How far should people be kept away or evacuated for safety? What kinds of personal protective gear (respirators, rubber suits, etc.) are in order? Arriving on the scene of an

FIGURE 11–14
The town of Times Beach, Missouri, had to be evacuated and destroyed when it was found that roads had been treated with oil contaminated with dioxin and a flood had spread it throughout the entire town. (Environmental Protection Agency.)

FIGURE 11–15
Emergency response. Police and fire departments have emergency response teams, teams trained to handle emergencies involving hazardous materials. Shown here is a drill, a team responding to a simulated accident involving a rail tanker of chlorine gas and a truck tanker of alcohol. (Photo by author.)

accident, the emergency response team immediately assesses the dangers by the placards, suits up, and goes into action as necessary.

Laws applying to hazardous wastes are summarized in Figure 11–16.

What You Can Do

Recognize that in a modern technological society, while we may reduce or even ban the use of certain materials, hazardous and toxic materials and wastes are not going to go away. The solution must be: *responsible management.* Responsible management begins with your recognition, use, and disposal of hazardous and toxic materials in your home, school, and workplace. Read labels and make yourself informed about the hazards of the various materials you use in the home and workplace. Heed any warnings. Eliminate or minimize your use of materials with heavy

metals or chlorinated hydrocarbons, and do not empty unused quantities of such materials down the drain. Remember that receiving an injury is not heroic; it simply adds to the problem rather than the solution.

How can you help to reduce discharges of toxic wastes that are in violation of the Clean Water Act or RCRA? The Clean Water Act amendments of 1977 specifically authorize citizen groups to bring lawsuits against violators in federal court. Such lawsuits will require the company to come into compliance, pay fines for past violations, and award attorney's fees and costs to the group bringing the suit. Public Interest Research Groups (PIRGs), who have an affiliation in about 20 states, have a program of "stream walks" in which they train citizens to notice and identify signs of toxic contamination (discolored water, oil films, and peculiar odors) as they walk along streams. On confirmation and identification of sources, they alert the responsible state agency to begin cleanup efforts, and if satisfaction is not forthcoming, they may bring suit. Join or form a PIRG group in your area (see Appendix A). Such groups can also bring suits against the state for allowing overly lax interim permits.

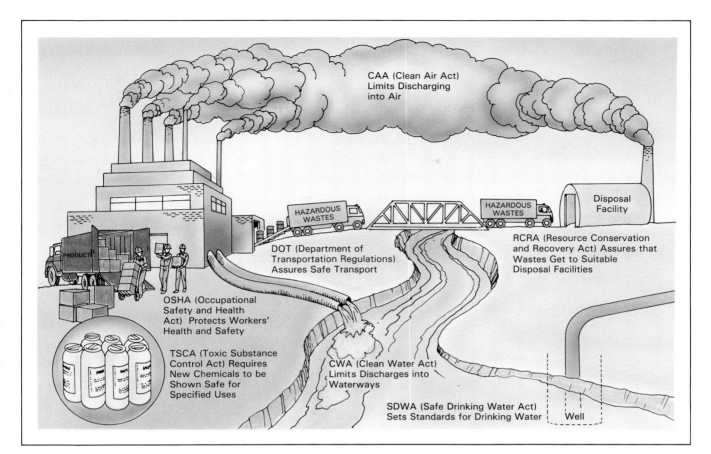

FIGURE 11–16
Laws applying to hazardous materials. The major laws pertaining to the protection of workers, the public, and the environment from hazardous materials are summarized in this illustration.

Investigate the practices of companies in your area regarding their handling of toxic wastes. Under RCRA each company must have a person who is trained and in charge of hazardous waste handling. Ask to see and get information from this person. Are disposal techniques/facilities safe to protect water supplies? Groups have been successful in bringing pressures to have companies upgrade their disposal standards. In this regard, be careful not to get automatically involved in opposing new hazardous waste incineration facilities because such opposition may simply prolong unsafe disposal. Your group should have an independent consulting engineer review the specifications of the new plant to be assured that there will be adequate air pollution control. However, in general, incineration of toxic organic wastes offers the potential for a much cleaner, safer environment than any form of land disposal.

Finally, it should not escape your notice that hazardous waste management and disposal, cleanup of toxic waste sites, groundwater remediation, and emergency response are growing areas in the economy that offer many job opportunities with the potential for rapid advancement. You may consider such a career for yourself.

12
Air Pollution and Its Control

CONCEPT FRAMEWORK

■ Outline

I. BACKGROUND OF THE AIR POLLUTION
   PROBLEM   296

II. MAJOR AIR POLLUTANTS AND
    THEIR EFFECTS   300
    A. Major Pollutants   300

    B. Adverse Effects of Air Pollution on
       Humans, Plants, and Materials   300

       1. Effects on Human Health   301
       2. Effects on Agriculture and Forests   301
       3. Effects on Materials and Aesthetics   305

CASE STUDY: HOW AIR POLLUTION IS AFFECTING
TREES   306

III. SOURCES OF POLLUTANTS AND CONTROL
     STRATEGIES   307
     A. Pollutants: Products of Combustion   307
        1. Direct Products   307
        2. Indirect Products   308

     B. Setting Standards   310

     C. Major Sources and Control Strategies   311
        1. Particulates   311
        2. Nitrogen Oxides, Hydrocarbons,
           Ozone and Other Photochemical
           Oxidants, and Carbon Monoxide   311
        3. Sulfur Dioxide and Acids   313
        4. Lead and Other Heavy Metals   314
     D. Taking Stock—What You Can Do   314

IV. INDOOR AIR POLLUTION   317

■ Study Questions

1. Define: *threshold level, dose.* How does the threshold level vary with concentration and time? Cite natural sources of air pollution. How does the biosphere handle them? How have humans changed the situation? Name and describe a weather condition that may aggravate the pollution problem. National action against air pollution was initiated by what law? List the four stages involved in improving air quality.

2. List the major air pollutants.

3. List three areas in which air pollution has adverse effects. Cite the adverse effects and name the pollutant(s) that are mainly responsible in each case.

4. Describe how smoking contributes to the adverse impact of air pollution.

5. Describe how each of the major air pollutants arises as a direct or indirect product of combustion.

6. Describe what is involved in setting air pollution standards.

7. For each of the pollutants listed in the outline opposite, cite its major source(s) and what has been done to control it. Discuss: Is present control sufficient? What changes or additional areas need to be addressed?

8. Discuss things you can do to contribute to maintaining or improving air quality.

9. Discuss: Is indoor air quality generally better or worse than outdoor air quality? Why? What are the sources of indoor air pollution? How may they be reduced?

oth plants and animals, as well as humans, are being adversely affected by poor air quality. The fact that sustaining the biosphere will depend on maintaining good air quality needs no elaboration. In this chapter, we will look at various aspects of this problem and what is being and still needs to be done to solve it.

BACKGROUND OF THE AIR POLLUTION PROBLEM

Organisms do have the capacity to deal with certain amounts or levels of pollutants without suffering ill effects. The level of a pollutant below which no ill effects are observed is called the **threshold level.** Above the threshold level effects begin to be observed. However, the effect caused by a pollutant depends on its concentration and the time of exposure. Higher levels may be tolerated if the time of exposure is short. Thus, the threshold level may be high for shorter exposures, but gets lower as exposure time increases (Fig. 12–1).

There are certain exceptions. Compounds that bioaccumulate (Chapter 11) have extremely low threshold levels, and radioactive compounds (Chapter 20), many feel, have a zero threshold level. That is, any exposure, regardless of how small, may cause

damage. However, the air pollution that we will be discussing in this chapter does not fall into these categories. Therefore it is important to keep in mind that it is not the absolute presence or absence of pollutants, but rather the *dose* that is important. **Dose** is defined as concentration (level) multiplied by the time of exposure.

Three factors determine the level of pollution:

☐ Inputs of pollutants into the air
☐ Amount of space into which pollutants are dispersed and diluted
☐ Outtakes—mechanisms that remove pollutants from the air.

Volcanoes, natural fires, and dust storms have sent smoke and other pollutants into the atmosphere for millions of years. But the biosphere does have means for removing, assimilating, and recycling these natural pollutants. They disperse and dilute in the atmosphere. Then they settle or come down to earth via precipitation, and poisonous gases are converted to harmless products by soil microorganisms (Fig. 12–2). For example, carbon monoxide and sulfur dioxide are converted to carbon dioxide and sulfate, which are plant nutrients. Thus natural inputs of pollution are kept well below threshold levels except, for example, in the immediate area of a volcanic eruption.

With the discovery of fire about 100,000 years ago, humans began adding to these natural pollutants. Our early ancestors probably did not consider the potential harmful effects of pollution, nor did their primitive technology provide any practical alternative. Thus, venting combustion and other fumes into the atmosphere became an accepted practice. It was assumed that natural processes would handle our additions and this assumption prevailed until the mid-1900s. Where air pollution occurred near certain industrial sources, it was considered to be simply a problem of dilution. "Dilution is the solution to pollution" was a commonly used phrase, and the attitude was "If you don't like it, move." This was not without some truth. A short distance away from pollution sources, air quality generally remained good.

In the 1950s, however, particularly with the mushrooming use of cars for commuting (see Chapter 22), whole metropolitan areas increasingly became enshrouded in a brownish haze on a daily basis (Fig. 12–3). This haze is called smog, or more properly, **photochemical smog,** because sunlight is involved in its formation as we shall see later.

Certain weather conditions intensify the smog levels, the most significant being *temperature inversions.* Normally, air temperature decreases with increasing elevation. In this situation, the warm air near

FIGURE 12–1
Threshold level. The threshold level of a pollutant is its concentration below which no ill effects will occur. However, the threshold level diminishes by a factor of 1000 or more with increasing exposure time. The threshold level differs for each pollutant. Also, different species and individuals may have very different threshold levels to a given pollutant, and the threshold level may differ depending on other pollutants or factors that may be causing stress.

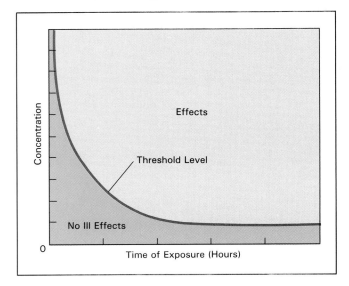

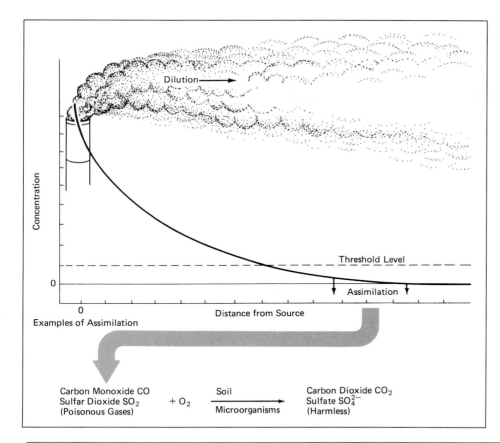

FIGURE 12–2
Dilution and assimilation of pollutants. As a pollutant mixes with a larger air mass, its concentration diminishes by dilution. With sufficient dilution, concentration is reduced to a threshold level, a level at or below which it is assumed to cause no ill effects. Further, soil microorganisms or other natural processes may absorb and assimilate pollutants, removing them from the system entirely. Thus, there is an assumption that "nature will take care of pollutants." Unfortunately modern civilization produces pollutants in such quantities or of such kinds that these assumptions do not hold.

FIGURE 12–3
A typical episode of photochemical smog. (a) Early in the morning, the air is clear. (Tom McHugh/Photo Researchers.) (b) Midmorning of the same day, the air very hazy with smog. (Georg Gerster/Photo Researchers.) The haze is the result of pollutants from the exhaust of rush-hour traffic reacting in the atmosphere. The reactions are promoted by sunlight. Photos are close to the same view over Los Angeles.

(a)

(b)

the ground rises, carrying pollutants upward and dispersing them at higher altitudes (Fig. 12–4a). In a **temperature inversion,** a layer of cold air near the ground is covered by warm air. This situation develops as an influx of cooler air moves in under the warm air. With a temperature inversion, the upward movement of air carrying pollutants is blocked and pollutants accumulate in the cool air near the ground (Fig. 12–4b). Local topography may also intensify the effects of a temperature inversion, as in Mexico City and Los Angeles where surrounding mountains prevent pollutants from moving horizontally.

Many people found smog the cause of headaches, nausea, irritation to eyes and throat, and aggravating to preexisting respiratory conditions such as asthma and emphysema. In some industrial cities under severe temperature inversions, air pollution reached such high levels that mortalities increased significantly (Table 12–1). These cases became known as *air pollution disasters.*

Even more, the ill effects were not limited to just cities. In one notorious episode in August 1969, a conspicuous buildup of smog covered almost the entire eastern United States from the Great Lakes to the Gulf Coast. Further, an increase in atmospheric turbidity (haze) was observed globally (Fig. 12–5). Many species of trees and other vegetation in cities began to die back, and farmers near cities such as Los Angeles

Table 12–1
Major Air Pollution Episodes

DATE		PLACE	EXCESS DEATHS
Feb.	1880	London, England	1000
Dec.	1930	Meuse Valley, Belgium	63
Oct.	1948	Donora, Penn., U.S.	20
Nov.	1950	Poca Rica, Mexico	22
Dec.	1952	London, England	4000
Nov.	1953	New York, U.S.	250
Jan.	1956	London, England	1000
Dec.	1957	London, England	700–800
Dec.	1962	London, England	700
Jan./Feb.	1963	New York, U.S.	200–400
Nov.	1966	New York, U.S.	168

Source: Wilfred Bach, *Atmospheric Pollution*, 1972, McGraw-Hill Book Co., reprinted by permission.

and New York suffered damage or even total destruction of their crops because of air pollution (Fig. 12–6). Indeed, pollution-induced damage became so routine that it forced the complete abandonment of citrus growing in some parts of California and vegetable growing in certain areas of New Jersey, which were among the most productive regions in the country. A conspicuous acceleration in the rate of metal corrosion and deterioration of rubber, fabrics, and other materials was also noted.

In short, it became conspicuous that human inputs of pollutants were overloading the natural system. Indeed, calculations showed that pollutants from human sources exceeded those from natural sources by 10 to 1,000-fold, depending on the particular pollutant. It was clear in the 1950s and 1960s that unrestricted discharge of pollutants into the atmosphere could no longer be tolerated.

Following demand by citizens for action, the U.S. Congress passed the Clean Air Act of 1970 and

FIGURE 12–4
A temperature inversion may cause episodes of high concentrations of air pollutants. (a) Normally air temperatures are highest at ground level and decrease at higher elevations. Since the warmer air rises, pollutants are carried upward and diluted in the air above. (b) A temperature inversion is a situation in which a layer of warmer air overlies cooler air at ground level. This blocks the normal updrafts and causes pollutants to accumulate like cigarette smoke in a closed room.

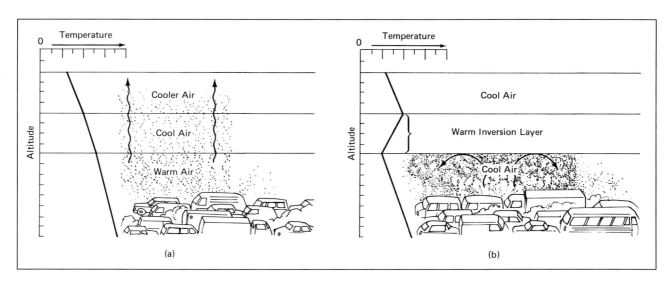

(a)

(b)

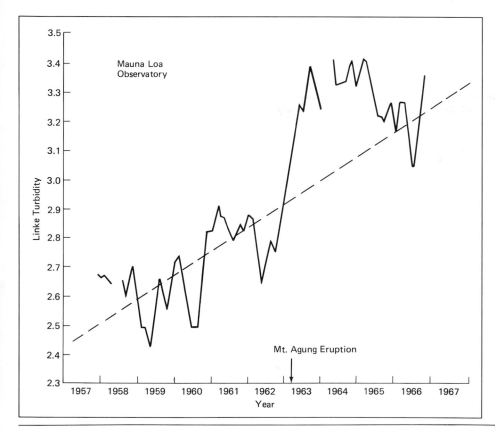

FIGURE 12–5
Increasing atmospheric turbidity. Measurements of atmospheric turbidity (haziness), taken at Mauna Loa, Hawaii, showed an increasing trend apart from natural events. Since Hawaii is far removed from human sources of pollution, this was taken to mean that human-produced pollution was affecting the entire atmosphere. (Redrawn by permission from Wilfred Back, *Atmospheric Pollution*, p 37, McGraw-Hill Book Company, 1972.)

its amendments of 1977. These laws remain the foundation of air pollution control efforts. They call for designating the most widespread pollutants, setting **standards**—levels that need to be achieved to protect environment and human health—and establishing timetables to meet these goals.

Four distinct stages are involved in meeting these mandates:

☐ Identifying the pollutants.

☐ Demonstrating which pollutants are responsible for particular adverse health and/or environmental effects, so that reasonable standards may

FIGURE 12–6
Sweet corn growing near Los Angeles about 1968. Injury was caused by the existing levels of air pollution, particularly photochemical smog. Such examples became increasingly common through the 1960s. (Photo courtesy of Ray Thompson, Statewide Air Pollution Research Center, Univ. of California, Riverside, Calif.)

be set. (If standards are set below threshold levels vast sums may be spent on control without benefit.)

☐ Determining their sources.

☐ Developing and implementing suitable controls.

All phases are ongoing processes. As new information regarding adverse effects of pollutants is discovered, standards may be made more stringent and new strategies for control must be developed and implemented.

As a result of progress under the Clean Air Act, air quality in most cities is markedly better now than it was in the early 1970s. However, not all problems have been solved and some have become worse. Also, there has been some backsliding in recent years. Let us turn our attention now to that progress and to the remaining problems in more detail.

MAJOR AIR POLLUTANTS AND THEIR EFFECTS

Major Pollutants

The following eight pollutants or pollutant categories have been identified as most widespread and serious:

1. **Particulates.** Particulates are tiny solid or liquid particles suspended in the air. We see these particles as smoke or haze. Other pollutants present as gas or vapor are not visible except in the case of nitrogen dioxide, which is a brownish gas. Particulates may carry any or all of the other pollutants dissolved in or adhering to their surfaces (Fig. 12–7).

2. **Hydrocarbons and other volatile organic compounds.** These include materials such as gasoline, paint solvents, and organic cleaning solutions that evaporate and enter the air in a vapor state.

3. **Carbon monoxide (CO).** Carbon monoxide is a highly poisonous gas.

4. **Nitrogen oxides (NO$_x$).** These include several nitrogen-oxygen compounds, all gases.

5. **Sulfur oxides, mainly sulfur dioxide (SO$_2$).** Sulfur dioxide is a poisonous gas to both plants and animals.

6. **Lead and other heavy metals.**

7. **Ozone and other photochemical oxidants.** You probably know that ozone in the upper atmosphere must be preserved to shield us from ultraviolet radiation (see Chapter 13). However, ozone is also a highly toxic gas to both plants

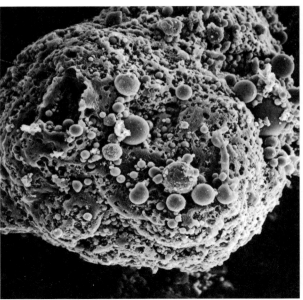

FIGURE 12–7
Fly ash from a coal-fired power plant (magnified 2000 times). A particle of ash is an adsorptive surface for many other pollutants, as shown by this photograph taken under a scanning electron microscope. (Photo by Roger J. Cheng, Atmospheric Sciences Research Center, The State University of New York at Albany.)

and animals. Therefore, *ground-level ozone* is a serious pollutant. This emphasizes the case that "a pollutant is a chemical out of place."

8. **Acids, mainly sulfuric and nitric acids.** These are commonly associated with liquid particles creating acid rain and acid fogs, and mists.

Adverse Effects of Air Pollution On Humans, Plants, and Materials

It is important to recognize that air pollution is not a single entity but an alphabet soup of the above materials mixed with the normal constituents of air (Fig. 2–1). Further, the amount of each pollutant present varies tremendously depending on proximity to the source and various conditions of wind and weather. Thus, we are exposed to a mixture that varies in makeup and concentration from day to day, even from hour to hour, and from place to place. Consequently, the effects we feel or observe are rarely if ever the effects of a single pollutant; they are the combined effect of the whole mixture of pollutants acting over the total life span, and frequently the effects are *synergistic* (see Chapter 11).

For example, both plants and animals may be stressed by pollution such that they become more vulnerable to other environmental factors such as drought or attack by parasites. Given the complexity of this situation, it is extremely difficult to determine the role of any particular pollutant in causing the observed result. Consequently, there is still a conspicuous lack of information concerning effects of various pollutants. Nevertheless, there are some general conclusions.

Effects on Human Health

During periods when pollution reaches high levels, many people complain of headaches, irritation of eyes, nose, and throat, nausea, and general ill-feeling. Ozone seems to be the predominant factor in irritation of mucous membranes. Acid particles, particularly those of sulfuric acid, correlate most closely with the increase in asthma attacks, and carbon monoxide may be responsible for weakened judgment, drowsiness, and headaches. High levels of particulates over prolonged periods correlate with respiratory disease and lung cancer. However, all the factors may contribute to various degrees.

In some cases, as shown in Table 12–1, air pollution has reached levels that caused death. Although this mortality increase is attributed to air pollution, it should be noted that these deaths occur among people already suffering from severe respiratory and/or heart disease. While the gases present in air pollution are known to be lethal in high concentrations, such concentrations are not reached in outside air. Therefore, deaths attributed to air pollution are not the direct result of simple poisoning. However, air pollution puts an additional stress on the body and, if a person is in an already weakened condition, this additional stress may be fatal.

The heavy metal and organic constituents of air pollution include many chemicals that are known to be carcinogenic in high doses. Therefore, there is a strong feeling that the presence of trace amounts of these chemicals in air may be responsible for a significant portion of the cancer observed in humans. However, the actual evidence to support this contention is far from clear. Initial studies did show a higher rate of lung disease among those living in cities with high levels of air pollution. However, as studies have progressed, the only pollution factor that clearly and indisputably correlates with serious lung disease is cigarette smoking (Fig. 12–8). When studies were repeated separating smokers and nonsmokers, it was found that nonsmokers living in polluted city air had little if any more lung disease than those living in clean air. However, smokers living in polluted air had a much higher incidence of lung disease than smokers living in clean air (Fig. 12–9). In other words, there is a strong synergistic effect between smoking and air pollution.

Certain diseases typically associated with oc-

cupational air pollution show the same synergistic relationship with smoking. For example, black lung disease is seen almost exclusively among coal miners who are also smokers. Lung disease among those exposed to asbestos predominates in smokers. In a 1983 article in *Science*, Dr. Bruce Ames, chairman of the Department of Biochemistry, University of California at Berkeley, stated, "despite numerous suggestions to the contrary, there is no convincing evidence of any generalized increase in the United States (or United Kingdom) of cancer rates other than what could plausibly be ascribed to the delayed effects of previous increases in tobacco usage."

The synergistic effect between smoking and air pollution may occur because smoking deadens the action of the cilia (tiny active hairs on the cells lining the lung passages), which normally serve to remove foreign particles (Fig. 12–10). Hence, in smokers, particles laden with other pollutants remain in the lungs much longer than in nonsmokers.

Some experts feel, however, that dismissing the trace amounts of heavy metals and organic compounds in air pollution as insignificant is not justified in view of the fact that these substances in higher concentrations are known to increase cancer risks. The final outcome of this long-term exposure is simply not known at present.

Lead is a pollutant that deserves special attention. Lead poisoning has been recognized for several decades as a cause of mental retardation. It was formerly thought that ingestion of peeling paint chips that contain lead was the main source of lead contamination. In the early 1980s, however, it was discovered that elevated lead levels were much more widespread than previously expected, and they were present in adults as well as children. Further, learning disabilities in children and high blood pressure in adults were found to be correlated with high levels of lead in the blood. The source of this widespread contamination was attributed to the use of leaded gasoline. The lead in exhaust fumes may be inhaled directly or may settle on food, water, or any number of items that are put in the mouth. This knowledge led the Environmental Protection Agency to mandate the phaseout of leaded gasoline, which was completed at the end of 1988.

Effects on Agriculture and Forests

The pollutants responsible for the damage to vegetation are experimentally determined by growing plants in chambers where they can be subjected to any desired concentration of pollutants and the results compared with field observations. Also, open-top chambers are set up in the field which enable plants in one chamber to receive filtered air while

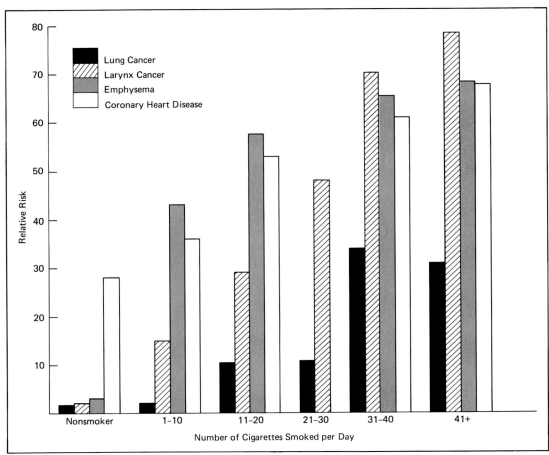

FIGURE 12–8
Many diseases and disease conditions are correlated with smoking, including cancer of the lung and larynx, emphysema, and coronary heart disease. (Note that data are not given for emphysema and heart disease at the 21–30 cigarettes-per-day level.) (Data from the U.S. Department of Health, Education and Welfare.)

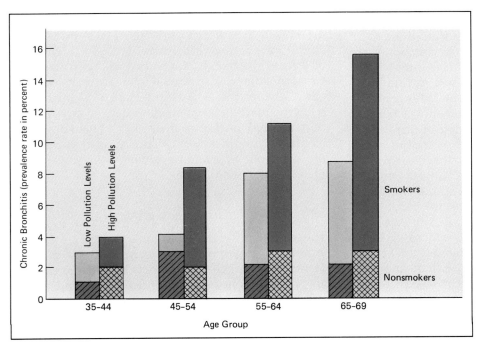

FIGURE 12–9
Incidence of chronic bronchitis seen among people living in "low pollution" rural areas and those living in "high pollution" city areas. Among nonsmokers no significant difference is seen between the two groups. But, the increased incidence caused by smoking is increased even more by air pollution, a clear synergistic effect. (From: P. M. Lambert and D. D. Reid, "Smoking, Air Pollution, and Health," *Lancet,* April 25, 1970.)

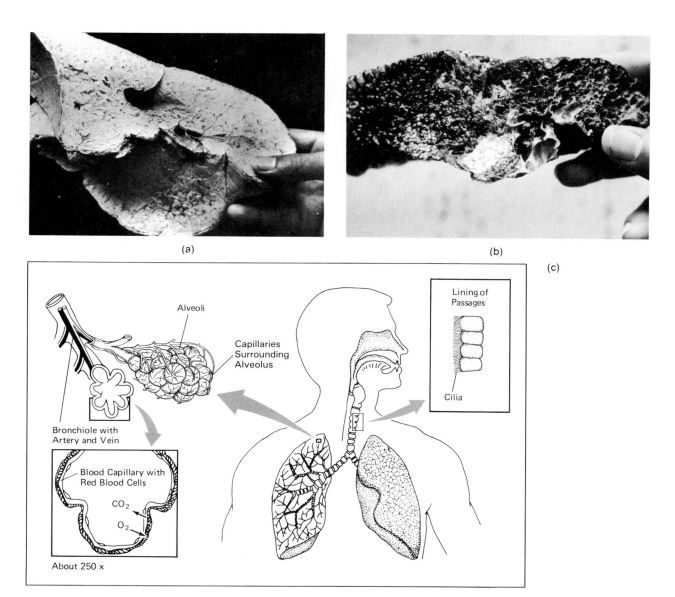

(a)

(b)

(c)

FIGURE 12–10
In the lungs, air passages branch and rebranch and finally end in millions of tiny sacs called alveoli. Alveoli are surrounded by blood capillaries. As blood passes through these capillaries, oxygen diffuses from the alveoli into the blood, and carbon dioxide diffuses in the reverse direction. Lung disease: (a) Normal lung tissue. (b) Lung tissue from person who suffered from emphysema, a chronic lung disease in which some of the structure of the lungs has broken down. Cigarette smoking is associated with the development of emphysema as well as other lung diseases. (EPA—Documerica, photos by Leroy Woodson; courtesy of the U.S. Environmental Protection Agency.) (c) Diagram of lung structure.

plants in an adjacent chamber receive unfiltered air, and pollutants are monitored (Fig. 12–11). Through such experiments it is possible to gain insight into which pollutants cause damage and to extrapolate from this information a broader picture of the effects of air pollution on agriculture, forests, and ecosystems in general.

Results show that plants are considerably more sensitive to gaseous air pollutants than humans. Before emission controls it was common to see wide areas of totally barren land or severely damaged vegetation downwind from smelters or coal-burning power plants (Fig. 12–12). The factor responsible was sulfur dioxide. The dieback of vegetation in cities and damage that farmers downwind of cities experienced was determined to be mainly due to ozone and other photochemical oxidants.

Even more insidious, however, open-chamber experiments show that plants in clean (filtered) air grow considerably larger than plants in unfiltered air (Fig. 12–13). This shows that existing levels of pollution in air are responsible for a general reduction of growth without conspicuous signs of damage or

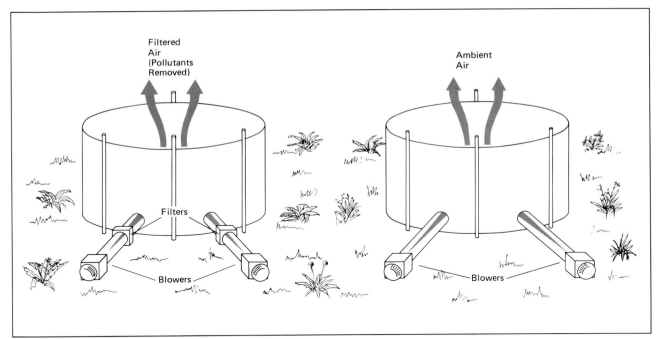

FIGURE 12–11
Testing the effects of pollutants in ambient (existing) air on plant growth. Open-top chambers as shown are placed around sample plots of vegetation, and filtered or unfiltered air is blown through. Comparing the growth of vegetation within the two chambers reveals the effect of ambient pollution levels under natural conditions. Many such experiments show that pollution-free air leads to a 10 to 40 percent increase in growth, which is to say that pollution levels in ambient air may be causing this degree of growth reduction. Ozone seems to be most significant in causing this effect.

FIGURE 12–12
Death of vegetation caused by air pollution. That air pollution can cause severe damage to vegetation has been long known. It is common to find vegetation killed back for considerable distances around various industrial operations. This forest was killed by various pollutants emanating from an open pit coal mine in British Columbia, Canada. (Paolo Koch/Photo Researchers.)

abnormality. Ozone is by far the most significant factor in this effect. A recent assessment of crop production in the United States estimated that without ozone pollution, crop yields would increase as follows: corn, 3 percent; wheat, 8 percent; soybeans, 17 percent; and peanuts, 30 percent. This represents about $5 billion worth of agricultural productivity or about 10 percent of total farm income. These figures represent an average across the United States. The effects in badly polluted areas are far more severe.

The negative impact of air pollution on wild plants and forest trees may be even greater than that noted for agricultural crops. Open-chamber experiments in the Blue Ridge Mountains in northwestern Virginia have shown that the growth of various wild plants was reduced between 32 and 57 percent by ozone even though it remained below standards set for human health. Significant damage to economically valuable ponderosa and Jeffrey pines is occurring

FIGURE 12-13
Pollution also causes a general reduction in growth without other conspicuous damage. Potato plant at right was grown in prevailing air at Beltsville, Maryland; plant at left was grown at the same location, but air was filtered to remove pollutants. (USDA photos.)

with very little if any noticeable effect. Then, just a small increase in concentration or duration of exposure may push the plant beyond its capacity to cope and cause drastic reduction of growth or dieback. This point is known as the **critical level** (Fig. 12–14).

Even without drastic dieoffs, the decrease in primary production must ultimately affect the rest of the ecosystem, including soils because they depend on primary production. As sensitive species die out, they are replaced by more resistant species in the process of ecological succession. Where this will lead is uncertain, but numerous foresters and ecologists have little doubt that large-scale biological changes are already underway in the landscape as a result of current levels of air pollution.

Effects on Materials and Aesthetics

Walls, windows, and all other surfaces turn grey and dingy as particulates settle on them. Paint and fabrics deteriorate more rapidly, and the sidewalls of tires

FIGURE 12-14
Critical level of pollution. (a) Control plants at left were grown in clean air; others were exposed to indicated concentration of sulfur dioxide (parts per 100 million). Note that the reduction in growth is not linear but that there is a sharp drop between 6 and 15 parts per 100 million. (USDA photo.) The concept of the critical level is shown graphically in (b).

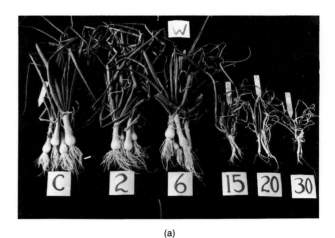

(a)

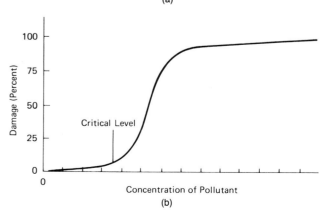

(b)

along the entire western slope foothills of the Sierra Nevada Mountains in California. In the San Bernardino Mountains of California, an area that receives air pollution from Los Angeles, production has been reduced by 75 percent. A study plot of white pine near a roadside in western Virginia showed a 40 percent decrease in wood production, and similar reductions are being observed throughout New England and other regions of the eastern United States. Again, a major factor seems to be ozone, although acid precipitation may also be a major factor (see Case Study 12 and Chapter 13).

Further, forests under pollution stress are more susceptible to damage by insects and other pathogens. For example, death of ponderosa and Jeffrey pines in California is generally attributed to western pine beetles, which invade pollution-weakened trees. Even normally innocuous insects may cause mortality when combined with pollution stress. If widespread pollution worsens, reductions in production could occur with disastrous suddenness. The effect may be swift because reduction in growth (or any other manifestation of pollution damage) is not a linear function of the pollution level. That is, one unit of pollution does not produce one increment of damage. Instead, plants will tolerate a certain level of pollution stress

Case Study 12

HOW AIR POLLUTION IS AFFECTING TREES

Across Europe and the United States, air pollution from burning fossil fuels is injuring and killing trees. Extensive damage and mortality from ozone pollution have been documented among ponderosa and Jeffrey pines in California and white pines in the eastern United States. Air pollution is also suspected in the decline of high-elevation red spruce in the Appalachian Mountains from Vermont to North Carolina. At polluted research sites on mountains in New York, Vermont, and New Hampshire, the red spruce by 1992 will have decreased in abundance by 50 to 80 percent since 1962. By 1987, almost half of the red spruce and Fraser fir on Mt. Mitchell's (North Carolina) west-facing slopes were dead. No insects or disease can account for these declines. Air pollution is also a leading suspect in the unexplained increased mortality among commercial yellow pines in the Southeast and sugar maples in the Northeast and Canada.

The air pollutants implicated in these declines are acid deposition and ozone. Levels of these pollutants are high where trees are dying. Along the Appalachian Mountain chain the *peak* acidity of cloud moisture is more than 1000 times greater than that of unpolluted precipitation (pH values of cloud moisture as low as 2.2 have been measured). Ozone levels on these remote mountains are often twice those at lower elevations.

Researchers in Europe and the United States have found that ozone and acid deposition injure trees both directly through damage to needles and leaves, and indirectly by leaching nutrients vital to tree growth from the soils. Pines are among the most sensitive trees to ozone. Ozone can enter the tree's needles and damage the cells containing chlorophyll. This impairs photosynthesis and leads to reduced tree growth. Photosynthetic activity can decline before

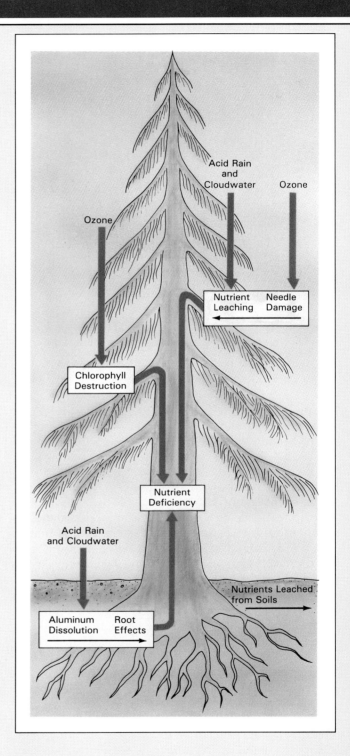

Acid rain and ozone together could lead to nutrient deficiency in a coniferous tree, according to a currently favored scenario for their possible role in forest decline. (From Mohnen, Volker A.: "The Challenge of Acid Rain," *Scientific American*, August, 1988. Copyright © 1988 by Scientific American, Inc. All rights reserved.)

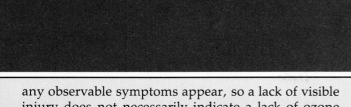

any observable symptoms appear, so a lack of visible injury does not necessarily indicate a lack of ozone effects. Ozone also accelerates the leaching by acid deposition of important nutrients from the tree's needles.

Experiments have also found that acid deposition can injure trees directly and through soil changes. Highly acidic rain and clouds can leach various nutrients such as magnesium, potassium, and calcium from both the needles of the trees and from the soils where the trees are growing. When nutrients are leached from the needles, the trees try to replace them by taking more from the soils. If the soils have been depleted by acid deposition, the trees are unable to compensate for the lost nutrients and develop symptoms of nutrient imbalance, such as yellowing of needles.

As pollution weakens the trees, they succumb to natural stresses such as insects, disease, or climate extremes (drought, frost, etc.) that they otherwise might have withstood. The significance of such injuries for these forest systems goes far beyond the immediate death of the damaged trees. As air pollution's effects on the ecosystem become progressively more severe, the whole ecosystem can deteriorate. Unlike a healthy ecosystem, which can normally weather bouts of fire and insects, a system exposed to chronic air pollution gradually loses the capacity to cope with changing natural stresses. Some scientists believe that ecosystems chronically exposed to air pollution will eventually collapse.

JAMES J. MACKENZIE
World Resources Institute

and other rubber products become hard and checkered with cracks because of oxidation by ozone. Corrosion of metals is dramatically increased by sulfur dioxide and acids derived from sulfur and nitrogen oxides (Fig. 12–15), as is weathering and deterioration of stonework. These and other effects of air pollutants on materials increase the costs for cleaning and/or replacement of products by hundreds of millions of dollars per year.

In addition, a clear blue sky and good visibility—in contrast to the haze of smog—have aesthetic value and a psychological impact. This has a direct impact on real estate values, since it has been seen that such values decrease with rising levels of air pollution.

SOURCES OF POLLUTANTS AND CONTROL STRATEGIES

Pollutants: Products of Combustion

Direct Products

In large measure air pollutants are direct and indirect byproducts of burning coal, gasoline, and other liquid fuels, and refuse (waste paper, plaster, etc.). These fuels and wastes are organic (carbon-hydrogen compounds). With complete combustion, the byproducts of burning them are carbon dioxide and water vapor ($CH_4 + 2O_2 \rightarrow CO_2 + 2H_2O$). However, oxidation is seldom complete.

Particles consisting mainly of carbon are emitted into the air; these are the **particulates** seen as smoke. In addition, there are various unburned or fragments of fuel molecules. These are the **hydrocarbon emissions.** Partially oxidized carbon is **carbon monoxide** (CO), in contrast to completely oxidized carbon, or carbon dioxide, CO_2. *Nitrogen-oxides* arise because burning takes place in the air, which is just 21 percent oxygen and 78 percent nitrogen. At high combustion temperatures, some of the nitrogen gas (N_2) is oxidized to form the gas, *nitric oxide* (NO). In the air, nitric oxide immediately reacts with additional oxygen to form *nitrogen dioxide* (NO_2) and/or *nitrogen tetroxide* (N_2O_4). These compounds are collectively referred to as **nitrogen oxides** (NO_x). Nitrogen dioxide absorbs light and is largely responsible for the brownish color of photochemical smog.

In addition, fuels or refuse contain impurities or additives and these are also emitted into the air when burned. Coal, in particular, contains from 0.2 to 5.5 percent sulfur. In combustion, this sulfur is oxidized

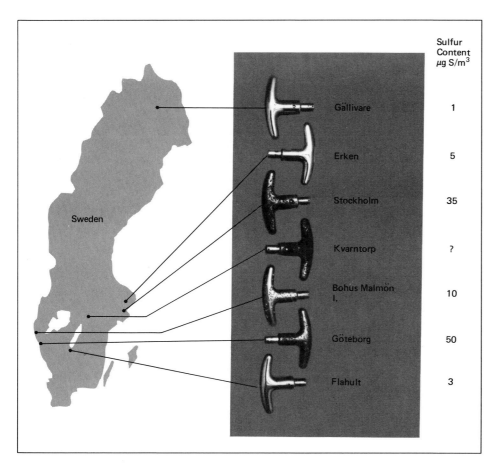

	Sulfur Content µg S/m³
Gällivare	1
Erken	5
Stockholm	35
Kvarntorp	?
Bohus Malmön I.	10
Göteborg	50
Flahult	3

FIGURE 12–15
Material corrosion and air pollution. The nickel-plated handles shown were exposed to the prevailing air conditions for 4.5 years. The association between corrosion and level of air pollution is conspicuous. (Swedish Corrosion Institute, Stockholm, Sweden.)

giving rise to the gas **sulfur dioxide** (SO_2). Coal may contain *heavy metal* impurities, and of course refuse contains an endless array of "impurities." Prior to the EPA-directed phaseout, lead was added to gasoline as an inexpensive way to prevent engine knocks. Emitted with the exhaust, it remained airborne and traveled great distances before settling (Fig. 12–16).

Indirect Products

The above direct products of combustion may undergo further reactions in the atmosphere and produce additional **indirect products** of combustion.

Ozone and numerous reactive organic compounds are formed as a result of chemical reactions between *nitrogen oxides* and *volatile hydrocarbons,* with *sunlight* providing the energy necessary to cause the reactions to occur. Since sunlight provides the energy, these products are collectively known as **photochemical oxidants.** The major reactions involved are shown in Figure 12–17. Light energy is absorbed by nitrogen dioxide and causes it to split to form nitric oxide and free oxygen atoms. The free oxygen atoms are spontaneously reversible; that is, if other factors were not involved, ozone and nitric oxide would react

FIGURE 12–16
Lead levels in Greenland glaciers. Age of ice samples is related to their depth in the glacier. Lead content of samples, while variable, is clearly correlated with amount of lead emitted into the air by industry and automobiles. (Reprinted with permission from M. Murozumi, T. J. Chow, and C. Patterson, "Chemical Concentrations of Pollutant Lead Aerosols, Terrestrial Dusts, and Sea Salts in Greenland and Antarctic Snow Strata," *Geochimica and Cosmochimica Acta,* 33. Copyright 1969, Pergamon Press, Ltd.)

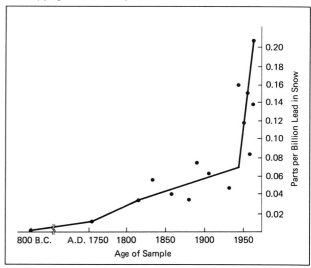

to reform the nitrogen dioxide and oxygen gas and there would be no appreciable accumulation of ozone (Fig. 12–17a).

But, when hydrocarbons are present, the nitric oxide reacts with them, and this has two diabolical effects. First, the reaction between nitric oxide and hydrocarbons leads to highly reactive and damaging synthetic organic compounds known as peroxyacetyl nitrates, or PANs (these are the "other" photochemical oxidants). Second, with the nitric oxide tied up in this way, the *ozone accumulates* (Fig. 12–17b).

Acids form as a result of sulfur dioxide and nitrogen oxides reacting with water vapor to form **sulfuric acid** and **nitric acid,** respectively.

Finally, fine particles also take on additional potency because they *adsorb* numerous other pollutants on their surface (Fig. 12–7). Adsorption means that the pollutants simply adhere to a surface, like flies sticking to flypaper as opposed to "soaking in" (absorption). Indeed, this is why activated charcoal makes an excellent chemical filter.

FIGURE 12–17
Formation of ozone and other photochemical oxidants. Ozone is the most injurious. (a) Nitrogen oxides, by themselves, would not reach damaging levels because reactions involving them are cyclic. (b) When hydrocarbons are also present, however, reactions occur which lead to the accumulation of numerous damaging compounds, most significantly ozone.

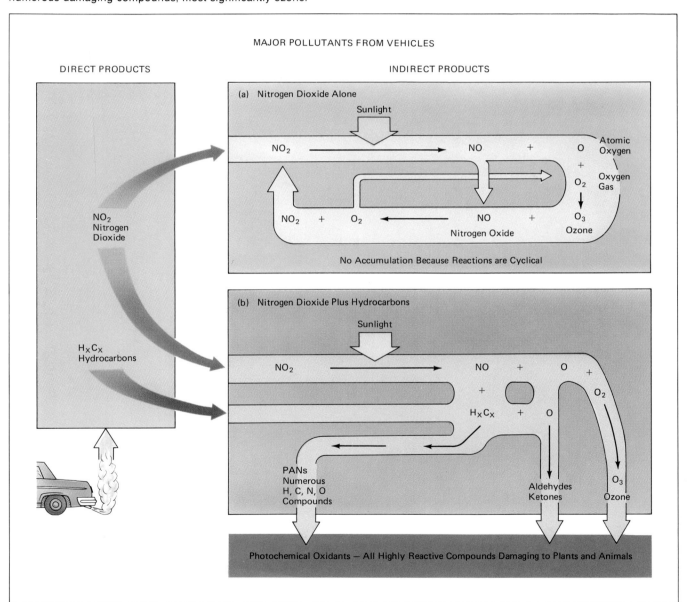

Sources of air pollution are summarized in Figure 12–18.

Setting Standards

The Clean Air Act of 1970 mandated setting standards for the following five pollutants, which, at the time, were recognized as the most widespread and objectionable: (1) total suspended particulates, (2) sulfur dioxide, (3) ozone, (4) carbon monoxide, and (5) nitrogen oxides. These are now known as **criteria pollutants.** The **primary standard** for each is based on the highest level that can be tolerated by humans

FIGURE 12–18
(a) The prime sources of the major air pollutants are depicted in this illustration. (b) Vehicles are the prime source of hydrocarbons and nitrogen oxides because in each cylinder of a gasoline engine a fuel-air mixture is (1) taken in, (2) compressed, (3) ignited, and (4) exhausted about 25 times each second during normal operation. The high-pressure combustion causes the production of nitrogen oxides; moreover, combustion is often incomplete, causing emissions of hydrocarbons and carbon monoxide.

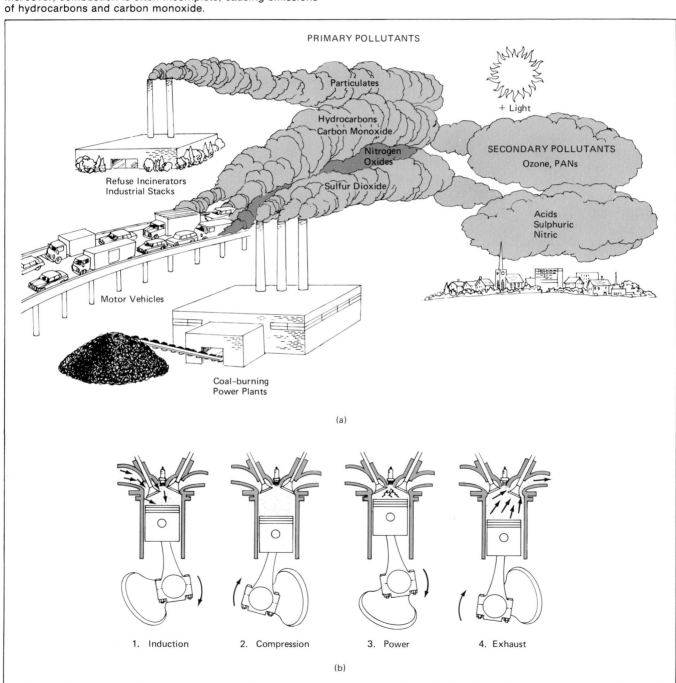

PRIMARY POLLUTANTS

Particulates

Hydrocarbons
Carbon Monoxide

Nitrogen
Oxides

Sulfur Dioxide

Refuse Incinerators
Industrial Stacks

Motor Vehicles

Coal-burning
Power Plants

+ Light

SECONDARY POLLUTANTS
Ozone, PANs

Acids
Sulphuric
Nitric

(a)

1. Induction   2. Compression   3. Power   4. Exhaust

(b)

Table 12–2
National Ambient Air Quality Standards for Criteria Pollutants

POLLUTANT	AVERAGING TIME[a]		PRIMARY STANDARD
Particulates (μg/m^3)[b]	1	year	75
	24	hours	260
Sulfur oxides (ppm)	1	year	0.03
	24	hours	0.14
	3	hours	—
Carbon Monoxide (ppm)	8	hours	9
	1	hour	35
Nitrogen Dioxide (ppm)	1	year	0.05
Ozone (ppm)[c]	1	hour	0.12

[a] Averaging time is the time period over which concentrations are measured and averaged.
[b] A (microgram) is one-millionth of a gram.
[c] Revised January 26, 1979.
Source: After Council on Environmental Quality, *Seventh Annual Report*, 1976, p. 215.

without noticing any ill effects, plus a 10 to 50 percent margin of safety (Table 12–2).

Air quality, which is now commonly given along with weather reports, is based on these standards. The pollutant having the highest concentration relative to its primary standard determines the index (Fig. 12–19). For example, if the highest of any of the criteria pollutants is 25 percent of its primary standard, the Pollution Standards Index (PSI) is 25 and air quality is in the "good" range; if the highest pollutant is 150 percent of its standard, the index is 150 and air quality is in the "unhealthful" range, as shown in Figure 12–19. If air pollution gets too high, factories and other pollution sources may be shut down until it clears.

Major Sources and Control Strategies

Particulates

Prior to the 1970s, the major sources of particulates were from the open burning of refuse (open burning dumps) and industrial stacks. The Clean Air Act of 1970 mandated the phaseout of open burning of refuse and required particulates from industrial stocks be reduced to "*no* visible emissions."

The alternative generally taken for refuse disposal was landfilling, a solution that has created its own set of environmental problems, as will be discussed further in Chapter 18. To reduce stack emissions industries installed filters, electrostatic precipitators, and other devices as shown in Figure 12–20. Here also, wastes must still go somewhere. Ironically, wastes removed from exhaust gases, which frequently contained heavy metals and other toxic substances, added to the burden of toxic substances to be disposed of (Chapter 11). Nevertheless, these measures resulted in a marked reduction of particulates over the 1970s (see Figure 12–23). However, the growing use of wood stoves and diesel cars has offset some of these gains. For this reason restrictions are now being placed on woodburning stoves in some regions and the sale of new diesel cars has been banned in California.

Nitrogen Oxides, Hydrocarbons, Ozone and Other Photochemical Oxidants, and Carbon Monoxide

The major source of hydrocarbons and nitrogen oxides, which lead to the ground-level ozone and PANs, and also carbon monoxide, was determined to be ve-

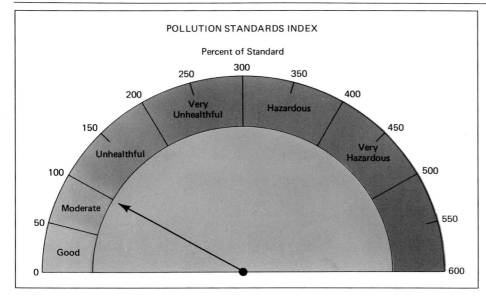

FIGURE 12–19
Pollution Standards Index (PSI) of air quality. Air quality is rated as shown according to the highest percentage of primary standards reached by any of the criteria pollutants.

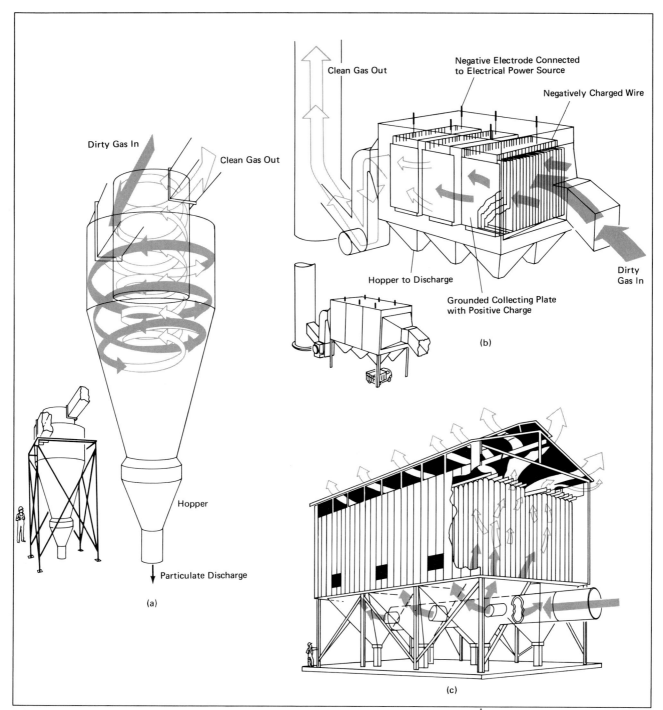

FIGURE 12–20
Devices to remove particles from exhaust gases. (a) Cyclone precipitator. Particles are removed by centrifugal force as exhausts are swirled. (b) Electrostatic precipitator. Particles are electrically (−) charged, then attracted to plates of the opposite charge (+). (c) Bag house. Exhaust gases are forced through giant vacuum cleaner bags. Note that none of these devices removes very fine particles or polluting gases such as sulfur dioxide. (American Lung Association.)

hicle exhaust; additional hydrocarbons were coming from evaporation of gasoline and oil vapors from the fuel tank and engine systems.

The Clean Air Act of 1970 mandated a 90 percent reduction in these emissions by 1975. Fuel tank evaporation became controlled by a system of hoses leading to a canister of activated charcoal—which held and rerouted fumes to be burned in the engine. Control of nitrogen oxide emissions proved problematic. The first control measures consisted of engine adjustments, which reduced combustion temperatures and pressures and, hence, resulted in less oxidation of nitrogen. Unfortunately, such adjustments also reduce engine efficiency and, consequently, increase fuel consumption. Thus, we see pollution control in conflict with another extremely important goal—energy conservation. However, hydrocarbons as well as nitrogen oxides are required for ozone production. It was considered that nitrogen oxides might not be a problem if hydrocarbons were well controlled. Therefore, efforts to control nitrogen oxide emissions from cars were essentially abandoned and efforts were focused on controlling hydrocarbon emissions.

Cars are now equipped with a considerable array of pollution control devices, including computerized control of fuel mixture and ignition timing, which give more complete combustion of fuel and decrease hydrocarbon emissions (Fig. 12–21). However, the most significant control device on cars remains the **catalytic converter** (Fig. 12–21b). As the exhaust passes through this device, a chemical catalyst consisting of platinum-coated beads causes the oxidation of hydrocarbons to carbon dioxide and water. The catalytic converter also completes the oxidation of carbon monoxide to carbon dioxide, thus giving excellent control of this product. It does nothing, however, to control nitrogen oxides.

Of course, the effectiveness of pollution control devices depends on proper maintenance. It is discouraging to note that a survey showed that one of five cars had pollution control devices that did not work because of improper maintenance or tampering. One of the most common problems was catalytic converters that had been disconnected in the mistaken belief that this improves fuel economy. To overcome this problem the 1977 Amendments to the Clean Air Act mandated that states have emission inspection programs in areas that fail to meet air quality standards (Fig. 12–22). If states don't comply they may lose federal highway funds. Note that the emissions controls of modern vehicles are so well integrated with engine systems that any tampering or lack of maintenance surely reduces engine performance.

Between emission controls and inspection programs, carbon monoxide, hydrocarbon, ozone, and smog levels declined in most U.S. cities from 1975 to 1983.

However, part of this improvement may represent a shift from city driving to rural driving as new shopping centers and industrial parks are built in ex-urban areas. Unfortunately, air pollution has not been carefully monitored outside cities, but judging from crop and forest damage, concentration of photochemical oxidants in rural areas did decrease as might be expected with the improvement in city air. Now, whatever initial reductions may have occurred are being offset again by the continuing increase in the number of vehicles and miles driven, as well as the many fleets of diesel trucks which comprise our goods distribution system. Also, there are very significant nonvehicle sources of hydrocarbons such as the evaporation from waste disposal facilities (e.g., surface impoundments), paint shops, service stations, and so on. In any case ozone levels are headed up again. In the summer of 1988, according to an Environmental Defense Fund researcher, ozone levels in Washington, D.C., were such that if it had been a factory it would have been shut down because of unsafe levels. Researchers are also finding increasing evidence of ozone damage to vegetation as well as forests (see Case Study 12–1).

Sulfur Dioxide and Acids

Measurements of sulfur dioxide in city air indicate that there has also been great success in controlling this pollutant (Fig. 12–23). Actually these results are highly misleading because control strategies have led to other problems that are as bad if not worse. The major source of sulfur dioxide is coal-burning electric power plants and smelters. A power plant may burn up to 10,000 tons of coal per day. Contaminated with 3 percent sulfur, this results in daily emissions nearing 1000 tons of sulfur dioxide.

Rather than reducing these emissions, these industries lobbied for and obtained permission under the Clean Air Act to use a "dilution solution." They shut down small inefficient plants in cities and built huge facilities in rural areas, with very tall smokestacks that ejected pollutants above the inversion layer. The idea was that if pollutants were ejected into the upper atmosphere, they would dilute and effectively disappear. Far from disappearing, however, the sulfur dioxide from the tall stacks of coal-burning electric power plants converts to sulfuric acid, and it has been shown to be the major source of acid precipitation (Chapter 13).

Thus, improvements in city air quality in this respect have been at the expense of declining rural air quality, and reducing sulfur dioxide at ground

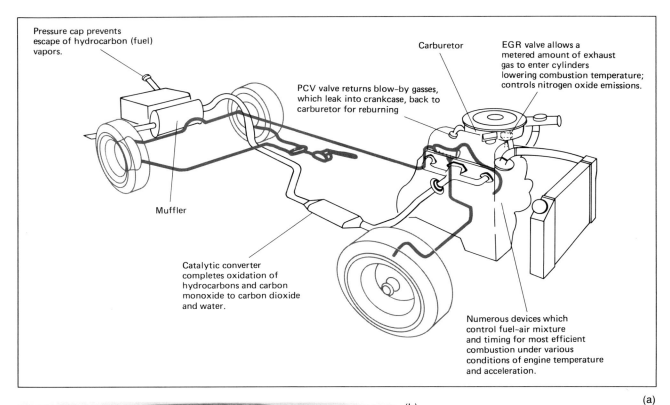

Pressure cap prevents escape of hydrocarbon (fuel) vapors.

Carburetor

EGR valve allows a metered amount of exhaust gas to enter cylinders lowering combustion temperature; controls nitrogen oxide emissions.

PCV valve returns blow-by gasses, which leak into crankcase, back to carburetor for reburning

Muffler

Catalytic converter completes oxidation of hydrocarbons and carbon monoxide to carbon dioxide and water.

Numerous devices which control fuel-air mixture and timing for most efficient combustion under various conditions of engine temperature and acceleration.

(a)

(b)

FIGURE 12–21
Pollution control features on cars. (a) New cars are now equipped with numerous pollution control devices. (b) The most significant device is the catalytic converter in the exhaust system. As exhausts pass through the converter, the catalyst, made of platinum-coated beads, causes hydrocarbons and carbon monoxide to react with more oxygen to form harmless carbon dioxide and water vapor. (General Motors photo.)

level has been at the expense of greatly worsening the acid rain problem. This has occurred despite the Clean Air Act stipulating that rural areas should not be allowed to decline.

Lead and Other Heavy Metals

A greater success has been achieved with lead, which, as noted previously, has been phased out of use as a gasoline additive (Fig. 12–23). However, other serious sources of heavy metals remain. In 1988 the state of Michigan issued a warning not to eat fish from its

inland lakes because they were bioaccumulating high levels of mercury, apparently getting into the lakes by way of air pollution. Sources had not been identified.

Taking Stock—What You Can Do

In the early 1970s, many cities were experiencing more than 100 days a year of pollution in the "unhealthful" range or above; Los Angeles was having in the order of 300 such days per year. The original

FIGURE 12–22
Auto emissions inspection station. A sample of gas is withdrawn from the tailpipe and automatically analyzed for carbon dioxide, carbon monoxide, and hydrocarbons. Each lane of the station can test up to 40 cars per hour. Repairs are required of cars not meeting standards. Where air pollution levels remain problematical, such inspections are mandated under the Clean Air Act. (Photo by author with permission of Systems Control, Inc.)

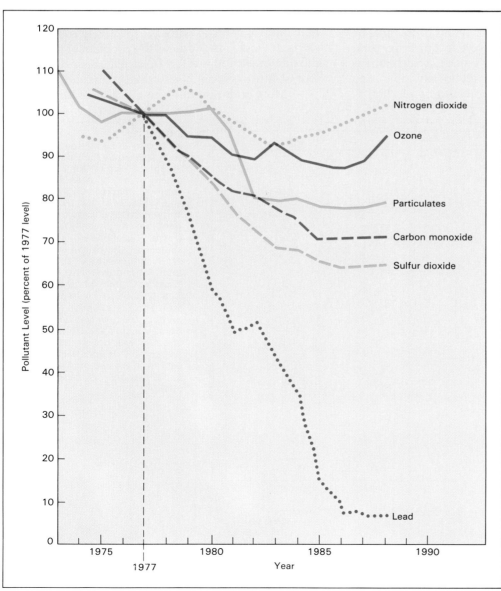

FIGURE 12–23
Trends in various constituents of air pollution compared to 1977 as a base of 100. Note that nitrogen oxides have not improved due to lack of attention to this pollutant. Ozone improved, but gains have nearly been erased as increasing numbers of vehicles and driving offset lower emissions per car. Apparent improvement in sulfur dioxide has been at the expense of increased acid rain because of the tall stacks that are used. Real improvements have been made in lead by phasing it out of gasoline. The data are averages of many stations; any specific location may differ greatly. (Data from the Environmental Protection Agency)

goal was to achieve "good" air quality on all days by 1975. It hardly needs to be said that this was not accomplished; however, marked progress did occur (Fig. 12–24). Certainly air quality is very much better than it would be if controls had not been initiated.

However, the backsliding of the last few years is very troubling. Basically, pollution control devices and greater fuel efficiency reduced a car's emissions by 95 percent from 1969 to 1989. As old polluting cars were replaced by new "cleaner" cars over this period, an overall reduction of emissions was achieved despite increasing numbers of cars and miles driven. Now, with most cars being cleaner models, this reduction-by-replacement has leveled out and the increasing number of cars is becoming the dominant factor, causing overall pollution to increase again. Clearly, renewed effort must be focused on controlling emissions, especially hydrocarbons and nitrogen oxides, which lead to ozone formation. However, further gains will be more costly and harder to achieve because the easy steps have already been taken.

In 1989, Los Angeles adopted a sweeping 20-year plan to control its notorious smog problem. The plan calls for most cars to convert to burning methanol (alcohol fermented from wood), which burns more cleanly than gasoline, and locally operating cars, trucks, and buses to be powered by electricity.

President George Bush proposed such measures on a nationwide scale. The problems with this are that production of methanol may produce more pollution and environmental side effects than the smog it is intended to abate, and electric vehicles still await some technological breakthroughs to be practical (see Chapter 21). Beyond reducing vehicle emissions, however, the plan will also require employers to encourage car pooling and it regulates over 100 other items, for example, banning aerosol hair sprays and deodorants and requiring companies to install state-of-the-art pollution control equipment regardless of cost.

A more general approach that would benefit the whole country would be to increase the fuel efficiency standard for cars, which has remained at 26 miles per gallon (mpg) since 1985. Technologies are available that could bring average mileage per gallon to 60 by the year 2000. Also, stricter fuel efficiency and emission control standards should be applied to light trucks and four-wheel-drive vehicles that many people are now using as private cars. Doubling the fuel efficiency of vehicles would respectively reduce emissions. It would also address the problem of the carbon dioxide greenhouse effect (Chapter 13) and the problem of future crude oil shortages (Chapter 19).

More attention must be given to reducing nitro-

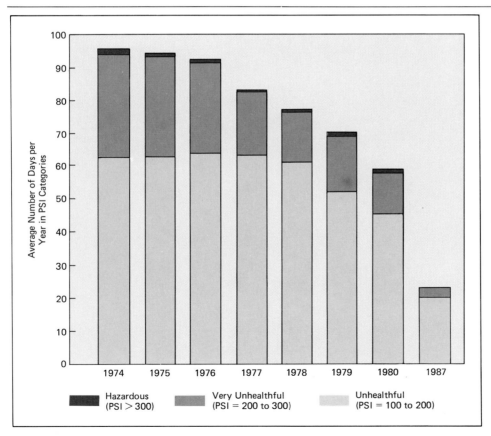

FIGURE 12–24
Air quality in 23 metropolitan areas as measured by the Pollution Standards Index (PSI). In general, air quality in metropolitan areas improved markedly in the late 1970s and 1980s, as shown by the decrease in the number of days that the PSI went into the unhealthful range or worse. This shows pollution abatement strategies have had a significant positive effect, although room for improvements still remains. (Source: Council on Environmental Quality Reprinted with permission from The Conservation Foundation, *State of the Environment: An Assessment at Mid-Decade.* 1984. The 1987 bar, not strictly comparable, is approximated from data from the *National Air Quality and Emissions Trends Report*, 1987, by EPA.)

gen oxide emissions because it has proven unfeasible to control ozone and other photochemical oxidants with controls on hydrocarbons alone. Additional reasons to reduce nitrogen oxide emission are its contributions to acid rain (Chapter 13) and to eutrophication (Chapter 9). Technologists are working to develop a catalytic converter that will turn nitrogen oxides back into nitrogen gas. This effort, which has near-term practicality, could use more support and pressure from a concerned public.

Write your Senators and Representative and ask that the next reauthorization of the Clean Air Act include requirements to:

☐ Set and enforce standards for ozone and other pollutants that will protect crops, forests, and all other aspects of the environment, not just human health.

☐ Control nitrogen oxides. Further delays are not tolerable.

☐ Set standards to increase average fuel efficiency of cars to 60 mpg by the year 2000.

☐ Set stricter standards for the fuel efficiency and emissions of trucks and buses, especially light trucks and four-wheel-drive vehicles used as private cars.

☐ Address nonvehicle sources of hydrocarbons.

☐ Reduce sulfur dioxide emissions from coal burning power plants. (The importance of this to control acid precipitation is discussed further in Chapter 13.)

Finally, even with an efficient car, pollution will be an inevitable byproduct of each mile you drive. Think of ways you can amend your lifestyle to drive fewer miles. Can you carpool? Use public transport? Live closer to school/work so that you could bicycle or walk? Are there ways you can increase the energy conservation in your home to use less fuel or electricity? Each bit of energy you save means less pollution.

INDOOR AIR POLLUTION

After recognizing the many pollutants in outside air one may be inclined to remain indoors to escape the hazards. However, air inside the home and workplace often contains much higher levels of hazardous pollutants than outdoor air. In a recent report, the National Academy of Sciences called indoor air pollution an issue that "has been largely overlooked" and a matter "of immediate and great concern." The overall indoor air pollution problem is threefold.

First, increasing numbers and kinds of products and equipment used in homes and/or offices give off potentially hazardous fumes. Second, buildings have become increasingly well-insulated and sealed; hence, pollutants are trapped inside where they accumulate to potentially dangerous levels. Third, exposure to indoor air pollution is longer in duration than exposure to outdoor pollution. The average person spends 70 to 80 percent of his or her time indoors, and the people who spend the most time inside are those who are most vulnerable to the harmful effects of pollution: small children, pregnant women, the elderly, and the chronically ill. The sources of indoor air pollution are numerous. They include:

☐ Formaldehyde and other synthetic organic compounds given off from plywood and particle board where they are used as binders, and from foam rubber and "plastic" upholstery where they are used as "softeners."

☐ A wide host of compounds from things burned on the stove or oven.

☐ Incomplete combustion and impurities from fuel-fired heating systems such as gas or oil furnaces, kerosene heaters, and woodstoves. Note that even complete combustion is not safe in well-insulated homes, as oxygen levels are lowered greatly.

☐ Fumes from oven and basin cleaners and other "strong" cleansing agents.

☐ Fumes from glues and hobby materials.

☐ Pesticides.

☐ Air fresheners and disinfectants. Most air fresheners work by either dulling the sense of smell so you don't notice noxious odors or by introducing "high intensity" smells that cover up odors.

☐ Aerosol sprays of all sorts.

☐ Radon. Radon is a radioactive gas resulting from the spontaneous breakdown of fissionable material in the earth's interior. It escapes naturally to the surface and forms part of the natural background radiation to which we are all exposed (see Chapter 20). However, as warm air escapes from the top of a house creating a partial vacuum, radon may be drawn through the basement floor and, trapped in the house, it may then reach hazardous levels.

☐ Asbestos. Asbestos is a natural mineral that has fiberlike crystals (Fig. 13–25). It is mined from the earth and was widely used as a heat-insulating and fire-retarding material for wrapping on steam heating pipes, ceiling coverings in many public buildings, ironing board covers, as

FIGURE 12–25
Asbestos. Asbestos is a natural mineral, the crystals of which are in the form of fine fibers. Since asbestos will not burn, it was used extensively in materials to insulate heating pipes and ducts and deaden sound throughout schools and other public buildings. Then it was discovered that asbestos fibers, when inhaled, may cause lung cancer. (Photo by author.)

well as some paints and roofing materials. In the 1960s, it was determined that the inhalation of **asbestos fibers** is associated with a unique form of lung cancer that develops as many as 20 to 30 years after exposure. The Environmental Protection Agency began regulating asbestos in the mid-1970s and initiated intensive campaigns to remove it from schools. However, this program has moved forward slowly and many buildings still contain asbestos, and many asbestos products remain in use.

☐ Smoking. Again, smoking carries a much higher health risk than the average exposure to any of the above materials, and it may act synergistically to increase the risk of exposure to other indoor (and outdoor) pollutants. Smoking has also been shown to increase the risks to those nonsmokers who are subjected to breathing "secondhand" smoke.

What can you do to decrease the risk of indoor pollution? If you are a smoker, stop, and insist that others not smoke in your presence or in your home. Minimize your use of strong, noxious chemicals and solvents. Replace lids securely and dispose of saturated rags in a sealed container. Use baking soda or vinegar for household cleaning. Also, properly maintain your furnace, woodstove, or kerosene heater, as well as your oven and other household appliances. Beyond this, ventilation is the key. Open a door or window whenever you are using a strong, noxious, smelly compound. Routinely air your house out, especially if it is well insulated, every few days, even in the winter. Most of the heat is in the solid walls and furnishings; relatively little is in the air. Thus, opening a door or window just long enough to change the air, usually 15–30 seconds, results in little heat loss.

13

Acid Precipitation, the Greenhouse Effect, and Depletion of the Ozone Shield

CONCEPT FRAMEWORK

▪ Outline

I. ACID PRECIPITATION   321
  A. Understanding Acidity   321
    1. Acids, Bases, and Water   321

    2. pH: The Measurement of Acidity   323

  B. Extent and Potency of Acid Precipitation   324

  C. The Source of Acid Precipitation   324

  D. How Acid Precipitation Affects Ecosystems   325
    1. Effects on Aquatic Ecosystems   326

    2. Effects on Forests   328
      a. Direct contact effects   328
      b. Removal of nutrients   329
      c. Mobilization of aluminum and other toxic elements   329
  E. Depletion of Buffering Capacity and Anticipation of Future Impacts   329

  F. Effects on Humans and Their Artifacts   332

  G. Strategies for Coping with Acid Precipitation   332
    1. Treating Symptoms   332

    2. Reducing Acid-Forming Emissions   333
      a. Fuel switching   333
      b. Coal washing   333
      c. Fluidized bed combustion   333
      d. Scrubbers   333
      e. Alternative power plants and conservation   333
    3. Prescription for Biosphere Collapse   333

    4. What You Can Do   335
II. THE GREENHOUSE EFFECT   336
  A. The Heat-Trapping Effect of Carbon Dioxide   336

▪ Study Questions

1. Define acid precipitation. Name the kinds.

2. Describe the chemical nature of *acids* and *bases*. Explain how an acid neutralizes a base. Why is water considered neutral?

3. Explain the *pH scale*. What numbers are *acid, base, neutral*? How much difference is there between each number?

4. How widespread is acid precipitation? How much more acid than normal is it?

5. Name the two major acids involved in acid precipitation and describe where they come from.

6. Describe how acid precipitation affects aquatic ecosystems. How does loss of the aquatic ecosystem affect terrestrial ecosystems?

7. Describe three ways in which acid precipitation affects forests. Give examples of forest diebacks being observed.

8. Define: *buffer, buffering capacity*. Explain how some ecosystems remain healthy while others have been destroyed under the same amount of acid precipitation.

9. Describe how statues and monuments are being affected by acid precipitation. Draw a relationship between this and loss of buffering capacity.

10. How do farmers maintain the pH of their soil? Is this a practical solution for ecosystems in general?

11. List and describe ways in which acid-forming emissions from coal-burning power plants may be reduced. Which seem most practical for the near-term? the long-term?

12. Give the arguments that the utility industry has used to block action on acid-precipitation. Point out the fallacies of these arguments. What has the public done to support the inaction?

13. Describe what you can do to promote positive action.

14. Describe the heat-trapping effect of carbon dioxide. Describe how the level of carbon dioxide is changing.

B. Sources of Carbon Dioxide Additions 336

C. Other Greenhouse Gases 337
D. Degree of Warming and Its Probable
Effects 338

E. Is the Greenhouse Warming Effect Here? 339

F. Strategies for Coping with Carbon Dioxide
Effect—What You Can Do 339

CASE STUDY: THE GREENHOUSE EFFECT 340

III. DEPLETION OF THE OZONE SHIELD 339
A. The Nature and Importance of the Ozone
Shield 339

B. The Formation and Breakdown of the
Ozone Shield 341

C. The Source of Chlorine Atoms Entering
the Stratosphere 342

D. The Ozone "Hole" 343

E. Coming to Grips with Ozone Depletion 343

F. What You Can Do 345

15. Where is the additional carbon dioxide coming from? How are you producing some of it yourself? Name and give sources of other gases involved.

16. Describe the degree of warming that will probably occur and the effects that it will have.

17. Describe the evidence that the carbon dioxide warming trend is already occurring.

18. Describe what can be done to mitigate the effect and how you can participate.

19. Describe the nature and importance of the ozone shield.

20. Describe how the ozone shield is formed and what is causing its breakdown.

21. List and describe the sources of chlorine entering the stratosphere. Define CFCs.

22. Describe when and where ozone depletion was first observed. Is depletion in other regions likely? Explain.

23. What is being done to cope with depletion of the ozone shield?

24. Discuss what you can do.

Thousands of lakes are lifeless and tens of thousands more are threatened, and forests are dying back because of acid precipitation. The climate is warming, threatening the world with unprecedented droughts and other climatic shifts because of increasing carbon dioxide and other "greenhouse gases" in the atmosphere. All life on earth may be threatened by increasing ultraviolet radiation due to depletion of the ozone shield. All three of these events have been clearly identified as stemming from human alterations, forms of pollution reaching global proportions and affecting the entire biosphere. In this chapter we shall examine the causes of these three events and what can be done to mitigate them.

ACID PRECIPITATION

Acid precipitation refers to any precipitation, rain, fog, mist, or snow, that is more acid than normal. It also involves the fallout of dry acid particles, referred to more specifically as **acid deposition.** In a nutshell, broad areas of North America, as well as most of Eu-rope and other industrialized regions of the world, are regularly experiencing precipitation that is between 10 and 1000 times more acid than normal. This is affecting ecosystems in diverse ways, as illustrated in Figure 13–1, and will be described in the following.

To understand the full extent of this problem, however, it is first necessary to understand some principles regarding the nature and measurement of acids.

Understanding Acidity

Acids, Bases, and Water

Acidic properties (sour taste and eating away of many materials) are due to the presence of **hydrogen ions** (H^+, hydrogen atoms without their electrons), which are highly reactive. Therefore, an **acid** is any chemical that will release hydrogen ions when it dissolves in water. Chemical formulas of a few common acids are shown in Table 13–1. Note that all of them ionize (the chemicals separate) to give hydrogen ions as well as the negative ion of the particular acid. The higher the

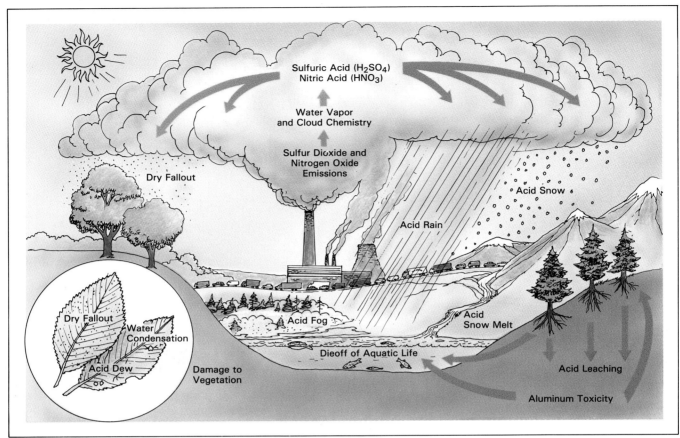

FIGURE 13–1
Acid precipitation. Emissions of sulfur and nitrogen oxides react with water vapor in the atmosphere to form their respective acids which come back down as dry acid deposition or mixed with water, causing the precipitation to be abnormally acidic. Various effects are noted.

concentration of hydrogen ions in a solution, the more acidic it is. In a similar way, the bitter taste and caustic properties of all alkaline or basic solutions are due to presence of **hydroxyl ions** (OH^-, oxygen-hydrogen groups with an extra electron). Hence, a **base**

is any chemical that will release OH^- ions (Table 13–1).

You are probably familiar with the fact that acids and bases neutralize each other. This occurs because the H^+ and OH^- ions simply come together to form HOH (i.e., H_2O) or water. Adding a base to an acid solution will gradually decrease the acidity as OH^- ions combine with H^+ until the neutral point associated with pure water is reached. Then there will be an increase in alkalinity. Thus, there is a continuum

Table 13–1
Common Acids and Bases

ACID	FORMULA	YIELDS	H^+ ION(s)	PLUS	NEGATIVE ION
Hydrochloric Acid	HCl	→	H^+	+	Cl^- chloride
Sulfuric Acid	H_2SO_4	→	$2H^+$	+	SO_4^{2-} sulfate
Nitric Acid	HNO_3	→	H^+	+	NO_3^- nitrate
Phosphoric Acid	H_3PO_4	→	$3H^+$	+	PO_4^{3-} phosphate
Acetic Acid	CH_3COOH	→	H^+	+	CH_3COO^- acetate

BASE	FORMULA	YIELDS	OH^- ION(s)	PLUS	POSITIVE ION
Sodium Hydroxide	NaOH	→	OH^-	+	Na^+ sodium ion
Potassium Hydroxide	KOH	→	OH^-	+	K^+ potassium ion
Calcium Hydroxide	$Ca(OH)_2$	→	$2OH^-$	+	CA^{2+} calcium ion
Ammonium Hydroxide	NH_4OH	→	OH^-	+	NH_4^+ ammonium ion

from acidity through neutrality toward increasing basicity or alkalinity (Fig. 13–2). The balance or neutral point between acid and base is water or, said the other way, water is neutral. The other ions associated with the acid and base remain in solution as a salt, but these do not affect acidity or alkalinity.

FIGURE 13–2
Adding a base to an acid. OH^- of the base combines with H^+ of the acid until the neutral point, pH 7, is reached. Further addition of base increases the OH^- concentration, making the solution basic.

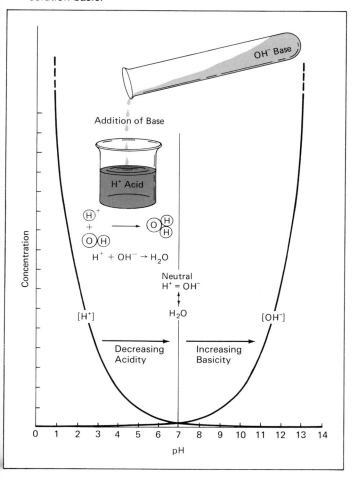

FIGURE 13–2
Adding a base to an acid. OH^- of the base combines with H^+ of the acid until the neutral point, pH 7, is reached. Further addition of base increases the OH^- concentration, making the solution basic.

pH: The Measurement of Acidity

The actual concentration of hydrogen ions along the continuum from acid to base is expressed as **pH**. The pH scale goes from 0 (highly acidic) through 7 (neutral) to 14 (highly basic) (Fig. 13–3). The numbers on the pH scale actually stand for the negative logarithm (powers of 10) of the hydrogen ion concentration in grams per liter. For example pH = 1 means that the concentrate of hydrogen ions in the solution is 10^{-1} or 0.1 g/L; pH = 2 means that the hydrogen ion concentration is 10^{-2} or 0.01 g/L, and so on.

At pH = 7 the hydrogen ion concentration is 10^{-7} (.0000001 g/L), but here the OH^- concentration is the same. Thus, this is the neutral point and expresses the tiny but equal amounts of H^+ and OH^- present in pure water. The pH numbers above 7 continue to express the negative exponents of hydrogen ion concentration, but more importantly they also represent an increase in OH^- ion concentration. For example, pH = 13 means that the hydrogen ion concentration is 10^{-13} (a decimal followed by 12 zeros and a one) grams per liter. However, the OH^- concentration at this point is equivalent to 10^{-1}

It is significant to note that since numbers on the pH scale represent powers of 10, there is a tenfold difference between each unit. For example, pH 5 is ten times more acidic (has ten times more H^+ ions) than pH 6; pH 4 is ten times more acidic than pH 5, and so on. There is a 1000-fold difference between pH 6 and pH 3 (tenfold for each unit: $10 \times 10 \times 10 = 1000$). The same is true for OH^- ion concentration as pH units progress above 7.

pH is easily measured by using pH indicator paper, which is available from any laboratory supply house. Such indicator papers contain pigments that readily absorb or release hydrogen ions depending on the pH of the solution and change color as a result. The pH is determined by simply dipping a strip of indicator paper in the solution and matching its color with a color chart provided with the paper. There are also more expensive electronic instruments for measuring pH.

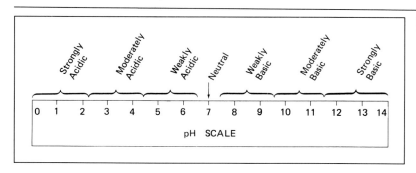

FIGURE 13–3
The pH scale.

Extent and Potency of Acid Precipitation

In the absence of any pollution, rainfall is normally slightly acidic, pH 5.6, because carbon dioxide in the air readily dissolves in and combines with water to produce a weak acid, carbonic acid (Fig. 13–4). *Acid precipitation*, then, is defined more precisely as any precipitation with a pH of 5.5 or less.

Unfortunately, this situation occurs over most of the industrialized world. The pH of rain and snowfall over large portions of the eastern United States and Canada, the West Coast, and most of Europe is typically about 4.5 (Fig. 13–5). Many areas within these regions regularly receive precipitation with a pH of 4.0. These are averages; the pH of particular events may be considerably lower. Fogs and dews have been found to be even more acidic. Researchers at the California Institute of Technology have found that fog in the Los Angeles area typically has a pH value of between 2.5 and 3. In forests in the mountains east of Los Angeles, scientists found fog water with a pH of 2.8 dripping from pine needles. In short, large portions of the industrialized world and other regions around urban centers typically receive precipitation that is 10 to 1000 times more acidic than normal, and values as low as pH 1.5, 10,000 times more acid than normal, have been recorded.

The Source of Acid Precipitation

Chemical analysis of acid precipitation reveals the presence of sulfuric acid (H_2SO_4) and nitric acid (HNO_3). In general about two-thirds of the acidity is due to sulfuric acid, one-third to nitric acid. The presence of sulfur and nitrogen, respectively, in these two acids says that the source of the problem must be sulfur and nitrogen emissions into the air. Knowing that burning fuels produce sulfur dioxide and nitrogen oxides (Chapter 12), we can surmise the source of the problem. Proof has been supplied by cloud sampling and chemical experiments which clearly document that both sulfur dioxide and nitrogen oxides gradually react with water vapor and, through a number of steps, become acids.

$$SO_2 + H_2O \xrightarrow{\hspace{3cm}} H_2SO_4$$

Cloud chemistry

$$NO_2 + H_2O \xrightarrow{\hspace{3cm}} HNO_3$$

Precipitation becomes acidic as it flushes these acids from the atmosphere. The pH of the precipitation depends on both the amount of acid and the amount of water that it is dissolved in. Heavy rains may be less acidic because there is relatively more water. Fogs and mists may be most acidic because the acid is dissolved in relatively little water.

It is now recognized that the acid may also simply settle out of the atmosphere by itself or on dust particles apart from any water. This dry **acid fallout** or *acid deposition* may accumulate on vegetation, and when it becomes wetted with a small amount of water, as occurs with the condensation of dew, the resulting solution may be strongly acidic. Hence, **acid dews** must be added to the list of acid precipitation.

Knowing the general amounts of emissions of sulfur dioxide and nitrogen oxides from various sources (Chapter 12), acid precipitation can be attributed to electrical utilities (coal-burning power plants), vehicles, and industries accordingly. For example, sulfur dioxide and nitrogen oxide emissions from various sources in the eastern United States are shown in Figure 13–6. Since about two-thirds of the acidity in precipitation is due to sulfuric acid and since three-fourths of the sulfur dioxide comes from coal-burning power plants, over 50 percent of acid precipitation falling over the eastern United States and Canada can be attributed to coal-burning power plants.

Furthermore, the locations of major offending power plants can be pinpointed by studying cloud tracks, recorded by weather satellites, preceding acid

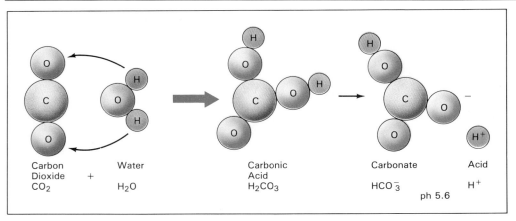

Carbon Dioxide CO_2 + Water H_2O → Carbonic Acid H_2CO_3 Carbonate HCO_3^- ph 5.6 Acid H^+

FIGURE 13–4
The pH of normal rainfall is 5.6. Normal precipitation is slightly acidic because carbon dioxide dissolved in water produces a weak acid as shown.

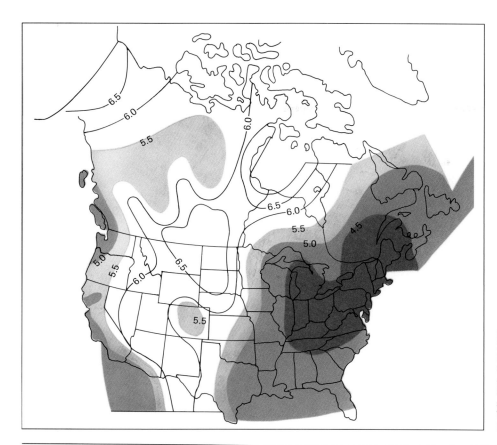

FIGURE 13-5
Regions receiving acid precipitation. Monitoring the pH of precipitation now reveals that acid precipitation is occurring over most of the United States and Canada. It is especially severe in the Northeast and along the West Coast. (Ontario Ministry of the Environment.)

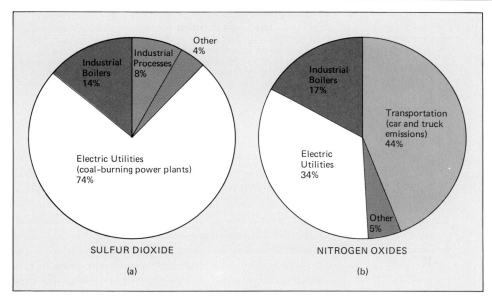

SULFUR DIOXIDE

(a)

- Electric Utilities (coal-burning power plants) 74%
- Industrial Boilers 14%
- Industrial Processes 8%
- Other 4%

NITROGEN OXIDES

(b)

- Transportation (car and truck emissions) 44%
- Electric Utilities 34%
- Industrial Boilers 17%
- Other 5%

FIGURE 13-6
Sources of acid-forming pollutants in the 31 eastern states. (a) Sulfur dioxide. (b) Nitrogen oxides. (Office of Technology Assessment; data for 1980.)

precipitation events. Thus, for the eastern United States the source of over 50 percent of the acid precipitation has been identified as the tall stacks of some 50 huge coal-burning power plants (Fig. 13-7). Recall (Chapter 12) that these tall stacks were built to alleviate the sulfur dioxide pollution at ground level. The unfortunate result of the "dilution solution" is that emitting sulfur dioxide and nitrogen oxides high in the air simply provides more opportunity for them to convert to acids and spread hundreds of miles from the source (Fig. 13-8).

How Acid Precipitation Affects Ecosystems

Acid precipitation has been recognized as a problem in and around industrial centers for over 100 years,

(a)

(b)

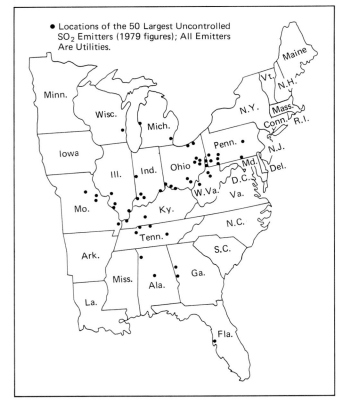

- Locations of the 50 Largest Uncontrolled SO$_2$ Emitters (1979 figures); All Emitters Are Utilities.

FIGURE 13-7

(a) Standard smokestacks of this coal-burning power plant were replaced by new 1000-foot stacks to aid in the dispersion of pollutants into the atmosphere. They may alleviate local problems, only to create more widespread ones. Shawville, Pennsylvania. (b) Locations of the 50 largest sulfur dioxide emitters, all of which are utility coal-burning power plants. They account for 74 percent of the sulfur dioxide and a large portion of the nitrogen oxide emissions as well. (a, Photo Researchers/© 1984 Grapes·Michaud; b, "Acid Rain, a Major Threat to the Ecosystem," Conservation Foundation Letter, December 1982.)

but effects on ecosystems were not noted until about 35 years ago when anglers started noting precipitous declines in fish populations in many lakes in Sweden, Ontario, Canada, and the Adirondack Mountains of upper New York State. Many hypotheses were suggested and tested to determine the reason. Scientists in Sweden were the first to identify the cause as increased acidity and to link this with abnormally low pH of precipitation. Since that time, while the ecological damage has continued to spread, studies have revealed many ways in which acid precipitation affects and may destroy ecosystems. We shall look at the main ways here.

Effects on Aquatic Ecosystems

pH is extremely critical because it affects the functioning of virtually all enzymes, hormones, and other proteins of the body which control all aspects of me-

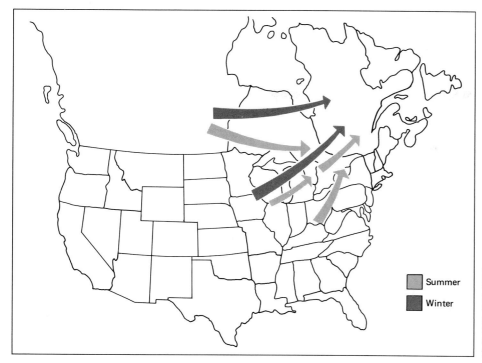

FIGURE 13–8
Major summer and winter storm paths. Acid-forming pollutants emitted from the tall stacks of power plants located in the Ohio Valley tend to fall out over New England and eastern Canada. (Ontario Ministry of the Environment.)

FIGURE 13–9
Effects of pH on the survival of various organisms. Some organisms are more tolerant than others to low pH, but very few can survive below pH 4.5, and upsets in ecosystems will occur even as the most sensitive organisms die off. (Office of Technology Assessment.)

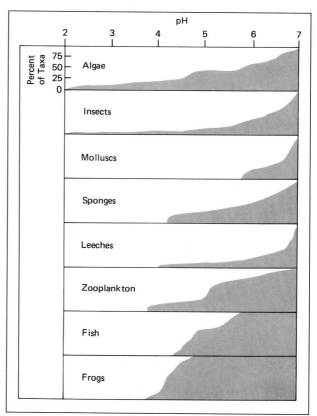

tabolism, growth, and development. Larger organisms may not be harmed by small changes in environmental pH because metabolism adjusts internal pH and skin offers a degree of protection. However, eggs, sperm, and developing young of aquatic organisms have little such protection. Most are severely stressed and many die if the environmental pH shifts as little as one unit from the optimum. Most freshwater lakes, ponds, and streams have a natural pH in the range of 6–7, and organisms are adapted accordingly.

As such aquatic ecosystems are acidified, there is a rapid dieoff of virtually all organisms because of failure to reproduce if not because of direct impact (Fig. 13–9). The impact of acid precipitation on aquatic ecosystems may be intensified by melting snow, which tends to bring all the acid precipitation accumulated over winter into a lake or river just at the time most organisms are reproducing.

Additional impact results from the fact that acid precipitation may leach aluminum and various heavy metals from the soil as it percolates through. Normally, the presence of these elements in the soil does not pose a problem because they are bound in insoluble mineral compounds and, therefore, are not absorbed by organisms. As these compounds are dissolved by low pH, however, they may be absorbed and they prove highly toxic to both plants and animals. For example, aluminum is quite abundant in most soils and when leached into lakes causes abnormal development and death of fish embryos.

In Norway and Sweden the fish have died in about 6500 lakes and 7 Atlantic salmon rivers. In Ontario, Canada, approximately 1200 lakes are now dead, and in the Adirondacks, a favorite recreational region for New Yorkers, more than 200 lakes are without fish, and many are devoid of all life save a few bacteria. The appearance of such lakes is deceiving. From the surface they are clear and blue, the outward signs of a healthy oligotrophic condition. However, a view under the surface is eerie. In spite of ample light shimmering through the clear water, there is not a sign of life; they are totally barren.

The loss of life does not stop with the dieoff of fish and other aquatic organisms. Numerous food chains supporting nearly all wildlife have origins in lakes and streams. For example, naturalists report that in areas of the Adirondacks where the lakes no longer support fish, there is a total absence of loons and other waterfowl. Populations of birds that feed on insects, many of which breed in the water, are also down, as are populations of raccoons and many other mammals. If the situation continues it is not hard to imagine what will happen to geese and other migrating birds that use lakes as places to stop and feed (Fig. 13–10).

FIGURE 13–10
Acidification of lakes and rivers has a profound effect on terrestrial wildlife because numerous birds and land animals are supported by food chains that have their origin in aquatic systems.

Effects on Forests

Along with dying lakes, decline of forests has also become conspicuous. Between 1963 and 1973, the growth rate of spruce trees in the Green Mountains of Vermont declined by 50 percent. More recently, scientists have observed similar slowdowns in the growth of forests over wide areas of the eastern United States and in California. An unprecedented number of trees is falling prey to insect and disease attack, and large-scale dieoff of a number of species is now occurring in New England, California, and other regions of high acid precipitation (Fig. 13–11). Most scientists are convinced that acid precipitation, along with ozone (Chapter 11), is a major factor in this because they have found that acid precipitation may affect vegetation in the following ways.

- ☐ Damaging the surface by direct contact
- ☐ Removing nutrients
- ☐ Mobilization of aluminum and other toxic elements.

In turn, trees stressed by any or all these factors fall victim more easily to pest attack.

Direct Contact Effects. Through simulated acid rain studies in greenhouses it has been shown that the acid damages the waxlike protective layer of leaves, making plants more vulnerable to attack by insects,

FIGURE 13–11
Dieback of forests. A retardation of growth and a dieback of a number of species are being observed in many areas impacted by acid rain. Acid precipitation, in addition to other factors of pollution, is apparently stressing forests to the critical point. Shown here: Dieback of forest on Mount Mitchell, western North Carolina (1985). (Courtesy of Dr. Dwight Billings, Duke Univ.)

fungi, or other plant pathogens as well as drought because water loss through the leaves increases. This effect may be particularly severe where there are intensely acid fogs or the accumulation of dry acid deposition.

Removal of Nutrients. Analysis of water draining from various natural areas under different conditions has shown that acid precipitation greatly increased the leaching of nutrients. Hydrogen ions effectively displace nutrient ions from their places in the soil and humus (Fig. 13–12). In addition, low pH also retards the activity of decomposers and nitrogen-fixing organisms, causing even further nutrient shortages. Finally acidic precipitation washing over vegetation has been shown to leach nutrients and other metabolites from leaves, especially when the leaf surfaces have been damaged by other factors. Any or all of these may cause nutrient deficiencies that result in decreased growth rate and increased vulnerability to natural enemies and drought.

Mobilization of Aluminum and Other Toxic Elements. Lastly, many plants are highly sensitive to aluminum. Aluminum is an extremely common element; substantial quantities are present in many rock and soil minerals. Normally, these minerals are very insoluble and thus harmless, but under the attack of acid they break down and release the aluminum into solution. The process is referred to as **mobilization;** in this case, mobilization of aluminum. Other toxic elements, such as lead and mercury, may be mobilized as well. Synergistic interactions between toxic elements and low pH may also occur. These toxic effects

may also account for reduced growth rates and diebacks.

Through these mechanisms, acid precipitation by itself may account for slowing growth rates and diebacks being observed in many forests, but ozone, as noted in Chapter 12, may have the same effects and both factors are generally present. Thus there is probably a synergistic interaction between ozone and acid precipitation (see Case Study 12–1, Chapter 12).

Depletion of Buffering Capacity and Anticipation of Future Impacts

Wide regions receive roughly equal amounts of acid precipitation, yet not all areas are equally affected. Many areas remain apparently healthy while others have acidified to the point of becoming lifeless. How is this possible? More importantly, how can we predict the future impacts? The key to both questions lies in understanding the concept of *buffering* and *buffering capacity.*

A system may be protected from pH change, despite addition of acid, by a *buffer.* A **buffer** is a substance that has a large capacity to absorb (or release) hydrogen ions at a given pH. If acid is added to a system containing a buffer, the additional hydrogen ions will be absorbed by the buffer and the pH will remain relatively constant.

Many, but not all natural bodies of water and soils are buffered by the presence of limestone, which chemically is calcium carbonate ($CaCO_3$). As shown in Figure 13–13, hydrogen ions react with the carbonate (CO_3^{2-}) to form water and carbon dioxide. This is the "fizzing" that is observed when a drop of

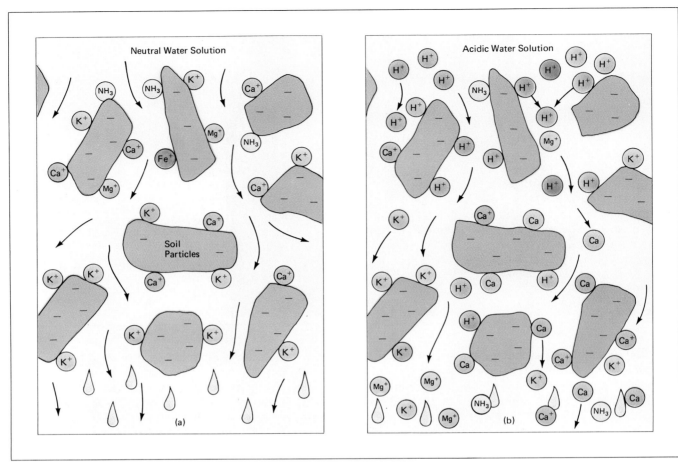

FIGURE 13–12
Acid precipitation leaches soil nutrients. Clay and humus particles tend to be negatively charged and hence bind and hold positively charged nutrient ions such as potassium (K^+), ammonium (NH_3^+), and calcium (Ca^{2+}). (a) The attractive force is strong enough to hold ions despite water percolating through the soil. (b) Acidic solutions cause leaching because hydrogen ions (H^+) displace nutrient ions.

FIGURE 13–13
Buffering. Acids may also be neutralized by certain nonbasic compounds called *buffers*. A buffer such as limestone (calcium carbonate) reacts with the hydrogen ions as shown. Hence the pH remains close to neutral despite the additional acid. Note, however, that the buffer is consumed by the acid. Limestone is the most widespread natural buffer.

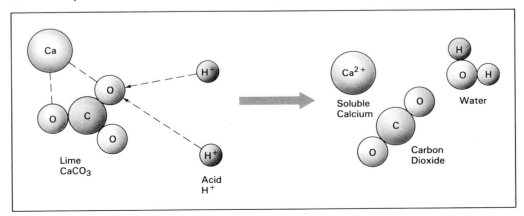

acid is put on limestone. Farmers have long used the addition of lime (pulverized limestone) to neutralize acid soils. Shells of clams and oysters, and eggshells, which are also calcium carbonate, are favored by organic gardeners for the same purpose.

Any buffer has limited capacity, however. Lime, for instance, is simply used up by the buffering reaction. Thus, we speak of the *buffering capacity* of a system. When the buffering capacity is exhausted, additional hydrogen ions will remain in solution and there will be a corresponding drop in the pH.

Under similar amounts of acid precipitation, ecosystems that have already acidified and collapsed are those that had very little buffering capacity. Those that remain healthy have more. But, what about the future if acid precipitation is allowed to continue? The unfortunate fact is that large portions of the United States and Canada that receive acid precipitation have rock and soil types that are low in buffering capacity (Fig. 13–14) and what capacity remains is being depleted rapidly. A recent study of the United States east of the Mississippi River conducted by the Congressional Office of Technology Assessment showed that of a total of 17,059 lakes, 2993 (about 16 percent) are already showing signs of damage, and another 6430 (about 30 percent) are endangered as indicated by very little remaining buffering capacity. Of a total 117,423 miles (187,876 km) of streams and rivers, 24,688 miles (21 percent) already show damage from acid precipitation and another 21 percent are endangered. Although the problem is worse in the Upper Midwest and Northeast where up to 80 percent of lakes and streams are threatened, nearly every state from Maine to Florida reports some damage. In Canada, Ontario alone stands to lose an additional 48,000 lakes within the next 15 years if acid precipitation continues as it is.

The rapid shift in pH that occurs as buffering capacity is exhausted deserves emphasis. It is graphically shown in Figure 13–15. Note that when a buffer is present, the addition of many units of acid has no appreciable effect on the pH. However, as the buffering capacity is exhausted, there is a precipitous drop in the pH with the addition of just a few more units of acid. Can you see the implications of this?

Indeed a major cause for concern is that all of these acid-related effects may intensify with precipitous suddenness as buffering capacities are exhausted. An example from Europe is noteworthy. In a region of high acid precipitation centered on the border between East Germany and Czechoslovakia, the woods died at an alarming rate. A 1980 study showed that 60 percent of the fir trees were healthy; two years later 98 percent were dead or dying. A similarly rapid dieback of forests occurred throughout West Germany. According to German government surveys, forest affected by damage as a result of pollution increased from 8 percent to 50 percent in just

FIGURE 13–14
Acid-sensitive regions. Large areas of North America, particularly in Canada, are especially sensitive to acid precipitation because these regions have granitic rock formations that have little buffering capacity. (Ontario Ministry of the Environment.)

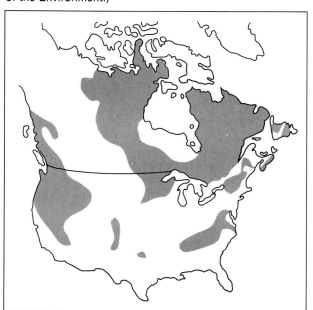

FIGURE 13–15
Exhaustion of buffering capacity. As the buffering capacity of a system is exhausted, the change in pH is *not* gradual. With the addition of very little more acid there is a sudden precipitous drop in the pH. Consequently, if we wait until buffers are depleted, disastrous consequences may occur before there is time for remedial action.

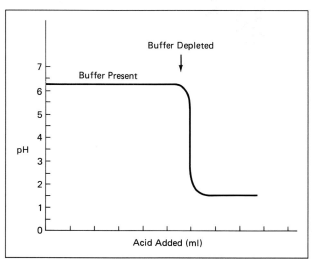

the two years between 1982 and 1984. Now about 50 percent of the forests throughout Northern Europe show damage. German researchers warn that symptoms observed in U.S. forests in 1985 resembled those seen in West Germany in 1980.

As changes occur in forests, you can surmise the effect on other wildlife populations. If a sudden collapse of the forest ecosystems occurs, the ramifying effects on soil erosion, sedimentation of waterways, flooding, and deterioration of water supplies will be catastrophic. At the very least, we can expect a succession in which the dying trees are replaced by acidloving species, but the variety of such plants is very limited. Most are mosses, ferns, and other scrubby plants that are economically worthless even for grazing.

Effects on Humans and Their Artifacts

From the viewpoint of the general public, one of the more noticeable effects of acid precipitation is the deterioration of artifacts. Limestone and marble (which is a form of limestone) are favored materials for the outside of buildings and monuments. The reaction of acid on limestone, shown in Figure 13–13, is causing these materials to weather and erode at a tremendously accelerated pace. Monuments and/or buildings that have stood for hundreds or even thousands of years with little change are now dissolving and crumbling away (Fig. 13–16).

While the decay of such artifacts is a tragic loss in itself, it should also stand as a grim reminder of how we are dissolving away the buffering capacity of ecosystems.

Additionally, some officials are concerned that acid precipitation's mobilization of aluminum and other toxic elements may result in contamination of both surface and groundwater supplies. It is noteworthy that aluminum has recently been implicated as a cause of Alzheimer's disease, a form of premature senility. Increased acidity of water also mobilizes lead used in some old plumbing systems and from the solder used in copper systems. Thus increasing acidity will very likely increase the toxic chemical problem (Chapter 11).

However, if acid precipitation is allowed to continue at the present rate, by far the greatest impact on humans will be from the deterioration and loss of lakes and forests and their associated economic, ecological, and aesthetic values, and from additional impacts occurring through soil erosion. Clearly, allowing acid precipitation to continue is contrary to a sustainable society.

Strategies for Coping with Acid Precipitation

Treating the Symptoms

As already noted, farmers maintain the pH of their soils by adding lime. Additionally, some lakes are being limed, although it is questionable whether this maintains a natural ecosystem. However, there are over 1 million square miles in the eastern United States and Canada alone that are being impacted by acid precipitation. The work and cost of maintaining this kind of area by artificial liming is ridiculous to even contemplate. Therefore, we must focus attention on reducing acid-forming emissions.

(a)

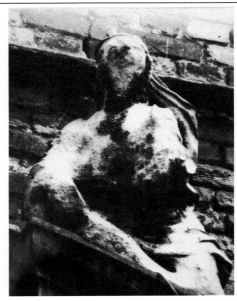
(b)

FIGURE 13–16
Corrosive effects of acid rain are highly destructive to stonework. The statue shown is in the Ruhr, an industrial district of West Germany. (a) Appearance in 1908; (b) appearance in 1969. Erected in 1702, it suffered more erosion in the last 60 years than it did in the previous 200. (From E. M. Winkler, *Stone: Properties, Durability in Man's Environment.* Photos supplied by Schmidt-Thomsen, Landesdenkmalamt, Westfalen-Lippe, Muenster, Germany.)

Reducing Acid-Forming Emissions

Scientists calculate that a 50 percent reduction in present acid-causing emissions would effectively prevent further acidification of the environment. This would not correct the already bad situations, but natural buffering processes are estimated to be capable of preventing further deterioration. Since we know that about 50 percent of the acid-producing emissions comes from the tall stacks of coal-burning power plants, control strategies center on these sources. Five main strategies have been proposed: (1) fuel switching, (2) coal washing, (3) fluidized bed combustion, (4) scrubbers, (5) alternative power plants, and (6) reducing consumption.

Fuel Switching. The present plants would not emit sulfur dioxide if the fuel were not contaminated with sulfur. Is it feasible to switch to low- or no-sulphur fuels? The United States does have abundant reserves of low-sulfur coal, but most are located in the western states (Montana, Wyoming, Colorado, and Utah). The logistics and economic implications of relocating miners and equipment, and transporting 200 million tons of coal per year to eastern power plants, are horrendous to contemplate. Also western low-sulfur coal has a lower heating value. Therefore, more coal would be needed to produce the same power. Thus, the net reduction in SO_2 emissions might be small. If low-sulfur fuel oil or natural gas were to be substituted, these fuels would have to come from increasing imports. This would be highly counterproductive in terms of U.S. efforts to become less dependent on foreign fuel supplies. Indeed, a significant factor in decreasing reliance on foreign oil in the 1970s was the switch by utilities from oil to coal.

Coal Washing. It is possible to pulverize coal and put it through a chemical wash to remove the sulfur before it is burned. But the economics of putting 10,000 tons of coal per day, the amount consumed by just one of these power plants, through such a wash is forbidding.

Fluidized Bed Combustion. In fluidized bed combustion, coal is burned in a mixture of sand and lime all of which is kept churning (fluid) by forced air coming from underneath (Fig. 13–17). In this process, the sulfur combines with the lime in the course of combustion and is removed with the ash. In the building of new plants this may be the preferred method of emission control, but to put this into existing plants would entail tearing down and rebuilding a major portion of the plant.

Scrubbers. *Scrubbers* are "liquid filters." They entail putting the exhaust fumes through a spray of water containing lime (Fig. 13–18). The sulfur dioxide reacts with the lime and is precipitated as calcium sulfate ($CaSO_4$). Plants in Europe, Canada, and a few in the United States have demonstrated that scrubbers can be added onto existing power plants in a relatively short period of time; they are highly effective in controlling emissions, and they are not prohibitively expensive.

Alternative Power Plants and Conservation. Sulfur dioxide is only one of several pollution problems inherent in coal-burning power plants. Others are massive land destruction and water pollution from mining the coal, pollution from disposal of the ash, and contribution to the carbon dioxide climatic warming effect. Even if acid-forming emissions are controlled these other problems will remain. Ultimately, we need alternative, cleaner methods of producing electrical power.

Interestingly, in the 1950s and 1960s it was considered that nuclear power would be this ultimate alternative power source. While nuclear power does not produce acid-forming compounds or carbon dioxide, it does present its own problems of radioactive wastes and the danger of dispersal of radioactive compounds into the environment in the event of an accident. Public concern regarding these dangers has been a major factor in limiting the building of nuclear power plants in the United States. In a very real sense, the huge coal-burning power plants that are the center of the acid precipitation problem were constructed as an alternative to nuclear power. Yet the technology exists for building fail-safe nuclear power plants. The option of using nuclear power as an alternative to coal is an alternative that should not be automatically ruled out. We shall explore this question in more depth in Chapter 20.

In the long run, various solar energy options, which will be discussed in Chapter 21, have even more potential for producing abundant, sustainable, clean, safe power. In the short term, increasing the efficiency of electricity use can reduce generation and delay the need for new power plants.

In summary, installing scrubbers on coal-burning power plants could provide a very significant reduction of acid-forming pollutants almost immediately. This could be followed by building alternative power plants for the long term. Of course, control of pollutants from other sources, particularly nitrogen oxide emissions from vehicles as discussed in Chapter 12, is still in order, as is energy conservation.

Prescription for Biosphere Collapse

The seriousness of the acid precipitation problem and the involvement of coal-burning power plants were clearly evident by the end of the 1970s. Indeed, in 1981 the National Academy of Sciences, after studying

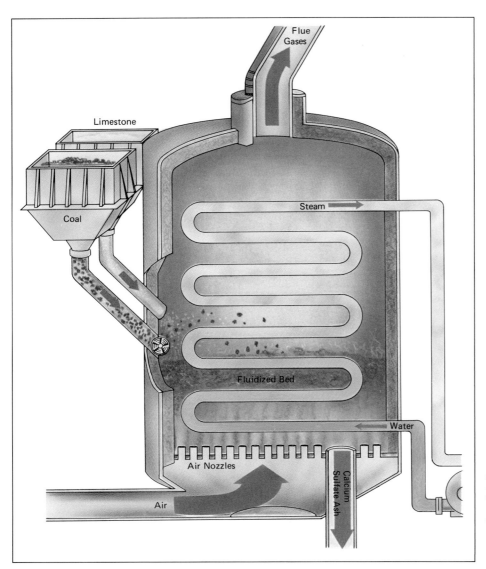

FIGURE 13–17
Fluidized bed combustion. Sulfur dioxide may be removed in the combustion process itself by burning the coal in a mixture of sand and lime, all of which is kept fluid by forcing air up from the bottom. (Redrawn from an illustration by Rick Farrell, copyright ©. All rights reserved.)

the problem, concluded that the circumstantial evidence linking power plant emissions to the production of acid precipitation was "overwhelming." Additional research and studies completed since have added to the already "overwhelming evidence." This analysis, combined with the fact that scrubbers are a proven technology, would lead one to expect swift action. Quite the opposite has been the case.

The electric power industry has raised every argument and excuse imaginable to confuse the issue, to disclaim any responsibility, and to avoid or delay emission controls. Many of the arguments they present are totally spurious. For example, they make the categorical statement, "All rain is acid," attempting to make the public believe that rainfall of pH 4.0 or less is no different from normal rain of pH 5.6. They also argue that acid precipitation has nothing to do with the situation because adjacent lakes receiving the same rainfall have different pH values. Of course, this ignores the differing buffering capacities that vary tremendously with limestone formations. They disclaim all evidence showing that acid precipitation is playing a role in forest damage.

One of their most common arguments is that scientists are unable to prove any case in which emissions from plant X have caused lake Y to become acid. Unless or until there is such definitive proof, they claim that they cannot and should not be held responsible. This argument is analogous to saying that we should not have pollution controls on cars because we cannot prove which car produced the smog that gave John Doe bronchitis. Clearly, this is absurd. Lakes and ecosystems are suffering from a general loading of the atmosphere with acid-producing pollutants, and it makes sense to reduce the major sources of this loading. It is both ridiculous and totally

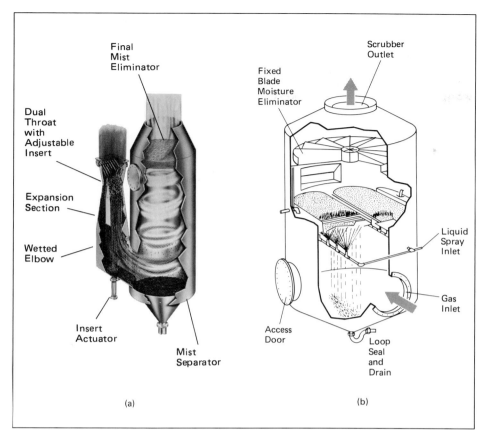

Final
Mist
Eliminator

Dual
Throat
with
Adjustable
Insert

Expansion
Section

Wetted
Elbow

Insert
Actuator

Mist
Separator

(a)

Fixed
Blade
Moisture
Eliminator

Scrubber
Outlet

Liquid
Spray
Inlet

Gas
Inlet

Access
Door

Loop
Seal
and
Drain

(b)

FIGURE 13–18
Scrubbers. Sulfur dioxide may be removed from flu gases by passing the furnace exhaust through a spray of lime and water. The sulfur dioxide reacts with the lime and a calcium sulfate precipitate is removed. Two designs for such devices, called *scrubbers*, are shown here. [(a) Courtesy of FMC Corporation-Air Quality Control. (b) A cutaway view of a SLY IMPINJET™ Gas Scrubber (The W. W. Sly Manufacturing Company, Cleveland, Ohio.)]

unnecessary to demand that lines be drawn from specific power plants to specific lakes.

Finally, the electric power industry claims the costs of installing scrubbers would make electricity unaffordable. But the industry does promote nuclear power plants as an answer. It is interesting to note that all 50 of the coal-fired power plants in question could be fitted with scrubbers for the cost of one or two nuclear power plants. In fact, actual experience from the Tennessee Valley Authority (TVA), a quasi-government utility company, shows that installing and operating a scrubber increases the costs by only 3 to 5 percent.

Blatantly absent from discussions are costs of environmental destruction that will be incurred if emissions are not controlled. In the United States, acid damage could be disastrous to the $49 billion forest and forest products industry, which employs 1.2 million people. This is not to mention the impacts on recreation, water supplies, wildlife, and general aesthetic values. The National Academy of Sciences warns that allowing emissions of sulfur and nitrogen oxides to continue unchecked *"in view of the known hazards is taking a grave risk"* with human health and biosphere protection. The academy goes on to strongly recommend prompt tightening of the controls on acid-forming emissions.

Unfortunately, much to the frustration of those who understand the seriousness of the problem, particularly the Canadians, who receive a large portion of acid precipitation from us, the Reagan Administration consistently echoed and supported electric power industry and consistently resisted any action toward emission controls. In early 1986 the administration did agree with Canadian officials to support a joint 5-billion-dollar program to determine the most economical way to clean up coal burning. This adds up to simply another delaying tactic. The $5 billion could have gone a long way toward installing scrubbers.

What You Can Do

We may feel justified in blaming the utility industry for inaction in facing up to the acid precipitation problem, but equal blame must be accepted by the public, the large majority of people who have remained strangely silent and apathetic on the issue. *Are you going to join the apathetic public in following the prescription for biosphere collapse, or are you going to take action?* To take action:

☐ Write or call your congressperson and senators and express the need for immediate legislation amending the Clean Air Act to require tight controls on sulfur dioxide emissions from power plants. Continuing to allow the "dilution solution" is unconscionable now that its linkage to the ecological destruction being caused by acid precipitation is clearly documented.

☐ Join an organization such as the Environmental Defense Fund or Natural Resources Defense Council. This will support their lobbying for effective controls.

☐ Inform others of the problem and what they can do.

THE GREENHOUSE EFFECT

Humans have tried to influence the weather for millennia. Suddenly we are now on the threshold of a major human-caused change in the climate. Unfortunately, the change was neither planned nor directed and may be catastrophic. It is being caused by the addition of carbon dioxide and certain other gases to the atmosphere, which will cause the earth's climate to warm resulting in rising sea levels and severe weather changes throughout the earth. Let's look at this problem in more detail.

The Heat-Trapping Effect of Carbon Dioxide

You are familiar with the way the interior of a car heats up when it is sitting in the sun with the windows closed. This occurs because sunlight (light energy) comes in through the windows and is absorbed by the seats and other interior objects. In being absorbed, the light energy is converted to heat energy, which causes the objects to become hot and give off heat energy in the form of **infrared** or heat radiation. Unlike light, infrared radiation is blocked by glass and trapped inside. The energy thus trapped causes the temperature to rise (Fig. 13–19). The same phenomenon occurs in a greenhouse giving rise to the term **greenhouse effect.**

On a global scale, carbon dioxide in the atmosphere plays a role analogous to the glass. Light energy comes through the atmosphere, is absorbed and converted to heat energy at the surface, and exits as infrared radiation. Carbon dioxide, however, absorbs infrared radiation; other natural gases in the atmosphere do not. As carbon dioxide absorbs the infrared radiation it becomes warm and this, in turn, warms the rest of the atmosphere. Consequently, it follows that the greater the amount of carbon dioxide, the greater the amount of infrared that will be absorbed and the warmer the atmosphere will be as a result.

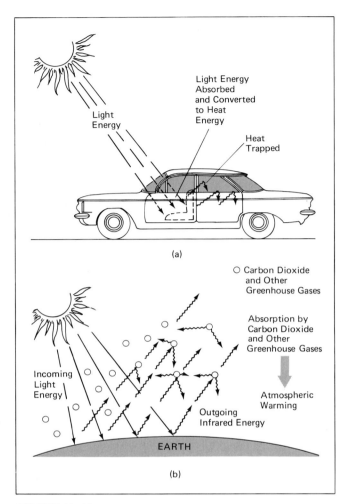

FIGURE 13–19
The greenhouse effect. A car or greenhouse in the sun gets hot because light energy entering through the glass is absorbed and converted to heat, infrared radiation which cannot escape through the glass. With the heat thus trapped, the temperature rises. Similarly, the earth's atmosphere is heated by light coming through, and outgoing infrared radiation is absorbed by carbon dioxide and certain other gases causing warming. The greater the concentration of carbon dioxide and/or other greenhouse gases, the greater will be the warming or greenhouse effect.

The temperatures/climates we are accustomed to on our Spaceship Earth have been given by a carbon dioxide concentration of just 0.03 percent or 300 parts per million. Now we are increasing this concentration as shown in Figure 13–20 and a warming trend is in prospect, if not already here.

Sources of Carbon Dioxide Additions

In the natural biosphere the level of carbon dioxide in the atmosphere is held steady by inputs equaling outtakes. We studied the greatest example of this in Chapter 2 in connection with the carbon cycle, in which carbon dioxide removed from the atmosphere

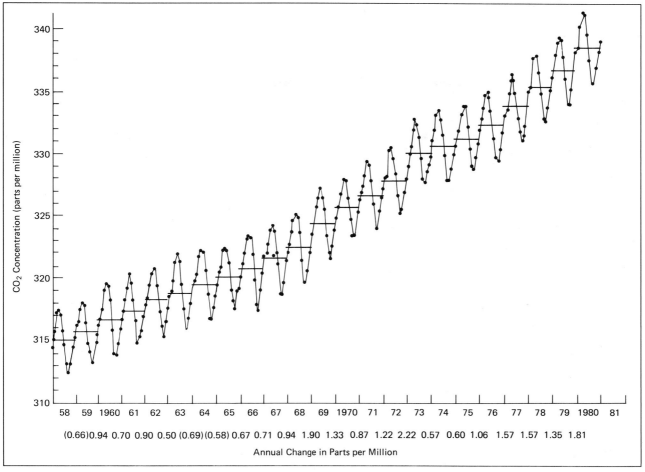

FIGURE 13-20
Atmospheric carbon dioxide concentration. Carbon dioxide concentration in the atmosphere fluctuates between winter and summer due to seasonal variation in photosynthesis. But the *average* concentration is gradually increasing owing to human activities, namely, burning fossil fuels and oxidation of soil organic matter. This trend may lead to an increase in global temperatures, which will result in other widespread effects. (M. C. MacCracken and H. Moss. "The First Detection of Carbon Dioxide Effects," *Bulletin of the American Meteorological Society* 63:1165, October 1982.)

by photosynthesis is replaced through respiration of organisms and burning.

Now humans are offsetting the natural balance by burning fuels and by deforestation. Every pound of fossil fuel (coal, liquid fuels derived from crude oil, and natural gas) that is burned results in the production of about three pounds or about 2 cubic meters of carbon dioxide. (The weight triples as each carbon atom in the fuel picks up two oxygen atoms in the course of burning and becoming carbon dioxide.) Close to 2 billion tons of fossil fuels are burned each year adding nearly 5.5 billion tons of carbon dioxide to the atmosphere. Another estimated 1.7 billion tons are released as tropical forests are cleared and burned and the organic material (humus) in the soil oxidizes.

As a result of these additions, the carbon dioxide level, which was about 290 parts per million at the turn of the century, is now up to 350 parts per million, an increase of 20 percent. Further, the rate of increase is accelerating as increasing population burns more fuel and clears more forests.

Other Greenhouse Gases

The greenhouse problem is being greatly intensified by certain other gases that humans are introducing into the atmosphere. In particular, methane, chlorofluorocarbons (CFCs), and nitrous oxide absorb infrared 50- to 100-fold more effectively than carbon dioxide. Consequently, while the amounts of these gases being emitted into the atmosphere are far less than the amounts of carbon dioxide, they have an effect on climatic warming which nearly equals that caused by the carbon dioxide additions. Thus, the term **greenhouse gases** is used to denote carbon dioxide and these other gases that absorb infrared radiation and lead to climatic warming.

Degree of Warming and Its Probable Effects

If current trends are allowed to continue, there will be a doubling of the equivalent carbon dioxide concentration by 2050. In turn, computer programs modeling various climatic parameters indicate that this will cause an overall warming of between 1.5° C and 4.5° C (3°–8° F). Warming is likely to be more pronounced in polar regions, as much as 10° C (18° F), and less pronounced in equatorial regions, 1°–2° C (2°–4° F). The reason for the considerable discrepancy is that exactly how the warming will affect cloud cover and how this will effect incoming radiation are not precisely known. However, there is no disagreement that warming will occur.

On first glance this seems modest. However, another 8–10 degrees added on top of a 100° F heat wave could be catastrophic. Furthermore, this warming will result in glaciers and the polar ice caps melting back enough to raise sea level by about 5 feet (1.5 meters). This will both flood and make many coastal areas much more prone to storms, forcing people to abandon properties and migrate inland. Are inland cities and communities ready to accommodate the millions of people that will be displaced? Are we ready to rebuild or modify all ports to accommodate the higher sea level?

The effects of the warming on rainfall and agriculture are likely to be even more serious. The difference in temperature between the poles and the equator is a major driving force for atmospheric circulation. The greater heating at the poles than at the equator will reduce this force. This will change atmospheric circulation patterns and, in turn, will affect the distribution of rainfall. Some regions of the world are likely to see an increase in rainfall; other areas, a decrease, as shown in Figure 13–21.

North Africa, which is largely desert at present, is likely to profit by increasing rainfall. However, the United States and Canada are likely to be losers. The central portion of North America is a major "breadbasket" of the world, producing huge amounts of surplus wheat and corn. Rainfall, already minimal for these crops, is likely to become much less. Quoting Dr. Walter Orr Roberts, former director of the National Center for Atmospheric Research,

> The Dust Bowl of the middle 1930s in the United States was the greatest climatic disaster in the history of our nation. . . . [However] the Dust Bowl of the 30s may seem like children's play in comparison to the Dust Bowl of the 2040s. Because of the effects of the warming . . . natural rainfall may decline by as much as 40%, and the summers will be hotter, increasing the evaporation of soil

FIGURE 13–21
Regions of the world that are likely to become wetter or drier than now under the influence of the carbon dioxide effect. Predictions may not be entirely accurate, but areas demarked by a dotted line have a higher probability of being correct. (William W. Kellogg and Robert Schware, 1981. *Climate Change and Society: Consequences of Increasing Atmospheric Carbon Dioxide.* Boulder, Colorado: Westview Press.)

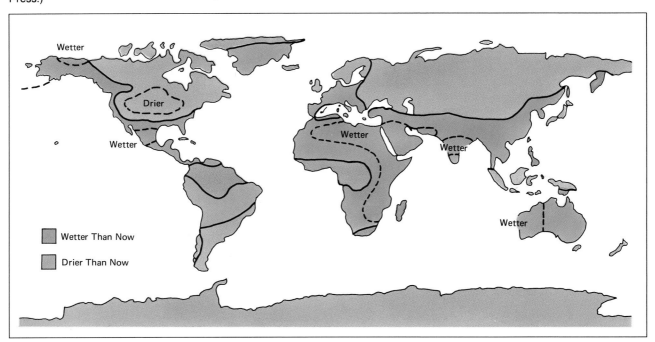

moisture. The soils will desiccate, and the winds will lift them to the skies.*

Nor can irrigation be expected to provide much relief. Recall that the water table is already being drawn down to support agriculture through much of the region; by 2040, and probably considerably before then, most of the groundwater for the region will be exhausted. Perhaps agriculture can be adapted to a different climate, simply moving it northward for example. The greatest difficulty, however, is in simply not knowing what to expect. Already farmers lose an average of one in five crops because of unfavorable weather. As the climate shifts the vagaries of weather will become more pronounced and crop losses due to "unfavorable weather" are likely to increase disastrously. Note the near 40 percent loss of the U.S. corn crop that occurred in 1988 as a result of drought.

Is the Greenhouse Warming Effect Here?

The projected warming because of increasing carbon dioxide and other infrared-radiation-absorbing (greenhouse) gases is based on extremely sound theory and no one can deny that the level of these gases in the atmosphere is rising. Until recently, however, it received relatively little attention, probably because it was considered to be too far in the future to be of concern. Also, there is so much natural variation in weather from year to year that a new trend is not immediately apparent. However, the future has a strange way of creeping into the present.

Accurate climate records have been kept for about 100 years. Six of the hottest years on record have occurred in the past decade: 1980, 1981, 1983, 1986, 1987, 1988. This and other data led noted climatologist James Hansen of the NASA Goddard Institute for Space Studies to testify before Congress in 1988 that what we are experiencing is indeed the beginning of the greenhouse effect. Many climatologists disagree with Hansen's analysis saying that the recent hot years may be nothing more than normal variations about the mean. However, data favorable to Hansen's interpretation continues to mount. In 1989, A. E. Strong of the National Atmospheric and Oceanic Administration reported that "satellite-derived sea surface temperatures for the period 1982–88 . . . [show] that the global ocean is undergoing a gradual but significant warming of about 0.1° C per

year." This is highly significant because the oceans have such thermal inertia that they respond little if at all to random fluctuations in climate. The fact that they show a warming trend is to say that the trend is real.

Thus, heat waves and droughts similar to or even more severe than that experienced in 1988 may become increasingly common occurrences leading to what Dr. Roberts described in the quote given above.

Strategies for Coping with the Greenhouse Effect—What You Can Do

The world's industry and transportation are so locked into the use of fossil fuels that massive additions of carbon dioxide to the atmosphere are bound to continue for the foreseeable future. However, there are ways we can lessen the additions and eventually bring about a sustainable balance. Most significant are:

☐ Increase fuel efficiency of vehicles and all other forms of energy conservation since energy production is nearly all based on the burning of fossil fuels.

☐ Develop and implement solar and other non-fossil-fuel energy alternatives.

☐ Halt further deforestation, particularly the clearing of tropical forests.

☐ Initiate, join, and support tree-planting programs. All of us can plant additional trees in our yards.

All of these actions are in keeping with other environmental goals. Energy conservation and development of energy alternatives are in keeping with reducing pollution. Maintaining and planting trees are in keeping with soil and water conservation and maintaining species diversity. They are all needed if we are to have a sustainable biosphere.

DEPLETION OF THE OZONE SHIELD

The Nature and Importance of the Ozone Shield

Radiation from the sun includes ultraviolet light along with visible light. **Ultraviolet light** is like light radiation but the wavelengths are slightly shorter than violet light, which are the shortest wavelengths seen

---

* Roberts, Walter Orr, "It is Time to Prepare for Global Climate Changes," *Conservation Foundation Letter*, April, 1983.

THE GREENHOUSE EFFECT

How will the coming greenhouse-gas induced climate change affect society as we know it? Shown in the figure is a computer model estimate of how we may expect the global mean temperature to change in the coming years, with three different scenarios of future trace gas growth. By the year 2000 all the curves indicate that the global mean temperature should be warmer than it has been during recorded history. Can we anticipate what effects this will have on the way we live?

Industrialized society has grown up in the past 200 years assuming that our climate is unchanging. Farmers plant the types of crops they know from experience will grow in a particular area. While weather varies from year to year, and so agricultural production does also, there has always been the expectation that in the long run, the weather will average out.

With a changing climate, this will no longer be the case. Hotter and perhaps drier summers should gradually become the norm, and crop production may well suffer. Choices will have to be made as to whether to grow different crops, or different strains of the same crop (say, more heat resistant varieties, if they are available), or use the land for some other purpose. If the land is no longer viable for agriculture, where will the food be grown? Perhaps in regions further north, say in Canada instead of in the Southern Great Plains, but this raises other issues. For one thing, food exports are a major source of income for the United States, and if this revenue is transferred to some other country it will represent a financial setback for this country. Also, the soils in Canada are not of the same quality as those further south, due to the action of the ice age glaciers which scraped the bedrock. It is not certain that the same amounts or quality of food could be grown there. Declining global food production, in conjunction with a continual growth in the world's population, would be a prescription for disaster.

If agriculture needs increased irrigation to cope with higher temperatures, this will require greater water availability. Warmer temperatures will also motivate consumers to use air conditioning more frequently, requiring greater electrical power consumption. In regions of the country where a significant portion of electrical supply comes from hydroelectric power plants, more water will be needed here too. And as a warmer climate will allow for more evaporation of moisture from the surface, there may well be less water available in the future. Water is one resource for which there is no adequate substitute. As the climate warms, we may expect that water shortages, and the ecological problems caused by declining water resources, will become more frequent.

These are but two of the fundamental ways in which climate affects society. We may expect almost everything we do to be influenced by the climate changes that are now being forecast. So what can we do? As the climate change is being driven by the emissions of carbon dioxide to the atmosphere due to the burning of fossil fuels for energy, we cannot realistically prevent the greenhouse warming from occurring without giving up our energy-dependent way of life. We must therefore prepare for climate warming

---

by the human eye (Fig. 13–22). While ultraviolet light is not visible, the rays are more energetic than those of visible light. On penetrating the atmosphere and being absorbed by biological tissues, they actually destroy protein and DNA molecules. This is what occurs when you get a sunburn. If the full amount of ultraviolet radiation falling on the upper atmosphere came through to the earth's surface, it is doubtful if any life could survive; plants and animals alike would simply be "cooked." Even the small amount (less than 1 percent) that does come through is responsible for all the sunburns and some 200,000–600,000 cases of skin cancer per year in the U.S.

We are spared more damaging effects from ultraviolet rays because most of it (over 99 percent) is absorbed and thus screened out by a layer of ozone in the stratosphere some 18 miles (25 km) above the earth. (See Fig. 13–23.) Thus, ozone in the stratosphere is commonly referred to as the **ozone shield.** The need to maintain it needs no elaboration. However, there are various human-made pollutants that are causing it to break down.

Yes, the ozone we are speaking of here is the same molecule (O_3) that we described in Chapter 12 as a serious air pollutant. Recall that one definition of pollution is a chemical out of place. There is negligible mixing between the lower atmosphere and the stratosphere. Therefore, the two situations, ozone

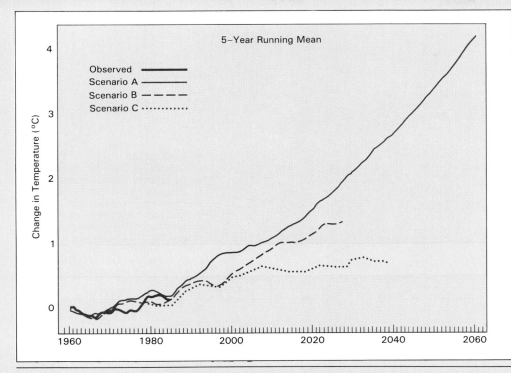

5–Year Running Mean

Observed
Scenario A
Scenario B
Scenario C

Computer model estimates of how we may expect the global mean temperature to change in the coming years given three different scenarios of future trace-gas growth. "A" assumes continued growth rates of emissions of greenhouse gases typical of the past 20 years (about 1.5% per year). "B" assumes such emissions are fixed at current rates, and "C" assumes drastically reduced emissions. By the year 2000 all the curves indicate that the mean temperature will be warmer than at any time in recorded history. The shaded range is an estimate of global temperature during the peak of interglacial periods 6000 and 120,000 years before present. The zero point for observations is the 1951–1980 mean. From Hansen et al. 1988, *Journal of Geophysical Research*, 93, 9341–9364.

of some magnitude by building more flexibility into our systems. If we expect water availability will be a problem in the future, start now to acquire the land and make plans for additional reservoirs and water storage capacity. If we think wheat growth in the Southern Great Plains will be affected, invest now in developing heat and drought resistant strains of wheat. If certain types of tree growth will be favored in a warmer climate, plant these trees instead. The processes just mentioned all take decades to bring to fruition. If decisions are made now, we may be able to minimize disruptions that are in store as climate warms.

DAVID RIND
NASA/Goddard Institute for Space Studies

being a pollutant in the lower atmosphere and its being essential in the stratosphere, are, for all practical purposes, separate issues.

The Formation and Breakdown of the Ozone Shield

Interestingly, ozone in the stratosphere is a product of ultraviolet (UV) radiation itself acting on oxygen molecules. The UV radiation causes some oxygen molecules (O_2) to split apart into free oxygen atoms, and these in turn may combine with other oxygen molecules to form ozone (O_3) as shown in Figure 13–23b. All the oxygen is not converted to ozone, how-

ever, because free oxygen atoms may also combine with ozone molecules causing them to break down to oxygen (Fig. 13–23c). Thus, the amount of ozone in the stratosphere is not static; it represents equilibrium or balance between these two reactions. In the 1970s, however, scientists hypotnesized that free chlorine atoms *catalyze* the breakdown of ozone. Humans had been unwittingly introducing such atoms into the stratosphere for decades.

A **catalyst** is defined as a chemical that promotes a chemical reaction without itself being used up by the reaction. The way a chlorine atom can catalyze the breakdown of ozone is shown in Figure 13–24. A relatively small amount of free chlorine can be ex-

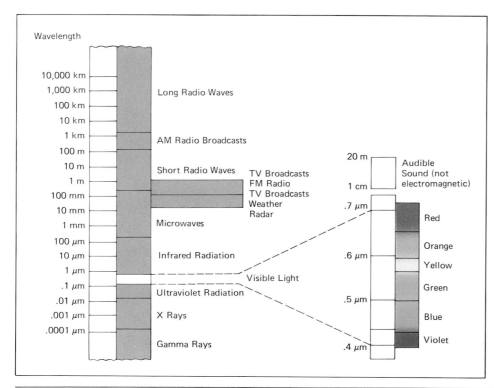

FIGURE 13–22
Ultraviolet, visible light, infrared, and many other forms of radiation are, in fact, just different wavelengths of the electromagnetic spectrum. (By permission from Joe R. Eagleman, *Meteorology: The Atmosphere in Action.* D. Van Nostrand copyright 1980 by Litton Educational Publishing, Inc. Reprinted by permission of Wadsworth Publishing Co.)

ceedingly damaging to the ozone shield, and the damage can persist for an indefinite period of time because the chlorine atoms are removed from the stratosphere very slowly.

The Source of Chlorine Atoms Entering the Stratosphere

Most of the chlorine used on earth for purposes such as water treatment is in the form of chlorine compounds or chlorine ions which are water soluble. Consequently, they are washed out of the atmosphere by the water cycle long before reaching the stratosphere. **Chlorofluorocarbons** (CFCs), which are simple hydrocarbon (carbon-hydrogen) molecules with both chlorine and fluorine atoms replacing some of the hydrogens, are an exception. They are both exceedingly volatile (gaseous) and water insoluble. Consequently, they are not flushed out of the atmosphere but continue to diffuse through the atmosphere until they finally reach the stratosphere. Here, they may be broken down releasing the free chlorine atoms, which proceed to destroy ozone. Thus, CFCs are damaging in that they act as transport agents to get chlorine atoms into the stratosphere.

Chlorofluorocarbons are relatively inert (non-reactive) chemically, nonflammable, and nontoxic. Fur-

thermore, while they are gases at room temperature, they liquify under modest pressure giving off heat in the process. When they revaporize, they reabsorb the heat and become cold. These attributes have led to their widespread use for the following purposes.

☐ Chlorofluorocarbons are used in virtually all refrigeration, air conditioners, and heat-pump units as the heat transfer fluid (Fig. 13–25). As these units break down or are ultimately scrapped, their CFC content generally escapes into the atmosphere.

☐ A second major use is in the production of plastic foams. The CFC is mixed into the liquid plastic under pressure. (It is soluble in organic materials.) When the pressure is released it causes the plastic to foam just as the carbon dioxide in a soda causes foaming when the pressure is released. From here, the CFCs escape to the air.

☐ A third major use is in the electronics industry for cleaning computer chips, which must be meticulously clean. Again, the CFC is allowed to escape into the air. Finally, CFCs are still used in most countries other than the United States as the pressurizing agent for aerosol cans, which, of course, release them into the air.

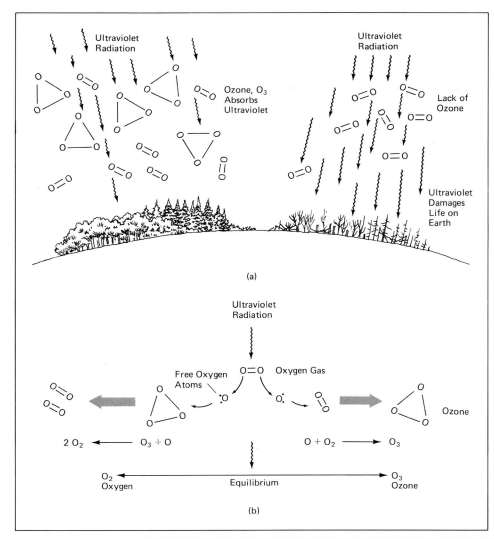

(a)

(b)

FIGURE 13–23
The ozone shield. (a) Ozone (O₃) in the stratosphere absorbs ultraviolet radiation (uv) from the sun. Without this protection uv could destroy most life on earth. (b) Ozone is formed in the stratosphere when uv causes oxygen molecules to split into free oxygen atoms that may then combine with another oxygen molecule to form ozone. However, a free oxygen atom may also combine with an ozone molecule causing it to break into two oxygen molecules as shown. Consequently an equilibrium or balance between oxygen and ozone is established and maintained. However, introduction of certain pollutants such CFCs, which catalyse the breakdown of ozone, are shifting the equilibrium toward less ozone.

FIGURE 13–24
Destruction of the ozone shield. Free chlorine atoms and certain other pollutants entering the stratosphere catalyze the breakdown of ozone as shown. Ozone will not be totally destroyed because it keeps being produced, but the oxygen-ozone balance is shifted to a lesser ozone concentration.

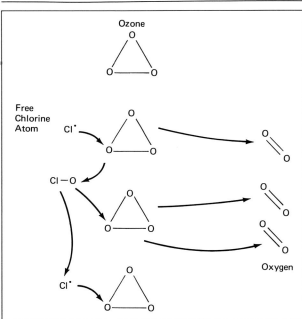

The Ozone "Hole"

The United States and a few other countries banned CFC use from aerosol cans soon after the ozone-CFC linkage was hypothesized in the early 1970s, but most did not. Nor was there any curtailing of the other uses of CFCs. Hence, worldwide, production and use of CFCs continued to grow, and since subsequent analyses indicated that ozone depletion would be relatively slight, (1–2 percent) the issue was largely ignored until 1985.

In the fall of 1985, however, satellite observations revealed a gaping "hole" in the ozone shield

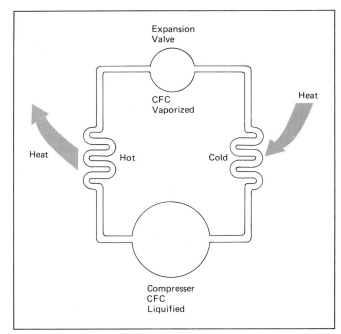

FIGURE 13-25
Freon is used for heat transfer in refrigerators, air conditioners, and heat pumps. When these are damaged or eventually scrapped, the freon escapes into the atmosphere.

over the South Pole (Fig. 13–26). In an area the size of the United States there was a 50 percent reduction in the amount of ozone. It had been assumed that the loss of ozone (the shift in the equilibrium) would be slow, gradual, and uniform over the whole earth. This hole came as a surprise, and if it had occurred anywhere but over the pole, the results in UV damage would have been catastrophic.

It would have been much easier if the issue had been taken more seriously when it was first discovered in the 1970s. This issue should be a warning that a "wait-and-see" attitude is exceedingly dangerous. Major biosphere upsets can occur with catastrophic suddenness. In 1987 the hole was worse than ever. What scientists had failed to foresee, but which they have since learned, is that cloud particles formed at very low temperatures are instrumental in releasing the active chlorine atoms from CFCs. Thus, considerable amounts of active chlorine are formed in the cold Antarctic winter. Then, sunlight in the spring promotes the destruction of ozone by active chlorine.

In February of 1989, researchers examined the stratosphere over the Arctic and found the same chemical factors existing. They concluded that the level of Arctic ozone is also poised for a fall. It will simply depend on certain weather conditions that may or may not occur in any given year. If an ozone hole develops over the Arctic, it will be more serious because there is more life there that may be damaged.

Even the recurring hole over the Antarctic may cause a significant destruction of marine phytoplankton. If it does, this will have a sever impact on virtually all the Antarctic wildlife from penguins to whales because it is the basis of nearly all the food chains. Scientists are currently examining this problem.

If current discharges of CFCs into the atmosphere continue, we can only anticipate that the polar ozone holes will become "deeper" and larger. Of course, ozone destruction over the poles will serve to dilute the ozone layer over the entire earth. Clearly, this is unacceptable.

Coming to Grips with Ozone Depletion

Under its Environmental Program, the United Nations convened a meeting in Montreal, Canada, in 1986. Member nations were able to reach an agreement, known as the Montreal Accord, to scale back production of CFCs by 50 percent by 1999, and chemical companies have mounted a search for substitutes. However, this was before CFCs were so clearly implicated in driving the ozone destruction and before the threat to the Arctic ozone was recognized. Under this accord, total CFC discharges would continue, and increase, and ozone destruction would as much as quadruple. This is clearly unacceptable. A second meeting in Helsinki in 1989 is pushing toward agreement for a total phaseout by 2000, but some are concerned that even this may not be strict enough.

FIGURE 13-26
Satellite map showing severe depletion or "hole" in the ozone layer over Antarctica in October, 1985. The hole was even more severe in 1987. It has been shown to be caused by CFCs released into the atmosphere. (Science Photo Library/Photo Researchers.)

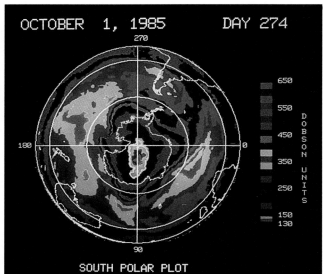

The problem is that there is so much CFCs already existing in refrigeration/air-conditioning units that normal breakdown of these will lead to increase of CFCs in the atmosphere for some years even if there is an immediate total ban for new production. Thus, many are calling for an immediate total ban.

What You Can Do

☐ Make sure that anyone servicing a refrigeration or air-conditioning unit for you will recapture and recycle the CFCs if the system needs to be opened.

☐ Initiate a campaign to see that all refrigeration/air-conditioning firms in your area adopt this practice.

☐ Promote local legislation that will require the above.

☐ Boycott all use of plastic foam until such time as substitute foaming agents come into use.

14
Risks and Economics
of Pollution

■ Outline

I. THE COST-BENEFIT ANALYSIS   347

II. PROBLEMS IN PERFORMING COST-BENEFIT
ANALYSIS   349
   A. Estimating Costs   349

   B. Estimating Benefits and Performing Risk
      Analysis   349

   C. Problems in Comparing Costs and
      Benefits   350
      1. Short-Term versus Long-Term View   350

      2. Who Bears the Costs and Who
         Receives the Benefits?   351

III. THE NEED FOR REGULATION AND
ENFORCEMENT   351

IV. CONCLUSION   351

■ Study Questions

1. What is a *cost-benefit analysis*? How is it used? List specific costs and benefits of pollution control. How do costs and benefits vary with increasing degree of control? Why is 100 percent cleanup not always necessary?

2. What expenses can be objectively estimated and what type of expenses must be projected? Why are pollution control costs higher when initiated than they are later on?

3. Give examples of the type of controls used to avoid or reduce risks. Why are these difficult to assess? What factors does risk assessment consider?

4. Discuss why different parties may come to different conclusions regarding cost-benefit analyses.

5. Discuss how and why the above conclusion varies with the time periods considered and parties doing the analyses.

6. Explain why economic pressures lead businesses to oppose pollution control. What steps have been taken to remedy this situation?

7. Discuss why it is important for you as a member of society to stand up for your interests.

Previous chapters have shown the many serious pollution problems we face. These chapters have also shown that we now have sufficient knowledge and technology to reduce, if not eliminate, many of these problems. The same holds true for environmental problems that will be discussed in subsequent chapters. Why, then, have solutions not been forcefully implemented?

In short, the answer is economics. Scientific information and the availability of control strategies are not enough to solve problems. Numerous competing economic interests often block action. The objective of this chapter is to examine these conflicts and discuss how the economic validity of a proposed project or action may be determined.

THE COST-BENEFIT ANALYSIS

A cost-benefit analysis is often used as a means of rationally deciding whether to go ahead with a given project. A cost-benefit analysis compares an estimate of the costs of the project to the value of the benefits that will be achieved by the project. A comparison of the costs and benefits is commonly referred to as the **benefit-cost** (or cost-benefit) ratio. A favorable benefit-cost ratio means that the benefits outweigh the costs, or that the project is **cost-effective** and there is economic justification for proceeding with the project. If costs are projected to outweigh the benefits, the project may be dropped.

In terms of pollution control, the costs include the price of purchasing, installing, operating, and maintaining pollution control equipment and/or implementing a control strategy. Even the banning of an offensive product costs money because jobs are lost, new products must be developed, and machinery may have to be scrapped. In some instances, controlling a pollutant may result in the discovery of a less expensive alternative. However, such money-saving controls are relatively rare. In most cases, any form of pollution control involves additional expenses. Further, costs generally increase exponentially with the percent of control to be achieved (Fig. 14–1a). That is, a small reduction in the level of pollution may be achieved by a few relatively inexpensive measures. However, further reductions gener-

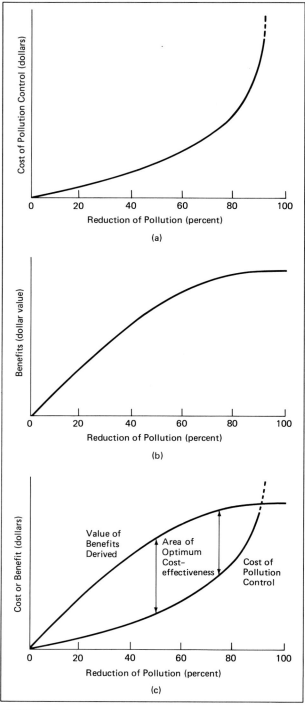

FIGURE 14–1
The cost-benefit ratio of pollution reduction. (a) The cost of pollution control increases exponentially with the degree of control to be achieved. (b) However, additional benefits to be derived from pollution control tend to level off and become negligible as pollutants are reduced to near or below threshold levels. (c) When the curves for costs and benefits are compared, we see that the optimum cost-effectiveness is achieved at less than 100 percent control. Expenditures to achieve maximum reduction may yield little if any additional benefit and, hence, may be cost-ineffective.

ally require increasingly expensive measures, and 100 percent control is likely to be impossible at any cost.

Benefits, on the other hand, include such things as improved public health, reduced corrosion and deterioration of materials, and/or reestablishing recreational use of a polluted area. The dollar value of these benefits is derived by estimating, for example, the reduction in health care costs, the reduction in maintenance and/or replacement costs, and the economic value generated by the enhanced recreational activity. Further examples of potential benefits are listed in Table 14–1.

The relationship between the percent reduction of pollution and value of benefits is very different from that for costs. Significant benefits are frequently achieved by modest degrees of cleanup. Yet, as cleanup approaches 100 percent, little if any additional benefits may be realized (Fig. 14–1b). This follows from the fact that organisms can tolerate a certain level of pollution without ill effect. It is only when pollutant levels exceed this "certain level," known as the **threshold level,** that harmful effects are observed, and then such effects may increase rapidly. Conversely, reducing a pollutant from its threshold to a still lower level will not yield an observable improvement.

When the relationship between costs and benefits is shown graphically, as in Figure 14–1c, it is clear that with modest degrees of cleanup, benefits outweigh costs. As cleanup efforts move toward the 100 percent mark, however, the lines cross and costs

Table 14–1
Benefits That May Be Gained by Reduction and Prevention of Pollution

1. Improved human health
   a. Reduction and prevention of pollution-related illnesses
   b. Reduction of worker stress caused by pollution
   c. Increased worker productivity
2. Improved agriculture and forest production
   a. Reduction of pollution-related damage
   b. More vigorous growth by removal of pollution stress
   c. Higher farm profits benefiting all agriculture-related industries
3. Enhanced commercial and/or sport fishing
   a. Increased value of fish and shellfish harvests
   b. Increased sales of boats, motors, tackle, and bait
   c. Enhancement of businesses serving fishermen
4. Enhancement of recreational opportunities
   a. Direct uses such as swimming and boating
   b. Indirect uses such as observing wildlife
   c. Enhancement of businesses serving vacationers
5. Extended lifetime of materials and cleaning
   a. Reduction of corrosive effects of pollution extending the lifetime of metals, textiles, rubber, paint, and other coatings
   b. Cleaning costs reduced
6. Enhancement of real estate values
   a. Real-estate values depressed in polluted areas
   b. Reduction of pollution will enhance them

exceed the value of benefits. Consequently, while it is tempting to argue that we should strive for 100 percent control, demanding upwards of 90 percent control may involve astronomical costs with little or no added benefit. At this point, it makes more sense to allocate dollars and effort to other projects where greater benefits may be achieved for the money spent. The optimum cost-effectiveness is achieved at the pollution reduction point where the benefit curve is the greatest distance above the cost curve.

It is important to emphasize that the cost-benefit analysis described here is only a general concept. Costs of controls and values of benefits achieved will differ with each pollutant in question. Each pollutant or each pollutant category must be analyzed separately and, in each case, there are many uncertainties. Consequently, ongoing questions include: How clean is clean? Is 50 percent clean, clean enough? Is 75 percent enough? Do we need 100 percent?

What has been the result of cost-benefit analyses to date? We noted that pollution of air and surface water reached critical levels in many areas and systems in the late 1960s and that huge sums of money were spent on pollution abatement. Cost-benefit analysis shows that, overall, these expenditures have more than paid for themselves in decreased health care costs and enhanced environmental quality. But does this mean that further expenditures on pollution control will prove equally cost-effective? Or, are we at the point where further expenditures will yield little if any benefit and the money will be effectively wasted? At the very least, industry, many economists, and government officials now demand more documentation of presumed benefits before consenting to further expenditures. Some believe that this demand represents real questions about the cost-effectiveness of additional expenditures for environmental protection. Others believe that it represents a more sophisticated method of protecting economic self-interests at the expense of environmental quality and society at large. To understand these views, we must take a more detailed look at the facets and problems of analyzing and comparing the costs and benefits.

PROBLEMS IN PERFORMING COST-BENEFIT ANALYSIS

The concept behind a cost-benefit analysis is relatively straightforward. Simply estimate costs that may be incurred and estimate the value of the benefits that may be derived from various degrees of cleanup. Then compare the two estimates. The difficulty in implementing this concept lies in obtaining realistic, objective estimates and in making objective comparisons.

Estimating Costs

In most cases, pollution control technologies and/or strategies are understood and available. Thus, equipment, labor, and maintenance costs can be estimated with a fair degree of objectivity. Unforeseen problems that cause costs to rise may occur, but as technology advances and experience is gained, lower-cost alternatives frequently emerge. Note also that pollution control itself provides jobs and hence leads to increased economic prosperity that can be considered a cost reduction or benefit over time. All told, the costs of pollution control are likely to be highest at the time they are initiated and tend to decrease as time passes (Fig. 14–2a). The importance of this will become evident when we consider the time span over which costs and benefits are compared.

Estimating Benefits and Performing Risk Analysis

The value of many of the benefits to be gained can also be estimated with a fair degree of objectivity. For example, it is well recognized that air pollution episodes cause increases in the number of people seeking medical attention. Since the medical attention provided has a distinct dollar value, eliminating air pollution episodes provides a health benefit of that value. Similarly, the rate at which materials corrode or deteriorate under polluted and nonpolluted conditions is known and we know how much money is spent to maintain and replace materials. In another example, consider a polluted lake that is upgraded to the point where it will again support water recreation. The benefits of this are estimated by assigning a value of $3 to $5 to each anticipated swimmer-day. This figure is based on the fact that most people will willingly pay this price for admission to a pool.

As we have seen in the preceding chapters, however, prime reasons for controlling present pollutants such as toxic chemicals, ozone, acid rain, and carbon dioxide do not involve protection to be gained as such. They involve protection against or avoidance of *future risks* of environmental degradation. For example, if toxic wastes are not disposed of properly, we risk widespread groundwater contamination and increased cancer rates. If acid precipitation is not controlled, we risk the widespread dieoff of aquatic life and forests. If fossil fuel consumption is not controlled, we risk climatic change due to the carbon dioxide effect, and so on. In such cases we are really considering reducing certain risks and placing a value on reducing such risks. This is not easy and there are

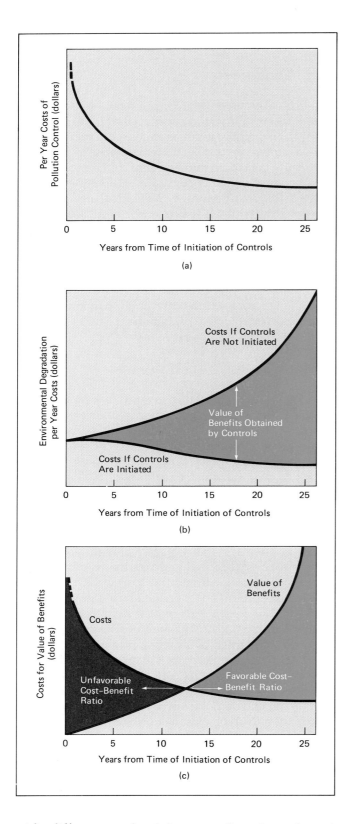

FIGURE 14–2
Evaluation of the optimum cost-effectiveness of pollution control expenditures changes with consideration of time. (a) Pollution control strategies generally demand high up-front costs. Costs generally decline as those strategies are absorbed into the overall economy. (b) Benefits may be negligible in the short term, but they increase and continue to accrue as environmental and human health recovers from the impacts of pollution or is spared increasing degradation. (c) When these two curves are compared, we see that what may appear as cost-ineffective expenditures in the short term (5–10 years) may, in fact, be very cost-effective expenditures in the long term.

sessment. A risk analysis includes consideration of factors such as the following:

- ☐ The number of people that may be affected.
- ☐ The geographic extent or area that may be affected.
- ☐ The nature and/or severity of the effects.
- ☐ The probability of the effects occurring. (Risks may range from "virtual certainty" to "little likelihood of occurring.")
- ☐ The immediacy of the threat.
- ☐ Indirect effects if the threat occurs.
- ☐ The reversibility of the threat.

By weighing these factors we can gain a more realistic appraisal of the values that may be attained by embarking on actions that will mitigate the risks. The value to be obtained is the *difference* between the cost of the damage that would probably occur in the absence of controls and the mitigated costs that may occur even with controls.

But here, the time span to be considered again becomes all-important. It is extremely important to recognize that benefits of risk reductions only begin to occur sometime after initial control strategies commence. However, they continue to accrue and become greater and greater as time passes (Fig. 14–2b).

Problems in Comparing Costs and Benefits

Even after valid cost and benefit estimates are obtained, the comparison is often a complicated matter.

Short-Term versus Long-Term View

We have seen that, while costs are high and observable benefits may be few if any during the initial stages of control, as time passes, costs generally moderate, whereas benefits increase and accumulate. Consequently, whether benefits outweigh costs or vice versa depends on whether one takes a short-term or a long-term point of view. A situation that appears

wide differences of opinion regarding the value of reducing any particular risk.

A new field of study has emerged to tackle this subject. It is referred to as **risk analysis** or **risk as-**

to be cost-ineffective in the short term may prove extremely cost-effective in the long term (Fig. 14–2c). This is particularly true in cases involving pollution problems like acid rain or groundwater contamination due to toxic wastes. In these instances, the consequences of delaying control may seriously affect large geographic areas and many millions of people, and may be irreversible.

Who Bears the Costs and Who Receives
the Benefits?

Those who bear the costs of pollution control and those who receive most of the benefits are frequently different groups of people. For example, industry and its shareholders may bear the costs of curtailing effluents into a river while fishermen gain the benefits. Obviously, the two parties are more than likely to reach different conclusions regarding whether benefits outweigh costs.

This problem is made more complex by the fact that the pollutants produced in one state or country may have their greatest negative impact in another state or country. This is particularly true of acid precipitation. Again, we find little agreement between the parties involved in the conflict over costs and benefits.

THE NEED FOR REGULATION
AND ENFORCEMENT

Many people feel that industry should control its pollutants out of good conscience. However, the economic pressures among competing businesses are often so severe that they preclude meaningful action. For example, consider two competing industries, *A* and *B*. Suppose *A* decides to undertake pollution control measures; it must either pass the costs on to customers in the form of higher prices, whereupon it may lose customers to *B*, who maintains lower prices, or it must accept lower profits, whereupon it begins to lose the financial support of investors. Thus, company *A* in its effort to be virtuous, loses competitive advantage to company *B* regardless of which road it takes. Consequently, it has proven necessary to institute laws and regulations to affect all offending companies equally. Nevertheless, we see many industries, individual companies, and special interest groups attempting to exempt themselves from regulations in order to gain an economic advantage.

CONCLUSION

It is important to recognize that ultimately all of society will receive the benefits of environmental protection and will suffer the costs of environmental degradation. Hence, a broad, long-term perspective of risks and benefits is in order. We can hardly afford to let groups with short-term economic and political interests prevent timely action. Whenever you are faced with the argument that near-term costs are too high, introduce the fact that much greater costs will be paid for environmental deterioration in the long-term if the near-term prices are not faced up to.

Part Five
PESTS AND PEST CONTROL

The dictionary defines *pest* as "any organism that is noxious, destructive, or troublesome." The term obviously includes a tremendous variety of organisms that interfere directly or indirectly with humans or their social and economic endeavors. The principal categories of pests are:

1. Organisms that cause disease in humans or domestic plants and animals. These pests include viruses, bacteria, and a wide variety of other parasitic organisms such as intestinal worms and flukes.

2. Organisms that harass people and domestic animals and that may transfer disease by biting or stinging. Common examples are flies, ticks, bees, and mosquitoes.

3. Organisms that feed on ornamental plants or agricultural crops, both before and after they are harvested. The most notorious of these organisms are various insects, but certain worms, snails and slugs, rats, mice, and birds also fit into this category.

4. Animals that attack and kill domestic animals, such as sheep-killing coyotes and chicken-killing foxes and weasels.

5. Organisms that cause wood, leather, and other materials to rot or deteriorate and cause food to spoil. Bacteria and fungi, especially molds, are largely responsible for this, but in warm, moist climates, termites are the primary culprit in the destruction of wood.

6. Plants that compete with agricultural crops, forests, and forage grasses for light and nutrients. Some may poison cattle; others simply detract

from the aesthetics of lawns and gardens. A plant in any of these roles is often referred to as a *weed*.

You may ask why humans seem virtually surrounded by pests while in a natural ecosystem the category of "pests" does not seem to exist. The answer lies in one's point of view. When an ecosystem is viewed as a whole, each organism is seen in its particular role as a producer, consumer, or decomposer and all play a role in the transfer of energy and nutrients. In considering pests, however, we adopt the viewpoint of a particular species within the system. For example, rabbits compete with woodchucks, insects, and other animals for food. They have their peculiar diseases and parasites and they suffer from biting insects. Other animals compete with them for living space, and they have numerous predators. From the rabbit's point of view, all these organisms would be considered "pests." Ecologically, however, they are considered as factors of natural environmental resistance that serve the important function of keeping the rabbit population in balance with the rest of the ecosystem. This allows the system as a whole to perpetuate itself. Without these "pests," the rabbit population would explode, often to the detriment of the ecosystem.

The prosperity of humans can be largely attributed to pest control. We would still live under extremely precarious conditions—our food supply and physical health at the mercy of such organisms—if it were not for our ability to control pests. Today, there are numerous ways to control pests, but two basic, yet extremely different, philosophies predominate.

One philosophy is based on a purely technological approach. It seeks the development of a "magic bullet," usually in the form of a human-made chemical, which will simply eradicate or exterminate the pest organism. This approach too often gives little if any consideration to the effects of this action on the ecosystem at large.

The second philosophy, which is now called **ecological pest management,** recognizes the importance of sustaining an overall ecological balance. It emphasizes the *protection* of people and domestic plants and animals from pest damage rather than *eradication* of the pest organism. Thus, the benefits of pest control can be obtained while the integrity of the ecosystem is maintained.

Traditionally, humans opted for the purely technological approach. As a result, huge quantities of chemical pesticides are sprayed on agricultural crops, forests, and our own lawns and gardens. Yet, for reasons which will be described in Chapter 15, the success of this approach is reaching its limit. Not only has it failed to eradicate pests, it has actually increased many pest problems. In addition, it has adversely affected wildlife. And, because residues from the chemicals are increasingly tainting water supplies, it poses potentially serious human health hazards. The situation reminds us that anything applied to the surface of the earth may be leached into groundwater or surface water. Some scientists believe that continued widespread use of chemical pesticides is the world's most severe pollution problem.

The futility of the technological approach, along with the pollution hazards it creates, has led more and more people involved in pest control to turn their thoughts and efforts toward ecological pest management. Chapter 16 describes various techniques which this method employs, and addresses some of the economic and social factors that are crucial to shifting from the philosophy of eradication by chemicals to that of ecological pest management.

15

The Pesticide Treadmill

CONCEPT FRAMEWORK

■ Outline

I. PROMISES AND PROBLEMS OF CHEMICAL
   PESTICIDES   357

   A. Development of Chemical Pesticides and
      Their Apparent Virtues   357

   B. Problems Stemming from Chemical
      Pesticide Use   357

      1. Development of Pest Resistance   358

      2. Resurgences and Secondary Pest
         Outbreaks   359

      3. Increasing Costs   362

      4. Adverse Environmental and Human Health
         Effects   362

   C. The Pesticide Treadmill   363

II. ATTEMPTING TO PROTECT HUMAN AND
    ENVIRONMENTAL HEALTH   365

   A. The Federal Insecticide, Fungicide, and
      Rodenticide Act (FIFRA)   365

   B. Shortcomings of FIFRA   365

      1. Inadequate Testing   366
      2. Bans Issued Only on a Case-by-Case
         Basis when Threats Are Proven   366
      3. Pesticide Exports   367
      4. Lack of Public Input   367
   C. Nonpersistent Pesticides: Are They the
      Answer?   367

   D. What You Can Do   369

■ Study Questions

1. Describe the need for early attempts to control pests.

2. Name the first synthetic organic chemical pesticide. Tell how it was discovered. Describe its apparent virtues.

3. List and define problems resulting from chemical pesticide use.

4. Describe how pest resistance develops as a result of pesticide use.

5. Describe how resurgences and secondary outbreaks occur as a result of pesticide use.

6. Explain why costs for pest control are increasing.

7. Describe adverse environmental and human health effects that are occurring as a result of pesticide use. Explain how the attributes of DDT that were considered "virtues" resulted in problems.

8. Explain what is meant by the statement: We are all on the pesticide treadmill. Discuss: Is this sustainable?

9. What is the law that regulates pesticide use?

10. Is FIFRA adequate to protect human and environmental health?

11. List and discuss the shortcomings of FIFRA.

12. What are nonpersistent pesticides? Are they the answer to pest control/environmental problems?

13. Discuss steps you can take to promote environmentally safer pest control.

Thousands of chemicals have been developed to eradicate pests. These chemicals are called **pesticides** (from *pest* and *cide*, "to kill"). Pesticides are categorized according to the group of organisms they aim to kill. Thus, there are insecticides (insect killers), rodenticides (mice and rat killers), fungicides (fungi killers), and so on. None of these chemicals, however, is entirely specific for the organisms it is designed to

control; they all pose hazards to other organisms and to humans. Therefore, they are also referred to as **biocides,** emphasizing that they may endanger many forms of life.

Our objective in this chapter is to become familiar with the problems inherent in the use of chemical pesticides. We shall focus on insecticides and the battle against insect pests, but the principles discussed can be applied to other categories of pesticides and organisms.

PROMISES AND PROBLEMS OF CHEMICAL PESTICIDES

Development of Chemical Pesticides and Their Apparent Virtues

Since the earliest times, humans have suffered the frustration and food losses brought on by destructive pests. For example, the Bible speaks of invasions of locusts that destroyed crops and caused famines. That certain chemicals would repel or kill pests was recognized at least 1000 years before the time of Christ. The Greek epic poet Homer wrote of "pest-averting sulfur with its properties of divine and purifying fumigation." Sulfur is still used, but it is hardly deserving of the term "divine"; by modern standards, its effectiveness is very limited. Therefore, devastating crop losses continued to occur intermittently well into this century, and still occur occasionally.

Finding effective materials to combat pests remains an ongoing endeavor. The early substances used included toxic heavy metals like lead, arsenic, and mercury. These *inorganic* compounds are frequently referred to as **first-generation pesticides.**

We now recognize that such toxic heavy metals may accumulate in soils and can inhibit plant growth. In some places, soils were so poisoned that now, 50 years later, they are still unproductive. Animal and human poisonings doubtlessly occurred, also. Further, these chemicals lost their effectiveness; pests became increasingly resistant to them. For example, in the early 1900s, citrus growers were able to kill 90 percent of injurious scale insects by placing a tent over the tree and piping in deadly cyanide gas for a short time. By 1930, this same technique killed as few as 3 percent of the pests. With agriculture expanding to meet the needs of a rapidly increasing population and first-generation (inorganic) pesticides failing, the farmers of the 1930s were begging for new pesticides.

These new **second-generation pesticides,** as they came to be called, were found in *synthetic organic chemicals.* The science of organic chemistry actually began in the early 1800s. Over the next century, chemists synthesized thousands of organic compounds but, for the most part, these compounds sat on shelves because uses for them had not been developed. In the 1930s a Swiss chemist, Paul Muller, began systematically testing some of these chemicals for their effect on insects. In 1938 he hit upon a chemical, **dichlorodiphenyltrichloroethane (DDT),** which had actually been synthesized some 50 years before.

DDT appeared to be nothing less than the long-sought "magic bullet," a chemical that was extremely toxic to insects and yet seemed relatively nontoxic to humans and other mammals. It was very inexpensive to produce. At the height of its use in the early 1960s, it cost no more than about 20 cents a pound. It was *broad-spectrum,* meaning that it was effective against a multitude of insect pests. It was *persistent,* meaning it did not break down readily in the environment and hence provided lasting protection. This attribute provided additional economy by eliminating both the material and labor costs of repeated treatments.

DDT was so effective, at least in the short run, that crop yields dramatically increased in many cases as a result of reduced pest damage. Growers could ignore other more painstaking methods of pest control such as crop rotation and destruction of old crop residues. They could grow less-resistant, but more productive varieties. They could grow certain crops at new locations under climatic conditions that had formerly been precluded because of the pest damage that would be incurred. In short, DDT gave growers more options for growing the most economically productive crop.

DDT also successfully controlled important insect disease carriers. During World War II, for example, the military used DDT to control body lice, which spread typhus fever among the men living in dirty battlefield conditions. As a result, World War II was the first war in which fewer men died of typhus than of battle wounds. The World Health Organization of the United Nations used DDT throughout the tropical world to control mosquitoes and thereby greatly reduced the number of deaths caused by malaria. There is little question that DDT saved millions of lives. In fact, the virtues of DDT seemed so tremendous that Muller was awarded the Nobel Prize in 1948 for his discovery. It is hardly surprising that DDT ushered in an unending parade of synthetic organic pesticides that continue to be used today in increasing amounts. Sales have increased from 2.7 billion in 1970 to 11.6 billion in 1980 and will probably exceed 18 billion in 1990.

Yet, problems associated with these pesticides soon became evident. DDT itself fell into disrepute and was banned from use in most industrialized countries in the early 1970s.

Problems Stemming from Chemical Pesticide Use

Problems associated with synthetic organic pesticides can be placed in four categories:

☐ Development of pest resistance

- ☐ Resurgences and secondary pest outbreaks
- ☐ Increasing costs
- ☐ Adverse environmental and human health effects.

Development of Pest Resistance

The most fundamental problem to growers is that chemical pesticides gradually lose their effectiveness; larger and larger quantities and/or new and more potent pesticides are required to obtain the same degree of control. Synthetic organic pesticides fared no better than first-generation pesticides in this respect. Numbers illustrate the situation. In 1946, 1 pound (0.45 kg) of pesticides provided enough protection to produce about 30,000 bushels of corn. In 1971, some 140 pounds (64 kg) were used for the same production, and losses due to pests actually increased during this period.

Why does this occur? Pest populations are not static; they represent a dynamic gene pool that is capable of relatively rapid evolution. Recall the process of natural selection described in Chapter 4. Pesticide treatments provide selective pressure, which leads to

resistance. This can be experimentally demonstrated. If you place 100 flies in each of several jars and treat them with different doses of pesticide, you can see that the flies show a high degree of variation in their sensitivity to the pesticide. As shown in Figure 15–1, low doses of pesticide will kill a few flies—those that are most sensitive to the pesticide. Higher doses kill higher percentages, and eventually all flies may be killed. However, note that a few flies will survive despite a significant dose of pesticide. These are the flies most resistant to the chemical. If you allow these few resistant flies to breed and repeat the experiment on the new population, you will find that a higher dose of pesticide is required to achieve the same kill-rate because the new population carries the genetic traits of resistance from their parents.

Such experiments may be used to interpret field observations. Pesticide treatments in the field destroy the sensitive individuals of the pest population while the more resistant individuals continue breeding, creating a new population of more resistant flies. Insects do this very rapidly because they have a phenomenal reproductive capacity. A single pair of houseflies, for example, can produce several hundred offspring,

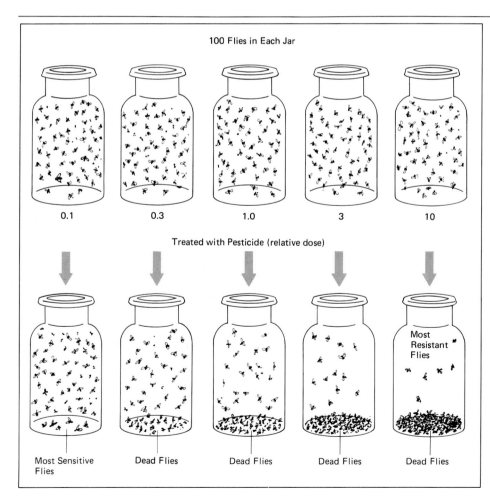

100 Flies in Each Jar

0.1   0.3   1.0   3   10

Treated with Pesticide (relative dose)

Most Resistant Flies

Most Sensitive Flies   Dead Flies   Dead Flies   Dead Flies   Dead Flies

FIGURE 15–1
Variation in resistance to pesticides. The kind of experiment shown here reveals that genetic variation in relative resistance to pesticides exists in all populations. Therefore, any pesticide treatment that kills fewer than all the individuals will invariably select those that are most resistant for survival and reproduction (i.e., survival of the fittest will occur). Thus, genes that enhance pesticide resistance are passed on to future generations.

which may mature and reproduce again in just 2 weeks. Consequently, repeated pesticide applications result in the unwitting selection and breeding of genetic lines that are highly, if not totally, resistant to the chemicals that were designed to eliminate them. Cases have been recorded in which the resistance of a pest population to a pesticide has increased as much as 25,000-fold.

Over the years of pesticide use, the number of resistant species has climbed steadily (Fig. 15–2). About 25 major pest species are resistant to all of the principal pesticides. Interestingly, as a pest population becomes resistant to one pesticide, it also may gain resistance to other unrelated pesticides even though it has not been exposed to the other chemicals.

Resurgences and Secondary Pest Outbreaks

The second problem with the use of synthetic organic pesticides is that after a pest outbreak has been virtually eliminated with a pesticide, the pests not only return, *they may return at higher and more severe levels.* This is known as a **resurgence.** To make matters worse, populations of insects that were previously of no concern because of their low numbers suddenly started multiplying, creating new problems. This phenomenon is called a **secondary pest outbreak.** For example, mites have become a serious pest problem, and the number of serious pests on cotton has increased from 6 to 16 with the use of synthetic organic pesticides.

At first, pesticide proponents denied that resurgences and secondary pest outbreaks had anything to do with the use of pesticides. However, careful investigations have shown otherwise. To prove the point, entomologist Paul DeBach has gone so far as to experimentally demonstrate pesticide "recipes" for fostering pests. For example, his recipe for growing California Red Scale on Citrus:

> Take four pounds of 50 percent wettable DDT and mix into 100 gallons of water. Using a three-gallon garden sprayer, spray one or two quarts of this mixture lightly over the tree, repeating at monthly intervals until the tree is defoliated or dying as a result of the increase in red scale. This may require from about six to twelve applications, depending upon the degree of initial infestation; the higher the numbers initially, the faster the explosion [of red scale population] occurs. Most growers do not like this sort of test and have usually insisted that we stop before trees were killed, but it is a wonderful way to raise red scale.*

By contrast, scale populations on untreated trees remain very low (Fig. 15–3).

Resurgences and secondary pest outbreaks occur because the insect world involves a complex food web just as in the world of higher animals. Populations of plant-eating insects are frequently held in check by other insects that are parasitic or predatory on them (Fig. 15–4). Pesticide treatments often have a *greater impact* on these natural enemies than on the plant-eating insects. Consequently, with natural enemies suppressed, populations of plant-eating insects not only recover but may "explode" (Fig. 15–5).

There are several reasons for pesticides having a greater impact on natural enemies than upon the target pest. The herbivore species may be intrinsically more resistant to the pesticide than is its predator. The predator may receive a higher dose because of biomagnification through the food chain. The predatory insects may be starved out by the temporary lack of prey as well as by the pesticide poisoning.

FIGURE 15–2
Number of species resistant to pesticides, 1940–1980. Using pesticides causes a selection (survival of the fittest) for those individuals that are resistant. Consequently, by using pesticides we are unwittingly breeding insects and other pests that are increasingly resistant to the pesticides used against them. (World Resources, 1986, World Resources Institute and from Georghiou, G.P., and R.B. Mellon, "Pesticide Resistance in Time and Space," in *Pest Resistance to Pesticides*. G.P. Georghiv and T. Saito, Eds. New York: Plenum Press, 1983, pp. 1–46.)

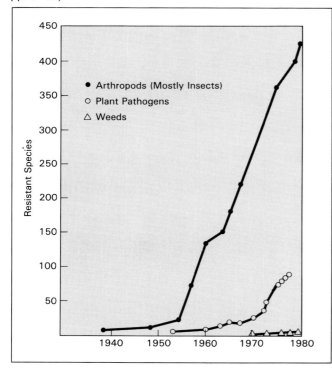

* Paul DeBach, *Biological Control by Natural Enemies* (London: Cambridge University Press, 1974), p. 2.

(a)

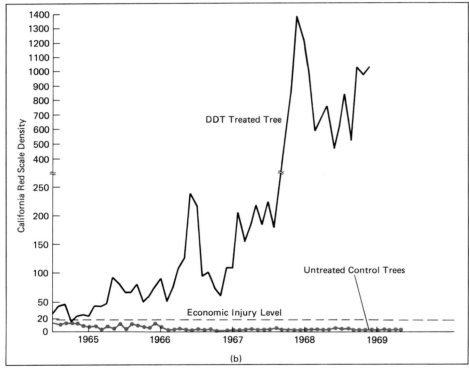
(b)

(b)

FIGURE 15–3
(a) Red scale on lemon. (USDA photo.)
(b) Increases in red scale infestation
caused by light monthly applications of
DDT spray compared to scale
populations on untreated trees under
biological control in the same grove.
Paul Debach, *Biological Control by
Natural Enemies*, p. 4. London:
Cambridge University Press, 1974.

Finally, another factor in the pesticide treatment may
subtly alter the chemistry of the plant so that it actually becomes more susceptible to pest attack.

To illustrate the seriousness of resurgences and
secondary pest outbreaks, a recent study in California
surveyed a sequence of 25 major pest outbreaks each
of which caused more than $1 million worth of dam-
age. All but one involved resurgences or secondary
pest outbreaks. Of course, the species involved in sec-

ondary outbreaks quickly became resistant to pesticides, thus compounding the problem.

The chemical approach fails because it is contrary to basic ecological principles. It assumes that the
ecosystem is a static entity in which one species, the
pest, can simply be eliminated. In reality, the ecosystem is a dynamic system of interactions, and a
chemical assault on one species will inevitably upset
the system and produce other undesirable effects.

FIGURE 15–3 (*cont.*)
(c) Scale insect. (Robert E. Pelham/Photo Researchers.)

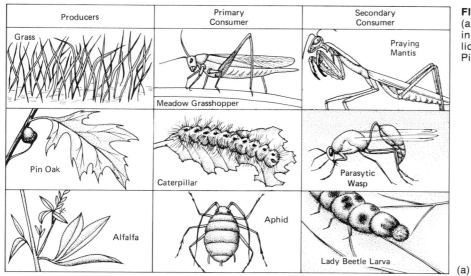

(a)

FIGURE 15–4
(a) Food chains exist among insects just as in other parts of the ecosystem. (b) Aphid lion feeding on aphids. (Courtesy of David Pimentel, Cornell University.)

Producers	Primary Consumer	Secondary Consumer
Grass	Meadow Grasshopper	Praying Mantis
Pin Oak	Caterpillar	Parasytic Wasp
Alfalfa	Aphid	Lady Beetle Larva

(b)

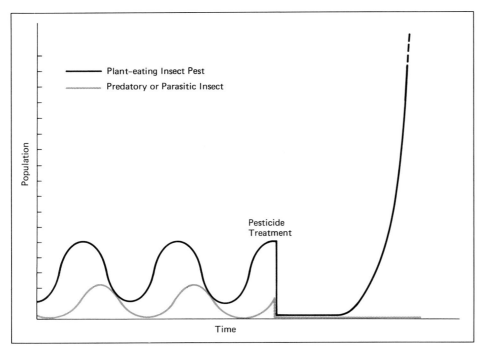

FIGURE 15–5
Populations of plant-eating pest insects may be held in check by natural enemies as in other predator-prey relationships. A chemical pesticide may affect the predator more than the pest. Freed from its natural enemy, the pest population increases rapidly.

Increasing Costs

We noted initially that DDT enabled growers to increase their yields and profits dramatically. However, the economic advantage is declining and in some cases has reached "zero return" as different and more expensive pesticides are applied more frequently and in larger quantities to counter increasing resistance, resurgences, and secondary pest outbreaks. Indeed, the growing of cotton in some regions is being abandoned because costs of pest control outweigh the value of cotton.

Adverse Environmental and Human Health Effects

Of greatest concern to society is the potential for adverse effects to human and environmental health from pesticides. The now classic story of DDT, used so widely during the 1940s and 1950s, illustrates the hazards.

In the 1950s and 1960s, ornithologists (people who study birds) observed drastic declines in populations of many species of birds that fed at the top of the food chains. Fish-eating birds such as the bald eagle and osprey were so affected that extinction was feared. Investigations showed that the problem was reproductive failure; eggs were breaking in the nest before hatching. In turn, it was shown that the fragile eggs had high concentrations of DDT and that DDT interferes with calcium metabolism, causing birds to lay thin-shelled eggs. Further study revealed that

birds were acquiring high levels of DDT by biomagnification (Chapter 11) through the food chain (Fig. 15–6). Indeed, it was these investigations that provided much of our understanding about the propensity of halogenated hydrocarbons to bioaccumulate. Recall from our discussion of DDT, page 275/Chap. 11, that it is a chlorinated hydrocarbon. Fish-eating birds were most affected because large amounts of DDT leached into waterways where long aquatic food chains provide more steps for biomagnification.

In addition, tissue assays showed that DDT was accumulating in the body fat of humans and virtually all other animals, including Arctic seals and Antarctic penguins even though these animals were far removed from any point of application (Fig. 15–7). While harmful effects on humans have not been substantiated, experimental testing has shown that related compounds are carcinogenic, mutagenic (mutation-causing), and teratogenic (birth defect-causing).

These findings led to the banning of DDT in the United States and most other industrialized countries in the early 1970s. Subsequently, several other related halogenated-hydrocarbon pesticides were banned when their propensity to bioaccumulate and cause adverse human health effects was documented.

In the years since the banning of DDT, observers have noted a marked recovery in the populations of birds that were adversely affected. However, this does not mean that the situation is under control.

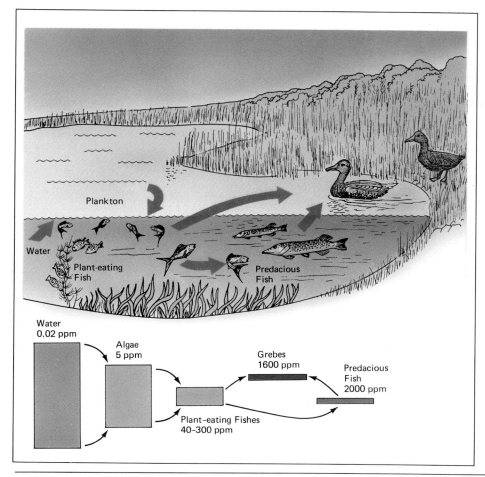

FIGURE 15–6
Biomagnification of a DDT-related pesticide. Figures are actual amounts added to or concentrations found in various parts of the ecosystem. Note that the pesticide concentration in animals at the top of the food chain is up to 100,000-fold higher than that of the water.

Within the figure:

Water 0.02 ppm

Algae 5 ppm

Plant-eating Fishes 40–300 ppm

Grebes 1600 ppm

Predacious Fish 2000 ppm

Plankton

Water

Plant-eating Fish

Predacious Fish

Because of increasing resistance, resurgences, and secondary pest outbreaks, the kinds and quantities of pesticides in use continue to grow. About 1.7 billion pounds (772 million kg) of pesticides, or nearly 5 pounds (2.72 kg) per person were used in the United States in 1982. Worldwide some half-million workers and others are injured (many fatally) each year through the production, use, and misuse of these toxic compounds, and from accidents involving their application.

The irony is that less than 1 percent of this huge amount of pesticides ever comes in contact with or is ingested by pest organisms. Because many pesticides are applied by aerial spraying, only about one-half of the chemicals even reach the crop being tested. The remainder drifts in the air and settles on surrounding ecosystems and bodies of water. Then, of the portion reaching the crop, less than 0.1 percent is ingested by the target pest. Residues may remain on foods. They may also leach from the soil into aquifers and waterways long after application. The potential for adverse environmental and human health effects is obvious.

Laws provide for withdrawing foodstuffs from the market and shutting wells if unsafe levels of con-

tamination are found. Further use of the pesticide may also be prohibited. In 1983 and 1984 numerous wells were closed and certain grain products were removed from store shelves because of EDB (ethylenedibromide) contamination, and use of EDB was banned. But do these measures provide adequate protection for human and environmental health? Just as important, do they protect natural ecosystems? We shall return to a fuller discussion of this controversy later in the chapter.

The Pesticide Treadmill

The late entomologist Robert van den Bosch coined the term **pesticide treadmill** to describe attempts to eradicate pests with synthetic organic chemicals. It is an apt term. The chemicals do not eradicate the pests. They increase resistance and secondary pest outbreaks, which lead to the use of new and larger quantities of chemicals, which in turn lead to more resistance and secondary outbreaks and so on. The process is an unending vicious cycle constantly increasing the risks to human and environmental health (Fig. 15–8). It is clearly not sustainable.

FIGURE 15—7
Bioaccumulation and magnification of DDT in virtually all organisms on earth. Tests revealed that DDT was accumulating in all organisms tested, not only those in or close to areas where DDT was used but also those far removed from the area of use, including Antarctic penguins. It was thus discovered that DDT vaporizes and, given its persistence, it was being carried to and contaminating all food chains on earth. All humans tested were also accumulating DDT in body tissues.

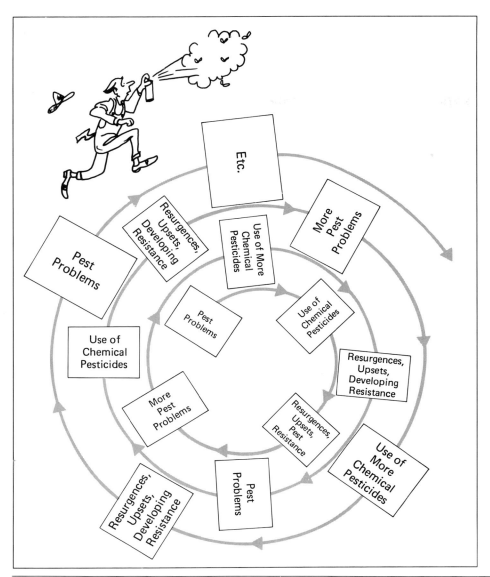

FIGURE 15–8
The pesticide treadmill. Use of hard chemical pesticides aggravates many pest problems; this demands the use of increasing amounts of pesticide, which further aggravates pest problems, which . . .

ATTEMPTING TO PROTECT HUMAN AND ENVIRONMENTAL HEALTH

The Federal Insecticide, Fungicide, and Rodenticide Act (FIFRA)

The key legislation to control pesticides in the United States is the Federal Insecticide, Fungicide, and Rodenticide Act, commonly known as **FIFRA.** This law requires manufacturers to register pesticides with the government before marketing them. The registration procedure involves testing to determine toxicity to animals (and by extrapolation, to humans). From the test results, standards regarding use are set. For example, highly toxic compounds such as chlordane, which is used to control termites, are not authorized for use on crops. For pesticides that are approved for food crops, standards are set regarding amounts of residues that may remain at harvest. Many foods have been withdrawn from the market because residues of certain pesticides were above the established standard.

If health hazards appear after a substance has been registered and marketed, the act provides for "deregistration," whereby the pesticide may be banned from one or more uses.

Shortcomings of FIFRA

Unfortunately FIFRA has many shortcomings, four of which are discussed below.

Inadequate Testing

The perils of bioaccumulation, biomagnification, and the potential for long-term exposure that may cause cancer, birth defects, mutations, and other physiological disorders were not fully appreciated until the late 1960s when bitter experience with DDT and other chemicals came to the forefront. This experience made officials and the public aware that many pesticides in use have never been adequately tested. Consequently, Congress amended FIFRA in 1972 to require the Environmental Protection Agency (EPA) to reevaluate and reregister all pesticide products then on the market.

The required testing involves feeding doses of the pesticide to a population of 100 to 200 mice over a 1- to 2-year period. At the conclusion of the testing period, the mice are examined for tumors and other physical disorders, and the findings are compared to a control group.

In 1972, when the amendment was adopted, there were already some 1400 chemicals and 40,000 formulations (mixtures of different chemicals) on store shelves. In addition, there was and still is tremendous pressure from the chemical industry to register numerous new pesticides. Needless to say, the EPA's office of pesticide programs has been overwhelmed. Ongoing budget restrictions have not helped the situation.

Since the law allows existing pesticides to remain in use unless proven hazardous, many products on the market still have not been subjected to adequate tests, and existing standards may actually allow hazardous levels of pesticide residues in food. In 1989, the Natural Resources Defense Council (NRDC) published the results of a 3-year study from which they concluded that many fruits and vegetables routinely contain levels of pesticide residues that, although within prescribed standards, "pose an increased risk of cancer, neurobehavioral damage, and other health problems" *to children.* The focus is on children because they both typically consume fruits and vegetables at a significantly greater rate than adults and because studies have shown that the young are frequently more susceptible than adults to carcinogens and neurotoxins. While the government disputes NRDC's assessment, they acknowledge that the risk to children in view of their eating habits and increased vulnerability has not been previously quantified in a comprehensive manner. Sadly, many of the toxic residues remaining on produce are from pesticides that are used only to improve its appearance. More will be said about this in Chapter 16.

Bans Issued Only on a Case-by-Case Basis when Threats Are Proven

DDT is just one of many chemically related, persistent, halogenated hydrocarbon pesticides. When the problems involving DDT were discovered, logic and prudence dictated banning the whole group because their chemical similarities made it very likely that they would all bioaccumulate and pose long-term human and wildlife health hazards.

Yet, FIFRA precludes such guilt by association. Each individual compound must be shown to pose a threat before it is subject to a ban. Even then the government must demonstrate that risks to humans are greater than the agricultural benefits derived from using the pesticide. This frequently involves litigations that take several years. Thus, in the years following the DDT ban, many related pesticides such as chlordane, heptachlor, aldrin, dieldrin, DBCP (dibromachloropane), kepone, mirex and, most recently, EDB (ethylenedibromide) were also banned, but only after serious incidents.

For example, kepone was banned in 1976, only after a number of workers in a kepone manufacturing plant on the James River in Virginia became seriously ill, a few fatally, with a variety of physical and nervous disorders, and after fish in the river were found to contain so much kepone that they were deemed unsafe to eat. The fish had bioaccumulated kepone from wash water discharged from the plant into the river. Therefore, in addition to banning the manufacture and use of kepone, the James River was closed to fishing (Fig. 15-9). Still kepone continued to recycle between the sediments and food chains. It was 1988 before levels of kepone had declined to low enough levels that commercial fishing in the James River was again permitted.

The story of EDB is perhaps more upsetting. It has been known since the mid-1970s that EDB causes cancer, birth defects, and other illnesses in laboratory animals. Yet its widespread use as a soil fumigant to control root nematodes (small worms that attack roots) continued. By 1982, 20 million pounds of EDB were pumped into the soil each year. It could be predicted that this practice would eventually result in groundwater contamination. In addition, EDB was increasingly used to fumigate and protect grains and other crops in storage. But EDB was not banned until 1984 after traces were found in hundreds of wells in Florida, California, Massachusetts, and other regions, and unacceptable amounts of residues of EDP were found in flour and other food products. Contaminated food products were taken off the market and well boring was stopped. But how long will the water remain unsafe? Unfortunately, since groundwater is

FIGURE 15-9
A fisherman removes his nets from the James River for the last time, out of business because of kepone contamination in 1976. The river remained closed until 1988. (United Press International photo.)

essentially a sealed system with little flushing action, contamination may linger indefinitely. To illustrate the situation, a study conducted in California in the early 1980s showed 50 different pesticides present in the groundwater of 23 separate counties. Some of these chemicals had not been used for more than a decade.

Clearly, waiting until after the consequences to ban such chemicals is hardly a prudent way to protect public health. Yet, this is precisely how the system operates. A few persistent halogenated hydrocarbon pesticides remain in use, and the use of some, notably lindane, is increasing. Perhaps most serious is the widespread and increasing use of **herbicides.** They now account for more than half of all pesticides used. Large numbers of people are being exposed to these compounds with unknown effects.

Pesticide Exports

Another ominous loophole in FIFRA is the lack of a requirement for registration of pesticides intended for export. The United States currently exports more than 100 million pounds of these chemicals to less-devel-

oped countries each year. Some 25 percent of this total consists of products banned in the United States. Chemical companies are free to promote their products in other countries; thus, less-developed countries hear only how pesticides will solve their pest problems. They are not informed about resurgences or secondary pest outbreaks, much less about hazards to human and environmental health. As a result, these chemicals are often overly and carelessly used, leading the World Health Organization to estimate that some 500,000 pesticide poisonings occur each year.

Ironically, a large portion of these pesticides are used by foreign countries on export crops, many of which are imported by the United States.

Lack of Public Input

Unlike the Clean Water Act and the Clean Air Act, FIFRA provides no mechanism for public input. Therefore, the legislation primarily reflects the chemical industry's interests, which are conveyed through intense lobbying efforts as opposed to public forums. The bias is obvious when one considers that the law imposes a $1000 penalty on those who misuse a pesticide and a $10,000 penalty on those who reveal a trade secret about a pesticide's formulation.

Nonpersistent Pesticides: Are They the Answer?

A key factor underlying the tendency of halogenated hydrocarbons to contaminate the environment and bioaccumulate is their persistence, that is, their slowness to break down. DDT, for example, has a half-life on the order of 20 years—half the amount applied is still present and active 20 years later; after 20 more years, half of that amount, or one-quarter of the amount applied, remains, and so on. Recognizing this fact, the agrochemical industry has, in large measure, substituted nonpersistent pesticides for the banned compounds.

These nonpersistent pesticides are also synthetic organic compounds, but are ones that break down into simple nontoxic products within a few days or weeks after application. Thus, there is no danger of their migrating long distances through the environment and affecting wildlife or humans long after being applied. These nonpersistent pesticides have been touted as "environmentally safe," but are they? For several reasons, nonpersistent pesticides are not as environmentally sound as their proponents claim.

First of all, total environmental impact is a function of persistence along with three important factors: toxicity, dosage applied, and location where applied.

Many of the less-persistent pesticides are actually more toxic than DDT. This, combined with the frequent applications needed to maintain control, presents a significant hazard to agricultural workers.

Second, nonpersistent pesticides may still have far-reaching environmental impacts. For example, to control outbreaks of the spruce budworm in New Brunswick, Canada, forests were sprayed with a nonpersistent organophosphate pesticide that was promoted as environmentally safe (Fig. 15–10). After spraying, however, an estimated 12 million birds died. These birds may have died by direct poisoning or by the loss of their food supply since a bird eats nearly its own weight in insects each day. In either case, visitors commented on the eerie silence and the numerous dead warblers littering the ground after spraying.

Third, nonpersistent pesticides may upset the ecosystems in treated areas. Insects are an integral part of many aquatic food chains. Population explosions of phytoplankton have occurred as insects that normally feed on it were killed. Likewise, soil ecosystems may be upset, affecting decomposition and release of nutrients. Populations of soil-dwelling pests may also increase due to the loss of natural enemies.

Fourth, desirable insects may be just as sensitive as pest insects to these substances. Bees, for example, which play an essential role in pollination, are highly sensitive to nonpersistent pesticides. Thus, use of

(a)

(b)

FIGURE 15–10
(a) Damage caused by spruce budworm. (William J. Jahoda/Photo Researchers.) (b) Spraying for spruce budworm. (U.S. Forest Service.)

these compounds creates an economic problem for beekeepers as well as jeopardizing pollination.

Finally, nonpersistent chemicals are just as likely to cause resurgences and secondary pest outbreaks as persistent pesticides, and pests become resistant to nonpersistent chemicals just as quickly.

What You Can Do

In conclusion, the mechanisms designed to control pesticides and protect human and environmental health leave much to be desired. The National Resources Defense Council (Appendix A) is spearheading a drive, ''Mothers and Others for Pesticide Limits,'' to make the public more aware of the hazards of pesticides and to get at least the most risky pesticides banned from use. You can contact NRDC to find out more about the problem and how you can support this effort.

Recognize, however, that banning one pesticide will accomplish little if another equally hazardous one is to be substituted. Fortunately, there is an alternative. To break the cycle of the pesticide treadmill we must recognize the dynamics of the ecosystem and shift our approach toward ecological pest management, which utilizes various natural techniques rather than synthetic chemicals to control pests. A substantial number of growers are beginning to do this and produce what are called *organically grown* fruits and vegetables. We will examine the techniques of this approach and what you can do to help implement them in Chapter 16.

16
Natural Pest Control Methods and Integrated Pest Management

CONCEPT FRAMEWORK

■ Outline

I. THE INSECT LIFE CYCLE AND ITS
VULNERABLE POINTS   372

II. METHODS OF NATURAL OR BIOLOGICAL
CONTROL   372
    A. Control by Natural Enemies   373
    B. Genetic Control   378

CASE STUDY: EXPERIENCE IN FINDING NEW
NATURAL ENEMIES   377

        1. Chemical Barriers   378
        2. Physical Barriers   378
    C. The Sterile Male Technique   378
    D. Cultural Controls   382
        1. Cultural Control of Human Pests   382
        2. Cultural Control of Pests Affecting
           Lawns, Gardens, and Crops   384
    E. Natural Chemical Control   384
    F. Economic Control versus Total
      Eradication   386

    G. Conclusion   386

III. INTEGRATED PEST MANAGEMENT   386

    A. Socioeconomic Factors Supporting
      Overuse of Pesticides   386
        1. "The Only Good Bug Is a Dead Bug"
          Attitude   386
        2. Aesthetic Quality and Cosmetic
          Spraying   387
        3. Insurance Spraying   387
        4. Chemical Company Profits   388
    B. Addressing Socioeconomic Issues   388
        1. Produce Labeling versus Spraying   388
        2. Advice from Trained Field Scouts and
          Pest-Loss Insurance   388
        3. Economics of Natural Controls   389
    C. Integration of Natural Controls   389

    D. Prudent Use of Pesticides   390

    E. What You Can Do to Promote Ecological
      Pest Management   390

■ Study Questions

1. Describe the life cycle of an insect and its vulnerable points.

2. Define *natural* or *biological control*. List five different methods of biological or natural control. Describe and give examples of each.

3. Is total eradication of a pest necessary? Explain. What is meant by the *economic threshold* for pest control?

4. Discuss: Is it possible to control pests by natural means?

5. Define *integrated pest management*.

6. List and describe economic and sociological factors that support continuing overuse of pesticides.

7. Describe how the economic and social factors supporting overuse of pesticides may be addressed.

8. Give an example of how several natural controls may be combined in a program of integrated pest management.

9. Describe when and why chemical pesticides remain a part of integrated pest management.

10. Tell what you can do to promote ecological pest management.

Numerous ecological and biological factors affect the relationship between a pest and its host (the plant or animal attacked). Ecological pest management seeks to manipulate one or more of these natural factors so that crop protection is achieved without upsetting the rest of the ecosystem or jeopardizing environmental and human health. Since ecological pest management involves working with natural factors instead of synthetic chemicals, the techniques are referred to as **natural control** or **biological control** methods.

Our objective here is to show the major categories of natural control methods and the opportunities they offer. We shall then focus on some of the economic and sociological factors that need to be overcome in moving from chemical to ecological control methods.

THE INSECT LIFE CYCLE AND ITS VULNERABLE POINTS

The ecological pest management approach, unlike the chemical biocide approach, depends on an understanding of the pest and its relationship with its host.

The more we know about the organisms involved, the greater our opportunities for natural control.

To illustrate, the life cycle of moths and butterflies is shown in Figure 16–1. Many groups of insects have a similarly complex life cycle. The development of each stage of the life cycle may be influenced by numerous abiotic factors such as temperature and humidity. One or another stage may be attacked by a parasite or predator. Additionally, proper completion of each stage depends on internal chemical signals provided by hormones. Locating mates, finding food, and other behaviors depend on external chemical signals. As will be seen in the following pages, all of these findings suggest vulnerable points or ways in which pest populations may be controlled without resorting to synthetic chemical pesticides.

METHODS OF NATURAL OR BIOLOGICAL CONTROL

The five general categories of natural or biological pest control are:

☐ Control by natural enemies

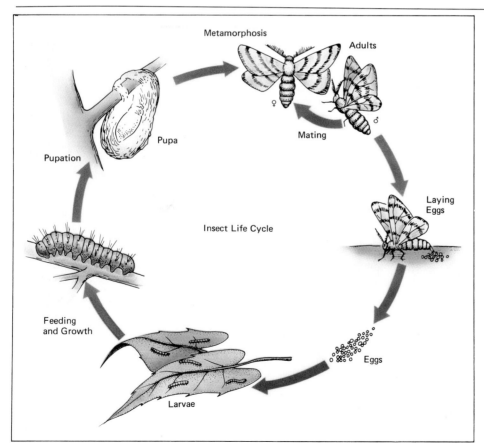

FIGURE 16–1
Insect life cycle. Most insects have a complex life cycle that includes a larval and adult stage. Each stage has special food and/or habitat requirements. Affecting any stage of the life cycle can prevent reproduction and affect control.

- Genetic control
- The sterile male technique
- Cultural control
- Use of natural chemicals.

Control by Natural Enemies

The following examples illustrate the range of possibilities for controlling pests with natural enemies:

- Scale insects, potentially devastating to citrus crops, have been successfully controlled by vedalia (ladybird) beetles, which feed on them (Fig. 16–2).
- Various caterpillars have been controlled by parasitic wasps (Fig. 16–3).
- Gypsy moths and Japanese beetles are controlled in part with bacteria (Fig. 16–4 and Fig. 16–5).
- Prickly pear cactus and numerous other weeds have been controlled by plant-eating insects (Fig. 16–6).
- Water hyacinths can be controlled by manatees (Fig. 16–7).
- Rabbits in Australia are controlled by an infectious bacterium.

The problem in using natural enemies is in finding and deploying organisms that provide control of the target species without attacking other desirable species. Entomologists estimate that only about 1 percent of about 50,000 known species of plant-eating insects that might be serious pests, actually are. The populations of the other 99 percent are held in check by one or more natural enemies such that they do not do significant amounts of damage. Therefore, the first step in using natural enemies for control should be *conservation*, preserving the natural enemies that already exist. This means avoiding the use of broad-spectrum chemical pesticides, which may affect natural enemies even more than the target pests. Elimination or considerable restriction of the use of broad-spectrum chemical pesticides will, in many cases, allow natural enemies to reestablish themselves and control secondary pests, those that became serious problems only after the use of pesticides.

However, effective natural enemies are not always readily available. In some cases, the lack of natural enemies is the result of accidentally importing the pest without its natural enemy. In several such cases effective natural enemies have been found by systematically combing the home region of an intro-

duced pest and finding various predatory or parasitic organisms. Also, entirely new relationships have been discovered (see Case Study 16).

Yet, the potential for utilizing natural enemies has barely been tapped. Of more than 2000 serious insect pest species, some 90 percent remain without effective natural enemies mainly because no one has looked. Entomologist Paul DeBach points out that this is because the use of natural enemies does not generate profits for industry the way that synthetic chemicals do. Without the profit incentive, the research necessary to find natural enemies has received very little financial backing, and this is likely to remain the case until enough of us recognize that a sustainable

FIGURE 16–2
(a) Cottony-cushion scale feeding on citrus trees. (Photo by Max E. Badgley.) (b) Vedalia (ladybird) beetle eating a scale insect. (Photo by Max E. Badgley.)

(a)

(b)

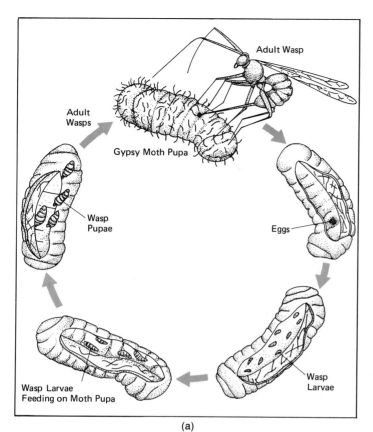

FIGURE 16–3
(a) A parasitic wasp depositing eggs in a gypsy moth pupa. (USDA photo.) (b) The life cycle of the parasitic wasp. (c) Another insect parasite, a braconid wasp lays its eggs on a tomato hornworm, the larva of the hawk moth. The wasp larvae feed on the caterpillar and shortly before the caterpillar's death they emerge and form cocoons seen here. (Scott Camazine/Photo Researchers, Inc.)

(a)

Adult Wasp

Adult Wasps

Gypsy Moth Pupa

Eggs

Wasp Pupae

Wasp Larvae

Wasp Larvae Feeding on Moth Pupa

(b)

(c)

(a)

(b)

(c)

FIGURE 16–4
Gypsy moth and larva. Control is possible by spraying disease-causing bacteria on the leaves which the larvae eat. (a) Adults laying egg masses on tree trunk. (Nobel Proctor/Photo Researchers, Inc.) (b) Larva feeding. (John M. Barnley/Photo Researchers, Inc.) (c) Larva stricken by bacteria. (USDA photo.)

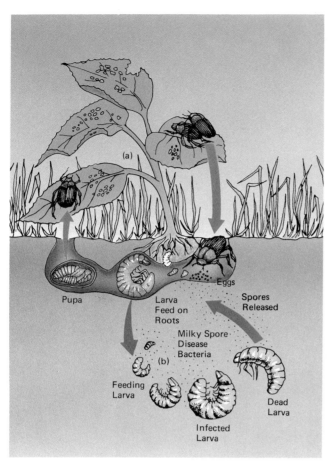

FIGURE 16–5
Milky spore disease is an effective control for Japanese beetles. (a) Normal life cycle. (b) Bacteria infect and multiply in the larva. As the larva dies, bacterial spores are released into the soil and the cycle is repeated.

(a)

(b)

FIGURE 16–6
Biological control of the weed prickly pear cactus by a cactus-eating moth larva in Queensland, Australia. (a) Homestead abandoned because of prickly pear infestation. (b) Homestead reoccupied following cleanup by the insect. (Queensland Department of Primary Industries; Australian Information Service.)

FIGURE 16–7
Manatees or sea cows might be used to maintain control of water hyacinths. Unfortunately poaching and injury by motor boats endanger their survival. (© Douglas Faulkner, Photo Researchers, Inc.)

Approximately 200 major pests in the world are effectively and permanently controlled by biological control parasites and predators, thus significantly reducing the need for pesticides. In addition to the environmental benefits, the economic benefits of biocontrol are enormous. For each dollar invested in biological control, at least $30 are returned in reduced losses of crops from pests.

Most of the successful cases of biological control in the past have involved the introduction of beneficial parasites and predators from the native home of the introduced exotic pest. This so-called "classical approach" in biocontrol required the introduction of about 20 parasites and predators before there was one case of success. In some cases, like the gypsy moth, nearly 100 beneficial parasites and predators have been introduced from Eurasia, the native home of the gypsy moth, but none of these biocontrol introductions has provided successful control of the moth. This has been discouraging and costly.

Convergent ladybug feeding on a black aphid. (J.H. Robinson/Photo Researchers.)

I discovered a new approach to improve biocontrol that focuses on "new associations" between parasite (and predator) and the host species. The "new association" approach provides greater efficiency and effectiveness in selecting natural enemies for biological control than the "classical approach." The new approach avoids the problem that in old associated parasite-host systems the host has evolved such resistance that its population is not greatly reduced by the parasite. When an introduced parasite has never interacted and evolved with its new host, often it severely attacks and greatly reduces a new pest-host population. For example, the myxoma-virus parasite was introduced for control of the European rabbit pest, which was destroying pastures and natural vegetation in Australia. The myxoma-virus was introduced from the South American tropical forest rabbit for biological control of the European rabbit (new association). The parasitic virus associated with the South American rabbit has minimal effects on its natural rabbit host, which suggests that the myxoma-virus parasite and South American rabbit had evolved some degree of commensalism or balance. Although there has been some evolution in the newly associated myxoma-virus parasite and European rabbit host in Australia, the European rabbit pest is still under good biological control from the newly associated virus parasite.

Employing the "new association" approach, we can reduce the number of biocontrol introductions by 75% to achieve successful biological control! At the same time, we can now control native pests, which was impossible with the "classical approach." This is also extremely important because from 60% to 80% of all pests are native organisms that moved from feeding on native vegetation to feeding on our introduced crops.

Employing biological control, especially the "new association" approach, offers a great many opportunities to control a wide array of pests more effectively and more efficiently than ever before possible with the "classical approach." This will help us reduce our dependence on pesticides and, thus, protect our environment and public health while reducing the economic costs of pest control.

DAVID PIMENTEL

biosphere may depend on such research and start demanding its support. Finding suitable natural enemies is not easy. After locating a potential natural enemy it must be propagated and carefully tested before it is released, to be sure that it will not harm other organisms. This often takes many years of painstaking research. When effective natural enemies are found and deployed, however, they may provide control indefinitely, providing savings of millions of dollars per year without further inputs. Pests generally don't develop "resistance" to the natural enemies because the natural enemy tends to evolve along with its host or prey.

Genetic Control

Most plant-eating insects and plant pathogens (bacteria, viruses, and other parasitic organisms) attack only one or a few closely related plant species. This implies a *genetic incompatibility* between the pest and species that are not attacked. The essence of **genetic control** is to develop genetic traits in the host species that provide the same incompatability, that is, resistance to attack. This technique has been extensively utilized in connection with plant "diseases," fungal, viral, and bacterial parasites. For example, in the years 1845–1847 the potato crop in Ireland was devastated by "late blight," a fungal parasite. Nearly a million people starved and another million people emigrated to escape the same fate. Now protection against such disasters is provided in large part by growing plant varieties that are resistant to attack. It is no overstatement to say that the world owes much of its production of corn, wheat, and other cereal grains to the painstaking work of plant geneticists who selected and bred these resistant varieties.

The same potential exists for breeding plants that are resistant to insect pests. Traits that provide resistance may be categorized in two groups: *chemical barriers* and *physical barriers*.

Chemical Barriers

A **chemical barrier** means that the plant produces some chemical that is lethal or at least repulsive to the would-be pest. The relationship between wheat and the Hessian fly provides an example. This fly lays its eggs on wheat leaves and the larvae move down the leaves and into the main stem as they feed. This weakens the stem so that it is killed or broken in the wind (Fig. 16–8). The Hessian fly was introduced to this country in the straw bedding of Hessian soldiers during the Revolutionary War. The fly eventually spread throughout much of the Midwest causing widespread devastation, until scientists at the University of Kansas developed a variety of wheat that causes the larvae to die as they feed on the leaves.

The phenomenon being exploited here? Through natural selection some plants have evolved the capacity to produce their own pesticidal chemicals. And through artificial selection plant breeders are enhancing this trait. However, some of these chemicals are also highly toxic to humans and some are known to be carcinogenic. Nicotine in tobacco is a notable example. In breeding plants for pest resistance, we must be careful that we do not also make them unsuitable for human or livestock consumption.

Increasing resistance through breeding may not provide 100 percent protection, but even partial protection can make the difference between profit or loss for the grower. In addition, any degree of resistance lessens the need for chemical pesticides.

Physical Barriers

Physical barriers are *structural traits* that impede the attack of a pest. For example, leafhoppers are significant worldwide pests of cotton, soybeans, alfalfa, clover, beans, and potatoes, but they can only damage plants with relatively smooth leaves. Hooked hairs on the leaf surfaces of some plants tend to trap and hold immature leafhoppers until they die (Fig. 16–9). Similarly, alfalfa weevil larvae are fatally entrapped by glandular hairs that exude a sticky substance. Such traits can be enhanced through selective breeding.

Unfortunately, pests may develop the ability to overcome genetic controls in the same way they develop resistance to pesticides. This means that breeders must continually develop new resistant varieties to substitute for old varieties as their resistance is overcome. This substitution process has occurred seven times in the case of wheat and the Hessian fly. Such substitution often takes place without the public ever knowing that a potential catastrophe is being averted. Note again the importance of maintaining biological diversity to make this possible.

The Sterile Male Technique

The **sterile male** technique involves inundating a natural population with sterile males that have been reared in laboratories.

Combating the screwworm fly provides a prime illustration. The screwworm fly, closely related and similar to an ordinary housefly, lays its eggs in open wounds of cattle and other animals. The larvae (maggots) feed on blood and lymph, keeping the wound open and festering (Fig. 16–10). Secondary infections frequently occur and often lead to the death of the animal.

Before World War II, this problem became so severe that cattle ranching from Texas to Florida and northward was becoming economically impossible.

(a)

(b)

APRIL	MAY	JUNE	JULY	AUGUST	SEPTEMBER
Eggs on Leaf Egg, Greatly Enlarged Fly Fly Emerges from Infested Wheat and Lays Eggs on Leaves of Healthy Wheat	Larva, or Maggot, Greatly Enlarged Maggot Hatches on Leaf and Goes to Stem	Flaxseed Damaged Stem Lodges and Maggot Becomes Pupa	Pupa Left in Stubble	Fly Emerges from Stubble and Lays Eggs on Young Wheat	Maggot Hatches and Feeds in Crown
OCTOBER	**NOVEMBER**	**DECEMBER**	**JANUARY**	**FEBRUARY**	**MARCH**
Maggot Becomes Pupa	Pupa Ready for Wintering	Weakened Plant Failing to Tiller	Pupa Wintering	Pupa Wintering	Pupa about to Become Fly

FIGURE 16–8
Genetic control by developing chemical resistance. (a) The Hessian fly, which is a serious
pest of wheat. (b) Its life cycle. Control has been gained by developing strains of wheat that
cause the larvae to die as they feed. (USDA.)

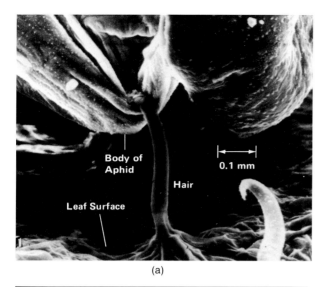

(a)

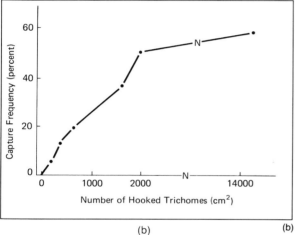

(b)

FIGURE 16-9
Genetic control by developing physical barriers. (a) The leafhopper is controlled by hairs on the leaf, a genetic trait, which hook immature leafhoppers. (b) Relationship between the number of hairs and the frequency of capture. (Courtesy of E. A. Pillemer and W. M. Tingey, New York State Agricultural Experiment Station, Ithaca, New York. *Science*, 193 (August 6, 1976), 482–84. © 1976 by the American Association for the Advancement of Science.)

In studying the situation, Edward Knipling, an entomologist with the U.S. Department of Agriculture, observed two essential features of screwworm flies: (1) their populations are never very high, and (2) the female fly mates just once, lays her eggs, then dies. Knipling reasoned that if the female mated with a sterile male, no offspring would be produced. To achieve this end huge numbers of screwworm larvae are grown on meat in laboratories. The pupae (see insect life cycle, Fig. 16–1) are collected and subjected to just enough high-energy radiation to render them sterile. These sterilized pupae are then air-dropped into the control area (Fig. 16–11). Ideally, 100 sterile

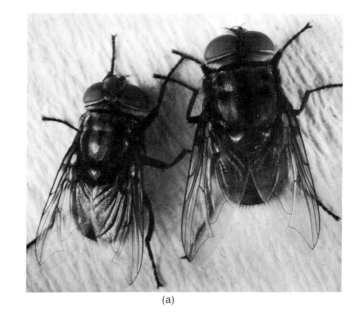

(a)

(b)

FIGURE 16-10
The screwworm fly, a deadly pest of cattle. (a) Adults. (b) Larvae.

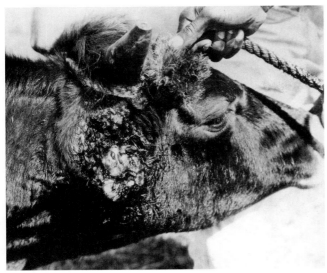

FIGURE 16-10 (cont.)
(c) Larvae of the screwworm fly feeding in a wound. Keeping the wound open allows the entry of other infections and frequently results in the death of the animal. (USDA photos.)

males are dropped for every normal female in the natural population, giving a 99 percent probability that wild females will mate with one of the sterile males.

This technique successfully eliminated the screwworm fly from Florida in 1958–1959, and it continues to be used to control it in the Southwest. The savings to the cattle industry is estimated at better than $300 million a year. It has also been used to eradicate infestations of imported pests before they gained a strong foothold. Populations of such insects are maintained in facilities around the world so that sterile males may be called up on very short notice if the need arises.

Again, pests may develop "resistance" to the sterile male technique. For example, females in the natural population may evolve the ability through natural selection to distinguish normal from sterile males. This means that entomologists must watch for the development of such traits in the natural population and breed the captive population accordingly.

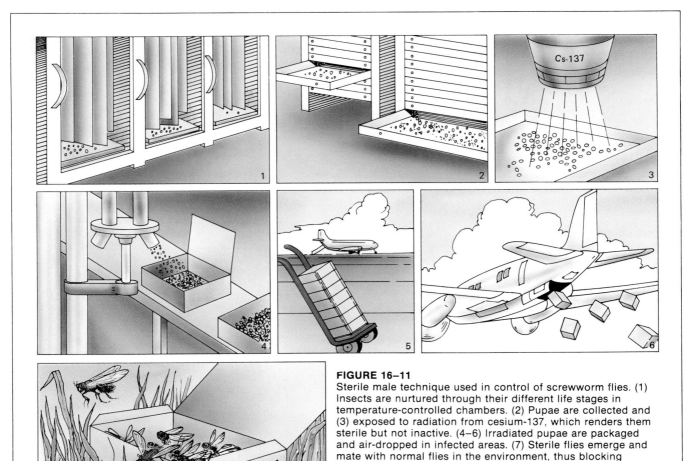

FIGURE 16-11
Sterile male technique used in control of screwworm flies. (1) Insects are nurtured through their different life stages in temperature-controlled chambers. (2) Pupae are collected and (3) exposed to radiation from cesium-137, which renders them sterile but not inactive. (4–6) Irradiated pupae are packaged and air-dropped in infected areas. (7) Sterile flies emerge and mate with normal flies in the environment, thus blocking reproduction. (Redrawn from photos from Comision Mexico Americana para la Erradicacion del Gusano Barrenador del Ganado and USDA.)

Cultural Controls

A **cultural control** is a nonchemical alteration of one or more environmental factors in such a way that the pest finds the environment unsuitable or is unable to gain access.

Cultural Control of Human Pests

We routinely practice many forms of cultural control against diseases and parasitic organisms. Some of these practices are so familiar and well entrenched in our culture that we no longer recognize them as such. For instance, proper disposal of sewage and the avoidance of drinking water from "unsafe" sources are cultural practices that protect against waterborne disease-causing organisms. Combing and brushing the hair, bathing, and wearing clean clothing are cultural practices that eliminate head and body lice, fleas, and other ectoparasites. Regular changing of bed linens protects against bedbugs. Proper and systematic disposal of garbage and keeping a clean, tight house with good screens are both effective methods in keeping down populations of roaches, mice, flies, mosquitoes, and other pests (Fig. 16–12). In turn, control of these pests protects us against numerous diseases that they may carry. Similarly, sanitation requirements in handling and preparing food as well as cleaning cooking and eating utensils are cultural controls designed to prevent the spread of diseases. Refrigeration, freezing, canning, and drying of foods are cultural controls that inhibit the growth of organisms that cause rotting, spoilage, and food poisoning.

If such practices of personal hygiene and sanitation are dropped, as they may be in any major disaster situation, there is very real danger of additional widespread mortality resulting from outbreaks of parasites and diseases. It is estimated that 80 percent of human illness in less-developed countries results from contaminated drinking water. The most economical way to improve human health in these areas is through cultural controls, namely providing clean water, sanitation facilities, and promoting personal hygiene.

Cultural Control of Pests Affecting Lawns, Gardens, and Crops

The following examples illustrate some of the major categories of cultural controls that can be used to manage lawn, garden, and agricultural pests.

Selection of What to Grow and Where to Grow It. Sun-loving plants are put in the shade; shade-loving plants are put in sunny locations; wet-loving plants are put in dry places, and so on. Plants (including trees and shrubs) grown in conditions to which they are not ecologically well adapted may survive, but they are stressed. The stress frequently renders them more susceptible to pests. Plants under their optimal conditions often have sufficient vigor that they resist attack, thus eliminating the need for any additional pest control. When these plants are attacked, growers are prone to use prodigious amounts of pesticides to maintain the ecological "mistakes." Cotton, for example, has a high profit potential so it is grown in many areas where it is particularly vulnerable to pests and requires huge amounts of pesticides, but the principle applies to ornamental plants around your

FIGURE 16–12
Cultural controls. Many practices of hygiene and sanitation are cultural controls against disease, parasites, and other pests. How many cultural controls can you find in this illustration?

home as well. A more practical alternative may be to substitute plants or crops that are better suited to the particular location.

Management of Lawns and Pastures. Homeowners are prone to use excessive amounts of pesticides to maintain a weed-free lawn. Weed problems in lawns are frequently a result of simply cutting the grass too short. If grass is not cut to less than 3 inches it will maintain a dense enough cover to keep out crabgrass and many other noxious weeds. Similarly, weed problems in pastures are frequently a problem of overgrazing, which allows noxious weeds to invade and thrive.

Water and Fertilizer Management. Plants stressed by lack of water or fertilizer may become more susceptible to certain pests; the same may occur with too much moisture or fertilizer. Therefore, water and fertilizer management is an important factor in controlling pests as well as in producing optimal growth.

Time of Planting. For a pest that emerges early in the spring, the planting of a susceptible crop may be delayed so that most of the pest population starves before the plants are available. On the other hand, if a pest population emerges late, early planting may be the answer. Early planting may allow the crop to grow and be harvested before the pest population multiplies to destructive levels.

Destruction of Crop Residues (Sanitation). Spores of plant disease organisms and insects may overwinter or complete part of their life cycle in the dead leaves, stems, or other plant residues that remain in the fields after harvest. Plowing under or burning the material may be very effective in keeping pest populations to a minimum. In gardens, a clean mulch of material such as grass clipping may be substituted to protect the soil from erosion.

Adjacent Crops and Weeds. Some plants are particularly attractive to certain pests; others are especially repugnant. In each case the effect may "spill over" to adjacent plants. Thus a gardener may control many pests by paying careful attention to eliminating plants that act as attractants and growing others that act as repellants. Some parasites require an alternate host. Control may be gained by eliminating the alternate host (Fig. 16–13).

Crop Rotation. Pest problems are exacerbated by growing the same crop year after year, which keeps the pest's food supply continuously available. Alternatively, crop rotation, the practice of alternating crops from one year to the next in a given field—for example, corn one year, beans the next—may provide control because corn pests cannot feed on beans and vice versa. Crop rotation is especially effective in controlling root nematodes, roundworms that live in

FIGURE 16–13
Wheat rust, a parasitic fungus that is a serious pest on wheat. (a) Injury to wheat caused by the rust. (b) The life cycle of the rust. Since part of the life cycle of wheat rust requires that it infest barberry, eliminating barberry in wheat-growing regions has been an important cultural control, along with developing resistant varieties of wheat. (USDA.)

(a)

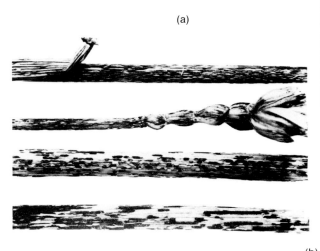

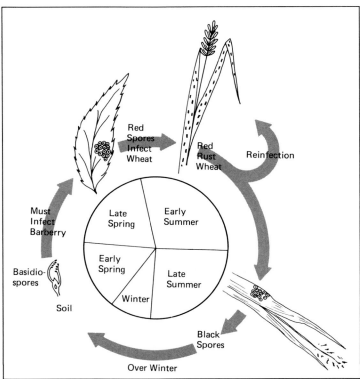

(b)

the soil and feed on roots, and other pests that do not have the ability to migrate appreciable distances (Fig. 16–14).

Polyculture. For economic efficiency agriculture has moved progressively toward *monoculture,* the growing of a single species or even a single variety over a wide area, for example, the Corn Belt, Cotton Belt, and so on.

Recall the ecological principle that diversity provides stability or, conversely, a simple system is ecologically *unstable.* When a pest outbreak occurs, monoculture is most conducive to its rapid multiplication and spread. Other natural controls, even if present, may be overwhelmed by the avalanche of spreading pests. Conversely, the spread of a pest outbreak is impeded and other natural controls may be more effective if there is a mixture of species, some of which are not vulnerable to attack. Indeed, much of our prodigious use of pesticides may be seen as a desperate attempt to maintain unstable monocultures.

Agriculturalists are experimenting with various systems of **polyculture** such as the growing of two or more species together or in alternate rows. One may harbor natural enemies of pests that attack the other.

Customs and Quarantines. Many of our most difficult-to-control pests are species that have been unwittingly imported from other parts of the world. Further, it is recognized that many species existing in other regions would be serious pests if they were introduced. Therefore emphasis is placed on keeping would-be pests out. This is a major function of the customs office. Biological materials that may carry pest insects or pathogens are either prohibited from entering the country or are subjected to quarantines, fumigation, or other treatments to assure that they are free of pests. The cost of such procedures is small in comparison to the costs that could be incurred if such pests gained entry and became established. It is extremely important that we recognize the need for customs checks and cooperate with procedures.

Natural Chemical Control

As in humans and other animals, each stage in insect development is controlled by **hormones,** chemicals produced by the organism which provide "signals" that control developmental processes and metabolic functions. Too much or too little hormone can cause severe abnormalities. In addition, insects produce many **pheromones,** a chemical secreted by one which influences the behavior of another.

The aim of **natural chemical control** is to isolate, identify, synthesize, and then use an insect's own hormones or pheromones to disrupt its life cycle. Advantages of natural chemicals are: (1) they are highly specific to the pest in question (they do not affect natural enemies to any appreciable extent), and (2) they are nontoxic. When eaten by another organism they are simply digested. Ways in which natural chemicals may be used are illustrated by the following.

Scientists have discovered that pupation of caterpillars is triggered by a decrease in the level of a chemical, which they named the **juvenile hormone.** If this chemical is sprayed on caterpillars, pupation does not occur. The larvae simply continue to feed and grow, become grossly oversized, and eventually die. Hormone sprays are now being used in part to control gypsy moths in some urban areas.

Adult female insects secrete pheromones that

FIGURE 16–14
Cultural control of nematodes, parasitic worms that feed on roots and cause serious damage to a number of plants. Populations of these pests may be kept to a minimum by crop rotation with alternate planting of a crop that does not support the nematodes. (USDA.)

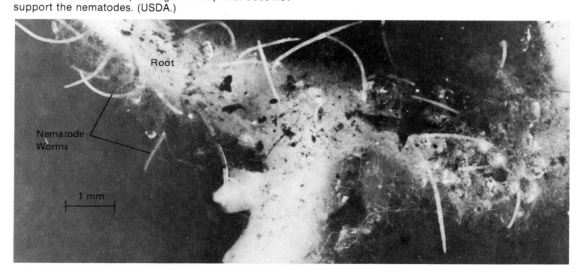

Root

Nematode Worms

1 mm

attract males for the function of mating. Once identified and synthesized, these **sex attractant** pheromones may be used in either of two ways: (1) the **trapping technique** or (2) the **confusion technique.** In the trapping technique, the pheromone is used to lure males into traps or to poison bait (Fig. 16-15). In the confusion technique, the pheromone is dispersed over the field in such quantities that males become confused, cannot find the females, and thus fail to mate. It has been found that insects develop resistance to specific natural chemicals through evolving variations of their hormones or pheromones. However, the process of identifying and synthesizing the compound can "evolve" in kind.

The theoretical potential of natural chemicals for controlling insect pests without causing other ecological damage or upsets is enormous and has been recognized for at least 30 years. However, the background research and necessary testing is just now reaching fruition. Tests are extremely promising and

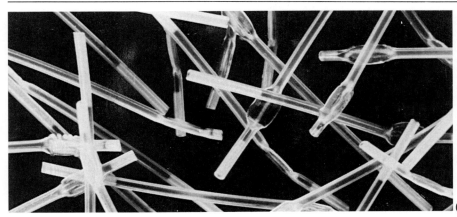

(a)

FIGURE 16–15
Natural chemical control. Control of pink bollworms using their sex pheromone. In this case, the pheromone is contained by tiny fibers, a microscopic close-up of which is shown in (a). The outside of the fiber is treated with a minute amount of insecticide. A single handful of fibers (b) is enough to treat two acres. The male bollworm moths are attracted to the fibers by pheromone evaporating from the open end (c). In attempting to mate with the fiber, the moth contacts the insecticide and dies. (Courtesy of Jack Jenkins, Scentry Inc. Buckeye, Arizona.)

(b)

(c)

it appears that the "great expectations" in this field are about to be realized as over 800 natural chemicals have been identified and more than 250 are being produced commercially.

According to many in the field, regulatory hurdles now represent the greatest barrier to deployment of these valuable, nontoxic compounds. The problem is that specific wording in FIFRA causes these natural or **biorational chemicals,** as they are sometimes called, to be treated the same as "hard" synthetic organic chemical pesticides. Thus, costly and time-consuming toxicity testing is required which is unnecessary for these natural compounds.

Economic Control versus Total Eradication

Throughout all natural control methods, it is important to keep in mind that a species only becomes a pest when its population multiplies to the point where it causes significant damage. Natural controls, in general, are aimed at keeping pest populations below damaging levels, not at total eradication. By keeping populations of pests down, natural controls avert significant damage while preserving the integrity and balance of the ecosystem.

Therefore, the question to be asked in facing any pest species is: Is it causing significant damage? If significant damage is not occurring, natural controls such as natural enemies and cultural conditions are already operating and the situation is probably best left as is. Spraying with synthetic chemicals at this stage is more than likely to upset the natural balance and make the situation worse through causing resurgences and upsets. On the other hand, if significant damage is occuring, a pesticide treatment may be in order.

While "significant damage" is open to value judgment, growers can put it in economic terms. Damage should not be taken as significant unless the cost of the damage considerably outweighs the cost of a pesticide application (Fig. 16–16). This point is called the **economic threshold.**

Conclusion

We see there is tremendous potential for controlling most, if not all, pest problems through natural means that would be sustainable. But, in large part, we continue to give relatively little support to development and deployment of natural controls. Instead we continue on the pesticide treadmill despite its declining effectiveness and environmental hazards. Why?

The answer is deeply rooted in sociological attitudes and economic interests. If we are to get off the pesticide treadmill these must be addressed as

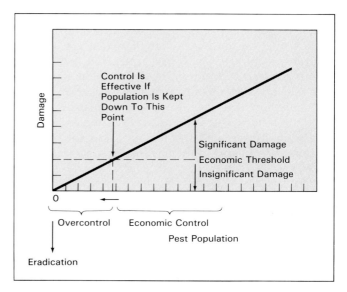

FIGURE 16–16
The objective need not be to eradicate the pest but only to keep population levels below the economic threshold.

well as the "technology" of natural controls. This brings us to the subject of *integrated pest management.*

INTEGRATED PEST MANAGEMENT

Integrated pest management (IPM) aims to minimize use of synthetic organic pesticides without jeopardizing crop protection through addressing all sociological, economic, and ecological factors involved. Importantly, IPM is not a technique in and of itself. Rather, it is an *approach that integrates many techniques.* We shall first examine the factors that lead to overuse of pesticides and see how these factors may be addressed.

Socioeconomic Factors Supporting Overuse of Pesticides

The economic and sociological factors that support the overuse of pesticides are closely linked, but they may be broken down into the following four categories.

"The Only Good Bug Is a Dead Bug" Attitude

The attitude that "The only good bug is a dead bug" is deeply ingrained in many people. They get great satisfaction watching insects literally drop dead as a result of spraying. Seldom do these people consider whether the "bug" is causing damage or is beneficial. Ignoring potential consequences, they apply pesticides more frequently and in quantities far greater than are needed to achieve adequate protection; because of resurgences and upsets, their efforts are often counterproductive (Fig. 16–17).

Aesthetic Quality and Cosmetic Spraying

A natural tendency is to pick over produce to find the best-looking fruits and vegetables. The not-so-good-looking items are frequently left and trashed. This approach to shopping can be called the "Snow White" syndrome. Like Snow White, we all tend to reach for the perfect, unblemished apple, orange, or head of lettuce and not consider that pesticide "potions" may have been used to produce it. In turn, supermarket chains, canneries, and others set aesthetic standards and reject anything less to create and maintain a "top quality" image. Thus, growers find that blemished produce brings a sharply reduced price—if it can be sold at all. This effectively forces them to indulge in **cosmetic spraying**—the use of pesticides to control pests that harm *only* the item's outward appearance. Indeed, millions of pounds of pesticides, with all their potential hazards to human and environmental health, are applied to protect cosmetic appearance alone with absolutely no benefit to yield, nutritional value, or any other practical aspect. Indeed, it is residues of such pesticides that frequently remain on produce creating a health risk.

The Snow White syndrome also applies to homeowners desiring a perfectly uniform, unblemished green lawn. Lawns are drenched two or three times a year with mixtures of herbicides, insecticides, and fungicides, as well as fertilizers. The short-term desire, a weed-free lawn, may be achieved, but the long-term costs to the environment may be considerable.

Insurance Spraying

It makes little difference whether the threat of loss to pests is real or imagined, close at hand or remote; what is important is how the grower perceives the threat. Even if there is no evidence of immediate pest

FIGURE 16–17

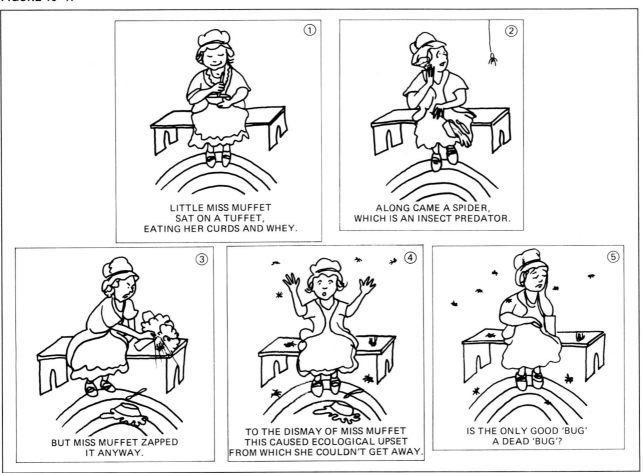

① LITTLE MISS MUFFET SAT ON A TUFFET, EATING HER CURDS AND WHEY.

② ALONG CAME A SPIDER, WHICH IS AN INSECT PREDATOR.

③ BUT MISS MUFFET ZAPPED IT ANYWAY.

④ TO THE DISMAY OF MISS MUFFET THIS CAUSED ECOLOGICAL UPSET FROM WHICH SHE COULDN'T GET AWAY.

⑤ IS THE ONLY GOOD 'BUG' A DEAD 'BUG'?

damage, a grower who believes his or her plantings are at risk is likely to succumb to **insurance spraying,** the use of pesticides "just to be safe." Few such insurance treatments actually prove necessary, but they all have negative impacts on the environment.

Chemical Company Profits

The production and sale of synthetic organic pesticides ($18 billion in 1989) is a highly profitable business for the huge and very powerful chemical industry. Thus chemical companies do all they can to promote and exploit the previously described attitudes to sell more pesticides. Through advertising and paid-on-commission field representatives, pesticide producers attempt to convince growers that the threat of pests is much greater than it is and that spraying pesticides is good insurance. Likewise, they capitalize on the Snow White syndrome by emphasizing the enhanced cosmetic quality that can be obtained with pesticides. The fact that unnecessary treatments can only aggravate pest problems through causing resurgences, secondary pest outbreaks, and increasing pest resistance only serves to benefit the chemical companies by creating a market for larger quantities and numbers of pesticides.

Addressing Socioeconomic Issues

Again, IPM must address these sociological and economic factors as well as providing alternative natural controls. Some ways of addressing these issues follow.

Produce Labeling versus Spraying

There is strong public feeling against the use of chemical pesticides. Cosmetic or other unnecessary spraying persists in large part because consumers are kept ignorant about the kinds or amounts of pesticides used. Experience shows that when consumers are informed many abandon the Snow White attitude. Food outlets selling "organically grown" produce (produce that is grown without synthetic chemical pesticides or fertilizers) are thriving, although the produce is not as cosmetically perfect and may be higher priced. As a result, many growers now find it more profitable to rely on natural controls and sell through these specialized markets.

A major supermarket in Switzerland, Migros, is exploiting this situation to benefit itself and growers. A special label that stands for health is placed on fruits and vegetables grown with minimum use of synthetic chemical pesticides. These products are marketed beside and at the same price as produce grown with traditional use of pesticides. The store explains the difference to the consumers on large wall posters proclaiming: "We grow our vegetables and fruits with fewer artificial fertilizers and pesticides and that means better soils, better energy use, and less pollution so that we can protect the health of our clients."[*]

Migros promotes its philosophy by employing farm advisors to instruct growers in the techniques of organic farming. If farmers comply with practices, Migros guarantees to buy their produce. An increasing percentage of Migros produce now bears the special label. As farmers become more experienced and natural controls become more established, standards for allowable synthetic chemicals are made stricter.

A similar scheme is possible in the United States because most supermarket chains contract directly with growers for produce. They can contract for produce grown without pesticides as well as for that grown with pesticides.

Advice from Trained Field Scouts and Pest-Loss Insurance

Devastating pest infestations generally occur only when certain weather and crop conditions coincide, making circumstances particularly favorable for the reproduction of pests. At other times pest populations, while present, may remain below the economic threshold.

Most growers are not trained entomologists and cannot be expected to recognize all the conditions that make a pest outbreak a real versus an imagined threat. Thus, they rely on information from others as to whether a pesticide application is necessary or not. Through the 1970s, at least 90 percent of such information that growers received was from pesticide producers or marketers. You can guess the bias of the information.

This situation is being remedied by *field scouts* employed by local agricultural extension services, farm cooperatives, or acting as independent consultants. **Field scouts** are persons trained in monitoring pest populations (traps baited with pheromones are used for this purpose) and other factors and in determining whether the pest population is exceeding economic threshold. Many growers have cut pesticide use dramatically without loss by using information from field scouts.

Additionally, providing **pest-loss insurance,** insurance that will pay the farmer in the event of loss due to pests, has enabled growers to get away from unnecessary insurance spraying.

---

* By permission: "All Things Considered," December 20, 1982, National Public Radio.

Economics of Natural Controls

Making the economic benefits of natural controls known to growers is another important aspect of integrated pest management. Initially, pesticides increased yields and profits. Many farmers still cling to pesticides, believing that they offer the only way to bring in a profitable crop. In some instances, however, the rising costs of pesticides, along with their tendency to aggravate pest problems, has eliminated their economic advantage.

The Center for the Biology of Natural Systems of Washington University compared 16 "organic" farms with 16 similar "conventional" farms in the Corn Belt. Organic farms relied on crop rotation and other cultural techniques to control pests; conventional farms used pesticides. Crop production per acre was virtually identical in the two cases. Total value of crops produced was slightly higher on conventional farms because pesticides permitted monocropping and hence a large percentage of the land was in high-value corn. However, this gain was offset by the cost of pesticides and fertilizer so that the eco-nomic return to the farmers was not significantly different. Productivity of crops on Amish farms, where pesticides were never used because of religious beliefs, is as high or higher than that of farms where pesticides are used.

Integration of Natural Controls

In addition to addressing social and economic issues, IPM may involve the integration of any number of natural controls and the prudent use of pesticides. The U.S. program to protect against the foreign Mediterranean fruit fly, or "Medfly," exemplifies how this may be done (Fig. 16–18).

Unlike our common fruit fly, which is attracted only to very ripe fruit, the Medfly lays its eggs on unripe fruits and vegetables in the field. The feeding maggots thus cause extensive damage before harvest and during storage and transport. If the Medfly established itself in the United States, huge amounts of pesticides would be needed to protect crops, and it would still cause millions of dollars worth of damage each year.

To prevent this, multiple lines of defense have been adopted. First, all imported produce is checked and if officials detect any trace of the insect it is fumigated. Second, a network of traps baited with a sex attractant is maintained. Monitoring these traps pro-

FIGURE 16–18
Integrated pest management (IPM). Integrated pest management involves the integration of a number of techniques, none of which would be completely effective by itself. The illustration depicts the techniques or "lines of defense" that are used to keep the Mediterranean fruit fly out of California.

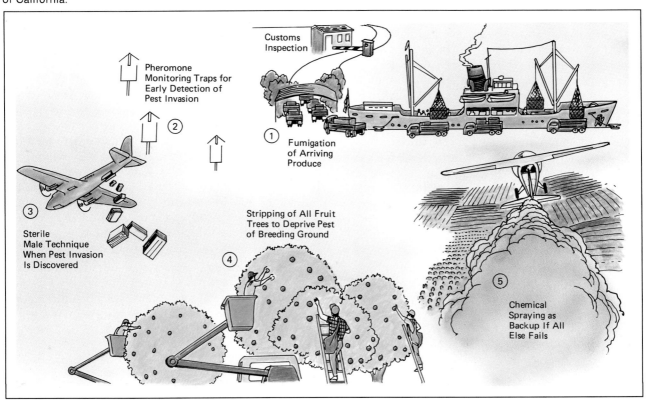

vides an early warning of the presence of Medflies that have slipped through "customs."

If Medflies are found, the sterile male technique is called on as the third line of defense. Stocks of Medflies are maintained in South America, and batches of sterile flies can be delivered on short notice and dropped over the infected area. In addition, fruit is stripped from the trees to prevent the Medfly from reproducing. Officials continue to use pheromone-baited traps to monitor the program's success.

If these efforts are not successful, they move to a final line of defense: the chemical malathion. Malathion has a relatively low toxicity to humans and animals but is very effective against Medflies. As an extra safeguard, however, the pesticide is applied to a Medfly bait, thus increasing its specificity to the pest and decreasing human and animal exposure.

Several Medfly invasions into the United States—in 1956 and 1961 in Florida, in 1966 in Texas, and in 1975, 1980–1981, and 1984 in California—have been eradicated by this IPM program, generally without having to use the last defense, malathion. But, because of an unfortunate accident, the 1980–1981 episode required widespread spraying and barely succeeded at that.

The accident was that a batch of supposedly sterile flies turned out to be fertile; they had been insufficiently irradiated. As a result, instead of curtailing the breeding of a few remaining natural Medflies, agents unwittingly introduced about 100,000 additional pairs of fertile Medflies. (Rather than tediously separate male and female insects in the laboratory, both males and females are irradiated and introduced. Normally the presence of the sterile females along with the sterile males does not influence the effectiveness of the technique.) With this magnitude of infestation and the status of "sterile" flies on hand in question, immediate spraying was called for. However, public protest against the use of pesticides delayed action and gave the flies another few weeks to establish themselves. Eventually, extensive spraying took place and the invasion was eradicated.

You should note here that public protest against the use of synthetic organic pesticides is an important element in promoting the movement toward natural controls. However, also note that such protests should be tempered with an understanding of the situation. In this case, public opposition to the use of pesticides resulted in the need for additional spraying and, had the invasion not been eradicated, the need for and use of pesticides would have increased enormously.

Prudent Use of Pesticides

The Medfly episode (above) also illustrates that while the aim of IPM is to reduce reliance of chemical pesticides, such pesticides still remain an integral part. Indeed, some entomologists feel that IPM is still overly reliant on the use of synthetic organic pesticides and is thus still contributing to driving the pesticide treadmill. These entomologists advocate renewed emphasis on the study of pest ecology and the development and implementation of natural controls—in short, a renewed emphasis on the concepts of ecological pest management.

What You Can Do to Promote Ecological Pest Management

It is estimated that about a third of the chemical pesticides in the United States are used in houses, on lawns, and in gardens. Furthermore, pesticides are most often overused here because homeowners are seldom constrained by costs in terms of how much pesticide they apply for "insurance" and in order to be sure all "bugs" are dead. Thus, urban and suburban dwellers are a very significant force in driving the pesticide treadmill. Likewise your actions toward pests in and around your home can be a significant factor in bringing it to a halt.

☐ When you have a pest problem in your own home or garden do not automatically use a synthetic chemical pesticide and do not get "advice" from a dealer who has a vested interest in selling them. Seek advice from your agricultural extension service agent and specifically ask: How much damage will this "bug" do? (It may be that the "bug" is actually harmless and will disappear by itself in a matter of days!) If it does present a real problem ask specifically about natural control methods. Cultural control methods are often most applicable. Finally, consider if a particular planting is really worth the risks of using pesticides. It may be much easier in the long run to substitute a planting that is more resistant. Encourage your friends and neighbors to do these things also.

☐ In any growing—small-scale gardening to large-scale farming—seek to use natural controls only. Remember that this starts with selection of resistant plants grown under conditions that are optimal. Rodale Press carries a wealth of publications on organic gardening. The *IPM*

Practitioner, a publication from the Bio Integral Resource Center (see Appendix A), also provides a wealth of information regarding biological control techniques.

☐ As a consumer, ask your grocery store manager to label produce for pesticides that have been used on it and to start carrying organically grown produce. As store managers perceive that a significant number of their customers want organically grown produce, they will start acquiring it. In turn, more growers will move into serving this market. Indeed, this is already beginning to occur. The number of organic farmers in the United States has more than doubled in recent years.

To attack the issue on a broader front:

☐ Initiate a drive to have your county require pesticide labeling on produce as described. Such legislation invariably starts at levels of local governments and works its way up.

☐ Write your congressperson asking that more support be given to research in finding and testing natural enemies for pests and developing other natural controls.

☐ Ask that FIFRA be rewritten to distinguish between "hard" synthetic organic chemical pesticides and natural chemicals so that the use of natural chemicals can be expedited.

☐ Ask that funding for training and employment of field scouts be increased.

☐ Consider a career for yourself in the field of biological pest control. Additional people, from field scouts to entomologists and organic chemists, are needed in this rapidly growing field.

Part Six

RESOURCES: BIOTA, REFUSE, ENERGY, AND LAND

Soil, water, air, and natural biota, which are the basis of all food and many other materials, are necessary for human survival. Civilization also depends on various *mineral resources*. **Minerals** are rocklike materials that comprise the earth's crust. Minerals that yield metallic elements such as iron, aluminum, and copper upon smelting are known as ores. Other minerals, such as phosphate rock, are valuable for the fertilizer nutrients they contain. Still others, such as asbestos, clays, and mica, are commercially useful in their existing state.

In the early 1970s there was considerable debate and concern that growing population and increasing demands would exhaust certain critical mineral resources and civilization might be brought to a halt as

a result. The problem is that rich ore deposits, those that can be mined and the desired element obtained at a reasonable cost, are limited and nonrenewable. However, the earth is a vast sphere made entirely of minerals and while some are rare outside of certain rich ore deposits, the theoretical quantities of even the most rare are huge. Second, in speaking of minerals we are generally speaking of particular elements such as iron, aluminum, lead, and mercury. Elements cannot be created, but they cannot be destroyed either. Therefore, even after any amount of use and throwing away, they are still theoretically available for rerefining and reuse. Consequently, the limiting factors in having adequate mineral resources are not in diminishing quantities of the elements *per se*.

The real limits in gaining mineral resources are in the environmental impacts we are willing to tolerate and in the energy we are willing to expend in the process of getting them. Another limit is in the environmental impacts of throwing them away. Recall from Chapter 11: The environmental and human health consequences of discharging heavy metals (mineral resources) into the air and water are far more pressing and urgent than any concern about running out of these minerals.

Therefore, the focus of this Part is on maintaining natural biota (Chapter 17), which are a critical resource in their own right, disposal of refuse (Chapter 18), and energy resources (Chapters 19–21). The way we use land and how this affects all other resources including soil and water and air are addressed in Chapter 22.

17
Biota: Biological Resources

CONCEPT FRAMEWORK

■ Outline

I. NATURAL BIOTA: TO PRESERVE OR
DESTROY?   397

  A. Values of Natural Biota   397
    1. Underpinnings for Agriculture and
      Forestry   397
    2. Resource for Medicine   399
    3. Providing Natural Services   400
    4. Recreational, Aesthetic, and Scientific
      Values   401
    5. Commercial Interests   402
  B. Assaults against Natural Biota   404
    1. Loss of Habitat to Alternative Land
      Uses   404
    2. Pollution   404
    3. Overuse   406
    4. Introduction of Foreign Species   407
    5. Combination of Factors and
      Environmental Degradation   407
    6. Conclusion and the Tragedy of the
      Commons   410

CASE STUDY: LAST STAND FOR AFRICA'S
ELEPHANTS   408

II. CONSERVATION OF NATURAL BIOTA AND
ECOSYSTEMS   412
  A. General Principles   412
    1. Definition and Objective of
      Conservation   412

    2. Concept of Maximum Sustainable Yield 414

    3. Specific Points   415

  B. Where We Stand with Regard to
    Protecting Biota   416
    1. Game Animals in the United States   416

    2. Exotic Species and the Endangered
      Species Act   417

    3. Aquatic Species   419

  C. Is Saving Species Enough—The Need to
    Focus on Ecosystems   421

■ Study Questions

1. Define: *natural biota*. Describe the historical relationship that has existed between humans and natural biota. How have humans become apparently independent from natural biota?

2. List and describe ways in which humans remain dependent on natural biota.

3. List and describe various ways in which natural biota are being diminished or destroyed.

4. Define and give examples that describe the concept of *the tragedy of the Commons*.

5. Define and give the objective of *conservation*.

6. Define and give examples illustrating the concept of *maximum sustainable yield*.

7. List specific points that must be addressed in order to prevent the tragedy of the commons and maintain a maximum sustainable yield.

8. Describe how regulations concerning game animals act to maintain a maximum sustainable yield.

9. Define *endangered species*. What laws/regulations have been enacted to protect endangered and aquatic species? What are the shortcomings?

10. What are the threats against aquatic species?

11. Discuss the questions: Is saving major exotic species enough? Why not? Why do whole ecosystems need to be protected?

D. Saving Tropical Forests	424	12. What is happening to tropical rainforests? Describe the economic and social pressures that are causing this to happen.
E. What You Can Do	424	13. Discuss what you can do to help save natural biota and ecosystems.

About 1.5 million species have been examined, named, and classified, but scientists estimate that there are at least 5 to 10 million additional species, particularly in tropical forests and marine ecosystems, which have not been systematically explored. Yet, around the world, these ecosystems are being destroyed or upset by clearing, exploitation, pollution, and other factors such that, if these trends are allowed to continue, millions of species will be lost to extinction over our lifetime.

All species are collectively referred to as **biota**. The fact that natural biota are being destroyed seems to say that, in general, society values what is gained from exploiting natural biota more than the natural biota. If this were not the case, we wouldn't be making the tradeoff, would we? Or, is it like the Indians who traded Manhattan Island for a few dollars worth of glass beads and other trinkets? Do we simply fail to recognize the relative values in the trades we are making? In this chapter we wish to examine the values of natural biota, the reasons behind sacrificing these values, and what needs to be done to preserve natural biota, if indeed, it is our choice to do so.

NATURAL BIOTA: TO PRESERVE OR DESTROY?

As primitive humans living in the context of natural ecosystems, we obtained all our food from the hunting and gathering of wild plants and animals; clothing was animal skins; many tools and implements, beyond clay and stone pots and implements, were made from wood and bone; shelters, outside of caves, were poles and leaf thatch bound with vines or strips of animal hide; the only fuel was firewood. In short, all food, clothing, fuel, and much material for building and implements came from natural biota. Thus, after air and water, natural biota were humans' most important resource.

Then, about 10,000 years ago, humans began learning to select certain species from the natural biota and to propagate these to the exclusion of others. This was the advent of agriculture. About 6000 years ago humans began learning to refine and fashion metals, and 250 years ago, the beginning of the

Industrial Revolution, we learned to build machines and run them with fuels other than firewood. Now, living in cities surrounded by modern technology and getting all our food from supermarkets, our connections to natural biota seem remote and perhaps irrelevant, and one can argue that without sacrificing natural biota to some extent, humans would still be living in caves. Thus the trend of sacrificing natural biota for alternative gains continues (Fig. 17-1).

Values of Natural Biota

But humankind's dependance on natural biota remains as crucial today and for our future survival as it ever did. Our accelerating pace of destroying natural biota is undercutting the sustainability of modern civilization and should be a factor of alarming concern. Indeed, it may be the most important factor as it acts as a "barometer." Preserving natural biota will indicate a balanced biosphere; nothing less is sustainable.

The values of natural biota can be categorized into five areas:

☐ Underpinnings for agriculture and forestry
☐ Resources for medicine
☐ Providing natural services
☐ Recreational, aesthetic, and scientific values
☐ Commercial values.

Underpinnings for Agriculture and Forestry

Since most of our food comes through agriculture, we tend to believe that it is independent of natural biota. This is not true. Recall that in nature both plants and animals are continuously subjected to the rigors of natural selection. Only the fittest survive. Consequently, wild populations have numerous traits for resistance to parasites, competitiveness, tolerance to adverse conditions, and other aspects of *vigor*. In addition, the gene pools of wild populations generally harbor the variations that enable adaptation to changing conditions (see Chapter 4).

Conversely, populations grown for many generations under the "pampered" conditions of agri-

FIGURE 17–1
"Graveyard" of extinct species at Bronx Zoo in New York City. Human impacts have already caused the extinction of numerous species. Only a small portion of the extinct species are depicted on these gravestones. If this graveyard were kept up to date, in the next few years the tomb stones might number in the millions. (New York Zoological Society photo.)

culture tend to lose these traits for vigor because, here, they are generally being selected for production as opposed to vigor. For example, a high-producing plant that lacks drought resistance is supplied with irrigation and its drought resistance is ignored. Also, in the process of breeding plants for maximum production, virtually all genetic variation is eliminated. Indeed, the cultivated population is commonly called a **cultivar** (for cultivated variety), indicating that it is a highly selected strain of the original species with a *minimum* of genetic variation. Such cultivars, when provided with optimal water and fertilizer, do give outstanding production under the specific climatic conditions to which they are attuned. However, with minimum genetic variation, they have virtually no capacity to adapt. If climatic conditions change from what a cultivar is adapted to, its production may drop to nil and, by itself, it cannot be adapted to the new conditions because its gene pool lacks the necessary genetic variations.

To maintain vigor in cultivars and/or to adapt them to different climatic conditions, plant breeders comb wild populations of related species for the desired traits. The desired traits, when found, are introduced into the cultivar through crossbreeding and reselection. If the specific gene for the trait can be identified, it may be possible to introduce it directly through the techniques of genetic engineering. In either case, however, the trait comes from a related wild population that is from natural biota. If natural biota with such wild populations are lost, agriculture will be severely crippled; indeed, it may not be sustainable.

Additionally, from the several million species existing in nature, our ancestors chose a handful for propagation, and modern agriculture has tended to focus on even fewer, much less considering new species. This limited diversity of agriculture makes it ill-suited to production under many environmental conditions. For example, we tend to consider that arid regions are unproductive unless irrigation is provided. However, there are may species of wild bean-related plants that produce abundantly under dry conditions (see Case Study 4). Scientists estimate that there are many thousands of wild species that might be brought into cultivation which would increase production in environments that are less than ideal. Also, such diversification is sorely needed to provide stability. Recall the principle that *simple systems are ecologically unstable; diversity provides stability*. Destroying natural biota will destroy such opportunities.

Finally, in Chapter 16 we discussed the tremendous and invaluable opportunities to control pests through introducing natural enemies and increasing genetic resistance. Natural enemies and genes for increasing resistance can only come from natural biota. Again, destroying natural biota will destroy such opportunities.

Essentially all the same arguments can be made in connection with forestry.

Another way of considering this issue is to look at natural biota as a reservoir or bank of genetic material, the sum of the gene pools of all the species involved. As long as natural biota are preserved we have a rich endowment of genes in the "bank," which we can draw upon as needed to sustain agriculture. Thus natural biota are frequently referred to as a **genetic bank.** Depleting this bank cannot help but deplete our future (Fig. 17–2).

Resource for Medicine

This genetic bank also serves medicine as the following example illustrates. For thousands of years, natives on the island of Madagascar used an obscure plant, the rosy periwinkle, in their folk medicine. Useless witchcraft? If the rosy periwinkle had become extinct before 1960 hardly anyone outside Madagascar would have noted the loss or cared. Then, in the 1960s, scientists discovered that a new chemical, which they called "vincristine," extracted from this plant did have medicinal properties. It revolutionized the treatment of childhood leukemia. Before the discovery of vincristine, leukemia was almost always fatal in children. With vincristine treatment, there is now a 95 percent chance of remission.

The rosy periwinkle is just one of hundreds of such stories. Innumerable drugs including all the an-

FIGURE 17–2
Natural biota serve as a "genetic bank." Wild species have many invaluable uses as depicted here. Hence, a healthy, diversified natural biota may be viewed as a genetic bank from which withdrawals for these uses can be made at any time. What will be the effects on the areas illustrated if natural biota are destroyed?

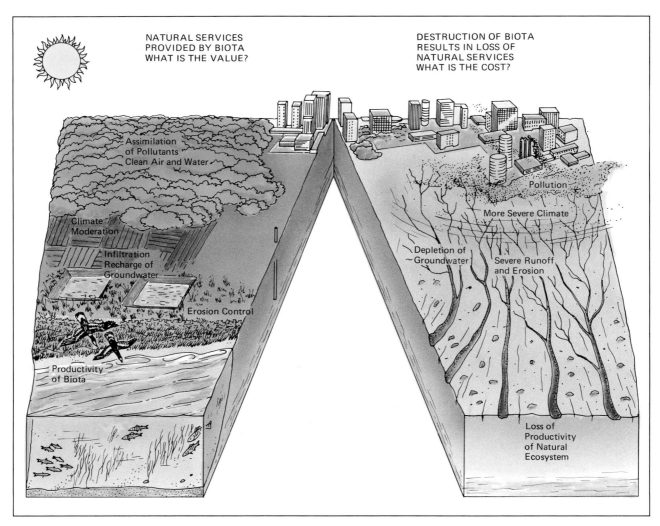

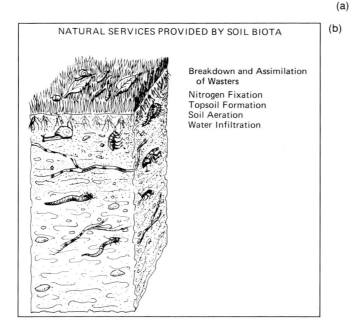

NATURAL SERVICES PROVIDED BY SOIL BIOTA

Breakdown and Assimilation
of Wasters

Nitrogen Fixation
Topsoil Formation
Soil Aeration
Water Infiltration

FIGURE 17–3

(a) Natural biota perform invaluable natural services, including water and air purification, control of the water cycle, propagation of fish, and climate control. (b) Soil biota provide numerous natural services as depicted. Destruction of natural biota results in the loss of these services. What are the costs to humans?

tibiotics were first discovered in natural plants, animals, or microbes. Yet, the search for such chemicals has barely scratched the surface. Given the opportunity to isolate and test the chemicals from natural biota, the potential discoveries are incalculable. But, pursuing the trends of destruction of natural biota, we may never have the opportunity.

Providing Natural Services

The vegetative cover provided by forest or grassland biota is crucial in preventing erosion, maintaining topsoil, allowing water to infiltrate and recharge

FIGURE 17–4
Tidal wetlands provide both the habitat and much of the food production for numerous fish and shellfish. They also serve to filter sediments and nutrients out of water. For each acre of wetlands that is destroyed by development or bulkheading shorelines, society stands to lose up to $100,000 per year in pollution control and wildlife propagation services.

groundwater supplies, reducing runoff and flooding, and holding and recycling nutrients. Wetland biota are particularly significant in stabilizing sediments, absorbing nutrients, and thus protecting water quality from eutrophication. Additionally, natural plant biota serve to maintain air quality as they fix carbon dioxide, release oxygen, and help to assimilate other air pollutants. By absorbing considerable solar radiation, and by releasing water vapor through transpiration, they moderate temperature and help to maintain the climate. Natural biota of various ecosystems provide a continuing production of wood, wildlife, fish, and other products of tremendous commercial and sport value. Soil biota are crucial in maintaining topsoil. Particular species may provide more specific functions; e.g., bees pollinate flowers.

All of these and other things that natural biota do may be called **natural services** (Fig. 17–3). We tend to take these services for granted, until the biota and thus the services are lost. For example, deforestation in India is largely responsible for the massive siltation and flooding that are causing human tragedy and suffering in Bangladesh. Loss of the wetlands, which provided sediment and nutrient control, is a major factor in the eutrophication and loss of productivity of Chesapeake Bay (Chapter 9). To put it in monetary terms, scientists have calculated that it would cost more than $100,000 a year to artificially duplicate the water purification and fish propagation capacity provided by a single acre (.4 ha) of natural tidal wetland (Fig. 17–4). Even then, the energy expenditures involved (producing and burning the requisite fuels) would probably lead to a net increase in pollution rather than a decrease. Thus, there is no real way that we can compensate for the losses of natural services that will be incurred as natural biota are destroyed. We will simply have to suffer the deterioration in quality of life that occurs.

Recreational, Aesthetic, and Scientific Values

The biota of natural ecosystems also provide the foundation for numerous recreational and/or aesthetic in-

FIGURE 17–5
Natural biota provide numerous recreational and aesthetic values, a few of which are depicted here.

terests ranging from sport fishing and hunting to hiking, camping, bird-watching, photography, and so on (Fig. 17–5). Interaction may range from casual aesthetic enjoyment through serious scientific study. Virtually all our knowledge and understanding of evolution and ecology has come through the study of natural biota and the ecosystems in which they live. Pleasure and satisfaction may even be indirect. For instance, one may never see a whale, but knowing that whales and other exotic species exist provides a certain aesthetic pleasure. Further, knowing that the earth and its biosphere continue to support and maintain such wildlife provides a sense of well-being.

Commercial Interests

The above recreational or aesthetic activities in turn support commercial interests, including a considerable share of the business of sporting goods outlets, tourist and travel accommodations, and so on. As leisure time increases, larger portions of the economy become connected to supporting activities related to the natural environment. Conversely, any loss or degradation of the natural environment affects such commercial interests. Examples abound of businesses folding as a lake or beach, for example, is polluted or despoiled (Fig. 17–6).

FIGURE 17–6
Pollution has a severe economic impact as seen by beaches being nearly deserted when medical wastes washed ashore. (Top, Photo Researchers/© 1988 Hank Morgan; bottom, Sygma/© 1988, Nola Tully.)

Finally, natural biota continue to support a number of commercial interests involved in direct exploitation. Commercial fishing, logging, and the trade in exotic "pets," uncommon species of fish, reptiles, mammals, birds, and plants, are the most conspicuous. Of course, it is these activities that are also, in many cases, destroying the natural biota.

Assaults against Natural Biota

Despite the values of natural biota, they are under assault from many sides, which may lead to extinction of all but a few remnants over the years of your lifetime. The major assaults are:

☐ Loss of habitat to alternative land uses
☐ Pollution
☐ Overuse
☐ Introduction of foreign species
☐ Combination of factors and environmental degradation.

Loss of Habitat to Alternative Land Uses

The greatest assault against natural biota by far is the destruction of habitats for alternative land-use. Clearing of forests and the draining or filling of wetlands for development or agriculture are prime examples (Fig. 17–7).

When a forest is cleared it is not just the trees that are destroyed. Every other plant and animal that occupies that ecosystem on a permanent or even a temporary basis, migrating birds for example, also suffers. The idea that this wildlife will simply move "next door" and continue to live in an undisturbed section is erroneous. Population balances, described in Chapter 3, produce a situation such that each area has all the wildlife it can support. Any loss of natural habitat can only result in a proportional reduction in all populations that require that habitat. For example, ornithologists report that populations of migratory song birds in the Eastern United States are down by 70 to 90 percent because of loss of their winter tropical forest habitat in Central and South America.

In many cases, the impact is *greater* than the proportion of land that is actually developed. For example, a highway that runs along a river may prevent animals from reaching a water source. Also, a certain minimum area is required to support a *critical number* of a species, the minimum number that can continue to interact and reproduce the population. This minimum area must be large enough to compensate for years of adverse weather. That is, more area will be required during a dry year than during a normal year. If development reduces suitable habitat to a point where it cannot support the critical number during an adverse year, the entire population will perish. Similarly, development such as a highway that splits a territory and prevents migration between the two sides will cause a population to perish if neither half can adequately support the critical number.

Likewise, ecosystems are not isolated from one another. Changes in one ecosystem are more than likely to affect adjacent ecosystems. A highway along a river, for example, in addition to blocking access to the river, will increase the runoff of water, sediments, and pollutants into the river, and temperature may be increased by reducing natural shading. Thus, the highway may render a river incapable of supporting certain species. Also, if a habitat used by migrating birds is destroyed, the birds will be eliminated as well, and this may have an impact on food webs of other ecosystems that they inhabit (Fig. 17–8).

The root cause underlying the destruction of natural ecosystems for alternative land uses is growing population with its demands for ever more land for developments, agriculture, and/or other resources. Typically, expanding human development consumes existing agricultural land and forces additional agricultural land to be carved out of natural ecosystems. However, the problem is made worse by inequitable land distribution, that is, some wealthy landlords keeping huge amounts of agricultural land out of food production. Also, huge amounts of agricultural land are unnecessarily lost to erosion and desertification (see Chapter 7).

Then, appetite for land for alternative uses is exacerbated by various economic factors and changing lifestyle. For example, development of coastal wetlands into recreational properties and the conversion of tropical forests to grassland to graze cattle are driven more by the profit motive of developers than by any actual need. This will be addressed in more depth later.

Pollution

Another major assault against natural biota is pollution. Of course, pollution is another form of habitat destruction or alteration. It is made a separate category, however, because the cause is not a wanton attack on the natural ecosystem as is the clearing of a forest, for example. Rather it is an "accidental" side effect of other activities.

Nevertheless, the consequences may be just as severe as occur in wanton destruction. Recall examples of forest dieback being caused by acid rain and air pollution, and the dieoff of species in Chesapeake Bay due to sediments and nutrients. The effects of pollution may be even more severe since they may be more widespread. Some scientists project that the

FIGURE 17–7
Tidal wetlands, in addition to their role in maintaining water quality, support a rich diversity of fish, shellfish, birds, and other wildlife. All this is destroyed as wetlands are converted to development. (a) Tidal wetlands at Wallops Island, Virginia. Inserts: Mussels; a willet nest. (Photos by author); (b) A similar area undergoing development. (EPA-Documerica photo by Flip Schulke; courtesy of U.S. Environmental Protection Agency.)

(a)

(b)

warming climate anticipated from the greenhouse heat-trapping effect may be the greatest catastrophe to hit natural biota in 60 million years. (Observations of the fossil record reveal that a massive episode of extinctions of plants and animals occurred at that time. The probable cause was that the earth was hit by a major asteroid, but there is still controversy regarding this hypothesis.) The reason for projecting catastrophic effects due to warming is that most species adapt only very slowly and, hence, they can adapt only to very gradual changes. The greenhouse effect may cause more warming in the next 50 years than would normally occur in the next 1000 years, a 40-fold increase in the rate of change. It is probable that numerous species will be unable to adapt so rapidly. Nor can they even migrate quickly enough. Tree species, for example, can only spread their seeds a mile or so at best. Then the tree must grow to maturity and shed seed before it can move the next mile, and so on. Consequently, scientists speculate that the rate of climatic warming will far outpace the ability of most tree species to migrate northward, thus trapping them in inhospitable climates. Then, every species that dies out doubtlessly will take others with it. If the forest trees can't migrate, neither can the rest of the wildlife that depend on them for food and habitat.

Overuse

It is obvious that taking whales, fish, trees, or any other species at a rate exceeding its capacity to reproduce and replace itself will lead to the ultimate extinction of the species. Yet, such wanton overuse is another major assault against natural biota. Such overuse is driven by a combination of economic greed and ignorance or insensitivity.

Some consumers are willing to pay exorbitant prices for such things as furniture made from tropical hardwoods like mahogany, exotic pets, furs from wild animals, and innumerable other "luxuries" including polar bear rugs, baskets made from elephant and rhinoceros feet, ivory-handled knives, and reptile-skin shoes and handbags. For example, some Indonesian and South American parrots sell for up to $5000. A panda-skin rug can bring $25,000. Such prices create a powerful economic incentive to exploit the species involved. For example, in 1983, more than 80,000 African elephants, about 8 percent of the continent's population, were killed. Most died to satisfy demand for ivory tusks by artisans who, in turn, are lured by customers willing to pay several thousand dollars for a carved knife handle, for example.

The prospect of extinction does not curtail the activities of exploiters because the prospect of a huge immediate profit outweighs other considerations. Even when the species is protected under the law the economic incentive is such that poaching and black market trade continue. Illicit trade in protected wildlife is second only to drug trafficking (Fig. 17–9 and Case Study 17.

It is easy to place the blame on the economic greed of the persons doing the actual killing. However, equal blame must be shared by the consumers who offer the "reward." It is their ignorance or total insensitivity to the fact that their money is fostering the extinction of invaluable wildlife that is fueling the situation. Unfortunately, the situation shows no sign of abatement. Increasing numbers of people want and are willing to pay the high price for rare furs or other such materials.

Particularly severe is the growing fad for exotic "pets," fish, reptiles, birds, and house plants. In virtually every case, these species are gathered, trapped, or hunted in the wild. Keeping these animals or plants for "pets" may seem acceptable since there is no intent to kill them. However, for every living specimen that makes it to the buyer, several are killed or die in the process of capture, transportation, and marketing. Few survive in captivity for any length of time. And regardless of survival, all are removed from the natural breeding population. In terms of maintaining the species, they are no better than dead. Numerous species of tropical fish, birds, reptiles, and plants are thus headed toward extinction because of the exploitation of the pet trade.

Unfortunately, as certain plants and animals become increasingly rare, the price some people are willing to pay goes up by the economic law of supply and demand. This, in turn, increases the fortune hunter's incentive to hunt or collect that animal or plant. Laws and regulations do little to curtail the exploitation because the potential profit is so great that many willingly accept the risks of illegal activities. The World Wildlife Federation reports that poaching and black market trade in endangered species have reached epidemic proportions and continue despite efforts to control them (Fig. 17–10).

Frequently exploiters will rationalize their activities by claiming that declines in populations are due to factors other than their exploitation. In some cases, other factors such as pollution or disease may be involved. However, it is significant to note that those with an economic interest in continuing exploitation often attempt to impede studies and raise every imaginable objection to any study that implicates their activities. In short, exploiters make extensive efforts to maintain lack of knowledge as a shield for callous insensitivity.

Even more severe examples of overuse are the overcutting of forests for firewood, the overcultivation of land, and overgrazing discussed in earlier chapters. These not only deplete the resource in question, they set into motion the cycle of erosion and desertification with effects far beyond the exploited area. Recall, for example, how sediments impact aquatic ecosystems (Chapter 9). The same may be said regarding the overuse of water resources (Chapter 8).

Introduction of Foreign Species

The destruction of the American chestnut by chestnut blight disease and the overgrowth of forests by kudzu described in Chapter 3 are prime examples of how the introduction of a foreign species can alter an ecosystem. You may recall other examples from Chapter 2 as well (Fig. 17–11).

Combination of Factors and Environmental Degradation

Depending on the particular situation, any of the above factors may be the most significant. Making matters worse, any combination of these factors may occur simultaneously. For example, a combination of several forms of pollution, elimination of wetlands that destroys breeding habitats, and overfishing are all involved in the decline of many fish and shellfish populations in the Chesapeake Bay (see Chapter 9). We see that we are in a vicious cycle of *environmental degradation*. That is, any of the factors may degrade the environment in such a way that further biota are adversely affected, causing still further environmental degradation (Fig. 17–12).

Case Study 17

LAST STAND FOR AFRICA'S ELEPHANTS
Record Ivory Prices Send Poachers after the Survivors

Striding majestically across the savanna, the African elephant is an unmistakable symbol of power and strength. As recently as the 1970s, its numbers were so great that some conservationists worried about overpopulation. Now the elephant is involved in a desperate struggle to survive, and the reason for its peril is one of its glories: the huge creature's magnificent tusks of ivory. Since the early 1980s, the price of ivory has surged from $25 per lb. to $80 per lb. As a result, growing bands of wily and ruthless poachers have taken to hunting down elephants illegally all across Africa, killing the animals with everything from automatic weapons to poison. About 10% of the remaining African elephants were killed last year, reducing their ranks to fewer than 750,000. If the slaughter continues at the present pace, the wild elephant could be close to extinction within a decade.

This week, to prevent such a tragedy, conservationists will unveil the most elaborate and costly plan in history to rescue a single species. Sponsored by the African Elephant Conservation Coordinating Group, a coalition of several international organizations, the plan calls for bolstering efforts to protect elephants against poachers, a study of ways to crack down on illegal trading of tusks, and a publicity campaign to alert people and governments to the relationship between the trade in ivory and the plight of the elephant. The AECCG hopes to raise at least $15 million in four years to finance its work.

The effort may be futile, though, unless demand for the animals' tusks is reduced sharply. Ivory is fashioned into everything from billiard balls and knife handles to necklaces and figurines. Craftsmen have even carved tusks into ornamental replicas of AK-47 assault rifles.

Theoretically, the business of taking ivory from animals alive or dead is highly regulated and ostensibly restricted by African governments. And under an international convention, there is a quota system that puts limits on the number of tusks each country can export.

So much for theory. In reality, the quota system

FIGURE 17–9
Tourists who buy elephant tusks like these on the Ivory Coast do not see the slaughter that supports the illegal trade and threatens a species. (© Valenti/Sygma.)

has been ineffective in controlling the trade. Up to 90% of the tusks that enter the marketplace have been taken illegally by poachers, and smugglers have little trouble getting the ivory out of Africa. Angolan rebel leader Jonas Savimbi has reportedly financed his insurrection with ivory taken from more than 100,000 elephants. Some countries seem to be conduits for the illegal trade. With roughly 4,500 elephants of its own, Somalia has still managed to export tusks from an estimated 13,800 elephants in the past three years, evidence that the country has been providing false documents for ivory poached elsewhere. In response, the U.S. is expected this week to announce a ban on imports of Somalian ivory.

The leading destination for legal and perhaps illegal ivory is Asia. Hong Kong is a major manufacturer and exporter of ivory jewelry, and 30% of the colony's output goes to Americans. "People in the U.S. just don't connect ivory with elephants," says Mark Stanley Price, a director of the African Wildlife Foundation, "but every bracelet represents a dead elephant." Another top consumer is Japan, where ivory has long been used for personalized seals called *hanko*. But under pressure from conservationists, Hong Kong and Japan have begun to check closely the documents on ivory imports to weed out illegal shipments. Japan's legal ivory imports, in particular, have dropped sharply in the past three years.

Unfortunately, the decline in ivory trade in Japan and elsewhere may not reflect a drop in demand so much as the decimation of adult elephants. As mature elephants are killed, it becomes harder to satisfy the world's appetite for ivory. Stephen Cobb, who leads an ivory study for the AECCG, says the reduction in trade "is a clear sign of the collapse of exploitable elephant populations."

Some conservationists would like to see a total ban on the ivory trade. But that would be no easier to enforce than the laws against selling cocaine and heroin. Dealers bold enough to defy the embargo could anticipate higher profits than ever. Moreover, poor African countries need the revenue from at least a limited amount of legal trading.

Realizing that if elephants vanish, so might tourists, some African nations are determined to slow down the killing. In addition, the animal is a vital part of Africa's unique ecosystem. For eons, elephants have knocked down trees, helping to give Africa its distinctive mix of forest and savanna and opening up the land for other big mammals.

Unwilling to let the elephant be wiped out, some governments have declared war on illegal killing. In Kenya armed patrols have orders to shoot poachers. Sometimes, though, the culprits are a formidable force themselves. At Kenya's Tsavo National Park, scores of poachers dressed in battle fatigues and armed with automatic weapons killed one policeman and wounded several others.

Besieged by armies of hunters, many herds are literally on the run. Conservationists use the phrase "refugee elephants" to describe animals fleeing Mozambique to crowd into protected areas in Zimbabwe. The killing of older animals with the biggest tusks threatens to reduce herds to what Tanzanian game manager Constantius Mlay describes as collections of naive teenagers without the wise old elephants needed as leaders in times of drought and food scarcity.

Conservationists cannot hope to protect elephants throughout their African homelands. For that reason, the AECCG, which includes such major conservation groups as the World Wildlife Fund, TRAFFIC and Wildlife Conservation International, envisions a triage approach. The group plans to concentrate its resources on about 40 populations that have the best chance of being guarded from poachers. That strategy would focus on saving about 250,000 elephants and would reluctantly leave another 500,000 to their fate.

This prospect is not so dismal as it sounds. If protected well, the remaining quarter-million elephants would be a large enough population to thrive and multiply again. In fact, David Western, director of WCI, asserts that if allowed to grow old and die naturally, the elephants in these herds could probably supply enough tusks to support an ivory market larger than today's illegal business.

—With reporting by Carter Coleman/Dar es Salaam and Roger Browning/Nairobi

EUGENE LINDEN
Time, February 20, 1989

FIGURE 17-10
"Products" from endangered species confiscated in a breakup of a black market ring. Some people will pay exorbitant prices for various items from wild animals. Despite the fact that trade in such products has been made illegal and the animals are supposedly protected under the law, poaching and black market trade to profit from this market are still causing merciless slaughter. Sadly, most of the products for which animals are slaughtered are trivial. (U.S. Fish and Wildlife Service photo. Steve Hillebrand.)

FIGURE 17-11
Introduced species can severely upset the natural balance. Semiarid grasslands throughout much of the southwest are now dominated by sagebrush due to overgrazing. Animals eat the grass but not the sagebrush, allowing it to spread. Wild burros seen here are one of many introduced animals responsible for the overgrazing. (Photo Researchers/© 1985 R. J. Erwin)

Conclusion and the Tragedy of the Commons

Given that we are destroying natural biota in the above ways, an alien visiting planet earth would logically conclude that humans value trinkets, freedom to pollute and despoil, and extravagant land use more than a sustainable biosphere. A phenomenon that perpetuates this absurdity is the *tragedy of the commons*.

As described by the biologist Garrett Hardin, in a now classic essay by the same title, the original "commons" involved areas of pastureland in England that were provided free by the king to anyone who wished to graze cattle. Despite the social virtue of the idea, farmers were quick to realize that whoever grazed the most cattle stood to benefit most. Even if he realized that the commons was being overgrazed, any farmer who withdrew his cattle simply sacrificed his own profits while others went on grazing. His loss

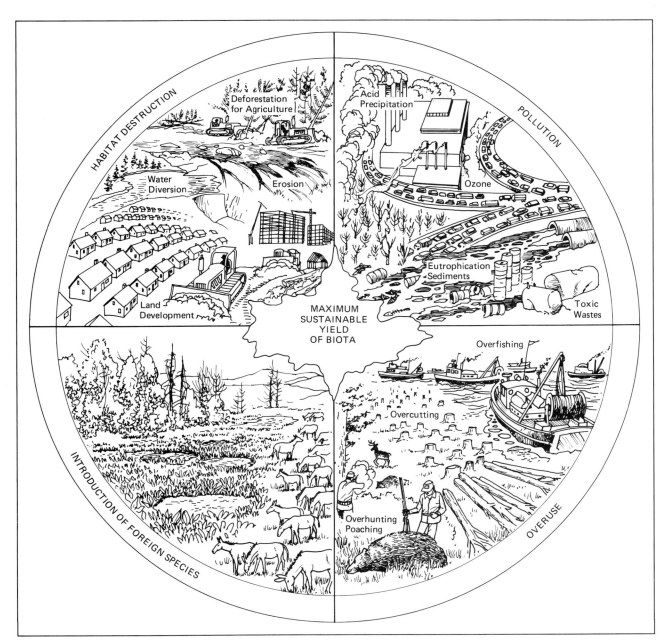

FIGURE 17–12
A summary of the ways in which sustainable yields of resources from natural biota are being undercut by human impacts.

became their gain, and the commons was overgrazed in any case. Consequently, farmers continued grazing until the commons was totally destroyed (Fig. 17–13).

Lobster trapping in New England provides a modern example. The lobster grounds are effectively a commons, free to be used by anyone with the means to do so. Although lobstermen may recognize that the lobsters are being overharvested, any lobsterman who curtails his own fishing diminishes his own in-

come while his competitors continue. His loss only becomes their gain, and the lobsters are depleted in any case. Thus, while catches diminish indicating overfishing, fishing continues. Increasing price brought on by shortage of supply continues to make the fishing profitable despite declining catches (Fig. 17–14). You can guess that poachers hunting endangered wildlife use the same logic of "If I don't, he will" to rationalize any feelings of guilt.

The concept of the tragedy of the commons extends to any situation where two or more independent people or groups are engaged in the exploitation of a resource. It also applies to: pollution—air and

Chap. 17: Biota: Biological Resources **411**

FIGURE 17–13
The tragedy of the commons. Whenever a resource is open to be exploited by numerous parties, it will tend to be overexploited because each party benefits maximally from its own exploitive efforts while the loss (destruction of the resource) is shared by all.

water comprise a commons into which polluters discharge their wastes; land development—land is a common resource exploited by competing developers; even wildlife trinkets on the market are a commons being exploited by competing collectors of such trivia. In each case, there is the factor of: "If I don't, he will; so I might as well be the one to get it." Thus the prices are paid and driven up and exploitation continues. Of course, in the long run, society at large will suffer the consequences of the losses. Perhaps the greatest tragedy of all is that we stand by apathetically and permit some to profit enormously at endeavors that will be to the detriment of us all.

CONSERVATION OF NATURAL BIOTA AND ECOSYSTEMS

How can natural biota be protected?

General Principles

Definition and Objective of Conservation

The beauty of natural biota is that it is a **renewable resource.** That is, it has the capacity to renew and replenish itself through its reproductive capacity despite certain quantities being taken, and this can go

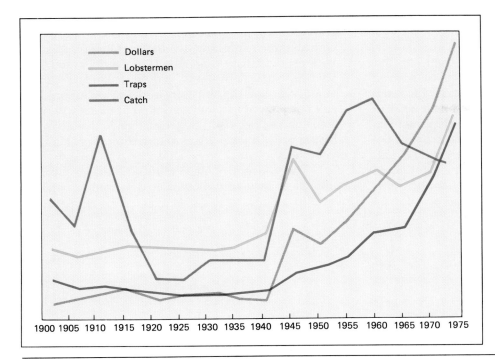

FIGURE 17–14
Economics may promote overexploitation, as shown in the case of lobster fishing in New England. As lobster catches have declined, price has increased owing to limited supply. The continued potential for profit fosters continued exploitation of already overharvested lobsters. (Copyright 1977 by the National Wildlife Federation. Reprinted from the April–May issue of *National Wildlife* magazine.)

on indefinitely. Recall from Chapter 3 that every species has the biotic potential to increase its numbers and that in a balanced ecosystem the excess numbers simply fall prey to parasites, predators, and other factors of environmental resistance. It is difficult to find fault with activities that effectively put some of this excess population to human use. Indeed, removal of excess numbers of herbivores is sometimes necessary to prevent overgrazing and the destruction of the ecosystem as a whole (Fig. 17–15). The tragedy occurs when users (hunters, fishermen, loggers, etc.) take more than the excess and deplete the breeding populations, threatening or causing extinction of the species.

Conservation of natural biota, then, does not, or at least should not, imply complete denial of any usage, although it may be temporarily expedient in a management program to allow a certain species to recover its population size. Rather the aim of conservation is to *manage or regulate the use* so that it does not exceed the capacity of the species or system to renew itself.

FIGURE 17–15
Preservation of wildlife requires a careful consideration of the habitat area required to support the animals. Photograph shows a preserve in Africa that has been largely destroyed by the overgrazing of elephants. (Mark Boulton/Photo Researchers, Inc.)

Concept of Maximum Sustainable Yield

The central question in conservation, then, is: *How much use can be sustained on a continuing basis without undercutting the capacity of the species or system to renew itself?* The maximum *use* a system can sustain without impairing its ability to renew itself is called the **maximum sustainable yield.** You may observe that the concept applies to more than just preservation of natural biota. It is also the central question in maintaining parks, air quality, water quality and quantity, soils, and, indeed, the entire biosphere. "Use" can refer to cutting timber, hunting, fishing, number of park visitations, discharge of pollutants into air or water, and so on. Natural systems can withstand a certain amount of use (or abuse in terms of pollution) and still remain viable. However, a point exists where increasing use begins to destroy regenerative capacity. Just short of that point is the maximum sustainable yield (Fig. 17–16).

The concept of maximum sustainable yield is quite clear in theory; however, determining what it

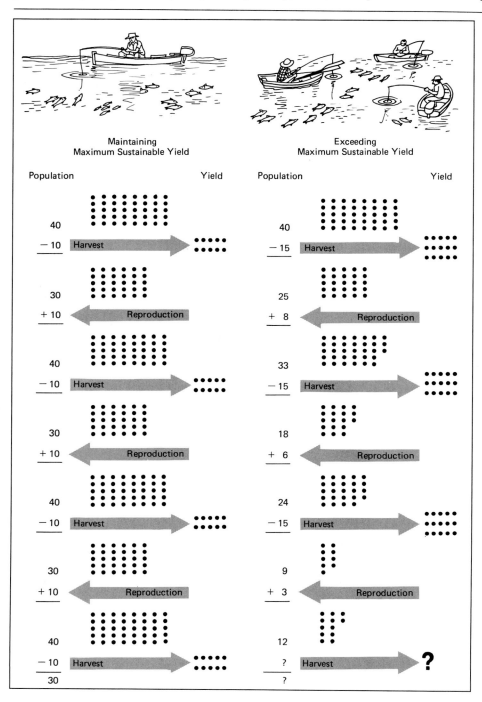

FIGURE 17–16
Maximum sustainable yield. Biological resources all have a maximum sustainable yield, a maximum harvest or use that can be sustained without impairing their productivity. The maximum sustainable yield may be exceeded in the short run only at the expense of undercutting the productivity of the system in the long run.

is in practice is much more difficult. An important consideration in determining the maximum sustainable yield is the **carrying capacity** of the system, the maximum population that the ecosystem can support. If a population is well within the carrying capacity of the ecosystem, allowing a population to grow will increase the number of reproductive individuals and, thus, the sustainable yield. However, as the population approaches the carrying capacity of the ecosystem, new individuals must compete for food and living space with those that are already present. As a result, recruitment may fall drastically. The net production in a mature climax forest, for instance, is zero because new growth is balanced by natural dieoff. If populations are at or near the carrying capacity of the ecosystem, production, and hence sustainable yield, can actually be increased by thinning the population so that competition is reduced and optimal growth and reproductive rates are achieved. Thus, maximum sustainable yield is obtained with *optimal*, not *maximum*, population size (Fig. 17–17).

The matter is further complicated, however, by the fact that carrying capacity and hence optimal population are not constant. They may vary from year to year as weather fluctuates. Replacement may also vary from year to year since some years are particularly favorable to reproduction and recruitment while others are not. Of course human impacts such as pollution and other forms of habitat alteration adversely affect reproductive rates, recruitment, carrying capacity, and consequently sustainable yields (Fig. 17–18).

Often, attempting to achieve maximum sustainable yield involves a process of alternating overuse until deterioration is observed and moratoriums enforcing no use until the system recovers.

Also, maximum sustainable yield of particular species may be increased by suitable management procedures. For example, certain fish are routinely propagated in hatcheries and released into rivers so that fishermen can continue to "overfish." Then, there is a gradation of management all the way from simple restocking to highly controlled agricultural systems. However, it should be obvious that in managing a system to increase the yield of one species, any number of other species may be lost in the process.

Specific Points

To protect natural biota, we must be fully aware of the concept of the maximum sustainable yield and those social and economic factors that tend to cause overuse and other forms of environmental degradation that diminish maximum sustainable yields. We must then exercise and cooperate with regulations that are necessary to protect natural biota. Specifically:

☐ Natural biota cannot be treated as a commons since, wherever it has been, the "tragedy of the commons" inevitably results. Biota must be put under an authority that is responsible for its sustainability and can regulate its use as necessary to meet that objective.

☐ There must be suitable enforcement of regulations.

☐ There must be a reduction in the economic incentives that promote violation of regulations.

☐ Suitable habitats must be preserved.

☐ Habitats must be protected from pollution.

With this background let us turn to a survey of where we stand with regard to protecting the earth's natural biota.

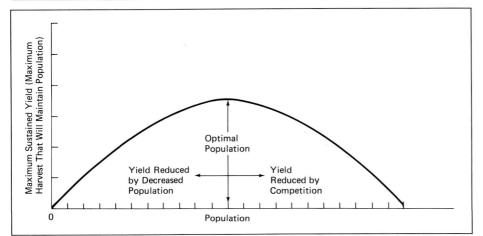

FIGURE 17–17
Maximum sustainable yield occurs at an optimum population level. At populations above the optimum, yield is diminished by competition between organisms. At levels below the optimum, yield is diminished by reduced stock.

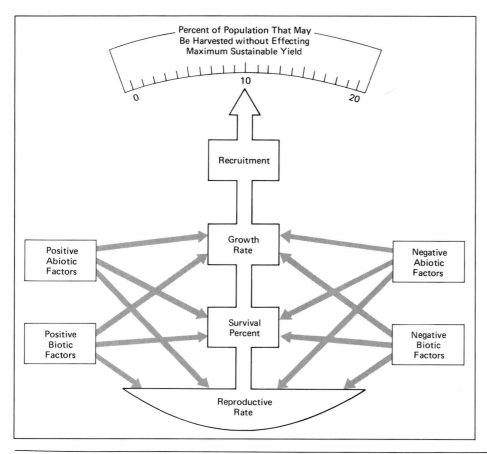

Percent of Population That May Be Harvested without Effecting Maximum Sustainable Yield

0   10   20

Recruitment

Growth Rate

Positive Abiotic Factors

Negative Abiotic Factors

Survival Percent

Positive Biotic Factors

Negative Biotic Factors

Reproductive Rate

FIGURE 17–18
The percentage of the population that can be harvested each year without depleting the stock depends on replacement capacity, which in turn depends on many other interconnected factors.

Where We Stand with Regard to Protecting Biota

Game Animals in the United States

Game animals are those animals traditionally sought by hunters for meat. In the early days of this country these animals were treated as a commons. A number of species including the American buffalo (bison) and the wild turkey were hunted to near extinction, and the great auk and the passenger pigeon did become extinct as a result. As the tragedy of the commons was recognized, regulations came into effect. Consider how current regulations address the points listed above.

State governments backed up by the federal government take over as the authority and set regulations and hire wardens for enforcement. Restrictions on hunting may include:

☐ A license is required.

☐ You are limited in the kinds of firearms you can use.

☐ You can only hunt within a specified season, or hunting of certain species may be banned.

☐ There may be size and/or sex limits to protect the young and breeding animals.

☐ There are limits on the number of a game animal you can take.

☐ Marketing is illegal, which undercuts hunting for profit.

By monitoring populations and adjusting seasons and take limits accordingly, game animal populations are maintained so that there is effectively a maximum sustainable yield. Additionally game preserves, parks, and other areas where hunting is prohibited serve to protect both habitat and certain breeding populations. Also, common game animals such as deer, rabbits, and pheasants are well adapted to the rural field/woods environment. Being thus adapted to the humanized environment and protected from overhunting, populations of game animals are being maintained. However, some problems are emerging.

Significant losses of habitat are now occurring as rural environments are increasingly pocketed by suburban or recreational developments and segmented by new and improved roadways. This process is fragmenting the environment into "ecological islands," many of which may be too small to support breeding populations. The number of animals killed on roadways now far exceeds that killed by hunters and is an added hazard to motorists as well. Draining wetlands is diminishing populations of many water-fowl, as mentioned before. A few animals such as opossums and raccoons have adapted and are thriving even in highly urbanized areas, but this creates hazards of a reverse sort. For example, a rabies epidemic among "urban" raccoons is considered a significant public health risk in several cities in the eastern United States.

Exotic Species and the Endangered Species Act

Numerous species have also suffered from commercial hunting for their furs, skins (alligators), plumage, tusks (elephants) and, more recently, as exotic animals to be simply sold as "pets." The snowy egret provides an example of how both laws and changing public attitudes have served to provide protection in some cases (Fig. 17–19).

FIGURE 17–19
Snowy egret. Hunted almost to extinction for its plumage in the late 1800s, egrets are now valued most for their living beauty. (Courtesy of Steve Simon, Catonsville Community College.)

In colonial days, huge flocks of snowy egrets inhabited coastal wetlands and marshes of the southeastern United States. In the 1800s, when women's fashion turned to fancy hats adorned with feathers, egrets and other birds were treated as a "commons" that was open to be exploited by any number of independent hunters seeking to profit from the lucrative plumage market. By the late 1800s, egrets were almost extinct. In 1886, the newly formed Audubon Society began an active campaign through the press to shame "feather wearers" and end this "terrible folly." The campaign caught on and, gradually, attitudes changed and laws followed. Florida and Texas were first to pass laws protecting plumed birds, and in 1900, Congress passed the Lacey Act, which forbid interstate commerce in illegally killed wildlife, making it more difficult for hunters to sell their kill. Finally, wildlife refuges have been established to protect the birds' breeding habitats. With millions of people visiting these refuges and gaining the enjoyment of seeing these birds in their natural locales, attitudes have changed so completely that today the thought of hunting these birds would be abhorrent to most of us, even if official protection were removed. Thus protected, populations were able to recover substantially. However, their numbers are headed down once again since habitats outside the very limited refuges continue to be destroyed and affected by pollution.

Thousands of other species throughout the world are even less well off as they continue to be hunted for paraphernalia or trapped for pets and lose their habitats (Fig. 17–20). Congress made some effort to protect species from extinction through the **Endangered Species Act** of 1966 (most recently reauthorized in 1988). An **endangered species** is defined as one which has been reduced to the point that it is in imminent danger of becoming extinct if protection is not provided. When a species is *officially recognized* as endangered, the law provides for substantial fines for any killing, trapping, uprooting (plants), or commerce of it or its parts. (Making such commerce illegal supports protection of the endangered species of other countries as well as those in the United States, because the United States is a major market.) Further, the act requires the mapping of habitats of endangered species and forbids any government agency from destroying such habitats in the course of construction projects such as highways, dams, or airports.

There are several shortcomings in the Endangered Species Act, however. A major one is that protection is not provided until a species is "officially recognized" as endangered by the Department of Interior, which is responsible for administering the act. Most ecologists are extremely concerned by the slow

FIGURE 17-20
A giraffe in Kenya killed by a poacher for the tuft from its tail, which some consider to be a prized ornament. (FAO photo by Thane Riney.)

pace at which species are being added to the official list. Hundreds of species in the United States, over 400 in Hawaii alone, may become extinct, or habitats will be so reduced that extinction is inevitable, before they gain recognition as endangered.

A second major problem is that, despite official protection, in some cases poaching and illegal trade remain rampant. This is to say, there is insufficient funding provided in the act to create adequate enforcement. This problem is worsening as increasing numbers of people, a result of the population explosion, are unable to find adequate employment and turn to illicit activities.

A few exotic species have gained exceptional public attention and heroic efforts have been mounted to save them. Efforts to save the whooping crane, for example, included virtually full-time monitoring and protection of the single remaining flock, which for years numbered only in the teens. To assure a higher recruitment rate, eggs were collected (the female normally lays two but only rarely does more than one chick survive) and artificially incubated. The chicks were then returned to nests of related cranes to be raised by foster parents (Fig. 17-21). This effort seems to have paid off; the whooping crane is now at least holding its own. Similar efforts to save the

FIGURE 17-21
A ten-day-old whooping crane chick being reared by sandhill crane foster parents in captivity. (Courtesy of U.S. Fish and Wildlife Service.)

FIGURE 17-22
Exotic animal "ranch" in Florida. Many feel that transferring small herds of exotic animals to ranches where they will be totally protected may be the only way to save them from extinction. (Gerald Davis Photography © All rights reserved.)

California condor, however, failed. It is now reduced to a few individuals in captivity, although attempts are still being made to breed it.

Most large zoos are beginning to take an active role in breeding other endangered animals such as the rhinoceros and elephant (Fig. 17-22). Likewise, seeds are being collected from wild plants and stored in "seed banks" maintained by international cooperative efforts among universities and governments. Many believe that this is the only way left to save endangered species. However, for every animal or plant that receives enough attention to be saved from extinction by breeding a few captive individuals in zoos or other facilities, hundreds of other lesser-known but perhaps no less valuable species will doubtlessly become extinct as habitat destruction and other factors take their toll. Finally, even at best, the few individuals maintained and bred in captivity are a very limited representation of the total genetic diversity that existed in the total wild population.

Aquatic Species

Open oceans have traditionally been considered an international commons, and by the end of the 1960s numerous species and areas were being seriously depleted by overfishing as a result (Fig. 17-23 and Fig. 17-24). In 1977, a number of nations including the United States extended their territorial limits from the previous 3 to 12 miles (5.8 to 19.3 km) to 200 miles (322 km) offshore. Since many prime fishing grounds are located between 12 and 200 miles from shores, this action effectively removed them from the international commons and placed them under the authority of a particular nation, allowing regulation of fishing. As a result of this action some fishing areas have recovered.

Whales, which inhabit the ocean outside the 200 mile limit, however, are still plagued with protection problems. An organization of nations with whaling interests, the International Whaling Commission, makes efforts to set quotas or moratoriums for various species but, lacking any real overall authority, the tragedy of the commons continues. Several whale species have been so depleted that many fear their extinction is imminent. Sufficient public outcry against whaling may make the difference.

Tuna is another species that continues to be overfished because U.S. commercial tuna fishermen managed to have themselves exempted from the 200 mile regulation so that they could continue to exploit tuna grounds around the world. Even more tragic, huge numbers of animals such as dolphins and endangered sea turtles get caught or entangled in the tuna nets and are needlessly killed as a result. The plastic nets continue to cause this problem even long after they are discarded.

For fish and shellfish of fresh water and estuaries, overfishing is generally controlled by regulations similar to those used to control hunting. The greatest threat now is pollution, sediments, and nutrients causing eutrophication as in Chesapeake Bay, and also toxic chemicals (Fig. 17-25). In some areas such as parts of the Great Lakes and the Hudson and Mississippi Rivers, some game fish continue to live, but they are contaminated with such levels of toxic chemicals, gained through bioaccumulation and biomagnification, that commercial fishing is banned and sportsmen are advised not to eat their catch. Notably, a few rivers such as the Willamette River in Oregon and the Detroit River in Michigan, which were little more than open sewers in the 1960s, have been cleaned up and support fishing again.

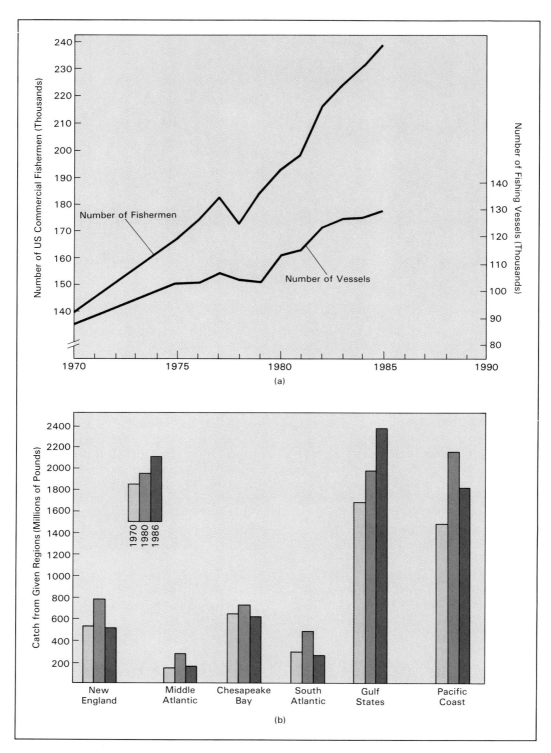

FIGURE 17–23
(a) The number of fisherman and fishing vessels has grown continuously from 1970 to 1986 as shown. (b) However, fish catches peaked around 1980 and have declined since in all areas except the Gulf states. Such declining catches in the face of increasing fishing indicates that fish populations are being depleted as a result of overfishing and/or environmental degradation. (U.S. Department of Commerce, *Statistical Abstracts of the United States, 1988.*)

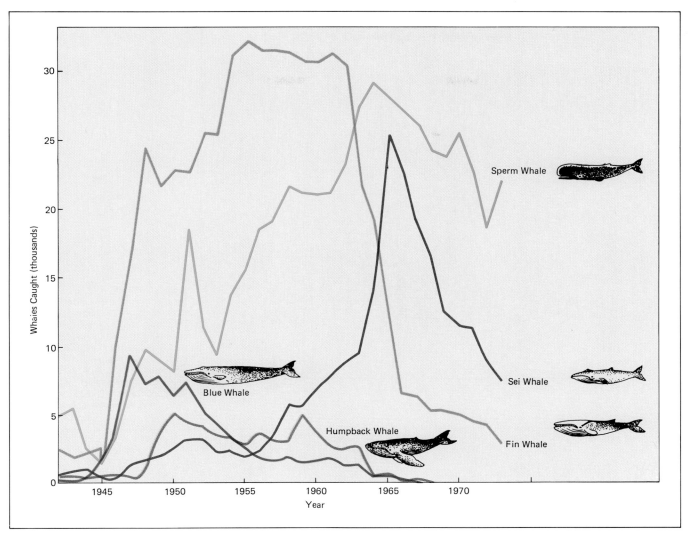

FIGURE 17–24
Catches reported for various species of whales from 1942 to 1973. Intensive whaling was resumed after World War II. Yields were temporarily high from overharvesting, but they declined steeply as stocks were depleted. Yet whaling continues. Can it be stopped before extinction occurs? (Data from International Whaling Commission.)

Is Saving Species Enough—The Need to Focus on Ecosystems

Increasingly, biologists and ecologists are concerned that the emphasis on saving species is distracting us from the broader picture of saving ecosystems. Even if we save a few exotic species through captive breeding, consider what will still be lost if their ecosystem-habitats are allowed to vanish in the process. Lost to us forever would be:

☐ Thousands—perhaps even millions in the case of tropical forests—of other species that may be of much more value in supporting the underpinnings of agriculture, forestry, and providing new medicines.

☐ All the natural services of maintaining the water cycle, water and air pollution control, and even climate.

☐ Scientific, recreational, and aesthetic values.

☐ Finally, the very commercial interests that are causing the overexploitation will come to an end as they ultimately destroy their last "victims."

Therefore, renewed emphasis must be put on saving the ecosystems. Saving the entire ecosystem is the only way that we can preserve the values of biota and protect the biosphere from ultimate collapse.

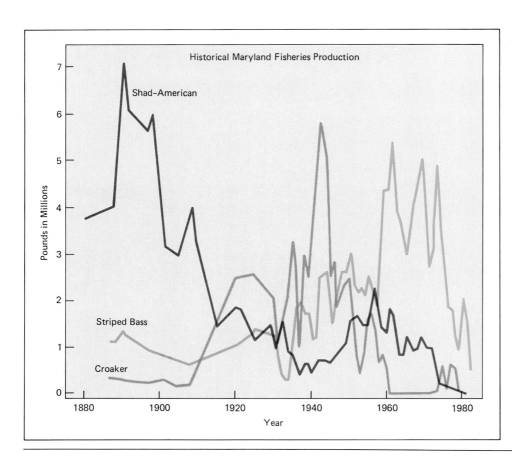

Historical Maryland Fisheries Production

Shad-American

Striped Bass

Croaker

Pounds in Millions

Year

FIGURE 17-25
Production of three important fish species from Chesapeake Bay. Declines observed early in the century were due to overfishing. However, further declines in the 70s and failure of recovery in the 80s are attributable to pollution factors and loss of habitat as well. (Maryland Dept. of Natural Resources, Tidewater Administration, Chris Bonzek.)

Saving Tropical Forests

Tropical rainforests are of greatest concern. Covering a broad, equatorial band across South America (mainly Brazil), Africa (mainly Zaire), and Indonesia, they are the habitat for millions of plant and animal species, a vast portion of which are still unidentified. They are also, many climatologists reason, crucial in maintaining the climate of the earth. At the very least, their destruction is contributing significantly to the loading of the atmosphere with carbon dioxide, which is currently warming the climate. Yet rainforests are being destroyed at a phenomenal rate; an area the size of Pennsylvania was destroyed in 1988. If the pace continues, and to date it has, tropical forests will be reduced to remnants in another 10 to 20 years (Fig. 17–26).

Destruction is being caused by a number of factors, all of which come down to the fact that the countries involved are both poor and have uncontrolled population growth. The huge numbers of young people in the newest generation cannot find jobs or live on the land that barely supports their parents. Therefore, to make a living they are burning the forests to obtain land for agriculture and cutting them down to get firewood or lumber to use or sell (Fig. 17–27).

FIGURE 17-27
Burning of tropical rainforest for agriculture in Amazonia, Brazil. Millions of acres of rainforest are being thus destroyed each year, undoubtedly causing the extinction of many species and sharply reducing genetic diversity of innumerable others and perhaps affecting global climate in the long run. (Earth Scenes. © 1985, Dr. Nigel Smith.)

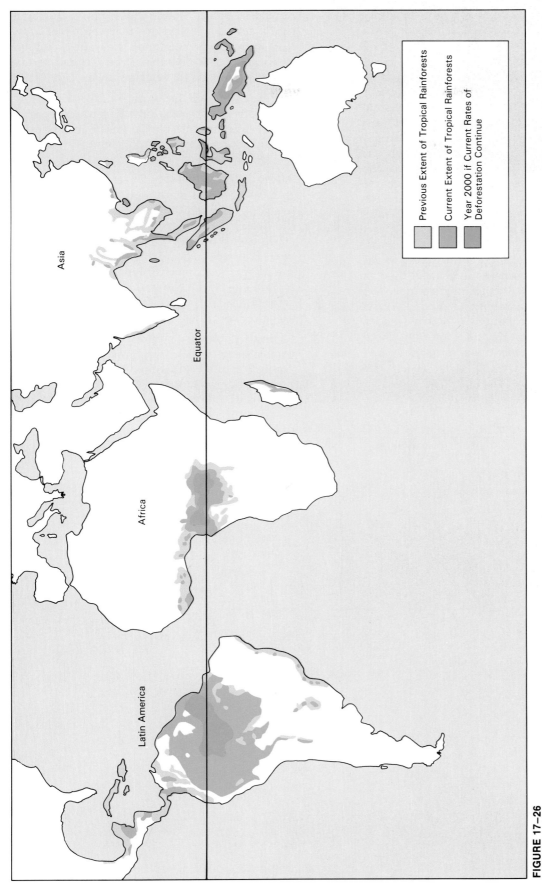

FIGURE 17–26

Tropical rain forests originally covered most of the equatorial regions of South America, Africa, and Indonesia. Their extent was already greatly reduced when recent exploitation and clearing for agriculture, which often proves unsustainable, accelerated vastly. At current rates of destruction little will remain in another 10 years. What may be the consequence for the biosphere of this loss? (Smithsonian Institution, Washington, D.C. Traveling Exhibition Service, 1988.)

Legend:
- Previous Extent of Tropical Rainforests
- Current Extent of Tropical Rainforests
- Year 2000 if Current Rates of Deforestation Continue

Asia

Equator

Africa

Latin America

Unfortunately tropical soils do not support much agriculture before they are exhausted of nutrients and mineralized into untillable clods. Thus, more of the forest is cut, leaving behind barren earth.

The problem is exacerbated by the governments involved. They have huge debts (over 100 billion dollars in the case of Brazil) incurred from past loans. The forests are their major "resource." To raise money to pay the interest on their loans they are selling the logging rights, generally to multinational companies, which come in and wastefully destroy the forests to obtain their hardwood logs for furniture and make no effort to replant or restore the forest. In other words they see the forest as a commons from which to get as *much* as they can *while* they can. They have no interest or concern in maintaining sustainable yield. Similarly, agricultural rights are sold to companies that come in and clear the forest and then replant it in grass for the grazing of cattle to supply fast-food chains with cheap hamburger. Again, we see how buying power in rich nations is driving the destruction of biota. From the destruction of this biota, however, we will all suffer in the end.

What You Can Do (to Help Save Natural Biota and Ecosystems)

☐ Do not buy any item that comes from an endangered species or indulge in the ownership of exotic "pets." Shame those who do. Animal rights activists would do much better to focus their attention on stemming the trade of wildlife items and exotic pets than on the use of certain common animals for research.

☐ Join and support The World Wildlife Fund, Nature Conservancy, Natural Resources Defense Council, Environmental Defense Fund, or other organizations committed to saving tropical forests and other ecosystems.

☐ Write your congressperson asking that saving tropical forests be made a priority in U.S. policy. Brazil's 100 billion dollar debt, which is causing the destruction of vast tracts of tropical forest, is less than a third of our military budget for one year. Is it worth sacrificing our future to collect the interest on this debt for a few more years? It would be possible to trade this debt for Brazil's commitment to save the forests. The same goes for other countries involved.

☐ Saving the biosphere must be an international commitment and this requires international organization and regulations. The United Nations Environmental Program (UNEP) is an organization with this potential if they receive the necessary support. Write your congressperson and the White House to increase support and commitment to UNEP.

18
Converting Refuse to Resources

■ Outline

I. THE SOLID WASTE CRISIS   427

  A. Background of Solid Waste Disposal   427
    1. Open Burning Dumps   428
    2. Landfills   428
  B. Problems of Landfills   429

    1. Leachate Generation and Groundwater
       Contamination   429
    2. Methane Production   429
    3. Settling   429
  C. Improving Landfills: Trying to Fix a Wrong
    Answer   430

  D. Escalating Costs of Landfilling   430

II. SOLUTIONS   430

  A. The Recycling Solution   430
    1. Impediments to Recycling   430
      a. Sorting   430
      b. Lack of standards   431
      c. Reprocessing   431
      d. Marketing   431
      e. Separation between government and
        private enterprise   431
      f. Vested interests in the status quo   431
      g. Hidden costs   432
    2. Addressing the Problems   432

      a. Forming partnerships between
        government and business   432
      b. Sorting   432
      c. Reprocessing and profits   432
    3. Promoting Recycling through Mandate   434
      a. Mandatory recycling laws   434
      b. Banning the disposal of certain
        items or constituents in landfills   434
      c. Mandating government purchase of
        recycled materials   434
      d. Advance disposal fees   436

CASE STUDY: VALUE ADDED: THE POT AT THE END
OF THE TRASH RAINBOW   435

  B. Composting   436

  C. Refuse-to-Energy Conversion   438

  D. Reducing Waste Volume   438
    1. Returnable versus Nonreturnable
       Bottles   438

■ Study Questions

1. Define and give the general contents of *municipal solid wastes.* What level of government is responsible for disposal of municipal solid waste?

2. Give a historical background of refuse disposal. What remains the most common method for disposal refuse?

3. List and describe problems stemming from landfilling of refuse.

4. Describe the features that new landfills must have in order to prevent the above problems.

5. Is landfilling sustainable? What are the costs and limitations?

6. List alternatives to landfilling some of the components of refuse.

7. List and describe the factors that impede recycling of refuse.

8. Describe how the factors impeding recycling can be and are being overcome.

9. Describe examples of laws that might be passed to promote recycling.

10. Describe what *composting* is and what components of refuse might be composted.

11. Describe how refuse may be converted to energy. Give the pros and cons of doing this.

12. Describe ways in which the total volume of refuse might be reduced.

    2. Other Measures   440
    E. Integrated Waste Management   440

III. CHANGING PUBLIC ATTITUDES AND
    LIFESTYLE—WHAT YOU CAN DO   440

13. Describe what is meant by *integrated waste management.*

14. Describe what you can do to promote converting refuse to resources.

The end use of countless materials and products ranging from old newspapers, emptied cans and bottles, food wastes, and packaging materials to wornout clothing, broken dishes, and unwanted appliances is in the home. From here, tradition and continuing practice is to *throw it away*, a clear violation of the basic ecological principle that all materials are recycled. Common sense dictates that a throw-away society is not sustainable, but it appears that we have had to prove this experimentally. The results: we are on the brink of a crisis as we are producing more trash than ever before and are rapidly running out of places to put it. Yet, solutions are available and are being implemented. In this chapter, we will look at the dimensions of the crisis and potential solutions in more detail.

THE SOLID WASTE CRISIS

Background of Solid Waste Disposal

The total of the above materials, thrown away from homes and commercial establishments, often called trash, refuse, or garbage, is technically referred to as **municipal solid waste.** This is to distinguish it from industrial wastes, agricultural wastes, and sewage wastes, which we have considered in previous chapters.

Over the years, the amount of municipal solid waste has grown steadily, in part because of increasing population, but more so because of changing lifestyles and the increasing use of disposable materials and containers (Fig. 18–1). It now amounts to some-

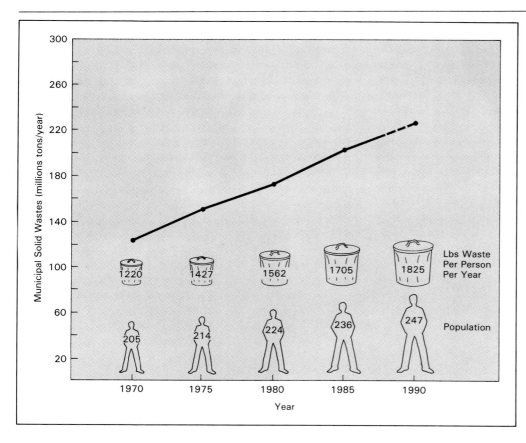

FIGURE 18–1
Output of solid wastes in the United States has grown much more rapidly than population and the trend is continuing. Can it be reversed? (Data from S. L. Blum, "Tapping Resources in Municipal Solid Waste," p. 47, in *Materials: Renewable and Nonrenewable Resources,* ed. P.H. Abelson and A. L. Hammond. © 1976 by American Association for the Advancement of Science, Washington, D.C.)

what over 4 pounds per person per day. Combining everyone in the United States, that amounts to enough to fill 63,000 garbage trucks each day, a total of 140 million tons per year.

Studies show that the refuse generated by a city is composed of roughly:

Paper	41%	Wood	5%
Food Wastes	21%	Rubber and Leather	3%
Glass	12%	Textiles	2%
Ferrous Metals	10%	Aluminum	1%
Plastics	5%	Other Metals	0.3%

However, the proportions vary greatly depending on the generator (commercial vs. residential), the neighborhood (affluent vs. poor), and the time of the year. During certain seasons, **yard wastes,** such as grass clippings, leaves, and other lawn wastes, add to the solid waste burden, often equaling all the other categories combined.

Traditionally, local governments have assumed the responsibility for collecting and disposing of municipal solid wastes. The local jurisdiction itself may own the trucks and employ workers or it may contract a private firm to provide the service. In either case, the service is paid for by local tax dollars, and the type and quality of the service ultimately depends upon what residents are willing to support and pay for.

Open Burning Dumps

Until the 1960s most municipal solid waste was disposed of in open, burning dumps. Burning was to reduce the volume and lengthen the lifespan of the dump site, but refuse does not burn well. Such smoldering dumps were the source of clouds of smoke that could be seen from miles away, of obnoxious smells, and created the breeding ground for flies, rats, and other vermin. Some cities used incinerators, but without air pollution controls these were also prime sources of air pollution. Public objection and air pollution laws forced open-burning dumps and most incinerators to be phased out during the 1960s and early 1970s.

Landfills

Landfills were generally adopted as the substitute. In a **landfill,** the raw trash is put on or in the ground and covered with earth. By not burning the refuse and covering each day's fill with a few inches of earth, air pollution and vermin are kept down (Fig. 18–2). Unfortunately, aside from those concerns and the minimizing of cost, no other factors were given real consideration. Municipal waste managers generally had no understanding or interest in ecology, the water cycle, or what products would be generated by decomposing wastes, and they had no regulations to guide them. Therefore, in general, any piece of cheap, conveniently located piece of land with an existing depression or hole became the site for a landfill. This was frequently a natural gully or ravine, an abandoned stone quarry, or a section of wetlands. With acquisition of the land, dumping commenced with no precautions taken. "A hole is to fill (period)," was the mentality. It was even planned that after filling, covering with earth, and seeding, the landfill site would become a park or playground, and there are many cases where this situation came to pass. Figure 18–3 is a photo of a golf course atop an old landfill. Thus,

FIGURE 18–2
Landfill in operation. Refuse is dumped (*left*), compacted by steel-wheeled vehicle (*left of center*) and covered with earth (*right*). All three operations proceed simultaneously so that there is never more than this amount of refuse uncovered. (Photo by author.)

FIGURE 18–3
Longview golf course, Baltimore County, Maryland. This area of the golf course is actually the completed top of a landfill. (Photo by author.)

landfilling was originally thought of as a means of upgrading "wasteland" to a higher use as well disposing of trash.

Problems of Landfills

Just with the understanding that you have gained from previous chapters, however, you should be able to predict the consequences of landfilling in this way. They include:

☐ Leachate generation and groundwater contamination
☐ Methane production
☐ Settling.

Leachate Generation and Groundwater Contamination

The most serious problem by far is groundwater contamination. Recall that as water percolates through any material, various chemicals may be dissolved and carried along, a process called *leaching,* and the water with various pollutants in it is called **leachate.** As water percolates through raw refuse, a particularly noxious leachate is generated, which consists of residues of decomposing organic matter combined with iron, mercury, lead, zinc, and other metals from rusting cans, discarded batteries and appliances, and generously spiced with discarded quantities of paints, pesticides, cleaning fluids, and other chemicals. Further, the siting of landfills and absence of precautionary measures noted above funnel this "witches brew" directly into groundwater aquifers.

All states have some municipal landfills that are or will be contaminating groundwater, but Florida is in a real crisis. Floridians located most of their landfills in wetlands, and they rely on groundwater for 92 percent of their fresh water. The result: they have over 200 municipal landfill sites on or to be added to the Superfund list. Recall that "Superfund" is the federal program to clean up sites that are in imminent danger of jeopardizing human health through groundwater contamination (see Chapter 11). It will cost between 10 and 100 million dollars to clean up each site. So much for cheap waste disposal!

Methane Production

A second problem is methane generation. Buried wastes do not have access to oxygen. Therefore, their decomposition is anaerobic and the waste resulting is biogas, which is about two-thirds methane, a highly flammable gas (see Chapter 10). Produced deep in a landfill, methane may seep horizontally through the earth, enter basements, and cause explosions as it accumulates and is ignited. Over 20 homes at distances up to 1000 feet (300 meters) from landfills have been destroyed and some deaths have occurred as a result of such explosions. Also, methane seeping to the surface kills vegetation through poisoning the roots. Without vegetation, erosion occurs, exposing the unsightly waste. A few cities resolved the problem by installing "gas wells" in older landfills. The wells trap the methane, which can be used as fuel.

Settling

Finally, as waste decomposes, it settles. This eventuality was recognized, so buildings were not put on landfills. However, it presents a problem in playgrounds, and golf courses as well, because the settling creates shallow depressions that collect and hold water, converting the playground back to a "wetland."

Improving Landfills:
Trying to Fix a Wrong Answer

Recognizing the above problems, the Environmental Protection Agency has upgraded the siting and construction requirements for new landfills. In accordance with current regulations:

☐ New landfills are sited on high ground well above the water table. Often the top of a hill is taken off to supply a source of cover dirt and create a floor that is still considerably above the water table.

☐ The floor is contoured to drain water into tiles, which will be a leachate collection system, and covered with at least 12 inches of impervious clay or a plastic liner. On top of this is a layer of coarse gravel and a layer of porous earth. This design is such that any leachate percolating through the fill will come to the impervious clay or plastic, then move through the gravel layer into the leachate collection system. Collected leachate can be treated as necessary.

☐ The gravel layer that eventually surrounds the entire fill also serves to vent methane.

☐ In the process of filling, layer upon layer of refuse is put down until the fill is built up in the shape of a pyramid. The finished landfill thus sheds water and minimizes infiltration, percolation, and the consequent formation of leachate.

☐ Finally, the entire site is surrounded by a series of groundwater monitoring wells that are checked periodically, and such checking must go on indefinitely.

These design features are summarized in Figure 18–4.

Escalating Costs of Landfilling

While new landfill design seems to address the problems observed, its costs are becoming prohibitive. Increasing costs are not just because of the new design features. Even more, they are expenses of acquiring a site and providing transportation.

No one wants a landfill in their community, and with spreading urbanization there is no area near cities not already dotted with high-priced suburban homes. Any site selection, then, is met with protests and often legal suits declaring, "Anywhere but here!" Delays and legal costs incurred in overcoming these objections, if they are overcome, are often as expensive as all other costs combined. Or, the process leads to selection of a very distant site, which involves inordinate hauling expense. There are also limits of state lines; no state wants to act as the dump site for another. The result is that in many cases costs for disposal in new landfills are going well over 100 dollars per ton.

With new landfills being held back by costs and/ or legal objections, the sad fact is that most municipal solid waste is still going into old landfills with inadequate safeguards. Of the 6000 municipal landfills in the United States which receive 75 percent (over 100 million tons per year) of municipal solid waste per year, 75 percent are unlined, 95 percent do not have leachate collection systems, and 75 percent do not monitor groundwater.

With about 1200 old landfills scheduled to close in the next five years because of reaching their capacity or environmental problems and new landfills being constructed at less than half this rate, the crisis is worsening. In the 1960s, many environmentalists thought that our throw-away society would meet limits in the form of shortages of resource materials. It is ironic to note that the actual limits are in terms of space to dump the garbage.

Coming back to common sense, you can see that even if sites for new landfills could be obtained, the system is not sustainable. The ultimate end, some have speculated, is a landscape covered with "pyramids" and most of society belonging to a "priesthood" that ritualistically samples surrounding wells, although they don't know why. Fortunately, there is a better way.

SOLUTIONS

The Recycling Solution

Recycling is the obvious solution. Of course, many have been advocating this, and various groups and individuals have been recycling paper, glass, and aluminum cans on a small scale for decades. What has prevented recycling from being implemented on a large scale? There are some real impediments to large-scale recycling that must be overcome. However, once these are recognized and understood, they can be and are being solved. Indeed, recycling is now a tremendous growth industry with prospects for a brilliant future.

Impediments to Recycling

Impediments to recycling that must be overcome are:

Sorting. We are used to the convenience of throw-

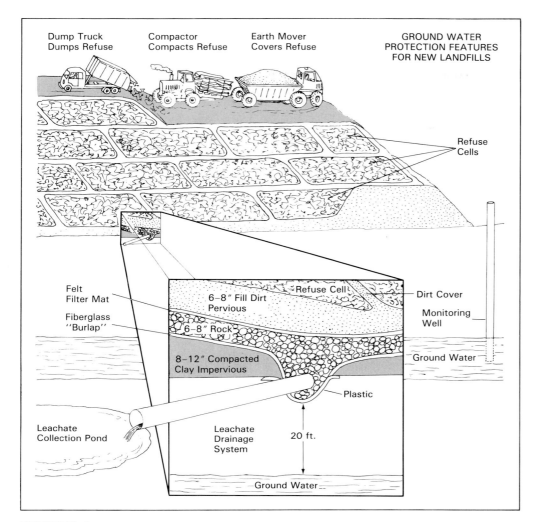

FIGURE 18–4
Features of a modern landfill with environmental safeguards.
Landfill is sited on a high location well above the water table.
Bottom is sealed with compacted clay and topped with a rock
layer as indicated to drain off leachate and vent methane.
Layer upon layer of refuse is built up so that the completed fill
has a pyramid-shape which sheds water. The fill is surrounded
with groundwater monitoring wells.

ing all our refuse into a single container and handling
it as one bulk mass. For recycling, the various constituents either must be separated in the home, or
there must be a means for separating them after collection.

Lack of Standards. Sorting is made even more difficult by lack of standards. That is, several kinds of
plastic or grades of paper may be used in similar products or even in the same product.

Reprocessing. There must be companies capable of
receiving the materials collected and converting them
into salable materials. Otherwise it is back to the landfill after sorting.

Marketing. There must be industrial or consumer
markets to buy the products made from recycled material. Otherwise the company is bankrupt and the
products become refuse before they are even sold.

**Separation between Government and Private Enter-
prise.** In general, refuse collection is done by local
government, and government is reluctant to (and
probably shouldn't) get into the business of producing and marketing materials, which is the realm of
private enterprise. Conversely, companies engaged
in production like to deal with clean, uniform raw
materials, which trash is not. Therefore, with few exceptions, they have been less than eager to mess with
refuse. Lack of cooperation between local governments and the private sector frequently impedes recycling.

Vested Interests in the Status Quo. There are tremendous profits that may be maintained indefinitely
in manufacturing and selling bottles, cans, and other

items that are used only once and then discarded. The vested interests who profit from the throw-away habit have been a potent force against implementing any form of recycling.

Hidden Costs. Since refuse disposal is financed out of tax revenues, people generally do not realize how much they are paying. Environmental costs such as groundwater contamination, cleaning up a hazardous site, or monitoring such a site forever and ever, are not tallied into the costs of disposal. With costs thus hidden, most people do not realize how much they are really paying for refuse disposal; indeed, it may seem like a free (and carefree) service. When the costs of alternatives are discussed they seem expensive by comparison, even though the long-term costs will be less.

Addressing the Problems

But problems should not be taken as an excuse for inaction; rather, they should be taken as an opportunity to develop creative solutions. Hundreds of communities across the nation are overcoming the aforementioned obstacles in one way or another and entering into recycling. A few of the basic ideas that are being tried follow.

Forming Partnerships between Government and Business. Companies that will provide full-service recycling—that is, collection through processing, including production of certain products from recycled materials—are forming and growing rapidly. Governments are forming partnerships with such companies. Basically the company is contracted to provide for collection and recycling of a certain minimum percentage of materials so that they won't end up back in landfills. In return, the government gives the company certain guarantees such as exclusive collection and marketing rights of certain recycled materials in their area. Further, the government may agree to purchase certain amounts of recycled paper, compost, and plastic. Finally the government agrees to continue to dispose of a certain quantity of material that cannot be recycled at the present time. Can you see the need for these guarantees? Without them the company could be forced out of business, to the detriment of both parties.

Sorting. Sorting may be done at the source (homes) by individuals or in facilities after collection. Sorting at the source requires the cooperation of a large portion of the population, but it is relatively inexpensive since the work is "volunteer." A system that is gaining acceptance in a number of communities is the issuance of colorcoded containers for plastic, metals, glass, paper, yard wastes, and "other" (Fig. 18–5). A

trailer with colored bins is drawn behind the regular trash truck and workers dump containers in the respective bins. Unsorted refuse continues to go into the regular trash truck for traditional disposal.

Alternatively, the pickup of trash can remain unchanged and separation can be conducted after collection. Refuse separation facilities have been built and are in operation. The general scheme for one such facility is shown in Figure 18–6. However, such equipment is very costly to purchase, and operation and maintenance costs are also high. The payback from sale of recycled materials comes nowhere close to offsetting these costs. Another scheme is to sort refuse by hand as it passes along conveyor belts.

In Third World countries many poor make their living by picking through dumps and reselling "garbage." This, however, is a sign of their desperate poverty, not a recommended solution to the garbage problem.

Reprocessing and Profits. There is an abundance of alternatives for reprocessing various components of refuse, and people are coming up with new ideas and techniques all the time. A few of the major established techniques follow:

☐ Paper can be repulped and reprocessed into recycled paper, cardboard, and other paper products; finely ground and sold as cellulose insulation; shredded and composted (see below).

☐ Glass can be crushed, remelted, and made into new containers; or crushed and used as a substitute for gravel or sand in construction materials such as concrete and asphalt.

☐ Plastic can be remelted and fabricated into "synthetic lumber" (Fig. 18–7). Such lumber, since it is not biodegradable, has tremendous potential for use in fence, sign, and guardrail posts, docks, decks, and other outdoor uses.

☐ Metals can be remelted and refabricated. Making aluminum from scrap aluminum saves up to 90 percent of the energy required to make aluminum from virgin ore.

☐ Food wastes and yard wastes (leaves, grass, and plant trimmings) can be composted to produce a humus soil conditioner.

☐ Textiles can be shredded and used to strengthen recycled paper products.

☐ Old tires can be remelted and made into a number of other products.

Additionally, literally hundreds of new processes are being developed and commercialized to make refuse components into more valuable end products. Thus recycling is becoming increasingly

FIGURE 18–5
In the Twin Cities area of Minnesota recycling much of household refuse is already a way of life. Residents separate cans, glass, plastics, paper, and "other" into separate containers. Workers dump containers into respective dumpsters aboard the truck and trailer shown. At strategic locations, dumpsters are emptied by forklifts into larger vehicles to be taken to recyclers. (Courtesy of Super Cycle Inc., St. Paul, Minnesota.)

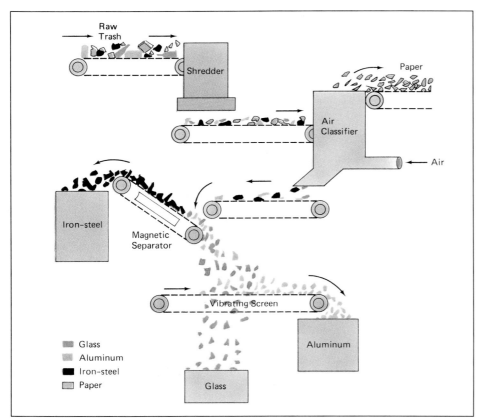

FIGURE 18–6
Schematic flow diagram for the separation of municipal solid waste. Separation can be achieved but is it cost-effective? Does the value of the separated materials justify the costs of separation?

(a)

(b)

profitable. (See Case Study 18) Consequently, the profit potential of recycling is bringing many new companies into the field despite the vested interests of some old established companies in maintaining the status quo.

Promoting Recycling through Mandate

A number of measures are being adopted by various states and/or local governments which mandate or at least support recycling. These include:

Mandatory Recycling Laws. A number of states have passed mandatory recycling laws. Such laws require that each county, under threat of loss of state funds, implement recycling of a certain percentage of their refuse by a certain date. Florida, for example, is aiming for recycling of all containers by 1992. The means which counties adopt to accomplish this is still up to them.

Banning the Disposal of Certain Items or Constituents in Landfills. Yard wastes are a good first candidate here because they take up considerable volume and they can easily be collected separately, composted into humus, and applied to park lands. To date, Florida, Illinois, Minnesota, New Jersey, Pennsylvania, and Wisconsin have banned yard wastes from landfills. Of course items that present a toxic or explosive hazard, such as car batteries, are already generally banned.

Mandating Government Purchase of Recycled Materials. A state can simply require that the purchases of all its agencies include a certain percentage of recycled paper. Requiring highway departments to use

This deck provided by the Chicago Park District as part of the 1988 Pilot Recycling Program. The deck material is manufactured from approximately *84,500 recycled plastic milk bottles* by Eaglebrook Profiles, Chicago, Illinois. Installation by Heritage Cabinet, Bedford Park, Illinois.

VALUE ADDED: THE POT AT THE END OF THE TRASH RAINBOW

Recycling with a big "R" has made it to the top of the agenda of the nation's solid waste management policy. The U.S. Environmental Protection Agency refers to recycling as the foundation of all future policies. In fact, the EPA is catching up with actual developments at the local level. There, basic economics are driving cities and counties to recycle.

In 1980, it cost about $3 per ton to put solid waste in the ground in the Northeast. Today, that cost is $65 and going up as new state and federal landfill requirements demand stricter management systems. Incineration, which had long been seen as the alternative to landfilling, is even more expensive, at over $100 per ton. Many environmental questions, particularly with ash and acid gas emissions, have restricted the application of incineration in major cities throughout the United States, including Austin, Texas; San Diego and Los Angeles, California; Philadelphia, Pennsylvania; Lowell and Holyoke, Massachusetts; Chattanooga, Tennessee; Seattle, Washington; Portland, Oregon; and Gainesville, Florida.

At the same time the cost of recycling is declining. This is so for two reasons. First, as more and more people in a community recycle, the cost of collection goes down on a per-ton basis because the same truck picks up from most households on the street instead of a few. Mandatory recycling laws have been passed at the state level (New Jersey, Rhode Island, and Connecticut), and at the local level (Islip, NY; Perkasie, PA; Woodbury, NJ) cities and towns are recycling from 20 to 50 percent of their solid waste on a regular basis.

Second, the raw materials that are available are attractive to industrial manufacturers so that the local economy is benefitting from recycling instead of using landfills or incineration.

The term "value added" explains how the community's economy benefits from recycling. If city *A* sells 1 ton of glass that its citizens recycle, it may earn from $30 to $70 per ton, depending upon how clean, or how close to "furnace ready," the glass is. But if city *A* has a small glass manufacturing plant, the glass material can be worth from $4000 to $7000 per ton, based on the value of the products produced. This is called "value added" because the increased value reflects the jobs, skills, and manufacturing tax base which enhance the local economy.

Scrap tires are a small but troublesome waste item. Every American throws away the equivalent of one 20-pound tire a year. The current disposal cost

Products made from recycled old tires. (Courtesy of Rubber Research Elastomerics, Minneapolis, Minnesota.)

is about $.10 per pound. A city the size of Newark, New Jersey, or Toledo, Ohio, or Tucson, Arizona, pays about $700,000 simply to bury scrap tires. This assumes landfills accept them. Many won't. Tires have a nasty habit of resurfacing years later. Some states, like Minnesota, ban land disposal of tires. Tires can be shredded and burned. But at today's depressed oil prices, this is economical only at about a penny a pound. Tires shredded into finer pieces can be added to road asphalt and earn a few pennies more. But the real benefit comes when the scrap is converted into a valuable final product.

Enter Fred Stark, a chemical engineer and president of the Minneapolis-based Rubber Research Elastomerics (RRE). Mr. Stark's patented liquid polymer, when added to pulverized tires, enables the material, called TireCycle, to compete both with virgin rubber and with plastics.

RRE's first plant opened in the northern Minnesota city of Babbitt in March 1987. Its primary customers are thermoplastic molders. Adding two parts TireCycle to one part polypropylene can "improve performance while retaining the easy molding characteristics of plastics," declares chief operating officer John Stark III. For this purpose, TireCycle sells for about $.50 per pound. The rubber/plastic compound is used in car door panels.

Babbitt's economy was strengthened even more when one manufacturer moved from Ohio to be near its raw material supplier. Whirlair Rubber Products uses TireCycle to make rubber mats for home and business.

Imagine that a city like Newark or Toledo or Tucson built a facility like Babbitt's and local companies processed the material into final products. The local economy would avoid most of the $700,000 a year it now pays for tire disposal and would add several million dollars in a new largely export-oriented activity to the local economy.

The key to TireCycle's success is twofold: First the company benefited from innovative financing packaged together by county and state economic development agencies. Second, the product it produces—crumb rubber—sells for $.35 per pound as compared to $.65 per pound for virgin rubber.

Thus, the enterprise helps the community rid itself of a disposal problem and contributes to the local economy.

In the plastic, metal, paper, and glass industries, firms are investing in recycled materials because they reduce energy costs, reduce pollution equipment requirement, reduce costs of raw materials and allow equipment to last longer in production. Finally, in this era of solid waste public awareness, it is good public relations to advertise use of recycled materials.

All of this means that there are "value added" opportunities for our cities.

NEIL N. SELDMAN
Director of Waste Utilization
Institute for Local Self-Reliance
Copyright, ILSR

plastic signposts and parks departments to use compost can extend recycling beyond paper.

Advance Disposal Fees. An **advance disposal fee** can be collected by adding one or two cents to the price of every item sold in a glass, metal, or plastic container. Everything from beverages to shampoo to canned dog food is included. The fee tells consumers that there is a cost for disposal. Revenues raised from this fee may be used to publicize recycling programs, provide grants for experimental recycling endeavors, give awards for recycling accomplishments, or fund any other means of promoting and supporting recycling.

Composting

One way of treating refuse that is rapidly growing in popularity is *composting*. Recall that composting involves the natural biological decomposition (rotting) of organic matter in the presence of air. The end product is a residue of humuslike material, which can be used as an organic fertilizer. Composting was one method of treating sewage sludge described in Chapter 10. Likewise, since refuse is usually 60 to 80 percent organic matter (paper and food wastes)—or more when yard wastes are added—it may be treated by composting (Fig. 18–8). A number of companies have entered into the business of selling equipment or actually building and running facilities for the composting of refuse. Glass, metals, and plastics may be removed either before or after composting occurs and recycled as desired. Also, raw sewage sludge may be mixed with the refuse to achieve a synergistic composting of both simultaneously. Paper helps to dewater the sewage sludge and provides better aeration, and the sludge supports better decomposition. There is a good market for compost from landscaping firms, and it may be used on city parks or agricultural fields.

(a)

(b)

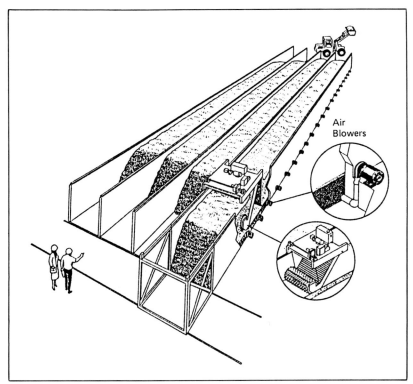

Air
Blowers

FIGURE 18–8
Composting of refuse. Keeping material moist but well aerated results in odor-free decomposition of paper, food, yard, and other organic wastes into humuslike material. (a) Windrow of refuse being turned and fluffed by machinery to aid composting. (Courtesy of Wildcat Manufacturing Co. Inc., Freeman, South Dakota.) (b) Where space is limited and odor control is critical, "in-vessel" composting may be the choice, because it allows more accurate control of moisture and aeration. The system shown enables a continuous input of refuse at one end and an output of compost at the other. (Enclosed Dynamic Composting System, courtesy of Royer Industries Inc., Kingston, Pennsylvania.)

Refuse-to-Energy Conversion

The organic content of refuse, including the plastics portion, also enables it to be burned, although it is a low-grade fuel. An alternative that falls somewhere between landfilling and ideal recycling is burning refuse and utilizing its energy output. A number of facilities are in operation and more are being built that will burn raw refuse and use the energy for producing electricity, which is always in demand. Air pollution may be controlled by the techniques noted in Chapter 12. Thus, problems of separation, producing, and marketing various products are largely avoided.

The most valuable materials in the waste—iron and aluminum—can still be recovered from the ash if desired. Other unburnable materials must be landfilled, but because these materials make up only about 10 to 20 percent of the original volume, the life of the landfill will be extended about five- to tenfold. More importantly, since the incinerated material is not subject to further decomposition and settling, it may be used as fill dirt in construction sites, road beds, and so forth. In other words, disposal of the incinerated material presents relatively few problems.

A refuse-to-energy plant that went into operation in Baltimore, Maryland, in 1984 is shown in Figure 18–9. This plant is capable of consuming 2000 tons of raw refuse per day. Steam produced in boilers drives a 60,000 kilowatt generator that produces enough power to service about 60,000 homes. Air pollution from the combustion is controlled by electrostatic precipitators.

One drawback of such facilities is that as refuse is consumed for energy, it largely precludes future options for recycling and/or composting.

Reducing Waste Volume

We noted earlier that the increased amount of waste produced over the last 30 years is largely a result of changing lifestyles, notably the growing use of disposable products. An option that receives too little attention is the potential for reducing the volume of wastes produced by keeping products in use longer. Reusing items in their existing capacity is the most efficient form of recycling. The use of returnable versus nonreturnable beverage containers is a prime case in point.

Returnable versus Nonreturnable Bottles

Prior to the 1950s, most soft drinks and beer were marketed by local bottlers and breweries in returnable bottles that required a deposit. Trucks delivered filled bottles and picked up empties to be cleaned and refilled. This procedure is efficient when the distance between the producer and the retailer is relatively short. However, as the distance increases, transportation costs become prohibitive because the consumer pays for hauling the bottles as well as for the beverage. In the 1950s, distributors, bent on expanding markets and growth, observed that transportation costs could be greatly reduced if they used lightweight containers that could be thrown out rather than shipped back. Thus, no-deposit, nonreturnable bottles and cans were introduced. The throwaway container is also an obvious winner for manufacturers who profit by each bottle or can they produce.

Through massive advertising campaigns promoting national brands and the convenience of throwaways, a handful of national distributors gained dominance and countless local breweries and bottlers were driven out of business during the 1950s and 1960s. At the same time, bottle and can manufacturing grew into a multibillion-dollar industry.

The average person drinks about a liter of liquid each day. For 245 million Americans, this daily consumption amounts to some 1.3 million barrels of liquid. That a significant portion of this fluid should be packaged in single-serving containers that are used once and then thrown away is bizarre. It is difficult to imagine a more costly, wasteful way to distribute fluids.

Indeed, beverages in nonreturnable containers appear to be priced competitively on the market shelf. But these containers constitute some 6 percent of all solid waste, and about 50 percent of the nonburnable portion; they constitute about 90 percent of the nonbiodegradable portion of roadside litter. The broken bottles are responsible for innumerable cuts and other injuries, not to mention flat tires. These containers are also environmentally undesirable because both the mining of the materials they are composed of and the manufacturing process create pollution. All of these are hidden costs that do not appear on the price tag. Consumers not only pay with taxes for litter cleanup, but also suffer the cost of treating injuries, flat tires, environmental degradation, and so on.

In an attempt to reverse the trend, environmental and consumer groups have promoted "bottle bills"—laws that encourage the use of returnable rather than throwaway beverage containers. Such bills generally call for a deposit on all beverage containers—returnable and throwaway. Since the deposit is wasted on nonreturnable containers, consumers gradually change their habits and buy returnables.

Bottle bills have been proposed in virtually every state legislature over the last decade. In every case, however, the proposals have met with fierce opposition from the beverage and container industries and certain other special interest groups. The

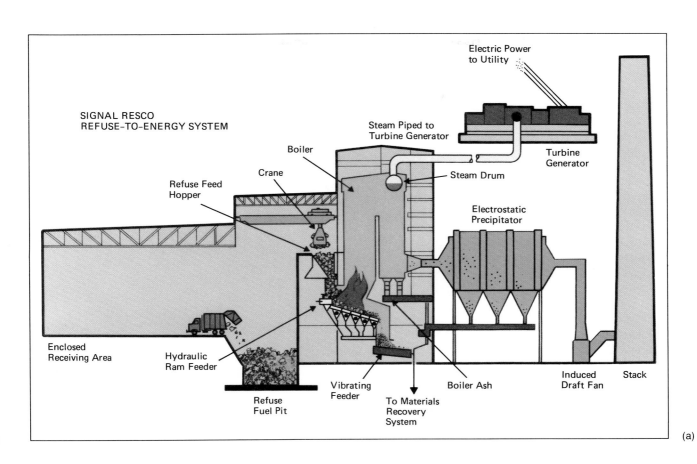

SIGNAL RESCO
REFUSE-TO-ENERGY SYSTEM

Electric Power to Utility

Steam Piped to Turbine Generator

Boiler

Crane

Refuse Feed Hopper

Steam Drum

Turbine Generator

Electrostatic Precipitator

Enclosed Receiving Area

Hydraulic Ram Feeder

Refuse Fuel Pit

Vibrating Feeder

To Materials Recovery System

Boiler Ash

Induced Draft Fan

Stack

(a)

(b)

FIGURE 18–9
Conversion of municipal solid waste to energy. This refuse-to-energy plant in Baltimore, Maryland, is capable of consuming 2000 tons of raw (unseparated) refuse per day to generate 60 megawatts (60 million watts) of on-line power, enough to supply some 60,000 homes. Air pollutants are removed by electrostatic precipitators. (a) Schematic diagram of plant. (b) The outside of the plant. Note the cleanliness and lack of smoke from the stack. This shows what modern "dumps" can be. (Courtesy of Signal Environmental Systems Inc.)

reason for their opposition is obvious—economic loss. But the arguments they put forth are more subtle. The industry contends that bottle bills will result in loss of jobs and higher beverage costs for the consumer. They also claim that consumers will not return the bottles and litter will not decline.

In most cases, the industry's well-financed lobbying efforts have successfully defeated bottle bills. However, some states—nine as of 1988—have adopted bottle bills despite industry opposition (Table 18–1). Their experience has proven the beverage and bottle industry's arguments false. More jobs are gained than lost, costs have not risen, a high percentage of bottles are returned, and there is a marked reduction in can and bottle litter. In some cases, local breweries and bottlers are making a comeback, thus improving the local economy.

A final measure of the success of bottle bills is the continued public approval of the program after it is initiated. Despite industry efforts to repeal bottle bills, no state that has one has done so. With some additional rise in public concern, a national bottle bill could be enacted in the near future.

Other Measures

Whenever items are reused rather than thrown away, the effect is a reduction in waste and better conservation of resources. In this respect, it is encouraging to see the growing popularity of yard sales, flea markets, and other "not new" markets. Other measures to reduce the amount of material going into the trash include reducing the amount of material in products, downsizing, and increasing durability.

Consideration of our domestic wastes and their disposal emphasizes what a mammoth stream of materials of all kinds flows in one direction from our resource base to disposal sites. Just as natural ecosystems depend upon recycling nutrients, the continuance of a technological society will also ultimately depend upon our learning to recycle or reuse not only nutrients but virtually all other kinds of materials as well.

Integrated Waste Management

Importantly, it is not necessary to fasten on a single method of waste handling. Almost any combination of recycling, composting, and reducing waste volume may be used. Further, recycling can be introduced gradually, pursuing a number of different options while phasing out landfilling. This system of having several alternatives in operation at the same time is called integrated waste management. Balancing the interests of all parties involved obviously requires skilled managers.

CHANGING PUBLIC ATTITUDES AND LIFESTYLE—WHAT YOU CAN DO

☐ Research how municipal solid waste is presently handled in your community and what are the future plans for its disposal. You can probably get this information with a call to your local department of sanitation or public works. If recycling is not at least in the planning stage, join or start a citizen group to promote it. Information and help are available from the Institute for Local Self-Reliance (see Appendix A). Also, the trade journal *BioCycle: Journal of Waste Recycling*, Box 351, Emmaus, PA 18049, is an invaluable source of information.

☐ Recognize the hidden costs of landfilling and the less-expensive long-term potential of recycling. Write or call your state delegates in support of the kinds of legislation suggested above, including a bottle bill.

☐ Start (if you have not already) taking your paper, bottles, and cans to recycling centers in your area.

☐ Buy and use durable products and minimize your use of disposables.

☐ Buy drinks in returnable bottles and return the empties.

☐ Be conscientious about not littering.

☐ Help spread the word.

☐ Consider a career for yourself in the rapidly growing area of waste recycling.

Table 18–1
States That Have Bottle Bills

STATE	DATE PASSED
Oregon	1972
Vermont	1973
Maine	1976
Michigan	1976
Connecticut	1978
Iowa	1978
Massachusetts	1978
Delaware	1982
New York	1983

19

Energy Resources and the Energy Problem

CONCEPT FRAMEWORK

■ Outline

I. ENERGY SOURCES AND USES IN THE
   UNITED STATES   444

   A. Background   444

      Electrical Power Production   446

   B. The Current Situation   449

II. THE ENERGY DILEMMA (OR CRISIS):
    DECLINING RESERVES OF CRUDE OIL   451
    A. Formation of Fossil Fuels   451

    B. Exploration, Reserves, and Production   452

    C. Declining U.S. Reserves and Increasing
       Importation of Crude Oil   453

    D. The Crisis   453

    E. Response to the Oil Crisis of the 1970s   454

    F. Victims of Success: Setting the Stage for
       Another Crisis   457

CASE STUDY: THE VALDEZ OIL SPILL   459

    G. Preparing for Future Oil Shortages   458

■ Study Questions

1. Describe how industrial development is correlated with energy consumption. Give examples of how sustaining development will depend on sustainable energy resources.

2. Describe how major sources of energy for the United States have changed from the year 1800 to the present. Why is electricity called a secondary energy source? What sources of energy are used to generate it? How has total consumption changed?

3. Name the major primary sources of energy now used in the United States. What percentage of the total does each supply? What is the main end use of each?

4. Name the *three fossil fuels* and describe their origin.

5. Define and distinguish between: *estimated reserves, proven reserves, production.* State the relationship between reserves and maximum rate of production; between declining reserves and cost of production. Explain why new reserves must be found to maintain production at a given level.

6. When and why did U.S. oil production start declining? What was happening to consumption at the same time? How was the "gap" filled?

7. Describe the nature of the "oil crisis" of the 1970s. What is OPEC and how did it control the price through the 1970s and early 1980s?

8. List the responses to the oil crisis of the 1970s. Describe how each is motivated and/or supported by higher oil prices. Describe how the shortage turned into a glut. What was the effect on oil prices?

9. Give the effects of lower oil prices on each of the responses to the oil crisis described above. Describe the trends for production, consumption, and imported oil for the latter 1980s. Are additional large oil reserves likely to be found in the United States? Give the argument as to why not? Are we likely to see OPEC controlling oil prices again in the near future? With what effects?

10. Should we be making preparations for shortages while oil supplies are abundant. Why? What preparations are in order?

1. Conservation 458

2. Development of Alternative Energy
   Sources 462
H. Promoting a Sound Energy Policy:
   What You Can Do 462

11. Define conservation. Contrast conservation with the discovery of a new oil field. List and describe areas where there are opportunities for conservation. Define cogeneration.

12. What can you do to help yourself and the nation prepare for future oil shortages?

Industrial development can be seen as a trend of increasing use of machines to perform labor. Agriculture provides an illustration. Using only hand labor to prepare soil, sow seed, pull weeds, and harvest provides a maximum yield of little more than enough for self-sufficiency. Therefore, before the mechanization of agriculture, more than 90 percent of the population was agrarian. A smaller percentage could not have produced enough for the others. With advancing mechanization, however, decreasing numbers of farmers could produce increasing quantities of foodstuffs. Now, only about 2 percent of the United States' population is involved in crop production. They not only feed the other 98 percent; they produce huge surpluses besides (Fig. 19–1).

However, machinery requires fuel. Therefore, as mechanization increased, fuel (energy) consumption per unit of food produced increased also (Fig. 19–1). From essentially no added energy, outside of photosynthesis and human labor, we have reached the point where about one barrel of oil is consumed (for tractor fuel, production of fertilizer and pesticides) for each ton of corn produced. Essentially the same can be said for other industries. In short, development has involved a substitution of nonhuman energy sources for human labor. This is seen also

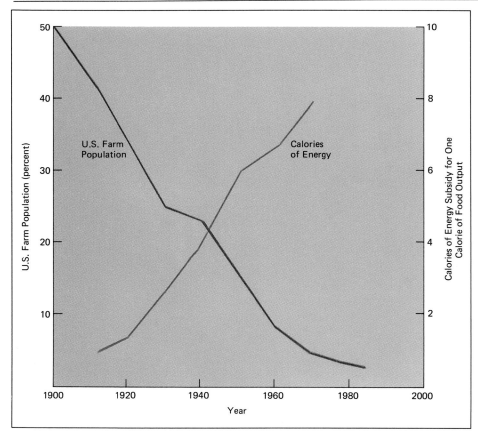

FIGURE 19–1
Industrialization has been basically a process of substituting machine labor for human labor. This is illustrated by the fact that a decline in farm population (largely farm labor) has been accompanied by an increase in energy consumed per calorie produced. (Population data from *Statistical Abstracts of the United States.* U.S. Dept. of Commerce, 1985. Calorie data from J. S. and C. E. Steinhart, "Energy Use in the U.S. Food System," *Science,* 184 [1974], p. 312.)

when level of development (as measured by gross national product per capita) is compared to energy consumption for various countries (Fig. 19–2). Although the correlation is not perfect, it is clear that the more-developed countries have a much higher level of energy consumption than less-developed nations.

It is true that much energy is used wastefully, and we can increase efficiency of our energy use; this is what is entailed in conservation. However, this can only go so far before we meet the limits of the laws of thermodynamics, which demand that more work requires more energy.

It should be clear, then, that *sustainable development depends on sustainable energy resources.* The fact that should cause considerable concern is that *the energy resources on which we currently depend are not sustainable.* The purpose of this chapter is to provide a clear picture of the nonsustainability of current energy resources, despite appearances to the contrary and to point out alternative courses of action which are discussed in Chapters 20 and 21.

ENERGY SOURCES AND USES IN THE UNITED STATES

Background

Throughout most of human history, the major energy source was human labor, doing things by hand. Some people lived in relative luxury by exploiting the labor of others—slaves, indentured servants, and minimally paid workers. Human labor was supplemented to some extent with domestic animals, water power (water wheels), and wind power (windmills). However, animals are not efficient in providing sustained power over long periods, and they, themselves, require a great deal of labor to feed, maintain, and clean up after. Water wheels are limited to certain locations and by the capacity of flowing streams and rivers; windmills are similarly limited as an energy source.

By the early 1700s designs for many kinds of machinery had already been invented. The limiting factor was a power source to run them. The break-

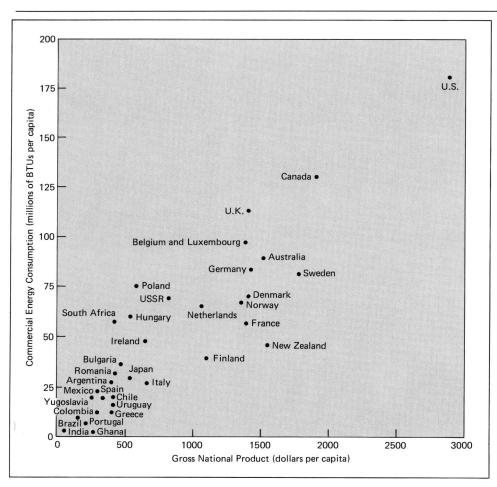

FIGURE 19–2
Correlation between energy consumption and gross national product. Gross national product per capita is one measure of standard of living. (Adapted from Earl Cook, "The Flow of Energy in an Industrial Society," *Scientific American,* Sept. 1971, pp. 134–144. Copyright © 1971 by Scientific American, Inc. All rights reserved.)

through that really launched the Industrial Revolution was the development of the steam engine in the late 1700s (Fig. 19–3). The steam engine rapidly became the power source for steamships, steam shovels, steam tractors, steam locomotives, and stationary engines to run sawmills, textile mills, and virtually all other industrial plants. The first major fuel for steam engines was firewood. Then, coal was substituted as demands for energy increased and forests were cut bare for fuel. By the end of the 1800s, coal became the dominant fuel. By 1920, coal provided 80 percent of all energy used in the United States; it was widely used for heating, cooking, and industrial processes as well as fuel for steam engines.

Coal, however, is hazardous to mine, messy to handle and transport, and polluting to burn. Air pollution in cities due to the innumerable coal fires was far worse than anything seen today. It also requires the disposal of quantities of ash, which is also messy and polluting. In addition, steam engines, which require starting the fire and heating the boiler before going into operation, are cumbersome to operate (Fig. 19–4).

In the late 1800s, simultaneous development of oil-well drilling technology, the ability to refine crude oil into fuels (gasoline, diesel fuel, and fuel oil), and the internal combustion engine provided an alternative. There is hardly a comparison between stoking up a coal fire and waiting for the boiler to heat, and pushing a button to start a gasoline engine. Compared to coal, gasoline and other oil-based fuels are relatively clean to handle and transport, they burn relatively cleanly, and there are no ashes to contend with. This is not to say that burning liquid fuels does not create pollution. Indeed, serious pollution problems from auto exhaust are now well recognized (see Chapter 12). Still, much less pollution is generated by gasoline than by coal burning.

Further, the gasoline internal-combustion engine provides a valuable power-to-weight advantage. A 100-horsepower gasoline engine weighs but a tiny fraction of a 100-horsepower steam engine and its boiler. In addition, a pound of gasoline has a higher energy content than a pound of coal, providing a further power-to-weight advantage. Automobiles and other forms of transportation would be cumbersome,

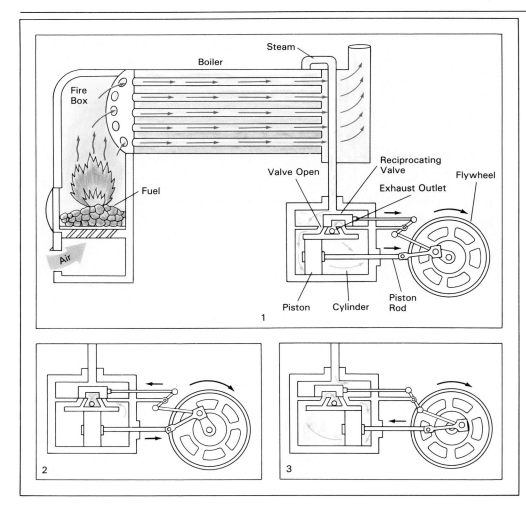

FIGURE 19–3
Steam engine. Steam pressure generated in boiler drives piston. Reciprocating valve shifts steam pressure and exhaust from one side of piston to the other. Steam engines permitting the physical conversions of fuels into useful work fostered the Industrial Revolution.

to say the least, without this power-to-weight advantage, and airplanes would be impossible. Given these advantages, it is not surprising that oil-based fuels running internal combustion engines became the dominant energy source by 1950. Oil was also substituted for coal, in large part, for heating of homes and other buildings.

Development of electrical power also came into its own during the first half of the 1900s. However, as described in the Box on pages 446–448, electricity is a **secondary energy** source. It requires a *primary* energy source to turn the generators. While coal was phased out of most of the above uses in favor of oil, it has remained and grown in use as the major fuel for generating electrical power. It is also still used and required in certain industrial processes such as smelting iron ore. In the 1960s nuclear power entered the picture as a means to generate electrical power.

ELECTRICAL POWER PRODUCTION

Electricity is produced by the phenomenon first described by Michael Faraday in 1831, namely, that when a coil of wire is passed through a magnetic field it induces a flow of electrons in the wire. The flow of electrons is synonymous with an electric current. An electric generator is basically a coil of wire that can be rotated within a magnetic field, or the magnetic field can be rotated within the coil of wire (Fig. 19–5). However, one does not get something for nothing. As an electric current begins to flow through the wire, it creates a magnetic field that is opposite to the existing field, and this resists the movement of the wire through the field. Therefore, an amount of energy proportional to the output must be expended in turning the generator. Since no energy conversion is 100 percent efficient, *more* energy must be put into turning the generator than one gets out in electricity.

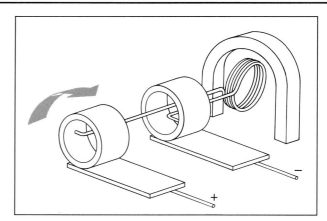

FIGURE 19–5
Principle of an electric generator. Rotating a coil of wire in a magnetic field induces a flow of electricity in the wire. But more energy must go into turning the generator than is gotten out in electricity.

Recognizing this principle, the main problem in producing electricity is finding a power source to turn the generator. The most common technique is to boil water to create high-pressure steam. The steam is passed over a turbine (a sophisticated paddle wheel) coupled to the generator (Fig. 19–6a). The combined turbine and generator are called a **turbogenerator.** Any heat source can be used to boil the water; coal, oil, and nuclear energy are most commonly used at present. Wood, wastepaper and garbage, solar energy, and geothermal energy (heat from the earth's interior) may be used more in the future.

In addition to steam turbines, gas and water (hydro-) turbines are also used. In a gas turbogenerator, the high-pressure gases produced by the combustion of a fuel, usually natural gas, drive the turbine directly (Fig. 19–6b). In a hydroturbogenerator, the high pressure of water from behind a dam or at the top of a waterfall is used to drive the turbine (Fig. 19–6c). The proportion of our electricity generated by different sources is indicated in each case.

The major drawback of electrical power is that despite its apparent cleanness in its final use, it has distinct environmental impacts related to the fuels or other energy source used to produce it. These impacts may be divided into three categories: (1) land destruction or alteration caused by mining the fuel—or building a dam and reservoir in the case of

FIGURE 19–6
Electricity is produced commercially by driving generators with (a) steam turbines, (b) gas turbines, and (c) water turbines. Percentage of electricity (United States) derived from each source is indicated. (d) Component of steam turbine being assembled. (Photo courtesy of Consulate General of Japan, N.Y.)

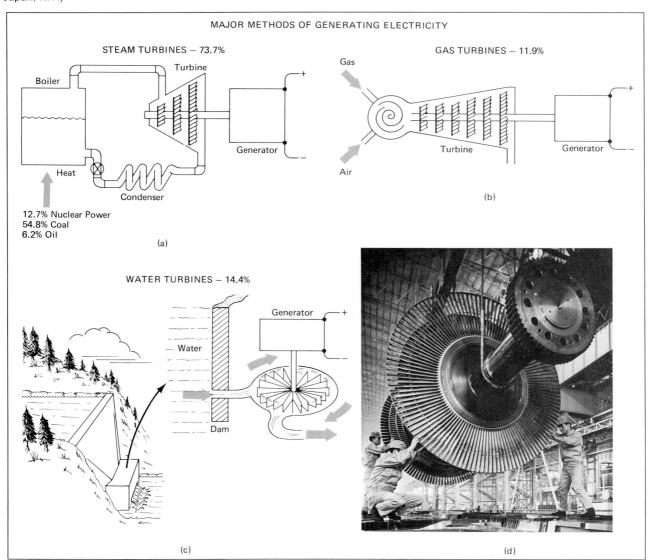

MAJOR METHODS OF GENERATING ELECTRICITY

STEAM TURBINES — 73.7%

12.7% Nuclear Power
54.8% Coal
6.2% Oil

(a)

GAS TURBINES — 11.9%

(b)

WATER TURBINES — 14.4%

(c)

(d)

hydroelectric power; (2) pollution or hazardous waste products resulting from consuming the fuel; and (3) thermal pollution from waste heat.

These problems take on even greater significance when the relatively low efficiency of electrical power production is considered. In order to drive a turbine, steam must pass from high pressure at the boiler side of the turbine to low pressure at the opposite side. In order to create low pressure, the steam leaving the turbine is passed through a condenser, which removes the heat and condenses it back to water. The heat removed is 60 to 70 percent of the energy that was used to produce the steam. Only 30 to 40 percent of the energy is actually converted into electricity; in terms of efficiency, this is only 30 to 40 percent efficient (Fig. 19–7a). The waste heat removed from condensers may be discharged into the atmosphere through cooling towers (see Fig. 19–7b), or it may be discharged into a body of water. Discharge of the waste heat into a natural body of water results in thermal pollution, which may have a severe impact on ecosystems.

(a) (b)

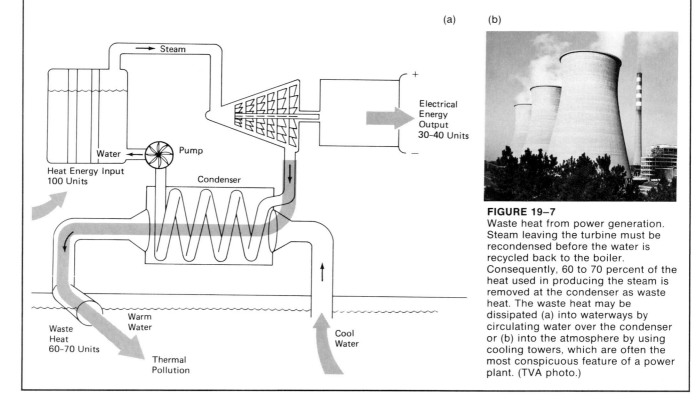

FIGURE 19–7
Waste heat from power generation. Steam leaving the turbine must be recondensed before the water is recycled back to the boiler. Consequently, 60 to 70 percent of the heat used in producing the steam is removed at the condenser as waste heat. The waste heat may be dissipated (a) into waterways by circulating water over the condenser or (b) into the atmosphere by using cooling towers, which are often the most conspicuous feature of a power plant. (TVA photo.)

Natural gas (methane), which is found in association with oil or in the process of looking for oil, burns even more cleanly than oil and is not subject to spillage. Thus, in terms of pollution, it is a most desirable fuel. At first, however, there was no practical means of transporting natural gas from wells to consumers. Natural gas was, and in many parts of the world still is, **flared,** that is, vented and burned in the atmosphere, a tremendous waste of valuable fuel. Gradually, however, the United States constructed a network of pipelines to carry the gas from wells to consumers. With the completion of these networks after World War II, the use of natural gas for heating, cooking, and industrial processes escalated rapidly because of its cleanliness, its convenience (no

storage bins or tanks are required on the premises), and, perhaps most importantly, its relatively low cost.

Along with the changing picture of fuels used, there has been a trend of increasing total amount consumed because of both increasing population and changing lifestyles. Particularly from the end of World War II to the mid-1970s there was a drastic increase in oil consumption (for gasoline) as much of the population shifted to suburban living and commuting by private cars. These trends of changing fuels and increasing total amount of fuel used are shown in Figure 19–8.

In the 1970s shortfalls between U.S. oil production and consumption occurred and these "energy crises" spawned efforts toward conservation and

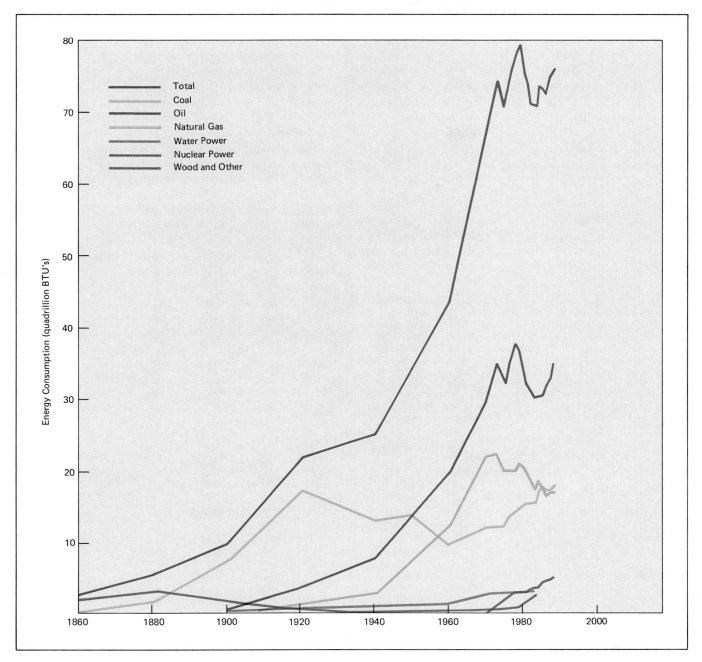

FIGURE 19–8
Energy consumption in the United States, 1860–1988; total consumption and major primary sources of energy. Note how the mix of primary sources has changed over the years while the total amount of energy consumed has continued to grow. Note skyrocketing increase in use of oil following World War II (1945) as society shifted to commuting by car. Conspicuous changes in the trends occurred in the 1970s with the recognition of limited oil reserves. (Data from the U.S. Department of Energy.)

finding alternative sources of energy. This dampened, for a time, the growth in energy consumption. Yet, the shortfall between production and consumption is considerably worse today than it was in the early 1970s; we are still using much more than we are producing.

The Current Situation

We remain dependent on oil-based fuels for 44 percent of our total energy consumption; natural gas provides about 21 percent, and coal provides another 22 percent. Nuclear power, water power, and various other sources provide about 13 percent. This energy is used for four major purposes as listed below and shown in Figure 19–9.

☐ *Transportation.* This includes trucks, buses, airplanes, trains, and boats, as well as cars, farm tractors, bulldozers, and similar machinery.

☐ *Industrial processes.* Production of all metals, synthetic chemicals, and other materials and fabrication of finished products.

☐ *Temperature control.* Space heating and cooling, and hot water heating in both homes and commercial buildings.

☐ *Generation of electrical power.* Electrical power is used for running electric motors that drive all kinds of appliances and machinery, provide lighting, and power electronic equipment in homes and commercial and industrial establishments.

It is very significant to note, as shown in Figure 19–9, that there is a high degree of matching between particular energy sources and specific end uses. In particular, transportation is essentially 100 percent dependent on liquid fuels refined from crude oil. Other energy sources are still impractical for this function. Similarly, it is impractical to use nuclear power, coal, and water power for anything but generating electricity. The significance of this matching is seen in addressing the question: Will developing more nuclear power alleviate a severe oil shortage? The answer is basically no, because nuclear power can only generate electricity. It cannot power transportation, which is the largest consumer of oil. Nuclear power can alleviate an oil shortage only insofar as electricity might substitute for oil in home or industrial heating or at such a time when electric vehicles become practical. The same applies to other alternative energy sources. A new source of energy is meaningless if

FIGURE 19–9

Energy flow from primary sources to end uses. This is a highly simplified diagram showing only major patterns, but it should emphasize that, given current technology, primary sources are connected to end uses in specific ways. Transportation is almost totally dependent on liquid fuels refined from crude oil, while coal and nuclear energy are useful only in producing electricity, almost none of which is used in transportation. Also, a large portion of energy is wasted at each conversion step. (Data from *Statistical Abstracts of the United States*, U.S. Department of Commerce, 1988.)

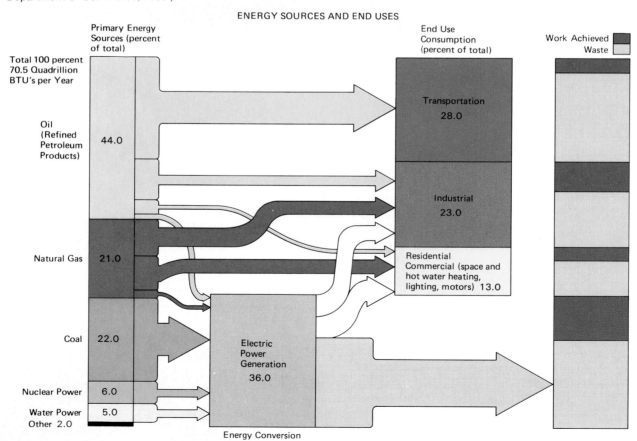

ENERGY SOURCES AND END USES

there is not a means of applying it to the end use that is energy short.

Herein lies the real dilemma for U.S. energy policy. We have created a lifestyle of commuting and transporting goods long distances which is dependent on oil-guzzling vehicles, while U.S. oil reserves are far short of meeting, much less sustaining, the demand. In the following section, we will describe this dilemma in more detail.

THE ENERGY DILEMMA (OR CRISIS): DECLINING RESERVES OF CRUDE OIL

To understand the limits of oil reserves, we must understand some basic concepts regarding their nature and exploitation.

Formation of Fossil Fuels

Early in the earth's biological history, photosynthesis outpaced the activity of consumers and decomposers. Consequently, large amounts of organic matter accumulated, especially on the bottoms of shallow seas and swamps. Gradually, this material was buried under sediments eroding from the land, and, over millions of years, it was converted to the materials we now recognize as coal, crude oil, and natural gas, the particular material being dependent on the specific conditions and time involved (Fig. 19–10). Since the origin of coal, crude oil, and natural gas is biological production, they are frequently called **fossil fuels.**

While fossil fuels are formed by ongoing, natural processes, two factors preclude any notion of their being replenished as we use them. First, biological conditions on earth have changed so that sig-

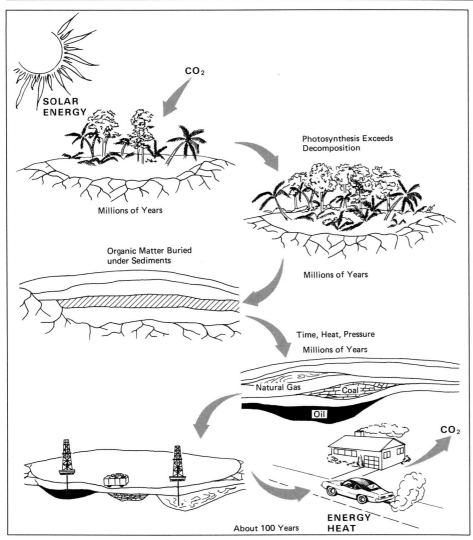

FIGURE 19–10
Energy flow through fossil fuels. Coal, oil, and natural gas are derived from photosynthesis of early geological times. Deposits are limited, and as they are used, the energy is gone forever.

nificant accumulations of organic matter no longer occur. Second, we are using them far faster than they ever formed. It is estimated that about 1000 years of natural production was required to produce the amount of crude oil that we now use each day. Because supplies are finite, there is no question that sooner or later we will run out of these fuels. The pertinent question is: How long will reserves last?

Exploration, Reserves, and Production

The science of geology provides some knowledge about the location and extent of ancient shallow seas. Based on this knowledge and past experience, geologists make educated guesses as to where oil and/or natural gas may be located and the amounts that may be found. These are the world's **estimated reserves.**

The next step is exploratory drilling. If oil is found, further exploratory drilling is conducted to determine the extent and depth of oil deposits in the area. From this exploratory drilling, a fairly accurate estimate can be made regarding how much oil can be recovered from the **oil field** at current prices. This oil now becomes **proven reserves** as opposed to estimated reserves. The final step, **production,** is the withdrawing of oil or gas from the field. Of course, "production" as used here is a euphemism. "It is production," in the words of ecologist Barry Commoner, "only in the sense that a boy robbing his mother's pantry can be said to be producing the jam supply."

But, production from a field cannot proceed at any rate desired since oil is contained in sedimentary rock like water in a sponge. Initially the field may be under so much pressure that the first penetration produces a gusher. But gushers are short-lived and the oil then seeps slowly from the rock into a well where it is withdrawn. A general rule is that *maximum production per year is limited to about 10 percent of the remaining reserves.* Consider this in terms of the maximum production from a field of 100 million barrels of recoverable reserves. The first year's maximum production will be 10 million barrels (10 percent of 100) leaving 90 million barrels. The second year's maximum production will be 9 million barrels (10 percent of 90) leaving 81. The third year's production will be 8.1 million barrels, and so on (Fig. 19–11).

Further, only about 30 percent of the oil present seeps from the rock into a well by itself. Additional oil recovery may be achieved by using secondary techniques such as injecting steam or other material into adjacent wells to force the oil into the producing

FIGURE 19–11
When will we run out of oil? The maximum yearly production from a given oil field is 10 percent of the remaining reserves. Therefore, although the maximum production from given reserves decreases, it may stretch out indefinitely as shown. However, the economic cutoff of production from a field occurs when the value of the oil obtained no longer justifies the costs of producing it. How is this cutoff point affected by the price of oil? Finally a net yield cutoff may occur when energy expended in obtaining oil outweighs the energy value of the oil obtained. The only way that production can be maintained or increased is to discover and develop new reserves at a rate equal to or exceeding consumption.

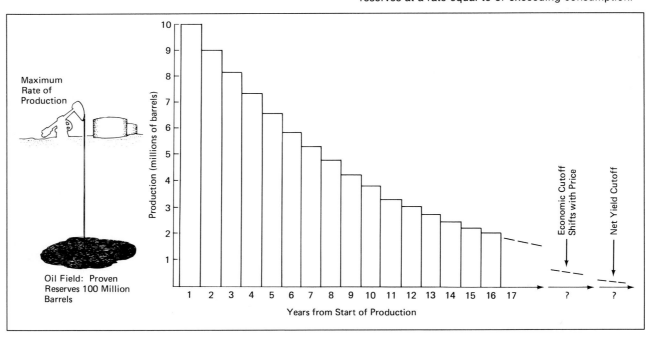

well. However, these techniques obviously involve more cost, and the 10 percent rule still applies.

Consequently, production from a given field both gradually declines and becomes more costly, but it may be stretched out indefinitely over time. The real cutoff of production from a field, its "running out," is an economic cutoff. When the cost of production reaches the value of the oil obtained, the well is shut down as uneconomical. However, note that if price of oil increases, such wells might become economical to operate again.

Of course, proven reserves are not only decreased by production; they may be increased by new discoveries. It follows from the 10 percent rule that proven reserves must be at least 10 times greater than annual production. Then, to hold oil production at a constant level, new discoveries (reserve additions) must, on the average, at least equal production (reserve subtractions).

Declining U.S. Reserves and Increasing Importation of Crude Oil

To maintain production, then, the important issue is whether new discoveries keep pace with production. If they do not, proven reserves will be drawn down and production will be diminished accordingly, although it may stretch out indefinitely depending on price. Having reserves ten times larger than annual production does *not* mean that we have a ten-year supply and then we will abruptly run out.

Here, then, is the energy problem in the United States. In 1970, U.S. oil production turned downward because new discoveries were no longer keeping pace with production. The last major discovery in the United States was made in 1968 in Alaska. Yet, the pace of oil consumption continued to increase, and the amounts involved are staggering. In 1970, the United States' consumption of crude oil was over 5 billion barrels per year (13.7 million barrels per day) and growing rapidly (Fig. 19–12). (One barrel = 42 gallons.)

To fill the "energy gap" between increasing consumption and falling production in the early 1970s, the United States turned to and became increasingly dependent upon foreign oil, primarily from the Arab countries of the Middle East (Fig. 19–12). European countries and Japan did likewise. Since imported oil cost only about $2.30 per barrel in the early 1970s and Middle East reserves were more than adequate to meet the demand, this course seemed to present few problems.

The Crisis

By the early 1970s the United States and most other Western nations were becoming increasingly dependent on oil imports from a small group of countries known as **OPEC,** the Organization of Petroleum Exporting Countries (Fig. 19–13). In 1973, OPEC, recognizing the dependence of industrialized nations on their oil, decided to take advantage of their situation. They temporarily suspended oil exports. This threw the world into a crisis. People lined up for blocks and waited for hours to get gasoline at the few stations that had it to sell. People were afraid to go anywhere, fearing that they would not have gas for emergencies. Motels, restaurants, and shopping malls became deserted.

Actually the panic was much greater than the actual shortage merited. To understand why this overreaction occurred, we must recognize that, compared with the amount of oil used (close to 20 billion barrels per year worldwide), relatively little is maintained in storage. We depend on a fairly continuous flow of oil from wells to the places of consumption. Consequently, if production is cut by just a few percent, supplies in storage are rapidly depleted, spot shortages occur, and, because our lifestyle is so dependent on using fuel for cars, the situation is perceived as a **crisis** and panic occurs.

We will see that the reverse situation also occurs. If production rises just a few percent and/or consumption drops, available storage tanks soon fill to capacity. There is literally no place to put the excess oil, and the situation is referred to as an oil glut (Fig. 19–14).

Anyhow, by curtailing supplies and causing the "oil crisis" of 1973, OPEC was able to raise the price from the previous $2.30 a barrel to about $10.50 a barrel. Further, by continuing to limit oil production through the 1970s, OPEC was able to keep supplies tight enough that they were able to continue to adjust the price upward. In the early 1980s a barrel of oil cost $30 to $35 and the United States was paying about $60 billion per year for imported oil (Fig. 19–15).

From a *world* view, then, running out of oil was not a real crisis in the 1970s. However, it was and it remains a real economic and political dilemma. U.S. reserves are far short of supporting our levels of consumption. As long as this is the case, we remain, in a sense, economically and politically hostage to OPEC. Our options, without risking a world war, are limited. Ultimately, the world will reach the limits of Middle East oil as well. If we and the rest of the world have not made suitable accommodations by the time

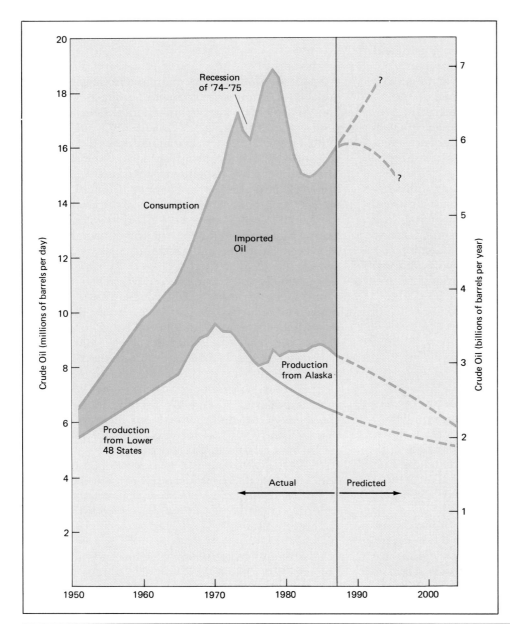

FIGURE 19–12
Oil production and consumption in the United States. Four stages can be seen: (1) Up to 1970, discovery of new reserves allowed production to parallel increasing consumption; (2) early in the 1970s, lack of new discoveries caused production to turn down while consumption continued to climb, causing a vast increase in oil imports and bringing on the oil crisis of the 1970s; (3) late in the 1970s/early in the 1980s, conservation decreased consumption, and higher prices simulated production, decreasing dependence on foreign oil; (4) late in the 1980s, resumption of trends of the early 1970s. Is it likely to bring on an "oil crisis of the 1990s?" (Data from *Statistical Abstracts of the United States, 1988*, U.S. Dept. to Commerce.)

that occurs, the "crisis" as seen above will be real and lasting.

Response to the Oil Crisis of the 1970s

Spurred by recognition of the need to become energy independent, and even more by the escalating price of fuel, the United States and other oil-dependent industrialized nations did take significant steps toward correcting the situation. Major steps were:

☐ Increasing exploratory drilling.
☐ Reopening production from old fields that had

been closed down because they were uneconomical when oil was getting only $2.30 a barrel.

☐ Conservation—most notably, the government-set standards for automobile fuel efficiency. In 1973 cars averaged 13 miles per gallon (mpg). The government mandated stepped increases so that new cars would average 27.5 mpg by 1985. Manufacturers achieved this goal mainly through reducing weight by downsizing, by going to front drive, and by improving aerodynamics. Also, notation of mpg was required in all car advertisements.

☐ Developing alternative energy sources. The gov-

FIGURE 19–13
The nations in OPEC, the Organization of Petroleum Exporting Countries.

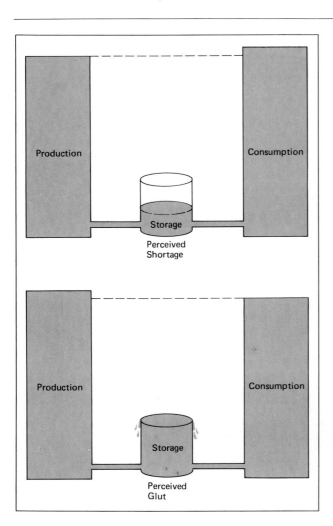

FIGURE 19–14
Shortages and gluts. The amount of oil in storage is relatively small compared to annual consumption. Therefore, production must be closely balanced with consumption. Any imbalance can rapidly give the impression of a shortage or glut.

ernment supported research and development efforts and gave tax breaks for the installation of alternative energy systems, thus providing additional economic incentive.

☐ A strategic oil reserve was created. We have "stockpiled" about 515 million barrels of oil (equivalent to 60 days of imports) in underground caverns in Louisiana (Fig. 19–16).

Again, note that all these steps, with the exception of the stockpile, only have appeal when the price of oil is high. If the price is low, the economic returns or savings do not justify the necessary investments and, consequently, they are not made.

Results were not immediate, nor can quick returns be expected. For example, it takes at least three years between design and start-up of production for a new car model. Then, it is another six to eight years before enough new models are sold and old ones retired to affect the overall "fleet." Therefore, through

455

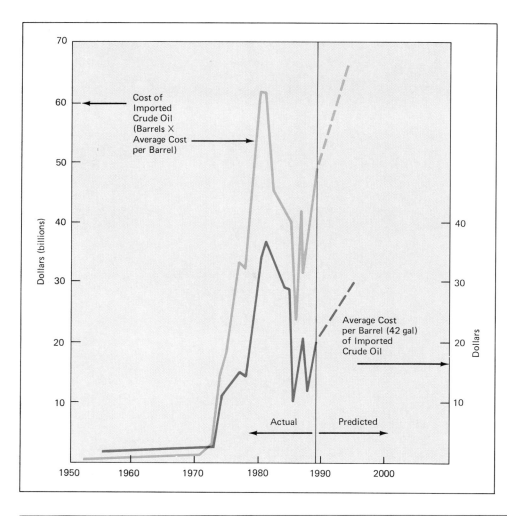

FIGURE 19–15
Cost of oil dependency. As the U.S. increased its dependency on foreign oil in the '70s, OPEC was able to increase prices, resulting in skyrocketing costs to the U.S. Conversely, decreasing oil dependency, enabled largely by conservation, brought about "oversupply" and a collapse of oil prices in 1986. Without continuing conservation the episode of the '70s will be repeated even more severely (dotted projections). (Data from *Statistical Abstracts of the United States,* U.S. Dept. of Commerce, 1988.)

FIGURE 19–16
Strategic petroleum reserves. To gain protection from future cutoffs of foreign oil supplies, the United States has been buying excess oil and putting it into storage in salt mines in Louisiana and Texas. The reserve contains about 60 days' supply of imported oil. (a) Aerial view of Morton Salt surface operation. (b) Underground at the work site. (U.S. Department of Energy.)

(a)

(b)

the 1970s OPEC was able to basically control the world oil market by restricting their own production, thereby keeping supplies tight and prices high. By the economic law of supply and demand, restricting supplies causes prices to increase.

As we came into the 1980s, however, the above efforts did pay off. The United States remained substantially dependent on foreign oil, but oil consumption was headed down, and through reopening old fields and building the Alaska pipeline, production was up (Fig. 19–12). Additionally, discovery and development of significant new fields in Mexico and the North Sea (between England and Scandinavia) made the world less dependent on OPEC oil. As a result of all these things and the inability of OPEC to restrain its own production sufficiently to keep supplies tight, oil production exceeded consumption in the mid-1980s and there was an oil glut. What happens when supply exceeds demand? Prices come down. In 1986, world oil prices crashed from about $30 per barrel to close to $10. Since that time they have fluctuated in the range of $14 to $18 per barrel.

Victims of Success: Setting the Stage for Another Crisis

The crash in oil prices was to the apparent benefit of consumers, industry, and oil-dependent Third World nations. But, does it mean that the underlying dilemma has been solved? It does not. In fact, it is simply serving to set the stage for another much more severe and lasting oil crisis which may occur as early as the mid-1990s unless significant steps are taken to avert it.

We noted above that high oil prices stimulated constructive responses. Conversely, the collapse in oil prices undercut these responses:

☐ Production from old fields, which was costing around $10 per barrel to pump, was terminated, causing economic devastation to the economies of Texas and Louisiana where most of it was occurring.

☐ Exploration, which had become more costly as wells were drilled deeper and in more remote locations, was sharply curtailed. In 1981, 4500 drilling rigs were operating. In 1989 the number was 740.

☐ Efforts toward further conservation have languished. Government standards for automobile fuel efficiency were decreased to 26 mpg with no intention to increase them further. The requirement to include the expected miles per gallon in car advertisements was dropped.

☐ Tax subsidies for installing alternative energy sources were terminated, destroying many new businesses engaged in solar and wind energy. Grants for research and development of alternative energy sources were sharply cut (except for nuclear power).

Therefore, since the oil price collapse of 1986, U.S. oil production has dropped, as has the exploration for and finding of new reserves, and consumption has started to climb again as people are buying less fuel efficient cars (Fig. 19–12). In particular, consumers are now buying more "muscle cars," four-wheel-drive vehicles, and light trucks, all of which get significantly lower mileage per gallon, and using them as cars for everyday commuting.

As a result, we are now dependent on foreign sources for 45 percent of our oil, more than in 1973 when the first oil crisis occurred, and this dependence is growing. You can easily see that following this trend will again lead to a situation in which OPEC can again control the market, tighten supplies, and raise prices, thus re-creating the situation we had in the 1970s.

The next time, however, the crisis is likely to be much more severe and long lasting because we have already exhausted the "easy options" that were available in the 1970s. For example, in 1970 the huge Alaskan reserve had just been found and was waiting to be exploited. No such proven reserves exist today, nor is it likely that any such new reserves will be found in U.S. territory.

The reason for this last statement is based upon what is called the "Easter egg" argument. When people first begin searching for hidden Easter eggs, the discovery rate is proportional to the searching effort. However, the searchers reach a point where, in spite of their continued effort, fewer and fewer eggs are found. This situation leads to the logical conclusion that most of the eggs have already been found.

The fact is that exploratory efforts of the late 1970s and early 1980s turned up remarkably little new oil. Indeed, multi-billion-dollar investments in exploratory drilling of the continental shelf off the east coast of the United States and in the Aleutian Islands off Alaska turned up nothing but dry holes (Fig. 19–17). Based on this decline in returns from exploratory drilling, the U.S. Geological Survey has lowered its forecasts of estimated reserves.

Thus, the United States in all likelihood is well into the period of irrevocably declining reserves and production. In the words of Earl Hayes, former chief scientist at the U.S. Bureau of Mines, "talk of rising petroleum (and gas) production for long periods is both immoral and nonsensical. Whatever slight gain might be achieved for a very few years will be at the

FIGURE 19–17
Despite intensive exploration, new finds have been disappointing. Exploratory wells in the Mukluk area of the Bering Sea, a 1.7 billion dollar venture, came up dry. (Lowell Georgia/Photo Researchers.)

expense of the youth of today.''* That is, forcing higher production now can only result in faster depletion and decline in production in the future.

At the very least we are undercutting our economic and political security. It was in the process of protecting oil shipments of the Persian Gulf that the United States accidentally shot down an Iranian Airliner and killed 290 people. This lead to the vengeful bombing of Pan Am flight 103, which killed 259 in the fall of 1988. These may be seen as additional prices being exacted for the renewed growth in our dependence on OPEC oil. As supplies become tight again, maintaining political stability will become more difficult. Assuming OPEC still has adequate reserves, we may get them but at OPEC's price and under its terms. Then the time will come when their reserves are also drawn down to the point of limiting production. Then, shortages will be more than economic and political. Transportation, industry, heating, and other needs will be curtailed without remission to a level supportable by current oil production. Moreover, there will be worldwide competition for dwindling supplies. The biggest and first losers are likely to be less-developed nations that do not have the purchasing power of industrialized nations. The potential for economic, political, and social chaos is severe. At the very least, there will be sharply higher prices and probably some form of rationing to achieve more equitable distribution of supplies.

The disastrous Alaskan oil spill of 1989 is another price we pay for oil (see Case Study 19).

---

* Hayes, Earl, "Energy Resources Available to the United States 1985–2000," *Science,* 203 (January 19, 1979), pp. 233–239.

Preparing for Future Oil Shortages

Instead of pretending that the oil shortages are history and embarking on increasing consumption and dependence again, we should be making every effort to accommodate ourselves to the decreasing supplies that will inevitably mark the future. Waiting until the crisis is upon us may well be too late because, as experience from the 1970s shows, it takes many years before measures that are started today are developed and implemented to the point of having a significant effect.

What measures should be taken? Basically, we need to reduce our consumption of oil-based fuels. This can be accomplished through any combination of:

☐ Conservation
☐ Development of alternative energy sources

Conservation

Consider the discovery of a new oil field with a potential production of at least 6 million barrels a day, perhaps more. This is three times the size of the Alaskan field, the last great U.S. discovery. Furthermore, this field is inexhaustible, it will go on producing indefinitely, and its exploitation will not adversely affect the environment. Of course, such an oil field sounds like a dream, but this is effectively what can be achieved through *conservation.*

Conservation is often promoted in terms of turning off lights, turning down thermostats, and car pooling. These activities can, and do, produce im-

THE VALDEZ OIL SPILL

On the evening of March 23, 1989, having completed the loading of 1.2 million barrels of crude oil from the terminus of the Alaska pipeline, the *Exxon Valdez* headed for refineries in California. Built in 1986, the tanker was equipped with state of the art navigational and safety equipment. In addition, shore-based Coast Guard radars were maintaining a continuous surveillance of shipping from the port of Valdez to the open ocean. The night was calm and clear. How could anything go wrong with conditions like that? Yet, just a short time later, at 12:04 a.m., the *Exxon Valdez*, more than a mile off course, plowed squarely into well-known rocks off Bligh Island that tore gashes in her hull, one of them 6 feet wide by 20 feet long. Eleven million gallons of oil poured out, the largest spill in North American waters in one of the most pristine and ecologically sensitive areas.

An oil spill can be contained and damage largely prevented by surrounding the oil slick with floating booms. But oil on water spreads rapidly; time is critical. Deplorably, it was 10 to 12 hours before a response was mounted. By this time the oil had already spread beyond control; there was inadequate equipment and some was rotted and unusable. Local fishermen deployed their own equipment to protect a few critical areas such as salmon hatcheries, but for the most part the spill ran its course. Less than 4 percent of the spilled oil was ever recaptured.

Most people who wanted to help were reduced to logging in the pathetic, oil-soaked, dead bodies of thousands of sea birds, hundreds of sea otters, and fish numbering in the millions. The actual counts will never be known as the slick covered over 900 square miles. Nor will the effects end with the immediate kills. Oil emulsifies with water and washes up and covers the shore line with a toxic, sticky, slimy goo that may last for many months. *Exxon* crews went to work to clean this mess, but after six weeks they had cleaned less than one of the more than 400 miles of shore line affected. Again it is clear the events will unfold with little mitigation by humans. What will happen to the flocks of birds, seals, and sea lions that come to breed and feed along these contaminated coasts? What will happen to the Sitka blacktailed deer feeding on contaminated kelp, or grizzlies and eagles feeding on the oiled-killed carcasses of dead birds and otters along the shore? What will happen to migrating whales and other marine animals as they find food chains decimated and contaminated by crude oil compounds that sink to the bottom affecting benthic organisms? Marine biologists who have studied the aftermath of other spills estimate that adverse ecological effects will persist for 10 to 20 years.

Can we learn something from this accident? One thing to recognize is that accidents with extremely low probability of occurring tend to lull people into

Workers cleaning up a section of oil-covered shoreline. Over 400 miles of coastline are affected by the spill. (Sygma.)

a state of incredible carelessness, making disasters almost inevitable. Environmentalists worried that such an accident would occur at the time the Alaska pipeline was being proposed, but they were out-voiced by oil people who said that such an accident couldn't happen. However, since it "couldn't happen," the captain of the tanker was below, apparently under the effects of alcohol while the ship was being piloted by an unauthorized third mate. The Coast Guard radar operator was apparently not watching the screens because the Coast Guard was unaware of the situation until word came from the *Exxon Valdez* that spillage was occurring, but this was almost half an hour after the accident occurred. Then, it was not clear who should be in charge of the emergency response, or even who knew where the containment equipment was located. When equipment was located, the barge which it was aboard was discovered to be out of order and it had to be transferred to another, and so on. Thus, all the way from attention and monitoring of the crews of its tankers through its ability to respond to an accident, the oil industry—there is no reason to believe that Exxon is a unique case—has shown itself to have a lackadaisical disregard if not utter disdain for maintaining any kind of

vigilance regarding protecting the environment. This attitude was demonstrated again four months later: when the *Exxon Valdez*, after presumably being thoroughly cleaned, as it was being towed into San Diego harbor for repairs it commenced exuding another slick.

Yet, with claims that it can be done without harming the environment, oil and mineral companies are pressing to be allowed to exploit even more ecologically sensitive areas of Alaska, the Antarctic, and elsewhere. To place our environment in such jeopardy while at the same time letting conservation efforts and development of alternative energy sources languish, and wasting at least 50 percent of the fuel we use is unconscionable. Indeed, the *Valdez* oil spill is but one aspect of the price we are paying for an automobile-based, oil-gluttonous lifestyle. More subtle but, in the long run, even more devastating will be the environmental degradation caused by air pollution and the greenhouse effect resulting from by burning these fuels. If the destruction of Prince William Sound can symbolize the total price of environmental degradation and prompt us toward a more rational and sustainable National Energy Policy, it will not have been in vain.

mediate fuel savings and they have helped get us through periods of limited supplies, such as those that occurred in the 1970s. However, such activities have distinct limitations. The actual savings are small, only 2 to 4 percent of overall energy use, and they do entail some inconvenience and/or discomfort. Therefore, they are difficult to sustain.

The real focus and potential of **conservation** is in the development of systems that use energy more efficiently so that the *same* or even improved transportation, lighting, heating, work, comfort, etc. are achieved but *less* energy is consumed in the process. There is no way around the fundamental Laws of Thermodynamics. However, note in Figure 19–9 that 60 to 80 percent of the energy consumed ends up as waste heat rather than performing the intended functions. The opportunity and potential of conservation is in reducing this waste.

As noted above, this "conservation reserve" has already been tapped to some extent, especially when the average efficiency of cars was doubled from 13 to 26 mpg. This by itself is already saving about 2 million barrels of crude oil per day. But other areas have contributed as well. As a result, during the late 1970s and early 1980s the economy expanded without a corresponding rise in energy consumption for the first time in history. If the same relationship between energy and gross national product that existed in the early

1970s had persisted, we would now be using about 25 percent more energy than we are (Fig. 19–18). Said another way, we are already obtaining about 25 percent of our energy from the conservation reserve at an annual savings approaching $100 billion. Efforts in conservation have been extremely cost effective.

Still the conservation reservoir has barely been touched. Simply by implementing currently available technologies energy consumption could be cut in half again. Cars are currently available which get at least double the current average mileage per gallon, and Peugeot, Toyota, Volkswagen, and Volvo all have prototypes which get from 70 to 100 mpg. Renault has a prototype which attained a remarkable 124 mpg in test runs. Unfortunately, all car companies are waiting for government mandates, consumer demand, or actual fuel shortages before putting these models into production so that high sales will be guaranteed.

Other conservation measures which are readily available but under-utilized include:

☐ Improved insulation and windows for homes and buildings which would reduce energy consumption for heating and cooling. At least another billion barrels of oil per year could be saved in this area.

☐ Substitution of fluorescent lights for conven-

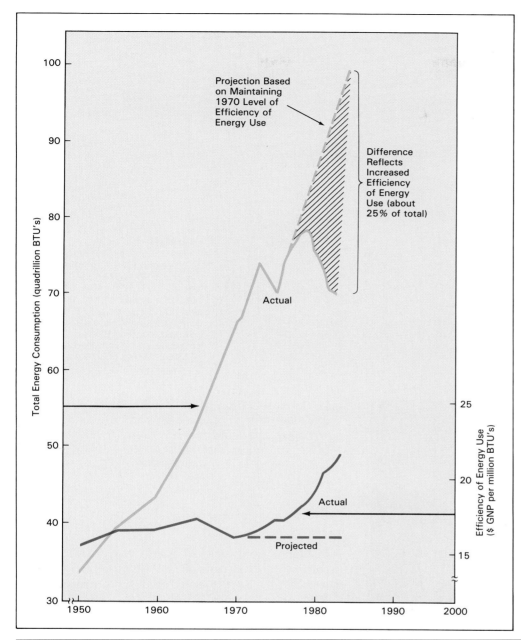

FIGURE 19–18
Increasing energy efficiency. If the connection between energy use and gross national product that existed in 1973 had continued, we would have used about 100 quadrillion BTUs in 1983. However, actual consumption was only about 77 quadrillion BTUs. The difference is attributable to increasing the efficiency of energy use. Hence, by far the largest contribution toward ameliorating the energy problem has come from conservation, i.e., increasing the efficiency of energy use. (Data from *Statistical Abstracts of the United States, 1985*. U.S. Department of Commerce.)

tional incandescent light bulbs. Traditional filament light bulbs are 5 percent efficient; 95 percent of the power ends up as waste heat. Fluorescent bulbs are about 95 percent efficient.

☐ Cogeneration. Virtually every building and factory requires both electricity and heat. (In the summer, heat can be used to drive air conditioning units.) Traditionally, electricity is obtained from a central power station where 60 to 70 percent of the heat from the fuel is wasted (see Fig. 19–9). Then heating is obtained by burning additional fuel. **Cogeneration** involves

placing an engine-driven generator in the building itself. While producing the necessary electricity, the engine's waste heat provides the necessary energy for temperature control and water heating. The utility company provides backup and buys surplus power. The result is a 30 percent savings in the overall amount of fuel required (Fig. 19–19). Cogeneration has the added advantage of providing protection against the wide-scale blackouts and brown-outs which may occur with highly centralized systems. Waste heat from other processes might similarly be put to use. Many European cities use waste

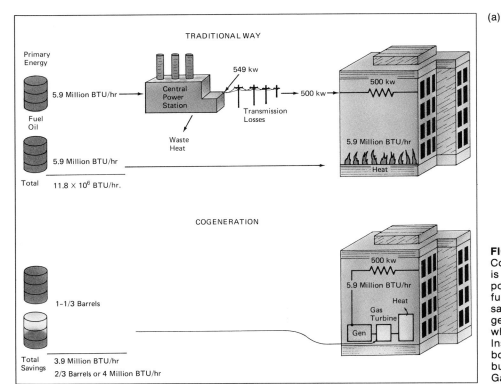

TRADITIONAL WAY

Primary Energy

5.9 Million BTU/hr — Central Power Station → 549 kw — Transmission Losses → 500 kw → 500 kw

Fuel Oil

Waste Heat

5.9 Million BTU/hr → 5.9 Million BTU/hr / Heat

Total   11.8 × 10⁶ BTU/hr.

COGENERATION

500 kw

5.9 Million BTU/hr

Heat

Gas Turbine

Gen

1-1/3 Barrels

Total Savings   3.9 Million BTU/hr
2/3 Barrels or 4 Million BTU/hr

FIGURE 19–19
Cogeneration: (a) Traditional means is to provide electricity from a central power plant and then burn additional fuel for heat. A 30 percent fuel savings may be realized by generating electricity in the building where waste heat is utilized. (b) Installed, cogeneration units provide both heat and electricity for an office building. (Photo courtesy of the Garrett Corporation.)

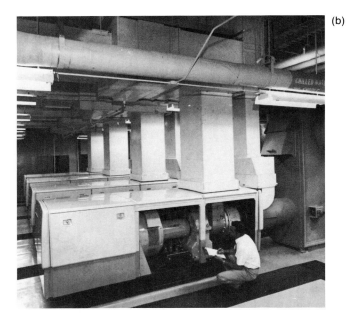

(b)

heat from power generation or waste incineration to heat surrounding homes and buildings through interconnecting steam lines, a concept known as **district heating.**

☐ Adjustments in lifestyle. Americans have developed a lifestyle that is energy extravagant. Further conservation may be attained by certain modifications in this area, which will be discussed in Chapter 22.

Conservation of crude oil and other fossil fuels will also serve to mitigate the carbon-dioxide climatic warming effect, acid rain, ground level ozone, and other air pollutants, because they all originate, in large part, from burning these fuels (See Chapters 12 and 13).

Development of Alternative Energy Sources

Insofar as nonfossil fuel energy may be substituted, consumption of crude oil and other traditional energy sources may be further reduced. Of course, such substitution will become essential in the long run because conservation can go only so far before the limits of thermodynamics are met. Various alternatives are discussed in Chapters 20 and 21. Solar alternatives are most promising, but despite these alternatives we will see that the need for conservation remains essential.

Promoting a Sound Energy Policy: What You Can Do

Economic and political security, the carbon dioxide effect, acid rain, ground level ozone, and other air pollutants: there are few occasions when so many serious concerns may all be addressed and mitigated with one single course of action with no undesirable side effects. Yet, conservation is one course of action that will moderate all these concerns. Since nearly 70 percent of the crude oil used in this country is for transportation and a large share of this is for cars, it is logical to focus attention on automobile fuel efficiency. As noted above, it could easily be doubled again bringing about a 30 percent reduction in overall crude oil demand.

☐ Write your congressperson and senators and ask that Congress:

Reinstate the policy of increasing fuel economy standards. Requiring new cars to have an average efficiency of 60 mpg by the year 2000 is realistic.

Establish a more stringent "gas guzzler" tax which increases the cost of inefficient vehicles.

Create a gasoline tax that would increase at the rate of 10¢ per year for the next five years to encourage conservation. Revenues from the tax should fund conservation, light rail-transit systems, and development of environmentally sound alternative energy sources.

☐ In buying a new or used car make high mileage per gallon a top priority.

☐ Try to establish or join a car or vanpool. Two rather than one in a car doubles the passenger miles per gallon.

☐ Suggestions regarding development of alternative energy sources and potential lifestyle changes will be considered in Chapters 21 and 22.

Moves to open up National Monuments, Parks and Wilderness Areas to oil exploration should be vigorously blocked. Sacrificing such areas in the hopes of finding more oil without first exploiting all the potential of conservation is unconscionable. Like an alcoholic selling family heirlooms to maintain his habit, it is the trading of national treasures simply to maintain profligate waste. What is not produced now will remain in the ground and may become invaluable resources later. Producing (and wasting) them now will only make future shortfalls more precipitous and severe as noted by Earl Hayes above.

20
Nuclear Power, Coal, and Synthetic Fuels

■ Outline

I. NUCLEAR POWER: A DREAM OR
DELUSION?   466

  A. Nuclear Power   468
    1. How It Works   468
      a. The fuel for nuclear power plants   468

      b. The nuclear reactor   469
      c. The nuclear plant   470

    2. Environmental Advantages of Nuclear
       Power   470

  B. Radioactive Materials and Their Hazard 474

    1. Disposal of Radioactive Wastes   475

    2. The Potential for Accidents   477

  C. Economic Problems with Nuclear Power   478

  D. More Advanced Reactors   479
    1. Breeder Reactors   479
    2. Fusion Reactors   480

II. COAL AND SYNTHETIC FUELS   483

  A. Coal   483
  B. Synthetic Fuels   483

■ Study Questions

1. Contrast the outlook for nuclear power in the 1960s with its present situation.

2. Define: *fission, fusion, uranium, isotopes.* Describe the basics of how nuclear power works. What is the fuel? How does it release energy? What is a *chain reaction?* How is it initiated, sustained, and controlled?

3. Describe the general design of a nuclear reactor, a nuclear power plant.

4. Nuclear power is basically an alternative to what conventional fuel? List the environmental advantages of nuclear power in contrast to coal. What are its disadvantages?

5. Define: *unstable isotope, radioactive emissions, radioactive substance.* Tell how radioactive wastes are produced. What are the hazards of radiation from radioactive substances? Are nuclear wastes the only source of such radiation?

6. Define: *radioactive decay, half-life.* Explain why there are two stages of disposal and what they are or what is proposed. What are the problems with achieving *long-term containment* and how may this affect the future of nuclear power? What is the current status of long-term containment?

7. Describe the scenario that occurred at Chernobyl. Discuss: Could it happen at a U.S. power plant? Define: *inherently safe nuclear power plants.*

8. Does/can nuclear power address our shortfall in crude oil production? Describe the economic balance between nuclear power and coal. Why are utilities opting for coal?

9. Briefly describe *breeder reactors, fusion reactors.* Does either offer promise for the projected near-term oil shortages?

10. Describe how liquid fuels might be obtained from coal and or oil shales. What are the drawbacks?

While U.S. reserves and production of crude oil are falling increasingly short of meeting demands, the United States does have abundant reserves of coal and uranium: the latter being the "fuel" for nuclear power. Can either of these energy sources be developed to fill in the growing gap between crude oil production and consumption? Our objective in this chapter is to investigate the potential of nuclear power and coal.

NUCLEAR POWER: A DREAM OR DELUSION?

From the time fossil fuels were first used, geologists recognized that they would not last forever. Sooner or later, other energy sources would be needed. The end of World War II, marked by the awesome power of the atom, was the time of decision. The U.S. government desperately wanted to show the world that the power of the atom could benefit humankind as well as destroy it and embarked on a course to lead the world into the "nuclear age."

It was anticipated that nuclear power could produce electricity in such large amounts and so cheaply that we would phase into an economy in which electricity would take over virtually all functions, including the generation of other fuels, at nominal costs. Thus, the government moved into research, development, and promotion of commercial nuclear power

plants (along with continuing development of nuclear weaponry) (Fig. 20–1). Utilizing this research, companies such as General Electric and Westinghouse construct the nuclear power plants that are ordered and paid for by utility companies (Fig. 20–2). The Nuclear Regulatory Commission (NRC), another arm of the federal government, sets and enforces safety standards for the operation and maintenance of these plants.

In the 1960s and early 1970s, utility companies embraced the concept and moved ahead with plans for numerous nuclear power plants (Fig. 20–3). By 1975, 53 plants were operating in the United States, producing about 9 percent of the nation's electricity, and another 165 plants were in various stages of planning or construction. Officials estimated that by 1990 several hundred plants would be on line, and by the turn of the century as many as 1000 plants would be operating. A number of other industrialized countries were in step with their own programs, and some less-developed nations were going nuclear as well by buying plants from industrialized nations.

However, since 1975, the picture has changed

FIGURE 20–1
U.S. government appropriations for research and development of commercial energy sources. The government's commitment to the development and promotion of nuclear power is demonstrated by the fact that nuclear power has consistently been given the lion's share of government support. Other energy sources and conservation received serious consideration only for a brief period under the Carter administration, 1976–1980.

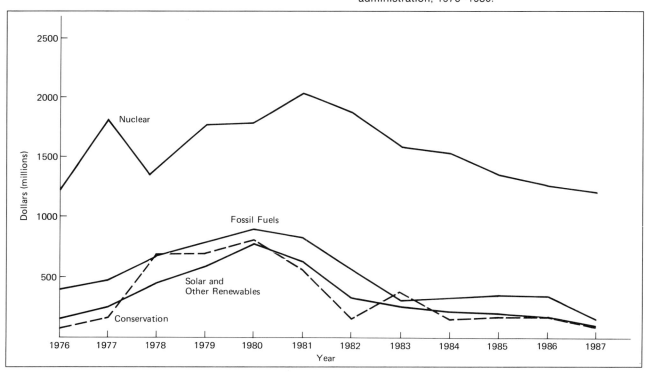

FIGURE 20-2
Arkansas Power and Electric's nuclear power plant near Russellville, Arkansas. The huge cone-shaped structure at left is a cooling tower to condense the waste steam back to water. About 70 percent of the heat produced is simply wasted into the environment here. Nuclear reactors are located in the two smaller cylindrical buildings. (U.S. Department of Energy.)

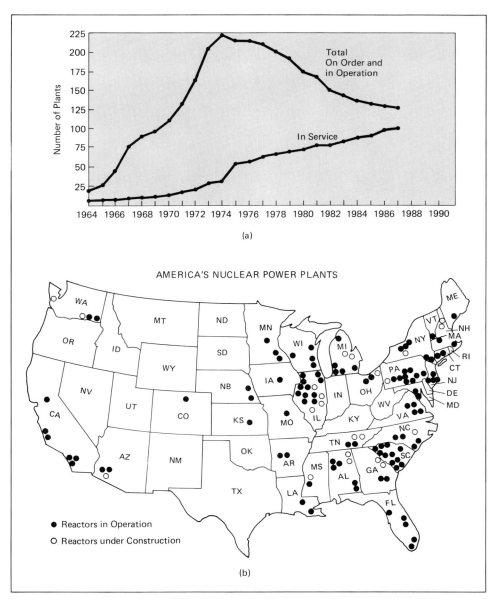

FIGURE 20-3
(a) Changing fortunes of nuclear power in the U.S. are evident in graphs of the number of nuclear plants on order and in operation. Since the early 1970s, when orders for plants reached a peak, few utilities have called for new plants and many have canceled earlier orders. (The last plant order not subsequently canceled was placed in 1974.) Nevertheless, the number of plants in service has increased steadily as plants under construction were completed. But, unless new orders are placed, 124 plants will be the peak. The figures for the number of plants operating in 1986 and beyond (*circles*) are estimates. (From Richard K. Lester, "Rethinking Nuclear Power," *Scientific American,* March 1986, p. 32. Copyright © 1986 Scientific American, Inc. All rights reserved.) (b) Locations of operating nuclear power plants (U.S. Department of Energy.)

dramatically. Utilities stopped ordering nuclear plants and numerous existing orders were canceled. The last plant ordered in the United States that was not subsequently canceled was ordered in 1974. In some cases construction was terminated even after billions of dollars had already been invested. Most striking, the Shoreham Nuclear Power Plant on Long Island, New York, after being completed and licensed at a cost of $5.5 billion was turned over to the state of New York in the summer of 1989 to be dismantled without ever having generated a single watt of power. The reason was that citizens and the State deemed that there was no possible way to evacuate people from the area should an accident occur. Similarly, just a few weeks earlier, California citizens voted to shut down and terminate the Rancho Seco Nuclear Power Plant located near Sacramento, which had had a 15-year history of troubled operation. Just 68 of those plants under way in 1975 have been completed, and another 3 are still under construction. Thus, it appears that nuclear power in the United States will top out at 124 operating plants in the early 1990s, generating about 16 percent of our electricity, and then head downward as older plants are "decommissioned." Beyond the United States, another roughly 400 plants are operating or under construction, mainly in France, the Soviet Union, Japan, West Germany, Canada, and the United Kingdom. But the story abroad is similar—drastic downward revisions if not total moratoriums on construction of new plants. Only France and Japan remain fully committed to pushing forward with nuclear programs.

After the catastrophic accident at Chernobyl in April 1986, it is not hard to see why nuclear power is being shunned. Yet, the impending oil shortages are real. Unless we develop alternative energy sources, the consequences of future shortages will be greatly exacerbated. By turning our backs on nuclear power, are we throwing away our best opportunity for supplying future energy needs? The nuclear industry is making every effort to have us believe so. They are advertising extensively regarding the promise and potential of nuclear power, and the lion's share of the Department of Energy's budget still goes to nuclear research and development (Fig. 20–1).

Public opinion and reaction will be the determining factor in whether we revive the nuclear dream or put it to rest. We need a clear understanding of what nuclear power is and its pros and cons so that we can act intelligently on this issue.

Nuclear Power

How It Works

The release of nuclear energy is a completely different phenomenon from the burning of fuels or other chemical reactions we have discussed. Recall our frequent observation up till now that material at the atomic level always remains unchanged, although the forms they constitute on the visible level undergo great transformation as atoms rearrange to form different compounds. Nuclear energy, however, does involve changing atoms through one of two basic processes, *fission* or *fusion*. In the process known as **fission,** a large atom of one element is split, resulting in two smaller atoms of different elements (Fig. 20–4a). In the process known as **fusion,** two small atoms combine to form a larger atom of a different element (Fig. 20–4b). In both fission and fusion, the mass of the products is less than the mass of the starting material, and the lost mass is converted into energy in accordance with the law of mass-energy equivalence ($E = mc^2$) first described by Albert Einstein. The amount of energy released by this conversion is tremendous. The sudden fission or fusion of about one kilogram (2.2 lb) of material releases the devastatingly explosive energy of a nuclear bomb.

The heart of current nuclear power plants is *controlled fission* so that energy is released gradually as heat. The heat is used to boil water and produce steam, which is then used to drive conventional turbogenerators.

The Fuel for Nuclear Power Plants. All current nuclear power plants utilize the fissioning (splitting) of uranium atoms, specifically uranium-235. Uranium is an element that occurs naturally in various minerals in the earth's crust. It exists in two primary forms or **isotopes:** uranium-238 (^{238}U) and uranium-235 (^{235}U).

The isotope numbers represent the number of subatomic particles present in the nucleus of the atom (e.g., ^{238}U has three more neutrons than ^{235}U). Most elements exist as a number of different isotopes, but since different isotopes of a given element are chemically identical, the phenomenon has little significance from a strictly chemical point of view. However, other characteristics of isotopes may differ profoundly. In this case, uranium-235 atoms will fission, or split, readily whereas uranium-238 atoms will not.

Occasionally, a uranium-235 atom fissions spontaneously, but it can be triggered to fission if it is struck by a neutron. This is the key to promoting the nuclear reaction. When one uranium-235 atom splits, it ejects two or three neutrons in addition to releasing energy. If one of these neutrons strikes another uranium-235 atom, fission occurs again releasing more energy and more neutrons, which repeat the process. A domino effect, known as a **chain reaction,** occurs (Fig. 20–5a).

A chain reaction does not occur in nature because uranium-235 atoms are too dispersed among

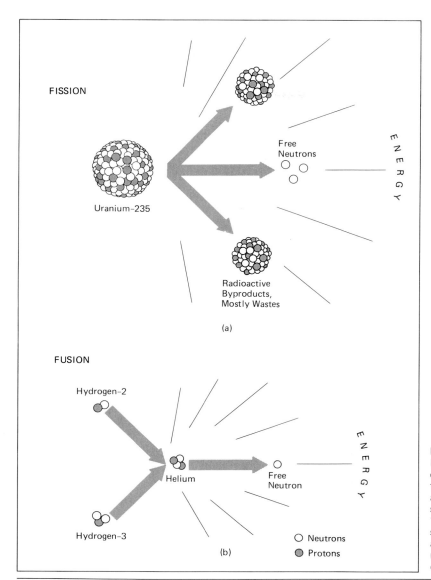

FISSION

Uranium-235

Free
Neutrons

Radioactive
Byproducts,
Mostly Wastes

(a)

FUSION

Hydrogen-2

Helium

Free
Neutron

Hydrogen-3

(b)

○ Neutrons
● Protons

FIGURE 20–4
Release of nuclear energy. Nuclear
energy is released from either (a)
fission, the splitting of certain large
atoms such as uranium-235 into
smaller atoms, or (b) fusion, the
"melting" together of small atoms
such as hydrogen to form a larger
atom. In both cases some of the
mass of the starting atom(s) is
converted to energy.

other elements and stable uranium-238 atoms. In fact, 99.3 percent of all uranium is ^{238}U; only 0.7 percent is ^{235}U. Hence when a ^{235}U atom spontaneously fissions in nature, it seldom triggers another atom, and the energy released by a single atom goes unnoticed without the aid of radiation detectors such as Geiger counters.

To make nuclear "fuel," uranium ore is mined, purified, and *enriched*. **Enrichment** involves separating ^{235}U from ^{238}U to produce a material with a higher concentration of ^{235}U. Since ^{238}U and ^{235}U are chemically identical, enrichment is based on the slight difference in mass. The difficulty of the enrichment procedure is the major technical hurdle that prevents less-developed nations from advancing their own nuclear capability.

When uranium-235 is highly enriched and placed in an appropriate mass, the spontaneous fissioning of an atom triggers a chain reaction. In nuclear weapons, small masses of virtually pure uranium-235 or other fissionable material are forced together so that the two or three neutrons from a spontaneous fission cause two or three more atoms to fission; each of these triggers two or three more, and so on. Thus, the whole mass undergoes fissioning in a fraction of a second, releasing all of the energy in one huge explosion (Fig. 20–5b).

The Nuclear Reactor. For a power plant, a nuclear reactor is designed to sustain a continuous chain reaction but not allow its amplification into a nuclear explosion (Fig. 20–5c). This is achieved by enrichment of the uranium to about 3 percent ^{235}U and 97 percent ^{238}U. This modest degree of enrichment will not sup-

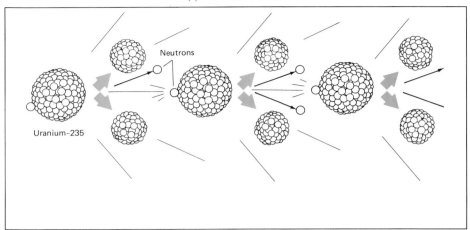

FIGURE 20–5
(a) A simple chain reaction. When a uranium atom fissions, it releases two or three high-energy neutrons in addition to the energy and the split "halves." If another uranium-235 atom is struck by a high-energy neutron, it fissions and the process is repeated, causing a chain reaction.

port the amplification of a chain reaction into a nuclear explosion. When a suitable mass is put together, however, it will support enough of a chain reaction so that the material will become intensely hot. Thus, the enriched uranium is then made into pellets that are loaded into long steel tubes. The loaded tubes are called **fuel elements** or **fuel rods.** When many fuel elements are placed close together in a **fuel assembly,** a chain reaction is initiated and sustained.

To control the chain reaction in the fuel assembly, rods of neutron-absorbing material referred to as **control rods** are inserted between the fuel elements. The chain reaction is started and controlled by withdrawing and inserting the control rods as necessary. The **nuclear reactor** is simply this assembly of fuel elements and movable control rods (Fig. 20–6). As the control rods are removed and a chain reaction is initiated, the fuel rods become intensely hot.

The Nuclear Power Plant. In a nuclear power plant, heat from the reactor is used to boil water to provide steam for driving conventional turbogenerators. Water may be boiled directly by circulating it through the reactor. However, in U.S. power plants, a double loop is employed. Water is intensely heated by circulating it through the reactor, which is mounted in a thick-walled vessel, the **reactor vessel,** but it is kept from boiling by keeping the system under extreme pressure. This super-heated water is circulated through a heat exchanger where it boils water to produce the steam used to drive the turbogenerator.

This "double loop" design aids in isolating hazardous materials in the reactor from the rest of the power plant. However, it has the drawback that if a break should occur, the sudden loss of water from around the reactor could result in its overheating and melting, an event called a **meltdown.** Then the mol-

ten material falling into the remaining water could cause a steam explosion with considerable consequences as will be discussed later. To guard against this, there are backup cooling systems to keep the reactor immersed in water should leaks occur, and the entire assembly is housed in a thick concrete **containment building** (Fig. 20–7).

The fissioning of about a pound (0.5 kg) of uranium fuel releases energy equivalent to burning 1000 tons of coal. Thus, one fueling of the reactor with about 3 tons of uranium is sufficient to run the power plant for as long as 2 years. The spent fuel elements are then removed and replaced with new ones.

Environmental Advantages of Nuclear Power

In canceling plans for nuclear power plants, we did not decide to do without electricity; effectively we opted to build coal-fired power plants instead. The United States does have abundant coal reserves, but is burning it the course of action we wish to pursue? We should note some decided environmental advantages of nuclear power in contrast to burning coal (Fig. 20–8). Comparing the environmental impacts of a 1000-megawatt nuclear plant with a coal-fired plant of the same capacity for one year, we find:

☐ Fuel need—The coal plant will consume about 3.5 million tons of coal obtained by stripmining with the resulting environmental destruction and acid leaching. The nuclear plant will require about 1.5 tons of uranium obtained from mining less than 1000 tons of ore.

☐ Carbon dioxide emission—The coal plant will emit over 10 million tons of carbon dioxide into

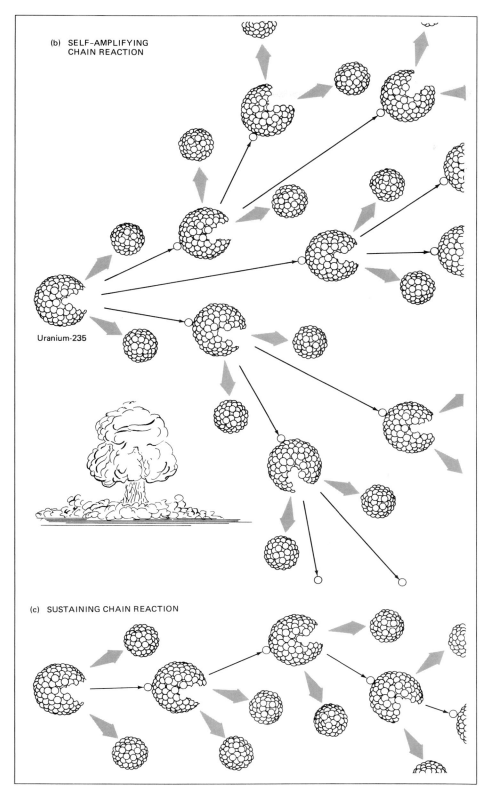

(b) SELF-AMPLIFYING
CHAIN REACTION

Uranium-235

(c) SUSTAINING CHAIN REACTION

FIGURE 20–5
(*cont.*) (b) A self-amplifying chain reaction leading to a nuclear explosion. Since two or three high-energy neutrons are produced by each fission, each fission may cause the fission of two or three additional atoms. Hence the entirety of a suitably concentrated mass of fissionable material may be caused to fission in a tiny fraction of a second, resulting in a nuclear explosion. (c) In a sustaining chain reaction the extra neutrons are absorbed in other materials so that amplification does not occur.

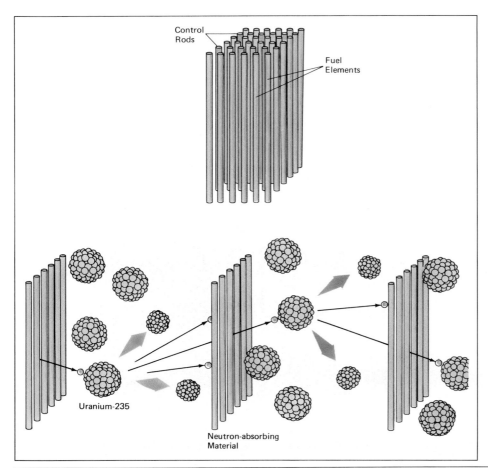

FIGURE 20–6
The nuclear reactor. In a nuclear reactor, a large mass of uranium is created by placing uranium in adjacent tubes, the fuel elements. The uranium is *not* sufficiently concentrated to permit a nuclear explosion, but it will sustain a chain reaction that will produce a tremendous amount of heat. The rate of the chain reaction is modulated by inserting or removing rods of neutron-absorbing material (control rods) between the fuel elements. Heat is removed from the fuel elements, and the reactor is kept from overheating, by circulating water through the reactor.

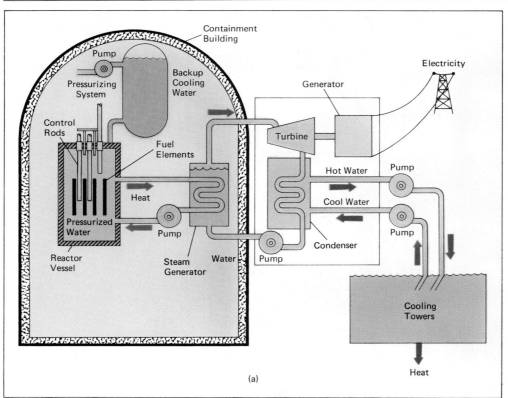

FIGURE 20–7a
(a) Schematic diagram of a nuclear power plant.

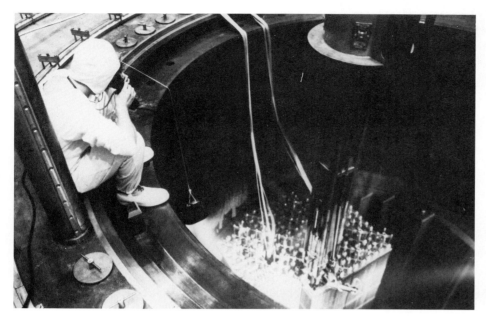

Figure 20–7b
(b) Technician sitting on edge of reactor vessel is monitoring the fuel-loading operation. Note the assemblies of fuel rods. (U.S. Department of Energy photo, courtesy of Combustion Engineering, Inc.)

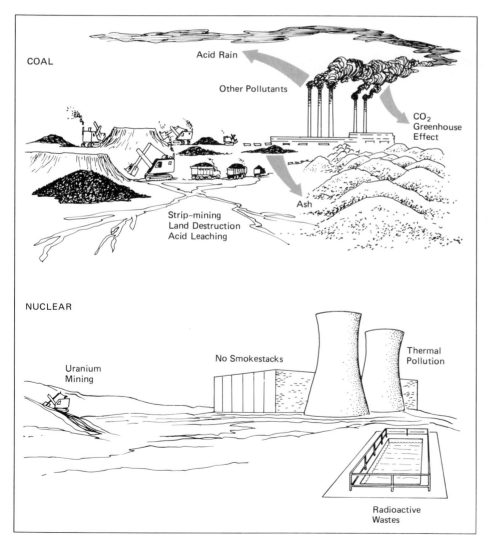

FIGURE 20–8a
(a) Diagram depicts the environmental impacts of nuclear power in contrast to those of coal. (*cont.*)

FIGURE 20–8b
(b) A nuclear power plant. Note the absence of smoke stacks and visible air pollution. (U.S. Department of Energy photo.)

the atmosphere, contributing to the greenhouse climatic warming effect. The nuclear plant will emit none.

☐ Sulfur dioxide and other acid-forming emissions—The coal plant will emit over 400,000 tons of sulfur dioxide and other acid-forming pollutants. The nuclear power plant will produce none.

☐ Solid wastes—The coal plant will produce about 100,000 tons of ash, requiring disposal. The nuclear plant will produce about 2 tons of radioactive wastes requiring disposal.

Of course it is these radioactive wastes and the potential for catastrophic accidents that have become the overriding concerns. Are these problems that cannot be overcome?

Radioactive Materials and Their Hazard

Assessing the hazards of nuclear power necessitates our understanding a little about radioactive substances and their danger.

When uranium or any other element undergoes fission, the split halves are atoms of lighter elements such as iodine, cesium, strontium, cobalt, or any of some 30 other different elements. These newly formed atoms, however, are generally *unstable isotopes* of their respective elements. Unstable isotopes gain stability by spontaneously ejecting subatomic particles and/or high-energy radiation, which are referred to as **radioactive emissions.** The unstable isotopes, which are emitting such radiation, are known as **radioactive substances.** In addition to the direct fission products, other materials in and around the reactor may be converted to unstable isotopes and become radioactive by absorbing neutrons from the fission

process (Fig. 20–9). These direct and indirect products of fission are the **radioactive wastes** of nuclear power. Radioactive fallout from nuclear explosions also consists of these direct and indirect fission products.

Radioactive emissions may penetrate through biological tissue in a manner analogous to tiny bullets. They leave no visible mark, nor are they felt, but they are capable of breaking molecules within cells. In high doses, radiation may cause enough damage to prevent cell division. Thus radiation can be focused on a cancerous tumor to destroy it. However, if the whole body is exposed to such levels of radiation, a generalized blockage of cell division occurs which prevents the normal replacement or repair of blood, skin, and other tissues. This result is called radiation sickness, which may result in death a few days or months after exposure. Very high levels of radiation may totally destroy cells, causing immediate death.

In low doses, however, radiation may damage DNA molecules, the genetic material inside the cell. Cells with damaged (mutated) DNA may then begin dividing and growing out of control, forming malignant tumors or cancer. If an egg or sperm cell is involved, the radiation may cause birth defects in offspring. These effects may go unseen until many years after the exposure. The greatest concern regarding nuclear power is that it may inadvertently cause large numbers of the public to be exposed to low levels of radiation, thus elevating the risk of cancer and/or birth defects.

But nuclear proponents are quick to point out that nuclear power is by no means the only source of such radiation. There is normal **background radiation** from radioactive materials, such as uranium itself and radon gas which occur naturally in the earth's crust, and from cosmic rays from outer space. In addition, we expose ourselves to radiation from X-rays, radium dial watches, and other sources. Thus, the argument

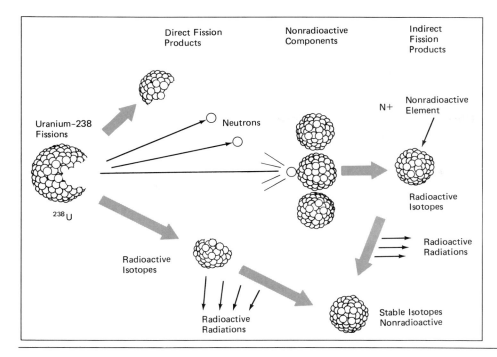

Direct Fission Products

Nonradioactive Components

Indirect Fission Products

Uranium-238 Fissions

Neutrons

^{238}U

Radioactive Isotopes

Radioactive Radiations

Nonradioactive Element

N+

Radioactive Isotopes

Radioactive Radiations

Stable Isotopes Nonradioactive

FIGURE 20–9
Radioactive wastes and radioactive emissions. Nuclear fission results in the production of numerous unstable isotopes as depicted. These unstable isotopes are the radioactive wastes. They give off potentially damaging radiations until they regain a stable structure.

becomes one of *relative risks*. Does or will the radiation stemming from nuclear power significantly raise the levels of radiation and elevate the risks of cancer to which we are already exposed?

During normal operation of a nuclear power plant, the radioactive fission products remain within the fuel elements and indirect or secondary fission products are maintained within the containment building that houses the reactor. No routine discharges of radioactive materials into the environment occur. Even very close to an operating nuclear power plant, radiation levels from the plant are less than 1 percent of normal background levels. A radiation detector will pick up more radiation from the earth and concrete on a basement floor than it will when held within 500 feet of a nuclear power plant.

However, the main concern is not in the *normal* operation. It is in how the radioactive wastes are ultimately disposed of and the potential for catastrophic accidents as occurred in the case of Chernobyl in 1986. We shall look at these two issues further.

Disposal of Radioactive Wastes

To understand the rationale of nuclear waste disposal, we must first understand the concept of *radioactive decay*. As unstable isotopes eject particles and radiation, they become stable and cease to be radioactive. This process is known as **radioactive decay.** As long as radioactive materials are kept isolated from humans and other organisms, the decay process proceeds harmlessly and radiation levels decrease accordingly.

The rate of radioactive decay is such that half of the starting amount of a given isotope will decay in a certain period of time. In the next equal period of time, half of the remainder (one fourth of the original) decays, and so on as shown in Figure 20–10. The "certain period of time" for half the amount of a radioactive isotope to decay is known as its **half-life.** The half-life of an isotope is always the same regardless of starting amount. Note that decay of a radioactive isotope is never 100 percent complete; it is only reduced by half in each half-life, so there is always an undecayed portion remaining. However, it is generally considered that after 10 half-lives, radiation will be reduced to insignificant levels.

Each particular radioactive isotope has a characteristic half-life; the half-lives for various isotopes range from a fraction of a second to many thousands of years. The fissioning of uranium results in a very heterogeneous mixture of radioactive isotopes, the most common of which are listed in Table 20–1. Note that the half-lives of these radioactive isotopes vary from a few days to many years; note especially plutonium with its half-life of 24,000 years. Thus, much of the radioactivity from fission wastes will dissipate in a period of months or a few years as the short-lived isotopes decay. However, to be safe, long-lived isotopes require isolation for up to 240,000 years (ten times the half-life of plutonium). Thus, the problem of nuclear waste disposal is divided into two areas:

☐ Short-term containment (a few years) to allow the radioactive decay of short-lived isotopes.

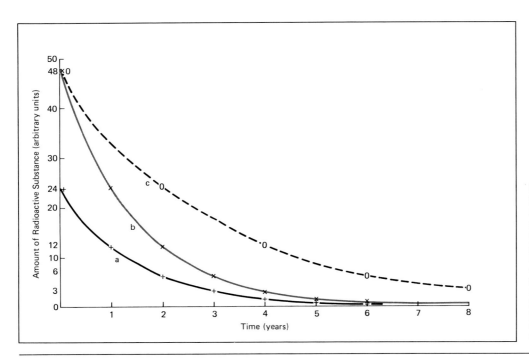

FIGURE 20–10
Radioactivity for any isotope declines as shown. Regardless of starting amount, one-half decays during each successive half-life. (a) A substance with a half-life of one year starting with 24 units; (b) the same substance starting with 48 units; (c) a substance with a half-life of two years. The half-life for different isotopes may vary from less than one second to many thousands of years.

Table 20–1
A Few of the Most Common Radioactive Isotopes Resulting from Uranium Fission and Their Half-lives

SHORT-LIVED FISSION PRODUCTS	HALF-LIFE (DAYS)
Strontium-89	54
Yttrium-91	59.5
Zirconium-95	65
Niobium-95	35
Molybdenum-99	2.8
Rubidium-103	39.8
Iodine-131	8.1
Xenon-133	15.3
Barium-140	12.8
Cerium-141	32.5
Praseodymium-143	13.9
Neodymium-147	11.1

LONG-LIVED FISSION PRODUCTS	HALF-LIFE (YEARS)
Krypton-85	10.27
Strontium-90	28.
Rubidium-106	1.0
Cesium-137	30
Cerium-144	.8
Promethium-147	2.6

ADDITIONAL PRODUCTS OF NEUTRON BOMBARDMENT	
Plutonium-239	24,000

Wastes can be handled much more easily and safely after this occurs.

☐ Ultimate long-term containment (tens of thousands of years) to provide protection from the long-lived isotopes

For short-term containment, the spent fuel rods are being stored in deep swimming pool–like tanks on the sites of nuclear power plants. The water serves to dissipate waste heat, which is still generated to some degree, and acts as a shield for radiation. Nuclear weapons facilities also maintain various tanks for short-term containment of radioactive wastes.

Curiously, however, the development and commitment to nuclear power went ahead without ever fully addressing and resolving the issue of ultimate long-term containment. It was generally assumed by nuclear proponents that the remaining long-lived nuclear wastes could be solidified, placed in sealed containers, and buried in deep stable rock formations as the need for such containment became necessary (Fig. 20–11). In the meantime, the short-term containment would suffice.

However, is there any rock formation that can be guaranteed to remain stable and dry for tens of thousands of years? Everywhere scientists look, there is evidence of volcanic or earthquake activity or groundwater leaching having occurred within the last 10,000 years or so, which is to say that it may occur again in a similar period of time. If such events did occur the still-radioactive wastes could escape into the

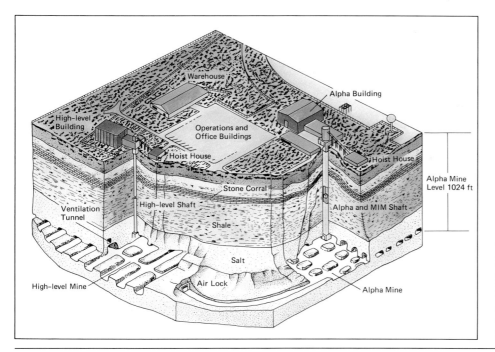

FIGURE 20–11
Disposal of radioactive wastes from nuclear power plants. These wastes must be isolated from the environment for thousands of years. Elaborate plans have been made for their disposal, but will this assure that they will stay where they are put? (U.S. Department of Energy.)

environment and contaminate water, air, or soil with consequent effects on humans and wildlife.

Therefore, efforts to locate a long-term containment facility, which have been going on for about 20 years, have been characterized by "NIMBY," which stands for "not in my back yard." A number of states, under pressure from citizens, have passed legislation categorically outlawing the disposal of nuclear wastes within their boundaries. In the meantime, the need for a long-term repository has become increasingly critical. Indeed, some nuclear power plants may be shut down in the next 5 to 10 years if this problem is not resolved because of lack of short-term containment facilities. Even more serious, it has been discovered that a number of military facilities are now leaking their radioactive wastes into groundwater because they have been kept in use far beyond their intended lifetimes.

Finally, at the end of 1987, Congress called a halt to debate and arbitrarily selected a remote site known as Yucca Mountain in southwestern Nevada to be the nation's nuclear dump. Intensive study and evaluation of this site are now under way. If it does check out as adequate, and many are skeptical that it will, it will begin receiving wastes from commercial and military facilities all over the country in 2003. It should not escape notice that this means that thousands of tons of radioactive wastes will be shipped by rail and truck through congested areas across the country to Yucca Mountain. Still, an even worse alternative is to let the problem remain unresolved. This is and will remain a serious problem for the foreseeable future

just with existing nuclear power plants. Building more plants can only intensify it.

The Potential for Accidents

Prior to 1986, the scenario for a worst-case nuclear power plant disaster was a matter of speculation. Then at 1:24 A.M. on April 26, 1986, events at Chernobyl in the Soviet Union made it a reality. In the process of shutting the plant down for maintenance, three engineers unfamiliar with the nuclear reactor had all safety systems turned off when they decided to conduct an unauthorized experiment. The result was that the reactor suddenly surged to full power, a meltdown occurred, a steam explosion blew the 1000-ton top off the reactor, and radioactive materials were ejected thousands of feet into the atmosphere.

Only 2 of the engineers were killed in the actual explosion, but 31 of the personnel brought in to contain the aftermath died of radiation sickness within the next 6 months. It is sobering to note that responding to a nuclear accident may be a suicidal mission, yet the escape of radioactive material would have been considerably greater without the efforts of these people.

As the radioactive fallout settled, 135,000 people, everyone in an 18-mile radius around the plant, were evacuated and eventually relocated. The area remains uninhabitable due to contamination of the soil with radioactive compounds.

Over a much broader area, efforts were taken to wash down buildings and roadways to flush away radioactive dust. Many tons of foodstuffs, caught in fallout areas in Europe, were banned from market. But even with these precautionary measures, many of those in or near the evacuation zone may have received levels of radiation that may lead to cancers and birth defects in future years. Indeed, increased levels of radiation resulting from the disaster were detected around the globe, including in the United States, but these increases were deemed to have no significance for human health because they were both temporary and only modestly above normal background levels.

Are we in danger of such disasters occurring at U.S. nuclear power plants? Nuclear proponents argue that the answer is no because U.S. power plants have a number of design features that should make a repeat of Chernobyl impossible. Specifically, there are more backup systems to prevent core overheating and reactors are housed in a thick concrete-walled *containment building* designed to withstand explosions such as the one that occurred at Chernobyl. The Russian reactor had no such containment building. However, humans seem to have an unlimited capacity to subvert safety systems through negligence or committing errors.

In 1979 the Three Mile Island nuclear power plant near Harrisburg, Pennsylvania, suffered a partial meltdown and came dangerously close to a major explosion because of sloppy maintenance procedures and the operators not responding properly to the initial malfunctions. It is not clear that the containment building could have withstood the resulting steam explosion had it occurred. The drama held the whole nation, particularly the 300,000 residents of Harrisburg poised for evacuation, in suspense for several days. The situation was eventually brought under control and no injuries occurred. The reactor, however, was so badly damaged and so much radioactive contamination had occurred inside the containment building that the cleanup and repair process won't be completed until the mid 1990s and is proving to be as costly as building a new power plant.

However, proponents point out that we have learned valuable lessons from Three Mile Island and other lesser incidents. As a result, the Nuclear Regulatory Commission has upgraded safety standards not only in the technical design of nuclear power plants, but also in maintenance procedures and the training of operators. Thus, proponents contend, nuclear plants were designed to be safe in the beginning and now they are safer than ever. It is estimated that as a result of new procedures instituted after the accident at Three Mile Island, nuclear plants are now six times safer than before. Even more, proponents of nuclear power point out that we now have the technology to build **inherently safe** nuclear reactors,

reactors designed in such ways that any accident would result in an automatic quenching of the chain reaction.

Economic Problems with Nuclear Power

Still, as noted, U.S. utilities have not ordered any new nuclear power plants that were not subsequently canceled, since 1974. Thus, U.S. utilities were turning away from nuclear power considerably before the disaster at Chernobyl. The reasons are mainly economic.

First, increasing safety standards for the construction and operation of nuclear power plants, which we would not want to do without, have served to escalate costs of nuclear power plants at least fivefold even after inflation is considered. Second, public objections have frequently served to delay construction and/or startup, which increases costs still more because the utility is paying interest on its investment of several billion dollars even when the plant is not producing power. As these costs are passed on, consumers become even more disillusioned with nuclear power (Fig. 20–12). The problem of long-term disposal of nuclear wastes will be unchanged. Finally, safety systems may protect the public, but they do not prevent an accident from being financially ruinous to the utility. Since radioactivity prevents straightforward cleanup and repair, an accident, as Three Mile Island demonstrated, can convert a multibillion-dollar asset into a multibillion-dollar liability in a matter of minutes. Thus, nuclear power involves a financial risk that utility executives are loath to take.

Another factor that promises to increase the cost of nuclear-generated electricity is a shorter-than-expected lifetime for nuclear power plants. Originally it was thought that nuclear plants would have a lifetime of about 40 years. It now appears that their lifetime will be considerably less, perhaps only 30 years. This difference substantially increases the cost of the power produced because the cost of the plant must be repaid in the shorter period.

The shorter-than-expected lifetime is due to a problem known as **embrittlement.** As well as maintaining a chain reaction, some of the neutrons from fission bombard the reactor vessel and other hardware. Gradually, this neutron bombardment causes the metals to become brittle such that they may crack under stress. When the reactor vessel becomes too brittle to be considered safe, the plant must be shut down or, in technological jargon, **decommissioned.**

Decommissioning itself may be an extremely costly process. By this time, the power plant components will have accumulated so much radioactivity from neutron bombardment that the only safe course of action will be to seal off the entire containment

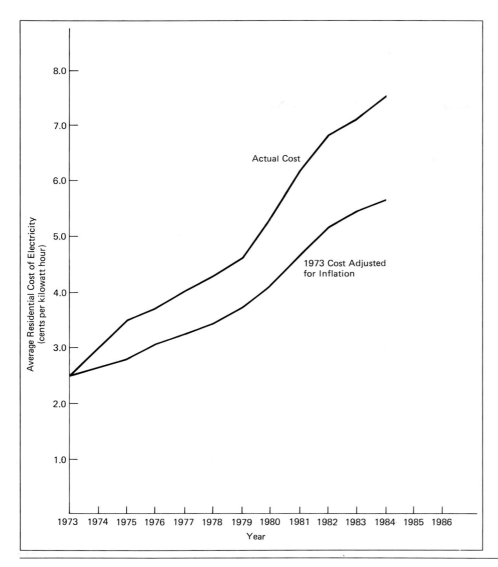

FIGURE 20–12
Over the past decade, the cost of electricity has risen even more sharply than inflation at large. This is, in part, attributable to the huge costs of nuclear power plants which consumers are obliged to assume as the plants are brought on line. Even so, there are additional costs hidden in tax moneys supporting nuclear research and development, regulation, and disposal of nuclear wastes. (*Statistical Abstracts of the United States*. U.S. Department of Commerce, 1985.)

building for an indefinite period of time and construct a new plant. Of the plants operating in the United States, it is estimated that at least ten will be decommissioned in the 1990s and most of the rest in the ten years thereafter.

Finally, there is the basic mismatch between nuclear power and the energy problem. As we have emphasized, our main energy problem involves a shortage of crude oil for transportation, yet nuclear power produces electricity, which is not used for transportation. If we were moving toward a totally electric economy including electric cars, nuclear-generated electricity could be substituted for oil-based fuels. Unfortunately, electric cars have not yet proved practical and the outlook for them in the near future remains dim. Consequently, nuclear power simply competes with coal-fired power plants in meeting limited demands for electricity. The simple fact is that, given the escalated costs and additional financial risks of

nuclear power plants, coal is cheaper, and the United States does have abundant coal reserves.

However, there are still the environmental problems of mining and burning coal, including acid precipitation and the carbon dioxide effect. If these were factored into the price of coal we would find its price considerably higher than it appears. Of course, costs such as long-term confinement of nuclear wastes and decommissioning of power plants have yet to be factored into the cost of nuclear power as well.

More Advanced Reactors

Breeder Reactors

Uranium, especially uranium-235, is not a highly abundant mineral on earth. At the height of nuclear optimism in the 1960s, when as many as 1000 plants were envisioned by the turn of the century, it was foreseen that shortages of uranium-235 would develop. Conversion to the **breeder reactor** was seen as the solution.

Recall that when a uranium-235 atom fissions, two or three neutrons are ejected. Only one hitting another ^{235}U atom is required to sustain the chain reaction; the others are absorbed by something else. The breeder reactor is designed so that these extra neutrons are absorbed by the nonfissionable uranium-238. When this occurs the ^{238}U is converted to plutonium-239, which is a fissionable. The plutonium-239 can be enriched and used as a nuclear fuel in the same manner as ^{235}U and such reactors can, in turn, breed more plutonium from ^{238}U. Thus, the breeder converts ^{238}U, which is nonfissionable and hence energy worthless, into a useful nuclear fuel. Since there are generally two neutrons in addition to the one needed to sustain the chain reaction, the breeder may actually produce more fuel than it consumes. Since 99.7 percent of all uranium is ^{238}U, converting this to plutonium-239 through breeder reactors effectively increases the nuclear fuel reserves over 100-fold. However, none of the other problems of nuclear power are alleviated.

Indeed, breeder reactors heighten the risks of nuclear power. They present all of the problems and hazards of standard fission reactors plus additional problems. If a meltdown occurred, the consequences would be much more serious because of the presence of large amounts of plutonium-239, which has the exceedingly long half-life of 24,000 years. In addition, because plutonium can be purified and fabricated into nuclear weapons more easily than uranium-235, the potential for the diversion of breeder fuel into weapons production is greater. Hence, the safety and security precautions needed for breeder reactors are greater. Can they be great enough?

With its scaled-down nuclear program, the United States currently has enough uranium stockpiled to fuel all reactors that are operating or under construction through their lifetimes. Thus, there is no urgency for the United States to develop breeders, and construction of the prototype breeder at Clinch River, Tennessee, has been discontinued. France, however, is proceeding with a breeder reactor development program.

Fusion Reactors

Fusion is another form of nuclear power. Rather than splitting a large atom to release energy, fusion "melts" together smaller atoms into a larger atom. As in fission, fusion results in a loss of mass, which is released as energy.

The vast energy emitted by the sun and other stars comes from this fusion process. The sun, as well as other stars, is composed mostly of hydrogen. Solar energy is the result of fusion of this hydrogen into helium. Scientists have duplicated this process in the hydrogen bomb; but hydrogen bombs hardly constitute a useful release of energy. The aim of fusion technology is to carry out this fusion in a controlled manner in order to provide a practical heat source for boiling water to power steam turbogenerators.

Since hydrogen is an abundant element on earth (two atoms in every molecule of water), and helium is an inert, nonpolluting, and nonradioactive gas, hydrogen fusion is promoted as the ultimate solution to all our energy problems—that is, pollution-free energy from a virtually inexhaustible resource, water. However, the dream is still a long way from reality. Indeed, a fusion plant has not yet been proven possible, much less a practical alternative.

At the present state of the art, fusion power is still an energy consumer rather than a producer. The problem is that it takes an extremely high temperature (some 100 million degrees Celsius) and pressure such as exist in the sun to get hydrogen atoms to undergo fusion. In the hydrogen bomb, the temperature and pressure are achieved by using a fission bomb as an igniter—hardly a practical alternative to sustained, controlled fusion.

A major problem is a means to contain the hydrogen while it is being heated to the tremendously high temperatures required for fusion. No material known can withstand these temperatures without vaporizing; however, two concepts are undergoing experimental testing. One is the Tokamak design, in which ionized hydrogen is contained within a magnetic field while it is heated to the necessary temperature (Fig. 20–13). The second concept is laser fusion, in which a tiny pellet of frozen hydrogen is dropped into a "bull's-eye" where it is hit simultaneously from all sides by powerful laser beams (Fig. 20–14). The laser beams simultaneously heat and pressurize the pellet to the point of fusion.

Some fusion has been achieved in these devices, but as yet, the break-even point has not been reached. More energy is required to run the magnets or lasers than is obtained by the fusion. The most optimistic workers in the field feel that with sufficient money for research—some $10 billion—the break-even point might be reached in the 1990s. Even if this goal is achieved, however, it is still a long way from a practical commercial fusion-reactor power plant. Developing, building, and testing such a plant would require at least another 20 to 30 years and many more billions of dollars. Additional plants would require additional years. Thus, fusion is, at best, a very long-

(a)

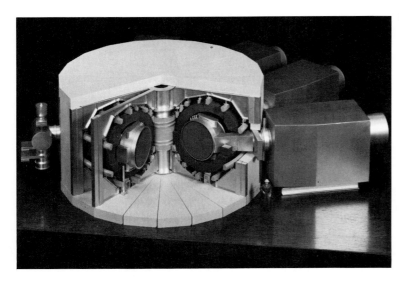

FIGURE 20–13
TOKAMAK. A magnetic
containment vessel for hydrogen
fusion. (a) Scale model. (b) The
real thing. (U.S. Department of
Energy photos.)

(b)

term option. It cannot solve the energy problems that
will probably grip the 1990s when world oil production begins to decline.

Many, then, believe that fusion power will continue to be the elusive pot of gold at the end of the
rainbow. If the break-even point is reached, fusion
energy still promises to be neither clean nor an unlimited resource. Current designs do not use regular
hydrogen but rather isotopes of hydrogen, namely

deuterium (^{2}H) and **tritium** (^{3}H). Fusion of regular
hydrogen (^{1}H) would demand even greater temperatures and pressures. Deuterium is a naturally occurring isotope that can be isolated in almost any desired amounts from the normal hydrogen in
seawater. Tritium, however, is an unstable radioactive isotope that must be produced. Current plans call
for the production of tritium by bombarding the element lithium with neutrons. The neutrons will be

FIGURE 20–14
Laser fusion. In this experimental instrument at Lawrence Livermore Laboratory, 30 trillion watts of optical power are focused onto a tiny pellet of hydrogen smaller than a grain of sand located in the center of this vacuum chamber. For less than a billionth of a second the fusion fuel is heated and compressed to temperatures and densities like those found on the sun. (U.S. Department of Energy photo.)

produced by the fusion of deuterium and tritium. The overall reaction is:

$$^2\text{H} \quad + \quad ^3\text{H} \quad \rightarrow \quad ^4\text{He} \quad + \quad n \quad + \quad \text{Energy}$$

^{2}H		^{3}H		^{4}He		n		Energy
Deuterium		Tritium		Helium		Neutron		
Isolated			↑					
from Water		^{3}H	+	^{4}He	←	n	+	^{6}Li
		Tritium		Helium		Neutron		Lithium

Lithium is not an abundant element and it could easily become the limiting factor in the wide-scale use of fusion reactors. Also, tritium is radioactive, hence very hazardous and difficult to contain. As a result, fusion reactors could easily become a source of radioactive tritium leaking into the environment. Finally, the hardware of the reactor itself will be embrittled and made highly radioactive by the constant bombardment from neutrons. Thus, there will be the cost of constantly replacing reactor components and the problem of disposing of components that have been made radioactive. Finally, fusion reactors promise to be the source of unprecedented thermal pollution. A steam turbogenerator itself is only 30 to 40 percent efficient and, if half the power produced must be fed back into the reactor to sustain the fusion process, the overall reactor is only 15 to 20 percent efficient. That is to say, 80 to 85 units of heat energy will be dissipated into the environment for every 15 to 20

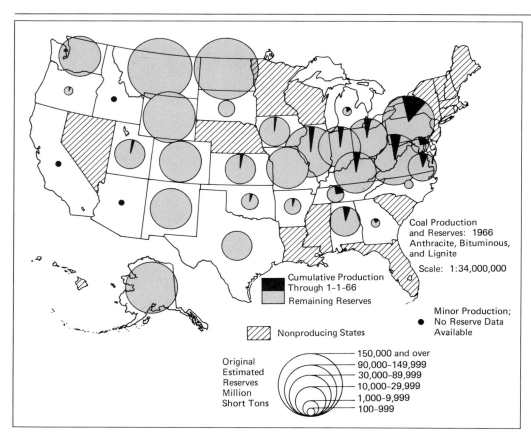

FIGURE 20–15
Major coal deposits in the United States: a solution or a potential ecological disaster? (U.S. Geological Survey.)

Coal Production and Reserves: 1966 Anthracite, Bituminous, and Lignite

Scale: 1:34,000,000

Cumulative Production Through 1-1-66
Remaining Reserves

Nonproducing States

Minor Production; No Reserve Data Available

Original Estimated Reserves Million Short Tons

150,000 and over
90,000–149,999
30,000–89,999
10,000–29,999
1,000–9,999
100–999

units of electrical energy produced. At best all this boils down to the fact that fusion power will be exceedingly expensive if it is achieved at all.

In the spring of 1989, there was great excitement regarding fusion power when two scientists announced that they had achieved the fusion of hydrogen atoms at room temperature. The claim was that electrolyzing heavy water (water in which the hydrogen atoms have been replaced by deuterium) using palladium electrodes, some of the deuterium diffused into the electrodes and fused, which generated heat in addition to that from the current applied for electrolysis (Fig. 21–17). If this proves to be the case, it will render the above negative arguments meaningless, and we may be on the way toward an exciting new power source. However, it is not yet clear whether the preliminary results can be confirmed.

In conclusion, fusion research may continue in an effort to increase human understanding of physical processes. However, it seems likely that solar technologies will be producing power less expensively and with fewer risks before fusion reactors become a reality.

COAL AND SYNTHETIC FUELS

Coal

In contrast to the rather limited reserves of crude oil, the United States is exceptionally well endowed with coal reserves. Huge reserves of coal remain unexploited in the United States (Fig. 20–15). These reserves could supply this country's needs for energy for over 100 years. Can this be the answer to the energy problem?

When the energy problem came to the forefront in the 1970s, the United States generated about 30 percent of its electrical power with oil and natural gas. Coal has since been substituted in many plants, and, to that extent, it has alleviated the problem of oil shortages. However, coal use is fraught with the environmental problems already discussed.

Most coal is in deposits that can be exploited practically only by stripmining. In underground mining, at least 50 percent of the coal must stay in place to support the mine roof. In stripmining, gigantic power shovels turn aside the rock and soil above the coal seam and then remove the coal (Fig. 20–16). It is evident that this procedure results in total destruction of the ecosystem. Although such areas may be reclaimed—that is, regraded and replanted—it takes many years before an ecosystem like the original is reestablished. In arid areas of the West, limits of water

are such that it is questionable whether the ecosystem could ever be reestablished. Consequently, strip-mined areas may be turned into permanent deserts. Furthermore, erosion and acid leaching from the disturbed earth may have numerous adverse effects on waterways and groundwater that drains from the site.

Also, coal-burning power plants have been identified as the major contributors to acid rain because sulfur in the coal converts to sulfur dioxide and then to acids in the atmosphere (Chapter 14). Acid-forming nitrogen oxides are also produced, along with particulates. In addition, burning coal generates many other wastes including trace elements such as arsenic, lead, and mercury, as well as small amounts of a number of radioactive substances. Finally, each ton of coal burned produces about three tons of carbon dioxide. Thus, burning coal is a significant contributor to the climatic warming effect.

Even with tighter regulations and controls regarding the mining and burning of coal, which seem extremely slow in coming, we will pay a high environmental and human health price for increasing coal consumption. Even then it may not solve the energy problem. Coal is practical for little more than the generation of electrical power. Now that coal is already substituting for oil in this area, further increasing its use only serves to substitute for nuclear power and other means of generated power. It will not help the oil situation.

Synthetic Fuels

A way in which coal might be used to help the oil shortage is through its chemical conversion to liquid compounds, which can then be substituted for oil-based fuels. Such coal-derived fuels are referred to as **synthetic fuels** or **synfuels.** With both government and corporate support, a great deal of research and development went into synfuel production in the late 1970s and early 1980s. While some government support remains, the major oil companies have now abandoned these projects because they proved too expensive, at least for the present. Profitable production of synthetic fuels would involve prices of $60 to $70 per barrel. In addition, if synfuel production becomes economically competitive, it promises to be an exceedingly "dirty" industry, which will generate numerous pollutants. It can only aggravate the negative impacts of stripmining and the greenhouse effect.

A second major potential source of synfuels is oil shale, a sedimentary rock that contains a tarlike organic material called **kerogen.** Vast formations of oil shale exist in several areas of the United States

FIGURE 20–16
(a) Coal being stripmined near Zanesville, Ohio. Forty-five feet of overlying rock and soil are being removed to get at a 3-foot seam of coal. (Fred McConnaughey/Photo Researchers, Inc.)
(b) Giant stripmine shovel located near Marissa, Illinois. (U.S. Department of Energy photo by Schneider.)

(Fig. 20–17). When oil shale is heated to about 600°C, the kerogen releases hydrocarbon vapors that can be recondensed to form a black, viscous, crude oil, which can in turn be refined into gasoline and other petroleum products. It is estimated that ultimately 600 billion barrels, about 100 years' supply of oil at current rates of use, might be recovered from oil shale; at least 50 billion barrels, 10 years' supply, could be recovered by current technology.

Nevertheless, oil shale is not the panacea for oil shortages either. In order to extract shale oil, the rock must be mined, crushed, and heated to distill off the kerogen, and then the rock waste must be disposed of. A ton of high-grade oil shale will yield little more

than half a barrel of oil. The mining, transportation, and disposal of wastes necessitated by an operation producing, for example, a million barrels a day would be a herculean task, to say nothing of its environmental impact. Perhaps for our environmental good fortune, oil shale, like oil from coal projects, has proven economically impractical for the present.

But in the long term, the most compelling reason to avoid all-out exploitation of every source of fossil carbon fuel is its inevitable exacerbation of the climatic warming caused by carbon dioxide.

In conclusion, we find that two of the most touted energy options, nuclear power and coal, are fraught with long-term risks to both human and en-

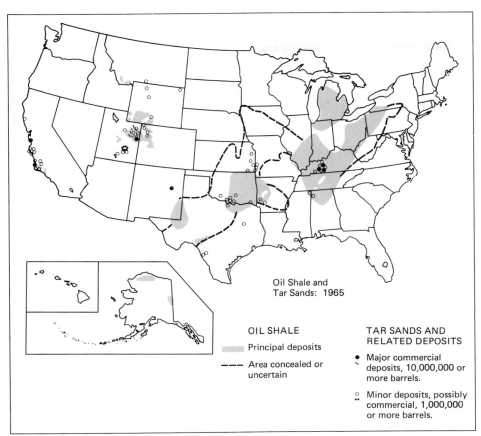

OIL SHALE

Principal deposits

- - - Area concealed or uncertain

Oil Shale and
Tar Sands: 1965

TAR SANDS AND
RELATED DEPOSITS

● Major commercial
deposits, 10,000,000 or
more barrels.

○ Minor deposits, possibly
commercial, 1,000,000
or more barrels.

(a)

(b)

FIGURE 20-17
(a) Map shows locations of major oil shale deposits in the United States. (U.S. Geological Survey.) (b) Self-sustaining combustion in high-grade "paper shale." Most shale will not burn but must be heated to remove oil-like material. (c) Anvil Points oil shale processing facility near Rifle, Colorado. (U. S. Department of Energy.)

(c)

vironmental health and neither are well-suited to meet the real problem—shortfalls in oil for transportation. Also, although the United States has relatively abundant reserves of coal and uranium, neither is a renewable resource. Opting to meet the energy problem by all-out exploitation of these resources would assuredly lead to another energy resources crisis within 100 years. Over the course of history, it is a very short period.

Recognizing these facts, we turn our attention to solar and other "renewable" energy options. Progress in these areas will be the subject of Chapter 21.

21

Solar and Other Renewable Energy Sources

CONCEPT FRAMEWORK

▪ Outline

I. SOLAR ENERGY   489

    A. Use of Direct Solar Energy   489

       1. Space and Water Heating   491

CASE STUDY: THE DAVIS EXPERIENCE   496

       2. Solar Production of Electricity   495
          a. Photovoltaic cells   497
          b. Power towers   500
          c. Solar ponds   500

       3. Solar Production of Hydrogen   501

    B. Use of Indirect Solar Energy   503

       1. Biomass Energy or Bioconversion   504
          a. Direct burning   504
          b. Methane (natural gas) production   505
          c. Alcohol production   505

       2. Hydropower   507

       3. Wind Power   508

II. GEOTHERMAL, TIDAL, AND WAVE POWER   510
    A. Geothermal Energy   510

    B. Tidal Power   512

    C. Wave Power   512

III. CONCLUSIONS   512

    A. Need for Conservation   514

▪ Study Questions

1. Define solar energy. What are its virtues? Drawbacks? Distinguish between direct and indirect solar energy.

2. List ways in which direct solar energy may be used.

3. Describe how solar energy may be used for heating water and space. Distinguish between active and passive heating systems. Which is more advantageous and why? Discuss how problems of heat storage may be overcome.

4. List and describe three methods for producing electricity by direct solar energy. How do these methods compare with nuclear or coal-generated power in terms of cost? Environmental impact?

5. Can hydrogen be used as a substitute fuel for vehicles? What is the drawback? Discuss the potential for solar production of hydrogen.

6. List and describe forms of indirect solar energy.

7. Define biomass energy and bioconversion. Describe three ways of converting biomass into useful energy. Discuss the potential and environmental drawbacks of each.

8. Discuss the potential and environmental drawbacks for developing more hydropower from large dams; from small dams.

9. Discuss the progress and potential in developing wind power. What kinds of wind turbines are most successful so far? Compare the cost-effectiveness and environmental impact of wind power with nuclear and coal-generated power.

10. Define geothermal energy. How is it or may it be obtained and used? Discuss its potential and environmental drawbacks.

11. Discuss the potential and environmental drawbacks for tidal and wave power.

12. Discuss what conclusions may be drawn from a study of alternative energy sources. Specifically, address: Is there a need for building more nuclear and/or coal-fired power plants? Is there a ready, environmentally sound substitute for crude oil to fuel vehicles?

13. Discuss why conservation should still be a top priority.

B. Energy Policy 514

C. What You Can Do 516

14. Cite and describe the requirements of certain laws that have aided the development of alternative energy sources or conservation. What is their current status? Outline the elements of a sound energy policy. What items should be given highest priority? What items currently receiving support could/should be defunded? Give arguments to support your position.

15. What can you do: To promote a sound energy policy? To conserve energy? To make yourself less vulnerable to energy shortages?

A major principle underlying the sustainability of natural ecosystems, as we emphasized in Chapter 2, is that they run on solar energy, which is abundant, everlasting, and nonpolluting. In contrast, we have seen in the last two chapters that reliance on fossil fuels and nuclear power is fundamentally nonsustainable because of resource depletion and/or polluting waste products. Progress toward a sustainable society will parallel progress in weaning ourselves from dependence on fossil and nuclear sources and developing our ability to use solar energy sources. The potential for meeting all our energy requirements through solar energy is real and promising. In this chapter we shall look at the range of technologies that are available and consider what is necessary to promote their development and implementation. We shall also consider some other "renewable" energy sources and the potential of conservation (Fig. 21–1).

SOLAR ENERGY

Solar energy is kinetic energy of radiation, mainly light, originating from thermonuclear reactions in the sun (Fig. 21–2). Since using solar energy will not deplete the sun in any way—astronomers estimate that the sun will "burn" for another several billion years—it is commonly referred to as a **renewable energy** source. In natural ecosystems, a small portion, less than 1 percent, is absorbed by the chlorophyll in plant leaves and used for photosynthesis, the production of organic matter from carbon dioxide and water. Energy is thus captured and stored as the high potential energy in organic matter. The organic matter then supplies the energy needs for the rest of the ecosystem.

It is calculated that a similarly small percentage of solar energy would be more than enough to meet all our energy demands for transportation, industry,

and residential purposes, not only now but for the foreseeable future. Furthermore, whether or not we use this energy to perform useful work before it is radiated back into space will not change the overall energy balance of the earth or upset the biosphere in any way.

The abundance of solar energy, however, stems from the fact that it falls over the entire earth; nowhere is it particularly intense. Thus, the hurdles to be overcome in using solar energy involve absorbing it over a broad area, and concentrating and converting it to forms that can power our transportation, residential, and industrial needs. Additionally, means for storage may be necessary to provide energy during dark or cloudy periods. The difficulties or costs of overcoming these hurdles have given solar energy the reputation of being impractical, at least for the present. However, in many cases these problems are overrated. The key to using solar energy is in applying it to situations where these costs are minimal or may be avoided altogether. As technological development lowers costs and traditional energy sources become more expensive, solar energy may be extended to additional applications.

Light energy may be captured and used directly as it comes from the sun; this is referred to as **direct solar energy**. Additionally, in the biosphere, winds, the water cycle, and the production of biomass are driven by solar energy. Thus, taking energy from these sources is, in effect, **indirect solar energy.**

Use of Direct Solar Energy

Direct solar energy is ideally suited for space heating of buildings and providing hot water. Through photovoltaic (solar) cells, it can be used to produce electrical power, and potentially it can be used to produce hydrogen gas, which may be substituted for natural gas and liquid fuels. We shall consider each of these areas in more detail.

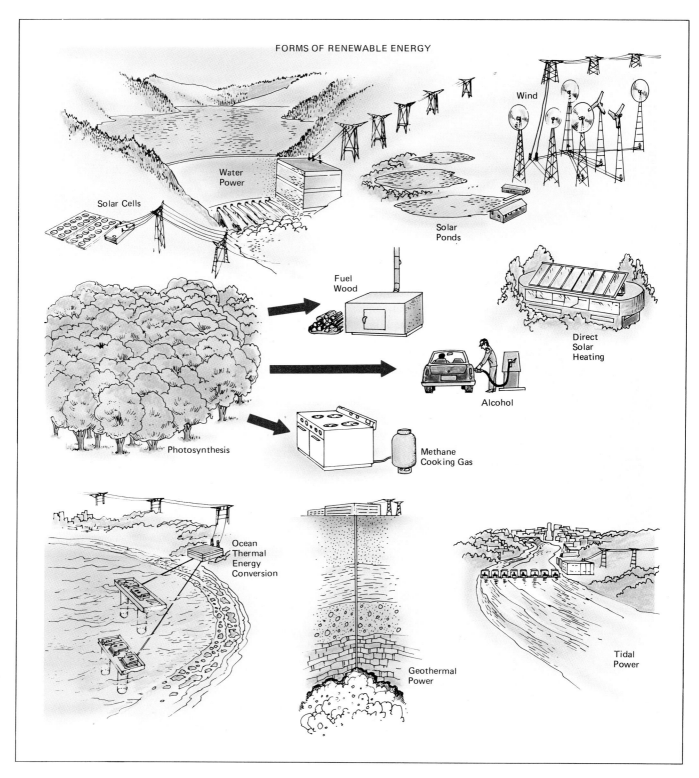

FORMS OF RENEWABLE ENERGY

Solar Cells

Water Power

Solar Ponds

Wind

Fuel Wood

Alcohol

Direct Solar Heating

Photosynthesis

Methane Cooking Gas

Ocean Thermal Energy Conversion

Geothermal Power

Tidal Power

FIGURE 21–1
Renewable energy resources. The natural energy sources shown will continue for hundreds of millions of years with little change, and they will not be altered appreciably whether or not humans utilize them. Therefore, these energy resources are referred to as *renewable*.

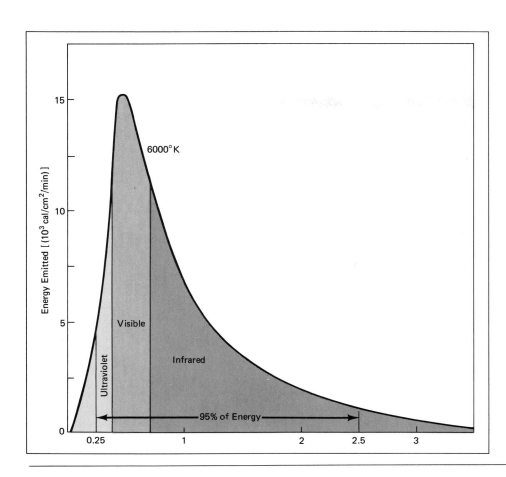

FIGURE 21–2
The solar energy spectrum. The greatest output of solar energy is in the visible light part of the spectrum. From: Joe R. Eagleman, *Meteorology: The Atmosphere in Action*, Copyright © 1980 by Litton Educational Publishing, Inc. Reprinted by permission of Wadsworth Publishing Co.

Space and Water Heating

About 25 percent of the total U.S. energy budget is used for space heating of homes and buildings in cold weather and for providing hot water for washing. The temperatures involved here are modest, 68° to 72° F (20°–22° C) for space and 120° to 140° F (50°–60° C) for hot water. Using a gas or oil flame with a temperature of over 1000° C to provide this *low-temperature heat* is an extremely inefficient use of this energy. Using electrical power for this purpose is even more wasteful. It is like using a rifle to kill a fly.

Solar energy, on the other hand, is ideally suited for this task because sunlight falling on any black surface is readily absorbed and converted to heat in this temperature range. You have probably experienced this in trying to walk barefoot on an asphalt pavement in the sun. Thus, complex collection or conversion equipment is not required. A *flat-plate collector* will suffice.

There are countless variations of **flat-plate collectors,** but all are basically composed of a black surface, with a clear plastic or glass "window" over it. The black surface absorbs the light and converts it to heat, and the window prevents the heat from escaping (Fig. 21–3); recall the greenhouse effect described

FIGURE 21–3
The principle of a flat-plate solar collector. Sunlight is converted to heat as it is absorbed by a black surface. A glass or clear plastic window over the surface allows the sunlight to enter but traps the heat. Air or water is heated by passing it over or through tubes in the surface.

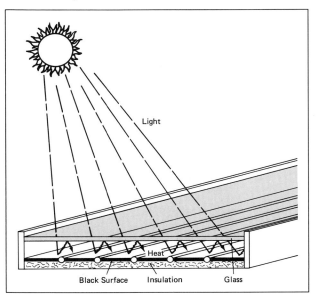

in Chapter 13. Air is heated by passing it between the window and the black surface and water may be heated by passing it through tubes in the surface. Thus, there is minimal cost in the collection and conversion of solar energy to heat.

Beyond the collector, however, heating systems may be *active* or *passive* and may or may not include a means of heat storage. An **active solar heating system** is defined as one that uses pumps or blowers to actively circulate the air or water from the collector to the desired location. A **passive solar heating system** relies on natural convection currents (the fact that hot air or water rises) to move the air or water. A schematic diagram of an active system with storage is shown in Figure 21–4. Note the number of pumps, blowers, valves, and plumbing features. Active systems may work, but they are costly to install and both troublesome and costly to maintain. Such systems have given solar heating the undeserved reputation of being economically impractical, but the problem is in the failure to follow what engineers call the "KISS" principle, which stands for *Keep It Simple, Stupid!*

Passive systems can be relatively inexpensive and maintenance-free. Innumerable plans are available for passive solar homes, but the basic concept is that shown in Figure 21–5. Relying on large south-facing windows, the building itself acts as the collector. In winter, sunlight beams through the window, heating the interior; at night, insulated drapes or

shades are pulled to trap the heat inside. To avoid excessive heatload in the summer months, an awning or overhang can shield the window from the high summer sun. Deciduous trees are also exceptionally efficient in providing cooling in summer and letting solar energy through in the winter. Thus, suitable architecture that may cost no more than conventional designs can provide heating and cooling that is virtually free.

A common criticism of solar heating, active or passive, is that a backup heating system is still required for periods of especially inclement weather. While this is true, it misses the point. Even if solar provides only 20 percent of the overall heating needs, this still reduces the demand for conventional fuels by 20 percent. Since our need is to accommodate to decreasing availability of oil-based fuels, every unit of savings is significant. Also, a 20 percent savings on your fuel bill may, depending on your solar investment, still be very cost-effective. Then, as solar technologies become more advanced and more widely used, we can expect the backup needs to decrease. Finally, in a passive solar home, a small wood stove, which is an indirect form of solar energy, often provides adequate backup.

The need for backup heating is also decreased by increasing heat-storage capacity. Early solar designs involved circulating air through a "reservoir" of rocks, which readily absorb excess heat and then

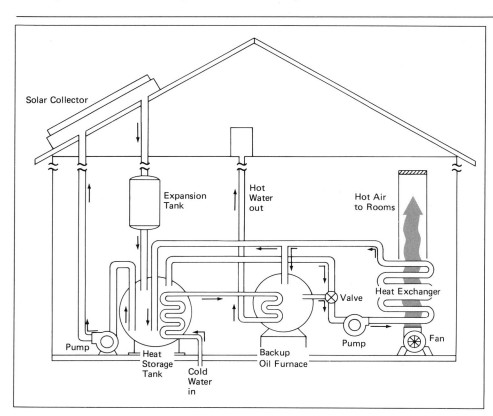

FIGURE 21–4
An active solar heating system with heat storage. As shown, a solar system that will provide all heating needs and function automatically like current central heating systems can be designed. In practice, however, such systems have generally proven to be cost-ineffective and fraught with maintenance problems. Note the amount of piping and number of pumps and blowers. (From Bruce Anderson, with Michael Riordan, *The Solar Home Book*, p. 118. Harrisville, N.H.: Brick House Publishing Co., 1976.)

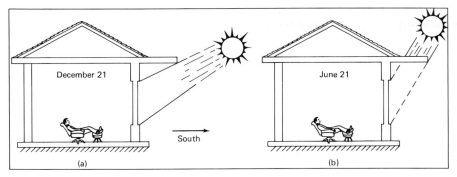

FIGURE 21-5
Passive solar heating. In contrast to expensive active solar systems, solar heating may be achieved by suitable architecture and orientation of the home at little or no additional cost. (a) The fundamental feature is large sun-facing windows that permit sunlight to enter during winter months. Insulating drapes are pulled to hold in the heat when the sun is not shining. (b) Suitable overhangs, awnings, or deciduous plantings will prevent excessive heating in the summer. (From Bruce Anderson, with Michael Riordan, *The Solar Home Book*, p. 87. Harrisville, N.H.: Brick House Publishing Co., 1976.)

FIGURE 21-6
Passive solar heating may include additional heat storage. Rocks are the preferred heat-storage material because they readily absorb and give off heat, and they are inexpensive. Still, heat storage such as this is generally not cost-effective. A more cost-effective plan is to incorporate heat storage into the architecture of the home in the form of interior stone or brick walls.

release it as the temperature drops (Fig. 21-6). However, in terms of the units of heat stored and later used it is the most expensive component of the solar installation. Experience has shown that such storage is not cost-effective. In contrast, efficient storage may be incorporated into the design of the building itself. By good insulation, insulating drapes that are closed to hold heat, and the use of interior brick or stone, the interior of the building itself is the heat-storage unit. It is ironic that in terms of solar energy, we traditionally construct buildings inside out. For solar efficiency, the insulation should be on the outside and the brick or stone on the inside.

The Center for Renewable Resources holds that a well-designed passive solar home can reduce en-

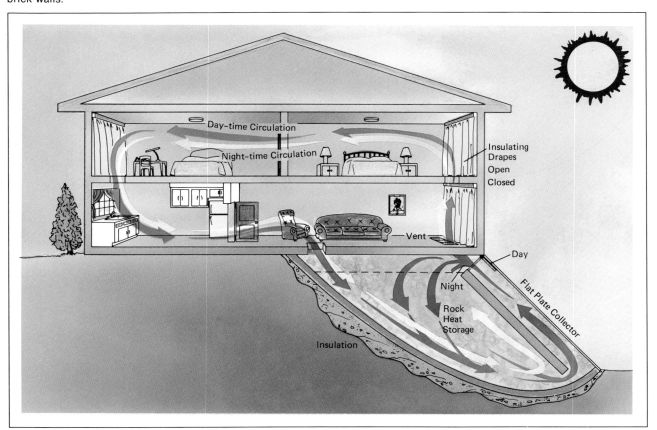

FIGURE 21–7
Many homeowners could gain heating economy by adding solar collectors to south-facing, exterior walls. (a) Do-it-yourself panels can be made from inexpensive materials as shown. (Bruce Anderson with Michael Riordan, *The Solar Home Book,* p. 118. Brick House Publishing Co., 1976.) (b) Home with retrofitted solar panels. (Photo by author)

(a)

(b)

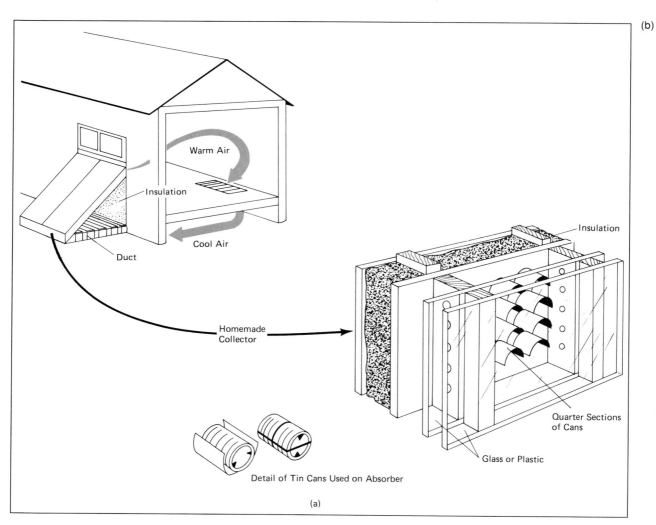

Warm Air

Insulation

Cool Air

Duct

Homemade Collector

Insulation

Quarter Sections of Cans

Glass or Plastic

Detail of Tin Cans Used on Absorber

(a)

ergy bills by 75 percent with an added construction cost of only 5 to 10 percent in almost any climate. This translates into a savings on fuel that is greater than the cost (interest) of borrowing the additional money. What about existing homes? In innumerable situations, passive solar features can be retrofitted into existing homes (Fig. 21–7). Such additions, especially if owner-installed, may pay for themselves and provide savings over and above the investment very quickly.

The Center for Renewable Resources estimates that over 1 million homes in the United States obtain at least part of their heating from direct sunlight, but this is less than 1 percent of the total number of homes. It is disappointing that more than 15 years after recognizing the energy crisis and the virtues of solar energy, we are still building and heating most homes in traditional ways and ignoring the potential benefits of the sun. The Center for Renewable Resources concludes, "The chief barrier to more widespread use of passive solar design is ignorance. Many builders and consumers are unaware of the potential benefits of passive solar design. . . . [Further,] they are often ignored by policy makers and have received meager government support." At least one factor in helping to maintain this "state of ignorance" has been intensive advertising campaigns promoted by utility and oil companies purporting that solar energy is not practical or cost-effective at present. It should not escape notice that these companies make a profit on each unit of energy regardless of its cost, where it comes from, or its future availability.

One place where heating and cooling efficiency standards achievable with solar-conscious designs have been incorporated into building codes is in Davis, California. This was due largely to the efforts of a group of graduate students at the University of California at Davis. (See Case Study 21.)

Much the same holds true for solar water heating. About one-third to one-half of the total energy consumed in an average household is for heating water. Solar water heating systems, which provide savings year-round as opposed to only in winter, may be even more cost-effective than space heating systems. In nonfreezing climates, passive systems suffice. In climates where freezing occurs, circulating antifreeze systems are preferred (Fig. 21–8). Again, a solar water heater may not provide the full heat desired, requiring traditional backup heating. However, a solar water heater that heats the water only halfway still saves half the conventional fuel because it takes the same amount of energy to heat water from 10° to 11° C as from 49° to 50° C.

On a per capita basis, Cyprus is the world's largest solar energy user; 90 percent of the homes and a large portion of the hotels and apartment buildings have solar water heaters. In Israel, solar energy provides some 65 percent of domestic hot water. In the United States, there are an estimated 800,000 solar hot water systems operating, but this represents only about one-half of 1 percent of the total.

In conclusion, there is tremendous opportunity to cost-effectively apply solar energy to space and water heating needs and thereby make the fuels that are now used for these purposes available for other purposes or future needs.

Solar Production of Electricity

Sunlight can be used to produce electrical power through photovoltaic cells or by generating high temperatures to produce steam to drive turbogenerators.

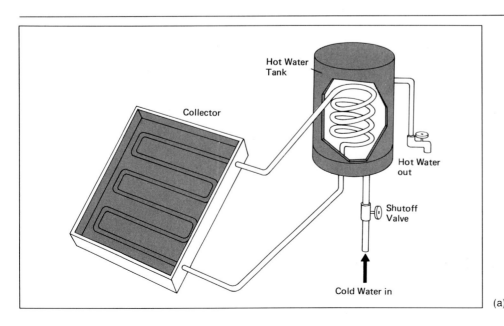

FIGURE 21–8
(a) Solar water heater. Since needs for hot water exist all year, solar water heaters may be very cost-effective. In nonfreezing climates, simple water convection systems may suffice. Where freezing occurs, an antifreeze fluid is circulated. (*cont.*)

Hot Water Tank

Collector

Hot Water out

Shutoff Valve

Cold Water in

(a)

FIGURE 21–8 (cont.)
(b) Solar panels supply 82 percent of the water heating required by this laundromat in Los Angeles, California. (Tom McHugh/Photo Researchers)

Case Study 21

THE DAVIS EXPERIENCE

I came to the University of California at Davis in 1970 as an Ecology graduate student. Committed to making positive environmental change, I began to acquaint myself with people involved in local politics. Simultaneously I organized a group of fellow graduate students to prepare a study on key Davis environmental problems including urban sprawl onto agricultural land and high household energy use. Working with a like-minded local politician, we formed our studies into a political platform and elected a sympathetic slate of city council officials.

One blustery fall day in 1972, while walking across the campus, I suddenly was struck by the inspiration that I could develop an energy conservation building code for the City of Davis. I realized that in the new political climate, the idea of a building code to encourage the use of energy conservation and solar energy had become viable.

A team of colleagues and I received funding to research and develop a building code appropriate for Davis. Being convinced that empirical data from actual houses would best demonstrate the benefits of such a code, we put recording thermographs in identical unoccupied apartments facing in various directions and measured their temperatures during the warmest and coldest months of the year. Not surprisingly, we found that the apartments with large south facing windows were warmer on sunny winter days and cooler on summer days than buildings with either east, west, or north facing windows. Energy bills from identical occupied apartments supported our findings.

Using engineering calculation techniques for heat loss and heat gain over a 24-hour "design day," we developed a code that required minimum standards of energy efficiency and that gave credit for elements that would maximize solar heat gains during the winter and prevent overheating in the summer. We also developed new neighborhood planning standards to encourage bicycle use and good solar access.

Just as we were releasing our report, we received an unexpected boost from a national crisis, the Arab oil boycott of 1973, which caused a rapid escalation of energy prices and a shortage of gasoline. We presented our findings to a variety of citizens' groups, and people began taking our suggestions very seriously. The Davis City Council asked us to work with a committee of local citizens including the head building inspector and prominent local builders to develop a building code. After over a year of negotiations and rewritings, our code was ready for submission to the city council for actual adoption.

Meanwhile Davis was becoming a center of environmental experimentation. Other designers and my company, Living Systems, began designing

Photovoltaic Cells. The most promising technology for solar production of electricity is the **photovoltaic** or **solar cell.** Solar cells consist of sophisticated materials in which light energy striking the materials causes a flow of electrons, that is, an electrical current, in the material. Thus, with no moving parts, they convert light energy directly to electrical power (Fig. 21–9). More specifically, the cell consists of two very thin layers. The lower layer has atoms with single electrons in the outer orbital; such electrons are easily lost. The upper layer has atoms lacking one electron from their outer orbital; such materials readily gain electrons. The kinetic energy of light striking this "sandwich" dislodges electrons from the lower layer and they go into the upper layer. This creates an electrical imbalance or voltage; the lower layer, which is missing electrons, is positive, and the upper layer with surplus electrons is negative. The circuit is completed by connecting a wire to the two sides so that the "surplus" electrons continue their flow from the upper side through a motor or other device and back to the lower side (Fig. 21–10). Current solar cells achieve a conversion efficiency of between 10 and 20 percent.

Under full sunlight, a single 2-inch (5-cm) diameter solar cell will provide about the same power output as a standard flashlight battery. However, solar cells can be connected together to obtain any amount of power desired (Fig. 21–11). Since there are no moving parts, solar cells do not wear out. What limits their life is deterioration and breakdown of the materials due to exposure to the weather. However,

passively heated solar houses. Local builders began making innovative neighborhoods. These ongoing projects both were encouraged by the code and helped to reinforce and assure its eventual passage. For example, it was effective to take people to a solar house on a day when the weather was 110 degrees and have them experience 72-degree inside temperatures achieved with no mechanical cooling system.

The most encouraging aspect of the Davis project was that what had once been the idea of a very few individuals became a creative force that helped a whole community to solve complex problems of energy utilization. Davis became a model for other communities as well, and eventually the State of California adopted an energy conservation building code inspired by the Davis experience.

JONATHAN HAMMOND
University of Illinois

Solar home in Davis, California. (Photo by Jonathan Hammond.)

FIGURE 21-9
The thin "wafer" of material with wires attached is a photovoltaic cell. Converting light to electrical energy, this cell provides enough energy to run a small electric motor. (Photo by author.)

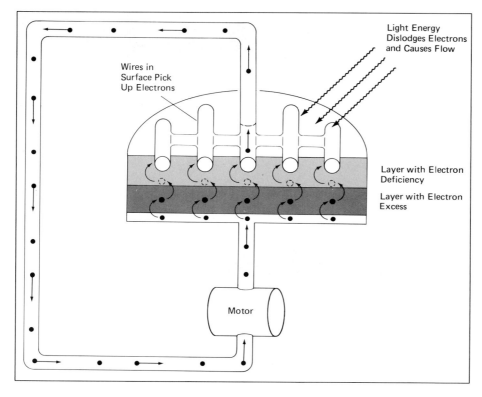

Light Energy Dislodges Electrons and Causes Flow

Wires in Surface Pick Up Electrons

Layer with Electron Deficiency

Layer with Electron Excess

Motor

FIGURE 21-10
The photovoltaic cell is actually a "sandwich" of two very thin layers of silicon treated in such a way that one layer tends to have a surplus of electrons while the other layer tends to have a deficiency. When light energy strikes the cell, electrons are dislodged and tend to pass from one layer to the other, producing the electric current.

average solar cells easily last 20 years, and this may be extended with development of more-resistant materials.

The photovoltaic effect was discovered more than 100 years ago but, until recently, the high cost of producing them kept their use limited to space satellites and signal devices in remote locations where other power sources were either unavailable or ex-ceedingly expensive. However, as less-expensive production techniques have been developed, costs have decreased dramatically, from 55 dollars per watt output in 1975 to about 5 dollars per watt currently, and production and use have increased correspondingly (Fig. 21-12).

Where relatively small amounts of power are required, photovoltaic cell assemblies are now more

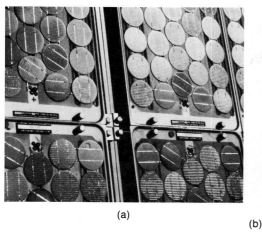

(a)

(b)

FIGURE 21–11
Solar cells as a power source. (a) A sufficient number of photovoltaic cells may be wired together to produce any desired amount of power. (b) An array of solar cells being used to power an irrigation project near Mead, Nebraska. (Department of Energy Technology Visual Collection)

cost-effective than replacing batteries or running power lines long distances. Therefore, their use is growing rapidly for such things as pocket calculators; telephone, television, and radio relay stations in remote areas; railroad signals; ocean buoys and distant lighthouses; offshore oil drilling platforms; and irrigation projects (Fig. 21–11). Rural electrification

FIGURE 21–12
World photovoltaic shipments and average market prices, 1975–1986. (From Shea, Cynthia Pollock, "Renewable Energy: Todays Contribution, Tommorrow's Promise," *Worldwatch Paper 81*, p. 32. Washington, D.C.: Worldwatch Institute, 1988.)

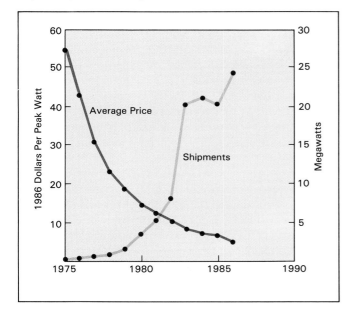

projects based on photovoltaic cells are beginning to spread throughout the Third World.

A 1000-megawatt (1 megawatt equals a million watts) nuclear power plant costs about 5 billion dollars now or 5 dollars per watt output. Thus photovoltaic power is already cost competitive with nuclear power and, when the hidden costs of nuclear power are considered, photovoltaic power already has a considerable advantage. With just a slight further reduction in production costs, which will probably be achieved in the near future, building photovoltaic power stations will be cost competitive with coal power.

Beyond the obvious environmental advantages, photovoltaic power stations are attractive to utilities because small units can be constructed and put into operation quickly and enlarged as the need develops. This eliminates the financial pitfalls associated with projecting needs many years in advance and making the huge upfront investments as are required for construction of nuclear power plants. The Pacific Gas and Electric Company in conjunction with ARCO Solar has already constructed a demonstration photovoltaic power plant near Bakersfield, California (Fig. 21–13). It has a capacity of 6.5 megawatts at peak, enough to power 2400 homes. Prospects for more such plants seem assured. It may also soon be cost-effective for consumers to mount photovoltaic panels on their rooftops to generate a portion of their own electrical needs. Particularly, solar powered air-conditioners, which could operate independently from the rest of the electrical system thus avoiding the costs of interconnecting systems, seem feasible.

FIGURE 21–13
World's first photovoltaic (solar cell) power plant located near Bakersfield, California. The array of 34-foot panels produces 6.5 megawatts at peak, enough for 2400 homes. (Photo courtesy of Pacific Gas and Electric Company)

Again, the problem that sunlight is not available during the nighttime is not as serious as might be supposed. Some 70 percent of electrical use occurs during daylight hours, when industries and offices are in operation. The demand for air conditioning, the largest single use of electrical power, correlates almost 100 percent with intensity of sunlight. If solar-generated electricity provided for the increased daytime load, existing facilities could carry the remaining load for many years to come. There are numerous options for storage, but there is no urgent need to develop them in the near future.

Power Towers. You have probably used a magnifying glass to focus sunlight onto a tiny spot to burn a hole through a piece of paper. A "power tower" is the ultimate expression of this concept. An array of sun-tracking mirrors is used to focus several acres of sunlight onto a boiler mounted on a tower (Fig. 21–14). The intense heat generates steam to drive a conventional turbogenerator. The facility shown in Figure 21–14, located in southern California, began operating in 1983, and some 17 similar facilities have been constructed since, 9 in California, others in Israel, France, Spain, Japan, Italy, and the Soviet Union. Such facilities are cost-competitive with nuclear power and have none of the environmental hazards.

Solar Ponds. Solar ponds are another even lower-cost method for collecting and storing solar energy.

An artificial pond is partially filled with brine (very salty water) and fresh water is placed over the brine. Because it is much denser than fresh water, the brine remains on the bottom and little or no mixing occurs. Sunlight passes through the fresh water but is absorbed and converted to heat in the brine. The fresh water then acts as an insulating "blanket," which holds the heat in (Fig. 21–15). In short, solar ponds produce a type of greenhouse effect with fluids. The hot brine solution can be circulated through buildings for heating, or it can be converted to electrical power by vaporizing fluids with low boiling points and using the vapors to drive low-pressure turbogenerators. Since the pond also acts as a very effective heat storage unit, a solar pond will supply power continuously. Israel has pioneered the development of solar ponds. A 48-megawatt solar pond is now under construction in California.

A drawback of all these methods of solar-produced power is that considerable area is required for collection, which must be located reasonably close (within 50 miles) to the place where the power is used. Otherwise, an unacceptable amount of power is lost in transmission. This handicap might eventually be overcome by development of superconducting transmission lines but, for the foreseeable future, finding suitable areas for solar ponds or power towers near many cities could pose a problem. Solar cells, on the other hand, can be dispersed on rooftops of buildings (Fig. 21–16).

FIGURE 21–14
The "power tower" method of producing electrical power from sunlight. Sun-tracking mirrors are used to focus a broad area of sunlight onto a boiler mounted on a tower in the center. The steam produced is used to drive a conventional turbogenerator. (a) A facility in southern California under construction. (U.S. Department of Energy.) (b) The completed facility. (Courtesy of Southern California Edison Co.)

(b)

Solar Production of Hydrogen

With crude oil and natural gas supplies diminishing, hydrogen has frequently been advocated as the "fuel of the future." Hydrogen gas (H_2) is highly flammable and can be used in place of natural gas with little change in distribution networks or furnaces. Additionally, cars have been run on hydrogen gas with only minor modifications of the carburetor. Also hydrogen is clean burning; the only waste product is water vapor.

$$2 H_2 + O_2 \rightarrow 2 H_2O + \text{kinetic energy}$$

Thus, pollution problems would be greatly reduced.

The only slight drawback (understatement) is that essentially *no free hydrogen exists on earth*. It has already been oxidized to water. Therefore, any hy-

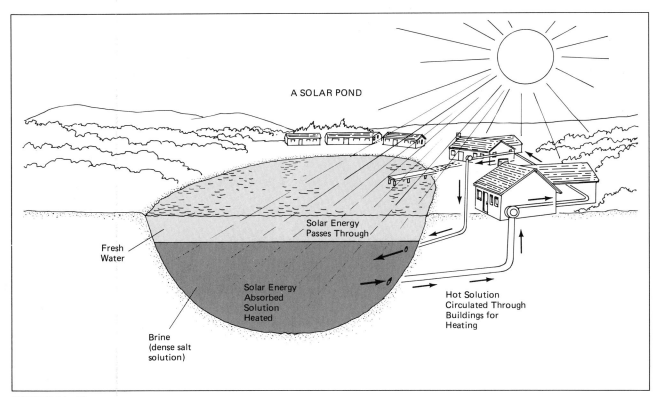

FIGURE 21–15
Solar ponds. Heat may be trapped and stored in brine (very salty water) covered with a layer of fresh water. Effectively the pond acts as a large flat-plate collector. The hot brine may be used for direct heating or to vaporize low-boiling-point fluids to drive low-pressure turbines.

FIGURE 21–16
Georgetown University's Intercultural Center (Washington, D.C.) supports an array of 4400 photovoltaic modules providing a power output of 300 kilowatts under full sun. Such rooftop arrays may become commonplace in the future as production costs of photovoltaic cells are reduced. (U.S. Department of Energy.)

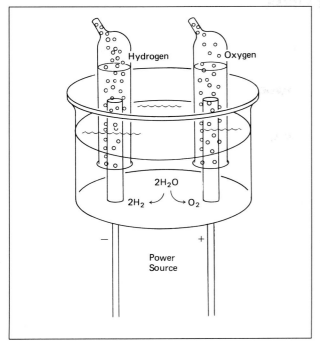

FIGURE 21–17
Electrolysis of water—one way to produce hydrogen gas. An electric potential applied to water causes its dissociation into hydrogen and oxygen. But the energy content of the hydrogen obtained only amounts to about 20 percent of the electrical energy used.

larger amounts of energy are needed to cause the disassociation than are released when the hydrogen and oxygen are recombined in burning. Indeed, this process is only about 20 percent efficient—100 calories of electrical energy are used for every 20 calories' worth of hydrogen produced. Consequently, hydrogen cannot become a widely used fuel unless there is an inexhaustible, nonpolluting energy source to produce it. Solar energy fits the bill.

In the initial reactions of photosynthesis, hydrogen atoms are disassociated from water molecules. It is the remaining oxygen atoms from water that are released into the atmosphere as oxygen gas. Laboratory experiments have shown that it may be possible to duplicate this initial step of photosynthesis in an artificial system (Fig. 21–18). But the process must be scaled up to commercial size at affordable costs. Unfortunately, the amount of money allocated to this research has been extremely limited, although the concept holds much greater promise than fusion nuclear power, for example, which continues to receive about 350 million dollars per year for research and development.

drogen to be had must be produced. This can be done through the **electrolysis** of water; electrical energy passed through water causes water molecules to disassociate into hydrogen and oxygen (Fig. 21–17). However, by the second law of thermodynamics,

Use of Indirect Solar Energy

Use of wind, water power, and biomass are considered indirect forms of solar energy because light provides the initial energy input. Likewise they are considered renewable energy resources.

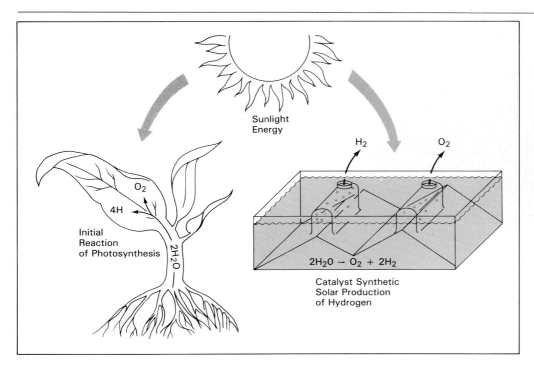

FIGURE 21–18
Solar production of hydrogen gas. (a) In photosynthesis, light energy causes the splitting of water into hydrogen and oxygen. (b) Research is being conducted to accomplish this in a stable artificial system. Some experiments have proven partially successful. If success is achieved, we may have solar production of a fuel (hydrogen) that can readily be substituted for gasoline.

Biomass Energy or Bioconversion

Biomass refers to all the organic material derived from photosynthetic production. **Biomass energy** or **bioconversion** refers to the use of this material as fuel or its conversion into fuels for external needs as opposed to food. A number of options are available (Fig. 21–19). It may be burned directly or it may be converted to methane (natural gas) or alcohol, which may be used as fuels.

Direct Burning. Wood stoves have enjoyed a tremendous resurgence in recent years. About 5 million homes in the United States rely entirely on wood for heating and another 20 million use it for partial heating, so much so that air pollution from wood stoves has become a significant problem in some communities, and restrictions are being applied.

In addition, many sawmills and woodworking companies burn wood wastes and a number of food processors, sugar producers for example, burn crop residues to supply all or most of their power. The burning of municipal refuse, which is largely wastepaper, for power production is another example (see Chapter 18). Some electric power plants that are fueled entirely by firewood are in operation. In these cases, air pollution is controlled by precipitators.

Some studies conclude that wood and wood wastes could supply as much as 20 percent of U.S. energy needs. However, the amount of cutting necessary to supply more than a small percent of the nation's energy needs could have severe environmental impacts. Access roads used in harvesting firewood, as well as the cutting itself, would greatly aggravate erosion. Managing forests for wood production would diminish the diversity of natural biota. Pest control problems might increase as a result of forest uniformity. Illicit cutting on both private and public lands might become a much greater problem than it already is.

Then there is the problem of keeping harvests within the limits of a maximum sustainable yield (see Chapter 18). As discussed in Chapter 7, over a billion people living in poverty in Third World countries do depend upon firewood as their only fuel for cooking—they have never used more sophisticated sources of energy. The deforestation from firewood cutting is causing severe erosion and desertification around many cities and villages in Africa and Asia. Indeed, Erik Eckholm, a senior researcher with Worldwatch Institute, says that "the famine in Africa has its origins in erosion and soil degradation resulting from deforestation in the search for firewood."

Therefore, burning wastes and wood where it is abundant and can be harvested without upsetting the forest ecosystem can be advocated, but this would

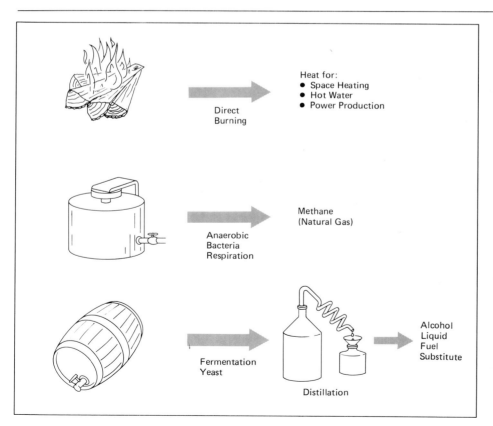

FIGURE 21–19
Bioconversion. As depicted, there are a number of ways of using biomass. It may be burned directly or converted into gas (methane) or liquid (alcohol) fuels.

supply no more than about 5 percent of our energy needs. A blanket recommendation to burn wood because in theory it is renewable, however, is environmentally irresponsible. Indeed, saving many Third World countries from ecological disaster will require finding alternative fuels to wood.

Methane (Natural Gas) Production. Bacteria feeding on organic matter under anaerobic conditions results in the production of *biogas,* which is about two-thirds methane gas. Recall the production of methane from sewage sludge described in Chapter 10 and the production of methane from landfills described in Chapter 18.

Using biogas for power production has considerable potential. For example, Richard Waybright, owner of the Mason-Dixon Dairy near Gettysburg, Pennsylvania, has adapted his operation so that manure from the cows (about 2000) is fed into anaerobic digesters and the biogas produced (purification of the biogas has proven unnecessary) fuels engines that drive generators supplying all the electrical power for the dairy and considerable excess, which is sold to the utility company. Waste heat from the engines is used to heat the digesters and also buildings. The nutrient-rich sludge remaining after digestion is complete is recycled back to maintain the fertility of the fields growing forage for the cows (Fig. 21–20). Between the savings on energy he does not have to buy and energy he sells, Mr. Waybright estimates that he makes close to a hundred thousand dollars a year on his energy system, and there are additional savings on fertilizer. He estimates that the system cost about 80 cents per watt to install, which compares most favorably with the 3 dollars per watt for coal and 5 dollars per watt for nuclear power.

If all dairy farms in the United States followed this example, we could be getting more electrical power from cows than we are getting from nuclear power at about 20 percent of the cost and the system would be sustainable and without adverse environmental consequences. Sewage sludges (human manure) could be processed and recycled similarly. However, so far very few have picked up on this concept. This is to say that there is a vast potential to be exploited in this area.

On a more individual level, millions of small farmers in China maintain a simple digester in the form of a sealed pit into which they put agricultural wastes. The biogas produced is used as a source of gas for cooking. This concept could be expanded throughout all the Third World providing an alternative to fuelwood.

Alcohol Production. Yeast feeding on sugars and/or starch under anaerobic conditions produces alcohol as a metabolic waste product; the process is known as *fermentation.* The alcohol can be further concentrated by *distillation,* boiling off and recondensing the alcohol. Fermentation and distillation have been used for millennia in the production of drinking alcohol. Here we are simply considering producing a sufficient quantity for fuel use as well.

Brazil has pioneered the development of large-scale fermentation plants that utilize sugarcane. A large percentage of Brazil's cars now run on alcohol or a mixture of alcohol and gasoline known as **gasohol.** In the United States, alcohol production from corn for use as gasohol increased rapidly in the early 1980s (Table 23–1). In 1984, the United States used about 160 million bushels of corn, about 2 percent of the total, to produce 420 million gallons of alcohol (about 0.5 percent of our liquid fuel).

However, alcohol production has severe limitations. It requires starting with sugary or starchy products such as sugarcane, corn, or potatoes. Therefore, a conflict between food and fuel is inevitable; if fuel producers offer a better price for grain than food markets will bear, the grain will go to fuel rather than food. The social consequences can be diabolical. It is reported that in Brazil, sugarcane is cultivated at record levels for alcohol production while food crops are down by 10 to 15 percent. This is occurring despite widespread malnutrition and a rapidly growing population. Until 1988, the United States had huge corn surpluses; therefore, conversion to alcohol helped alleviate the surplus problem. However, the drought of 1988 virtually eliminated surpluses. We will have to see where alcohol production goes from here.

Another problem is pollution. While alcohol is promoted as clean burning, its production generates much pollution because inexpensive, dirty-burning fuels such as soft coal are used for distillation and an amount equivalent to at least one-half gallon of fuel is used for every gallon of alcohol produced. This makes alcohol expensive—roughly double the cost of gasoline. The apparent competitive pricing of gasohol is due to tax subsidies amounting to $.60–$1.35 per gallon of alcohol.

A suggestion that is made in connection with all forms of biomass fuels is that crops be grown specifically for fuel purposes (Fig. 21–21). With few exceptions such as growing water hyacinths to remove nutrients from waste water (Chapter 10) such endeavors will be in competition with land, water, and fertilizer resources used for food production. Also, fuel culture would invariably increase erosion and the resulting impacts of sediments on waterways and the water cycle.

In summary, significant energy contribution with positive environmental impact can be achieved through bioconversion of manure, sewage sludges, and other organic wastes to biogas with recycling of

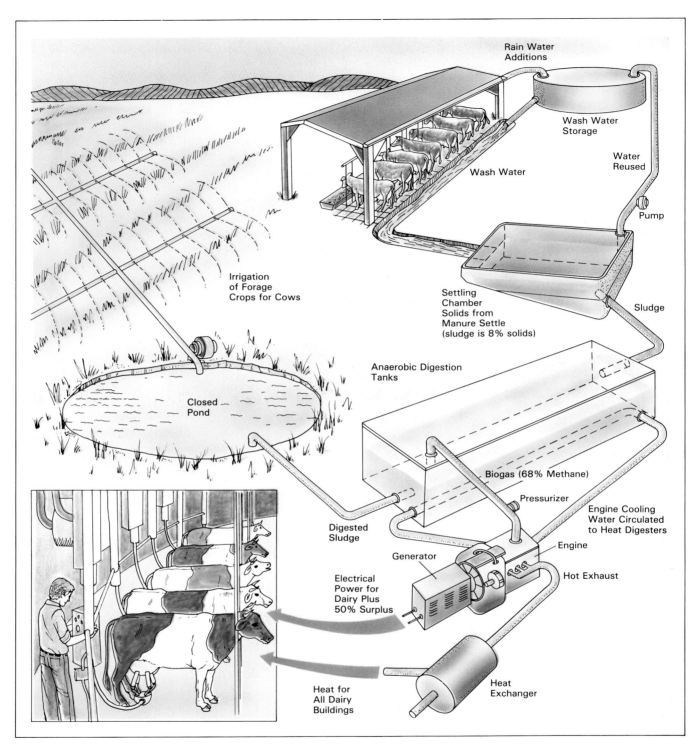

FIGURE 21–20
Dairy farm operated on cow manure. The total power needs for the Mason-Dixon Dairy located near Gettysburg, Pennsylvania, are obtained as a byproduct of cow manure as shown. Excess power, nearly half of what is produced, is sold to the local utility.

FIGURE 21–21
Sweet sorghum being grown for alcohol production. Such ''energy crops'' will inevitably compete with food crops for land and other agricultural inputs. (U.S. Department of Energy.)

the nutrient-rich residues. However, exploitation of other biomass energies involves environmental risks that are, potentially, as or even more severe than those connected with our traditional fuels.

Hydropower

Because sunlight drives the water cycle, hydro or water power is another indirect form of solar energy. Water power has been used for millennia by diverting water from natural falls over various kinds of paddle wheels or turbines (Fig. 21–22). But relatively few nat-

ural falls with significant volume exist in the United States. Therefore, over the last century or so, the trend has been to build huge dams to create artificial falls that will generate substantial **hydroelectric power** (Fig. 21–23). About 13.5 percent of the overall electrical power, 5.5 percent of total energy, in the United States currently comes from **hydroelectric dams,** most of it from about 300 major dams concentrated in the Northwest and Southeast. Can the use of hydroelectric power be increased?

While hydroelectric power is basically a nonpolluting, renewable energy source, it still involves tremendous tradeoffs.

FIGURE 21–22
Utilizing the force of falling water was one of the earliest sources of power. Mill near Pigeon Forge, Tenn., built in 1831. (Dick Foster/Stockpile.)

FIGURE 21–23
Hoover Dam. About 13.5 percent of the electrical power used in the United States comes from large hydroelectric dams such as this. Water flowing through the base of the dam drives turbines. (U.S. Department of Energy.)

Dams have drowned out some of the most beautiful stretches of river in North America as well as wildlife habitat; productive farmlands; forests; and areas of historic, archaeological and geological value. Glen Canyon Dam [on the border of Arizona and Utah] drowned one of this world's most spectacular canyons. Tellico Dam in Tennessee eliminated an ancient Cherokee village and the site of the oldest continuous habitation on the North American continent.*

The reservoir behind the Aswan High Dam in Egypt has caused the spread of the parasitic worm, *Shistosoma*, which causes a serious debilitating dis-

---

* Blackwelder, Brent. "Dams: A Change of Course." *National Parks,* 58 (July/Aug.), 1984, pp. 8-13.

ease. The dam has also increased humidity, which is now causing rapid deterioration of ancient monuments and artifacts that stood virtually unchanged for many centuries.

Since water flow is regulated according to the need for power, dams also play havoc downstream because water levels may go from near flood levels to virtual dryness and back in a single day. Other ecological factors are also affected as smaller amounts of nutrients and sediments reach the river's mouth.

Consequently, proposals for more dams are embroiled in increasing controversy over whether the projected benefits justify the ecological and sociological tradeoffs. In any case, few sites conducive to large dams remain in the United States. It is notable that a proposal to construct a dam on the New River in Virginia was defeated in 1976 after years of litigation between environmentalists and the utility wishing to build the dam. The New River was the *last* large undammed river on the East Coast. Similarly, dam proposals on the Buffalo River were defeated, preserving it as the *only* large undammed stream in the Arkansas Ozarks.

However, the potential for increasing hydroelectric power with small dams still exists. In the early days of hydroelectric power, numerous small dams were constructed on small streams and rivers to generate local power. As power production went to large, centralized facilities, these small dams were abandoned (Fig. 21–24). Now, groups of independent investors are renovating some of these small dams and selling the power to local utility companies. Also, many dams built for flood control and water supplies could be retrofitted for power generation. If all of these options were taken, production of hydroelectric power could be doubled. The amount of energy thus generated would equal that of 50 to 60 new nuclear or coal-fired power plants. Many less-developed countries have great potential for developing water power, but the tradeoffs noted above must be kept in mind.

Wind Power

Since winds result from unequal solar heating of the atmosphere, wind power is also a form of indirect solar energy. Along with water power and domestic animals, windmills have been used throughout history. Until the 1930s, most farms in the United States used windmills for pumping water and/or generating small amounts of electricity (Fig. 21–25). In the 1930s and 1940s, windmills fell into disuse as transmission lines brought low-cost power from central generating plants. However, in the 1970s and 1980s the use of wind energy for generating electrical power has rebounded. The modern machines for generating elec-

FIGURE 21–24
This dam on the Patapsco River near Baltimore, Maryland, was used for power generation in the early 1900s but was abandoned as large centralized facilities came into use. If the thousands of such dams were renovated, as some have been, production of hydropower could be nearly doubled. (Photo by author.)

FIGURE 21–25
Windmills like this were widely used on American farms for pumping water until the 1940s. (USDA photo.)

tricity are called **wind turbines** as opposed to windmills, which were used for mechanical energy for such work as pumping water and milling grain.

Since the energy to be gained increases with the area transected by the blades, doubling the length of blades increases the potential power output by a factor of four. Seeing this theoretical economy in size, government-funded programs focused on promoting huge wind turbines, machines with blades as much as 300 feet (about 100 meters) from tip to tip mounted on 200-foot towers (Fig. 21–26). Such machines in optimal wind conditions can generate as much as 2.5 megawatts (1 megawatt = 1 million watts), enough to provide for about 2500 homes. (For comparison, a nuclear power plant produces about 1000 megawatts.) About a dozen such wind turbines were built, seven in the United States and others in other countries, but results have proved disappointing; the stresses are tremendous and they suffered from repeated breakdowns. Canada, Denmark, the Netherlands, Sweden, the United Kingdom, and West Germany are still pursuing such programs, but they have been essentially abandoned by the United States.

On the other hand, setting up arrays of more

FIGURE 21-26
MOD 2 wind turbine in Washington State; blades, 300 ft from tip to tip; tower, 200 ft tall; capacity, 2.5 megawatts in winds 14–45 mph. The future of such wind turbines is in doubt, however, because severe stresses cause maintenance problems. (U.S. Department of Energy photo.)

17,000 machines with a total capacity of about 1500 megawatts, enough to eliminate the need for 1.5 nuclear power plants.

Best of all, the cost of installing such machines has come down to about $1.25 per watt, more than competitive compared to coal or nuclear power at $3 and $5 per watt respectively. When environmental concerns are considered there is hardly a comparison at all. It is hardly surprising that the world wind-turbine industry, from modest beginnings in 1980, grew to annual sales of over 500 megawatts valued at over a billion dollars in 1985. Electricity-producing wind turbines are now installed in 95 countries from the tropics to the Arctic, but still the potential has hardly been tapped. Most regions have areas where winds are constant enough to make wind turbines practical.

Again the problem of storage is not a present problem as there are adequate conventional sources to provide backup.

GEOTHERMAL, TIDAL, AND WAVE POWER

Geothermal Energy

Because of the energy released in the decay of naturally occurring radioactive materials, the earth's interior is a huge ball of molten rock, which occasionally erupts to the surface in volcanoes. This heat of the earth's interior is referred to as **geothermal** energy. If it can be obtained at a practical cost, geothermal energy is virtually infinite and everlasting. Some geothermal energy may be obtained where natural groundwater comes in contact with hot rock. Such

modestly-sized wind turbines, machines with a blade diameter of about 50 ft (17 m) capable of generating about 100 kilowatts, have proved practical (Fig. 21–27). Arrays of 50 to several thousand such wind turbines are called **wind farms.** As of 1987, California, the world's largest user, had installed a total of nearly

FIGURE 21-27
Wind farm located 40 miles east of San Francisco. Large arrays of modestly sized wind machines are proving more practical than a few very large wind machines. (Photo by author.)

FIGURE 21–28
Geothermal well on the island of Hawaii. (U.S. Department of Energy photo.)

heated water may come to the surface in natural steam vents like those in Yellowstone National Park, or the steam may be obtained by drilling into superheated aquifers (Fig. 21–28). The steam is used to drive turbogenerators. As of 1988 there were about 5000 megawatts of such geothermal energy facilities installed, about half of them in California, and others in the Philippines, Mexico, Italy, and Japan (Fig. 21–29). This amount could be increased several fold in the next 10 years.

However, large-scale development of geothermal power presents many problems. Hot steam and water brought to the surface are frequently heavily laden with salt and other contaminants, particularly sulfur compounds. These contaminating compounds are highly corrosive to turbines and other equipment, and they result in both air and water pollution as they are finally released into the environment. Sulfur pollution from a geothermal plant may be equivalent to that from a plant burning high-sulfur coal, and the hot brines released into streams or rivers may be ecologically disastrous. Finally the locations of natural steam geothermal sources are limited to relatively few areas, and many of these are not close enough to where the power is needed.

The real potential for geothermal energy lies in development of *hot dry rock*. Hot dry rock can be found anywhere by drilling deeply enough. In theory, two parallel holes can be drilled into hot rock and then fractures created between the two holes. Water forced down one hole will be heated as it seeps through the fractures and will come up the other hole as steam to drive turbines (Fig. 21–30). Recycling the water would avoid the pollution problems noted above. Unfortunately, the technological problems of drilling deeply enough (3–4 miles) into hot rock and then causing fractures to occur as desired have proven difficult to overcome. While work is proceeding, the ultimate success of these endeavors is still speculative.

FIGURE 21–29
One of 11 geothermal units operated by the Pacific Gas and Electric Company at The Geysers in Sonoma and Lake counties, California. The field is currently generating 2000 megawatts. (Photo courtesy of Pacific Gas and Electric Company.)

Chap. 21: Solar and Other Renewable Energy Sources **511**

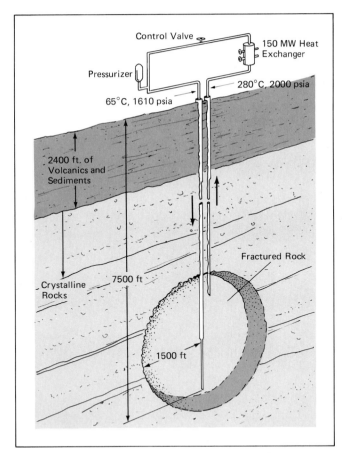

FIGURE 21-30
A dry geothermal well. Water is pumped into hot rock, where it is heated. The hot water is returned through a second well. (U.S. Department of Energy.)

Tidal Power

A great deal of energy is inherent in the twice-daily rise and fall of the tides, and many imaginative schemes have been proposed for capturing this eternal, pollution-free source of energy. The most straightforward is to build a dam across the mouth of a bay and mount turbines in the structure (Fig. 21-31). The incoming tide flowing through the turbines generates power. As the tide shifts, the blades are reversed so the outflowing water continues to generate power. Two such tidal plants are presently in operation—one in France and the other in the Soviet Union.

Developing tidal power requires a fluctuation between high and low tide of at least 20 feet (about 6 meters) in order to produce enough "head" of water pressure to make the tidal dam worthwhile. Otherwise, more energy is required in construction and operation of the facility than is produced. There are approximately 15 locations in the entire world that have

tides of this magnitude, only one of which is in North America, the Bay of Fundy in Nova Scotia where a major tidal power plant is under consideration.

But tidal power, is not without adverse environmental impact. In addition to the loss of unique aesthetic and recreational pleasures in these areas, there would be far-reaching environmental effects due to the dam's trapping of sediments, impeding of the migration of marine organisms, and from changing circulation and the mixing of fresh and salt water. In conclusion, tidal power does not have the potential to contribute more than a minute amount to overall energy use and only at the sacrifice of unique bays and estuaries. Indeed, potential impacts may extend over much broader areas; some scientists project that the Bay of Fundy project may affect tides as far south as Boston.

Wave Power

Waves are generated by wind, but they are mentioned here because they have much in common with tidal power. The energy inherent in waves holds the same general allure as the energy inherent in tides. It also presents the same difficulties with respect to harnessing it. Along coasts of the United States, consistent wave action is not great enough to develop a head of water pressure high enough to generate a significant amount of power with any device of practical size. Again, the net energy output of the wave generators conceived today appears to be zero or less. Therefore, wave power also has little, if any, potential for contributing to the solution of our energy problem in the United States. England and Ireland, however, have access to consistently higher waves and are experimenting with various wave generators. Again, all such schemes must be economically competitive to be practical. It appears likely that these alternatives will be surpassed by one or more of the solar technologies already discussed.

CONCLUSIONS

In surveying these various alternative energy options we see that three in particular, wind turbines and photovoltaic cells and biogas production, are at or very near the stage of large-scale commercialization. These, along with conservation, can probably handle growth; therefore, there is probably no need to build additional nuclear or coal-fired power plants. However, existing power plants will need to be kept in operation to provide backup and stability—at least for the short term. In the longer term, storage systems must be developed to complement solar and wind power.

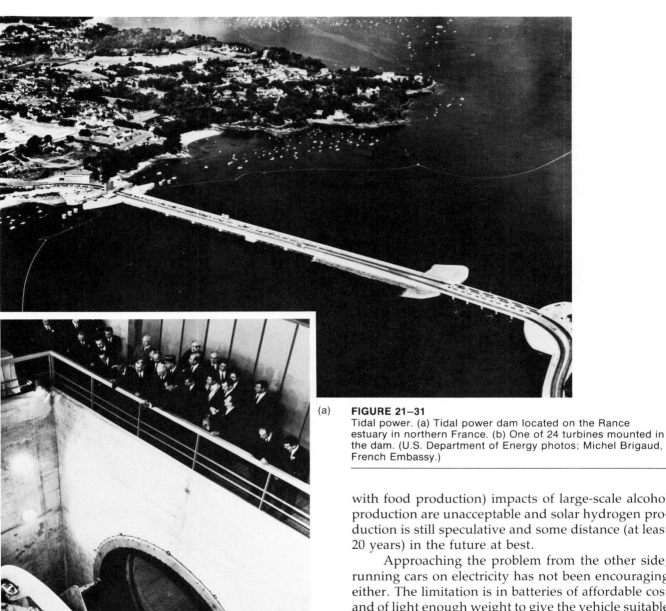

(a)

FIGURE 21–31
Tidal power. (a) Tidal power dam located on the Rance estuary in northern France. (b) One of 24 turbines mounted in the dam. (U.S. Department of Energy photos; Michel Brigaud, French Embassy.)

(b)

with food production) impacts of large-scale alcohol production are unacceptable and solar hydrogen production is still speculative and some distance (at least 20 years) in the future at best.

Approaching the problem from the other side, running cars on electricity has not been encouraging either. The limitation is in batteries of affordable cost and of light enough weight to give the vehicle suitable performance and range. Electric vehicles to date, with a heavy load of batteries, have a maximum range of about 50 miles at 30 miles per hour. The most advanced prototype, shown in Figure 21–32, charges its battery with photovoltaic cells as it travels, but it still has averaged only a little over 40 miles per hour under ideal conditions and has little range without sunlight. Consequently, without a radical breakthrough in battery development, which seems unlikely, electric cars are not in prospect for the foreseeable future except for situations where short distances and low speeds suffice. Consequently, following current trends, the shortfalls of crude oil will present us with a crisis situation despite photovoltaic cells, wind power, and other alternative energy sources.

Passive solar heating of space and hot water is also economically practical and it would be prudent

The prospects are bright, then, for environmentally sustainable alternative ways of generating electrical power. Unfortunately, this does not address the problem of declining availability of crude oil, which is still required for transportation. There were two options, discussed earlier, for producing fuel for vehicles, alcohol production and solar production of hydrogen. The environmental and social (competing

FIGURE 21–32
The SunPacer built by General Motors is a prototype car powered by solar energy. Limitations in speed and maximum range of solar cars prevent them at this time from offering a viable alternative to internal combustion engines. (Courtesy of General Motors Corp.)

to use passive solar architecture for all new homes and buildings. However, as long as existing buildings remain in use and unchanged, this will not diminish demand for traditional fuels; it will only mitigate the growth in demand. What could diminish demand for traditional fuels would be a massive program of increasing insulation and retrofitting solar space and water heating into existing homes and buildings. This would allow some fuel oil and natural gas to be transferred to the transportation sector. But, even at best, this would not offset the prospective shortfalls of crude oil for transportation.

This demands refocusing our attention specifically on the transportation sector.

Need for Conservation

Despite the oil shocks of the 1970s, numbers of cars and distance of commuting have continued to expand and not just in the United States. There are now some 400 million cars in use worldwide, 110 million (27 percent) in the United States, and the number is growing rapidly. Therefore, while other sectors of the economy have generally cut back on use by substituting alternative fuels, the total amount and the percentage going to transportation have continued to increase. This has occurred despite increasing the average miles per gallon. About 70 percent of the oil consumed in the United States in 1988 went for transportation, the bulk of it for private cars.

It is conspicuous that this trend is on a collision course with shortages that lie ahead. The United

States will be particularly hard hit because Americans have developed a lifestyle in which, for most people, cars are no longer a luxury but a necessity with virtually no alternatives. Over 80 percent of the miles driven (and fuel consumed) is for commuting to work and shopping for necessities, as opposed to trips for pleasure or visiting. In cities such as Houston and Los Angeles about 90 percent of the people get to work by cars, and even in cities such as New York, which have good mass transit systems, the figure is close to 70 percent. This contrasts with Tokyo, for example, where just 15 percent of people commute to work by car.

In an oil shortage, essentials such as agriculture, emergency vehicles, and buses will have the priority; private transportation will be crippled. With dim prospects for alternative energy sources for vehicles in the near future, the best we can achieve is to prolong the time before shortages become critical. In this way we can buy time to develop alternatives such as solar production of hydrogen and electric vehicles. Fortunately, there is a straightforward way to accomplish this. It is through conservation, in particular increasing the miles per gallon for cars and other vehicles.

Energy Policy

The progress that has been made in developing alternative energy sources has been aided by certain laws passed by Congress. The Public Utilities Regulatory and Policies Act of 1978 (PURPA) has been a key factor in the development of alternatives that produce electricity, such as wind farms, renovation of small hydroelectric dams, and cogeneration. Without the ability to tie into the electrical grid and sell power to the major utilities, such projects would be impossible. Major utilities are generally reluctant to construct such projects or to buy power from other producers because this runs counter to their making profits on their huge investments for nuclear power plants. PURPA, however, requires utilities to buy power from independent producers at "full cost," a price defined as the same price it would cost the utility to produce it. Since a number of the alternatives can produce power more cheaply than the utility can, independent business people who set up a wind farm, for example, stand to make a handsome return on their investments. PURPA, in effect, guarantees the production of power in the most inexpensive ways, which will also turn out to be the most efficient, environmentally sound, and sustainable ways. PURPA is a law that definitely should be maintained.

From 1975 to 1985, investments in alternative energy were made tax deductible. For example, an individual homeowner could write off the cost of a solar hot-water heater against income and thus receive a considerable tax break. This made such investments even more economically attractive and was a large factor in spurring both the solar and wind industries. Unfortunately Congress terminated this tax break in 1986, and with the fall of oil prices consumer interest in alternative energy all but died. Hundreds of solar and other alternative energy businesses went bankrupt and tens of thousands of workers in these industries became unemployed. There has been very little growth of alternative energy since that time. The loss in income taxes from these businesses and workers probably more than offset the gains for the government. The Center for Renewable Resources cites this sudden withdrawal of federal incentives as an "enormous mistake, crippling an industry that the government [had] carefully nurtured up to the point of economic viability."

Finally, development of alternative energy technologies was promoted by federal money allocated to research, development, and demonstration projects. Funding for renewable energy alternatives, as well as traditional sources of energy, increased steadily through the late 1970s. Then, during the early 1980s, despite the progress that had been made under previous funding, the government slashed budgets for conservation and development of renewable resources while maintaining budgets for nuclear power (both fission and fusion) (Fig. 21–33; Table 21–1). Dennis Hayes, chairman of the Solar Lobby, notes

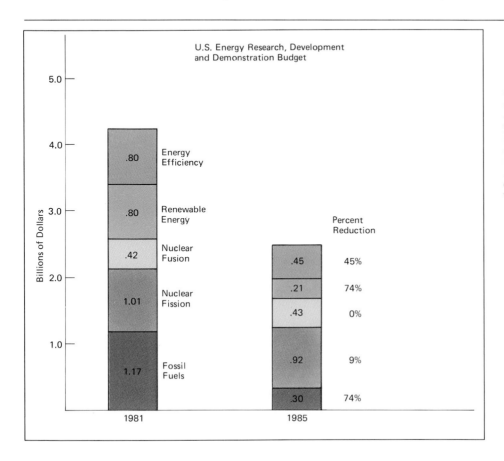

FIGURE 21–33
Energy research, development, and demonstration budget in the United States. Despite the remarkable contributions from the areas of energy efficiency and renewable energy sources, funding for these areas has been slashed in recent years while that for nuclear power has remained constant or increased. (Source: U.S. Department of Energy. Redrawn with permission from *Renewable Energy at the Crossroads*, Center for Renewable Resources, Washington D.C., 1985.)

Table 21–1
United States Government
Spending on Renewable Energy
(millions)

ENERGY SOURCE	1981	1985
Alcohol Fuels	21	31
Other Biomass	46	—
Hydropower	22	—
Solar Thermal	219	46
Photovoltaics	160	57
Geothermal Energy	199	32
Windpower	86	29
Other	44	17
Total	797	212

Source: U.S. Department of Energy.
Reprinted with permission from *Renewable Energy at the Crossroads,* Center for Renewable Resources, Washington, D.C. 1985.

that the largest recipients of tax breaks and supports are oil, coal, and nuclear industries due to their well-established, powerful lobbies.

What You Can Do

Informed and active citizens can do much to promote and support a policy that is more equitable and environmentally sound. All the suggestions for increasing the efficiency of vehicles listed in Chapter 19 should be reemphasized. In addition, write your congressperson in support of the following measures:

☐ Research and development funding for solar production of hydrogen, low-cost production of photovoltaic cells, and light-weight, inexpensive, high-capacity storage batteries should be given priority over nuclear power.

☐ A higher percentage of the funds currently going for highway construction, which only encourages more fuel consumption, should be shifted to construction of electric light rail systems.

☐ Some economic incentives for conservation and adding solar space for water heating should be reestablished. We will gain more long-term economic and political security by spending money in these areas than by spending it to secure safe passage of oil out of the Persian Gulf.

☐ Congress is to be commended for its foresight in enacting PURPA.

Note that none of these steps involves spending more money; they simply involve shifting of priorities to endeavors that will lead to a sustainable society.

Investigate, study, and implement ways in which you can conserve energy in your own home (without suffering discomfort) through weatherizing, adding insulation, and adding solar space and hot-water heating where possible.

Finally, to protect yourself from the almost certain fuel shortages that lie ahead, select a place to live where it is possible to meet your transportation needs without a car.

22

Lifestyle, Land Use, and Environmental Impact

■ **Outline**

I. URBAN SPRAWL: ITS ORIGINS AND
   CONSEQUENCES   519

   A. The Origin of Urban Sprawl   521

   B. Environmental Consequences of Urban
      Sprawl   522
      1. Depletion of Energy Resources   522
      2. Air Pollution, Acid Rain, and the
         Greenhouse Effect   524
      3. Degradation of Water Resources and
        Water Pollution   524
      4. Loss of Recreational and Scenic Areas,
        Natural Services, and Wildlife Habitat 525
      5. Loss of Agricultural Land   525

   C. Social Consequences of Urban Sprawl 526

      1. Exurban Migration and Economic and
        Ethnic Segregation   526

      2. Declining Tax Base and Deterioration
        of Public Services   528

      3. Segmentation and Desocialization
        Caused by New Highways   529

      4. Loss of Businesses and Employment 529

      5. Vicious Cycle of Urban Sprawl and
        Inner City Decay   530

   D. Conclusion   532
II. MAKING CITIES SUSTAINABLE   532
   A. Cities Can Be Beautiful   533
   B. Cities Can Be Sustainable   533
   C. Slowing Urban Sprawl and Aiding Urban
      Redevelopment   536
      1. Slowing Urban Sprawl   536
        a. Clustered versus detached housing 536
        b. Zoning laws   536
        c. Agricultural and conservation
          districts   538
        d. Growth management initiatives 538
        e. Land trusts   538

■ **Study Questions**

1. Define *urban sprawl.*

2. Describe the transportation and structure of cities prior to World War II and how these began to change after the war. Describe factors contributing to unplanned development. How did governments become involved in promoting urban sprawl? Define and describe the role of the Highway Trust Fund.

3. List the environmental consequences of urban sprawl. Describe how urban sprawl is, in large part, responsible for each of the factors listed in the outline.

4. As urban sprawl started, where were most people moving from? What is the term that describe this movement?

5. Describe how *exurban migration* resulted in economic and ethnic segregation.

6. List the public services that cities provide and tell how they raise money for these services. Describe how the exurban migration affects tax revenues and services.

7. How do commuters affect the quality of life in the city?

8. Describe why cities have much higher rates of unemployment than suburbs.

9. Describe how suburban sprawl and urban decay become a vicious cycle. Has recent urban renewal eliminated the problem? Cite evidence.

10. Can cities be made to provide a high quality of life which is resource and energy efficient and sustainable? Explain.

11. List and describe ways in which urban sprawl may be controlled.

2. Stemming Urban Blight and Aiding
   Urban Redevelopment   538
   a. Changing the property tax   538
   b. Aiding residents to own buildings   539
   c. Urban homesteading   539
3. The Integral Urban House   539

D. What You Can Do   540

CASE STUDY: AN INTEGRAL URBAN HOUSE   541

12. List and describe ways in which urban decay may be halted and reversed.

13. Describe how urban living may be compatible with ecological living and environmentalism. Describe the concept and some of the specifics of an *integral* urban house.

14. Tell what you can do to reduce or control urban sprawl. How might you alter your own lifestyle to aid the process?

All the environmental problems we have been discussing, ranging from degradation of agricultural soil and water resources through the various forms of pollution to depletion of traditional energy resources, have their origins with *people* and *lifestyle*—that is, us and the way we live. Since World War II, Americans have pursued a lifestyle marked by living in spread-out suburbs and meeting all transportation needs with private cars. Satisfying as this lifestyle is, it greatly intensifies the negative impacts we bring to bear on the environment. Understanding how this is so and exploring possibilities for moderating our lifestyle so that we have less environmental impact are the subject of this chapter.

URBAN SPRAWL: ITS ORIGINS AND CONSEQUENCES

Farms and natural areas surrounding cities are continually giving way to new housing developments, shopping centers, industrial parks, parking lots, and other facilities interconnected with new highways carrying streams of traffic (Fig. 22–1). This rampant growth is accepted and promoted as an indication of a strong economy. However, it is largely unplanned and uncontrolled. Therefore it is also commonly referred to as **urban sprawl,** and it lies at the root of many of our environmental problems (Fig. 22–2).

Most of you reading this text have grown up with cars and urban sprawl. Therefore it may be difficult to imagine that things were ever different. However, urban sprawl really only started in the late 1940s when, following World War II, private ownerships of cars became common. Of course, cars were developed around the turn of this century, but they were unaffordable to most people until Henry Ford developed assembly line production in the 1920s. However, ownership of cars was cut short by the Great Depression and then by World War II, at which time all production facilities were given over to the war effort.

FIGURE 22–1
Somehow escaping the initial phase of land clearing, a jack-in-the-pulpit gives its last bloom. In the United States, some 2 million acres of land per year goes into development. (Photo by author.)

FIGURE 22–2
Urban sprawl (Detroit, Michigan). Complexes of housing developments, highways, and shopping centers spread over the countryside in a largely unplanned, helter-skelter fashion, consuming prime agricultural land at the rate of over 2 million acres per year. (Photri, Inc.)

Prior to the end of World War II, therefore, a relatively small percentage of people owned cars.

Without cars, cities had grown in a way that accommodated people getting to school, work, and shopping for necessities by the means of available transport, namely walking. (For most, ownership of a horse in a city was as untenable then as now because of the expense, work, and mess of maintaining one.) But walking distances were generally short; nearly every block had a small grocery, hardware, drug, and other stores and had professional offices integrated with residences. Often a building had a store below at the street level with the residences above (Fig. 22–3). Cities were provided with parks that were heavily

FIGURE 22–3
Before the advent of widespread use of cars, cities had an integrated structure. A wide variety of small stores and offices on ground floors with residences above placed everyday needs within walking distance. This structure still exists in certain areas of some cities. This is a location in North Baltimore. (Photo by author.)

used (Fig. 22–4). Public transportation, horse-drawn trolleys, cable cars and, later, buses, provided for longer-distance commuting within cities, but they did not change the compact structure because people still needed to walk to the transit line. The small towns and villages surrounding cities—the suburbs—were compact for the same reasons. At the end of the transit lines, cities gave way abruptly to farms and open country. This pattern held until the end of World War II; then it began to change dramatically.

The Origin of Urban Sprawl

Despite their virtues, walking and public transit did not make cities desirable places to live. Especially in industrial cities, poor housing, inadequate sewage systems, inadequate refuse collection, pollution from home furnaces and industry, and generally congested, noisy conditions combined to make cities trying places to live. A decrease in services during the war aggravated these problems. Hence, many people had the desire, the American dream if you will, to live in their own house on their own piece of land away from the city. Furthermore, because of lack of consumer goods during the war, many civilians as well as returning veterans had accumulated considerable savings.

As mass production of cars resumed at the end

of the war, people flocked to buy them. With private cars, people were no longer limited in where they lived by the distance they could walk or the location of the transit line. The car could make any commuting possible.

Developers responded quickly to the new demand. They bought farms wherever they could and put up houses, each with its own well and septic system. The government aided this trend by providing low-interest mortgages through the Veterans Administration and the Federal Housing Administration, and interest payments on mortgages were made tax deductible. In addition, property taxes in the suburbs were much lower than in the city. These financial factors meant that, for the first time in history, making monthly payments on one's own home in the suburbs was actually cheaper than paying rent for equivalent or less living space in the city. Motivated by these factors, people bought the new homes as fast as they could be put up. Proximity to work, shopping, and schools was of little consequence since the car could provide all transportation needs.

Thus, development proceeded according to the whims of individual developers, governed only by where they could acquire land as opposed to any overall planning (Fig. 22–5). Indeed, overall planning was not merely neglected; it was actually prevented by the fact that large cities were surrounded by a maze of more or less autonomous local jurisdictions (towns, townships, counties, municipalities). No governing body existed to come up with an overall plan, much

FIGURE 22–4
Central Park, New York City, about 1900. Before the advent of the automobile, city parks were heavily used because there was not ready access to other open areas. (From the collection of the Library of Congress.)

FIGURE 22–5
Scattered groups of homes shown here reflect which farms developers were able to buy. An overall plan which would include schools, shopping centers, parks, offices, and other facilities does not exist. (USDA—Soil Conservation Service photo.)

less enforce it. Local governments were simply thrown into a "catch-up" role of trying to provide schools, roads, sewers, water systems, and other public facilities to accommodate the uncontrolled growth. The result is urban sprawl (Fig. 22–6).

As governments (local, state, and federal) tried to accommodate urban sprawl, they actually became caught up in its growth and perpetuation. In particular, the influx of commuters into a previously rural area soon creates congestion that makes the need for new and larger roads conspicuous. To meet this need, Congress created the **Highway Trust Fund,** a fund financed by a tax on gasoline to be used exclusively for building new roads.

But a new highway not only alleviates existing congestion. It also perpetuates further development and sprawl because *time, not distance,* is the limiting factor for commuters. The average person is willing to spend about 20 to 40 minutes commuting to a destination. If he/she has to walk, this time limit means living a maximum of 1 to 2 miles from work. But, with a car on an expressway, this time limit means that the commuter can live 20 to 40 miles from work and still get there in the same time. Paradoxically, then, new highways do not get people to work faster; they simply lead people to live further from where they work. Average commuting distance has doubled since 1960, while time to get to work has remained about the same. Thus, you can see how the Highway Trust Fund creates a vicious cycle. A new highway stimulates more farflung development, which produces more commuting and congestion, and also raises money to build more roads that stimulate further development, and so on (Fig. 22–7).

Farmers were and still are caught in the middle of this process. A farmer may abhor the sprawling development, but when the time comes for him to sell his farm, it is logical for him to seek the highest price, and developers can invariably offer the best price. Farmers who do not initially plan to sell are often forced to do so because the potential for development boosts land values and increases property taxes to the point that the farmer must sell to pay the taxes. The problems of farmers may be further aggravated by harassment from suburbanites who walk across their fields and molest their livestock. Also, as farming in an area declines so do all the agricultural support services, so the remaining farmers are forced to sell. Some farmers or land speculators choose to hold land, waiting for market prices to increase further. Again, you can see how this contributes to a "crazy quilt" pattern of urban sprawl.

Residential developments are followed (or sometimes led) by shopping centers, shopping malls and, more recently, industrial parks and office complexes. In contrast to original integrated cities and towns, where all needs could be easily met by walking or public transit, the overall result of urban sprawl is spread-out residential areas with highly centralized shopping and work places many miles apart with no access except by car.

What are the environmental and social consequences that have come with urban sprawl?

Environmental Consequences of Urban Sprawl

Depletion of Energy Resources

The urban-sprawl lifestyle translates into an increasing demand for energy resources, especially crude oil needed to fuel motor vehicles as the numbers of cars and commuting miles have increased. Thus, the problems and consequences of oil shortages that we face (Chapter 19) are a direct consequence of urban-sprawl lifestyle (Fig. 22–8). Likewise, individual suburban homes require 1.5 to 2 times more energy for heating and cooling than comparable attached city dwellings.

FIGURE 22-6

A large number of environmental problems and conflicts arising from lack of sufficient planning are evident in this area. Indicated on the photograph: (1) Storm runoff from parking lots causes severe flooding problems along the stream. (2) The stream is badly polluted by sewage overflows from the office complex. (3) Park versus highway conflict has kept this interstate a dead end for more than ten years. Highways block convenient routes between many areas. (Baltimore County, Department of Planning.)

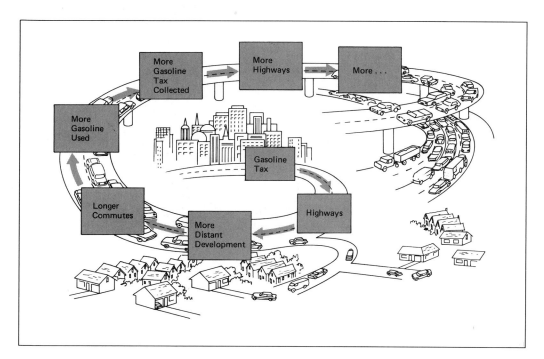

FIGURE 22–7
The gasoline tax, which is specifically designated to build new highways, creates a self-perpetuating cycle of increasing urban sprawl.

FIGURE 22–8
The problems and consequences of oil shortages which we face are a direct result of our urban sprawl lifestyle with its demand for cars and commuting. (Jack Sullivan/Photo Researchers.)

Therefore, a large part of the growth in demand for fuel oil, natural gas, and electrical power, which is largely produced from coal or nuclear power plants, is a result of the shift to more spacious suburban living.

Air Pollution, Acid Rain, and the Greenhouse Effect

Cars are the major contributors to photochemical smog, ground level ozone, carbon dioxide, and the nitrogen oxide portion of acid rain, and coal-burning power plants are the major contributors of the sulfur dioxide portion of acid rain and additional carbon dioxide. Consequently, we can see that urban sprawl with its increasing use of cars and high electrical and fuel demands of suburban homes underlies a major portion of air pollution problems.

Degradation of Water Resources and Water Pollution

Urban sprawl, with all its roadways, parking lots, and rooftops, results in increasing runoff and decreasing infiltration over massive areas. Even suburban lawns increase runoff because they are more compacted than the original soils. Recall the consequences of increasing runoff and decreasing infiltration: increased flooding, streambank erosion, depletion of groundwater, drying of springs and waterways, saltwater encroachment, to name a few (Chapter 8). Further, exorbitant quantities of water used by suburbanites for watering lawns, gardens, washing cars, and so on, may seriously deplete supplies.

About a third of the fertilizer nutrients and pesticides polluting waterways are from these materials being overused or misused on suburban lawns and gardens. To this extent, suburban sprawl contributes to eutrophication (Chapter 9) and pesticide pollution.

Loss of Recreational and Scenic Areas, Natural Services, and Wildlife Habitat

New highways to relieve the congestion spawned by urban sprawl are often placed through parks or along stream valleys because such open areas seem to provide the least expensive rights of way. The result is the sacrifice of aesthetic, recreational, and wildlife values in the metropolitan areas where they are so important. Highway planners argue that a highway through a relatively large park will take a relatively small portion of the total area. If a park is to provide humans with a measure of peace and tranquility, however, it is dubious that two halves split by a pollution- and noise-generating highway still add up to the whole. Certainly wildlife must find their two halves incomplete if they are blocked from their source of drinking water, as frequently happens when highways are placed along rivers or stream valleys.

In addition, highway construction is responsible for more erosion and sedimentation of waterways than any other form of building. After completion, highways are responsible for runoff of oil, grease, and numerous other pollutants generated by traffic.

As natural areas are sacrificed for development, virtually all the wildlife they supported and the natural services they performed such as infiltration and recharge of groundwater, air purification, and so on are likewise sacrificed (Chapter 17).

Loss of Agricultural Land

Finally, and perhaps most serious, is the loss of prime agricultural land. You may have heard this problem minimized by arguments that the world has plenty of land. However, some 70 percent of the world's land is tundra, rugged mountains, severe deserts, or wetlands, all of which preclude agriculture. Another 19 percent of the land is considered marginal; it is excessively dry, wet, hilly, or cold, but given suitable inputs, it might be used for agriculture. Only the remaining 11 percent has the climate, relatively level terrain, and good soil necessary for prime agricultural land.

Because food has always been the most basic resource, it was natural to locate cities in the same prime agricultural regions. Consequently, sprawling development extending out from cities involves the sacrifice of about 2.5 million acres (1 million hectares) per year of the best agricultural land in the United States alone (Fig. 22–9). Consequently, there is a serious and growing conflict between urbanization and the loss of prime agricultural land.

FIGURE 22–9
Land that goes into urban development is often prime agricultural land. The same is true around all major American cities. (USDA—Soil Conservation Service photo.)

The loss of agricultural land is partially offset by the development of some new agricultural land. However, the "new" land is virtually all marginal land; there is hardly any prime land that is not already in production. Development of this land takes place at the expense of natural ecosystems, the biota they support, and the natural services they perform. Then, development may require extensive water diversion projects for irrigation or drainage. Such projects have a further adverse impact upon the natural environment. Finally, agriculture on marginal lands may not be sustainable because of salinization or waterlogging of the soils or depletion of water supplies caused by irrigation. Or, the land may be too hilly resulting in intolerable amounts of erosion. Another compensation for the loss of agricultural land is that productivity per acre is increased through higher-yielding crops and better farming methods. Still, increasing production per hectare cannot compensate for the loss of land beyond a certain point.

Food shortages in the United States are extremely unlikely because we are so abundantly endowed with agricultural land in proportion to our population. However, an increasing number of nations do rely on our surpluses and they are a major factor in our balance of trade. Thus, squandering ag-

ricultural land for urban sprawl is economically fool-hardy and unconscionable in view of the developing world population pressures.

Social Consequences of Urban Sprawl

Urban sprawl has high social costs as well as environmental costs. The social costs stem from the fact that urban sprawl is largely a phenomenon of **exurban migration,** people moving from the central city into outlying suburbs and **exurbs,** communities beyond the suburbs. This has a profound economic and social impact because it also involves a segregation of the population along economic and/or ethnic lines.

Exurban Migration and Economic and Ethnic Segregation

Historically, cities included a wide diversity of people with different economic and ethnic backgrounds. However, in general, moving to the suburbs demands the ability to own and drive a car and to obtain a home mortgage. This immediately excludes the poor, the elderly, and the handicapped. Generations of discrimination in education and jobs made blacks, and certain other minority ethnic groups, disproportion-

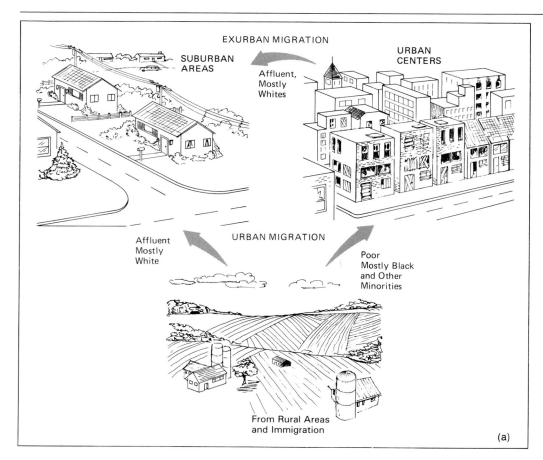

FIGURE 22–10
(a) Major migrations from rural areas to metropolitan areas and from cities to suburbs have also tended to segregate people along economic and racial lines. Such separation is called gentrification.

(b)

(c)

FIGURE 22–10(cont.)
(b) an area of suburbia and (c)
an area of inner-city Baltimore
contrast the extremes. (Photos
by author.)

ately poor. Therefore, the economic segregation was also, in large part, a racial and ethnic segregation (Fig. 22-10a). Then discriminatory lending practices on the part of banks and sales practices on the part of real estate companies served to exclude blacks and other minorities even when money was not a factor. Civil rights laws passed in the 1960s made such practices illegal, but there is much evidence that they still occur.

As a result, urban sprawl continues to reflect a **gentrification** of society—a segregation in which communities are composed of people with similar economic, social, and cultural backgrounds. New suburban and exurban developments continue to be occupied by the economically advantaged, while many areas that comprised the original cities and older suburbs are left to the poor, and now have populations that are comprised of predominantly economically depressed blacks, Hispanics, and other minorities, as well as by poorly educated whites (Fig. 22-10b and c).

A migration of people from rural areas to metropolitan areas has also occurred. More recently, there has been a tremendous migration from the

Northeast to the southern "sunbelt" states. The same pattern of gentrification emerges as these people settle in new areas.

Declining Tax Base and Deterioration of Public Services

By its economic segregation, exurban migration created a downward spiral of urban blight and decay that remains today as one of society's most vexing problems. To understand this downward spiral, we must note several points concerning local government.

☐ First, the local government (usually city or county, though the particular entities vary from state to state) is responsible for providing:
Schools
Local roads
Police and fire protection
Public water and sewer systems
Delivery of welfare services
Library and information services

☐ Second, close to 90 percent of the revenue for local governments to pay for these services comes from **property taxes,** taxes on all buildings and property proportional to their value. If values increase or decrease, the property tax is adjusted accordingly. The values are determined by the marketplace, what the properties—or similar properties in the same neighborhood—actually sell for.

☐ Third, in most cases, the central city is a gov-

ernmental jurisdiction separate from the surrounding suburbs.

These factors produce the following consequences. By the economic law of supply and demand, property values in the suburbs escalate with the influx of affluent newcomers. Thus, suburban jurisdictions enjoy increasing amounts of tax revenue with which they can improve and expand local services. By the same token, however, property values in the city crash to the levels that the remaining poor are able to afford. In all too many cases, owners, unable to find buyers at any price, simply abandon the properties that they left. In these situations, the government takes ownership in lieu of unpaid taxes, and the property, which was once a source of tax revenue, becomes a liability for the government (Fig. 22–11). The declining tax revenue resulting from declining property values is referred to as **declining** or **eroding tax base.**

As a result of an eroding tax base, city governments are forced to cut the quality and quantity of services provided and/or to increase the tax rate. Hence, the property taxes on a home in the city are often two to three times greater than on a comparably priced home in the suburbs, while schools, refuse collection, street repair, and other services are often far inferior. The services most neglected tend to be maintenance of roads and bridges and water and sewer systems because the money available goes to more visible and politically sensitive areas. Because of lack of maintenance, these things, which comprise the city's **infrastructure,** are approaching a state of imminent collapse in many older cities.

FIGURE 22–11
Growing effects of exurban migration. While suburban areas receive the economic benefits of an active home construction industry, cities receive the burden of dealing with the problems of abandoned buildings. (EPA-Documerica photo; courtesy of the U.S. Environmental Protection Agency.)

Segmentation and Desocialization Caused by New Highways

Many people who move to the suburbs maintain jobs in the city so each morning and evening the city is swamped and often choked in a tide of commuter traffic. To alleviate the congestion, new highways are constructed. A city's parks and its poorer areas are usually seen as the least expensive rights of way. Untold numbers of parks have been destroyed and millions of city residents have been crudely displaced to make way for new highways. Even when new freeways do not displace people, they frequently create barriers between formerly close neighborhoods and between where they live and work. Many cities have become more devoted to moving and parking commuter cars than to serving their own residents. In some cities as much as two-thirds of the land area is devoted to highways, parking lots, service stations, and other car-related facilities (Fig. 22–12).

Thus, building more highways to serve commuters getting in and out and through the city also serves to segment and depersonalize the city and further erodes it with noise and pollution. Further, because new highways tend to attract additional commuting, traffic congestion in cities is ever growing. The average commuting speed on Los Angeles freeways is now 30 miles per hour and declining. Manhattan often finds itself in complete gridlock and has decided to impose a heavy commuter tax to alleviate traffic.

Loss of Businesses and Employment

It is not just people who leave in the process of exurban migration. Stores, restaurants, theaters, professional offices, and other commercial establishments close in the cities and move to the new shopping malls in the suburbs because that is where their major customers and clients are. More recently, new

(a)

(b)

FIGURE 22–12
A large portion of the land area of many American cities has been used to accommodate automobiles rather than people. (a) Dallas, Texas. (EPA-Documerica, Bob W. Smith; courtesy of the U.S. Environmental Protection Agency.) (b) Map is an area of downtown Los Angeles in which about two-thirds of the land, indicated in black, is used for automobiles. (By permission from the Victor Gruen Center for Environmental Planning.)

and growing industries have been locating production, office, and warehouse facilities in new exurban **industrial parks** (Fig. 22–13).

The exurban migration of business causes further erosion of the city's tax base, and it makes the city a less desirable place to live because shopping opportunities and professional services become increasingly limited. The most serious consequence, however, is the loss of employment opportunities for city residents. Between 1960 and 1980, two-thirds of all new job openings were in the suburbs, while cities lost at least two jobs for every new job created. The poor do not own reliable cars, and the new outlying locations are, in large part, not serviced by public transit systems. Hence, inner-city residents cannot take advantage of job opportunities in the suburbs. As a result, unemployment in cities has grown steadily. The unemployment rate among young urban blacks has risen and remained near 50 percent, which means that a person finishing or dropping out of high school has about an equal chance of working or hustling to survive. Hustling, of course, involves petty crimes, drugs, and then larger crimes generally committed against other city residents. This atmosphere of high unemployment and crime breeds contempt for education; then, lack of education, of course, further undercuts opportunities for employment. Welfare and other social services aimed at combating these problems place greater stress on the eroding tax base and lead to further deterioration of the city's quality of life.

At the same time, employers in suburbs find it difficult to find workers for entry-level and unskilled jobs. Recently, steps have been taken in some cities to provide public transportation to enable city residents to get to jobs in outlying shopping malls and industrial parks. While this is a positive step in terms of employment, this reverse commuting consumes additional resources and energy and only small areas are served so that many jobs still remain inaccessible.

Vicious Cycle of Urban Sprawl and Inner City Decay

You can see how exurban migration and urban blight become a vicious cycle. Deteriorating quality of life in the city causes more people to leave, which causes more businesses to leave, which causes more unemployment and consequent crime, which causes more people to leave, and so on. These factors can force even the most fervent city dwellers to leave if they can muster the means to do so. Thus, the downward spiral persists (Fig. 22–14).

In recent years, a conspicuous rebirth of the core-areas in cities across America has occurred and is spreading. Redeveloped city centers include a wide variety of shops and restaurants, hotels, convention centers, theaters and other places for art, music, and entertainment, office buildings, and residences, and places to just walk or sit by fountains and relax (Fig. 22–15). This is bringing a new spirit of life and hope, as well as the practical assets of taxes and employment, back into cities. It is an encouraging new trend, so far as it goes.

However, we should not let the redevelopment overshadow urban blight, which is still all too real and in many cases worsening. While most of the country was enjoying economic prosperity during the 1980s, another 10 million people trapped between the suburbs and the growing centers sank below the pov-

FIGURE 22–13
One of many plants in an industrial park being developed in the rural countryside north of Baltimore, Maryland. Such industrial parks are the latest but accelerating trend in exurban migration. By transferring employment opportunities from urban to exurban areas, they will cause further hardships for city dwellers. They will also promote the sacrifice of huge amounts of additional surrounding land for residential and commercial uses. (Photo by author.)

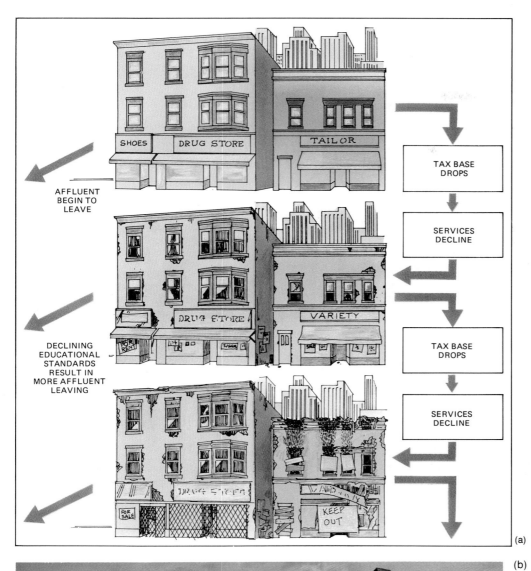

AFFLUENT BEGIN TO LEAVE

SHOES DRUG STORE TAILOR

DECLINING EDUCATIONAL STANDARDS RESULT IN MORE AFFLUENT LEAVING

DRUG STORE VARIETY

DRUG STORE KEEP OUT FOR SALE

TAX BASE DROPS

SERVICES DECLINE

TAX BASE DROPS

SERVICES DECLINE

(a)

(b)

FIGURE 22–14
(a) Migration of affluent people from cities to suburbs creates a vicious cycle leading to further urban decay. (b) Boarded-up shops in an inner-city area, the result of exurban migration. (Photo by author.)

FIGURE 22–15
In recent years there has been a conspicuous rehabilitation of the core areas of many cities. Shown here is the new inner harbor area of Baltimore, Maryland. Left foreground: harbor. Green-roofed building: a complex of small variety shops and eating places surrounded by a brick promenade, small amphitheater, park, and fountain. White building at left: McCormick Spice factory (not new) with new and renovated housing to the back and left. Center: new hotels and convention center. Right: new office buildings. Note elevated pedestrian walkway from shopping center to hotels. But the basic problems remain. Huge areas affected by urban blight are present in the background.

erty line. Walk two blocks in any direction from the sparkling, revitalized core of most American cities and you will find urban blight and decay continuing unabated. In the words of James Rouse, renowned city planner and developer,

> But right alongside [of the redeveloped center] in almost every city is a second city of those who are not making it—the city of the poor. It is like a Third World city—the city of people who are struggling to survive in miserably unfit housing in wretched, disorderly neighborhoods—with too little food, too little health care, too little work, and too little training for work, too little education, too little happiness, too little hope. It is like another nation where we are growing people who feel left out, abandoned, separated from the opportunities for the good life that abounds all around them. . . .
>
> We (may) have read the dismal figures, seen pictures of dilapidated housing in derelict neighborhoods, but most of us have not walked those streets, stepped inside those houses, climbed the stairs in those apartments, have not seen good people with clean, decent families huddled in that miserable housing, paying outrageous rents; have not looked into the saddened, sullen faces; felt the hopelessness, the distrust, suspicion, and separation that pervades their life and all around them.*

---

* James Rouse. "Suffering in the Second City," *EPA Journal* 14(4) (May, 1988), pp. 25–26.

Conclusion

Seventy-five percent of us in America live and work in metropolitan areas, which include everything from the outer edges of suburban sprawl to the revitalized urban cores. Yet, a substantial portion of this complex, as it presently exists, is not sustainable.

The suburban sprawl is not sustainable because of its insatiable demand on limited fossil fuel energy resources, its consumption of prime agricultural land, and the air and water pollution that result. Even if these environmental limits did not exist, "[t]he alternative," writes Luther Popst of the Conservation Foundation, "is final victory for urban sprawl, as accidental cities fill in the spaces between traditional cities, devouring natural beauty and cultural assets while creating an unsightly urban mass with no sense of place and little worth caring about." Likewise, the inner city with its decaying housing and lack of employment and opportunity is not only unsustainable and morally unconscionable; the antisocial behaviors that it breeds threaten the rest of society.

The redeveloped urban cores, however, demonstrate that cities can work.

MAKING CITIES SUSTAINABLE

With 75 percent of us living in metropolitan areas, there is no way that we can abandon present cities and start over; limits of land and other resources simply won't permit it. Present cities and suburbs must be made to provide the quality of life we desire and at the same time they must be sustainable.

Cities Can Be Beautiful

Given the blight and decay existing in most American cities and the fact that most Americans have been pursuing the ever more distant suburbs for the last 40 years, the thought of turning back to the city may seem little short of abhorrent. Yet, cities can be beautiful aesthetically, socially, culturally, and emotionally. Famous architect-planner Victor Gruen conveys this eloquently in the following quotation, which is based on his experience of many old European cities.

> The city is the sum total of countless features and places, of nooks and crannies, of vast spaces and intimate spots, an admixture of the public and private domain, of rooms for work and rooms for living, of rooms for trade, where money and wares change hands, and rooms where music and drama lift the soul, of churches and night spots, of landmarks, expressing the spirit of the community and homes for the comfort of the individual.
>
> It is the fountains and flower beds, the trees shading streets and boulevards, the sculptures and monuments, the rest benches placed in thousands of spots. The city is the little merchants who make their living on the streets, the vendors of balloons and pretzels, of newspapers, chestnuts, ice cream, flowers, lottery tickets and souvenirs. And the city, of course, is also the buildings . . . the relationship of these buildings to each other and, most important of all, to the spaces created between them. [Fig. 22-16 a–d]
>
> The city is the countless cafes . . . where a person with little money may spend hours over a cup of coffee and a newspaper. . . .
>
> The city is the crowded sidewalks, the covered galleries in Italy, the arcades and colonnades and the people on them and in them, some bustling, some walking for pleasure . . . The city is the parks: the tiny green spots with benches, the middle-sized ones that interrupt rhythmically the sea of stone and brick, the big ones that act as lungs and recreational places. . . .
>
> The city is the community of soul and spirit rising from an audience in a theater or at the opera or from those attending services in a church. The city is the daily chat (with the butcher, the newsman, the baker, and numerous other merchants with whom one shops).
>
> A real city is full of life, with ever-changing moods and patterns: the morning mood, the bustling day, the softness of evening and the mysteries of night,

the city on workdays so different from the city on Sundays and holidays.*

Unfortunately, most American cities never achieved the aesthetic qualities or the sociocultural development described by Victor Gruen. Driven by rapidly expanding commerce and industrialization, where leaders were generally seeking increasing profits, social amenities that make cities pleasant places to live were, by and large, overlooked or given minimum attention (Figs. 22–17 and 22–18). Consequently, American cities generally became places that people desired to leave rather than places to stay and enjoy.

However, the rebirth of the central areas of cities, which is patterned after the concepts Gruen described for European cities, is finally showing millions of Americans that cities and city living can be beautiful. It can also be sustainable.

Cities Can Be Sustainable

Basically, cities with an integrated, heterogeneous structure of homes, stores, offices, schools, places for entertainment and culture, and open spaces for relaxation and recreation are sustainable because they are, or can be, energy and resource efficient. The proximity of residences, shops, and workplaces enables everyday needs to be met by walking. Other transportation needs within the city, being relatively short, could be met with electric vehicles. Reduction in driving miles and use of electric vehicles would virtually eliminate air pollution. Heating and cooling of clustered buildings is also more energy efficient than that of separate homes, providing additional energy conservation and reduction of pollution. Most of the electrical power for buildings could be provided with exteriors of photovoltaic cells.

Provision of water, sewer, and other utilities is more efficient with the close proximity of users in a city. Likewise collection, processing, and recycling of wastes can be more efficiently handled in cities. Each city can provide thousands of low-skilled jobs in the recycling industries, jobs that don't exist now as wastes are simply landfilled or incinerated. Thousands of additional jobs will be created in maintaining the infrastructure of the city. The social ills of the inner city will not be solved by continuing to isolate ourselves from it. They can only be solved by providing a framework in which all segments of society

---

* *The Heart of Our Cities—The Urban Crisis: Diagnosis and Cure.* Copyright © 1964 by Victor Gruen. Reprinted by permission of Simon & Schuster, a Division of Gulf & Western Corporation.

FIGURE 22–16
Various aspects of a ''real'' city.

STREET SCENE IN PARIS

STREET SCENE IN NEW YORK

FIGURE 22–17
''A Humiliating Contrast.'' Historically, the cities of the United States, growing rapidly, neglected the creation of urban quality in their early stages. *Harper's Weekly*, in 1881, criticized New York by printing two contrasting drawings with this original caption.

FIGURE 22–18a
The inhumanity of many urban areas persists (*cont.*). (Photos by author.)

FIGURE 22–18b

can become integrated socially and economically. Integrated cities can provide this framework. Finally, by locating ourselves more in cities, the open places beyond will continue to exist and can be gotten to and enjoyed all the more so without the encroachment of more highways, shopping malls, and housing developments.

Slowing Urban Sprawl and Aiding Urban Redevelopment

To make cities sustainable, urban sprawl must be contained and development must be focused on the city.

Slowing Urban Sprawl

Clustered versus Detached Housing. Separate or detached homes spread over the landscape are the most land- and resource-extravagant ways of providing housing and assure dependence on cars. While it does not eliminate urban sprawl, **clustered development** enables housing the same number of people while preserving a major portion of the natural landscape. For example, Figure 22–19 shows two alternative plans for the same tract of land. In one plan, it is subdivided into 134 half-acre lots making the entire tract a uniform housing development. In the second plan, the same number of housing units are clustered, preserving the rest of the tract for nature or recreational uses. In the clustered plan the amount of road surface including driveways is reduced by 80 percent, allowing many additional acres to remain open as opposed to being covered with blacktop. Distances, costs, and the disturbance to the land of running water, sewer, and utility lines are similarly reduced.

A study of the Baltimore metropolitan region determined that a more centralized pattern of housing, including increased clustering as opposed to detached houses on separate lots, would:

- ☐ Reduce the burden of stormwater pollution from development by about one-half.
- ☐ Reduce space heating requirements and resulting air pollution by 25 percent.
- ☐ Reduce motor vehicle fuel consumption and resulting air pollution by 17 percent.
- ☐ Reduce air pollution from space heating by 24 percent.
- ☐ Reduce farmland and forest losses by 76 and 65 percent respectively.
- ☐ Reduce transportation costs for solid waste collection and haulings.
- ☐ Reduce student transportation costs by 38 percent.
- ☐ Provide more affordable housing.
- ☐ Make public transportation more cost-effective, if it is provided, because there are fewer stops and more ridership per mile.

At the very least, then, further suburban development should emphasize logical clustering of housing.

Zoning Laws. Zoning laws specify and limit the kind of development that can occur in given areas.

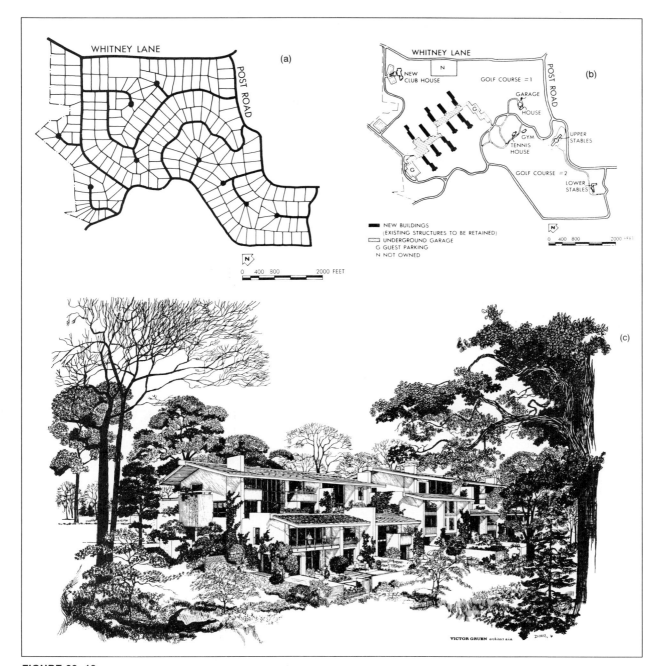

FIGURE 22–19
Alternative plans for developing the same tract of land. (a) Typical subdivision into half-acre lots. (b) The same number of units are clustered into groups of attached dwellings. The rest of the land is retained for open space and other amenities, and is cooperatively owned, used, and maintained by the residents. (c) Attached dwellings planned for clustered development. (By permission from the Victor Gruen Center for Environmental Planning.)

For example, one region may be designated for residential housing only with minimum lot size specified, while other areas are zoned for commercial development and apartment buildings. In theory, zoning laws could lead to planned, efficient devel- opment, but more often they have caused the reverse. Historically, zoning laws originated as a means of keeping "undesirable" development (undesirable according to the present constituents) from coming into an area. For example, counties have zoned major portions of their land for residential use only with a minimum lot size of three acres to assure that newcomers will be of considerable economic means. This effectively excludes and prevents gaining the efficiencies of clustered development noted above. It assures sacrificing huge amounts of agricultural land and open space to house modest numbers of people, and it guarantees continuing dependence on cars and commuting long distances to work or shopping. Thus it exacerbates urban sprawl and intensifies gentrification. Zoning laws frequently stand in the way and must be amended to permit more coherent logical development.

Agricultural and Conservation Districts. Since development contributes a major portion to the economy of states and counties, it is probably unrealistic to expect that states or local governments can or will enact legislation that will stop urban sprawl. However, they can make more order of it. Recognizing that suburban development tends to drive out farming and recognizing the need to preserve agriculture, a number of states have enacted legislation creating **agricultural** or **conservation districts.** Subdivision is greatly restricted in these districts while it is supported by building roads and other facilities in other areas. Although this does not stop urban sprawl, it tends, along with suitable amendments to zoning ordinances, to focus and cluster development so that it is somewhat more efficient in use of land and other resources.

Growth Management Initiatives. Through petitions, citizens can bring an issue to a vote as an initiative on state or local ballots and, if it passes, thus force the government to take certain actions. Citizens are beginning to use such initiatives to control growth. In 1987, 57 citizen initiatives requiring the tightening of growth controls came to a vote in various communities of California, and over 70 percent passed. Similarly, citizens have been approving bond issues to buy and preserve open space in a number of states.

Land Trusts. Increasingly, citizens are becoming dismayed by cherished open areas being traded for more development, and they are taking action themselves. They are forming groups called **land trusts.** The group raises money and negotiates with the owner to buy the land and put it into an open space trust, a legal designation that means that the land can't be developed. Frequently, the owner doesn't really want the land developed either but finds it necessary to sell for a variety of reasons. Therefore, he/she is willing to negotiate a bargain price with the land trust group. The Nature Conservancy, the largest national land trust organization, has saved 3.5 million acres, some in every state. However, there are some 700 land trusts working in local areas that have protected an additional 2 million acres nationwide. For example, a group in Pinellas County, Florida, worked with other community activists to protect a tract zoned for condominium development. "The site

held a nest where a pair of Bald eagles had produced 16 fledglings. The area is now a preserve for the eagles and wading birds such as wood storks and herons. Its wetlands play a critical watershed role by filtering stormwater runoff."* (See Fig. 22–20.)

Steps to control suburban growth meet considerable resistance from people arguing that it will cripple the economic future of the region. However, evidence is increasingly showing the reverse. In making their decision to remain or locate in an area, businesses rank environmental quality or "livability" high on the list. In turn, they rank open space for aesthetic and recreational enjoyment as a prime factor of environmental quality. This is particularly true for the new, rapidly growing computer and electronics industries. In short, preserving environmental quality will be decisive in our economic future.

Stemming Urban Blight and Aiding Urban Redevelopment

There is tremendous irony in the fact that high unemployment and decrepit housing exist hand in hand. It bespeaks a tragic failing in our economic system that facilities crying out for maintenance and repair and people crying out for jobs are not put together for the betterment of both. In fact, with a little attention and imagination, they can be.

Changing the Property Tax. Urban decay and blight are actually encouraged by an antiquated property tax code that taxes a property according to the market value of the house or building on it. This effectively discourages poor people from improving their homes because any improvements simply bring higher taxes. For example, suppose you put in a new bathroom. The tax assessor notes this and decides that this increases the value of your home by $3000, and "rewards" you with another $150 per year in taxes. Conversely, people are effectively rewarded for letting their properties deteriorate because, with decreasing values, their tax bills are lowered. This applies to landlords as well, and it may be used as a form of land speculation. A landlord can hold slum properties indefinitely, receiving substantial rents and paying virtually no property taxes. At some point he can sell the land for redevelopment, making a huge profit.

Pittsburgh, Pennsylvania, and some other cities have brought urban decay to a halt simply by changing the tax code so the bulk of the property tax is assigned to the land rather than to the building on it. Thus there is no financial penalty for improving dwellings and no reward in letting them deteriorate.

---

* *The Trust for Public Land* newsletter, summer 1988.

FIGURE 22–20
The Big Bend coast, which includes some 200 miles at the extreme northeast corner of the Gulf of Mexico, embraces immense wet hummocks and stream-veined marshlands that sustain one of the nations most productive fisheries and shelter a plethora of species from globally endangered West Indian manatees to federally protected Kemp's Ridley and green sea turtles. Yet the area is under intense pressure for development. Over the last 20 years the Nature Conservancy has worked with its agency partners in Florida to establish five federally or state-protected natural areas along the Big Bend coast. The Lower Suwanne River National Wildlife Refuge, shown here, is one of the five. (Susan Bournique, The Nature Conservancy)

Aiding Residents to Own Buildings. Many inner-city apartment buildings are sadly neglected and abused by their tenants, making high rents coupled with miserable conditions almost inevitable. However, this abuse stems more from frustration of helplessness than from actually not caring. Tremendous turnarounds have been achieved where such buildings have been converted into cooperatives owned, managed, and maintained by the residents themselves. Rents are lower, conditions are improved, and people gain a new sense of pride in themselves and where they live. Notably, the system provides a substantial number of full- or part-time jobs for residents who may have been unemployed or underemployed as they take take up new roles as plumbers, electricians, painters, carpenters, and groundskeepers. Habitat for Humanity (see Appendix A), using donated funds and materials, and volunteer labor, is aiding thousands of both urban and rural families build themselves decent housing.

Urban Homesteading. Because of exurban migration, most cities have large numbers of abandoned residences that they acquired in lieu of taxes. Under a program known as urban homesteading, originated in Wilmington, Delaware, and now being used in cities nationwide, such properties are turned over to individuals for the sum of one dollar. In signing the contract for the home, the individual agrees to fix it up and live in it for at least 3 years. To aid him/her in this effort, the city provides low-interest, long-term loans that effectively become a mortgage on the property.

This has proven to be a very successful program in which everyone wins. The "homesteaders" end up with a nice home at reasonable cost, or achieve a con- siderable profit on selling after the three years. The city gets the property back on the tax rolls. Most importantly, as affluent consumers come back into the city so do stores and other businesses, which provide employment and additional taxes, helping to reverse the entire cycle of urban blight (Fig. 22–21).

The Integral Urban House

Many environmentalists aspire to have their own piece of land where they live more self-sufficiently and "ecologically" by growing their own food, recycling wastes, utilizing solar energy, and so on. During the 1960s and 1970s many retreated to far reaches to do this. Living in the city and environmentalism seemed contradictory. However, there were also many who realized that this was no solution. There is simply not enough wilderness to accommodate more than a minute fraction of the population, and those who retreat there to carve out their own homestead are simply adding their impact on open space, which is already under extreme pressure. Then, there are many social and cultural ties to the city that we do not want to break. Thus, certain groups began considering: Is it possible to live more ecologically within the confines of the urban environment? Out of this was born the concept of the **integral urban house,** a typical urban residence that, by following all the ecological principles, is made environmentally sound and nearly self-sufficient in virtually all respects including food production (see Case Study 22). The integral urban house is a concept that can be pursued by nearly all city and suburban residents and which would reduce our impact on the environment and lead toward a sustainable biosphere.

(a)

FIGURE 22–21
"Dollar homes" being renovated in Baltimore, Maryland. (a) Note the still-abandoned structures on the left which are typical of the "before" condition. In the center are homes nearing completion of renovation. (b) Front side of homes. (Photos by author.)

(b)

What You Can Do

- [ ] Investigate land-use planning and policies in your region. Is urban sprawl and/or urban blight a problem? What kinds of policies or controls are being used to restrain them?
- [ ] Join your local neighborhood association to learn more about land use and zoning concerns in your area. Introduce concepts of ecologically sustainable land use to the group.
- [ ] Become involved in zoning issues in your neighborhood by attending zoning hearings. Contribute your understanding in support of zoning that will support rather than hinder ecologically sound land use.
- [ ] Contact the Nature Conservancy, The Trust for Public Land (see Appendix A), or local conservation organizations to identify particular areas in your region that are under pressure from development and which conservationists feel should be saved.
- [ ] Investigate why conservationists feel these areas should be protected and what is being done to save them. Join and contribute to one or more of these efforts.
- [ ] If there is a lot (large or small) in your area which is not receiving protection but which you feel should be, form a land trust group to protect it. Information on how to form a land trust can be obtained from The Trust for Public Land (see Appendix A).
- [ ] Write your city or county legislators to change the property tax code so that it does not penalize home improvements.
- [ ] Consider where you live and your lifestyle in terms of urban sprawl and its consequences. Consider moving to a location where your commuting needs and environmental impacts will be less.
- [ ] Investigate the principles of an integral urban house and begin to make additions to your home that will move it toward being an integral urban house.

Earth day, a media event in the Spring of 1969, inspired the creation of a very popular "Man and Environment" class at the University of California (UCB). My husband and I team-taught three subsections. This gave us ample opportunity to examine information on the impact of humans upon the planet, from both a physical and an ethical perspective.

The reaction of many aquaintances was "back to the land," i.e. find a piece of land in the country and homestead it. This was not an option for us. My husband, William Olkowski, was completing his Ph.D. in Biological Control at UCB. I was an M.S. student in Urban Pest Management at Antioch College West. We had oriented our academic efforts toward reducing the use of toxic pesticides. The dominating concern of our personal life became the daily, minute examination of our own lifestyle.

We were intensely focused on the kinds of materials that entered our home each day and what happened to them when they magically transmuted

themselves from being valuable to being "wastes." As part of changing our own lives, Bill started a recycling center, and inspired the inauguration of several others.

Because we were biologists, our investigations tended to emphasize the cycles of energy and minerals that support life. What captured the attention of our students was that we produced our own food right in the city.

So, we started a student food garden on University land, and a class to go with it, "Urban Garden Ecosystems." We used this class as a way to examine a wide range of issues, from population growth and spread of cities onto farmland to the many other ways that urban areas affect the countryside. Most of the people in this country live in cities. How can you live in the city so that you don't destroy the country? Our answer was the idea of the integral urban house. A way in which everyday activities in your own home can be integrated into the preservation of the planet. We worked on our own lives, our students worked on theirs.

When an elegant hillside home was donated to the small, nonprofit Farallones Institute, we proposed it be sold and the proceeds used to buy and renovate an old house in the inner city. Thus began a public demonstration project called The Integral Urban House.

For approximately a decade this house was open to the public for tours and classes that explored its many features and their potential for adoption elsewhere. Sun energy was captured through a vegetable garden, greenhouse, solar oven, and rooftop and window water collectors. Minerals were cycled through a composting toilet, and gray water and aerobic composting systems. Chickens and rabbits, raised partially on garden products, honeybees, and crayfish raised in a wind-aerated pond, added to the diet of the students who maintained the house. Use of a variety of water and other resource-saving appliances, as well as benign pest management methods, were illustrated through the daily activities of the residents. A book written by ourselves, students, and colleagues, published by Sierra Club Books, gives details on all the systems, and the thinking behind them.

The house was closed when the funds to support it were needed for other purposes by the parent institution. Although the public demonstration project ended, the idea of the integral urban house is alive and well. If you want to find out more about it, start creating one where you are living now.

HELGA OLKOWSKI

EPILOGUE

As environmental issues came to the forefront in the 1960s, a common question was, How long do we have? How long will it be before we reap the tragic consequences of our environmental disregard? The answer was 30 to 35 years. Now, approaching 30 years later, this answer haunts me. As we are beginning to experience a hotter climate, holes in the protective ozone shield, toxic chemicals in groundwater, food contaminated with pesticide residues, dieback of forests, increasing extinctions, and famines, I think the prediction was not far off. We simply do not have another 30 years to carry on life as usual while we debate and study the issues. Over the next 30 years, we must either create a sustainable society or bear the consequences. The consequences will be a deteriorating quality of life as air and water, natural environments, and wild species are increasingly despoiled or destroyed by our polluting wastes and lust for exploitation. Also, more and more people will become environmental refugees, depending on help from others, as depleted soil and water resources no longer support a productive livelihood.

The last 30 years have not been wasted: a firmer basis of scientific understanding has been gained; regulatory agencies at all levels of government have been formed; numerous citizens' groups dedicated to protecting the environment have been organized; many environmental laws and regulations have been passed and some international agreements have been reached.

However, remedial measures have tended to focus on "end-of-pipe" solutions rather than getting at the root causes. For example, we have tried to make landfills leak proof rather than questioning the practice of dumping in contrast to recycling. We have tried to make nuclear power plants acceptable by adding on more and more safety features, which makes them exorbitantly expensive, rather than developing solar alternatives, which are environmentally benign and exploit an energy source that is everlasting. We are putting more and more pollution control devices on cars and desperately seeking more oil rather than questioning the inordinate energy demand of urban sprawl. We are desperately trying to save a few species from extinction while not really addressing the population explosion which is overwhelming natural systems.

The basic point made in Part I should be abundantly clear now: Systems which are contrary to natural principles and laws are not sustainable. Attempts to sustain them only become increasingly expensive and complex and are destined for ultimate failure in any case. Even the most elaborate landfill will not contain wastes forever, and finally there is no place to put more landfills. The safest nuclear plant will still produce long-lived radioactive materials. Carbon dioxide outputs from burning fuels cannot be eliminated by any pollution control device. It is impossible to maintain most species apart from their natural ecosystems. Thus, we can bankrupt ourselves with end-of-pipe remedies and still not achieve sustainable solutions. Environmentalism has frequently been cast in the light of being counterproductive to progress and a strong economy. In should be abundantly clear that exactly the opposite is the case.

In order to achieve lasting solutions, we must refocus our attention and priorities on the principles underlying sustainability. These are:

- ☐ stabilizing population.
- ☐ attaining a sustainable agriculture, one which does not deplete soil and water resources or contaminate the earth and food supplies with pesticides.
- ☐ recycling wastes.
- ☐ developing benign solar energy alternatives.
- ☐ adopting more energy- and resource-conservative lifestyles.

I recognize that the problems seem so enormous and forces with vested interests in the status quo seem so entrenched that you may often feel that there is nothing you can do which will make any difference. Yet, everything you do already does make a difference. You cannot avoid the facts: the car you drive, the products you use, the wastes you throw away, and everything else you do contribute to the problems in a greater or lesser amount. Whatever you may do to lessen impacts on the environment is part of the solution. Consequently, it is not a matter of having an effect. It is simply a matter of asking yourself, *Will I be part of the problem or part of the solution?*

There are four levels on which you may participate to bring about solutions (Fig. E–1):

☐ individual lifestyle change.
☐ political involvement.
☐ membership and participation in environmental organizations.
☐ career choices.

Lifestyle changes may involve such things as switching to a more fuel-efficient car, recycling waste paper and bottles, adapting your home to solar energy, or composting and recycling food and garden wastes into your own lawn or garden. Political involvement means supporting and voting for candidates with strong environmental positions, and writing or making calls in support of environmental legislation. Environmental candidates don't get elected and environmental bills don't get passed unless constituents have *expressed* support for them.

Membership in environmental organizations can enhance both lifestyle change and political involvement. As a member of an environmental organization you will receive and may help disseminate information making you and others more aware of particular environmental problems and things you can do to help. Specifically, you will be informed re-

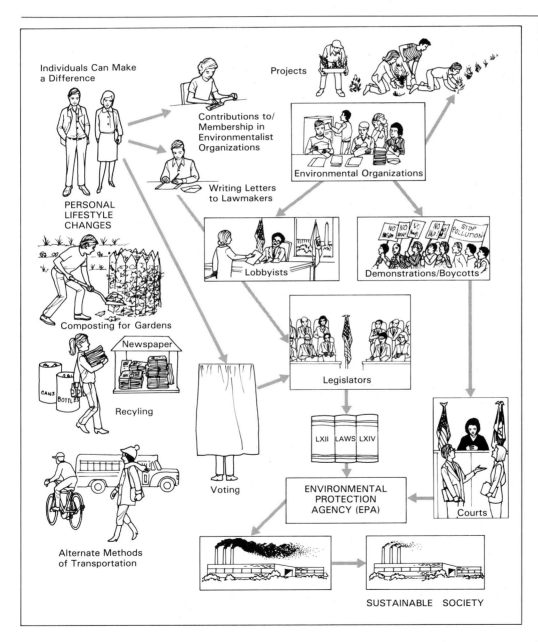

FIGURE E–1
Avenues of environmental action.

garding environmentally significant legislation so that you may focus your political efforts at the most effective time and place. Also, your membership and contributions serve to support lobbying efforts of the organization. A lobbyist representing only him or herself has relatively little impact on legislators. On the other hand, if the lobbyist represents a million-member organization that can follow up with that many phone calls and/or letters (and ultimately votes) the impact is considerable. Finally, where enforcement of existing law has been the weak link, some organizations such as Natural Resources Defense Council and Environmental Defense Fund have been very influential in bringing polluters or the government to court to see that the law is upheld. Again, this can only be done with the support of members.

Finally you may chose to devote your career to promoting solutions to environmental problems. In this way environmentalism may be your vocation as an avocation. Environmental careers go far beyond the traditional occupation of wildlife and park management. There are any number of lawyers, doctors, nurses, social workers, teachers, journalists, scientists, and others focusing their talents and training on environmental issues or risks. There are innumerable business opportunities in pollution control, recycling, waste management, environmental monitoring and analysis, nonchemical pest control, production and marketing of organically grown produce, and so on. Some music groups have gained renown for songs promoting environmental awareness. Some developers concentrate on rehabilitation and reversal of urban blight as opposed to contributing more to urban sprawl. Indeed, it is difficult to think of a vocation that cannot be focused on promoting solutions to environmental problems.

There are many more chapters of environmental science to be written, but they will be written by you, perhaps not in a formal way on paper, but by the things you do and accomplish toward bringing us all toward a sustainable society.

BIBLIOGRAPHY

GENERAL REFERENCES

ALLEN, JOHN, ED.   Annual editors: *Environment 88/89*. Guilford, Connecticut: Dushkin Publishing Group, Inc., 1988. An annual anthology of recent articles from respected periodicals and magazines.

CONSERVATION FOUNDATION. *State of the Environment: A View Toward the Nineties*. Washington, D.C.: The Conservation Foundation, 1988. Excellent coverage of selected environmental issues.

FUND FOR RENEWABLE ENERGY AND THE ENVIRONMENT. *The State of the States*. Washington, D.C.: Fund for Renewable Energy and the Environment, 1987. An overview of environmental problems and how states are coping with them.

GOUDIE, ANDREW. *The Human Impact*. Cambridge, Mass.: The M.I.T. Press, 1982.

MYERS, NORMON, ED. *Gaia: An Atlas of Planet Management*. Garden City, N.Y.: Anchor Press, 1984.

NATIONAL GEOGRAPHIC SOCIETY.   "Can Man Save This Fragile Earth?" *National Geographic*, 174 (December, 1988). Special environmental issue, excellent look at population and lifestyles, biological diversity, wildlife, conservation.

NATIONAL WILDLIFE FEDERATION.   "The Annual Environmental Quality Index," *National Wildlife* (February issue of each year). An annual assessment of various environmental quality indicators.

"Planet Earth: How It Works, How To Fix It." *U.S. News and World Report*, (Oct 31, 1988). Special issue confronting global warming and ozone depletion, deforestation, and overpopulation.

"Planet of the Year: Endangered Earth," *TIME*, (Jan 2, 1989). Special issue devoted to environmental problems facing the planet, covering everything from rainforests to toxic waste to the greenhouse effect. Looks at international solutions.

SOUTHWICK, CHARLES H., ED. *Global Ecology*. Sunderland, Mass.: Sinaver Associates, Inc., 1985.

U.S. EXECUTIVE OFFICE, COUNCIL ON ENVIRONMENTAL QUALITY. *Annual Environmental Quality Report*. Washington, D.C.: U.S. Government Printing Office.

WORLD RESOURCES INSTITUTE AND INTERNATIONAL INSTITUTE FOR ENVIRONMENT AND DEVELOPMENT. *World Resources 1988/89*. New York: Basic Books, Inc., 1988. An annual compilation of data and reports covering a wide range of environmental issues.

WORLDWATCH INSTITUTE. *State of the World 1989: A Worldwatch Institute Report on Progress Toward a Sustainable Society*. Washington, D.C.: Worldwatch Institute. An annual anthology of comprehensive studies, covering an array of environmental issues.

WORLDWATCH INSTITUTE. *World Watch*. A periodical published by the Worldwatch Institute, Washington, D.C. Excellent articles covering the whole range of environmental issues.

INTRODUCTION

BAKER, JEFFREY J.W., AND GARLAND E. ALLEN. *Hypothesis, Prediction, and Implication in Biology*. Reading, Mass.: Addison-Wesley Publishing Co., 1968.

BRADY, DONALD. *Logic of the Scientific Method*. New York: Irvington Publishers, 1973.

GIEVE, RONALD N. *Understanding Scientific Reasoning*, 2nd ed. New York: Holt, Rinehart & Winston, Inc., 1984.

LASTRUCCI, CARLO. *Scientific Approach*. Cambridge, Mass.: Schenkman Bks., Inc., 1967.

McCAIN, GARVIN, AND ERWIN M. SEGAL. *The Game of Science*. 4th ed. Florence, Ky.: Brooks-Cole Publishing Co., 1981.

UNITED NATIONS WORLD COMMISSION ON ENVIRONMENT AND DEVELOPMENT. *Our Common Future*. New York: Oxford University Press, 1987.

WALKER, M. *The Nature of Scientific Thought*. Englewood Cliffs, N.J.: Spectrum Book/Prentice-Hall, Inc., 1963.

CHAPTER 1. ECOSYSTEMS: WHAT THEY ARE

ALLEN, ROBERT. "Margin of Life," *International Wildlife*, 7 (Mar/Apr 1977) pp. 20–28.

CLOUD, PRESTON. "The Biosphere," *Scientific American*, 249 (Sep 1983), pp. 176–189.

CLOUDSLEY-THOMPSON, J.L. *Terrestrial Environments*. New York: Halsted Press, 1975.

CURRY-LINDAHL, K. *Wildlife of the Prairies and Plains*. New York: Harry N. Abrams, Inc., 1981.

EHRLICH, PAUL AND JOHNATHAN ROUGHGARDEN. *The Science of Ecology*. New York: Macmillan Publishing Company, 1987.

HUGHES, CAROL AND DAVID. "Teeming Life of a Rain Forest," *National Geographic*, 163 (Jan 1983), pp. 49–64.

HUTCHINSON, G. EVELYN. "The Biosphere," *Biosphere*, A Scientific American Book, pp. 1–11. San Francisco: W.H. Freeman and Company, 1970.

Kormondy, E.J. *Concepts of Ecology*, 3rd ed. Englewood Cliffs, N.J.: Prentice-Hall, Inc., 1984.

Margulis, Lynn, et al. "Microbial Communities," *Bio-Science*, 36 (Mar 1986), pp. 160–170.

Pimm, S.L. "Properties of Food Webs," *Ecology*, 61 (1980), pp. 219–225.

Smith, R.L. *Ecology and Field Biology*, 3rd ed. New York: Harper and Row, 1980.

Sutton, A., and M. Sutton. *Wildlife of the Forests*. New York: Harry N. Abrams, Inc., 1979.

Wagner, F.H. *Wildlife of the Deserts*. New York: Harry N. Abrams, Inc., 1980.

Whittaker, R.H. *Communities and Ecosystems*, 2nd ed. Toronto, Ontario: The Macmillan Co., 1974.

Whittaker, R.H., and G.E. Likens, eds. "The Primary Production of the Biosphere," *Human Ecology*, 1 (1973), pp. 299–369.

CHAPTER 2. ECOSYSTEMS: HOW THEY WORK

Berner, Robert A. and Antonio Lasaga. "Modeling the Geochemical Carbon Cycle," *Scientific American*, 260 (Mar 1989), pp. 74–81.

Bolin, B., and R.B. Cook. *The Major Biogeochemical Cycles and Their Interactions*. New York: John Wiley and Sons, 1983.

Brill, W.J. "Biological Nitrogen Fixation," *Scientific American*, 236 (Mar 1977), pp. 68–74, 79–81.

Detweiler, R.P., and Charles A.S. Hall. "Tropical Forests and the Global Carbon Cycle," *Science*, 239 (Jan 1, 1988), pp. 42–46. Many new estimates show a balance in the carbon cycle.

Frieden, E. "The Chemical Elements of Life," *Scientific American*, 227 (July 1972), pp. 52–60.

Gosz, J.R., R.T. Holmes, G.E. Likens, and F.H. Bormann. "The Flow of Energy in a Forest Ecosystem," *Scientific American*, 238 (Mar 1978), pp. 92–102.

Janick, Jules. "Cycles of Plant and Animal Nutrition," *Scientific American*, 235 (Sept 1976), pp. 74–84. Excellent overview of nutrient cycles and energy flow.

Jordan, Carl F. *Nutrient Cycling in Tropical Forest Ecosystems*. Somerset, N.J.: John Wiley & Sons, Inc., 1985.

Oort, A.H. "The Energy Cycle of the Earth," *Scientific American*, 223 (Sept 1970), pp. 54–63.

Pimentel, David, et al. "Environmental Quality and Natural Biota," *BioScience*, 30 (Nov 1980), pp. 750–755.

Rounick, J.S., and J.J. Winterbourn. "Stable Carbon Isotopes and Carbon Flow in Ecosystems," *Bioscience*, 36 (Mar 1986), pp. 171–177.

Scientific American, 223 (Sept 1970). Issue devoted to articles describing nutrient cycles and energy flow in ecosystems.

Sunquist, Fiona. "Zeroing In On Keystone Species," *International Wildlife*, 18 (Sept/Oct 1988), pp. 18–23.

Wood, Tim, et al. "Phosphorous Cycling in a Northern Hardwood Forest: Biological and Chemical Control," *Science*, 223 (Jan 27, 1984), pp. 391–393.

CHAPTER 3. ECOSYSTEMS: WHAT KEEPS THEM THE SAME? WHAT MAKES THEM CHANGE?

Brainhardt, Wilton. "The Death of Ducktown," *Discover*, 8 (Oct 1987), pp. 34–43. Area in Tennessee killed by fumes from a smelter.

Carey, John. "Science in a Tub," *National Wildlife*, 25 (Jun/July 1987), pp. 32–35. Population studies in artificial ponds. Excellent.

Chen, Allan. "Unraveling Another Mayan Mystery." *Discover*, 8 (Jun 1987), pp. 40–49.

Conant, Shiela. "Saving Endangered Species by Translocation," *Bioscience*, 38 (Apr 1988), pp. 254–257.

Ehrlich, Gretel. "In New Mexico: Desert Healer," *TIME*, 130 (Dec 7, 1987). Restoration of desertified lands by breaking surface and increasing infiltration with livestock trampling.

Ehrlich, Paul R. "Habitats in Crisis," *Wilderness*, 50 (Spring 1987), pp. 12–15. Damage of overgrazing. Excellent.

Golgel, Monica, and Susan Bratton. "Exotics in the Parks," *National Parks*, 57 (Jan/Feb 1983), pp. 25–29.

Holland, Robert. "Too Many Mallards," *Audubon*, 89 (Jan 1987), pp. 64–67. Overpopulation of mallards on domestic ponds. Good example of population upset of wild species caused by humans.

Kurlansby, Mark. "Haiti's Environment Teeters on the Edge," *International Wildlife*, 18 (Mar/Apr 1988), pp. 35–38.

Lewin, Roger. "Ecologists' Opportunity in Yellowstone's Blaze," *Science*, 241 (Sept 30, 1988), pp. 1762–63. Yellowstone, and the role of fire in ecosystems.

Lugo, A.E. "The Future of the Forest: Ecosystem Rehabilitation in the Tropics," *Environment*, 30 (Sept 1988), pp. 17–20, 41–45.

May, R.M. "Parasitic Infections as Regulators of Animal Populations," *American Scientists*, 71 (1983), pp. 36–45.

Naiman, Robert J. "Animal Influences on Ecosystem Dynamics," *BioScience*, 38 (Dec 1988), pp. 750–752.

Nash, Stephen. "The Blighted Chestnut," *National Parks*, 62 (July/Aug 1988), pp. 14–19.

Odum, Eugene P. "Trends Expected in Stressed Ecosystems," *BioScience*, 35 (July/Aug 1985), pp. 419–422.

Pastor, John, et al. "Moose, Microbes, and the Boreal Forest," *BioScience*, 38 (Dec 1988), pp. 770–777.

Power, J.F. and R.F. Follert. "Monoculture," *Scientific American*, 256 (Mar 1987), pp. 79–86.

Stevens, Larry. "King Kong Kudzu, Menace to the South," *Smithsonian*, 7 (Dec 1976), pp. 93–99.

Strohm, John. "Who's Afraid of the Big Bad Wolf," *National Wildlife*, 25 (Aug/Sept 1987), pp. 4–11. Controversy over reintroducing wolves into Yellowstone.

Waters, Tom. "Return to Krakatau," *Discover*, 9 (Oct 1988), pp. 64–67. Account of natural succession on Indonesian archipelago destroyed by volcanic eruption in 1883.

WATT, KENNETH E.F. "Seep Question about Shallow Seas," *Natural History*, 96 (July 1987), pp. 60–65. Population balances among the many fish species in shallow seas. Excellent.

WEST, RICHARD F. "Tolling the Chestnut," *American Forests*, 94 (Sept/Oct 1988), pp. 10, 76–77. A history of the blight and current research.

WILLIAMS, TED. "Incineratrion of Yellowstone," *Audubon*, 91 (Jan 1989), pp. 38–45.

WRIGHT, HENRY A., AND ARTHUR W. BAILEY. *Fire Ecology— United States and Southern Canada.* Somerset, N.J.: John Wiley & Sons, Inc., 1982.

ZURICK, DAVID. "A Question of Balance," *Sierra*, 72 (July/ Aug 1987), pp. 47–50.

CHAPTER 4. ECOSYSTEMS: ADAPTATION AND CHANGE OR EXTINCTION

BATTEN, MARY. "How Plants Fight Back," *International Wildlife*, 18 (July/Aug 1988), pp. 40–43. Chemical protection and its role in plant evolution.

BISHOP, J.A., AND L.M. COOK. "Moths, Melanism, and Clean Air," *Scientific American*, 232 (Jan 1975), pp. 90–99.

CAIRNS-SMITH, A.G. "The First Organisms," *Scientific American*, 252 (Jun 1985), pp. 90–100.

CLARKE, B. "The Causes of Biological Diversity," *Scientific American*, 233 (Aug 1975), pp. 50–60.

COHN, JEFFREY P. "Gauging the Biological Impacts of the Greenhouse Effect: How Might Species Cope With a Warmer World?" *BioScience*, 39 (Mar 1989), pp. 142–46. A look at genetic changes and behavioral adaptations.

DIAMOND, JARED. "The Golden Age that Never Was," *Discover*, 9 (Dec 1988), pp. 71–79. An interesting look at pre-industrial humans and their often devastating effect on the environment.

DIAMOND, JARED. "The Great Leap Forward," *Discover*, 10 (May 1989), pp. 50–60.

ISAAC, GLYNN. "The Food-Sharing Behavior of Protohuman Hominids," *Scientific American*, 238 (Apr 1978), pp. 90–108.

LEWIN, ROGER. "How Did Vertebrates Take to the Air?" *Science*, 221 (July 1, 1983), pp. 38–39.

LEWIS, THOMAS A. "Will Species Die Out as the Earth Heats Up?" *International Wildlife*, 17 (Nov/Dec 1987), pp. 18–21.

MYERS, NORMAN. "Extinction Rates Past and Present," *BioScience*, 39 (Jan 1989), pp. 39–42.

NABHAN, GARY PAUL. "Seeds of Renewal," *World Monitor*, 2 (Jan 1989), pp. 17–20.

NAYLOR, BRUCE G., AND PAUL HANDFORD. "In Defense of Darwin's Theory," *BioScience*, 35 (Sept 1985), pp. 478–484.

SCHWARTZ, ANNE. "Banking on Seeds to Avert Extinction," *Audubon*, 1 (Jan 1988), pp. 22–27.

Scientific American, 249 (Sept 1983). This issue is devoted to various aspects of change in and on the planet Earth over time.

STRINGER, C.B. AND P. ANDREWS. "Genetic and Fossil Evidence for the Origin of Modern Humans," *Science*, 239 (Mar 11, 1988), pp. 1263–68.

TIERNEY, JOHN, LYNDA WRIGHT, AND KAREN SPRINGDEN. "The Search for Adam and Eve," *Newsweek*, (Jan 11, 1988), pp. 46–52. Fascinating evidence in DNA.

WILSON, ALLAN C. "The Molecular Basis of Evolution," *Scientific American*, 253 (Oct 1985), pp. 164–173.

WOLF, EDWARD C. "Avoiding a Mass Extinction of Species," *State of the World 1988*, Chapter 6. Washington, D.C.: Worldwatch Institute, 1988.

PART TWO POPULATION—GENERAL REFERENCES

MURPHY, ELAINE M. *World Population: Toward the Next Century.* Washington, D.C.: Population Reference Bureau, Inc., 1985. Seventeen-page booklet. Great overview of population for classroom.

Population Bulletin, published by the Population Reference Bureau, Inc., Washington, D.C. Comprehensive, timely reviews written by experts on world and national demographic trends and their implications. Average length, 48–52 pages.

Population Today, a monthly newsletter published by the Population Reference Bureau, Inc., Washington, D.C. Contains latest information and interesting articles/statistics worldwide.

WORLD POPULATION NEWS SERVICE, *Popline*, published by the Population Institute, Washington, D.C. A news and feature service provided to more than 2100 daily newspapers worldwide, with correspondents in 159 countries.

CHAPTER 5. THE POPULATION PROBLEM: ITS DIMENSIONS AND CAUSES

BIRDSALL, NANCY. "Population Growth and Poverty in the Developing World," *Population Bulletin* 35. Washington, D.C.: Population Reference Bureau, 1980.

BOUVIER, LEON F. "Planet Earth 1984–2034: A Demographic Vision," *Population Bulletin* 39. Washington, D.C.: Population Reference Bureau, 1984.

BROWN, LESTER R. AND CHRISTOPHER FLAVIN. "The Earth's Vital Signs," *State of the World 1988*, Chapter 1. Washington, D.C.: Worldwatch Institute, 1988, pp. 1–21.

CROSSON, PIERRE. "Agricultural Land: Will There Be Enough?" *Environment*, 26 (Sept 1984), pp. 16–20, 40–45.

EHRLICH, PAUL R. AND ANNE H. EHRLICH, "Population, Plenty, and Poverty," *National Geographic*, 174 (Dec 1988), pp. 914–941.

EHRLICH, PAUL R. AND ANNE H. EHRLICH. "World Population Crisis," *Bulletin of the Atomic Scientists*, 42 (Apr 1986), pp. 13–19.

FORNOS, WERNER. *Gaining People, Losing Ground.* Washington, D.C.: Population Institute, 1987.

McDOWELL, BART. "Mexico City: An Alarming Giant," *National Geographic*, 166 (Aug 1984), pp. 138–172.

MELLOR, JOHN W. "The Intertwining of Environmental

Problems and Poverty," *Environment,* 30 (Nov 1988), pp. 8–13, 28–30.

ODELL, RICE, ED. "Egypt: A Case of Ecological Vulnerability," *Conservation Foundation Letter,* (Aug 1983), pp. 1–8.

POPULATION REFERENCE BUREAU, WASHINGTON, D.C. A source of current data and information concerning human population, both U.S. and world. Numerous publications are available on an individual or subscription basis.

PORTER, GARETH WITH DELFIN J. GANAPIN, JR. *Resources, Population, and the Philippines' Future: A Case Study.* Washington, D.C.: World Resources Institute, 1988.

REPETTO, ROBERT. "Population, Resources, Environment: An Uncertain Future," *Population Bulletin,* 42. Washington, D.C.: Population Reference Bureau, July 1987.

U.S. CONGRESS COUNCIL ON ENVIRONMENTAL QUALITY AND THE U.S. DEPARTMENT OF STATE. *The Global 2000 Report to the President: Entering the Twenty-First Century,* vol. 1, *The Summary Report.* Washington, D.C.: U.S. Government Printing Office, 1981.

WESTLAKE, MELVYN. "Trouble in the Third World's Environment," *World Press Review,* 34 (July 1987), pp. 53–54. Interconnection between population explosion, poverty, and pollution.

WESTOFF, C.F. "Fertility in the United States," *Science,* 234 (Oct 1987), pp. 554–59. Focuses on low fertility since the baby boom.

WORLD BANK. *World Development Report 1984.* New York: Oxford University Press, 1984. See also World Development Reports for other years.

WORLD RESOURCES INSTITUTE. "Population and Health," *World Resources 1988–89,* Chapter 2. New York: Basic Books, Inc., 1988.

WRIGHT, SUZAN. "Europe Faces Population Decline," *Futurist* (July/Aug 1987), p. 50.

CHAPTER 6. ADDRESSING THE POPULATION PROBLEM

BARDACH, JOHN. "Aquaculture: Moving from Craft to Industry," *Environment,* 30 (Mar 1988), pp. 6–11, 36–40.

BROWN, LESTER R. "The Changing World Food Prospect: The Nineties and Beyond," *Worldwatch Paper 85.* Washington, D.C.: Worldwatch Institute, October, 1988.

BROWN, LESTER. "Growing Grain Gap," *World Watch,* 1 (Sept/Oct 1988), pp. 10–18.

BROWN, LESTER R., AND EDWARD C. WOLF. "Reversing Africa's Decline," *Worldwatch Paper 65.* Washington, D.C.: Worldwatch Institute, June 1985.

CHANDLER, WILLIAM V. "Improving World Health: A Least Cost Strategy," *Worldwatch Paper 59.* Washington, D.C.: Worldwatch Institute, July 1984.

CREWS, KIMBERLY A. "Human Needs and Nature's Balance: Population, Resources, and the Environment," *A Population Learning Series.* Washington, D.C.: Population Reference Bureau, Inc., 1987.

DOVER, MICHAEL J. AND LEE M. TALBOT. "Feeding the Earth: An Agroecological Solution," *Technology Review,* 91 (Feb/Mar 1988), pp. 27–35.

EDDYT, WILLIAM. "Rhythms of Survival," *National Parks,* 61 (Sept/Oct 1987), pp. 9–10, 21–23. Competition, conflict between population growth needs and wildlife preservation in Kenya.

HAUB, CARL. "Selecting a Starting Fertility Level," *Population Bulletin,* 42 (Dec 1987), pp. 12–15.

HOLDEN, CONSTANCE. "U.S. Anti-abortion Policy May Increase Abortions," *Science,* 238 (Nov 27, 1987), p. 1222.

JACOBSON, JODI. "Planning the Global Family," *State of the World 1988,* Chapter 9. Washington, D.C.: Worldwatch Institute, 1988, pp. 151–169.

KENT, MARY. "Thailand's Achievement," *Population Today,* 16 (Nov 1988), pp. 3–4.

KINSINGER, ANNE E. *A Fund for Sustainable Agriculture.* Washington, D.C.: Natural Resources Defense Council, October 1987.

NATIONAL SCIENCE FOUNDATION. *Appropriate Technology,* A report prepared for the Committee on Science and Technology, U.S. House of Representatives. Washington, D.C.: National Science Foundation, 1979.

PLUCKNETT, DONALD L., AND NIGEL J.H. SMITH. "Sustaining Agricultural Yields," *BioScience,* 36 (Jan 1986), pp. 40–45.

REPETTO, ROBERT AND MALCOLM GILLIS, EDS. *Public Policies and the Misuse of Forest Resources.* Washington, D.C.: World Resources Institute, 1988.

SERRILL, MICHAEL S. "Famine," *TIME,* 130 (Dec 21, 1987), pp. 34–43.

TALBOT, LEE. *Helping Developing Countries Help Themselves.* Washington D.C.: World Resources Institute, 1985.

VAN DUIVENDRJK, HANS AND PABLO BARTHOLOMEW. "They Stopped the Sea: A Tide of Human Muscle Dikes a River," *National Geographic,* 172 (July 1987), pp. 92–95. Bangladesh, population, self-help. Daming a river for irrigation increases crop production.

VIETMEYER, NOEL. "Roast Rodent?," *International Wildlife,* 18 (Nov/Dec 1988), pp. 14–16.

WORLD BANK. *World Development Report 1988: Opportunities and Risks in Managing the World Economy.* New York: Oxford University Press, 1988.

WORLD RESOURCES INSTITUTE. "Food and Agriculture," *World Resources 1988–89,* Chapter 4. New York: Basic Books, Inc., 1988.

CHAPTER 7. SOIL AND THE SOIL ECOSYSTEM

ALLEN, EDITH B. "Stopping Desertification," *American Land Forum,* 7 (Mar/Apr 1988), pp. 14–16.

BATIE, SANDRA S., AND ROBERT G. HEALY. "The Future of American Agriculture," *Scientific American,* 248 (Feb 1983), pp. 45–53.

BLAIKIE, PIERS AND HAROLD BROOKFIELD. *Land Degradation and Society.* Corelo, Calif.: Island Press, 1987.

BREMAN, H., AND C.T. DE WIT. "Rangeland Productivity

and Exploitation in the Sahel," *Science*, 221 (Sept 30, 1983), pp. 1341–1347.

BROWN, LESTER R., AND EDWARD C. WOLF. "Soil Erosion: Quiet Crisis in the World Economy," *Worldwatch Paper* 60. Washington, D.C.: Worldwatch Institute, Sept. 1984.

CARTER, VERNON G., AND T. DALE. *Topsoil and Civilization.* Norman, Oklahoma: University of Oklahoma Press, 1974.

COHEN, ELLEN. "This Fall Grow Soil," *Rodale's Organic Gardening*, 32 (Sept 1985), pp. 55–61.

EL-SWAIFY, S.A., ET AL., EDS. *Soil Erosion and Conservation.* Soil Conservation Society of America. Corelo, Calif.: Island Press, 1985.

GRANATSTEIN, DAVID. *Reshaping the Bottom Line—On-Farm Strategies for a Sustainable Agriculture.* Land Stewardship Project. Corelo, Calif.: Island Press, 1988.

LEWIS, THOMAS. "Cropland Erosion Nears Crisis Levels in Many Areas of the Nation," *Crops and Soil*, 2 (Feb/Mar 1989), pp. 37–39.

LOWE, MARCIA D. "The Sahara Swallows Mauritania," *World Watch*, 1 (Sept/Oct 1988), pp. 38–39.

MING, LU. "Fighting China's Sea of Sand," *International Wildlife*, 18 (Nov/Dec 1988), pp. 38–45.

NORMAN, C. "Expanding Deserts, Shrinking Resources." *Science*, 235 (Feb 27, 1987), p. 963.

ORAM, PETER A. "Moving Toward Sustainability: Building the Agroecological Framework," *Environment*, 30 (Nov 1988), pp. 14–17, 30–36.

PIMENTEL, DAVID, ET AL. "World Agriculture and Soil Erosion," *BioScience*, 37 (Apr 1987), pp. 277–283.

POSTEL, SANDRA. "Halting Land Degradation," *State of the World 1989*, Chapter 2. Washington, D.C.: Worldwatch Institute, 1989.

POSTEL, SANDRA, AND LORI HEISE. "Reforesting the Earth," *State of the World 1988*, Chapter 5. Washington, D.C.: Worldwatch Institute, 1988.

SEARS, PAUL B. *Deserts On The March.* Corelo, Calif.: Island Press, 1988.

TAYLOR, MARCIA ZARLEY. "Uncle Sam Makes Conservation Compulsory," *Farm Journal*, 111 (Nov 1987), p. 22. Laws and regulations being imposed on farmers regarding soil conservation/erosion control.

TODD, DENNIS. "Mycorrhizae: The Beneficial Root Fungi," *Organic Gardening*, 33 (Sept 1986), pp. 49–52.

TRIPLETT, GLOVER B., JR., AND DAVID M. VAN DOREN, JR. "Agriculture Without Tillage." *Scientific American*, 236 (Jan 1977), pp. 28–33.

WALD, JOHANNA AND DAVID ALBERSWERTH. *Our Ailing Public Rangelands: Condition Report—1985.* Washington D.C.: National Wildlife Federation and Natural Resources Defense Council, December 1985.

WARD, JUSTIN R. *Ending Tax Subsidies for Sodbusting and Swampbusting.* Project on Agricultural Conservation and Tax Policy: A Complement to the Farm Bill. Washington, D.C.: Natural Resources Defense Council, 1987. (Aug 1987).

WARD, JUSTIN R., ET AL. *Dollars and Sense in Irrigated Agriculture: The need for a consistent tax policy.* Project on Agricultural Conservation and Tax Policy. Washington, D.C.: Natural Resources Defense Council, 1987. (Jun 1987).

WOLF, EDWARD C. "Mimicking Nature: The Sustainability of Many Traditional Farming Practices Lies in the Ecological Models They Follow," *The FAO Review*, 20 (Jan/Feb 1987), pp. 20–24.

WORLD RESOURCES INSTITUTE. "Forests and Rangelands," *World Resources 1986*, Chapter 5. New York: Basic Books, Inc., 1986.

YORK, E.T., JR. "Improving Sustainability with Agricultural Research," *Environment*, 30 (Nov 1988), pp. 18–20, 31–39.

CHAPTER 8. WATER, THE WATER CYCLE, AND WATER MANAGEMENT

ADLER, JERRY. "Where Will the Cranes Go?," *Newsweek*, (Apr 3, 1989), pp. 62–64.

CAUFIELD, CATHERINE. "Fowl Play," *OMNI*, 9 (Jun 1987), pp. 22–29. Description of irrigation drainage disaster to Kesterson Reservoir National Wildlife Refuge.

CHRISTOPHER, THOMAS. "Drip Irrigation," *Horticulture*, 66 (Apr 1988), pp. 28–31.

CLIFTON, MERRITT. "Water From the North?" *Environmental Action*, 20 (Jan/Feb 1989), pp. 25–27.

CUMMINGS, RONALD G., VICTOR BRAJER, JAMES W. MCFARLAND, JOSÉ TRAVA, AND MOHAMED T. EL-ASHRY. *Improving Water Management in Mexico's Irrigated Agriculture.* Washington, D.C.: World Resources Institute, 1989.

EDELSON, B. DAVID. "Water Recycling in Buildings," *Business Week*, 23 (Nov 1987), p. 29. Treatment and recycling of toilet water for flushing only.

FRANCKO, DAVID A., AND ROBERT G. WETZEL. *To Quench Our Thirst: The Present and Future Status of Freshwater Resources of the United States.* Ann Arbor: University of Michigan Press, 1983.

GOWERS, ANDREW, AND TONY WALKER. "Water War in the Middle East," *World Press Review*, 36 (May 1989), pp. 57–58.

HANSEN, KEVIN. "South Florida's Water Dilemma: A Trickle of Hope for the Everglades," *Environment*, 26 (Jun 1984), pp. 14–20, 40–42.

JOHNSGARD, PAUL A. "The Platte: A River of Birds," *The Nature Conservancy News*, 33 (Sept/Oct 1983), pp. 6–10.

LEVEEN, DR. PHILLIP, E., AND LAURA B. KING. *Turning Off the Tap on Federal Water Subsidies.* San Francisco: Natural Resources Defense Council, Inc., Aug 1985.

MICKLIN, PHILIP P. "The Vast Diversion of Soviet Rivers," *Environment*, 27 (Mar 1985), pp. 12–20, 40–45.

NATIONAL WILDLIFE. "Water: A Special Report," *National Wildlife*, 22 (Feb/Mar 1984), pp. 6–21. Several authors/articles.

PIMENTEL, DAVID. "Water Resources in Food and Energy Production," *BioScience*, 32 (Dec 1982), pp. 861–867.

Postel, Sandra. "Conserving Water: The Untapped Alternative," *Worldwatch Paper 67*. Washington, D.C.: Worldwatch Institute, Sept 1985.

Postel, Sandra. "North China Exceeds Its Water Budget," *World Watch*, 1 (Sept/Oct 1988), pp. 40–41.

Postel, Sandra. "Water: Rethinking Management in an Age of Scarcity," *Worldwatch Paper 62*, Washington, D.C.: Worldwatch Institute, Dec 1984.

Raloff, J. "U.S. River Quality: Not All Signs are Good," *Science News*, 131 (Apr 4, 1987), p. 214.

Reisner, Marc. "Water Folly: Dam the Rivers and Damn the Taxpayers: How Power Politics Is Costing You Billions of Dollars," *Common Cause Magazine* (Nov/Dec 1986), pp. 12–17.

Scudder, Thayer. "Conservation vs. Development: River Basin Project in Africa," *Environment*, 31 (Mar 1989), pp. 4–9/27–31.

Sheldon, Martha O. "Make Your Wash Water Do Double," *The Mother Earth News*, 88 (July/Aug 1984), pp. 102–104.

Sidey, Hugh. "The Big Dry," *TIME*, 131 (July 4, 1988), pp. 12–15.

Sloggett, Gordon and Clifford Dickason. *Ground-Water Mining in the United States*. Natural Resources Economics Division, Economic Research Service, U.S. Department of Agriculture, Agricultural Report No. 555, August. Washington, D.C.: U.S. Gov. Printing Office.

Splinter, William E. "Center-Pivot Irrigation," *Scientific American*, 234 (Jun 1976), pp. 90–99.

Staff. "Drought and Wetlands Drainage Take a Heavy Toll on Many Species," *National Wildlife*, 27 (Feb/Mar 1989), p. 34.

Stegner, Wallace. "Water and the Dimensions of the Crisis," *Wilderness*, 51 (Fall 1987), pp. 14–18.

Steinhart, Peter. "A Vision of Lakes," *Audubon*, 88 (July 1987), pp. 8–11.

Stokes, Bruce. "Water Shortages: The Next Energy Crisis," *The Futurist*, 17 (Apr 1983), pp. 37–41, 45–47.

Turner, Fredrick. "Slowly Sinking in the West," *Wilderness*, 5 (Fall 1987), pp. 47–50. Land subsidence due to water withdrawal.

Wesely, Edwin F. Jr. *Easy Ways to Save Water, Money, & Energy at Home*. Frederick, Md.: Potomac River and Trails Council, 1983.

World Resources Institute. "Freshwater," *World Resources 1988–89*, Chapter 8. New York: Basic Books, Inc., 1988.

Young, Gordon. "The Troubled Waters of Mono Lake," *National Geographic*, 162 (Dec 1982), pp. 504–515.

PART FOUR POLLUTION—GENERAL REFERENCES

Chesapeake: A Quarterly Report from Maryland Governor William Donald Schaefer. Provides general information regarding projects and activities to advance the restoration of the Chesapeake Bay.

Speth, James Gustave. *Our Polluted Environment: A Long-Term Perspective.* Washington, D.C.: World Resources Institute, 1988.

U.S. Department of Agriculture. Soil Conservation Service. *Water Quality Field Guide.* SCS-TP-160. Washington, D.C.: U.S. Government Printing Office, 1988.

CHAPTER 9. SEDIMENTS, NUTRIENTS, AND EUTROPHICATION

American Society of Limnology and Oceanography. *Nutrients and Eutrophication: The Nutrient Limiting Controversy*, Lawrence, Kans.: American Society of Limnology and Oceanography, 1972.

Clark, Edwin H. II, et al. *Eroding Soils: The Off-Farm Impacts*. Washington, D.C.: The Conservation Foundation, 1985.

DeVore, R. William, ed. *Proceedings 1983 International Symposium on Urban Hydrology, Hydraulics and Sediment Control*. Lexington, Ky.: O.E.S. Publications, 1983.

Environmental Law Institute. *Wetlands of the Chesapeake: Protecting the Future of the Bay*. Corelo, Calif.: Island Press, 1985.

Kusler, Jon A. *Our National Wetland Heritage: A Protection Guidebook*. Environmental Law Institute. Corelo, Calif.: Island Press, 1983.

Lowrance, Richard, et al. "Riparian Forests as Nutrient Filters in Agricultural Watersheds," *BioScience*, 34 (Jun 1984), pp. 374–377.

McCloskey, William. "Hard Times Hit the Bay," *National Wildlife*, 22 (Apr/May 1984), pp. 6–14.

National Academy of Sciences. *Eutrophication: Cause, Consequence, Corrections*. Washington, D.C.: National Academy of Sciences, 1969.

Officer, Charles B., et al. "Chesapeake Bay Anoxia: Origin, Development, and Significance," *Science*, 223 (Jan 6, 1984), pp. 22–27.

Organization for Economic Cooperation and Development (O.E.C.D.). *Eutrophication of Waters: Monitoring, Assessment, and Control*. Washington, D.C.: O.E.C.D. Publications, 1982.

Orth, Robert J., and Kenneth A. Moore. "Chesapeake Bay: An Unprecedented Decline in Submerged Aquatic Vegetation," *Science*, 222 (Oct 7, 1983), pp. 51–53.

St. Onge, Julie. "Runoff Runs Amok," *Sierra* 73, (Nov/Dec 1988), pp. 28–30.

Towsend, Colin R. *The Ecology of Streams and Rivers*. Baltimore, Md.: Edward Arnold Publishers, 1980.

Wetzel, R.G. *Limnology*. Philadelphia: Saunders, 1983. A textbook that covers the physical, chemical, and biologic aspects of aquatic ecosystem function.

Williams, John Page. "Living Upstream," *Chesapeake Bay Magazine*, 18 (Oct 1988), pp. 80–90.

CHAPTER 10. WATER POLLUTION DUE TO SEWAGE

Aberley, Richard C., and Susan Berg. "Finding Uses for

Sludge," *American City and County,* 101 (Jun 1986), pp. 38–46.

BARBER, LARRY. "Long Term Fate of Organic Micropollutants in Sewage-Contaminated Ground Water," *Environmental Science and Technology,* 22 (Feb 1988), pp. 205–211.

BASTIAN, ROBERT K. *Land Application of Municipal Sludge.* EPA Process Design Manual. Washington D.C.: U.S. Environmental Protection Agency, 1984.

CALLBURN, MELISSA, AND PATRICK HUNT. "Seeking the Sludge Solution," *American City American County,* 104 (Jan 1989), pp. 44–52.

DOERSAM, JIM. "Treating Wastewater with Hyacinths," *Biocycle,* 88 (Aug 1987), pp. 30–32.

ECKENFELDER, W. WESLEY, JR. "Biological Phosphorus Removal: State of the Art Review," *Pollution Engineering,* 18 (Sept 1987), pp. 88–93.

GOLDSTEIN, JEROME. *Sensible Sludge: A New Look at a Wasted Natural Resource.* Emmaus, Penna.: Rodale Press, 1977.

LOWE, MARCIA D. "Down the Tubes," *World Watch,* 2 (Mar/Apr 1989), pp. 22–29.

MARSHAL, ELIOT. "The Sludge Factor," *Science,* 242 (Oct 28, 1988), pp. 506–508.

MARX, WESLEY. "Swamped by Our Own Sewage," *Environmental Protection Agency Journal,* 14 (Mar 1988), pp. 37–39.

McEVEDY, COLIN. "The Bubonic Plague," *Scientific American,* 258 (Feb 1988), pp. 118–123.

REED, SHERWOOD C., ET AL. *Natural Systems For Waste Management And Treatment.* New York: McGraw-Hill Book Company, 1988.

UNITED STATES ENVIRONMENTAL PROTECTION AGENCY. *Land Application of Sludge; A Viable Alternative.* Washington, D.C.: Environmental Protection Agency, 1983.

VALDES-COGLIANO, SALLY J. "International Drinking Water Decade," *Environment,* 27 (Oct 1985), pp. 41–42.

WILSON, DR. THOMAS. "Chlorination vs. Alternative Disinfecting: Which is the Best Choice?" *Water Engineering and Management,* 135 (Oct 1988), pp. 42–45.

WOLFF, ANTHONY. "Boston's Toilet: The True Story," *Audubon,* 90 (March 1989), pp. 26–32.

CHAPTER 11. TOXIC CHEMICALS AND GROUNDWATER POLLUTION

ATHESON, TIMOTHY. "The Unrealized Potential of SARA," *Environment,* 29 (May 1987), pp. 6–11, 40–43. Superfund Amendments and Reauthorization Act of 1986.

ATKESON, TIMOTHY B., AND ROGER C. DOWER. "Mobilizing New Protection for Natural Resources, *Environment,* 29 (May 1987), pp. 6–44.

BISHOP, JIM. "Remediation: An Overview," *Hazmat World,* 1 (Sept 1988), pp. 30–38.

BROWN, MICHAEL H. "Toxic Wind," *Discover,* 8 (Nov 1987), pp. 42–49.

BURTIS, BILL. "Emergency Responses," *Hazmat World,* 1 (Sept 1988), pp. 39–41.

COBB, CHARLES. "The Great Lake's, Troubled Waters," *National Geographic,* 172 (July 1987), pp. 2–6, 14–30.

CUMMINS, JOSEPH. "Extinction: The PCB Threat to Marine Mammals," *The Ecologist,* 18 (Nov–Dec 1988), pp. 193–195.

DUFOUR, JEAN-PAUL, AND CORINNE DENIS. "The North's Garbage Goes South: The Third World Fears It Will Become the Global Dump," *World Press Review,* 35 (Nov 1988), pp. 30–32.

ELKINGTON, JOHN AND JONATHAN SHOPLEY. *Cleaning Up: U.S. Waste Management Technology and Third World Development.* Washington, D.C.: World Resources Institute, 1989.

EPSTEIN, S., L. BROWN, AND C. POPE. *Hazardous Waste in America.* San Francisco, CA: Sierra Club Books, 1982.

FORTUNA, RICHARD C. AND DAVID J. LENNETT. *Hazardous Waste Regulation—The New Era.* New York: McGraw-Hill Book Company, 1987.

GORDON, WENDY. *A Citizen's Handbook on Groundwater.* New York: Natural Resources Defense Council, 1984.

GORDON, WENDY AND JANE BLOOM. *Deeper Problems: Limits to Underground Injection as a Hazardous Waste Disposal Method.* New York: Natural Resources Defense Council, Inc., 1985.

Hazmat World. A trade journal covering Hazardous Materials Management Issues, Technology, and People.

JAFFE, MARTIN AND FRANK DiNOVA. *Local Groundwater Protection.* American Planning Association. Corelo, Calif.: Island Press, 1987.

KEITH, SUSAN. "Ground Water Quality Protection: State and Local Strategies," *Environment,* 29 (May 1987), pp. 25–27.

LAVE, LESTER B., AND ARTHUR C. UPTON, EDS. *Toxic Chemicals, Health, and the Environment.* Baltimore: The Johns Hopkins University Press, 1987.

LAVO, CARL. "Not in My Backyard," *National Wildlife,* 26 (Jun 1988), pp. 24–27.

LIPSKE, MIKE. "Are You Throwing Poisons into the Trash?" *National Wildlife,* 24 (Aug/Sep 1986), pp. 20–23.

MONASTERSKY, R. "Waste Wells Implicated in Ohio Quake," *Science News,* 134 (Aug 1988), p. 132.

MONMANCY, T. "Poison in the Plumbing?" *Newsweek,* 110 (Dec 21, 1987), p. 56.

MORELL, VIRGINIA. "Fishing for Trouble—A Cancer Epidemic in Fish is Warning Us: You May Be Next," *International Wildlife,* 14 (July/Aug 1984), pp. 40–43.

NATURAL RESOURCES DEFENSE COUNCIL. *Citizen's Handbook on Water Quality Standards.* Washington, D.C.: National Resources Defense Council, 1987.

POSTEL, SANDRA. "Controlling Toxic Chemicals," *State of the World 1988,* Chapter 7. Washington, D.C.: Worldwatch Institute, 1988.

ROBERTS, LESLIE. "Discovering Microbes with a Taste for PCBs," *Science,* 237 (Aug 1987), pp. 975–977.

SCHWARTZ, ANNE. "Poisons in Your Home: A Disposal Dilemma," *Audubon,* 89 (May 1987), pp. 12–17.

SHIELDS, DENNIS. "Agricultural Chemicals in Groundwater: Suggestions for EPA Strategy," *Environmental Sciences*, 30 (May 1987), pp. 23–27.

SIERRA CLUB LEGAL DEFENSE FUND. *The Poisoned Well—New Strategies for Groundwater Protection*. Corelo, Calif.: Island Press, 1989.

SIMON, ANNE W. AND PAUL HAUGE. *Contamination of New England's Fish and Shellfish:* A Report to the Governors and the Public. Conservation Law Foundation of New England. Washington, D.C.: Coast Alliance, June 1987.

STAMPS, D. "The Real Price of Road Salt," *National Wildlife*, 27 (Dec 1988/Jan 1989), p. 28.

THOMAS, JOHN. "Toxic Wastes and Their Problems," *The Futurist*, 29 (May/Jun 1989), pp. 17–20.

U.S. CONGRESS OFFICE OF TECHNOLOGY ASSESSMENT. *Superfund Strategy*. Washington, D.C.: U.S. Government, 1985.

VASS, DAVE. "Microbes May Aid Toxic Cleanups," *Science*, 241 (Sept 1988), pp. 46, 48–49.

WEISSKOPF, MICHAEL. "Lead Ashtray: The Poisoning of America," *Discover*, 8 (Dec 1987), pp. 68–74. Severity and frequency of low-level lead poisoning.

WINTON, JOHN M. AND RICH, LAURIE A. "Hazardous Waste Management: Putting Solutions into Place," *Chemical Week*, 143 (Aug 24, 1988), pp. 26–58.

CHAPTER 12. AIR POLLUTION AND ITS CONTROL

ALLEN, ALEXANDRA. "Poisoned Air," *Environmental Action*, 20 (Jan/Feb 1988), pp. 19–21.

AMERICAN LUNG ASSOCIATION. A wide assortment of pamphlets addressing specific air pollution problems. Washington, D.C.: American Lung Association.

AMERICAN MEDICAL ASSOCIATION. *Journal of the American Medical Association*, 225 (Feb 28, 1986). Entire issue devoted to effects of smoking tobacco.

ARTIS, THERESA. "Radon—Nature's Own Toxic Waste," *Discover*, 9 (Oct 1988), pp. 85–87/91.

BORMANN, F.H. "The Effects of Air Pollution on the New England Landscape," *Ambio*, 11 (5) 1982, pp. 338–346.

BROWN, MICHAEL H. "Toxic Wind," *Discover*, 8 (Nov 1987), pp. 42–49.

EDWARDS, J.W. "Sounding Taps for the Sugar Maple." *National Wildlife*, 25 (Oct/Nov 1987) p. 20. Sugar maples in the Northeast are deteriorating apparently because of pollution/acid rain stresses.

FEDER, WILLIAM A. "Cumulative Effects of Chronic Exposure of Plants to Low Levels of Air Pollutants," *Air Pollution Damage to Vegetation*, pp. 21–30. Washington, D.C.: American Chemical Society of Washington, 1973.

GROVE, NOEL. "Air, An Atmosphere of Uncertainty," *National Geographic*, 171 (Apr 1987), pp. 503–536.

HENSCHEL, BRUCE D., ET AL. *Radon Reduction Techniques for Detached Houses*. Washington, D.C.: Environmental Protection Agency, 1988.

LEGGE, ALLAN H., AND SAGAR V. KRUPA, EDS. *Air Pollutants And Their Effects On The Terrestrial Ecosystem*. Corelo, Calif.: Island Press, 1986.

LIPSKE, MIKE. "How Safe is the Air Inside Your Home?," *National Wildlife*, 25(3) (May 1987), pp. 34–39.

MACKENZIE, JAMES J., AND MOHAMED T. EL-ASHRY. *Airsick Crops and Trees: The Science and Policy Implications of Multiple Air Pollutants*. Washington, D.C.: World Resources Institute, 1988.

MACKENZIE, JAMES J., AND MOHAMED T. EL-ASHRY. *Ill Winds: Airborne Pollution's Toll on Trees and Crops*. World Resources Institute, Sept 1988.

MEHR, CHRISTIAN. "Are the Swiss Forests in Peril?," *National Geographic*, 175 (May 1989), pp. 637–651.

PETERS, SUSAN. "The ABCs of Asbestos Cleanup," *Sierra*, 73 (Sept/Oct 1988), pp. 27–29.

PETERSON, CASS. "Scenic Sites Under Siege: Air Pollution is Spreading a Vail of Haze over Some of the Country's Most Breathtaking Property, the National Parks," *National Wildlife*, 25(4) (Jun/July 1987), pp. 44–45.

PETERSON, I. "Searching for a Breath of Clean Air," *Science News* 132(22) (Nov 28, 1987), p. 340. American Forestry Association report stating that "air pollution poses a significant threat to the health and productivity of U.S. forests."

SEINFELD, J.H. "Urban Air Pollution: State of the Science," *Science*, 243 (Feb 10, 1989), pp. 745–752.

SEINFELD, JOHN. *Atmospheric Chemistry and Physics of Air Pollution*. New York: John Wiley & Sons, Inc., 1986.

SMITH, KIRK R. "Air Pollution: Assessing Total Exposure in the United States," *Environment*, 30 (Oct 1988), pp. 10–15/33–38.

SPETH, JAMES GUSTAVE. *Environmental Pollution: A Long-Term Perspective*. Washington, D.C.: World Resources Institute, 1988.

STRAIT, DONALD S. AND RICHARD E. AYRES, "High Noon For Smog Control," *Environment*, 29 (Sept 1987), pp. 43–45.

SUN, MARJORIE. "Tighter Ozone Standard Urged by Scientists," *Science*, 240 (Jun 24, 1988), pp. 1724–1725.

UEHLING, D. "Missing the Deadline on Ozone," *National Wildlife*, 25(6) (Oct–Nov 1987), pp. 34–37. Ground level ozone pollution.

WARK, KENNETH, AND CECIL F. WARNER. *Air Pollution: Its Origin and Control*, 2nd ed. New York: Harper and Row Publs., Inc., 1981.

WELLBORN, STANLEY N. "The New Soldier in the Clean Air War: You," *U.S. News and World Report*, 103(96) (Aug 10, 1987), pp. 50–51. Meeting clean air standards demands more citizen cooperation in individual activities such as fireplaces and outdoor grills.

CHAPTER 13. ACID PRECIPITATION, THE GREENHOUSE EFFECT, AND DEPLETION OF THE OZONE SHIELD

BALZHISER, RICHARD E. AND KURT E. YEAGER. "Coal-fired

Power Plants for the Future," *Scientific American,* (Sept 1987), pp. 100–107.

BJERKLIE, DAVID. "The Heat is On," *TIME,* 130(16) (Oct 19, 1987), pp. 58–64. Ozone and CO_2

BLUESTONE, MIMI. "Long-term Lime Aid for Ailing Lakes," *Business Week* (Jun 15, 1987), p. 65.

CONRAD, JIM. "An Acid-Rain Trilogy," *American Forests,* 93(11–12) (Nov/Dec 1987), pp. 21–23/77–79. Addresses philosophical problem of inaction regarding acid rain.

EL-SAYED, SAYED Z. "Fragile Life Under the Ozone Hole," *Natural History,*97 (Oct 1988), pp. 72–80.

ENVIRONMENT CANADA. *Downwind: The Acid Rain Story.* Ottawa: Ministry of Supply and Services, 1981.

HEKSTRA, G.P. "Global Warming and Rising Sea Levels: The Policy Implications," *The Ecologist,* 19 (Jan and Feb 1989), pp. 4–14.

HORDIJK, LEEN. "A Model Approach to Acid Rain," *Environment,* 30 (Mar 1989), pp. 17–20/40–41.

HOUGHTON, RICHARD A. AND GEORGE M. WOODWELL. "Global Climatic Change," *Scientific American,* 260 (Apr 1989), pp. 36–44.

JOHNSON, ARTHUR H. "Acid Deposition: Trends, Relationships, and Effects," *Environment,* 28 (May 1986), pp. 6–11/34–39.

JONES, ROBIN RUSSELL. "Ozone Depletion and Cancer Risk," *Lancet,* 8556 (Aug 1987), pp. 443–446.

KAHAN, ARCHIE M. *Acid Rain: Reign of Controversy.* Colorado: Fulcrum, Inc., 1986.

KERR, RICHARD A. "Arctic Ozone is Poised for a Fall," *Science,* 243 (Feb 24, 1989), pp. 76–78.

KERR, RICHARD A. "Stratospheric Ozone is Decreasing," *Science,* 239 (Mar 25, 1988), pp. 1489–1491.

KIESTER, EDWIN JR. "A Deadly Spell is Hovering above the Black Forest," *Smithsonian,* (Nov 1985), pp. 211–230.

LEWIS, THOMAS A. "Will Species Die Out as the Earth Heats Up?" *International Wildlife* 17(6) (Nov/Dec 1987), pp. 18–21.

LUOMA, JON R. "Acid Murder No Longer A Mystery," *Audubon,* 90 (Nov 1988), pp. 126, 128–129/131–135.

LUOMA, JON. "Black Duck Decline: An Acid Rain Link," *Audubon,* 89 (May 1987), pp. 19–24.

MACKENZIE, JAMES J. *Breathing Easier: Taking Action on Climate Change, Air Pollution, and Energy Insecurity.* Washington, D.C.: World Resources Institute, 1988.

MACKENZIE, JAMES J. AND MOHAMED T. EL-ASHRY, EDS. *Air Pollution's Toll on Forests and Crops.* Washington, D.C.: World Resources Institute, 1989.

MACKERRON, CONRAD B. "A Nontoxic Ozone Protector," *Chemical Week,* 142(3) (Jan 20, 1988), pp. 8–11. Substitutes for chlorofluorocarbons.

MELLO, ROBERT A. *Last Stand of the Red Spruce.* Washington, D.C.: Island Press, 1987.

MINTZER, IRVING M. *A Matter of Degrees: The Potential for Controlling the Greenhouse Effect.* Washington, D.C.: World Resources Institute, 1987.

MINTZER, IRVING M. *Can We Save the Sky? Probing the Prospects for a Stable Environment.* Washington, D.C.: World Resources Institute, 1989.

MINTZER, IRVING M., WILLIAM R. MOOMAW, AND ALAN S. MILLER. *Protecting the Ozone Shield: Strategies for Phasing Out CFC's During the 1990's.* Washington, D.C.: World Resources Institute, 1989.

MOHNEN, VOLKER A. "The Challenge of Acid Rain," *Scientific American,* 259 (August 1988) p. 30–38.

MONASTERSKY, R. "Acid Dew: What it Does," *Science News,* 132(16) (Oct 17, 1987), p. 247.

MOOMAW, WILLIAM R. AND IRVING M. MINTZER. *Strategies for Limiting Global Climate Change.* Washington, D.C.: World Resources Institute, 1989.

NATIONAL ACADEMY OF SCIENCES. *Acid Deposition: Long-Term Trends.* Washington, D.C.: National Academy Press, 1986.

NILSSON, STEW. "Acid Rain in Europe: The Extent of Forest Decline in Europe," *Environment,* 29 (Nov 1987), pp. 4–15.

NYLANDER, CARL. "S.O.S. for Ancient Monuments," *Archaeology,* 41 (July/Aug 1988), pp. 54–75.

PARRY, MARTIN, ET AL. "Climactic Change: How Vulnerable Is Agriculture?" *Environment,* 27 (Jan/Feb 1985), pp. 4–5, 43.

POSTEL, SANDRA. "A Green Fix to Global Warming," *World Watch,* 1 (Sept/Oct 1988), pp. 29–36.

POSTEL, SANDRA. "Air Pollution, Acid Rain, and the Future of Forests," *Worldwatch Paper* 58. Washington, D.C.: Worldwatch Institute, Mar 1984.

REUKIN, ANDREW C. "Cooling Off the Greenhouse," *Discover,* 10 (Jan 1989), pp. 30–32.

ROBERTS, LESLIE. "California's Fog Is Far More Polluted Than Acid Rain," *BioScience,* 32 (Nov 1982), pp. 778–779.

ROBERTS, LESLIE. "How Fast Can Trees Migrate?," *Science,* 243 (Feb 10, 1989), pp. 735–737.

ROBERTS, WALTER ORR. "It Is Time to Prepare for Global Climate Changes," *Conservation Foundation Letter* (Apr 1983), pp. 1–7.

SCHINDLER, D.W. "Effects of Acid Rain on Freshwater Ecosystems," *Science,* 239(4836) (Jan 8, 1988), pp. 149–153.

SCHNEIDER, STEPHEN H. "Doing Something About the Weather," *World Monitor,* 1 (Dec 1988), pp. 28–37.

SCHWARTZ, S.E. "Acid Deposition: Unraveling a Regional Phenomenon," *Science,* 243 (Feb 10, 1989), pp. 753–761.

SHELL, ELLEN R. "Solo Flights into the Ozone Hole Reveal its Causes," *Smithsonian,* 18 (Feb 1988), pp. 142–155.

STAFF. "Atmosphere and Climate," *World Resources 1988–89.* Chapter 10. New York: The World Resources Institute and The International Institute for Environment and Development, 1988.

STOLARSKI, RICHARD S. "The Antarctic Ozone Hole," *Scientific American,* 258(1) (Jan 1988), pp. 30–36.

TAUBES, GARY. "Made in the Shade? No Way," *Discover,* 8(8) (Aug 1987), pp. 62–71. (Chlorofluorocarbons and the ozone shield.)

U.S. Congress Office of Technology Assessment. *The Regional Implications of Transported Air Pollutants: An Assessment of Acidic Deposition and Ozone.* Washington, D.C.: U.S. Government, 1982.

Woods, F.W. "The Acid Rain Question." *Futurist,* 21 (Jan–Feb 1987), pp. 34–37.

CHAPTER 14. RISKS AND ECONOMICS OF POLLUTION

Ames, Bruce N. "Dietary Carcinogens and Anticarcinogens," *Science,* 221 (Sept 23, 1983), pp. 1256–1264.

Bell, Lauren. "The High Cost of Neglecting Wildlife," *National Wildlife,* 27 (Mar 1989), pp. 4–8.

Borrelli, Peter, et al. *Crossroads: Environmental Priorities for the Future.* Washington, D.C.: Island Press, 1989.

Carpenter, Richard A., and John A. Dixon. "Ecology Meets Economics: A Guide to Sustainable Development," *Environment,* 27 (June 1985), pp. 6–11, 27–32.

Hohenemser, C., et al. "The Nature of Technological Hazard," *Science,* 220 (22 Apr 1983), pp. 378–384.

Kneese, Allen V. *Measuring the Benefits of Clean Air and Water.* Baltimore, Md.: Resources for the Future, 1984.

Koshaland, Daniel E. Jr., ed. "Immortality and Risk Assessment," *Science,* 236 (Apr 17, 1987), p. 241.

Martin, Larry. "The Case for Stopping Wastes at their Source," *Environment,* 28(3) (Apr 1986), pp. 32–34.

Mills, E., and P. Graves. *Economics of Environmental Quality.* Washington, D.C.: Sidney Kramer Books, 1985.

Morgernster, Richard, and Stuart Sessions. "Weighing Environmental Risks: EPA's Unfinished Business," *Environment,* 30 (July/Aug 1988), pp. 15–17/34–39.

Repetto, Robert, William Magrath, Michael Wells, Christine Beer, and Fabrizio Rossini. *Wasting Assets: Natural Resources in National Income Accounts.* Washington, D.C.: World Resources Institute, 1989.

Russel, Milton. "Environmental Protection for the 1990s and Beyond," *Environment,* 29(7) (Sept 1987), pp. 12–15/34–38.

Stavins, Robert. "Harnessing Market Forces to Protect the Environment," *Environment,* 31 (Jan/Feb 1989), pp. 4–7/28–38.

Steinhart, Peter. "Respecting the Law," *Audubon,* 89(6) (Nov 1987), pp. 10–13.

Summers, Emmanuel. "Environmental Hazards Show no Respect for National Boundaries," *Environment,* 29(5) (Jun 1987), pp. 7–9/31–33.

Summers, Emanuel. "Transboundry Pollution and Environmental Health," *Environment,* 29(5) (Jun 1987), pp. 6–9/31–33.

Wann, David. "Environmental Crime: Putting Offenders Behind Bars," *Environment* 29(8) (Oct 1987), pp. 5/44–45. Illegal dumping of toxics continues. EPA trys to stop it.

CHAPTER 15. THE PESTICIDE TREADMILL

Aeppel, Timothy. "Echoes From the 'Silent Spring'," *The Christian Science Monitor,* (July 20, 21, and 22, 1987), pp. 1, 6; 3, 6, 7.

Barrons, Keith C. "How Risky are Pesticides?" *Science of Food and Agriculture,* 5(1) (Jan 1988), pp. 21–25.

Boralko, Allen A. "The Pesticide Dilemma," *National Geographic,* 157 (Feb 1980), pp. 144–183.

Bull, D. *Growing Problem: Pesticides and the Third World Poor.* Washington, D.C.: Sidney Kramer Books, 1984.

Carson, Rachel. *Silent Spring.* Boston: Houghton Mifflin Company, 1962.

Dover, Michael J., and Brian A. Croft. "Pesticide Resistance and Public Policy," *BioScience,* 36 (Feb 1986), pp. 78–85.

Edmonson. "Hazards of the Game," *Audubon,* 88 (Nov 1987), pp. 25–37. Environmental impacts of pesticides and lawn chemicals used on golf courses.

Landsberg, Hans. "Reducing Pesticide Use," *Environment,* 29 (July/Aug 1987), pp. 52–53.

Marshall, Eliot. "The Murky World of Toxicity Testing," *Science,* 220 (Jun 10, 1983), pp. 1130–1132.

May, Elizabeth E. "Canada's Moth War," *Environment,* 19 (Aug/Sept 1977), pp. 16–23.

McEvedy, Colin. "The Bubonic Plague," *Scientific American,* 258(2) (Feb 1988), pp. 118–123.

Natural Resources Defense Council. *A Report on Intolerable Risk: Pesticides in our Children's Food.* Washington, D.C.: Natural Resources Defense Council, 1989.

Norman, Colin. "EPA Sets New Policy on Pesticide Cancer Risks," *Science,* 242 (Oct 21, 1988), pp. 366–367.

Pimentel, David. *Ecological Effects of Pesticides on Non-Target Species.* Washington, D.C.: U.S. Government Printing Office, 1971.

Pimentel, David, and Lois Levitan. "Pesticides: Amounts Applied and Amounts Reaching Pests," *BioScience,* 36 (Feb 1986), pp. 86–91.

Postel, Sandra. "Controlling Toxic Chemicals," *State of the World 1988,* Chapter 7. Washington, D.C.: Worldwatch Institute, 1988.

Postel, Sandra. "Defusing the Toxics Threat: Controlling Pesticides and Industrial Waste," *Worldwatch Paper 79.* Worldwatch Institute, 1987.

Revkin, Andrew C. "March of the Fire Ants," *Discover,* 10 (Mar 1989), pp. 71–76.

Schneider, Keith. "Faking It: The Case Against Industrial Bio-Test Laboratories," *The Amicus Journal,* 4 (Spring 1983), pp. 14–26.

Van den Bosch, Robert. *The Pesticide Conspiracy.* Garden City, N.Y.: Doubleday & Co., Inc., 1978.

Vietmeyer, Noel D. "We Haven't Zapped the Boll Weevil Yet," *Smithsonian,* 13 (Aug 1982), pp. 60–68.

Walsh, John. "Locusts in Africa: A Plague is Possible," *Science,* 242 (Dec 23, 1988), pp. 1627–1628.

Weir, David, and Mark Schapiro. "The Circle of Poison," *The Nation,* 231 (Nov 15 1980), pp. cover, 514–516.

CHAPTER 16. NATURAL PEST CONTROL METHODS AND INTEGRATED PEST MANAGEMENT

BARNETT, MARY. "A Better Way to Fight Bugs: Integrated Pest Management," *The Amicus Journal*, 4 (Spring 1983), pp. 27–31.

BioScience, 30. (Oct 1980). Special issue on integrated pest management.

Common Sense Pest Control Quarterly. Bio Integral Resource Center, Berkley, Calif. Quarterly Journal giving numerous tips regarding the natural control of common house and garden pests.

DAHLSTEN, DONALD L. "Pesticides in an Era of Integrated Pest Management," *Environment*, 25 (Dec 1983), pp. 45–54.

DEBACH, PAUL. *Biological Control by Natural Enemies*. London: Cambridge University Press, 1974.

DETHIER, VINCENT. "Smart Strategies for Insect Control," *Horticulture*, 61 (Jan 1983), pp. 48–59.

DOVER, MICHAEL J. *A Better Mousetrap*. Washington, D.C.: World Resources Institute, 1985.

EDWARDS, PETER J., AND STEPHEN D. WRATTEN, PH.D. *Ecology of Insect-Plant Interactions*. Baltimore, Md.: Edward Arnold Publishers, 1980.

HORWITH, BRUCE. "A Role for Intercropping in Modern Agriculture," *BioScience*, 35 (May 1985), pp. 286–291.

LARSON, ERIK. "A Close Watch on U.S. Borders to Keep the World's Bugs Out," *Smithsonian*, 18(3) (Jun 1987), pp. 106–117.

MAXWELL, FOWDEN G., AND PETER R. JENNINGS, EDS. *Breeding Plants Resistant to Insects*. New York: John Wiley and Sons, Inc., 1980.

MCWILLIAMS, PATRICIA G. "Companion Planting," *Country Journal*, 14(6) (Jun 1987), pp. 43–45. Pest control through companion planting.

METCALF, ROBERT L., AND WILLIAM H. LUCKMANN, EDS. *Introduction to Insect Pest Management*. Somerset, N.J.: John Wiley & Sons, Inc., 1982.

MILLER, LOIS K., ET AL. "Bacterial, Viral, and Fungal Insecticides," *Science*, 219 (Feb 11, 1983), pp. 715–721.

OLKOWSKI, WILLIAM. "Update: Great Expectations For Non-Toxic Pheromones," *The IPM Practitioner*, 10 (Jun/July 1988), p. 1.

PILLEMER, ERIC A., AND WARD M. TINGEY. "Hooked Trichomes: A Physical Plant Barrier to a Major Agricultural Pest," *Science*, 193 (Aug 6, 1976), pp. 482–484.

POWER, J.F. AND R.F. FOLLERT. "Monoculture," *Scientific American*, 256 (Mar 1987), pp. 79–86.

ROSENTHAL, GERALD A. "The Chemical Defenses of Higher Plants," *Scientific American*, 254 (Jan 1986), pp. 94–99.

THE IPM Practitioner—Monitoring the Field of Pest Management. Bio Integral Resource Center, Berkley, Calif. A journal covering biological pest control methods.

WALSH, J. "Cosmetic Standards: Are Pesticides Overused for Appearance's Sake?" *Science*, 193 (Aug 27, 1976), pp. 744–747.

WOOD, DAVID L., ET AL., EDS. *Control of Insect Behavior By Natural Products*. New York: Academic Press, 1970.

CHAPTER 17 BIOTA: BIOLOGICAL RESOURCES

ALLARD, WILLIAM ALBERT, AND LOREN MCINTYRE. "Rondonia: Brazil's Imperiled Rain Forest," *National Geographic*, 174 (Dec 1988), pp. 772–799.

ALTIERI, MIGUEL A., ET AL. "Developing Sustainable Agroecosystems," *BioScience*, 33 (Jan 1983), pp. 45–49.

BERRY, WENDELL. "Preserving Wilderness," *Wilderness*, 50(176) (Spring 1987), pp. 39–40/50–54. Discusses realistic reasons for preserving wilderness.

BURLEY, F. WILLIAM. *Saving Critical Ecosystems*. Washington, D.C.: World Resources Institute, 1988.

BUSCHBACHER, ROBERT J. "Tropical Deforestation and Pasture Development," *BioScience*, 30 (Jan 1989), pp. 22–29.

CAREY, JOHN. "Trouble in Paradise," *National Wildlife* 25(6) (Oct/Nov 1987), pp. 42–44. Adverse effect of development on Florida Keys deer. Species is endangered by loss of habitat from development.

COHN, JEFFREY P. "Halting the Rhino's Demise," *BioScience*, 38 (Dec 1988), pp. 740–744.

CROTHERS, ANDRE. "And Then There Were None," *Green Peace*, 12(1) (Jan/Mar 1987), pp. 13–15. Exploitation of attempts to save the sea turtles.

CROTHERS, ANDRE. "Cry of the Beluga," *Greenpeace*, 12(2) (Apr/Jun 1987), pp. 14–17. White whale inhabiting the Gulf to St. Lawrence is endangered due to loss of habitat and toxics. Good example of bioaccumulation.

D'AULAIRE, EMILY, AND PER OLD. "Pangolins Are All the Rage: These Artichokes on Legs Are Victims of the Latest Fad in Footware," *International Wildlife*, 13 (Jan/Feb 1983), pp. 14–16.

DISILVESTRO, R.L. "A Saltmarsh in L.A.! and Other Good News," *Audubon*, 89(6) (Nov 1987), pp. 98–99.

DISILVESTRO, ROGER L. "Saga of AC-9, The Last Free Condor," *Audubon*, 89(4) (July 1987), pp. 12–14.

DOHERTY, JIM. "Hail, Lobsterman . . . and Farewell," *National Wildlife*, 15 (Apr/May 1977), pp. 42–49.

DREW, L. "Are We Loving the Panda to Death?," *National Wildlife*, 27 (Dec 1988, Jan 1989), pp. 14–17.

EHRLICH PAUL R. "Habitats in Crisis," *Wilderness*, 50(176) (Spring 1987), pp. 12–15. Damage of overgrazing. Excellent.

GRADWOHL, JUDITH AND GREENBERG, RUSSELL. *Saving the Tropical Forests*. Washington, D.C.: Smithsonian Institution, 1988.

HAZLEWOOD, PETER T. *Cutting Our Losses: Policy Reform to Sustain Tropical Forest Resources*. Washington, D.C.: World Resources Institute, 1989.

HINCK, JON. "The Tangled Web: Driftnets and The Decline of the North Pacific," *Greenpeace*, 12(2) (Apr/Jun 1987), pp. 6–9.

HYMAN, RANDALL. "Fall of the Oyster Kings," *International Wildlife*, 18 (Nov/Dec 1988), pp. 18–23.

KARR, JAMES R., ET AL. "Fish Communities of Midwestern Rivers: A History of Degradation," *BioScience*, 35 (Feb 1985), pp. 90–95.

KERASOTE, TED. "Is Nepal Going Bald?" *Audubon*, 89(5) (Sept 1987), pp. 28–37.

LAWREN, BILL. "The High Cost of Neglecting Wildlife," *National Wildlife*, 30 (Apr/May 1989), pp. 4–8.

LEWIS, THOMAS. "Searching for Truth in Alligator Country," *National Wildlife*, 25(6) (Nov 1987), pp. 12–18. Comeback of alligators. Were they ever really endangered?

LUGO, ARIEL E. "The Future of the Forest: Ecosystem Rehabilitation in the Tropics," *Environment*, 30 (Sept 1988), pp. 16–20/41–45.

MARES, M.A. "Conservation in South America: Problems, Consequences, and Solutions," *Science*, 233 (Aug 1986), pp. 734–739.

MARGOLIS, MAC. "Amazon Ablaze: An Eyewitness Report," *World Monitor*, 2 (Feb 1989), pp. 20–29.

McINTYRE, J.W., AND M.S. QUINTON. "The Loon Cries for Help," *National Geographic*, 175 (Apr 1989), pp. 512–524.

McLARNEY, WILLIAM O. "Still a Dark Side to the Aquarium Trade." *International Wildlife*, 18 (Mar/Apr 1988), pp. 46–51. Losses in capturing aquarium pets.

McNEELY, JEFFREY A., KENTON R. MILLER, WALTER V.C. REID, AND RUSSELL MITTERMEIER. *Conserving the World's Biological Resources: A Primer on Principles and Practice for Development Action.* Washington, D.C.: World Resources Institute, 1989.

MISRA, HERMANTA RAJ, AND ERIC DIMERSTAIN. "New zip codes for resident rhinos in Nepal," *Smithsonian*, 17 (Sept 1987), pp. 67–72. Saving an endangered species by breeding on ranches in the U.S.

MLOT, CHRISTINE. "The Science of Saving Endangered Species," *BioScience*, 39 (Feb 1989), pp. 68–70.

MORGAN, JOSEPH R. "Large Marine Ecosystems: An Emerging Concept of Regional Management," *Environment*, 29 (Dec 1987), pp. 4–9/29.

OBEE, BRUCE. "Seal-Salmon Controversy Escalates in B.C.," *Oceans*, 20 (Oct 1987), pp. 8–10. Controversy between reestablishment of seal population, which eats salmon, and salmon fishermen.

OMANG, JOANNE. "In the Tropics, Still Rolling Back the Rain Forest Primeval," *Smithsonian*, 17(12) (Mar 1987), pp. 56–60/62/64–67. Deforestation for cheap export meat production.

PETERS, ROBERT. "Wildlife Reserves: Can They Do the Job?" *Conservation Foundation Letter*, (Jan/Feb 1984), pp. 1–7.

RAVENS, PETER. *"Disappearing Species: A Global Tragedy,"* *The Futurist*, 19 (Oct 1985), pp. 8–14.

REID, WALTER V.C. AND KENTON R. MILLER. *Keeping Options Alive: The Scientific Basis for Conserving Biodiversity.* Washington, D.C.: World Resources Institute, 1989.

REPETTO, ROBERT. *The Forest for the Trees? Government Policies and the Misuse of Forest Resources.* Washington, D.C.: World Resources Institute, 1988.

REPETTO, ROBERT, AND MALCOLM GILLIS, EDS. *Public Policy and the Misuse of Forest Resources.* Washington, D.C.: World Resources Institute, 1988.

ROBOTHAM, ROSEMARIE. "Paradise in Peril," *Life*, 10(12) (Nov 1987), pp. 93–96. Logging controversy in Tongass National Forests. Nature vs. Economic Interests.

RUSSELL, DICK, AND BRIAN KEATING. "Sushi Today, Gone Tomorrow: The Plight of the Bluefin Tuna," *The Amicus Journal*, 6 (Winter 1984), pp. 38–46.

SHEN, SUSAN. "Biological Diversity and Public Policy," *BioScience*, 37(10) (Nov 1987), pp. 709–712.

TANGLEY, LAURA. "Studying (and Surveying) the Tropics," *BioScience*, 38 (Jun 1988), pp. 375–385.

TENNESEN, MICHAEL. "Putting the Sting on Poachers," *National Wildlife*, 25(6) (Oct/Nov 1987), pp. 27–28.

TUTTLE, LIZA. "End of the Old-Growth Canopy: The Decline of the Spotted Owl Signals the Loss of our Last Virgin Timber," *National Parks*, 61(5–6) (May/Jun 1987), pp. 16–21.

VOLLERS, MARYANNE. "Healing the Ravaged Land," *International Wildlife*, 18 (Jan/Feb 1988), pp. 4–11.

WEXLER, MARK. "Modern Mission To Save An Ancient Mariner," *National Wildlife*, 26 (Jun/July 1988), pp. 4–10.

WHITE, CHRISTOPHER P. *Endangered and Threatened Wildlife in the Chesapeake Bay Region: Delaware, Maryland, and Virginia.* Baltimore, Md.: Tidewater Publishers, 1982.

WHITE, PETER T. "Nature's Dwindling Treasures—Rain Forests," *National Geographic*, 163 (Jan 1983), pp. 2–9, 20–46.

WILSON, E.O., ED. *Biodiversity.* Washington, D.C.: National Academy Press, 1988.

WOLF, EDWARD C. *On the Brink of Extinction: Conserving the Diversity of Life.* Worldwatch Paper 78. Washington, D.C.: Worldwatch Institute, 1987.

WORLD RESOURCES INSTITUTE. "Wildlife and Habitat," *World Resources 1988–89*, Chapter 6. New York: Basic Books, Inc.

WORLD WILDLIFE FUND. World Wildlife Fund Letter, Washington, D.C., World Wildlife Fund. Periodical provides current news about worldwide species' endangerment and conservation efforts, 1988.

CHAPTER 18. CONVERTING REFUSE TO RESOURCES

BioCycle: The Journal of Waste Recycling. Emmaus, Penna.: The JG Press, Inc. A monthly periodical devoted to waste recycling. Excellent coverage of current technology and progress.

ENVIRONMENTAL PROTECTION AGENCY. "The Garbage Crisis: Understanding it; Finding Answers." *EPA Journal*, 15 (Mar–Apr 1989). Entire issue devoted to the "garbage crisis".

HOGAR, BARBARA. "All Baled Up and No Place to Go," *The Conservationist*, 42(4) (Jan/Feb 1988), pp. 36–39.

LANGONE, JOHN. "A Stinking Mess," *TIME*, 133 (Jan 2, 1989), pp. 44–45, 47.

LAWREN, BILL. "Getting into a Heap of Trouble," *National Wildlife*, 26 (Aug/Sept 1988), pp. 18–24.

LIPSKE, MIKE. "Are You Throwing Poisons into the Trash?" *National Wildlife*, 24 (Aug/Sep 1986), pp. 20–23. Household output of toxics.

LYHUS, RANDY. "Composting at the Landfill," *Environment*, 30 (Oct 1988), pp. 21–22.

MORGANTHAU, TOM, ET AL. "Don't Go Near the Water," *Newsweek*, (Aug 1, 1988), pp. 40–47.

NATURAL RESOURCES DEFENSE COUNCIL. *A Solid Waste Blueprint for New York State*. New York: Natural Resources Defense Council, 1988.

NEAL, HOMER A., AND J.R. SCHUBEL. *Solid Waste Management and the Environment—The Mounting Garbage and Trash Crisis*. Englewood Cliffs, N.J.: Prentice Hall, 1987.

NOLL, KENNETH E., ET AL. *Recovery, Recycle, and Reuse of Industrial Waste*. Chelsea, Minn.: Lewis Publs., Inc., 1985.

O'LEARY, PHILIP R., ET AL. "Managing Solid Waste," *Scientific American*, 259 (Dec 1988), pp. 36–42.

PLATT, BRENDA. *Wasted Material, Wasted Youth*. Washington, D.C.: Institute for Self-Reliance, 1989.

POLLOCK, CYNTHIA. "Mining Urban Wastes: The Potential For Recycling." *Worldwatch Paper 76*. Washington, D.C.: Worldwatch Institute, 1987.

SCHWARTZ, ANNE. "Poisons in Your Home: A Disposal Dilemma," *Audubon*, (May 1987), pp. 12–17.

SCHWARTZ, JOHN, ET AL. "Turning Trash Into Hard Cash," *Newsweek*, (Mar 14, 1988), pp. 36–37.

SHEA, CYNTHIA POLLOCK. "Building a Market for Recyclables," *World Watch*, 1 (May/Jun 1988), pp. 12–18.

SMITH, EMILY T. "On the Jersey Shore, Business Crumbles Like a Sand Castle," *Business Week* (Oct 12, 1987), pp. 104–106.

VOGEL, SHAWNA. "Waste is a Terrible Thing to Mind," *Discover*, 9(1) (Jan 1988), pp. 76–78. Story of the 1987 garbage barge fiasco.

WEISSKOPF, MICHAEL. "Plastic Reaps a Grim Harvest in the Oceans of the World," *Smithsonian*, 18 (November 1988), pp. 58–66.

WHITE, PETER T. "The Fascinating World of Trash," *National Geographic*, 163 (Apr 1983), pp. 424–440, 447–456.

CHAPTER 19. ENERGY RESOURCES AND THE ENERGY PROBLEM

ABELSON, PHILIP. "Energy Futures," *American Scientist* (Nov/Dec 1987), pp. 584–592.

BURNETT, W.M. AND S.D. BAN. "Changing Prospects for Natural Gas in the United States," *Science* 244 (April 1989), pp. 305–310.

CHANDLER, WILLIAM U. "Energy Productivity: Key to Environmental Protection and Economic Progress," *Worldwatch Paper 63*. Washington, D.C.: Worldwatch Institute, 1985.

COWLEY, GEOFFRY. "Dead Otters, Silent Ducks," *Newsweek* (Apr 24, 1989), pp. 70–73.

CRAWFORD, MARK. "The Mixed Blessings of Inexpensive Oil," *Science*, 242 (Dec 2, 1988), pp. 1242–1243.

FLAVIN, CHRISTOPHER, AND ALAN DURNING. "Raising Energy Efficiency," *State of the World 1988*, Chapter 3. Washington, D.C.: Worldwatch Institute, 1988.

FLOWER, ANDREW. "World Oil Production," *Scientific American*, 238 (Mar 1987), pp. 42–49.

GEVER, JOHN, ET AL. *Beyond Oil: The Threat to Food and Fuel in the Coming Decades*. Cambridge, Mass.: Carrying Capacity, Inc., 1986.

GODLEMBER, JOSE, ET AL. *Energy For Development*. Washington, D.C.: World Resources Institute, 1987.

HIRSCH, ROBERT L. "Impending United States Energy Crisis," *Science*, 235 (Mar 20, 1987), pp. 1467–1473.

MARSHALL, ELIOT. "Fill the Oil Reserve, Academy Report Says," *Science*, 232 (Apr 25, 1986), pp. 441–442.

NORMAN, COLIN. "Interior Slashes Offshore Oil Estimates," *Science*, 228 (May 24, 1985), p. 974.

OURISSON, GUY, ET AL. "The Microbial Origins of Fossil Fuels," *Scientific American*, 251 (Aug 1984), pp. 44–51.

RENNER, MICHAEL. "Rethinking Transportation," *State of the World 1989*, Chapter 6. Washington, D.C.: Worldwatch Institute, 1989.

ROBERTS, LESLIE. "Long Slow Recovery Predicted for Alaska," *Science*, 244 (Apr 7, 1989), pp. 22–24.

ROBERTS, LESLIE. "Valdiz: The Predicted Oil Spill," *Science*, 244 (Apr 7, 1989), pp. 20–24.

ROSS, MARC. "Improving the Efficiency of Electricity Use in Manufacturing," *Science*, 244 (April 1989), pp. 311–317.

U.S. DEPARTMENT OF ENERGY/ENERGY INFORMATION ADMINISTRATION. *Annual Energy Review*. Washington, D.C.: U.S. Government Printing Office. Annual publication of U.S. Energy Statistics.

WORLD RESOURCES INSTITUTE. "Energy," *World Resources 1988–89*, Chapter 7. New York: Basic Books, Inc., 1988.

CHAPTER 20. NUCLEAR POWER, COAL, AND SYNTHETIC FUELS

BROWN, GEORGE E. "U.S. Nuclear Waste Policy: Flawed but Feasible," *Environment*, 29(8) (Oct 1987), pp. 6–7/25.

EDWARDS, MIKE. "Chernobyl—One Year After," *National Geographic*, 171(5) (May 1987), pp. 632–653.

EISENBUD, MERRIL. "Sources of Ionizing Radiation Exposure," *Environment*, 26 (Dec 1984), pp. 6–11, 30–33.

ENVIRONMENTAL ACTION FOUNDATION. *Rate Shock: Confronting the Cost of Nuclear Power*. Washington, D.C.: Environmental Action Foundation, 1984.

FLAVIN, CHRISTOPHER. "How Many Chernobyls?," *World Watch*, 1 (Jan/Feb 1988), pp. 14–18.

FLAVIN, CHRISTOPHER. "Ten Years of Fallout," *World Watch*, 2 (Mar/Apr 1989), pp. 30–37.

GLASSTONE, SAMUEL. *Nuclear power and its environmental effects*. Chicago: Wesley Publishing Co., 1987.

GROSSMAN, DAN, AND SETH SHULMAN. "A Nuclear Dump:

The Experiment Begins," *Discover*, 10 (Mar 1989), pp. 49–56.

HOHENEMSER, CHRISTOPH, AND ORTWIN RENN. "Chernobyl's Other Legacy," *Environment*, 30 (Apr 1988), pp. 5–11/40–45.

HUBBARD, H.M. "Photovoltaics Today and Tomorrow," *Science*, 244 (April 1989), pp. 297–304.

KAKU, MICHIO, AND JENNIFER TRAINER, EDS. *Nuclear Power: Both Sides—The Best Arguments For and Against the Most Controversial Technology.* New York: W.W. Norton and Co., 1983.

LANDSBERG, HANS. "The Case for Coal in the United States," *Environment*, 29(6) (July/Aug 1987), pp. 18–20/38–43.

LESTER, RICHARD K. "Rethinking Nuclear Power," *Scientific American*, 254 (Mar 1986), pp. 31–39.

LUMPKIN, ROBERT E. "Recent Progress in the Direct Liquefaction of Coal," *Science*, 239 (Feb 19, 1988), pp. 873–877.

MANNING, RUSS. "The Future of Nuclear Power—Where Do We Go From Here?" *Environment*, 27 (May 1985), pp. 12–17, 31–37.

POLLOCK, CYNTHIA. "Decommissioning: Nuclear Power's Missing Link," *Worldwatch Paper 64*. Washington, D.C.: Worldwatch Institute, 1986.

TAYLOR, JOHN J. "Improved and Safer Nuclear Power," *Science*, 244 (April 1989), pp. 318–325.

WEINBERG, ALVIN M., AND IRVING SPIEWAK. "Inherently Safe Reactors and a Second Nuclear Era," *Science*, 224 (Jun 29, 1984), pp. 1398–1402.

CHAPTER 21. SOLAR AND OTHER RENEWABLE ENERGY SOURCES

ANDERSON, BRUCE, WITH MICHAEL RIORDAN. *The Solar Home Book: Heating, Cooling and Designing with the Sun.* Harrisville, N.H.: Brick House Publishing Co., 1976.

BROWN, LESTER R. "Food or Fuel: New Competition for the World's Cropland." *Worldwatch Paper 35*. Washington, D.C.: Worldwatch Institute, 1980.

DEUDNEY, DANIEL. "Rivers of Energy; The Hydropower Potential." *Worldwatch Paper 44*. Washington, D.C.: Worldwatch Institute, 1981.

FLAVIN, CHRISTOPHER. "Electricity For A Developing World: New Directions." *Worldwatch Paper 70*. Washington, D.C.: Worldwatch Institute, 1986.

FLAVIN, CHRISTOPHER. "Electricity's Future: The Shift to Efficiency and Small-Scale Power." *Worldwatch Paper 61*. Washington, D.C.: Worldwatch Institute, 1984.

FLAVIN, CHRISTOPHER. "Photovoltaics: A Solar Technology for Powering Tomorrow," *The Futurist*, 17 (Jun 1983), pp. 41–50.

FLAVIN, CHRISTOPHER. "Selling Solar Cells," *World Watch*, 1 (Sept/Oct 1988), pp. 42–43.

FLAVIN, CHRISTOPHER. "Wind Power: A Turning Point." *Worldwatch Paper 45*. Washington, D.C.: Worldwatch Institute, 1981.

GALLIGAN, MARY. "In Solar Village, Sunshine Is Put In Harness," *U.S. News & World Report*, 98 (Feb 11, 1985), pp. 72–73.

GOLDEMBER, JOSE, ET AL. *Energy For A Sustainable World.* Washington, D.C.: World Resources Institute, 1987.

GREENBAUM, E., ET AL. "Biological Solar Energy Production With Marine Algae," *BioScience*, 33 (Oct 1983), pp. 584–585.

HELLER, ADAM. "Hydrogen-Evolving Solar Cells," *Science*, 223 (Mar 16, 1984), pp. 1141–1148.

HOLDEN, CONSTANCE. "Hawaiian Rainforest Being Felled," *Science*, 228 (May 31, 1985), pp. 1073–1074.

KERR, RICHARD A. "Hot Dry Rock: Problems, Promise," *Science*, 238 (Nov 27, 1987), pp. 1226–1228.

MUNSON, RICHARD. "International: Israel's Solar Ponds Grow Larger," *Environment*, 26 (Jan/Feb 1984), pp. 41–42.

OGDEN, JOAN M., AND ROBERT H. WILLIAMS. *Solar Hydrogen: Moving Beyond Fossil Fuels.* Washington, D.C.: World Resources Institute, 1989.

PIMENTEL, D., ET AL. "Environmental and Social Costs of Biomass Energy," *BioScience*, 34 (Feb 1984), pp. 89–94.

POOL, ROBERT. "Solar Cells Turn 30," *Science*, 241 (Aug 19, 1988), pp. 900–901.

SHEA, CYNTHIA POLLOCK. "Harvesting the Wind," *World Watch* 1 (Mar/Apr 1988), pp. 12–17.

SHEA, CYNTHIA POLLACK. "Renewable Energy: Today's Contribution, Tomorrow's Promise." *Worldwatch Paper 81*. Washington, D.C.: Worldwatch Institute, 1988.

SHEA, CYNTHIA POLLOCK. "Shifting to Renewable Energy," *State of the World 1988*, Chapter 4. Washington, D.C.: Worldwatch Institute, 1988.

VOGEL, SHAWNA. "Wind Power," *Discover*, 10 (May 1989), pp. 46–49.

WILSON, HOWARD G., ET AL. "Lessons of Sunracer," *Scientific American*, 260 (Mar 1989), pp. 90–97.

CHAPTER 22. LIFESTYLE, LAND USE, AND ENVIRONMENTAL IMPACT

BALDWIN, MAL. "Wetlands: Fortifying Federal and Regional Cooperation," *Environment*, 29 (Sept 1987), pp. 6–10/39–42.

BENDICK, ROBERT L. JR. "Saving America's Open Space," *USA Today*, 116 (Jan 1988), pp. 48–51.

CASTELLS, MANUEL. *The City and the Grassroots.* Berkeley: University of California Press, 1983.

CLARK, W.C. AND R.E. MUNN. "Sustainable Development of the Biosphere," *Environment*, 29(9) (Nov 1987), pp. 25–27.

DeGEORGE, GAIL AND ANTONIO N. FINS. "Is the Sun Setting on Florida's Overbuilding?" *Business Week* (Oct 19, 1987) pp. 102–105.

ELFRING, CHRIS. "Preserving Land Through Local Land Trusts," *BioScience*, 39 (Feb 1989), pp. 71–74.

HARGROVE, EUGENE. *Beyond Spaceship Earth: Environmental Ethics.* San Francisco: Sierra Club Books, 1986.

Kohl, L. "Heavy Hands on the Land," *National Geographic,* 174 (Nov 1988), pp. 632–651.

Krasemann, Stephen J. "Quietly Conserving Nature," *National Geographic,* 174 (Dec 1988), pp. 818–845.

Lalo, Julie "The Problem of Roadkill," *American Forests,* 93 (Sept/Oct 1987), pp. 50–52/72.

McGrath, Susan. "They've Been Walking on the Railroad," *National Wildlife,* 26 (Aug/Sept 1988), pp. 40–43.

Mohlenbrock, Robert H. "This Land," *Natural History,* 97 (Feb 1988), pp. 12–15.

O'Riordan, Timothy. "The Earth as Transformed by Human Action: An International Symposium," *Environment,* 30(1) (Jan/Feb 1988), pp. 25–28.

Pimental, D., et al. "Land Degradation: Effects on Food and Energy Resources," *Science,* 194 (Oct 8, 1976), pp. 149–155.

Poole, Daniel A. "The Public Lands: Bust or Boon for Wildlife," *American Forests,* 93(9 and 10) (Sept/Oct 1987), pp. 17–20. Overview of history of public lands.

Propst, Luther. "Problems on the Urban Frontier," *EPA Journal,* 14 (May 1988), pp. 16–18.

Quinn, Barbara, "Country Congestion," *American City and County,* 103(2) (Feb 1988), pp. 52–59. Rural areas are becoming as congested as cities as businesses/industries move plants and offices to suburbs.

Richardson, David B. "To Rebuild America—$2.5 Trillion Job," *U.S. News & World Report,* 93 (Sept 27, 1982), pp. 57–61.

Rouse, James W. "Suffering in the Second City," *EPA Journal,* 14 (May 1988), pp. 24, 25.

Steinhart, Peter. "Mitigation Isn't," *Audubon,* 89(3) (May 1987), pp. 8–11. Suggested requirement that developers mitigate the impact of their development.

Stokes, Bruce. "Recycled Housing," *Environment,* 21 (Jan/Feb 1979), pp. 6–14.

United States Environmental Protection Agency. *EPA Journal,* 14 (May 1988). Washington, D.C.: United States Environmental Protection Agency. Entire issue devoted to cities and the environment.

Ward, Justin R. "The Conservation Reserve and the Rural Environment," *Environment,* 28 (Sept 1986), pp. 3–5.

EPILOGUE

Brown, Lester R., Christopher Flavin, and Sandra Postel. "Outlining a Global Action Plan," *State of the World 1989,* Chapter 10. Washington, D.C.: Worldwatch Institute, 1989.

Brown, Lester R., and Pamela Shaw. "Putting Society on a New Path," *Environment,* 24 (Sept 1982), pp. 29–33.

Brown, Lester R., and Pamela Shaw. "Six Steps to a Sustainable Society," *Worldwatch Paper,* 48 (Mar 1982). Washington D.C.: Worldwatch Inst, 1982.

Brown, Lester R. and Edward C. Wolf. "Reclaiming the Future," *State of the World 1988,* Chapter 10. Washington, D.C.: Worldwatch Institute, 1988.

Cobb, R.W., and C.D. Elder. *Participation in American Politics.* 2nd ed. Baltimore: Johns Hopkins University Press, 1983.

Durning, Alan B. "Mobilizing at the Grassroots," *State of the World 1989,* Chapter 9. Washington D.C.: Worldwatch Institute, 1989.

Norman, Colin. "Soft Technologies, Hard Choices," *Worldwatch Paper 21.* Washington, D.C.: Worldwatch Institute, Jun, 1978.

Schumacher, E.F. *Small Is Beautiful: Economics As If People Mattered.* New York: Perennial Library, Harper and Row, 1973.

World Resources Institute. *The Crucial Decade: The 1990's and the Global Environmental Challenge.* Washington, D.C.: World Resources Institute, 1989.

Appendix A
ENVIRONMENTAL ORGANIZATIONS

This is a list of organizations active in environmental matters. Included here are national organizations as well as some small, specialized ones. Many of these have internship positions available for those wishing to do work for an environmental group. A more complete listing can be found in the *Conservation Directory* put out by National Wildlife Federation (address below). This directory includes local, regional and national organizations. The cost is $15.00 plus $2.00 for postage.

AMERICAN LUNG ASSOCIATION,   1740 Broadway, New York, N.Y. 10019. (212) 315-8700. Research, education: air pollution effects and means of control.

AMERICAN RIVERS CONSERVATION COUNCIL,   801 Pennsylvania Avenue, S.E., Suite 303, Washington, D.C. 20003. (202) 547-6900. Lobbying: wild and scenic rivers.

CENTER FOR SCIENCE IN THE PUBLIC INTEREST,   1501 16th Street, N.W., Washington, D.C. 20036. (202) 332-9110. Research and education: food, nutrition, health.

CHESAPEAKE BAY FOUNDATION,   162 Prince George St., Annapolis, MD 21401. (301) 268-8816. Research, education, and litigation: environmental defense and management of Chesapeake Bay and surrounding land.

CLEAN WATER ACTION PROJECT,   317 Pennsylvania Ave., S.E., Washington, D.C. 20003. (202) 547-1196. Lobbying: water quality.

COMMON CAUSE,   2030 M Street, N.W., Washington, D.C. 20036. (202) 833-1200. Lobbying: government reform, energy reorganization, clean air.

CONCERN INC.,   1794 Columbia Road, N.W., Washington, D.C. 20009. (202) 328-8160. Research and education: environmental education.

CONGRESS WATCH,   215 Pennsylvania Avenue, S.E., Washington, D.C. 20003. (202) 546-4996. Lobbying: consumer health and safety, pesticides.

CONSERVATION FOUNDATION,   1250 24th St., N.W., Washington, D.C. 20037. (202) 293-4800. Research and education: land use, energy conservation, air and water quality.

CONSUMER FEDERATION OF AMERICA,   1424 16th Street, N.W., Suite 604, Washington, D.C. 20036. (202) 387-6121. Lobbying: energy policy, telecommunications, public product liability, banking and insurance reform.

CRITICAL MASS ENERGY PROJECT,   215 Pennsylvania Avenue, S.E., Washington, D.C. 20003. (202) 546-4996. Research: nuclear power, alternative energy.

DEFENDERS OF WILDLIFE,   1244 19th Street, N.W., Washington, D.C. 20036. (202) 659-9510. Research, education and lobbying: endangered species.

ENVIRONMENTAL ACTION, INC.,   1525 New Hampshire Avenue, N.W., Washington, D.C. 20036. (202) 745-4870. Lobbying: transportation, solid waste, water quality, solar energy, energy conservation, toxic substances, deposit legislation.

ENVIRONMENTAL DEFENSE FUND,   257 Park Avenue South, New York, N.Y. 10010. (212) 505-2100. Research, litigation, and lobbying: cosmetic safety, drinking water, energy, transportation, pesticides, wildlife, air pollution, cancer prevention, radiation.

ENVIRONMENTAL LAW INSTITUTE,   1616 P Street, N.W., Suite 200, Washington, D.C. 20036. (202) 328-5150. Research and education: institutional and legal issues affecting the environment.

ENVIRONMENTAL POLICY INSTITUTE,   218 D Street, S.E., Washington, D.C. 20003. (202) 544-2600. Lobbying: all aspects of energy development, preservation, restoration, and rational use of earth.

GREENPEACE, USA, INC.,   1436 U Street, N.W., Washington, D.C. 20009. (202) 462-1177. Non-violent direct action: whales, ocean waste disposal, acid rain, nuclear weapons testing, seal pups, and Antarctica.

HABITAT FOR HUMANITY INTERNATIONAL,   Habitat and Church Streets, Americus, Georgia 31709-3498. (912) 924-6935. Build homes, educate on housing issues.

INSTITUTE FOR LOCAL SELF-RELIANCE,   2425 18th Street, N.W., Washington, D.C. 20009. (202) 232-4108. Research and education: appropriate technology for community development.

IZAAK WALTON LEAGUE OF AMERICA, INC.,   Box 824, Iowa City, Iowa 52244. (319) 351-7037. Research and education: conservation, air and water quality, streams.

LEAGUE OF CONSERVATION VOTERS,   1150 Conn Avenue, N.W., Suite 201, Washington, D.C. 20036. (202) 785-8683. Political action: evaluation of environmental records of public officials.

LEAGUE OF WOMEN VOTERS OF THE U.S.,   1730 M Street, N.W., Washington, D.C. 20036. (202) 429-1965. Education and lobbying: general environmental issues.

MONITOR CONSORTIUM OF CONSERVATION AND ANIMAL WELFARE ORGANIZATIONS,   1506 19th Street, N.W., Washington, D.C. 20036. (202) 234-6576. Lobbying: endangered marine species.

NATIONAL AUDUBON SOCIETY,   950 Third Avenue, New

York, N.Y. 10022. (212) 546-9100. Research, lobbying, education: wildlife, wilderness, public lands, endangered species, water resource management.

NATIONAL PARK FOUNDATION, P.O. Box 57473, Washington, D.C. 20037. (202) 785-4500. Lobbying, education, and acquisition of land: National Parks.

NATIONAL PARKS AND CONSERVATION ASSOCIATION, 1015 31st Street, N.W., Washington, D.C. 20007. (202) 944-8530. Research and education: parks, wildlife, forestry, general environmental quality.

NATIONAL RESOURCES DEFENSE COUNCIL, 40 West 20th St., New York, N.Y. 10011. (212) 727-2700. Research and litigation: water and air quality, land use, energy, pesticides, toxic waste.

NATIONAL WILDLIFE FEDERATION, 1400 16th Street, N.W., Washington, D.C. 20036. (202) 797-6800. Research, education, lobbying: general environmental quality, wilderness and wildlife.

NATURE CONSERVANCY, 1815 North Lynn Street, Arlington, VA 22209. (703) 841-5300. Research and education: identification, protection and management of natural areas.

NEW DIRECTIONS FOR POLICY, 1101 Vermont Avenue, Suite 400, N.W., Washington, D.C. 20005. (202) 289-3907. Lobbying: international energy policy and self-reliant economic development.

PLANNED PARENTHOOD FEDERATION OF AMERICA, 810 Seventh Avenue, New York, N.Y. 10019. (212) 541-7800. Education, services, and research: fertility control, family planning.

THE POPULATION INSTITUTE, 110 Maryland Avenue, N.E., Washington, D.C. 20002. (202) 544-3300. Education, research, and lobbying: population control.

RACHEL CARSON COUNCIL, INC., 8940 Jones Mill Road, Chevy Chase, MD 20815. (301) 652-1877. Publication distribution and educational conferences: pesticides, toxic substances.

RAIN FOREST ACTION NETWORK, 301 Broadway, Suite A, San Francisco, CA 94133. (415) 398-4404.

RENEW AMERICA, 1001 Connecticut Avenue, N.W., Suite 719, Washington, D.C. 20036. (202) 466-6880. Research and education: energy policy community organizing on issues concerning renewable resources, solar energy.

RESOURCES FOR THE FUTURE, 1616 P Street, N.W., Washington, D.C. 20036. (202) 328-5000. Research and education: conservation of natural resources, environmental quality.

RURAL AMERICA, 725 15th Street, N.W., Suite 900, Washington, D.C. 20005. (202) 628-1480. Research and education: health, community development, housing, environmental quality.

SCIENTISTS INSTITUTE FOR PUBLIC INFORMATION, 355 Lexington Avenue, New York, N.Y. 10017. (212) 661-9110. Education: scientists of full range of disciplines provide information for public.

SIERRA CLUB, 730 Polk Street, San Francisco, CA 94109. (415) 776-2211. Education, lobbying: water quality, energy, offshore energy development, wilderness areas, urban recreation, wilderness trips.

TRUST FOR PUBLIC LAND, 116 New Montgomery, 4th Floor, San Francisco, CA 94105. (415) 495-4014. Works with citizen groups and government agencies to acquire and preserve open space.

U.S. PUBLIC INTEREST RESEARCH GROUP, 215 Pennsylvania Avenue, S.E., Washington, D.C. 20003. (202) 546-9707. Research and education: alternative energy, utilities regulation, environmental and consumer lobbying and public interest.

WATER POLLUTION CONTROL FEDERATION, 601 Wythe Street, Alexandria, VA 22314. (703) 684-2400. Research, education, and lobbying.

WILDERNESS SOCIETY, 1400 I Street, N.W., 10th Floor, Washington, D.C. 20005. (202) 842-3400. Research, education and lobbying: wilderness, public lands.

WORLD WILDLIFE FUND (AND THE CONSERVATION FOUNDATION), 1250 24th Street, N.W., Washington, D.C. 20037. (202) 293-4800. Research and education: endangered species.

WORLDWATCH INSTITUTE, 1776 Massachusetts Avenue, N.W., Washington, D.C. 20036. (202) 452-1999. Research and education: energy, food, population, health, women's issues, technology, the environment.

ZERO POPULATION GROWTH, INC., 1400 16th Street, N.W., Suite 320, Washington, D.C. 20036. (202) 332–2200. Research, education, lobbying: population.

Appendix B
UNITS OF MEASURE

The Metric System and Equivalent English Units

LENGTH	1 centimeter (cm) × 10 = 1 cm = 0.39 inches 1 inch = 2.54 cm	1 decimeter (dm) × 10 = 1 dm = 3.94 inches 1 foot = 3.05 dm	1 meter (m) × 1000 = 1 m = 1.09 yards 1 yard = .91 m	1 kilometer (km) 1 km = 0.62 miles 1 mile = 1.61 km
AREA	a square 1 cm on each side is 1 square centimeter (cm²) 1 cm² = 0.155 square inches 1 square inch = 6.45 cm²	a square 1 m on each side is 1 square meter (m²) 1 m² = 10.8 square feet 1 m² = 1.20 square yards 1 square yard = .836 m²	a square 100 m on each side is 1 hectare (ha) 1 ha × 100 = 1 km² 1 ha = 2.47 acres 1 acre = 0.405 ha	a square 1 km on each side is 1 square kilometer (km²) 1 km² = 0.39 square miles 1 square mile = 2.59 km²
VOLUME	a cube 1 cm on each side is 1 cubic centimeter (cm³) or 1 milliliter (ml) 1 ml × 1000 = 1 L 1 ml = .203 teaspoons 1 teaspoon = 4.9 ml	a cube 1 dm on each side is 1 cubic decimeter (dm³) or 1 liter (L) 1 L × 1000 = 1 m³ 1 L = 1.06 quarts 1 quart = .95 L	a cube 1 m on each side is 1 cubic meter (m³) 1 m³ = 264.2 gallons 1 m³ = 36.5 cubic feet 1 m³ = 28.4 bushels (dry) 1 m³ = 1.31 cubic yards 1 cubic yard = .76 m³	
MASS (WEIGHT)	1 ml of water at 4°C weighs 1 gram (g) 1 g × 1000 = 1 kg 1 g = .035 ounces 1 ounce = 28.4 g	1 liter of water at 4°C weighs 1 kilogram (kg) 1 kg × 1000 = 1 metric ton 1 kg = 2.2 pounds 1 pound = 0.45 kg	1 cubic meter of water at 4°C weighs 1 metric ton (t), also called a long ton 1 t = 2200 pounds 1 t = 1.1 short tons 1 short ton (2000 pounds) = .91 t	

Energy Units and Equivalents

1 Calorie, food calorie, or kilocalorie—The amount of heat required to raise the temperature of one kilogram of water one degree Celsius (1.8°F).

1 BTU (British Thermal Unit)—The amount of heat required to raise the temperature of one pound of water one degree Fahrenheit.

1 Calorie = 3.968 BTU's
1 BTU = 0.252 calories

1 therm = 100,000 BTU's
1 quad = 1 quadrillion BTU's

1 watt standard unit of electrical power

  1 watt-hour (wh) = 1 watt for 1 hr. = 3.413 BTU's

1 kilowatt (kw) = 1000 watts

  1 kilowatt-hour (kwh) = 1 kilowatt for 1 hr. = 3413

BTU's

1 megawatt (Mw) = 1,000,000 watts

  1 megawatt-hour (Mwh) = 1 Mw for 1 hr. = 34.13 therms

1 gigawatt (Gw) = 1,000,000,000 watts or 1,000 megawatts

  1 gigawatt-hour (Gwh) = 1 Gw for 1 hr. = 34,130 therms

1 horsepower = .7457 kilowatts; 1 horsepower-hour = 2545 BTU's

1 cubic foot of natural gas (methane) at atmospheric pressure = 1031 BTU's

1 gallon gasoline = 125,000 BTU's

1 gallon No. 2 fuel oil = 140,000 BTU's

1 short ton coal = 25,000,000 BTU's

1 barrel (oil) = 42 gallons

Appendix C
SOME BASIC CHEMICAL CONCEPTS

ATOMS, ELEMENTS, AND COMPOUNDS

All matter, whether gas, liquid, or solid, living or non-living, organic or inorganic, is comprised of fundamental units called **atoms.** Atoms are extremely tiny. If all the world's people, about 5,000,000,000 of us, were reduced to the size of atoms, there would be room for all of us to dance on the head of a pin. In fact, we would only occupy a tiny fraction (about $\frac{1}{10,000}$) of the pin's head. Given the incredibly tiny size of atoms, even the smallest particle which can be seen with the naked eye consists of billions of atoms.

The atoms comprising a substance may be all of one kind; or they may be of two or more different kinds. If the atoms are all of one kind, the substance is called an **element.** If the atoms are of two or more different kinds bonded together, the substance is called a **compound.**

Through countless experiments, chemists have ascertained that there are only 96 distinct kinds of atoms which occur in nature. They are listed in Table C–1 with their chemical symbols. By scanning Table C–1, you can see that a number of familiar substances such as aluminum, calcium, carbon, oxygen, and iron are elements; that is, they are a single distinct kind of atom. However, most of the substances with which we interact in every-day life, such as water, stone, wood, protein, and sugar, are not on the list. Their absence from the list is indicative that they are not elements; rather, they are compounds, which means they are actually comprised of two or more different kinds of atoms bonded together.

ATOMS, BONDS, AND CHEMICAL REACTIONS

In chemical reactions, atoms are neither created, nor destroyed, nor is one kind of atom changed into another. What occurs in chemical reactions, whether mild or explosive, is simply a rearrangement of the ways in which the atoms involved are bonded together. An oxygen atom, for example, may be combined and recombined with different atoms to form any number of different compounds, but a given oxygen atom always has been, and always will be, an oxygen atom. The same can be said for all the other kinds of atoms. In order to understand how atoms may bond and rearrange to form different compounds, it is necessary to first have some concepts concerning the structure of atoms.

Structure of Atoms

In every case, an atom consists of a central core called the nucleus (not to be confused with the cell nucleus). The nucleus of the atom contains one or more **protons** and, except for hydrogen, one or more **neutrons** as well. Surrounding the nucleus are particles called **electrons.** Each proton has a positive (+) electric charge and each electron has an equal but opposite negative (−) electric charge. Thus, the charge of the protons may be balanced by an equal number of electrons making the whole atom neutral. Neutrons have no charge.

Atoms of all elements have this same basic structure consisting of protons, electrons, and neutrons. The distinction among atoms of different elements is in the number of protons. The atoms of each element have a characteristic number of protons which is known as the **atomic number** of the element (see Table C–1). The number of electrons characteristic of the atoms of each element also differs corresponding to the number of protons. The general structure of the atoms of several elements is shown in Figure C–1.

The number of protons and electrons, i.e., the atomic number of the element, determines the chemical properties of the element. However, the number of neutrons may also vary. For example, most carbon atoms have six neutrons in addition to the six protons as indicated in Figure C–1. But some carbon atoms have eight neutrons. Atoms of the same element which have different numbers of neutrons are known as **isotopes** of the element. The total number of protons plus neutrons is used to define different isotopes. For example, the usual isotope of carbon is referred to as carbon-12 while the isotope noted above is referred to as carbon-14. The chemical reactivity of different isotopes of the same element is identical. However, certain other properties may differ. Many isotopes of various elements prove to be radioactive as is carbon-14.

Table C–1
The Elements

ELEMENT	SYMBOL	ATOMIC NUMBER	ELEMENT	SYMBOL	ATOMIC NUMBER
Actinium	Ac	89	Neodymium	Nd	60
Aluminum	Al	13	Neon	Ne	10
Americium	Am	95	Neptunium	Np	93
Antimony	Sb	51	Nickel	Ni	28
Argon	Ar	18	Niobium	Nb	41
Arsenic	As	33	Nitrogen	N	7
Astatine	At	85	Nobelium	No	102
Barium	Ba	56	Osmium	Os	76
Berkelium	Bk	97	Oxygen	O	8
Beryllium	Be	4	Palladium	Pd	46
Bismuth	Bi	83	Phosphorus	P	15
Boron	B	5	Platinum	Pt	78
Bromine	Br	35	Plutonium	Pu	94
Cadmium	Cd	48	Polonium	Po	84
Calcium	Ca	20	Potassium	K	19
Californium	Cf	98	Praseodymium	Pr	59
Carbon	C	6	Promethium	Pm	61
Cerium	Ce	58	Protoactinium	Pa	91
Cesium	Cs	55	Radium	Ra	88
Chlorine	Cl	17	Radon	Rn	86
Chromium	Cr	24	Rhenium	Re	75
Cobalt	Co	27	Rhodium	Rh	45
Copper	Cu	29	Rubidium	Rb	37
Curium	Cm	96	Ruthenium	Ru	44
Dysprosium	Dy	66	Samarium	Sm	62
Einsteinium	Es	99	Scandium	Sc	21
Erbium	Er	68	Selenium	Se	34
Europium	Eu	63	Silicon	Si	14
Fermium	Fm	100	Silver	Ag	47
Fluorine	F	9	Sodium	Na	11
Francium	Fr	87	Strontium	Sr	38
Gadolinium	Gd	64	Sulfur	S	16
Gallium	Ga	31	Tantalum	Ta	73
Germanium	Ge	32	Technetium	Tc	43
Gold	Au	79	Tellurium	Te	52
Hafnium	Hf	72	Terbium	Tb	65
Helium	He	2	Thallium	Tl	81
Holmium	Ho	67	Thorium	Th	90
Hydrogen	H	1	Thulium	Tm	69
Indium	In	49	Tin	Sn	50
Iodine	I	53	Titanium	Ti	22
Iridium	Ir	77	Tungsten	W	74
Iron	Fe	26	Unnilennium	Une	109
Krypton	Kr	36	Unnilhexium	Unh	106
Lanthanum	La	57	Unniloctium	Uno	108
Lawrencium	Lr	103	Unnilseptium	Uns	107
Lead	Pb	82	Uranium	U	92
Lithium	Li	3	Vanadium	V	23
Lutetium	Lu	71	Xenon	Xe	54
Magnesium	Mg	12	Ytterbium	Yb	70
Manganese	Mn	25	Yttrium	Y	39
Mendelevium	Md	101	Zinc	Zn	30
Mercury	Hg	80	Zirconium	Zr	40
Molybdenum	Mo	42			

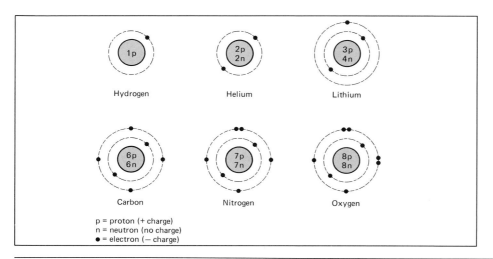

FIGURE C-1
Structure of atoms. All atoms consist of fundamental particles: protons (P), which have a positive electric charge, neutrons (n), which have no charge, and electrons, which have a negative charge. Protons and neutrons are located in a central core, the nucleus. The positive charge of the protons is balanced by an equal number of electrons, which occupy various levels or orbitals around the nucleus. The uniqueness of each element is given by its atoms having a distinct number of protons, its atomic number.

Bonding of Atoms

The chemical properties of an element are defined by the ways in which its atoms will react and form bonds with other atoms. By examining how atoms form bonds, we shall see how the number of electrons and protons determines these properties. There are two basic kinds of bonding: (1) **covalent bonding,** and (2) **ionic bonding.**

In both kinds of bonding, it is first important to recognize that electrons are not randomly distributed around the atom's nucleus. Rather, there are, in effect, specific spaces in a series of layers, or **orbitals,** around the nucleus. If an orbital is occupied by one or more electrons but not filled, the atom is unstable; it will tend to react and form bonds with other atoms to achieve greater stability. A stable state is achieved by having all the spaces in the orbital filled with electrons. But, it is also important to keep the charge neutral, i.e., the total number of electrons equal to that of the protons.

Covalent Bonding

These two requirements, filling all the spaces and keeping the charge neutral, may be satisfied by adjacent atoms sharing one or more pairs of electrons as shown in Figure C-2. The sharing of a pair of electrons holds the atoms together in what is called a **covalent bond.**

Covalent bonding, by satisfying the charge-orbital requirements, leads to discrete units of two or more atoms bonded together. Such units of two or more covalently bonded atoms are called **molecules.** A few simple but important examples are shown in Figure C-2.

A chemical formula is simply a shorthand description of the number of each kind of atom in a given molecule. The element is given by the chemical symbol and a subscript following the symbol gives the number present, no subscript being understood as one. A molecule with two or more different kinds of atoms may also be called a compound, but a molecule comprised of a single kind of atom, oxygen (O_2) for example, is still defined as an element.

Only a few elements, namely carbon, hydrogen, oxygen, nitrogen, phosphorus and sulfur, have configurations of electrons which lend readily to the formation of covalent bonds. But, carbon specifically, with its ability to form four covalent bonds, can produce long, straight or branched chains, or rings (Fig. C-3). Thus, an infinite array of molecules can be formed by using covalently bonded carbon atoms as a "backbone" and filling in the sides with atoms of hydrogen or other elements. Thus, it is covalent bonding among atoms of carbon and these few other elements that produces all natural organic molecules, those molecules that comprise all the tissues of living things, and also synthetic organic compounds such as plastics.

Ionic Bonding

Another way in which atoms may achieve a stable electron configuration is to gain additional electrons to complete the filling of an orbital, or lose electrons which are over a completed orbital. In general, the maximum number of electrons that can be gained or lost by an atom is three. Therefore, an element's atomic number determines whether one or more electrons will be lost or gained. If an atom's outer orbital is one to three electrons short of being filled, it will always tend to gain additional electrons. Conversely, if an atom has one to three electrons over its last complete orbital it will always tend to give them away.

Of course gaining or losing electrons results in the number of electrons being greater or less than the

FIGURE C–2

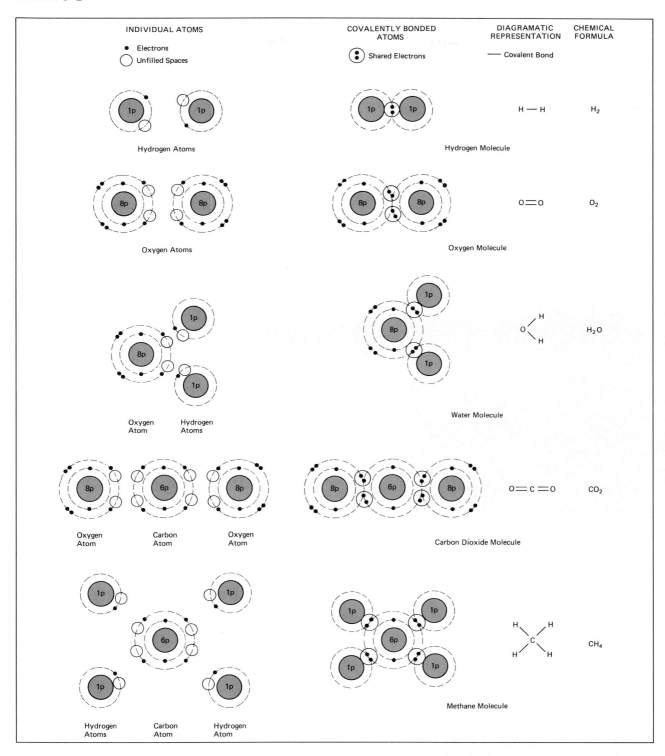

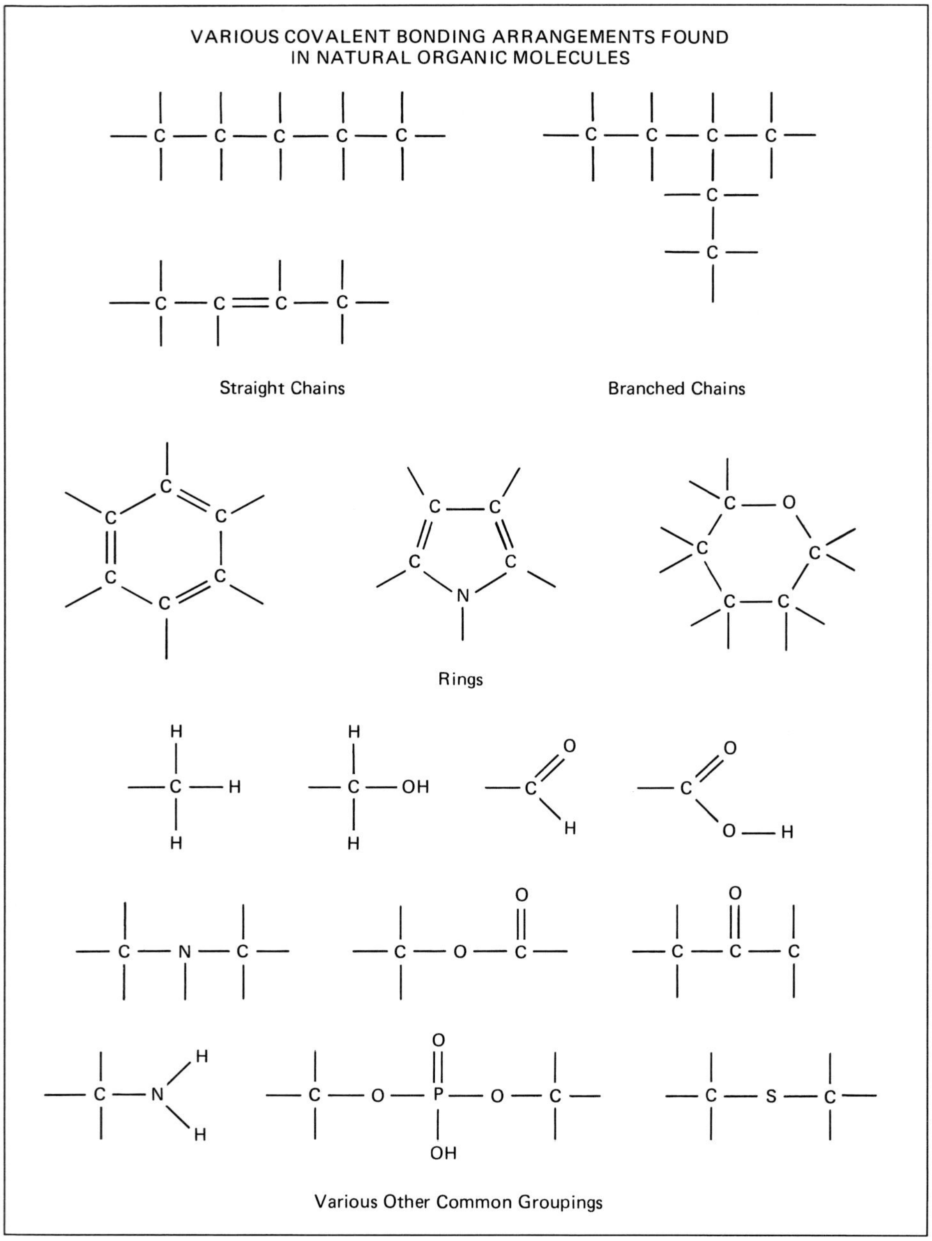

VARIOUS COVALENT BONDING ARRANGEMENTS FOUND IN NATURAL ORGANIC MOLECULES

Straight Chains

Branched Chains

Rings

Various Other Common Groupings

FIGURE C–3
Covalent bonding and organic molecules. The ability of carbon and a few other elements to readily form covalent bonds leads to an infinite array of complex molecules, organic molecules, which constitute all living things. A few major kinds of groupings are shown here. Note that each element forms a characteristic number of bonds: carbon, 4; nitrogen, 3; oxygen, 2; hydrogen, 1; sulfur, 2; phosphorus, 5. Bonds (dashed lines) left "hanging" indicate attachments to other atoms or groups of atoms.

number of protons, and the atom consequently having an electric charge. The charge will be one negative for each electron gained or one positive for each electron lost (Fig. C–4). A covalently bonded group of atoms may acquire an electric charge in the same way. An atom or group of atoms which has acquired an electric charge in this way is called an **ion,** positive or negative. Ions are designated by a superscript following the chemical symbol giving the number of positive or negative charges. Absence of superscripts indicates that the atom or molecule is neutral. Some important ions are listed in Table C–2.

Since unlike charges attract, positive and negative ions tend to join and pack together in dense clusters in such a way as to neutralize the overall electric charge. This joining together of ions through the attraction of their opposite charges is called **ionic bonding.** The result is the formation of hard, brittle, more or less crystalline substances of which all rocks and minerals are examples (Fig. C–5).

It is significant to note that whereas covalent bonding leads to discrete molecules, ionic bonding does not. Any number and combination of positive and negative ions may enter into an ionically bonded

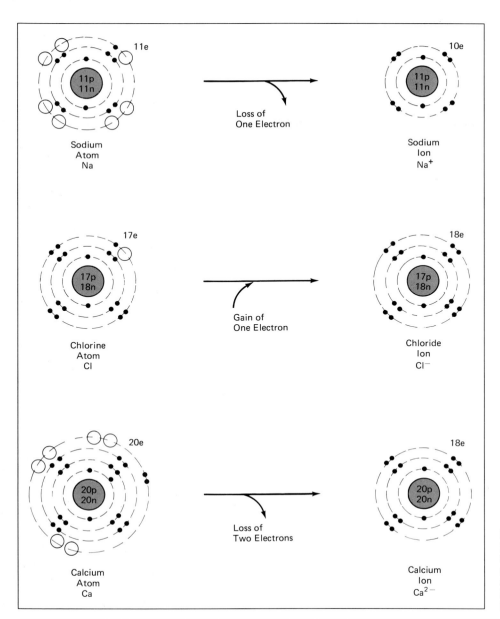

FIGURE C-4
Formation of ions. Many atoms will tend to gain or lose one or more electrons in order to achieve a state of complete (electron-filled) orbitals. In doing so they become positively or negatively charged ions as indicated.

Table C-2
Ions of Particular Importance to Biological Systems

NEGATIVE (−) IONS		POSITIVE (+) IONS	
Phosphate	PO_4^{3-}	Potassium	K^+
Sulfate	SO_4^{2-}	Calcium	Ca^{2+}
Nitrate	NO_3^-	Magnesium	Mg^{2+}
Hydroxyl	OH^-	Iron	Fe^{2+}, Fe^{3+}
Chloride	Cl^-	Hydrogen	H^+
		Ammonium	NH_4^+
		Sodium	Na^+

FIGURE C-5
Positive and negative ions bond together by their mutual attraction.

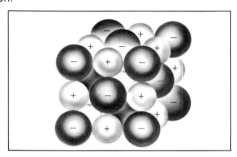

cluster to produce crystals of almost any size. The only restriction is that the overall charge of positive ions is balanced by that of negative ions. Thus, ionicly bonded substances are properly called compounds but not molecules. When chemical formulas are used to describe such compounds, they define the ratio of various elements involved, not specific molecules.

Chemical Reactions and Energy

While atoms themselves do not change, the bonds between atoms may be broken and reformed with different atoms producing different compounds and/or molecules. This is essentially what occurs in all chemical reactions. What determines whether a given chemical reaction will occur or not? We noted above that atoms form bonds because they achieve a greater stability by doing so. But some bonding arrangements may provide greater overall stability than others. Consequently substances with relatively unstable bonding arrangements will tend to react to form one or more different compounds which have more stable bonding arrangements. Common examples are the reaction between hydrogen and oxygen to produce water, and the reaction between carbon and oxygen to produce carbon dioxide (Fig. C–6).

Additionally, energy is always released in the process of gaining greater overall stability as indicated in Figure C–6. Thus, energy being released from a chemical reaction is synonomous with the atoms achieving more stable bonding arrangements. Thus, it may be said that chemical reactions always tend to go in a direction that releases energy as well as one which gives greater stability.

However, chemical reactions can be made to go in a reverse direction. With suitable energy inputs and under suitable conditions, stable bonding ar-

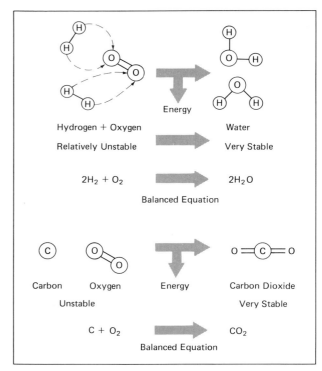

FIGURE C–6
Some bonding arrangements are more stable than others. Chemical reactions will go spontaneously toward more stable arrangements, releasing energy in the process. But, reactions may be driven in the opposite direction with suitable energy inputs.

rangements may be broken and less stable arrangements formed. As described in Chapter 2, this is the basis of photosynthesis occurring in green plants. Light energy is brought to bear on splitting the highly stable hydrogen-oxygen bonds of water and forming less stable carbon-hydrogen bonds thus creating high-energy organic compounds.

GLOSSARY

A

Abiotic. Pertaining to factors or things that are separate and independent from living things; nonliving.

Acid. Any compound that releases hydrogen ions when dissolved in water. Also, a water solution that contains a surplus of hydrogen ions.

Acid deposition. Any form of acid precipitation and also fallout of dry acid particles. (See **acid precipitation**)

Acid dew. Acidic dew; results from water vapor condensing on dry acid fallout.

Acid fallout. Molecules of acid formed from reactions involving nitrogen and sulfur oxides and water vapor settling out of the atmosphere without additional water.

Acid precipitation. Includes acid rain, acid fog, acid snow, and any other form of precipitation that is more acidic than normal, i.e., less than pH 5.6. Excess acidity is derived from certain air pollutants, namely sulfur dioxide and oxides of nitrogen.

Activated charcoal. A form of carbon that readily adsorbs organic material. Therefore it is frequently used in air and/or water filters to remove organic contaminants. It does not remove ions such as those of the heavy metals.

Activated sludge. Sludge comprised of clumps of living organisms feeding on detritus which settles out and is recycled in the process of secondary wastewater treatment.

Activated sludge system. A system for removing organic wastes from water. The system uses microorganisms and active aeration to decompose such wastes. The system is most used as a means of secondary sewage treatment following the primary settling of materials.

Active solar heating system. Solar heating system using pumps and/or blowers to transfer the heat from the collector to the place of use.

Adaption (ecological or evolutionary). A change in structure or function that produces better adjustment of an organism to its environment, and hence enhances its ability to survive and reproduce.

Adsorption. The process of chemicals (ions or molecules) sticking to the surface of other materials.

Advance disposal fee. A fee added to the purchase price of all products in glass, metal, and plastic cans or bottles to cover the cost of disposing of the container.

Advanced treatment (sewage treatment). Any of a variety of systems that follow secondary treatment and that are designed to remove one or more nutrients, such as phosphate, from solution.

Aeration. *Soil:* The property of a soil relating to its ability to allow the exchange of oxygen and carbon dioxide, which is necessary for the respiration of roots. *Water:* The bubbling of air or oxygen through water to increase the dissolved oxygen.

Agricultural district. A region in which the state or county limits subdivision and development in order to preserve the agricultural viability of the region.

Air. The mixture of gases, namely 78 percent nitrogen, 21 percent oxygen, and .035 percent carbon dioxide, making up the atmosphere. Water vapor and various pollutants may also be present.

Alga, pl. **algae.** Any of numerous kinds of photosynthetic plants that live and reproduce entirely immersed in water. Many species, the planktonic forms, exist as single or small groups of cells that float freely in the water. Other species, the "seaweeds," may be large and attached.

Algal bloom. A relatively sudden development of a heavy growth of algae, especially planktonic forms. Algal blooms generally result from additions of nutrients, whose scarcity is normally limiting.

Alleles. The two or more variations of a gene for any particular characteristic, e.g., blue and brown are alleles of the gene for eye-color.

Anaerobic digestion. The breakdown of organic material by microorganisms in the absence of oxygen. The process results in the release of methane gas as a waste product.

Anaerobic respiration. Respiration carried on by certain bacteria in the absence of oxygen. Methane, which can be used as fuel gas (it is the same as natural gas), may be a byproduct of the process.

Annuals. Plants which grow from seed, flower, set seed, and die, thus completing their life cycle in a single year.

Appropriate technology. Technology which seeks to increase the efficiency and productivity of hand labor without displacing workers. That is, it seeks to enable people to improve their well-being without disrupting the existing social and economic system.

Aquaculture. A propagation and/or rearing of any aquatic (water) organism in a more or less artificial system.

Aquifer. An underground layer of porous rock, sand, or other material that allows the movement of water between layers of nonporous rock or clay. Aquifers are frequently tapped for wells.

Artesian aquifer. An aquifer in which the groundwater is under such pressure that it comes to the surface when a well is drilled into it.

Artificial selection. Plant and animal breeders' practice of selecting individuals with the greatest expression of desired traits to be the parents of the next generation.

Asbestos fibers. Crystals of asbestos, a natural mineral, which have the form of minute fibers.

Assimilate. To incorporate into the natural working or functioning of the system as, for example, natural organic wastes are assimilated (broken down and incorporated) into the nutrient cycles of the ecosystem.

Atom. The fundamental unit of all elements.

Autotroph. Any organism that can synthesize all its organic substances from inorganic nutrients, using light or certain inorganic chemicals as a source of energy. Green plants are the principal autotrophs.

B

Background radiation. Radioactive radiation which comes from natural sources apart from any human activity. We are all exposed to such radiation.

Bacteria. Any of numerous kinds of microscopic organisms which exist as simple, single cells that multiply by simple division. Along with fungi, they comprise the decomposer component of ecosystems. A few species cause disease.

Bar screen. A set of iron bars about an inch apart used to screen debris out of waste water.

Base. Any compound that releases hydroxyl (OH^-) ions when dissolved in water. A solution that contains a surplus of OH^- ions.

Basic science. Science conducted purely for the purpose of gaining understanding as opposed to applied science where a particular application is the purpose.

Bedload. The load of coarse sediment, mostly coarse silt and sand, that is gradually moved along the bottom of a river bed by flowing water rather than being carried in suspension.

Benefit-cost. Used to describe an analysis and/or comparison of the value benefits in contrast to the costs of any particular action or project.

Benthic plants. Plants which grow underwater attached to or rooted in the bottom. For photosynthesis, they depend on light penetrating the water.

Best management practice. Farm management practices which serve best to reduce soil and nutrient runoff and subsequent pollution.

Bioaccumulation. The accumulation of higher and higher concentrations of potentially toxic chemicals in organisms. It occurs in the case of chemicals such as heavy metals and chlorinated hydrocarbons that may be absorbed or ingested with food but can neither be broken down nor excreted. Consequently, organisms act as strainers, accumulating increasing amounts. Through a food chain, organisms at higher trophic levels may accumulate concentrations as much as a millionfold higher than those present in the environment. Also called **biomagnification.**

Biocide. Applies to any pesticide or other chemical which is more or less toxic to many, if not all, kinds of living organisms.

Bioconversion (energy). The use of biomass as fuel. Burning materials such as wood, paper, and plant wastes directly to produce energy, or converting such materials into fuels such as alcohol and methane.

Biodegradable. Able to be consumed and broken down to natural substances such as carbon dioxide and water by biological organisms, particularly decomposers. Opposite: **nonbiodegradable.**

Biogas. The mixture of gases, about two thirds methane, one third carbon dioxide and small portions of foul smelling compounds, resulting from the anaerobic (without air) digestion of organic matter. The methane content enables biogas to be used as a fuel gas.

Biological control (pest control). Control of a pest population by introduction of predatory, parasitic, or disease-causing organisms.

Biological oxygen demand (BOD). A measure of water quality; the amount of dissolved oxygen that will be consumed by biological organisms in the process of decomposing organic material present. The greater the BOD, the lower the water quality.

Biological treatment (sewage treatment). (See **secondary treatment**)

Biomagnification. (See **bioaccumulation**)

Biomass. Mass of biological material. Usually the total mass of a particular group or category; for example, biomass of producers.

Biomass energy or **biomass fuels.** Energy, or fuels such as alcohol and methane, produced from current photosynthetic production of biological material. (See **bioconversion**)

Biomass pyramid. Refers to the structure that is obtained when the respective biomasses of producers, herbivores and carnivores in an ecosystem are compared, i.e., producers are found to have the largest biomass followed by herbivores and then carnivores.

Biome. A group of ecosystems that are related by having a similar type of vegetation governed by similar climatic conditions. Examples include prairies, deciduous forests, arctic tundra, deserts, and tropical rainforests.

Biosphere. The overall ecosystem of the earth. It is the sum total of all the biomes and smaller ecosystems, which are ultimately all interconnected and interdependent through global processes such as water and atmospheric cycles.

Biota. Refers to any and all living organisms and the ecosystems in which they exist.

Biotic. Living or derived from living things.

Biotic potential. The potential of a species for increasing its population and/or distribution. The biotic potential of every species is such that, given optimum conditions, its population will increase. Contrast environmental resistance.

Biotic structure. The organization of living organisms in an ecosystem into groups such as producers, consumers, detritus feeders and decomposers.

Birth control. Any means, natural or artificial, which may be used to reduce the number of live births.

BOD. (See **biological oxygen demand**)

Borrowed time. Time preceding a predictable and inevitable collapse or failure of a system during which nothing is done to avert the end result despite awareness of it.

Breeder reactor. A nuclear reactor that in the course of producing energy also converts nonfissionable uranium-238 into fissionable plutonium-239, which can be used as fuel. Hence, a reactor that produces as much nuclear fuel as it consumes, or more.

Broad-spectrum pesticides. Chemical pesticides that kill a wide range of pests. They also kill a wide range of nonpest and beneficial species; therefore, they may lead to environmental upsets and resurgences. Contrast narrow spectrum pesticides and biorational pesticides.

BTU (British Thermal Unit). A fundamental unit of energy in the English system. The amount of heat required to raise the temperature of one pound of water one degree Fahrenheit.

Buffer. A substance that will maintain the pH of a solution by reacting with the excess acid. Limestone is a natural buffer which helps to maintain water and soil at a pH near neutral.

Buffering capacity. Refers to the amount of acid that may be neutralized by a given amount of buffer.

C

Calorie. A fundamental unit of energy. The amount of heat required to raise the temperature of 1 gram of water 1 degree Celsius. All forms of energy can be converted to heat and measured in calories. Calories used in connection with food are kilocalories or "big" calories, the amount of heat required to raise the temperature of 1 liter of water 1 degree Celsius.

Capillary water. Water that clings in small pores, cracks, and spaces against the pull of gravity, like water held in a sponge.

Carbon dioxide effect. (See **greenhouse effect**)

Carbon monoxide. A highly poisonous gas, the molecules of which consist of a carbon atom with one oxygen attached. Contrast with nonpoisonous carbon dioxide, a natural gas in the atmosphere.

Carcinogenic. Having the property of causing cancer, at least in animals and by implication in humans.

Carnivore. An animal that feeds more or less exclusively on other animals.

Carrying capacity. The maximum population of a given animal or humans that an ecosystem can support without being degraded or destroyed in the long run. The carrying capacity may be exceeded, but not without lessening the system's ability to support life in the long run.

Catalyst. A substance that promotes a given chemical reaction without itself being consumed or changed by the reaction. Enzymes are catalysts for biological reactions. Also catalysts are used in some pollution control devices, e.g., the **catalytic converter.**

Catalytic converter. The device used by American automobile manufacturers to reduce the amount of carbon monoxide and hydrocarbons in the exhaust. The converter contains a catalyst that oxidizes these compounds to carbon dioxide and water as the exhaust passes through.

Cell (biological). The basic unit of life, the smallest unit that still maintains all the attributes of life. Many microscopic organisms consist of a single cell. Large organisms consist of trillions of specialized cells functioning together.

Cell respiration. The chemical process that occurs in all living cells wherein organic compounds are broken down to release energy required for life processes. Higher plants and animals require oxygen for the process as well and release carbon dioxide and water as waste products, but certain microorganisms do not require oxygen. (See **anaerobic respiration**)

Cellulose. The organic macromolecule that is the prime constituent of plant cell walls and hence the major molecule in wood, wood products, and cotton. It is composed of glucose molecules, but since it cannot be digested by humans its dietary value is only as fiber, bulk, or roughage.

Cell wall. A more or less rigid wall, composed mainly of cellulose, which surrounds plant cells and provides the supporting structure of plant tissues.

Center pivot irrigation. An irrigation system consisting of a spray arm several hundred meters long supported by wheels pivoting around a central well from which water is pumped.

Chain reaction (nuclear). Reaction wherein each atom that fissions (splits) causes one or more additional atoms to fission.

Channelization. The straightening and deepening of stream or river channels to speed water flow and reduce flooding.

Chemical barrier. In reference to genetic pest control, a chemical aspect of the plant that makes it resist pest attack.

Chemical energy. The potential energy that is contained in certain chemicals; most importantly, the energy contained in organic compounds such as food and fuels, which may be released through respiration or burning.

Chemosynthesis. The ability of some microorganisms to utilize the chemical energy contained in certain inorganic chemicals such as hydrogen sulfide for the production of organic material. Such organisms are producers.

Chlorinated hydrocarbons. Synthetic organic molecules in which one or more hydrogen atoms have been replaced by chlorine atoms. They are extremely hazardous compounds, because they tend to be nonbiodegradable, they tend to bioaccumulate, and many have been shown to be carcinogenic. Also called organochlorides.

Chlorination. The processes of adding chlorine to drinking water or sewage water in order to kill microorganisms that may cause disease.

Chlorofluorocarbons. Synthetic organic molecules that contain one or more of both chlorine and fluorine atoms.

Chlorophyll. The green pigment in plants, responsible for absorbing the light energy required for photosynthesis.

Clean Water Act of 1972. The cornerstone federal legislation addressing water pollution.

Clearcutting. Cutting every tree, leaving the area completely clear.

Climate. A general description of the average temperature and rainfall conditions of a region over the course of a year.

Climax ecosystem. The last stage in ecological succession. An ecosystem in which populations of all organisms are in balance with each other and existing abiotic factors.

Clustered development. The development pattern in which homes and other facilities are arranged in dense clusters on a relatively small portion of the land considered for development, allowing the rest of the land to remain open.

Cogeneration. The joint production of useful heat and electricity. For example, furnaces may be replaced with gas turbogenerators which produce electricity while the hot exhaust still serves as a heat source. An important avenue of conservation, it effectively avoids the waste of heat that normally occurs at centralized power plants.

Compaction. Packing down. *Soil:* Packing and pressing out air spaces present in the soil. Reduces soil aeration and infiltration and thus reduces the capacity of the soil to support plants. *Trash:* Packing down trash to reduce the space that it requires.

Composting. The process of letting organic wastes decompose in the presence of air. A nutrient-rich humus or compost results.

Composting toilet. A toilet that does not flush wastes away with water but deposits them in a chamber where they will compost. (See **composting**)

Compound. Any substance (gas, liquid, or solid) that is made up of two or more different kinds of atoms bonded together. Contrast **element.**

Condensation. The collecting together of molecules from the vapor state to form the liquid state, as, for example, water vapor condenses on a cold surface and forms droplets. Opposite: **evaporation.**

Confusion technique (pest control). Applying a quantity of sex attractant to an area so that males become confused and are unable to locate females. The actual quantities of pheromes applied are still very small because of their extreme potency.

Conservation. The management of a resource in such a way as to assure that it will continue to provide maximum benefit to humans over the long run. Conservation may include various degrees of use or protection, depending on what is necessary to maintain the resource over the long run. *Energy:* Saving energy. It not only entails cutting back on use of heating, air conditioning, lighting, transportation, and so on, but also entails increasing the efficiency of energy use. That is, developing and instigating means of doing the same jobs, e.g., transporting people, with less energy.

Conservation district. A region in which the state or county limits subdivision and development in order to preserve/conserve the natural environment and its values.

Consumers. In an ecosystem, those organisms that derive their energy from feeding on other organisms or their products.

Consumptive (water use). Use of water for such things as irrigation, where it does not remain available for potential purification and reuse.

Containment building (of nuclear power plant). Reinforced concrete building housing the nuclear reactor. Designed to contain an explosion should one occur.

Contour farming. The practice of cultivating land along the contours across rather than up and down slopes. In combination with strip cropping it reduces water erosion.

Contraceptive. Any device or drug that is designed to allow normal sexual intercourse but prevent unwanted pregnancies from occurring.

Control group. The group in an experiment that is the same as and is treated like the experimental group in every way except for the particular factor being tested. Only by comparison to a control group can one gain specific information concerning the effect of any test factor.

Controlled experiment. An experiment with adequate control groups. (See **control group**)

Control rods (nuclear power). Part of the core of the reactor, the rods of neutron-absorbing material that are inserted or removed as necessary to control the rate of nuclear fissioning.

Convection currents. Wind or water currents promoted by the fact that warming causes expansion, decreases density, and thus causes the warmer air or water to rise. Conversely, the sinking of cooler air or water.

Cooling tower. A massive tower designed to dissipate waste heat from a power plant (or other industrial process) into the atmosphere.

Cooperative energy use. Utilizing the waste heat from one process as the source of heat for another process which requires a lesser temperature.

Cosmetic damage (of fruits and vegetables). Damage to the surface that affects appearance but does not otherwise affect taste, nutritional quality, or storability.

Cosmetic spraying. Spraying of pesticides which is done to control pests which only damage the surface appearance.

Cost-benefit ratio or **benefit-cost ratio.** The value of the benefits to be gained from a project divided by the costs of the project. If the ratio is greater than 1, the project is economically justified; if less than 1, it is not economically justified.

Cost-effective. Pertaining to a project or procedure that produces economic returns or benefits that are significantly greater than the costs.

Covalent bond. A chemical bond between two atoms, formed by sharing a pair of electrons between the two atoms. Atoms of all organic compounds are joined by covalent bonds.

Criteria pollutants. Certain pollutants the level of which is used as a gage for the determination of air (or water) quality.

Critical level. The level of one or more pollutants above which severe damage begins to occur and below which few if any ill effects are noted.

Critical number. Refers to the minimum number of individuals of a given species that is required to maintain a healthy, viable population of the species. If a population falls below its critical number its extinction will almost certainly occur.

Crop rotation. The practice of alternating the crops grown on a piece of land. For example, corn one year, hay for two years, then back to corn.

Crude birth rate. Number of births per 1000 individuals per year.

Crude death rate. Number of deaths per 1000 individuals per year.

Crystallization. The joining together of molecules or ions from a liquid (or sometimes gaseous) state to form a solid state.

Cultivar. A cultivated variety of a plant species. All individuals of the cultivar are genetically highly uniform.

Cultural control (pest control). A change in the practice of growing, harvesting, storing, handling, or disposing of wastes that reduces the susceptibility or exposure to pests. For example, spraying the house with insecticides to kill flies is a chemical control; putting screens on the windows to keep flies out is a cultural control.

D

DDT (dichlorodiphenyltrichloroethane). The first and most widely used of the synthetic organic pesticides belonging to the chlorinated hydrocarbon class.

Debt crisis. Refers to the fact that many less-developed nations are so heavily in debt that they may not be able to meet their financial obligations, e.g., interest payments. Failing to meet such obligations could have severe economic impacts on the entire world.

Debt-for-nature swap. Trading portions of the foreign debt which Third World countries owe us for their saving portions of their natural environment, e.g., tropical rain forests.

Declining tax base. The loss of tax revenues that occurs as a result of affluent taxpayers and businesses leaving an area and a subsequent decline of property values. It has been especially severe in inner cities as a result of migration to suburbs and exurbs.

Decommissioning (of nuclear power plants). Refers to the inevitable need to take nuclear power plants out of service after 25–35 years because the effects of radiation will gradually make them inoperable.

Decomposers. Organisms the feeding action of which results in decay, rotting, or breakdown of complex to simpler components. The primary decomposers are fungi and bacteria.

Degrade. To lower the quality or usefulness of.

Demographic transition. The transition from a condition of high birth rate and high death rate through a period of declining death rate but continuing high birth rate finally to low birth rate and low death rate. This transition may result from economic development.

Density dependent (factors). In reference to population balance, factors such as parasitism which increase and decrease in intensity corresponding to population density.

Deoxyribonucleic acid. (See **DNA**)

Desertification. Declining productivity of land due to mismanagement. Overgrazing and overcultivation allowing erosion and salinization are the major causes.

Desertified. Land for which productivity has been significantly reduced (25 percent or more) due to human mismanagement. Erosion is the most common cause.

Detritus. The dead organic matter, such as fallen leaves, twigs, and other plant and animal wastes, that exists in any ecosystem.

Detritus feeders. Organisms such as termites, fungi, and bacteria that obtain their nutrients and energy mainly by feeding on dead organic matter.

Deuterium (^{2}H). A stable, naturally occurring isotope of hydrogen. It contains one neutron in addition to the single proton normally in the nucleus.

Developed countries. Industrialized countries, United States, Canada, Western Europe, Japan, Australia, and New Zealand, in which the gross national product exceeds $7000 per capita.

Developing countries. All free market countries in which the gross national product is less than $7000 per capita.

Development rights. Legal documents that grant permission to develop a given piece of property. They must be owned by the developer before development can occur. They can be bought and sold apart from the property itself.

Digest. The biological breakdown of organic material into simpler molecules.

Dioxin. A synthetic organic chemical of the chlorinated hydrocarbon class. It is one of the most toxic compounds known to humans, having many harmful effects, including induction of cancer and birth defects, even in extremely minute concentrations. It has become a widespread environmental pollutant because of the use of certain herbicides which contain dioxin as a contaminant.

Direct solar energy. (See **solar energy**)

Disinfection. The killing (as opposed to removal) of microorganisms in water or other media where they might otherwise pose a health threat. For example, chlorine is commonly used to disinfect water supplies.

Dissolved oxygen (DO). Oxygen gas molecules (O_2) dissolved in water. Fish and other aquatic organisms are dependent upon dissolved oxygen for respiration. Therefore concentration of dissolved oxygen is a measure of water quality.

Distillation. A process of purifying water or other liquids by boiling the liquid and recondensing the vapor. Contaminants remain behind in the boiler.

District heating. The heating of an entire community or

city area through circulating heat (e.g., steam) from a central source; particularly, utilizing waste heat from a power plant or from incineration of refuse.

Diversion (of water). Taking some or all of the flow of a natural waterway and carrying it to other places for uses such as municipal water supplies or irrigation.

DNA (deoxyribonucleic acid). The natural organic macromolecule that carries the genetic or hereditary information for virtually all organisms.

DO. (See **dissolved oxygen**)

Domestic solid wastes. Wastes that come from homes, offices, schools, and stores, as opposed to wastes that are generated from agricultural or industrial processes.

Dose (of exposure to a hazardous material). A consideration of the concentration times the length of exposure. For any given material or radiation, effects correspond to the product of these two factors.

Doubling time. The time it will take a population to double in size assuming the continuation of current fertility rate and no change in survival rate.

Drip irrigation. Supplying irrigation water through tubes which literally drip water onto the soil at the base of each plant.

E

Ecological pest management. Control of pest populations through understanding the various ecological factors that provide natural control and so far as possible utilizing these factors as opposed to using synthetic chemicals.

Ecological regard. Taking into consideration the environmental impact, direct and indirect, of one's actions and lifestyle. Adjusting actions and lifestyle to minimize the impact as much as possible.

Ecological upset. A drastic, relatively sudden change in an ecosystem, some species benefiting and/or becoming much more abundant while others are eliminated. Such upsets are most often caused by humans altering some biotic or abiotic factors.

Ecologists. Scientists who study ecology, i.e., the ways in which organisms interact with each other and their environment.

Ecology. The study of any and all aspects of how organisms interact with each other and/or their envionment.

Economic threshold (in reference to pest management). A certain level of pest damage which, to be reduced further, would require an application of pesticides that is more costly than the economic damage caused by the pests.

Ecosystem. A grouping of plants, animals, and other organisms interacting with each other and their environment in such a way as to perpetuate the grouping more or less indefinitely. Ecosystems have characteristic forms such as deserts, grasslands, tundra, deciduous forests, and tropical rain forests.

Ectoparasite. (See **parasites**)

Electrolysis (of water). The use of electrical energy to split water molecules into their constituent hydrogen and oxygen atoms. Hydrogen gas and oxygen gas result.

Electrons. Fundamental atomic particles that have a negative electrical charge but virtually no mass. They surround the nuclei of atoms and thus balance the positive charge of protons in the nucleus. A flow of electrons in a wire is synonymous with an electric current.

Element. A substance that is comprised of one and only one distinct kind of atom. Contrast **compound**.

Embrittlement. Becoming brittle. Pertains especially to the reactor vessel of nuclear power plants gradually becoming prone to breakage or snapping as a result of continuous bombardment by radiation. It is the prime factor forcing the decommissioning of nuclear power plants.

Emergency response teams. Teams of people, generally associated with police or fire departments, specially trained to handle accidents involving hazardous materials.

Endangered species. A species the total population of which is declining to relatively low levels, a trend which if continued will result in extinction.

Endangered Species Act. The federal legislation which mandates protection of species and their habitats that are determined to be in danger of extinction.

Endoparasite. (See **parasites**)

Energy. The ability to do work. Common forms of energy are light, heat, electricity, motion, and chemical bond energy inherent in compounds such as sugar, gasoline, and other fuels.

Enrichment. With reference to nuclear power, signifies the separation and concentration of uranium-235 so that, in suitable quantities, it will sustain a chain reaction.

Entomologist. A scientist who studies insects, their life cycles, physiology, behavior, and so on.

Environment. The combination of all things and factors external to the individual or population of organisms in question.

Environmental impact. Effects on the natural environment caused by human actions. Includes indirect effects through pollution, for example, as well as direct effects such as cutting down trees.

Environmental impact statement. A study of the probable environment impacts of a development project. The National Environmental Policy Act of 1968 (NEPA) requires such studies prior to proceeding with any project receiving federal funding.

Environmental movement. Refers to the upwelling of public awareness and citizen action regarding environmental issues that occurred during the 1960s.

Environmental resistance. The totality of factors such as adverse weather conditions, shortage of food or water, predators, and diseases which tend to cut back populations and keep them from growing or spreading. Contrast **biotic potential.**

Environmental science. The branch of science concerned with environmental issues.

EPA. United States Environmental Protection Agency. The federal agency responsible for control of all forms of pollution and other kinds of environmental degradation.

Epidemiological study. Determination of causes of disease conditions (e.g., lung cancer) through the study and comparison of large populations of people living in different locations or following different lifestyles and/or habits (e.g., smoking versus non-smoking).

Erosion. The process of soil particles being carried away by wind or water. Erosion moves the smaller soil particles first and hence degrades the soil to a more coarse, sandy, stony texture.

Estimated reserves. (See **reserves**)

Estuary. A bay open to the ocean at one end and receiving freshwater from a river at the other. Hence, mixing of fresh and salt water occurs (brackish).

Euphotic zone. The layer or depth of water through which there is adequate light penetration to support photosynthesis.

Eutrophication. Nutrient enrichment of a body of water which leads, in turn, to excessive growth of algae and then depletion of dissolved oxygen as dead algae is consumed by decomposers.

Evaporation. Molecules leaving the liquid state and entering the vapor or gaseous state as, for example, water evaporates to form water vapor.

Evolution. The theory that all species now on earth are descended from ancestral species through a process of gradual change brought about by natural selection.

Evolutionary succession. The succession of different species that have inhabited the earth at different geological periods, as revealed through the fossil record. The process of new species coming in through the process of speciation while other species pass into extinction.

Experimental group. The group in an experiment which receives the experimental treatment in contrast to the control group, used for comparison, which does not receive the treatment.

Exponential growth. The growth produced when the base population increases by a given *percentage* (as opposed to a given amount) each year. It is characterized by doubling again and again, each doubling occurring in the same period of time. It produces a J-shaped curve.

Extinction. The death of all individuals of a particular species. When this occurs, all the genes of that particular line are lost forever.

Exurban migration. Refers to the pronounced trend since World War II of people relocating homes and businesses from the central city and older suburbs to more outlying suburbs.

F

FAO. Food and Agriculture Organization of the United Nations.

Fecal. Refers to the solid excretory wastes of animals. Consists of undigested material passing through the gut and bacteria that have begun to feed on it.

Fecal coliform test. A test for the presence of *E. coli,* the bacterium that normally inhabits the gut of humans and other mammals. A positive test indicates sewage contam-

ination and the potential presence of disease-causing microorganisms carried by sewage.

Fermentation. A form of respiration carried on by yeast cells in the absence of oxygen. It involves a partial breakdown of glucose (sugar) which yields energy for the yeast and the release of alcohol as a byproduct.

Fertility rate. The number of live births occurring per 1000 women aged 15–44. Specifically referred to as the general fertility rate. (See also **total fertility rate**)

Fertilization. The application of fertilizer. (See **fertilizer**)

Fertilizer. Material applied to plants or soil to supply plant nutrients, most commonly nitrogen, phosphorus, and potassium but may include others. Organic fertilizer is natural organic material such as manure, which releases nutrients as it breaks down. Inorganic fertilizer, also called chemical fertilizer, is a mixture of one or more necessary nutrients in inorganic chemical form.

Field capacity. A measure of the maximum volume of water that a soil can hold by capillary action, i.e., against the pull of gravity.

Field scouts (regarding pest management). Persons trained to survey crop fields and determine whether applications of pesticides or other procedures are actually necessary to avert significant economic loss.

FIFRA. Federal Insecticide, Fungicide and Rodenticide Act.

Filtration. The passing of water (or other fluid) through a filter to remove certain impurities.

First basic principle of ecosystem function. Resources are supplied and wastes are disposed of by recycling all elements.

First-generation pesticides. Toxic inorganic chemicals that were first used to control insects, plant diseases and other pests. Included mostly compounds of arsenic and cyanide, and various heavy metals such as mercury and copper.

First Law of Thermodynamics. The fact based on irrefutable observations that energy is never created nor destroyed but may be converted from one form to another, e.g. electricity to light. (See also **Second Law of Thermodynamics**)

Fission. The splitting of a large atom into two atoms of lighter elements. When large atoms such as uranium or plutonium fission, tremendous amounts of energy are released.

Flare (verb, to flare). The burning of natural gas and/or other gaseous, flammable byproducts when there is no economical means available for utilizing it.

Flat-plate collector (solar energy). A solar collector that consists of a stationary, flat, black surface oriented perpendicular to the average sun angle. Heat absorbed by the surface is removed and transported by air or water (or other liquid) flowing over or through the surface.

Food chain. The transfer of energy and material through a series of organisms as each one is fed upon by the next.

Food web. The combination of all the feeding relationships that exist in an ecosystem.

Fossil fuels. Mainly crude oil, coal, and natural gas,

which are derived from prehistoric photosynthetic production of organic matter on earth.

Fuel assembly (nuclear power). The assembly of many rods containing the nuclear fuel, usually uranium, positioned close together. The chain reaction generated in the fuel assembly is controlled by rods of neutron-absorbing material between the fuel rods.

Fuel elements (nuclear power). The pellets of uranium or other fissionable material that are placed in tubes, which, with the control rods, form the core of the reactor.

Fungus, pl. **fungi.** Any of numerous species of molds, mushrooms, brackets, and other forms of nonphotosynthetic plants. They derive energy and nutrients by consuming other organic material. Along with bacteria they form the decomposer component of ecosystems.

Fusion. The joining together of two atoms to form a single atom of a heavier element. When light atoms such as hydrogen are fused, tremendous amounts of energy are released.

G

Gasohol. A blend of 90 percent gasoline and 10 percent alcohol, which can be substituted for straight gasoline. It serves to stretch gasoline supplies.

Gene pool. Refers to the sum total of all the genes which exist among all the individuals of a species.

Genes. The chemical units, passed from parents to offspring through the egg and sperm, that determine hereditary features. They include physical, physiological, and, to some extent, behavioral characteristics. Genes may be changed by mutations, and new combinations of genes in the offspring may lead to characteristics not observed in parents.

Genetic bank. The recognition that natural ecosystems with all their species serve as a tremendous repository of genes that is frequently drawn upon to improve domestic plants and animals and to develop new medicines, among other uses.

Genetic control (pest control). Selective breeding of the desired plant or animal to make it resistant to attack by pests. Also, attempting to introduce harmful genes—for example, those that cause sterility—into the pest populations.

Genetic engineering. The artificial transfer of specific genes from one organism to another.

Genetic makeup (of an individual). Refers to all the genes that an individual posesses and which determine all his/her/its inherited characteristics.

Genetics. The study of heredity and the processes by which inherited characteristics are passed from one generation to the next.

Gentrification. The trend seen in modern society of people moving into more or less isolated communities with others of similar economic, ethnic, and social backgrounds.

Geothermal. Refers to the naturally hot interior of the earth. The heat is maintained by naturally occurring nuclear reactions in the earth's interior.

Geothermal energy. Useful energy derived from the naturally hot interior of the earth.

Glucose. A simple sugar, the major product of photosynthesis. Serves as the basic building block for cellulose and starches and as the major "fuel" for the release of energy through respiration in both plants and animals.

Gravitational water. Water that is not held by capillary action in soil but percolates downward by the force of gravity.

Gray water. Wastewater, as from sinks and tubs, that does not contain human excrements. Such water can be reused for many purposes since it does not pose a health problem.

Greenhouse effect. An increase in the atmospheric temperature because of increasing amounts of carbon dioxide and certain other gases absorbing and trapping heat radiation that normally escapes from the earth.

Green revolution. Refers to the development and introduction of new varieties of wheat and rice (mainly) which increased yields per acre dramatically in some countries.

Grit chamber (of waste water treatment plants). Part of preliminary treatment, a swimming-pool-like tank where the velocity of the water is slowed enough to let sand and other gritty material settle.

Grit settling tank. Same as **grit chamber.**

Gross national product per capita. The total value of all goods and services exchanged in a year in a country divided by its population. A common indicator for the average level of development and standard of living for a country.

Groundwater. Water that has accumulated in the ground, completely filling and saturating all pores and spaces in rock and/or soil. Groundwater is free to move more or less readily. It is the reservoir for springs and wells, and is replenished by infiltration of surface water.

Groundwater remediation. The repurification of contaminated groundwater by any of a number of techniques.

Growth momentum (of the human population). Refers to the fact that the human population will continue to grow for some time even after the fertility rate is reduced to 2.0 because there is currently such an excessive number of children moving into the reproduction age brackets.

Gully erosion. Gullies, large or small, resulting from water erosion.

H

Habitat. The specific environment (woods, desert, swamp) in which an organism lives.

Half-life. The length of time it takes for half of an unstable isotope to decay. The length of time is the same regardless of the starting amount. Also refers to the amount of time it takes compounds to break down in the environment.

Halogenated hydrocarbon. Synthetic organic compound containing one or more atoms of the halogen group, which includes chlorine, fluorine, and bromine.

Hard water. Water that contains relatively large amounts of calcium and/or certain other minerals that cause soap to precipitate.

Heavy metals. Any of the high atomic weight metals such as lead, mercury, cadmium, and zinc. All may be serious pollutants in water or soil because they are toxic in relatively low concentrations and they tend to bioaccumulate.

Herbicide. A chemical used to kill or inhibit the growth of undesired plants.

Herbivore, adj., **herbivorous.** An organism such as rabbit or deer that feeds primarily on green plants, or plant products such as seeds or nuts. Synonym: **primary consumer.**

Herbivory. Refers to the feeding on plants that occurs in an ecosystem. The total feeding of all plant-eating organisms.

Heterotroph, adj., **heterotrophic.** Any organism that consumes organic matter as a source of energy.

Highway Trust Fund. The monies collected from the gasoline tax designated for construction of new highways.

Hormones, and **pheromones.** Natural chemical substances that control development, physiology, and/or behavior of an organism. Hormones are produced internally and affect only that individual. Pheromones are secreted externally and affect the behavior of other individuals of the same species, as, for example, attracting mates. Both hormones and pheromones are coming into use in pest control. (See **natural chemical control**)

Host. In feeding relationships, particularly parasitism, refers to the organism which is being fed upon, i.e., supporting the feeder.

Host-parasite relationship. The combination of a parasite and the organism it is feeding on.

Human ecosystem. The system involving humans and their agricultural plants and animals.

Humidity. The amount of water vapor in the air. (See also **relative humidity**)

Humus. A dark brown or black, soft, spongy residue of organic matter that remains after the bulk of dead leaves, wood, or other organic matter has decomposed. Humus does oxidize, but relatively slowly. It is extremely valuable in enhancing physical and chemical properties of soil.

Hunter-gatherers (referring to early humans). Humans surviving by hunting, and gathering seeds, nuts, berries, and other edible things from the natural environment.

Hybrid. A plant or animal resulting from a cross between two closely related species which do not normally cross.

Hybridization. Cross-mating between two more or less closely related species.

Hydrocarbon emissions. Exhaust of various hydrogen-carbon compounds due to incomplete combustion of fuel.

Hydrocarbons. *Chemistry:* Natural or synthetic organic substances that are composed mainly of carbon and hydrogen. Crude oil, fuels from crude oil, coal, animal fats, and vegetable oils are examples. *Pollution:* A wide variety of relatively small carbon-hydrogen molecules that result from incomplete burning of fuel, especially in internal combustion engines and that are emitted into the atmosphere

through the exhaust. They are a prime factor in the formation of photochemical smog.

Hydroelectric dam. A dam and associated reservoir used to produce electrical power by letting the high-pressure water behind the dam flow through and drive a turbogenerator.

Hydroelectric power. Electrical power that is produced from hydroelectric dams or, in some cases, natural waterfalls.

Hydrogen bonding. A weak attractive force which occurs between a hydrogen atom of one molecule and, usually, an oxygen atom of another molecule. It is responsible for holding water molecules together to produce the liquid and solid states.

Hydrogen ions. Hydrogen atoms that have lost their electrons. Chemical symbol, H^+.

Hydrological cycle. (See **water cycle**)

Hydroponics. The culture of plants without soil. The method uses water with the required nutrients in solution.

Hypothesis. An educated guess concerning the cause behind an observed phenomenon which is then subjected to experimental tests to prove its accuracy or inaccuracy.

I

Indirect products (of air pollution). Air pollutants that are not contained in emissions but which are formed as a result of such compounds undergoing various reactions in the atmosphere.

Indirect solar energy. See **solar energy.**

Infiltration. The process of water soaking into soil as opposed to its running off the surface.

Infiltration-runoff ratio. The ratio between the amount of water soaking into the soil and that running off the surface. (The ratio is given by dividing the first amount by the second.)

Infrared radiation. Radiation of somewhat longer wavelengths than red light, the longest wavelengths of the visible spectrum. Such radiation manifests itself as heat.

Infrastructure. The sewer and water systems, roadways, bridges, and other facilities which underlie the functioning of a city and which are owned, operated, and maintained by the city.

Inherently safe (nuclear power plants). Nuclear power plants of such a design that any mishap or accident will result in the nuclear reaction extinguishing itself, thus averting any potential for a meltdown and explosion regardless of the nature of the accident or the actions (or inactions) of the operators.

Inorganic. *Classical definition:* All things such as air, water, minerals, and metals that are neither living organisms nor products uniquely produced by living things. *Chemical definition:* All chemical compounds that do not contain carbon atoms as an integral part of their molecular structure. Contrast **organic.**

Insecticide. Any chemical used to kill insects.

Insurance spraying (in reference to use of pesticides). Spraying that is done when it is not really needed in the belief that it will insure against loss due to pests.

Integral urban house. A house in an urban setting that utilizes ecological principles, including water and materials conservation and recycling, solar energy, and intensive cultivation of food plants in so far as possible.

Integrated pest management (IPM). Two or more methods of pest control carefully integrated into an overall program designed to avoid economic loss from pests. The objective is to minimize the use of environmentally hazardous, synthetic chemicals. Such chemicals may be used in IPM, but only as a last resort to prevent significant economic losses.

Interim permits (referring to discharge of waste materials into air and water). Permission to discharge undesirable amounts of certain wastes into air and/or water until pollution control facilities can be installed. Such permits are granted to prevent forcing the closure of certain industrial plants.

Inversion. (See **temperature inversion**)

Ion. An atom or group of atoms that has lost or gained one or more electrons and consequently acquired a positive or negative charge. Ions are designated by + or − superscripts following the chemical symbol.

Ion-exchange capacity (soil). The ability of a soil to bind, hold, and release plant nutrients.

Ionic bond. The bond formed by the attraction between a positive and a negative ion.

IPM. (See **integrated pest management**)

Irrigation. Any method of artificially adding water to crops.

Isotope. A form of an element in which the atoms have more (or less) than the usual number of neutrons. Isotopes of a given element have identical chemical properties, but they differ in mass (weight) as a result of the additional (or lesser) neutrons. Many isotopes are unstable and give off radioactive radiation. (See **radioactive decay, radioactive emissions,** and **radioactive substances**)

J

Juvenile hormone. The insect hormone sufficient levels of which preserve the larval state. Pupation requires diminished levels; hence artificial applications of the hormone may block development.

K

Kerogen. A hydrocarbon material contained in oil shale that vaporizes when heated and can be recondensed into a material similar to crude oil.

Kinetic energy. The energy inherent in motion or movement including molecular movement (heat) and movement of waves, hence radiation including light.

L

Landfill. A site where wastes (municipal, industrial, or chemical) are disposed by burying them in the ground or placing them on the ground and covering them with earth.

Land subsidence. The phenomenon of land gradually sinking in elevation. It may result from removing groundwater or oil, which is frequently instrumental in supporting the overlying rock and soil.

Land trusts. Land that is purchased and held by various organizations specifically for the purpose of protecting its natural environment and biota that inhabit it.

Larva, pl. **larvae,** adj. **larval.** A free-living immature form that occurs in the life cycle of many organisms and that is structurally distinct from the adult. For example, caterpillars are the larval stage of moths and butterflies.

Law of conservation of matter. In chemical reactions, atoms are neither created, changed nor destroyed. They are only rearranged.

Law of limiting factors. Also known as Liebig's law of minimums. A system may be limited by the absence or minimum amount (in terms of that needed) of any required factor. (See **limiting factor**)

Leachate. The mixture of water and materials that are leaching.

Leaching. The process of materials in or on the soil gradually dissolving and being carried by water seeping through the soil. It may result in valuable nutrients being removed from the soil, or it may result in wastes buried in the soil being carried into and contaminating groundwater.

Legumes. The group of land plants that is virtually alone in its ability to fix nitrogen (see **nitrogen fixation**). The legume group includes such common plants as peas, beans, clovers, alfalfa, and locust trees but no major cereal grains.

Liebig's law of minimums. (See **law of limiting factors**)

Life cycle. The various stages of life, progressing from the adult of one generation to the adult of the next.

Limiting factor. A factor primarily responsible for limiting the growth and/or reproduction of an organism or a population. The limiting factor may be a physical factor such as temperature or light, a chemical factor such as shortage of a particular nutrient, or a biological factor such as a competing species. The limiting factor may differ at different times and places.

Limitists. Those who believe that our present economic-social system may be sharply curtailed in the future by resource shortages. Consequently, those who believe that changes in our system must be instigated now in order to accommodate future resource shortages.

Limits of tolerance. The extremes of any factor, e.g., temperature, which an organism or a population can tolerate and still survive and reproduce.

Lipids. A class of natural organic molecules that includes animal fats, vegetable oils, and phospholipids, the latter being an integral part of cellular membranes.

Litter. In an ecosystem, the natural cover of dead leaves, twigs, and other dead plant material. This natural litter is subject to rapid decomposition and recycling in the ecosystem, whereas human litter, such as bottles, cans, and plastics, is not.

Loam. A soil consisting of a mixture of about 40 percent sand, 40 percent silt, and 20 percent clay.

Longevity. The average lifespan of individuals of a given population.

M

Macromolecules. Very large, organic molecules such as proteins and nucleic acids which comprise the structural and functional parts of cells.

Malnutrition. Improper nutrition. It may consist of too much as well as too little of particular nutrients or calories, or it may be the lack of a proper balance among nutrients and calories.

Mariculture. The propagation and/or rearing of any marine (saltwater) organism in more or less artificial systems.

Matter. Anything that occupies space and has mass. Refers to any gas, liquid, or solid. Contrast with **energy.**

Maximum sustainable yield (renewable resources). The maximum amount that can be taken year after year without depleting the resource. It is the maximum rate of use or harvest that will be balanced by the regenerative capacity of the system—for example, the maximum rate of tree cutting that can be balanced by tree regrowth.

Meltdown (nuclear power). The event of a nuclear reactor getting out of control or losing its cooling water so that it melts from its own production of heat. The melted reactor would continue to produce heat and could melt its way out of the reactor vessel and eventually down into groundwater, where it would cause a violent eruption of steam which could spread radioactive materials over a wide area.

Metabolism. The sum of all the chemical reactions that occur in an organism.

Methane. A gas, CH_4. It is the primary constituent of natural gas. It is also produced as a byproduct of anaerobic respiration carried on by certain bacteria. Hence, it may be commercially produced by decomposing organic wastes.

Microbe. A term used to refer to any microscopic organism, primarily bacteria, viruses, and protozoans.

Microclimate. The actual conditions experienced by an organism in its particular location. Due to numerous factors such as shading, drainage and sheltering, the microclimate may be quite distinct from the overall climate.

Microfiltration or **reverse osmosis.** A process for purifying water, in which water is forced under very high pressure through a membrane that filters out ions and molecules in solution.

Microorganism. Any microscopic organism particularly bacteria, viruses, and protozoa.

Midnight dumping. The wanton illicit dumping of materials, particularly hazardous wastes, frequently under the cover of darkness.

Minamata disease. A "disease" named for a fishing village in Japan where an "epidemic" was first observed. Symptoms, which included spastic movements, mental retardation, coma, death, and crippling birth defects in the next generation, were found to be the result of mercury poisoning.

Mineral. Any hard, brittle, stonelike material that occurs naturally in the earth's crust. All consist of various combinations of positive and negative ions held together by ionic bonds. Pure minerals or crystals are one specific combination of elements. Common rocks are comprised of mixtures of two or more minerals. (See also **ore**)

Mineralization (soil science). The process of gradual oxidation of the organic matter (humus) present in soil which leaves just the gritty mineral component of the soil.

Mobilization. In soil science, the bringing into solution of normally insoluble minerals. Presents a particular problem when the elements of such minerals have toxic effects.

Molecule. A specific union of two or more atoms held together by covalent bonds. The smallest unit of a compound that still has the characteristics of that compound.

Monocropping. The practice of growing the same crop year after year on the same land. Contrast **crop rotation.**

Monoculture. The practice of growing a single crop over very wide areas, for example, thousands of square kilometers of wheat, and only wheat, grown in the Midwest.

Municipal solid waste. The entirety of refuse or trash generated by a residential and business community. The refuse that a municipality is responsible for collecting and disposing of, distinct from agricultural and industrial wastes.

Mutagenic. Causing mutations.

Mutation. A random change in one or more genes of an organism. Mutations may occur spontaneously in nature, but their number and degree are vastly increased by exposure to radiation and/or certain chemicals. Mutations generally result in a physical deformity and/or metabolic malfunction.

Mutualism. Refers to a close relationship between two organisms in which both organisms benefit from the relationship.

Mycelia. The thread-like feeding filaments of fungus.

Mycorrhizae. The mycelia of certain fungi that grow symbiotically with the roots of some plants and provide for additional nutrient uptake.

N

NASA. National Aeronautics and Space Administration.

Natural (adjective to describe a substance or factor). Occurring or produced as a normal part of nature apart from any activity or intervention of humans. Opposite of artificial, synthetic, human-made, or caused by humans.

Natural chemical control (pest control). The use of one or more natural chemicals such as hormones or pheromones to control a pest.

Natural control methods (pest control). Any of many techniques of controlling a pest population without resorting to the use of synthetic organic or inorganic chemicals. (See **biological control, genetic control, cultural control, hormones** and **pheromones**)

Natural enemies. All of the predators and/or parasites that may feed on a given organism are referred to as its natural enemies. Organisms used to control a specific pest through predation or parasitism.

Natural increase (in populations). The number of births minus the number of deaths. It does not consider immigration and emigration.

Natural laws. Derivations from our observations that matter, energy and certain other phenomena apparently always act (or react) according to certain "rules."

Natural organic compounds. (See **organic compounds**)

Natural rate of change (for a population). The percent of growth (or decline) of a population during a year. It is found by subtracting the crude death rate from the crude birth rate and changing the result to a percent. It does not include immigration or emigration.

Natural selection. The process whereby the natural factors of environmental resistance tend to eliminate those members of a population that are least well adapted to cope and thus, in effect, select those best adapted for survival and reproduction.

Natural services. Functions performed by natural ecosystems such as control of runoff and erosion, absorption of nutrients, and assimilation of air pollutants.

NEPA. National Environmental Policy Act.

Net yield (resources). The amount of a resource obtained minus the amount that must be expended in mining, refining, and delivery of the resource to consumers.

Neutron. A fundamental atomic particle found in the nuclei of atoms (except hydrogen) and having one unit of atomic mass but no electrical charge.

Niche (ecological). The total of all the relationships that bear on how an organism copes with both biotic and abiotic factors it faces.

Nitric acid (HNO_3). One of the acids in acid rain. Formed by reactions between nitrogen oxides and the water vapor in the atmosphere.

Nitrogen dioxide. (See **nitrogen oxides**)

Nitrogen fixation or **nitrogen fixing.** The process of chemically converting nitrogen gas (N_2) from the air into compounds such as nitrates (NO_3^-) or ammonia (NH_3) that can be used by plants in building amino acids and other nitrogen-containing organic molecules.

Nitric oxide. (See **nitrogen oxides**)

Nitrogen oxides (NOx). A group of nitrogen-oxygen compounds formed as a result of some of the nitrogen gas in air combining with oxygen during high-temperature combustion, they are a major category of air pollutants. Along with hydrocarbons, they are a primary factor in the production of ozone and other photochemical oxidants which are the most harmful components of photochemical smog. They also contribute to acid precipitation (see **nitric acid**).

Major nitrogen oxides are: nitric oxide, NO; nitrogen dioxide, NO_2; nitrogen tetroxide, N_2O_2;

Nitrogen tetroxide. (See **nitrogen oxides**)

NOAA. National Oceanic and Atmospheric Administration.

Nonbiodegradable. Not able to be consumed and/or broken down by biological organisms. Nonbiodegradable substances include plastics, aluminum, and many chemicals used in industry and agriculture. Particularly dangerous are nonbiodegradable chemicals that are also toxic and tend to accumulate in organisms. (See **biodegradable, bioaccumulation**)

Nonconsumptive (water use). Refers to the use of water for such purposes as washing and rinsing where the water, albeit polluted, remains available for further uses. With suitable purification, such water may be recycled indefinitely.

Nonpersistent (with respect to pesticides and other chemicals). Chemicals that break down readily to harmless compounds, as, for example, natural organic compounds break down to carbon dioxide and water.

Nonpoint sources (of pollution). Pollution from general runoff of sediments, fertilizer, pesticides, and other materials from farms and urban areas as opposed to pollution from specific discharges (contrast with **point sources** of pollution).

Nonrenewable resources. Resources such as ores of various metals, oil, and coal that exist as finite deposits in the earth's crust and that are not replenished by natural processes as they are mined.

No-till agriculture. The farming practice in which weeds are killed with chemicals (or other means) and seeds are planted and grown without resorting to plowing or cultivation. The practice is very effective in reducing soil erosion.

Nuclear power. Electrical power that is produced by using a nuclear reactor to boil water and produce steam which, in turn, drives a turbogenerator.

Nuclear Regulatory Commission. Independent governmental body charged with assuring and upholding safety standards for nuclear power plants.

Nuclear winter. A pronounced global cooling that would occur as a result of a large-scale nuclear conflict. It is based on theoretical projections concerning the amount of dust and smoke that would be ejected into the atmosphere and the resulting decrease in solar radiation.

Nucleic acids. The class of natural organic macromolecules that function in the storage and transfer of genetic information.

Nucleus. *Physics:* The central core of atoms, which is made up of neutrons and protons. Electrons surround the nucleus. *Biology:* The large body contained in most living cells that contains the genes or hereditary material, DNA.

Nutrient. *Plant:* An essential element in a particular ion or molecule that can be absorbed and used by the plant. For example, carbon, hydrogen, nitrogen, and phosphorus are essential elements; carbon dioxide, water, nitrate (NO_3^-), and phosphate (PO_4^{-3}) are respective nutrients.

Animal: Materials such as protein, vitamins, and minerals that are required for growth, maintenance, and repair of the body and also materials such as carbohydrates that are required for energy.

Nutrient cycles. The repeated pathway of particular nutrients or elements from the environment through one or more organisms back to the environment. Nutrient cycles include the carbon cycle, the nitrogen cycle, the phosphorus cycle, and so on.

Nutrient-holding capacity. The capacity of a soil to bind and hold nutrients (fertilizer) against their tendency to be leached from the soil.

O

Observations. Things or phenomena which are perceived through one or more of the basic five senses in their normal state. In addition, to be accepted as factual, the observations must be verifiable by others.

Oil field. The area in which exploitable oil is found.

Oil shale. A natural sedimentary rock that contains a material, kerogen, which can be extracted and refined into oil and oil products.

Oligotrophic. Refers to a lake the water of which is nutrient poor. Therefore, it will not support phytoplankton, but it will support submerged aquatic vegetation, which get nutrients from the bottom.

Omnivore. An animal that feeds more or less equally on both plant material and other animals.

OPEC. Organization of Petroleum Exporting Countries.

Optimum. The condition or amount of any factor or combination of factors that will produce the best result. For example, the amount of heat, light, moisture, nutrients, and so on that will produce the best growth. Either more or less than the optimum is not as good.

Optimum population (resources). The population that will provide the maximum sustainable yield. The yield is reduced at higher or lower populations.

Ore. A mineral rich in a particular element such as iron, aluminum, or copper which can be economically mined and refined to produce the desired metal. High-grade ore contains a relatively high percentage and low-grade ores contain a relatively low percentage of the desired element.

Organic. *Classical definition:* All living things and products that are uniquely produced by living things, such as wood, leather, and sugar. *Chemical definition:* All chemical compounds, natural or synthetic, that contain carbon atoms as an integral part of their structure. Contrast **inorganic.**

Organic compounds or molecules. Chemical compounds, the structure of which is based on bonded carbon atoms with hydrogen atoms attached. Natural organic compounds: those which are made by organisms and which comprise their bodies, the most common examples being cellulose making up wood, proteins, fats, and starches and sugars. Synthetic organic compounds: those which are man-made. Many of these are toxic.

Organic fertilizer. (See **fertilizer**)

Organic gardening or farming. Gardening or farming without the use of inorganic fertilizers, synthetic pesticides, or other human-made materials.

Organic molecules. (See **organic compounds**)

Organic phosphate. Phosphate (PO_4^{-3}) bonded to an organic molecule.

Organism. Any living thing—plant, animal, or microbe.

Osmosis. The phenomenon of water diffusing through a semipermeable membrane toward where there is more material in solution. (Where there is more material in solution there is a relatively lower concentration of water.) Has particular application regarding salinization of soils where plants are unable to grow because of osmotic water loss.

Outbreak (of pests). A population explosion of a particular pest. Often caused by an application of pesticides which destroys the pest's natural enemies.

Overgrazing. The phenomenon of animals grazing in greater numbers than the land can support in the long run. There may be a temporary economic gain in the short run, but the grassland (or other ecosystem) is destroyed and its ability to support life in the long run is vastly diminished.

Overreproduction. Refers to the observation that all species produce far more offspring, eggs, or seeds than would seem to be necessary to sustain their population.

Oxidation. A chemical reaction process that generally involves breakdown through combining with oxygen. Both burning and cellular respiration are examples of oxidation. In both cases, organic matter is combined with oxygen and broken down to carbon dioxide and water.

Ozone. A gas, O_3, which is a pollutant in the lower atmosphere, but necessary to screen out ultraviolet radiation in the upper atmosphere. May also be used for disinfecting water.

Ozone shield. The layer of ozone gas (O_3) in the upper atmosphere that screens out harmful ultraviolet radiation from the sun.

P

PANs (peroxyacetylnitrates). A group of compounds present in photochemical smog which are extremely toxic to plants and irritating to eyes, nose, and throat membranes of humans.

Parasites. Organisms (plant, animal, or microbial) that attach themselves to another organism, the host, and feed on it over a period of time without killing it immediately but usually doing harm to it. Commonly divided into *ectoparasites*, those that attach to the outside, and *endoparasites*, those that live inside their hosts.

Parent material. The rock material, the weathering and gradual breakdown of which is the source of the mineral portion of soil.

Particulates. A category of air pollutants consisting of visible particles appearing as smoke or mist.

Parts per million (ppm). A frequently used expression of concentration. It is the number of units of one substance present in a million units of another. For example, one gram of phosphate dissolved in one million grams (= 1 ton) of water would be a concentration of 1 ppm.

Passive solar heating system. A solar heating system that does not use pumps or blowers to transfer heated air or water. Instead natural convection currents are used or the interior of the building itself acts as the solar collector.

Pathogen, adj., pathogenic. An organism, usually a microbe, that is capable of causing disease.

PCBs (polychlorinated biphenyls). A group of very widely used industrial chemicals of the chlorinated hydrocarbon class. They have become very serious and widespread pollutants, contaminating most food chains on earth, because they are extremely resistant to breakdown and are subject to bioaccumulation. They are known to be carcinogenic.

Percolation. The process of water seeping through cracks and pores in soil or rock.

Perennials. Plants which survive and grow year after year as opposed to annuals which die and must be restarted from seed each year.

Permafrost. The ground of arctic regions which remains permanently frozen. Defines tundra since only small herbaceous plants can be sustained on the thin layer of soil that thaws each summer.

Persistent (with respect to pesticides or other chemicals). Nonbiodegradable and very resistant to breakdown by other means. Such chemicals therefore remain present in the environment more or less indefinitely.

Pesticide. A chemical used to kill pests. Pesticides are further categorized according to the pests they are designed to kill—for example, herbicides to kill plants, insecticides to kill insects, fungicides to kill fungi, and so on.

Pesticide treadmill. Refers to the fact that use of chemical pesticides simply creates vicious cycle of "needing more pesticides" to overcome developing resistance and secondary outbreaks caused by the pesticide applications.

Pest-loss insurance. Insurance that a grower can buy and which will pay in the event of loss of his crop due to pests.

Petrochemical. A chemical made from petroleum (crude oil) as a basic raw material. Petrochemicals include plastics, synthetic fibers, synthetic rubber, and most other synthetic organic chemicals.

pH. Scale used to designate the acidity or basicity (alkalinity) of solutions or soil. pH 7 is neutral; values decreasing from 7 indicate increasing acidity; values increasing from 7 indicate increasing basicity. Each unit from 7 indicates a tenfold increase over the preceding unit.

Pheromone. A chemical substance secreted by certain members which affect the behavior of other members of the population. The most common examples are sex attractants which female insects secrete to attract males.

Phosphate. An ion composed of a phosphorus atom with four oxygen atoms attached. $PO_4{}^{3-}$. It is an important plant nutrient. In natural waters it is frequently the limiting factor. Therefore, additions of phosphate to natural water are frequently responsible for algal blooms.

Photochemical oxidants. A major category of air pollu-

tants including ozone which are highly toxic and damaging especially to plants and forests. Formed as a result of interreactions between nitrogen oxides and hydrocarbons driven by sunlight.

Photochemical smog. The brownish haze that frequently forms on otherwise clear sunny days over large cities with significant amounts of automobile traffic. It results largely from sunlight-driven chemical reactions among nitrogen oxides and hydrocarbons, both of which come primarily from auto exhausts.

Photosynthesis. The chemical process carried on by green plants through which light energy is used to produce glucose from carbon dioxide and water, and oxygen is released as a byproduct.

Photosynthetic organism. An organism capable of carrying on photosynthesis. Opposite: nonphotosynthetic.

Photovoltaic cells. (See **solar cells**)

Physical barrier (regarding genetic control of pests). The presence of a genetic feature on a plant such as sticky hairs that physically blocks attack by pests.

Phytoplankton. Any of the many species of algae that consist of single cells or small groups of cells that live and grow freely suspended in the water near the surface. Given abundant nutrients, they may become so numerous as to give the water a green "pea soup" appearance and/or form a thick green scum over the surface.

Plankton, adj., planktonic. Any and all living things that are found freely suspended in the water and that are carried by currents as opposed to being able to swim against currents. It includes both plant (**phytoplankton**) and animal (**zooplankton**) forms.

Plant community. The array of plant species, including numbers, ages, distribution, etc., that occupies a given area.

Point sources (of pollution). Pollutants coming from specific points such as discharges from factory drains or outlets from sewage treatment plants. (see and contrast with **nonpoint sources**.)

Pollutant. A substance the presence of which is causing pollution.

Pollution. Contamination of air, water, or soil with undesirable amounts of material or heat. The material may be a natural substance, such as phosphate, in excessive quantities, or it may be very small quantities of a synthetic compound such as dioxin which is exceedingly toxic.

Polyculture. The growing of two or more species together in contrast to monoculture, the usual practice of growing only one species in a field.

Poor. Economically unable to afford adequate food and/or housing.

Population. A group within a single species, the individuals of which can and do freely interbreed. Breeding between populations of the same species is less common because of differences in location, culture, nationality, and so on.

Population density. The numbers of individuals per unit of area.

Population explosion. The exponential increase observed to occur in a population when or if conditions are such that a large percentage of the offspring are able to survive and reproduce in turn. Frequently leads, in turn, to overexploitation, upset, and eventually collapse of the ecosystem.

Population profile. A bar graph that shows the number of individuals at each age or in each five-year age group.

Population structure. Refers to the proportion of individuals in each age group. For example, is the population predominantly made up of young people, old people, or a more or less even distribution of young and old?

Potential energy. The ability to do work that is stored in some chemical or physical state. For example, gasoline is a form of potential energy; the ability to do work is stored in the chemical state and is released as the fuel is burned in an engine.

ppm. (See **parts per million**)

Practical availability (nonrenewable resources). The fraction of a material such as copper that can be obtained from the earth's crust given restraints of unacceptable pollution, expenditure of energy, and other costs associated with production from low-grade ores or remote sources.

Precipitation. Any form of moisture condensing in the air and depositing on the ground.

Predator. An animal that feeds upon another.

Predator-prey relationship. A feeding relationship existing between two kinds of animals. The predator is the animal feeding upon the prey. Such relationships are frequently instrumental in controlling populations of herbivores.

Pretreatment (of sewage). Passing sewage water through a coarse screen to remove large pieces of debris and slowing the water flow enough to let coarse grit and sand settle out.

Prey. In a feeding relationship, the prey is the animal that is killed and eaten by the other.

Primary consumer. An organism such as a rabbit or deer that feeds more or less exclusively on green plants or their products such as seeds and nuts. Synonym: **herbivore.**

Primary standard. The maximum tolerable level of a pollutant that is believed to protect human health with some margin of safety.

Primary succession. (See **succession**)

Primary treatment (of sewage). The process that follows pretreatment. It consists of passing the water very slowly through a large tank, which permits 30–50 percent of the organic material in the water to settle out. The settled material is raw sludge.

Principle of ecosystem stability. Ecosystem diversity provides ecosystem stability. Wide population swings, "booms and busts," are common in simple ecosystems, rare in diverse ecosystems.

Principle of population change. The population of a species is the result of a dynamic balance between its biotic potential and the environmental resistance it faces.

Producers. In an ecosystem, those organisms, mostly green plants, that use light energy to construct their organic constituents from inorganic compounds.

Production (of oil). The withdrawing of oil reserves.

Profligate growth. Growth characterized by extravagant and wasteful use of resources.

Property taxes. Taxes which the local government levies on privately owned properties, generally a few dollars per hundred of property value. This is the major source of revenue for local governments.

Protein. The class of organic macromolecules that is the major structural component of all animal tissues and that functions as enzymes in both plants and animals.

Proton. Fundamental atomic particle with a positive charge, found in the nuclei of atoms. The number of protons present equals the atomic number and is distinct for each element.

Protozoan, pl. **protozoa.** Any of a large group of microscopic organisms that consist of a single, relatively large complex cell or in some cases small groups of cells. All have some means of movement. Amoebae and paramecia are examples.

Proven reserves. Amounts of mineral resources (including oil, coal, and natural gas) which have been discovered and surveyed and are available for exploitation.

Puddling (soil science). The phenomenon of bare soil being beaten down and compacted by falling rain. It reduces soil aeration and infiltration, and this reduces the ability of the soil to support plants.

Pure science. Same as basic science. (See **basic science**)

R

Radioactive decay. The reduction of radioactivity that occurs as an unstable isotope (radioactive substance) gives off radiation and becomes stable.

Radioactive emissions. Any of various forms of radiation and/or particles that may be given off by unstable isotopes. Many such emissions have very high energy and may destroy biological tissues or cause mutations leading to cancer or birth defects.

Radioactive substances. Substances that are or that contain unstable isotopes and that consequently give off radioactive emissions. (See **isotope, radioactive emissions**)

Radioactive wastes. Waste materials which are or contain or are contaminated with radioactive substances. Many materials used in the nuclear industry become wastes because of their contamination with radioactive substances.

Rain shadow. The low-rainfall region that exists on the leeward (downwind) side of mountain ranges. It is the result of the mountain range causing the precipitation of moisture on the windward side.

Range of tolerance. The range of conditions within which an organism or population can survive and reproduce, for example, the range from the highest to lowest temperature which can be tolerated. Within the range of tolerance is the optimum or best condition.

Raw sewage. The sum total of all the waste water collected from homes and buildings before any treatment has occurred. Also called raw wastewater.

Raw sludge. The untreated organic matter that is removed from sewage water by letting it settle. It consists of organic particles from feces, garbage, paper, and bacteria.

Raw wastewater. (See **raw sewage**)

Reactor vessel. Steel-walled vessel that contains the nuclear reactor.

Recharge area. With reference to groundwater, the area over which infiltration and resupply of a given aquifer occurs.

Recruitment. With reference to populations, the maturation and entry of young into the adult breeding population.

Recycling. The practice of processing wastes and using them as raw material for new products as, for example, scrap iron is remelted and made into new iron products. Contrast **reuse.**

Relative humidity. The measure in percent of how much moisture is in the air compared to how much the air can hold at the given temperature.

Renewable energy. Energy sources, namely solar, wind, and geothermal, which will not be depleted by use.

Renewable resources. Biological resources such as trees that may be renewed by reproduction and regrowth. Conservation to prevent overcutting and protection of the environment are still required, however.

Replacement capacity (biological resources). The capacity of a system to recover to its original state after a harvest or other form of use.

Replacement fertility. The fertility rate that will just replace the reproducing population, given the average survivorship from birth through the preproductive period. Replacement fertility equals 2 divided by the survivorship from birth through reproductive years.

Reproductive rate. Refers to the rate at which offspring, eggs, or seed are produced.

Reserves. With reference to a mineral resource. The amount remaining in the earth that can be exploited using current technologies and at current prices. Usually given as *proven reserves,* those which have been positively identified, and *estimated reserves,* those which have not yet been discovered but which are estimated to exist.

Respiration, cellular respiration. The chemical process through which organic molecules are broken down in cells to release energy required for the functioning of the cell. In most organisms, respiration involves the breakdown of glucose in the presence of oxygen and the release of carbon dioxide and water as waste products. Certain microbes are able to derive sufficient energy from the partial breakdown of organic molecules which can occur in the absence of oxygen, and in this case other waste products result. (See **fermentation** and **anaerobic respiration**)

Resurgence (with respect to populations, especially of pests). The rapid comeback of a population after a severe dieoff, especially populations of pest insects, after being largely killed off by pesticides, returning to even higher levels than before the treatment.

Reuse. The practice of reusing items as opposed to throwing them away and producing new items, as, for example, bottles can be collected and refilled. Compare **recycling.**

Risk analysis or risk assessment. A study seeking to identify the exact nature and probability of risks involved in pursuing any particular project, or policy.

Rivulet erosion. A treelike pattern of numerous tiny (less than 15 cm) gullies caused by water erosion.

Runoff. That portion of precipitation which runs off the surface as opposed to soaking in.

S

Salinization. The process of soil becoming more and more salty until finally the salt prevents the growth of plants. It is caused by irrigation because salts brought in with the water remain in the soil as the water evaporates.

Saltwater intrusion, saltwater encroachment. The phenomenon of seawater moving back into aquifers or estuaries. It occurs when the normal outflow of fresh water is diverted or removed for use.

Sand. Mineral particles 0.2–2.0 mm in diameter.

Scientific fact. An observation regarding some object or phenomenon that has been repeated and confirmed by the scientific community.

Scientific method. The methodology by which scientific information is generated. Involves observations, formulating specific questions and hypotheses regarding the question's answer, then testing the hypotheses through experimentation.

SCS. U.S. Department of Agriculture—Soil Conservation Service.

Secondary consumer. An organism such as a fox or coyote that feeds more or less exclusively on other animals that feed on plants.

Secondary energy. A form of energy such as electricity that must be produced from another primary energy source such as burning coal or nuclear power.

Secondary pest outbreak. The phenomenon of a small, and therefore harmless, population of a plant-eating insect suddenly exploding to become a serious pest problem. Often caused by the elimination of competitors through pesticide use.

Secondary succession. (See **succession**)

Secondary treatment (of sewage). Also called biological treatment. A process that follows primary treatment. Any of a variety of systems that remove most of the remaining organic matter by enabling organisms to feed on it and oxidize it through their respiration. Trickling filters and activated sludge systems are the most commonly used methods.

Second basic principle of ecosystem function. Ecosystems run on solar energy, which is exceedingly abundant, nonpolluting, constant, and everlasting.

Second-generation pesticides. Synthetic organic compounds used to kill insects and other pests. Started with the use of DDT in the 1940s.

Second Law of Thermodynamics. The fact based on irrefutable observations that in every energy conversion, e.g., electricity to light, some of the energy is converted to heat and some heat always escapes from the system because it always moves toward a cooler place. Therefore, in every energy conversion, a portion of energy is lost. Therefore, since energy cannot be created (First Law) the functioning of any system requires an energy input.

Secured landfill. A landfill with suitable barriers, leachate drainage and monitoring systems such that there is deemed adequate security against hazardous wastes in the landfill contaminating groundwater.

Sediment. Soil particles, namely sand, silt, and clay, being carried by flowing water. The same material after it has been deposited. Because of different rates of settling, deposits are generally pure sand, silt, or clay.

Sediment trap. A device for trapping sediment and holding it on a development or mining site.

Sedimentation. The filling-in of lakes, reservoirs, stream channels, and so on with soil particles, mainly sand and silt. The soil particles come from erosion, which generally results from poor or inadequate soil conservation practices in connection with agriculture, mining, and/or development. Also called **siltation.**

Seep. Where groundwater seeps from the ground over some area as opposed to a spring which is the exit as a single point.

Selection pressure. With reference of evolution, an environmental factor which results in individuals with certain traits, which are not the norm for the population, surviving and reproducing more than the rest of the population. This results in a shift in the genetic makeup of the population. For example, the presence of insecticides provides a selection pressure toward increasing pesticide resistance in the pest population.

Selective breeding. The breeding of certain individuals because they bear certain traits and the exclusion from breeding of others.

Selective cutting. The cutting only of particular trees in a forest. Contrast **clearcutting.**

Sex attractant. A natural chemical substance (pheromone) secreted by the female of many insect species which serves to attract males for the function of mating. Sex attractants may be used in traps or for the **confusion technique** to aid in the control of insect pests.

Sexual reproduction. Reproduction involving segregation and recombination of chromosomes such that the offspring bear some combination of genetic traits from the parents. Contrast with asexual reproduction where all the offspring are exact genetic copies of the parent.

Sheet erosion. The loss of a more or less even layer of soil from the surface due to the impact and runoff from a rainstorm.

Shelter belts. Rows of trees around cultivated fields for the purpose of reducing wind erosion.

Silt. Soil particles between the size of sand particles and clay particles; namely, particles 0.002 to 0.2 mm in diameter.

Siltation. (See **sedimentation**)

Sinkhole. A large hole resulting from the collapse of an underground cavern.

Slash-and-burn agriculture. The practice, commonly exercised throughout tropical regions, of cutting and burning jungle vegetation to make room for agriculture. The process is highly destructive of soil humus and may lead to rapid degradation of soil.

Sludge cake. Treated sewage sludge which has been dewatered to make it a moist solid.

Sludge digesters. Large tanks in which raw sludge (removed from sewage) is treated through anaerobic digestion by bacteria.

Smog. (See **photochemical smog**)

Snow White syndrome. The human attribute that expects or demands cosmetic perfection in produce even though this is often obtained only by using huge quantities of environmentally hazardous pesticides.

Soft water. Water with little or no calcium, magnesium, or other ions in solution that will cause soap to precipitate (form a curd that makes a "ring around the bathtub").

Soil erosion. The loss of soil due to particles being carried away by wind and/or water.

Soil profile. A description of the different, naturally formed layers within a soil.

Soil structure. The phenomenon of soil particles (sand, silt, and clay) being loosely stuck together to form larger clumps and aggregates, generally with considerable air spaces in between. Structure enhances infiltration and aeration. It develops as a result of organisms feeding on organic matter in and on the soil.

Soil texture. The relative size of the mineral particles that make up the soil. Generally defined in terms of the sand, silt, and clay content.

Solar cells. Technically known as photovoltaic cells, devices which enable a direct conversion of light energy to electrical energy. Generally constructed from a "sandwich" of two thin layers of silica which are treated such that light striking one causes electrons to fall to the other.

Solar energy. Energy derived from the sun. Includes direct solar energy (the use of sunlight directly for heating and/or production of electricity) and indirect solar energy (the use of wind, which results from the solar heating of the atmosphere, and biological materials such as wood, which result from photosynthesis).

Solid waste. The total of materials that are discarded as "trash" and handled as solids, as opposed to those that are flushed down sewers and handled as liquids.

Solubility. The degree to which a substance will dissolve and enter into solution.

Solution. A mixture of molecules (or ions) of one material in another. Most commonly, molecules of air and/or ions of various minerals in water. For example, seawater contains salt in solution.

Specialization. With reference to evolution, the phenomenon of species becoming increasingly adapted to exploit one particular niche but, thereby, becoming less able to exploit other niches.

Speciation. The evolutionary process whereby populations of a single species separate and, through being exposed to differing forces of natural selection, gradually develop into distinct species.

Species. All the organisms (plant, animal, or microbe) of a single kind. The "single kind" is determined by similarity of appearance and/or by the fact that members do or potentially can mate together and produce fertile offspring. Physical, chemical, or behavioral differences block breeding between species.

Splash erosion. The destruction and compaction of soil structure that results from rainfall impacting bare soil.

Spores. Reproductive cells produced by fungi, some bacteria, and lower plants.

Springs. Natural exits of groundwater.

Standards (air or water quality). Set by the federal or state governments, they are the maximum levels of various pollutants that are to be legally tolerated. If levels go above the standards, various actions may be taken.

Standing biomass. That portion of the biomass (population) which is not available for consumption but which must be conserved to maintain the productive potential of the population.

Starvation. The failure to get enough calories to meet energy needs over a prolonged period of time. It results in a wasting away of body tissues until death occurs.

Sterile male technique (pest control). Saturating the area of infestation with sterile males of the pest species that have been artificially reared and sterilized by radiation. Matings between normal females and sterile males render the eggs infertile.

Stomas. Microscopic pores in leaves, mostly in the undersurface, that allow the passage of carbon dioxide and oxygen into and out of the leaf and that also permit the loss of water vapor from the leaf.

Stormwater. In cities, the water that results directly from rainfall, as opposed to municipal water and sewage water piped to and from homes, offices, and so on. The extensive hard surfacing in cities creates a vast amount of stormwater runoff, which presents a significant management problem.

Stormwater management. Policies and procedures for handling stormwater in acceptable ways to reduce the problems of flooding and erosion of stream banks.

Stormwater retention reservoirs. Reservoirs designed to hold stormwater temporarily and let it drain away slowly in order to reduce problems of flooding and stream bank erosion.

Stream bank erosion. The eating away of the stream bank by flowing water. Is greatly aggravated by flooding.

Strip cropping. The practice of growing crops in strips alternating with grass (hay) at right angles to prevailing winds or slopes in order to reduce erosion.

Strip mining. The mining procedure in which all the earth covering a desired material such as coal is stripped away with huge power shovels in order to facilitate removal of the desired material.

Submerged aquatic vegetation (SAV). Aquatic plants rooted in bottom sediments growing under water. Depend on light penetrating through the water for photosynthesis.

Subsistance farming. Farming that meets the food needs of the farmers but little more.

Subsoil. In a natural situation, the soil beneath topsoil. In contrast to topsoil, subsoil is compacted and has little or no humus or other organic material, living or dead. In many cases, topsoil has been lost or destroyed as a result of erosion or development and subsoil is at the surface.

Succession (ecological). The gradual, or sometimes rapid, change in species that occupy a given area, with some species invading and becoming more numerous while others decline in population and disappear. Succession is caused by a change in one or more abiotic or biotic factors benefitting some species but at the expense of others. *Primary succession.* The gradual establishment, through a series of stages, of a climax ecosystem in an area that has not been occupied before, e.g., a rock face. *Secondary succession.* The reestablishment, through a series of stages, of a climax ecosystem in an area from which it was previously cleared.

Sulfur dioxide (SO_2). A major air pollutant, this toxic gas is formed as a result of burning sulfur. The major sources are burning coal (coal-burning power plants) that contains some sulfur and refining metal ores (smelters) that contain sulfur.

Sulfuric acid (H_2SO_4). The major constituent of acid precipitation. Formed as a result of sulfur dioxide emissions reacting with water vapor in the atmosphere. (See also **sulfur dioxide**)

Superfund. The popular name for the Comprehensive Environmental Response, Compensation, and Liability Act of 1980. This Act provides the mechanism and funding for the cleanup of potentially dangerous hazardous waste sites.

Surface water. Includes all bodies of water, lakes, rivers, ponds, etc., that are on the surface of the earth in contrast to groundwater which lies below the surface.

Survival of the fittest. The concept that individuals best adapted to cope with both biotic and abiotic factors in their environment are the "fittest" and most likely to survive and reproduce.

Survival rate. (See **survivorship**)

Survivorship. The proportion of individuals in a specified group alive at the beginning of an interval, e.g., a five-year period, who survive to the end of the period.

Survivorship graph. A graph that shows the probability of an individual's surviving from birth to any particular age. Survivorship graphs are constructed (and may differ) for specific categories, such as sex, race, nationality, economic status, and so on.

Suspension. With reference to materials contained in or being carried by water, materials kept "afloat" only by the water's agitation and which settle as the water becomes quiet.

Sustainable. Able to continue indefinitely.

Sustainable agriculture. Agriculture that maintains the integrity of soil and water resources such that it can be continued indefinitely. Much of modern agriculture is depleting these resources and is, hence, not sustainable.

Sustainable development. Development that provides people with a better life without sacrificing/depleting resources or causing environmental impacts that will undercut future generations.

Symbiosis. The intimate living together or association of two kinds of organisms, especially in a way that provides a mutual benefit to both organisms.

Synergism, synergistic effect, synergistic interactions. The phenomenon in which two factors acting together have a very much greater effect than would be indicated by the sum of their effects separately—as, for example, modest doses of certain drugs in combination with modest doses of alcohol may be fatal.

Synfuels, synthetic fuels. Fuels similar or identical to those that come from crude oil and/or natural gas, produced from coal, oil shale, or tar sands.

Synthetic. Human-made as opposed to being derived from a natural source. For example, synthetic organic compounds are those produced in chemical laboratories, whereas natural organic molecules are those produced by organisms.

Synthetic organic compounds. (See **organic compounds**)

T

Taxonomy. The science of identification and classification of organisms according to evolutionary relationships.

Technological assessment. A study aimed at projecting the environmental and social impacts of introducing a new technology into a less-developed nation. Choices between alternative development projects may be made accordingly.

Technology. The application of scientific information to solve practical problems or achieve desired goals.

Temperature inversion. The weather phenomenon in which a layer of warm air overlies cooler air near the ground and prevents the rising and dispersion of air pollutants.

Teratogenic. Causing birth defects.

Terracing. The practice of grading sloping farmland into a series of steps and cultivating only the level portions in order to reduce erosion.

Territoriality. The behavioral characteristic seen in many animals, especially birds and mammalian carnivores, to mark and defend a given territory against other members of the same species.

Tertiary treatment. Third stage of sewage treatment, following primary and secondary, designed to remove one or more of the nutrients, usually nitrogen and/or phosphate, from the wastewater. Necessary to reduce the problem of eutrophication.

Test group. Synonym for experimental group. (See **experimental group**)

Texture. With reference to soils, the sizes of the particles, sand, silt, and/or clay, which make up the mineral portion.

Theory. A conceptual formulation which provides a rational explanation or framework for numerous related observations.

Thermal pollution. The addition of abnormal and undesirable amounts of heat to air or water. It is most significant with respect to discharging waste heat from electric generating plants, especially nuclear power plants, into bodies of water.

Third basic principle of ecosystem function. Large biomasses cannot be supported at the end of long food chains. Increasing population means moving closer on the food chain to the source of production.

Threatened species. A species the population of which is declining precipitously because of direct or indirect human impacts.

Threshold level. The maximum degree of exposure to a pollutant, drug, or other factors that can be tolerated with no ill effect. The threshold level will vary depending on the species, the sensitivity of the individual, the time of exposure, and the presence of other factors that may produce synergistic effects.

Topsoil. The surface layer of soil, which is rich in humus and other organic material, both living and dead. As a result of the activity of organisms living in the topsoil, it generally has a loose, crumbly structure as opposed to being a compact mass. In many cases, because of erosion, development, or mining activity, the topsoil layer may be absent.

Total fertility rate. The average number of children that would be born alive to each woman during her total reproductive years, assuming she follows the average fertility at each age.

Total watershed planning. A consideration of the entire watershed and planning development and other activities so as to maintain the overall water flow characteristics of the area.

Trace elements. Those essential elements that are needed in only very small or trace amounts.

Trade offs. The things that are given up to get or achieve something else that is valued.

Tragedy of the commons. The overuse or overharvesting and consequent depletion and/or destruction of a renewable resource that tends to occur when the resource is treated as a commons, that is, when it is open to be used or harvested by any and all with the means to do so.

Trait (genetic). Any physical or behavioral characteristic or talent that an individual is born with.

Transpiration. The loss of water vapor from plants. Evaporation of water from cells within the leaves and exiting through stomas.

Trapping technique (of pest control). The use of sex attractants to lure male insects into traps.

Treated (sewage) sludge. Solid organic material that has been removed from sewage and treated so that it is non-hazardous.

Trickling filters. Systems where wastewater trickles over rocks or a framework coated with actively feeding microorganisms. The feeding action of the organisms in a well-aerated environment results in the decomposition of organic matter. Used in secondary or biological treatment of sewage.

Tritium (^{3}H). An unstable isotope of hydrogen that contains two neutrons in addition to the usual single proton in the nucleus. It does not occur in significant amounts naturally but is human-made.

Trophic level. Feeding level with respect to the primary source of energy. Green plants are at the first trophic level, primary consumers at the second, secondary consumers at the third, and so on.

Turbid. Refers to water purity; means cloudy.

Turbine. A sophisticated "paddle wheel" driven at a very high speed by steam, water, or exhaust gases from combustion.

Turbogenerator. A turbine coupled to and driving an electric generator. Virtually all commercial electricity is produced by such devices. The turbine is driven by gas, steam, or water.

Turnover rate. The rate at which a population is replaced by the next generation.

U

Ultraviolet radiation. Radiation similar to light but with wavelengths slightly shorter than violet light and with more energy. The greater energy causes it to severely burn and otherwise damage biological tissues.

Upset (ecological). A vast shift in the relative size of one or more populations within an ecosystem. It may result in tremendous damage or even total destruction of the original ecosystem. It may be caused by any factor that changes the normal balances between species, such as introduction of a foreign species or eliminating a predator that was instrumental in controlling the population of an herbivore. Especially the phenomenon of an economically insignificant insect becoming a serious threat when its predators are killed off by pesticide treatment.

Urban decay. General deterioration of structures and facilities such as buildings and roadways, and also the decline in quality of services such as education, that has occurred in inner city areas as growth has been focused on suburbs and exurbs.

Urban sprawl. The rapid expansion of metropolitan areas through building housing developments and shopping centers farther and farther from urban centers and lacing them together with more and more major highways. Widespread development that has occurred without any overall land-use plan.

USDA. U.S. Department of Agriculture.

Utterly Dismal Theorem. The fact that supplying free food to food-poor nations undercuts the recipients' agricultural systems and worsens the problem it is intended to solve.

V

Verification. The checking of observations and/or experiments by others to determine their accuracy.

Vitamin. A specific organic molecule that is required by the body in small amounts but that cannot be made by the body and therefore must be present in the diet.

W

Waste inventories. Rosters of kinds and quantities of various waste products produced by various industries in a region. The information is compiled and distributed to facilitate recycling.

Water. Chemically, H_2O. All naturally occurring water contains additional materials in solution.

Water balance. Refers to the capacity and necessity of all organisms to control the relative volume of water inside vs. outside their cells.

Water cycle. The movement of water from points of evaporation through the atmosphere, through precipitation, and through or over the ground, returning to points of evaporation.

Water-holding capacity (soil). The property of a soil relating to its ability to hold water so that it will be available to plants.

Waterlogging (of soil). The total saturation of soil with water. Results in plant roots not being able to get air and dying as a result.

Watershed. The total land area that drains directly or indirectly into a particular stream or river. The watershed is generally named from the stream or river into which it drains.

Water table. The upper surface of groundwater. It rises and falls with the amount of groundwater.

Water vapor. Water molecules in the gaseous state.

Weathering. The gradual breakdown of rock into smaller and smaller particles, caused by natural, chemical, physical, and biological factors.

Wetlands. Areas that are constantly wet and are flooded at more or less regular intervals. Especially marshy areas along coasts that are regularly flooded by tide.

Wind farms. Arrays of numerous, modestly sized wind turbines for the purpose of producing electrical power.

Windrows. Piles of organic material extended into long rows to facilitate turning and aeration to enhance composting.

Wind turbines. "Wind mills" designed for the purpose of producing electrical power.

Work (physics). Any change in motion or state of matter. Any such change requires the expenditure of energy.

Workability. With reference to soils, the relative ease with which a soil can be cultivated.

Y

Yard wastes. Grass clippings and other organic wastes from lawn and garden maintenance.

Z

Zones of stress. Regions where a species finds conditions tolerable but suboptimal. Where a species survives but under stress.

Index

A

Abiotic factors, 35–39
  biotic interaction with, 41
  fire, 78-81
  as limiting factors, 37–39, 40
Abiotic structure, 25, 35–39
Abortion, 146, 150–51
Accidents:
  impact on population of, 129
  nuclear, potential for, 477–78
  of toxic chemical exposure, reducing, 289–92
Acid(s):
  common, 322
  defined, 321
  effects on trees, 306–7
  as indirect product of combustion, 309
  as major air pollutant, 300
  sources of and control strategies for, 313–14, 315
Acid deposition, 321, 324
Acid dews, 324
Acid fallout, 324
Acid-forming emissions:
  of coal-fired vs. nuclear power plant, 474
  reducing, 333, 334, 335
Acidity, 321–23
Acid leaching, stripmining and, 483
Acid precipitation, 109, 241, 313, 321–36, 351
  defined, 321
  depletion of buffering capacity and anticipation of future impacts, 329–32
  effects on ecosystems, 325–29
  effects on humans and their artifacts, 332
  extent and potency of, 324
  sources of, 324–25, 326, 483
  strategies for coping with, 332–36
  synergistic effect between ozone and, 329
  understanding acidity, 321–23
  urban sprawl and, 524
Activated sludge, 254
Activated sludge system, 254, 257
Active solar heating system, 492
Adaptation, 103
  development of ecosystems and, 100–102
  to different climates, 101
  evolutionary succession, 102–4
  examples of, 97
  gene pools and their change, 94–100
  genetic variation, 91–94, 107
  "heirloom vegetables," 109–10
  implications for humans, 106–9
  limits of change, 104–6
  natural selection and, 96–100
  to niches and habitats, 74–77
  to other species in ecosystem, 100
Adirondack Mountains, acid precipitation in, 326, 328

Adsorption, 309
Advance disposal fee, 436
Advanced treatment of sewage, 255–59
Aeration:
  to combat eutrophication, 233–34
  in composting, 260
  soil, 160
Aeration tank for sewage treatment, 254, 257
Aesthetic quality of produce, cosmetic spraying for, 387
Aesthetics, air pollution's effects on, 305–7, 308
Aesthetic values of natural biota, 401–2
African Elephant Conservation Coordinating Group, 408
African elephants, killing of, 408–9
Agrarian society, fertility rates in, 132
Agricultural districts, 538
Agricultural Extension Service, 168
Agriculture:
  air pollution's effect on, 301–5
  appropriate technology and, 144
  best management practices on farms, 242–44
  centralized development projects in developing countries and, 142
  crop rotation, 169, 383–84
  cultural control of pests affecting, 382–84
  development of, 41
  expansion of production, 137–39
  greenhouse effect on, 338–39, 340, 341
  heirloom vegetables, 109–10
  monoculture, 88
  narrow genetic base of modern, 110
  natural biota as underpinnings for, 397–99
  no-till, 182–84, 185, 186
  organic farming, 388, 389
  subsistence farming, 176
  sustainable, 142, 144
  techniques for preventing erosion, 182–88
  urban sprawl and loss of land, 525–26
  (see also Soil)
Air:
  interrelationships between water, minerals and, 47, 48
  molecules in, 45, 46
Air fresheners, 317
Air pollution, 294–318
  adverse effects of, 300–307
    on agriculture and forests, 301–7
    on human health, 301, 302, 303
    on materials and aesthetics, 305–7, 308
  background of problem, 296–300
  dilution and assimilation of pollutants, 296, 297
  disasters, 298
  factors determining level of, 296
  indoor, 317–18

in Los Angeles, 314, 316
major pollutants, 300
sources of pollutants and control strategies, 307–17, 428, 445, 504
standards, 299–300, 310–11
trends in various constituents of, 315
urban sprawl and, 524
Alaska, Valdez oil spill (1989), 459–60
Alaska pipeline, 457
Alcohol production, 505–7, 513
Aldrin, 366
Algae:
  harvesting, 234
  mutual relationship with fungus in lichens, 35
  planktonic, 233
Alkalinity (basicity), 160, 322, 323
Alleles, 91, 93, 98
Aluminum, mobilization of, by acid precipitation, 329, 332
Alzheimer's disease, 332
American Chestnut, demise of, 75–76
American Farmland Trust, 187
Ames, Bruce, 301
Anaerobic body of water, 250
Anaerobic digestion, 259–60
Anaerobic respiration, 259
Animals:
  in different biomes, 18–24
  limiting factors and distribution of, 39–41
  specialization to habitats and niches, 74–76
Apathy about sewage treatment, public, 268
Appropriate technology, 143–44
Aquaculture, 262–63
Aquatic ecosystems, acid precipitation and, 326–28
Aquatic plants, types of, 228–29
Aquatic species, protection of, 419–21
Aquifers, 198, 199, 204
Aral Sea, 204
ARCO Solar, 499
Arizona, Groundwater Management Act of 1980 in, 219
Artesian aquifer, 198, 199
Artifacts, acid precipitation's effect on, 332
Asbestos, 317–18
Aswan High Dam (Egypt), 508
Atmospheric circulation patterns, 338
Atmospheric turbidity (haze), 298, 299
Atomic number of element, 564
Atoms, 564
  bonding of, 566–70
  structure of, 564, 566
Audubon Society, 417
Automobiles:
  fuel efficiency, 316, 454, 460, 463
  fuel requirements for, 513–14
  pollution control devices, 313–16
  urban sprawl and, 519–520, 521
Avril, Prosper, 188

B

Background radiation, 474
Bacteria:
  cell respiration, 57
  coliform, 269, 270
  decomposition caused by, 28, 31
Bag house, 312
Bakersfield, California:
  nutrient-rich wastewater used for
    irrigation in, 262
  photovoltaic power plant near, 499,
    500
Balance (see Population balance)
Balanced diet, 56
Baltimore, Maryland, refuse-to-energy
  plant in, 438, 439
Baltimore Harbor, dredging of, 238–40
Bangladesh:
  deforestation of India and, 401
  family planning and health care
    programs in, 148–49
Bar screen, 253
Base, defined, 322
Basicity, 160, 322, 323
Basic science, 9
Bedload, 238, 239
Beland, Pierre, 289
Benefit-cost ratio, 347–48
  (see also Cost-benefit analysis of
    pollution control)
Benefits of pollution control, estimating,
  349–50
Benthic plants, 228, 229–230
Benzo-a-pyrene (BaP), 289
Best management practices, 242–44
Bhopal, India disaster, 287, 291
Bibliography, 547–60
Bioaccumulation:
  of DDT, 362, 364
  problem of, 275–77
Biocides, 356
  (see also Pesticides)
Bioconversion, 504–7
Biodegradation of toxic chemicals, 289
Biogas, 259, 268, 429, 505, 506
Biological oxygen demand (BOD), 250,
  251
Biological pest control (see Natural pest
  control)
Biological treatment of sewage, 254–55,
  256
Biomagnification, 275, 277, 362, 363, 364
Biomass, 17, 504
  decreasing, at higher trophic levels,
    62–64, 65
  standing, 62–63
Biomass energy, 504–7
Biomass pyramid, 33, 34
Biomes:
  climatic characteristics supporting, 18–
    24, 37–38
  coniferous forest, 21
  defined, 17
  desert, 20
  grassland, 19
  savanna, 23
  temperate forest, 18
  tropical rainforest, 24
  tundra, 22
  world distribution of major, 16
Biorational chemicals, 386
Biosphere, 25
  collapse, prescription for, 333–35
  defined, 15

Biosphere II, 66, 67
Biota, 25, 396–424
  assaults against, 404–12
  conservation of natural, 412–24
    aquatic species, 419–21
    exotic species and Endangered
      Species Act, 417–19
    game animals in U.S., 416–17
    general principles of, 412–16
    need to focus on ecosystems, 421–24
  values of natural, 397–404
Biotic factors:
  abiotic interactions with, 41
  as limiting factors, 39–41
Biotic potential, environmental
  resistance vs., 70–72, 87
Biotic structure, 25–35, 39–41
Birth control, 88, 124–25, 145–49
Birth defects, radiation and, 474
Birth rates, 125–27
  (see also Fertility)
Black lung disease, 301
Black market trade in endangered
  species, 407, 410
Blue Ridge Mountains, air pollution's
  effect on, 304
Bonding:
  of atoms, 566–70
  hydrogen, 192, 193
Borrowed time, 4
Bottles, returnable vs. nonreturnable,
  438–40
Boulding, Kenneth, 141
Brazil:
  agricultural use of land in, 140
  destruction of tropical rainforests in,
    124–25, 177, 242, 422
  family planning in, 124–25
  fermentation plants in, 505
Breastfeeding, 146
Breeder reactors, 479–80
Breeding, selective, 94–96
Broadleaf evergreen tree species, 38
Buffer, 329, 330
Buffering capacity, 334
  depletion of, 329–32
Buildings, acid precipitation's effect in,
  332
Bulkheading of shorelines, 241–42
Burning dumps, open, 428
Bush, George, 316
Businesses, loss of urban, 529–30, 531

C

Calcium carbonate as buffer, 329–31
California:
  acid precipitation in, 328
  solar energy in, 500
  wind farms in, 510
  (see also Los Angeles)
California Institute of Technology, 324
California Red Scale on citrus, 359, 360
Calories, 51, 56
Canada:
  acid precipitation in, 326, 328
  beluga whales of St. Lawrence River,
    288–89
  greenhouse effect on, 338
Cancer, 301, 474
Capillary water, 197, 199
Carbon, covalent bonding of, 566
Carbon cycle, 58–59

Carbon dioxide:
  additions, sources of, 336–37, 484
  atmospheric concentration of, 337
  emissions of coal-fired vs. nuclear
    power plant, 470–74
  heat-trapping effect of, 336
Carbon monoxide:
  as direct product of combustion, 307
  as major air pollutant, 300
  major sources and control strategies
    for, 311–13, 314, 315
Carcinogens, 275, 366
Career opportunities in industrial vs.
  agrarian societies, 132
Careers, environmental, 545
Carnivores, 28
Carrying capacity of ecosystem, 173, 415
Cars (see Automobiles)
Cash crops, 140
Catalyst, defined, 241
Catalytic converter, 313
Caterpillars, natural control of, 373, 374
Cell respiration, 55–56, 57, 160
Cellulose, 56
Center for Renewable Resources, 493–95,
  515
Center for the Biology of Natural
  Systems of Washington
  University, 389
Center-pivot irrigation, 177, 202, 203,
  206
Centralized projects for economic
  development, 142–43
Centre for Diarrhoeal Disease Research,
  Bangladesh (ICDDR,B), 148
Channelized stream, 217, 218
Channels, stream:
  blockage of, 212, 215
  sediments filling, 238–40
Chemical barriers against pests, 378
Chemical company profits, pesticide
  promotion for, 388
Chemical concepts, 564–70
Chemical control of pests, natural, 384–
  86
Chemical energy, 50, 53
Chemical fertilizers, 158
Chemical reactions, 564
  energy and, 570
Chemical treatments to combat
  eutrophication, 233, 234
Chemical wastes (see Toxic chemicals)
Chemosynthesis, 55
Chernobyl, accident at (1986), 468, 477–
  78
Chesapeake Bay, 255
  bulkheading of shorelines, 241–42
  eutrophication in, 227, 401
  production of three important fish
    species from, 422
  sources of polluting nutrients in, 241
Children:
  as economic asset or liability, 132
  pesticide's effects on, 366
  (see also Family planning; Fertility)
China, family planning in, 149–50
Chlordane, 365, 366
Chlorinated hydrocarbons, 259, 275–77,
  282, 289, 362
Chlorine, disinfection with, 259
Chlorine atoms:
  as catalyst for ozone breakdown, 341–
    42, 343
  entering stratosphere, source of, 342–
    43

Chlorofluorocarbons (CFCs):
  banning some uses of, 343, 344–45
  as greenhouse gas, 337
  low temperatures and release of
    chlorine from, 344
  ozone shield depletion by, 342
  sources of, 342
Chlorophyll, 25, 26
Cities:
  declining tax base and deterioration of
    public services, 528
  infrastructure of, 528, 533
  integrated urban house, 539, 541–42
  prior to World War II and mass
    production of cars, 519–21
  stemming urban blight and aiding
    redevelopment, 538–39
  sustainable, 532–42
  urban sprawl, 519–32
    environmental consequences of, 522–
      26
    origin of, 521–22
    slowing, 536–38
    social consequences of, 526–32
Classical approach in biocontrol, 377
Claton County, Georgia, nutrient-rich
    wastewater used for irrigation in,
    262
Clean Air Act of 1970, 278, 298, 310,
    311, 313
  amendments of 1977, 299
Clean Water Action Project, Inc., 220
Clean Water Act of 1972, 242–43, 268,
    278, 285, 286
  amendments of 1977, 292
  reauthorized in 1987, 243
Climate, 37–38
  adaptation to different, 101
  defined, 37
  greenhouse effect and, 338–41
  increasing food production and, 139
  tropical rainforests' contribution to,
    424
Climax, 86
Climax ecosystem, 86
Clivus Multrum, 267
Cloning, 94
Clustered development, 536, 537
Coal, 445, 449, 482, 483, 484
  formation of, 451–52
  major deposits in U.S., 482
Coal-fired power plants, 470
  acid precipitation and, 324–25, 326
Coal washing, 333
Cobb, Stephen, 409
Cogeneration, 460–61
Coliform bacteria, fecal coliform test to
    detect, 269, 270
Collapse of ecosystem, 87
Colloidal material in raw sewage, 253
Colonialism, legacy of, 132–33, 140
Colorado River, 202, 203
Combustion:
  air pollutants as products of, 307–10
  direct products of, 307–8
  fluidized bed, 333, 334
  indirect products of, 308–10
  indoor air pollution from, 317
Commercial interests in natural biota,
    402–4
Commoner, Barry, 452
Commons, tragedy of the, 410–12, 416
Commuters, 521, 522, 529
  (see also Automobiles)

Compaction, 160, 161
Competition among plants, 77–78
Competitive relationships, 35
Composting:
  of sludge, 260–62
  as solid waste solution, 436–37
Composting toilet, 267
Compounds, 570
  defined, 564
  inorganic, 47–48
  organic, 47, 48
Comprehensive Environmental Response,
    Compensation, and Liability Act of
    1980, 284–85
Condensation, 192–93, 195
Confusion technique of using sex
    attractant pheromones, 385
Congressional Office of Technology
    Assessment, 331
Coniferous forest biome, 21
Conservation:
  definition and objective of, 412–13
  of energy, 454, 458–62, 514
  of natural biota and ecosystems, 412–
    24
    aquatic species, 419–21
    exotic species and Endangered
      Species Act, 417–419
    game animals in U.S., 416–17
    general principles of, 412–16
    need to focus on ecosystems, 421–24
  preservation of natural enemies of
    pests, 373
  to prevent acid precipitation, 333
  of water, 209–10, 219–20
Conservation districts, 538
Conservation Foundation, 187
Conservation of Matter, Law of, 8, 48,
    64–65
Construction sites, sediments from, 235,
    236, 237
  control of, 244–46
Consumers, 27–28, 29, 30
  matter and energy changes in, 55–56,
    57
Consumers, wildlife extinction due to
    overuse by human, 407
Consumptive use of water, 202
Containment building in nuclear power
    plant, 470, 478
Containment of radioactive wastes, 475–
    77
Contour farming, 182
Contraception, 124–25
  availability of, 132
  family planning for developing
    countries, 145–49
  technology, 151
  trends in prevalence, 145
Controlled experiments, 7–8
Control rods, 470
Coral reefs, 35
Cosmetic spraying, 387
Cost–benefit analysis of pollution
    control, 347–51
Cost-effectiveness, 347
  of landfills, 430
  of pesticides, increasing, 362
  of pollution control, evaluation of
    optimum, 349, 350
  of recycling, 435–36
Covalent bonding, 566, 567, 568
Criteria pollutants, 310, 311
Critical level of pollution, 305

Critical number, 72
  habitat to support, 404
Croplands as source of sediments, 235
  (see also Agriculture)
Crop residues, destruction of, 383
Crop rotation, 169, 383–84
Crown fire, 80–81
Crude birth rate (CBR), 125–27
Crude death rate (CDR), 125–27
Cultivar, 398
Cultural controls against pests, 382–84
Cultural eutrophication, 232
Cummings, R., 141
Cummins, Joseph, 289
Customs, pest control via, 384
Cuyahoga River fire, 278
Cyclone precipitator, 312
Cyprus, solar energy usage in, 495
Czechoslovakia, acid precipitation's effect
    in, 331

D

Dams, hydroelectric, 507–8, 509
Darwin, Charles, 91, 99
Davis, California, energy conservation
    experience in, 496–97
DBCP (Dibromachloropane), 366
DDT, 275, 277, 357, 362–63, 364, 367
Death rates, 125–27
DeBach, Paul, 359, 373
Debris in raw sewage, 252–53
Debt crisis of developing countries, 123,
    143, 424
Debt-for-nature swaps, 189–190
Decentralized projects for economic
    development, 143–44
Deciduous forests and trees, 38, 77
  secondary succession and development
    of, 83–84, 85
Declining or eroding tax base, 528
Decommission of nuclear power plant,
    478–79
Decomposers, 28–31
  composting and, 260
  eutrophication and, 230
  matter and energy changes in, 56–57
Deep well injection, 278–79, 282
Deforestation, 174–77
  carbon dioxide increases due to, 337
  destruction of tropical rainforests in
    Brazil, 124–25, 177, 242, 422
  in developing countries, 176–77, 504,
    505
  from firewood cutting, 504
  in India, 401
  as source of sediments, 235
Degradation, environmental, 407, 420
Demographic transition, 129–31
  (see also Population)
Density dependence, 72
Deoxyribonucleic acid (DNA), 91–94, 95,
    96
Department of Energy, 468
Department of Interior, 417
Department of Transportation, 291
Desert biome, 20
Desertification, 173
  areas subject to, 181
  causes of, 173–78
  preventing, 182–89
  rehabilitation of desertified lands,
    185–86
  salinization, 178
  (see also Erosion)

Deserts of world, 181
Detergents containing phosphate, 241, 246–47
Detritus, 28
  reduction to humus, 165–67
Detritus-based food web, 32
Detritus feeders, 28–31
  composting and, 260
  matter and energy changes in, 56–57
  secondary sewage treatment and, 254, 256
Detroit River, 419
Deuterium, 481–82
Developed countries, 116
  crude birth rate and death rate in, 126
  distribution of wealth among, 118
  fertility rates, 131–33
  flood aid from, 140–42
  food production and population in, 137
  population profile for, 119, 120
Developing countries, 116
  agricultural policies in, 141–42
  crude birth rate and death rate in, 126
  debt crisis, 123, 143, 424
  debt-for-nature swaps, 189–90
  deforestation in, 176–77, 504, 505
  distribution of wealth among, 118
  dumping of toxic wastes in, 287
  economic development of, 142–44
  fertility rates, 131–33
    reducing, 144–51
  food production and population in, 137
  growth momentum in, 119
  pesticide exports to, 367
  population profile for, 119, 120
  sewage problems in, 267–68
Development:
  sediment control and, 244–46
  stream flow and effect of, 212, 213, 215, 216
  sustainable, 4, 42, 66, 124
  (see also Urban sprawl)
Diarrheal disease, 148
Dibromachloropane (DBCP), 366
Dichlorodiphenyltrichloroethane (DDT), 275, 277, 357, 362–63, 364, 367
Dieldrin, 366
Diet, balanced, 56
Digestion, 55–56
  anaerobic, 259–60
Dioxin, 275
Dip irrigation systems, 186
Direct burning of biomass fuel, 504–5
Direct solar energy, 489–503
  solar production of electricity, 495–501
  solar production of hydrogen, 501–3, 513
  space and water heating, 491–95, 513–14
Disease:
  epidemic, 127
  from untreated sewage, 250
Disinfectants, 317
Disinfection of wastewater, 259
Dissolved materials in raw sewage, 253
Dissolved oxygen:
  aeration to maintain deep, 233–34
  depletion of, 227, 250, 251
  in oligotrophic vs. eutrophic conditions, 229, 230
Distillation, 193, 195
Distribution of wealth, disparity of, 116–18
District heating, 461

Diversity, stability through, 73, 78, 88, 398
DNA, 91–94, 95, 96
Dodder, 28, 29
Dose, 296
Doubling time, 127
Drinking–water supplies, assuring safe, 284
Drip irrigation, 210
Droughts, 139, 186
Dry waste treatment system, 267
Dry wells and trenches, 217
Dumping, midnight, 280, 282, 287
Dust Bowl, 186–87
Duvalier, Jean–Claude, 188

E

E. coli (Escherichia coli), 270
Earth day (1969), 541
East Germany, acid precipitation's effect in, 331–32
Eckholm, Erik, 504
Ecological pest management, 354, 372, 390–91
  (see also Integrated pest management (IPM); Natural pest control)
Ecological regard, 112
Ecological upset, 86
Ecologists, defined, 25
Ecology, 4–5
  defined, 25
Economic categories of nations, 115–16
Economic development of developing countries, 142–44
Economic growth, real, 123
Economic incentives for reduced fertility, 149–50
Economics:
  of ivory trade and destruction of elephants, 408–9
  of natural pest controls, 389
  nuclear power and problems of, 478–79
  overuse of natural biota and, 406–7
  of pollution control, 347–51
  of recycling, 435–36
  tragedy of the commons, 410–12, 416
Economic threshold, 386
Economy, local, effect of food aid on, 140–42
Ecosystem(s), 15–42
  acid precipitation and, 325–29
  balance, 70–82, 100, 102
  carrying capacity of, 173, 415
  change, 82–87
  climax, 86
  collapse of, 87
  defined, 15
  development of, 100–102
  as dynamic system of interactions, 360
  evolution of, 103
  examples of, 15–25
  flow of energy through, 61–64
  function, principles of, 57–66
  human, 25, 41–42, 60, 64–66
  interaction between, 404
  need to focus on saving, 421–24
  reasons for different regions supporting different, 37–41
  soil, 162–69
  structure of, 25–37
    abiotic, 25, 35–39
    biotic, 25–35, 39–41
  transitional regions between, 17–25
  upset by nonpersistent pesticides, 368

Ecosystem stability, principle of, 73, 78, 88
EDB (ethylenedibromide), 363, 366–67
Education:
  family planning and, 146–49
  in industrial vs. agrarian societies, 132
Einstein, Albert, 468
Electrical power:
  energy usage for, 450
  production, 446–48
  running cars on electricity, 513, 514
  solar production of, 495–501
Electric power industry, arguments against emissions control, 334–35
Electrolysis, 503
Electromagnetic spectrum, 342
Electrons, 564
  bonding of atoms and, 566–68
Electrostatic precipitator, 312
Elements, 564
  organization of, living and nonliving systems, 45–48, 49
  table of, 565
Elephants, African, 408–9
Embrittlement, 478
Emergency response teams, 291–92
Emissions controls on cars, 313–16
Emmitsburg, Maryland, overland flow wastewater treatment system in, 265
Employment, loss of urban, 529–30, 531
Endangered species, 407, 410, 417
  poaching and, 407, 408, 409, 418
Endangered Species Act of 1966, 417–19
Endothia parasitica, 75
Enemies, natural (see Natural enemies)
Energy, 48–57
  categories of, 50–51
  changes in organisms, 54–57
  chemical, 50, 53
  chemical reactions and, 570
  common forms of, 50
  conservation of, 454, 458–62, 514
  defined, 50
  developing alternative sources of, 454–55, 457, 462, 514–16
  efficiency increasing, 460–61
  electrical power production, 446–49
  flow through ecosystems, 61–64
  geothermal, 510–12
  kinetic, 50, 51, 53, 55, 192, 193
  Laws of Thermodynamics, 8, 51–53, 61, 63
  matter vs., 50–53
  organic matter and, 53–54
  potential, 50, 53
  refuse-to-energy conversion, 438, 439
  units and equivalents, 563
Energy resources:
  coal, 445, 449, 482, 483, 484
  consumption of, 444, 449
  crisis, 453
    responses to, 454–57
    setting stage for another, 457–58
  crude oil, 451–62
    declining U.S. reserves and increasing importation of, 453, 456
    exploration, reserves and production, 452–53, 457
    preparing for future shortages, 458–62
  energy policy, 462–63, 514–16
  formation of fossil fuels, 451–52
  secondary energy, 446

sources and uses in U.S., 444–51
synthetic fuels, 483–86
U.S. government appropriations for
  research and development of, 461
urban sprawl and depletion of, 522–24
(see also Nuclear power; Solar energy)
Enrichment of uranium, 469–70
Environment:
  pesticide's effects on, 362–63
  population, poverty and, 123–24
  sustainable, 3, 5
Environmental Defense Fund, 241, 313,
  545
Environmental degradation, 407, 420
Environmental impact, 3, 112
Environmental movement, 3–4, 5
Environmental organizations:
  list of, 561–62
  membership in, 544–45
Environmental Pollution Panel of the
  President's Science Advisory
  Committee, 221
Environmental Protection Agency (EPA),
  3, 227, 282, 284–85, 287, 301, 318,
  366, 435
Environmental resistance, biotic
  potential and, 70–72, 87
Environmental science, 5, 10
Epidemic diseases, 127
Epidemiological studies, 8
Erie, Lake, eutrophication of, 232–33
Eroding tax base, 528
Erosion, 153–54, 170–89
  causes of, 173–78
  dimensions of problem, 178–82
  gully, 171, 172, 236
  natural biota preventing, 400
  preventing, 182–88
  process of, 170–73
  rate of, 179
  rivulet, 171, 172
  sheet, 171, 172
  as source of sediments, 236
  splash, 171, 172, 173
  from storm drain outlets, 212, 214
  stream bank, 212, 213, 236, 238
  stripmining and, 483
  (see also Sediments)
Estimated reserves, 452
Estuaries, overdraft of surface waters
  and effect on, 203
Ethiopia, soil erosion and famines in, 180
Ethylenedibromide (EDB), 363, 366–67
Euphotic zone, 228
Eutrophication, 227–35, 419
  advanced sewage treatment to prevent,
    255–59
  in Chesapeake Bay, 227, 401
  combating, 232–35
  in Everglades, 206–7
  of Lake Erie, 232–33
  natural vs. cultural, 232
  nutrient enrichment, upsetting balance
    by, 229–31
  process of, 230, 231
  sediments and nutrients and, 235–47
    controlling, 242–47
    loss of wetlands and bulkheading of
      shorelines, 241–42
    sources of, 235–41
  untreated sewage and, 250
  urban sprawl and, 525
Evaluation of information, 5–10
Evaportion, 159, 192, 193, 194

Evaporation-transpiration loop, 198
Evapotranspiration, 194, 197
Everglades, upsetting the, 206–7
Evolution, 91, 99, 100, 106–7, 108
Evolutionary succession, 102–4, 106
Exotic species, protection of, 417–19
Experiments, controlled, 7–8
Exploration, oil, 452, 457
Extinction, 72, 100, 103, 398
  criteria for, 104–6
  Endangered Species Act and, 417–19
  human causes of, 87–88, 407
Exurban migration:
  economic and ethnic segregation and,
    526–28
  effect on city tax base and public
    services, 528
  loss of urban businesses and
    employment due to, 529–30, 531
Exurbs, 526
Exxon Valdez oil spill, 459–60

F

Family planning, 119, 145–49
  in Brazil, 124–25
  costs of, 151
  promotion of, 150–51
Family Planning International
  Assistance, 151
Famines, 140–41, 180
Faraday, Michael, 446
Farallones Institute, 542
Farmers, urban sprawl and, 522
  (see also Agriculture)
Fecal coliform test, 269, 270
Fecal wastes, 56
Federal housing administration, 521
Federal Insecticide, Fungicide, and
  Rodenticide Act (FIFRA), 365–67
Feeding relationships, 30, 31–33
Fermentation, 57, 505
Fertility:
  education and, 146, 147
  factors in, 131–32
    income and, 146, 147
    reducing, 144–51
    replacement, 119–22
Fertility rates, 118–19
  in Brazil, 124
  disparity between developed and
    developing countries in, 131–33
  lower, as solution to population
    explosion, 129–33
  total, 118–19, 146, 147
  in U.S., 131
Fertilization, 91, 93
Fertilizers:
  increasing food production through,
    138
  inorganic (chemical) vs. organic, 158,
    167–69
  leaching of, 240
  management for pest control, 383
  treated sludge as, 260
Field scouts, advice on pests from, 388
FIFRA, 365–67
Figueiredo, Joao Baptista, 124
Fire, population balance through, 78–81
First–generation pesticides, 357
First Law of Thermodynamics, 51, 52
Fish, protection of endangered, 419
Fish harvests, 139, 140
Fission, 468, 470

Flared natural gas, 448
Flat–plate collectors, 491–92
Flooding, increased runoff and, 212
Flood irrigation, 177
Florida, groundwater contamination in,
  429
Florida Everglades, 206–7
Flowers and insects, relationship
  between, 35
Flow of energy through ecosystems, 61–
  64
Fluidized bed combustion, 333, 334
Fluorescent lights, 460
Food, roles of, 55–56
Food aid, 140–42
Food and Drug Administration, 8
Food-buying capacity, problem of, 139–40
Food chain, 31, 32, 33, 58
  acid precipitation and, 328
  biomagnification through, 275, 362,
    363, 364
  eating high on, 140
  among insects, 361
Food First (Lappe), 140
Food production, increasing, 137–39
Food Security Act of 1985, 187
Food web, 31–33, 41, 73
Ford, Henry, 519
Foreign species, introduction of, 407, 410
Forest biomes, 18, 21, 24
Forestry, natural biota as underpinnings
  for, 397–99
Forests:
  acid precipitation and, 328–29
  air pollution's effect on, 301–7
  deciduous, secondary succession and
    development of, 83–84, 85
Formaldehyde, 317
Fossil fuels, 59, 65
  air pollution from, 306
  carbon dioxide produced by, 337
  formation of, 451–52
  greenhouse effect and, 340, 341
Freon, 344
Fresh water, sources and uses of, 200–
  202
Fuel assembly, 470
Fuel efficiency, automobile, 316, 454,
  460, 463
Fuel elements or rods, 470
Fuel switching, 333
Fuller, Buckminster, 11
Fungi, 27
  mutual relationship with algae in
    lichens, 35
  parasitic, 28
  rotting or decomposition caused by,
    28–31
Fungicides, 356
  (see also Pesticides)
Fusion, 468
Fusion reactors, 480–83

G

Galapagos Islands, 102
Game animals in U.S., 416–17
Garbage (see Solid waste crisis)
Gardens:
  best management practices on, 242–44
  cultural control of pests affecting, 382–
    84
Gasohol, 505
Gasoline, 445

Gasoline tax, 524
Gas turbogenerator, 447
Gene pools, 94–100, 103
    change through natural selection, 96–100
    change through selective breeding, 94–96
    defined, 94
General Electric, 466
Genes, 91
Genetic bank, 399
Genetic control, 378, 379, 380
Genetic diversity, 105
Genetic engineering, 96, 138–39, 398
Genetic makeup, 91
Genetics, 91–94
Genetic variation, 91–94, 107
    geographic distribution and, 105
    of natural biota, 398
    in relative resistance to pesticides, 358
Gentrification of society, 527–28
Geothermal energy, 510–12
Glen Canyon Dam, 508
Glossary, 571–90
Glucose:
    digestion and, 55
    produced in photosynthesis, 54, 55
Government:
    city, eroding tax base of, 528
    purchase of recycled materials, 434–36
    spending on energy research and development, 466, 516
    subsidies, waste of water and, 210
    urban sprawl and, 522
Grain production, 137–38, 139, 140
Grapes of Wrath, The (Steinbeck), 187
Grassland biome, 19
Gravitational water, 197–98, 199
Gravity, law of, 8
Gray water, 267
    reuse of, 210, 211
Grazing:
    limitation of, 184–85
    overgrazing, 173–74, 235
Great Lakes, phosphate ban in, 247
Great Plains region, 204
Greenhouse effect, 109, 110, 139, 316, 336–39
    degree of warming and its probable effects, 338–41
    possible catastrophic effect of, 406
    strategies for coping with, 339
    urban sprawl and, 524
Greenhouse gases, 337
Green revolution, 139
Grit chamber or grit settling tank, 253, 254
Grit in raw sewage, 252–53
Gross national product (GNP):
    centralized projects for economic development and, 142
    per capita, 122
        nations of world grouped according to, 117
Ground fires, 80
Ground-level ozone, 300
Groundwater, 197–198, 199, 200, 210–11
    overdraft of, 204–9
Groundwater loop, 198
Groundwater pollution, 271–93
    sources of, 274, 366–67, 429
    by toxic chemicals, 273–93
        cleanup and management of, 284–93
        components of, 274–75

environmental contamination with, 277–84
    problem of bioaccumulation, 275–77
    synergistic effects, 277
Groundwater recharge, 204, 241
Groundwater remediation, 285, 286
Growth management initiatives, 538
Growth momentum, 119–22
Gruen, Victor, 533
Gully erosion, 171, 172, 236
Gypsy moths, natural control of, 373, 375

H

Habitat:
    defined, 74
    loss of, to alternate land uses, 404
        development, 416–17
        urban sprawl, 525
    specialization to, 74–77
Habitat for Humanity, 649
Haiti, 189–90
Half-life, 475, 476
Halogenated hydrocarbons, 275, 277, 362
Handler, Philip, 111
Hansen, James, 339
Hardin, Garrett, 410
Hawaiian Islands, 102, 104
Hayes, Dennis, 515
Hayes, Earl, 457, 463
Hazardous wastes (see Toxic chemicals)
Haze, 298, 299
Health, human:
    air pollution's effect on, 301, 302, 303
    pesticide's effects on, 362–63
Health care, family planning and, 146
Heart disease, air pollution and, 301
Heat radiation, 336
Heavy metals, 275–278
    as major air pollutant, 300
    as pesticides, 357
    reclamation and recycling of, 287–89
    sources of and control strategies for, 314, 315
Heirloom vegetables, 109
Heptachlor, 366
Herbicides, 88, 367
    to combat eutrophication, 233, 234
    no-till agriculture and, 182–84, 185, 186
Herbivores, 28, 77–78
Herbivory, 77
Hessian fly, 378, 379
Hidden costs:
    of nonreturnable bottles, 438
    of refuse disposal, 432
High-income countries, 115
    (see also Developed countries)
Highly developed countries (HDCs), 116
    (see also Developed countries)
Highway construction, sediment control on, 244
Highways, segmentation and desocialization caused by new, 529
Highway Trust Fund, 522
High-yielding plant varieties, 138–39
Homesteading, urban, 539
Hoover Dam, 508
Hormones, natural pest control with, 384–86
Horse, evolution of, 99, 100
Host, 28
Host-parasite relationship, 28, 72–74
Host species, genetic pest control in, 378

Hot dry rock, 511, 512
Hot water heating, solar, 513–514
Housing:
    aiding residents to own buildings, 539
    clustered vs. detached, 536, 537
    urban homesteading, 539
    (see also Cities)
Houston-Galveston Bay Region, 208
Hubbard Brook Forest, 174–75
Human ecosystem, 25, 41–42, 60
    adaptation vs. extinction, implications for, 106–9
    balance and change and, 87–88
    limiting factors and, 42
    principles of ecosystem function and, 64–66
Human excrements, polluting nutrients from, 241
    (see also Sewage; Solid waste crisis)
Humans, acid precipitation's effect in, 332
    (see also Health, human)
Humidity, 192
    relative, 195, 196
Humus, 165–67
    from sewage, 259, 260, 262, 266, 267
Hungry, plight of the, 139–140
Hunter-gatherers, 41
Hybrid, 95–96
Hybridization, 95
Hydrocarbons:
    chlorinated, 259, 275–77, 282, 289, 362
    as direct product of combustion, 307
    halogenated, 275, 277, 362
    as major air pollutant, 300
    reaction between nitric oxide and, 308–9
Hydroelectric dams, 507–8, 509
Hydroelectric power, 507–8, 509
Hydrogen, solar production of, 501–3, 513
Hydrogen bomb, 480
Hydrogen bonding, 192, 193
Hydrogen fusion, 480–81
Hydrogen ions, 321–22, 329
Hydrological cycle (see Water cycle)
Hydroponic culture, 167
Hydroponics, 162
Hydropower, 507–8, 509
Hydroturbogenerator, 447
Hydroxyl ions, 322
Hypotheses, 6–7
    defined, 6
    testing, 7–8

I

Ice, 192
Incineration of toxic chemicals, 289, 290
Income, fertility and, 146, 147
India:
    deforestation in, 401
    population in, 129
Indian pipe, 27, 28
Indirect products of combustion, 308–10
Indirect solar energy, 489, 503–10
    biomass energy or bioconversion, 504–7
    hydropower, 507–8, 509
    wind power, 508–10
Indoor air pollution, 317–18
Industrialized countries, 115
    reforestation in, 175–76
    (see also Developed countries)

Industrial parks, 530
Industrial Revolution, 445
Industrial society, fertility rates in, 132
Industrial wastes, contamination of
 treated sewage with, 268
Industry, energy usage for, 450
Infant and childhood mortality, 127
Infiltration, 159, 197–98, 199
 effects of decreased, 215, 217
Infiltration/runoff ratio, 197, 211, 217
Information, evaluation of, 5–10
Infrared radiation, 336
Infrastructure of cities, 528, 533
Injection wells, 278–79, 282
Inorganic chemicals, 27
Inorganic compounds, 47–48
Inorganic fertilizer, 158
Insecticides, 356
 (see also Pesticides)
Insects:
 food chain among, 361
 life cycle and its vulnerable points, 372
 (see also Pesticides)
Institute for Local Self-Reliance, 440
Instruments, role in science of, 9
Insurance, pest-loss, 388
Insurance spraying, 387–88
Integral urban house, 539, 541–42
Integrated pest management (IPM), 386–
 91
Integrated waste management, 440
Interim permits, 285
Internal-combustion engine, 445
International Child Health Foundation,
 148
International Monetary Fund, 143
International Planned Parenthood
 Federation, 151
International whaling commission, 419
Intestinal tapeworm, 29
Ion-exchange capacity of soil, 157–58
Ionic bonding, 566–70
Ions, 568, 569
IPM Practitioner, 390–91
Irrigation:
 center-pivot, 177, 202, 203, 206
 as consumptive use of water, 202
 dip irrigation systems, 186
 drip, 210
 flood, 177
 increasing food production through,
  138
 potential for water conservation in,
  209–10, 211
 projects in developing countries, 142–
  43
 salinization and desertification from,
  178
 using nutrient-rich water for, 262–66
Isotopes, 468, 475, 476, 564
 unstable, 474, 475
Israel, 186
 solar energy usage in, 495, 500
Ivory, elephant, 408–9

J

Japanese beetles, natural control of, 373,
 376
Junk foods, 56
Juvenile hormone, 384

K

Kenya, 123, 127, 409
Kepone, 366, 367
Kerogen, 483–84
Kinetic energy, 50, 51, 53, 55, 192, 193
Knipling, Edward, 380
Kormondy, Edward, 223
Kudzu, 78, 79

L

Labeling of produce for pesticides, 388,
 391
Lacey Act (1900), 417
Lakes:
 death of, acid precipitation and, 328
 gradual invasion by forest ecosystems,
  83, 84
Lamprey, 29
Land area devoted to grain production,
 137–38
Landfills, 428–30
 banning disposal of certain items or
  constituents in, 434
 for disposal of toxic chemicals, 278–83
  management problems, 280–82, 283
  methods, 278–80
  non-secure, 280–82
  secure, 280
 escalating costs of, 430
 groundwater pollution and, 274
 improving, 430, 431
 problems of, 429–30
Land subsidence, dropping water table
 and, 207–8
Land trusts, 538
Land use:
 effects caused by changing, 210–15
 loss of habitat to alternate, 404, 416–
  17, 525
 (see also Soil; Soil erosion)
Lappe, Frances Moore, 140
Laser fusion, 480, 482
Law, natural, 8
Lawns:
 best management practices on, 242–44
 cultural control of pests affecting, 382–
  84
Law of Conservation of Matter, 8, 48,
 64–65
Law of gravity, 8
Law of Limiting Factors, 36–37
Laws of Thermodynamics, 8, 51–53, 61,
 63
Leachate, 280
 collection system, 430
 generation by landfill, 429
Leaching, 157, 158, 199–200, 274
 acid, stripmining and, 483
 as nonpoint pollution, 242
 of nutrients, acid precipitation and,
  329, 330
 as source of polluting nutrients, 240
 (see also Groundwater pollution)
Lead:
 as major air pollutant, 300, 301
 mobilization of, by acid precipitation,
  332
 sources of and control strategies for,
  314, 315
Legumes, 61
Less-developed countries (see Developing
 countries)

Lichens, 33–35
Liebig, Justus von, 36–37
Liebig's Law of Minimums, 37
Lifestyle:
 human impact on environment and,
  111–12
 need to moderate, 462, 514, 519, 544
 solid waste disposal and changes in,
  440
 (see also Cities)
Limestone as buffer, 329–31
Lime to neutralize acid soils, 331, 332
Limiting factors, 36–37, 70
 abiotic, 37–39, 40
 biotic, 39–41
 environmental resistance, 70–71
 human ecosystem and, 42
Limits of change, 104–6
Limits of tolerance, 35–36
Lithium, 481–82
Living Systems, 496
Loam, 162, 167
Lobster trapping in New England, 411,
 413
Longevity, effect on population growth,
 129, 130
Long Island, phosphate ban on, 247
Long-term vs. short-term view of
 pollution control, 350–51
Los Angeles:
 air pollution in, 314, 316
 groundwater pollution in, 282
 pH of fog in, 324
 water conservation in, 219
Loss of habitat to alternate land uses,
 404
 development, 416–17
 urban sprawl, 525
Love Canal, 281–82
Low-income countries, 116
 (see also Developing countries)
Lung disease, air pollution and, 301, 303

M

Madagascar, 399
Madison (Wisconsin) Metropolitan
 Sewage District, 260
Malathion, 390
Malnutrition, 56, 139
Malthus, Thomas, 137
Manatees, 373, 376
Mandate, promoting recycling through,
 434–36
Mandatory recycling laws, 434
Mariculture in Philippines, 263, 264
Marketing of recycled materials, 431
Marsupials, 102
Maryland:
 phosphate ban in, 247
 sediment control regulation in, 245
 water conservation regulation in, 219–
  20
Maryland, University of, 227
Mason-Dixon Dairy, 505, 506
Materials, air pollution's effects on, 305–
 7, 308
Matter:
 changes in organisms, 54–57
 defined, 50
 energy vs., 50–53
Maximum sustainable yield, concept of,
 414–15, 416, 504
Measure, units of, 563

Meat consumption, 140
Medicine, natural biota as resource for, 399–400
Mediterranean fruit fly, U.S. program to protect against, 389
Meltdown, 470, 480
Mercury poisoning, 277
Merkel, W.H., 75
Metals, heavy (see Heavy metals)
Methane, 448, 449
  from biogas, 259
  as greenhouse gas, 337
  production, 429, 505, 506
Metric system, 563
Microbes, 17
Microclimate, 39
Middle-income countries, 115
  (see also Developing countries)
Midnight dumping, 280, 282, 287
Migration, 102
Milky spore disease, 376
Minamata disease, 277
Mineral(s):
  atoms of rock and soil, 45, 47
  defined, 393
  interrelationships between air, water and, 47, 48
  nutrients and nutrient-holding capacity, 157–58, 163, 164
  particles in soil, size of, 162–65
Mineralization, 167, 168
Mineral resources, 393–94
  (see also Energy resources)
Mining sites, sediments from, 235–36, 237
  control on, 244–46
Minorities, exurban migration and segregation of, 526–28
Mirex, 366
Mlay, Constantius, 409
Mobilization, 329, 332
Moderately developed countries, 115
  (see also Developing countries)
Molecules:
  in air and water, 45, 46, 47
  defined, 45
  organic, 47, 48
    energy and, 53
Monoculture, 88
Mono Lake, 203–4, 205
Montreal Accord (1986), 344
Monuments, acid precipitation's effect on, 332
Moss, primary succession by, 82–83
Mosses, 33
Muller, Paul, 357
Municipal solid waste, 427–28
Municipal water use, 200–202
  potential for conservation in, 210
Mutagenic effects of synthetic organic chemicals, 275
Mutations, 91–94, 105
Mutualism, 33–35, 56–57, 60
Mycelia, 28
Myocorrhizae, 167, 168

N

Nabhan, Gary Paul, 109
National Academy of Sciences, 5, 333, 335
National Audubon Society, 187
National Resources Defense Council, 369
Native Seeds/SEARCH, 109

National disasters, impact on population of, 129
Natural enemies, 73–74, 399
  experience in finding new, 377
  pest control by, 373–78
  suppression by pesticides, 359–60
Natural eutrophication, 232
Natural gas, 448, 449
  formation of, 451–52
  production, 429, 505, 506
Natural increase in population, 125
Natural law, 8
Natural organic compounds, 48
Natural pest control, 372–86
  control by natural enemies, 373–78
  cultural controls, 382–84
  economic control vs. total eradication, 386
  genetic control, 378, 379, 380
  integration of, 389–90
  natural chemical control, 384–86
  physical and chemical barriers, 41, 378, 380
  sterile male technique, 378–82, 390
Natural Resources Defense Council (NRDC), 366, 545
Natural selection, 102
  gene pool change through, 96–100
  natural pesticidal chemical barriers through, 378
  pest resistance development and, 358
Natural services:
  natural biota providing, 400–401
  urban sprawl and loss of, 525
Nature Conservancy, 538, 539, 540
Needwood, Lake, flood control project on, 220
Nematodes, 167, 168
Neurotoxins, 366
Neutrons, 564
New association approach to biocontrol, 377
New England:
  acid precipitation in, 328
  lobster trapping in, 411, 413
New York City, raw sludge dumping by, 268
Niche:
  defined, 74
  specialization to, 74–77, 98
NIMBY ("not in my back yard") attitude, 477
Nitric acid, 309
  in acid precipitation, 324
  as major air pollutant, 300
Nitric oxide, reaction between hydrocarbons and, 308–9
Nitrogen cycle, 60, 61
Nitrogen fixation, 61
Nitrogen oxide pollution, 300
  as direct product of combustion, 307
  major sources and control strategies for, 311–13, 314, 315
  need to reduce, 316–17
Nitrous oxide, as greenhouse gas, 337
Nonconsumptive use of water, 202
Non-feeding relationships, 33–35
Nonpersistent pesticides, 367–69
Nonpoint pollution, 242–43
Norplant (contraceptive), 151
North Africa, greenhouse effect on, 338
Norway, acid precipitation in, 328
No-till agriculture, 182–84, 185, 186
Nuclear power, 333, 450, 466–83
  economic problems with, 478–79

environmental advantages of, 470–74
how it works, 468–70
more advanced reactors, 479–83
radioactive materials and their hazard, 474–78
Nuclear power plants, 335, 468–69, 470, 472, 474, 478
Nuclear reactors, 469–70, 472, 479–83
Nuclear Regulatory Commission (NRC), 466, 478
Nuclear weapons facilities, 476, 477
Nutrient cycling, 57–61, 64, 65, 157
Nutrient enrichment, upsetting balance by, 229–31
Nutrient-holding capacity of soil, 157–58, 163, 164
Nutrient-poor (oligotrophic) condition, 229, 230
Nutrient-rich humus from treated sludge, 259, 260
Nutrient-rich wastewater, use in irrigation of, 262–66
Nutrients:
  controlling, 242–47
  leaching of, acid precipitation and, 329, 330
  recycling of, 157
  sources of polluting, 240–41
Nutrition, increasing food production and, 137–39
Nutritive role of food, 56

O

Observations, 5–6
Occupational exposure to toxic chemicals, reducing, 289–92
Occupational Safety and Health Administration (OSHA), 290
Ocean, warming of, 339
Ogallala aquifer, 204
Oil and oil-based fuels, 449
  background on use of, 445–46
  declining reserves of crude, 451–63
  exploration, reserves and production of, 452–53, 457
  formation of, 451–52
  future shortages of, preparing for, 458–62
  urban sprawl and depletion of, 522–24
Oil crisis of 1970s, 453–54
  response to, 454–57
  setting stage for another, 457–58
Oil field, 452
Oil glut, 453, 457
Oil shale, 483–84, 485
Oil spill, Valdez (1989), 459–60
Oligotrophic conditions, 229, 230
Olkowski, William, 541
Omnivores, 28
Ontario, Canada, acid precipitation in, 326, 328
OPEC (Organization of Petroleum Exporting Countries), 453, 455–58
Open burning dumps, 428
Optimal range, 35–36
Optimum, 35–36
Orbitals, 566
Ores, 393
Organic chemicals, 27
  (see also Synthetic organic chemicals)
Organic compounds, volatile, as major air pollutants, 300

Organic farming, 388, 389
Organic fertilizers, 158, 167–69
Organic gardening, 390–91
Organic material in raw sewage, 253
Organic molecules, 47, 48, 53
Organic phosphate, 60
Organisms:
  categories of, 25–31
  common feeding relationships among, 30
  matter and energy changes in, 54–57
Organization of elements in living and nonliving systems, 45–48, 49
Organophosphate pesticide, 368
Osmosis, 160
Osmotic pressure, 160–62
Outbreak, 77
Overcultivation, 173
Overdraft of water resources, 202–9
  groundwater, 204–9
  surface waters, 202–4
Overfishing, 419, 420
Overgrazing, 173–74, 235
Overland flow wastewater treatment system, 265
Overuse of natural biota, 406–7
Oxidation, 48, 50
Oxygen:
  aeration and, 160
  biological oxygen demand, 250, 251
  in oligotrophic vs. eutrophic body of water, 229, 230
  (see also Dissolved oxygen)
Ozone:
  as disinfecting agent, 259
  effect on agriculture and forests, 303, 304, 306–7
  increase in, 313
  as major air pollutant, 300, 301
  major sources and control strategies for, 311–13, 314, 315
  synergistic interaction between acid precipitation and, 329
Ozone shield, 339–45
  depletion, 109, 343
  formation and breakdown of, 341–42, 343
  nature and importance of, 339–41
  ozone "hole," 343
  source of chlorine atoms entering stratosphere, 342–43

P

Pacific Gas and Electric Company, 499
PANs, 309
Parasites, 28, 29
  balance in host-parasite relationships, 72–74
  (see also Natural enemies)
Parasitic wasps, 373, 374
Parent material, 157
Particulates:
  as direct product of combustion, 307
  as major air pollutant, 300
  major sources and control strategies for, 311, 312
Passive solar heating of space and hot water, 492–95, 513–14
Pasteur, Louis, 127, 251
Pathogens, 250, 254
PBB poisoning, 291
PCBs, 275
Pennsylvania, phosphate ban in, 247

Percolation of water, 197
Perennial crops, 184
Permafrost, 38
Peroxyacetyl nitrates (PANs), 309
Pest control (see Integrated pest management (IPM); Natural pest control; Pesticides)
Pesticides, 88, 354, 355–69
  categorization of, 356
  development and apparent virtues of, 357
  export of, 367
  Federal Insecticide, Fungicide, and Rodenticide Act (FIFRA), 365–67
  first-generation, 357
  increasing food production through, 138
  nonpersistent, 367–69
  problems stemming from, 274, 357–63, 525
  produce labels describing use of, 388, 391
  prudent use of, 390
  second-generation, 357
  socioeconomic factors supporting overuse of, 386–89
  treadmill, 363–65
Pest-loss insurance, 388
Pest resistance, development of, 358–59
Pests:
  defined, 353
  principal categories of, 353–54
pH, 160, 323
  buffering capacity and, 329–32
  importance of, 326–27
  of normal rainfall, 324
  scale, 323
  (see also Acid precipitation)
Pheromones, natural chemical control through, 384–86
Philippines, mariculture in, 263, 264
Phosphate:
  detergents containing, 241, 246–47
  organic, 60
  removal in sewage treatment, 255
Phosphorous cycle, 59–60
Photochemical oxidants, 308
  effect on agriculture and forests of, 303
  as major air pollutant, 300
  major sources and control strategies for, 311–13, 314, 315
Photochemical smog, 296–98, 313, 316
Photosynthesis, 25–27, 54–55, 158, 503, 570
  carbon cycle and, 58, 59
  defined, 25
  glucose produced in, 54, 55
  reverse of, 55
  sediments in streams and rivers preventing, 238
Photovoltaic cells, 497–500, 533
Physical barriers against pests, 41, 378, 380
Phytoplankton:
  in Antarctic, ozone "hole" and threat to, 344
  pollution of Chesapeake Bay by, 227
  turbid water and, 228–29
Pippard, Leone, 288
Planktonic algae, 233
Plant community, 17
Plant-herbivore balance, 77–78
Plants:
  aquatic, 228–29

competition among, 77–78
soil and, 157–70
  growing plants without soil, 162
  mineral nutrients and nutrient-holding capacity, 157–58, 163, 164
  mutual interdependence of, 169–70
  oxygen and aeration, 160
  relative acidity (pH), 160
  salt and osmotic pressure, 160–62
  soil ecosystem, 162–69
  water and water-holding capacity, 158–59
  specialization to habitats and niches, 76–77
Plastics reprocessed into synthetic lumber, 432, 434
Plutonium, 475
Poaching, 407, 408, 409, 418
Point sources, 242–43
Political involvement, individual, 544–45
Pollution, 221–23
  as assault against natural biota, 404–6
  categories of, 221–23
  control, cost-benefit analysis of, 347–51
  critical level of, 305
  defined, 221, 222
  dose of, 296
  economic impact of, 403
  from electrical power production, 448
  increased runoff and increase in, 212–15
  natural inputs of, 296
  need for regulation of, 351
  nonpoint vs. point sources, 242–43
  by sediments and nutrients, 235–47
  thermal, 482
  threshold level for, 296, 348
  Valdez oil spill (1989), 459–60
  (see also Air pollution; Eutrophication; Water pollution)
Pollution Standards Index (PSI), 311
Polycropping techniques, 144
Polyculture, 384
Ponds, solar, 500, 502
Popst, Luther, 532
Population:
  disparity in growth rates, 118–19
  doubling time, 127
  efforts to improve lives of people, 137–44
  environment and, 123–24
  growth momentum, 119–22
  impact on tropical forests, 124–25
  poverty and, 118–24
  rate of growth, 115, 137
  reduction of fertility, 144–52
  in rich and poor nations, 115–18
Population balance, 70–82
  biotic potential vs. environmental resistance, 70–72, 87
  case study of American Chestnut, 75–76
  density dependence and critical numbers, 72
  mechanisms of, 72–82
Population balances, 404
Population change, principle of, 70–72
Population density, 37, 72
Population explosion, 70, 115, 116, 124–33
  birth rates, death rates, and population equation, 125–27
  cause of, 127–29
  solution to, 129–33

Population profile, 119, 120–21
Postreproductive death, change from prereproductive to, 127–29
Potato blight in Ireland (1845–1847), 378
Potential energy, 50, 53
Potomac River, 255
Poverty:
  centralized development projects and, 142
  the poor in developed vs. developing countries, 118
  population and, 118–24
Power plants:
  alternative, 333
  coal-fired, 324–25, 326, 470
  nuclear, 335, 468–69, 470, 472, 474, 478
Power towers, 500, 501
Precipitation, 195–97
  (see also Acid precipitation)
Predator, 28
  (see also Natural enemies)
Predator-prey relationship, 28, 35
  balance in, 72–74
  natural selection and, 100
Preliminary treatment of sewage, 253, 254
Prereproductive death, change to postreproductive death from, 127–29
Preservation of wetlands, 246
Prey, 28
Price, Mark Stanley, 409
Prickly pear cactus, natural control of, 373, 376
Primary clarifiers, 253, 255
Primary consumers, 28
Primary detritus feeders, 28
Primary standard for pollutant, 310–11
Primary succession, 82–83, 84
Primary treatment of sewage, 253–54, 255
Principles, 8
"Pro-choice" groups, 150–51
Produce labeling for pesticides, 388
Producers, 25–27
  matter and energy changes in, 54–55
Production oil, 452–53
Property taxes, 528, 538
Protons, 564
Proven reserves, 452, 453
Public apathy about sewage treatment, 268
Public attitudes about solid wastes, 440
Public interest research groups (PIRGs), 292
Public transportation, 521
Public Utilities Regulatory and Policies Act of 1978 (PURPA), 514–16
Puddling, 171
Pure science, 9

Q

Quarantines, pest control via, 384

R

Rabbits, control of, 373, 377
Raccoons, rabies epidemic among "urban," 417
Racial segregation, exurban migration and, 526–28

Radiation, background, 474
Radiation sickness, 474
Radioactive decay, 475
Radioactive emissions, 474, 475
Radioactive materials, 474–78
Radioactive wastes, 333, 474
  disposal of, 475–77
Radon, 317
Rainfall:
  in different biomes, 18–24
  greenhouse effect on, 338–39
  separation of biomes and, 37
  (see also Acid precipitation)
Rain shadow, 195–97, 198
Rancho Seco Nuclear Power Plant (Sacramento), 468
Range of tolerance, 35–36
Raw sewage, 252–53
Raw sludge, 253, 259
Raw wastewater, 252
RCRA, 285–87, 292, 293
Reactor vessel, 470, 473
Reagan administration, 285, 335
Recharge, groundwater, 204, 241
Recharge area, 198
Reclamation of toxic chemical waste, 287–89
Recreational areas, urban sprawl and loss of, 525
Recreational values of natural biota, 401–2
Recruitment, 70
Recycling, 144, 430–36
  alternative sewage treatment and, 262–67
  in cities, 533
  impediments to, 430–32
  nutrient, 57–61, 64, 65, 157
  promotion through mandate, 434–36
  of toxic chemical waste, 287–89
Reforestation, 185–86
Refuse (see Solid waste crisis)
Refuse-to-energy conversion, 438, 439
Regulation(s):
  of game animals in U.S., 416–17
  of pollution, need for, 351
  for sediment control, 245
  of toxic waste management, 285–87, 293
  for water conservation, 219–20
  (see also specific regulations)
Relative acidity (pH), 160
  (see also pH)
Relative humidity, 195, 196
Religious beliefs in industrial vs. agrarian societies, 132
Renewable energy sources, 489, 490
  (see also Geothermal energy; Solar energy; Tidal power; Wave power)
Renewable resources, 412–13
Replacement fertility, 119–22
Reprocessing of recycled materials, 431
  profits from, 432–34
Reproduction, sexual, 91, 93
Reproductive capacity, survival of species and, 105
Reproductive rate, 70
Reserves, oil, 452–53
  strategic, 455, 456
Reservoirs:
  sediments filling, 238–40
  stormwater, 218
Resistance, pest, 358–59
Resource Conservation and Recovery Act of 1976 (RCRA), 285–87, 292, 293

Resources:
  mineral, 393–94
  renewable, 412–13
  (see also Biota; Energy resources; Water)
Respiration:
  anaerobic, 259
  cell, 55–56, 57, 160
Respiratory disease, air pollution and, 301
Restoring the Quality of Our Environment, 221
Resurgence, 359–62
Retention ponds, stormwater, 217–19
Returnable vs. nonreturnable bottles, 438–40
Reuse of water, potential for, 209–10, 211
Rhizobium, 60
Right-to-Know legislation, 285, 287, 290
"Right-to-life" groups, 150–51
Risk analysis of pollution, performing, 349–50
Rivers, impact of sediments on, 236–40
Rivulet erosion, 171, 172
Roberts, Walter Orr, 338
Rock Creek, Maryland, 217
Rodale Press, 187, 390
Rodenticides, 356
  (see also Pesticides)
Roman Empire, 178
Roots, symbiotic relation between soil fungi and, 167, 168
Rouse, James, 532
Rubber Research Elastomerics (RRE), 436
RU486 (drug), 151
Runoff, 197, 211–15
  effects of increasing, 212
  as nonpoint pollution, 242

S

Safe Drinking Water Act of 1974, 284
St. Petersburg, Florida, nutrient-rich wastewater used in irrigation in, 262
Salinization, 178
  countering, 186, 187
  retarding, 210
Salt, soil, 160–62
Saltwater intrusion (encroachment), 208–9
San Bernardino Mountains, effect of air pollution on, 305
Sanitary sewer system, 252
San Joaquin Valley (California), 207
SARA (Community Right-to-Know legislation), 287, 290
Savanna biome, 23
Save Our Streams, 246
Savimbi, Jonas, 409
Scale insects, natural control of, 373
Scenic areas, urban sprawl and loss of, 525
Science, 301
Science:
  basic or pure, 9
  defined, 5
  role of instruments in, 9
  technology and, 9
  unresolved questions and controversies, 9–10
  value judgments and, 10

Scientific fact, 6
Scientific method, 5–8, 9
Scientific values of natural biota, 401–2
Screwworm fly, combating, 378–81
Scrubbers, 333, 335
Secondary consumers, 28
Secondary detritus feeders, 28
Secondary energy, 446
Secondary pest outbreak, 359–62
Secondary succession, 83–86
Secondary treatment of sewage, 254–55, 256
Second-generation pesticides, 357
Second Law of Thermodynamics, 51–53, 63
Secure landfill, 280
Sediments:
    controlling, 242–47
    impacts on streams and rivers, 236–40
    pollution of Chesapeake Bay by, 227
    sources of, 235–40
Sediment trap, 245
Seep, 198
Segregation, exurban migration and, 526–28
Selection pressures, 98, 99, 102
Selective breeding, 94–96
Septic systems, individual, 266–67
Settling of landfills, 429–30
Sewage, 248–70
    handling and treatment, 251–67
        alternative systems, 262–67
        conventional, 252–62
        progress and lack of progress, 267–69
    hazards of untreated, 250–51
    monitoring for contamination, 269–70
    raw, 252–53
    sludge:
        activated, 254
        composting, 436
        methane production from, 505
        raw, 253, 259
        treatment, 259–62
    as source of polluting nutrients, 241
    treatment, upgrading, 247
Sex attractant pheromones, 385, 389
Sexual reproduction, 91, 93
Sheet erosion, 171, 172
Shellfish, protection of endangered, 419
Shelter belts, 182
Shoreham Nuclear Power Plant (Long Island), 468
Shorelines, bulkheading of, 241–42
Short-term vs. long-term view of pollution control, 350–51
Sierra Club, 187
Sierra Nevada Mountains, effect of air pollution on, 305
Sinkhole, 208
Slash-and-burn methods of ground clearing, 124
Sludge (see Sewage)
Sludge cake, 260
Sludge digesters, 259, 260
Smog, 296–98, 313, 316
Smoking:
    as indoor air pollutant, 318
    synergistic effect between air pollution and, 301, 302
Snow White syndrome, 387, 388
Snowy egret, protection of, 47
Socialist countries, 116
Socioeconomic factors supporting overuse of pesticides, 386–89

Soil, 153–54, 157–70
    deterioration, 169–70
    ecosystem, 162–69
    growing plants without, 162
    lime to neutralize, 331, 332
    microclimate and type of, 39
    mineral nutrients and nutrient-holding capacity, 157–58, 163, 164
    mutual interdependence of plants and, 169–70
    oxygen and aeration, 160
    particles, USFA classification of, 163
    relative acidity (pH), 160
    salt and osmotic pressure, 160–62
    water and water-holding capacity, 158–59
Soil Conservation Act of 1935, 187
Soil Conservation Service (SCS), 187, 240, 246
Soil erosion, 153–54, 170–89
    causes of, 173–78
    dimensions of problem, 178–82
    preventing, 182–88
    process of, 170–73
Soil organisms, 165, 166, 167
Soil profile, 165, 167
Soil structure, 165, 166
Soil texture, 162–65
Soil texture triangle, 162, 164
Solar cells, 497–500, 533
Solar energy, 62, 65, 66, 333, 489–510, 513–14
    Davis experience with, 496–97
        direct, use of, 489–503
        for production of electricity, 495–501
        for production of hydrogen, 501–3, 513
        for space and water heating, 491–95, 513–14
        indirect, use of, 489, 503–10
            biomass energy or bioconversion, 504–7
            hydropower, 507–8, 509
            wind power, 508–10
        spectrum, 491
Solar ponds, 500, 502
Solid waste crisis, 426–40
    background of solid waste disposal, 427–29
    changing public attitudes and lifestyle and, 440
    emissions of coal-fired vs. nuclear power plant, 474
    landfills, 428–30
        escalating costs of, 430
        improving, 430
    problems of, 429–430
    output in U.S., 427–28
    solutions to, 430–40
        composting, 436–37
        recycling, 430–36
        reducing waste volume, 438–40
        refuse-to-energy conversion, 438, 439
Sorting of solid wastes, 430–32
South Pole, ozone "hole" over, 344
Space Biospheres Ventures, 67
Space heating solar, 491–95, 513–14
Specialization to niches and habitats, 74–77, 98
Speciation, 98–99, 101
Species:
    basic genetics of, 91–94
    defined, 15
    diversity, ecosystem stability and, 73, 78, 88

gene pools and their change, 94–100
Splash erosion, 171, 172, 173
Spores, 31
Spring, 198
Stability, diversity and, 73, 78, 88, 398
Standard of living, population growth and, 122–23
    (see also Lifestyle)
Standards:
    air pollution, 299–300, 310–11
    recycling, lack of, 431
Standing biomass, 62–63
Stark, Fred, 436
Starvation, 56, 139
Steam engine, 445
Steinbeck, John, 187
Sterile male technique of pest control, 378–81, 390
Storm drains, 212, 214
    streams incorporated into, 215, 217
Stormwater control legislation, 246
Stormwater management, 215–19, 220
Strategic petroleum reserves, 455, 456
Stratosphere:
    chlorine atoms entering, source of, 342–43
    ozone shield in, 340–41
Stream:
    blockage of channel, 212, 215
    channelized, 217, 218
    effect of development on flow of, 212, 213, 215, 216
    impact of sediments on, 236–40
Stream bank erosion, 212, 213, 236, 238
Stress, zones of, 35–36
Strip cropping, 182, 183
Stripmining, 483, 484
Strong, A.E., 339
Submerged aquatic vegetation (SAV), 228, 229–30
Subsidies, government, waste of water and, 210
Subsistence farming, 176
Subsoil, 165–67
Suburbs, 521–22
    (see also Urban sprawl)
Succession, 82–87:
    climax ecosystem, 86
    evolutionary, 102–4, 106
    primary, 82–83, 84
    secondary, 83–86
    (see also Eutrophication)
Sulfur dioxide:
    as direct product of combustion, 308
    effect on agriculture and forests of, 303
    emissions, reducing, 333, 334, 335
    emissions of coal-fired vs. nuclear power plant, 474
    as major air pollutant, 300
    sources of and control strategies for, 313–14, 315
Sulfuric acid, 309
    in acid precipitation, 324
    as major air pollutant, 300
Sulfur oxides, as major air pollutant, 300
Sulfur pollution from geothermal plant, 511
Superfund, 284–85, 429
Superfund Amendments and Reauthorization Act of 1986 (SARA), 285, 287
    Title III, 290
Surface impoundments, 279–80, 282
Surface mining sites, sediments from, 235–36, 237

Surface runoff loop, 198
Surface water, 197
  diminishing, 207
  overdraft of, 202–4
Survivorship, 127, 128
Sustainability, principles underlying, 543
Sustainable agriculture, 142, 144
Sustainable cities, 532–42
Sustainable development, 4, 42, 66, 124
Sustainable environment, 3, 5
Sweden, acid precipitation in, 326, 328
Symbiosis, 35
Synergistic effect:
  of air pollutants, 300–301
    with smoking, 301, 302
  between ozone and acid precipitation, 329
  of toxic chemicals in groundwater, 277
  between toxic elements and low pH, 329
Synthetic fuels (synfuels), 483–86
Synthetic organic chemicals, 275, 276, 357
Synthetic organic compounds, 48, 289
  indoor air pollution from, 317
Synthetic organic pesticides, problems associated with, 357–63
Synthetic organic wastes, 278

T

Taxes, property, 528, 538
Technology:
  appropriate, 143–44
  contraceptive, 151
  defined, 9
  science and, 9
Tellico Dam, 508
Temperature forest biome, 18
Temperature:
  control, energy usage for, 450
  in different biomes, 18–24
  ecosystems and, 37–38
Temperature inversion, 296–98
Tennessee Valley Authority (TVA), 335
Teratogenic effects of synthetic organic chemicals, 275
Terracing, 182, 184
Terrain, microclimate and, 39
Territoriality, population balance through, 81–82
Thailand, family planning program in, 146–49
Theories, defined, 8
Thermal pollution, 482
Thermodynamics, Laws of, 8, 51–53, 61, 63
Third World countries, 116
  (see also Developing countries)
Three Mile Island nuclear power accident (1979), 478
Threshold, economic, 386
Threshold level of pollution, 296, 348
Tidal power, 512, 513
Tidal wetlands, 241, 401, 404, 405
Times Beach, Missouri, 291
Time span of risk analysis, 350
TireCycle, 436
Tobacco Institute, 10
To Feed This World (Wortman and Cummings), 141
Tokamak design, 480, 481
Tolerance, limits and range of, 35–36

Topsoil, 165–67:
  formation or mineralization, 168
  loss of, 179–80
  (see also Erosion)
Total fertility rate, 118–19, 146, 147
Toxic chemicals:
  cleanup and management of wastes, 284–93
    current wastes, 285–87
    drinking-water supplies, assuring safe, 284
    existing waste sites, 284–85
    future wastes, 287–89
    groundwater remediation, 285, 286
    occupational and accidental exposures, reducing, 289–92
  components of, 274–75
  environmental contamination with, 277–84
    background on problem, 278
    land disposal management problems, 280–82, 283
    land disposal methods and, 278–80
    major sources of wastes, 278
    scope of problem, 282–83
  in groundwater, 273–93
    bioaccumulation, 275–77
    cleanup and management of, 284–93
    components of, 274–75
    synergistic effects, 277
  threat to aquatic species, 419
  transport of, 291
Tradeoffs, 3
Traffic, 409
Tragedy of the commons, 410–12, 416
Traits:
  defined, 91
  natural selection and, 96
  selective breeding on basis of, 95
Transitional regions between ecosystems, 17–25
Transpiration, 158–59, 194
Transportation:
  fuel requirements for, 450, 513–14
  public, 521
Trapping technique of using sex attractant pheromones, 385
Trash (see Solid waste crisis)
Treated sludge, 259–60
Trees, air pollution's effect on, 301–7
  (see also Forests)
Trenches, dry, 217
Trickling filter systems, 254, 256
Tritium, 481–82
Trophic levels, 33
  flow of energy and decreasing biomass at higher, 62–64, 65
Tropical rainforests:
  biome, 24
  in Brazil, destruction of, 124–25, 177, 242, 422
  burning of, 422
  extent of, 423
  impact of population on, 124–25
  saving, 422–24
Trout Unlimited, 246
Trust for Public Land, 540
Tsavo National Park (Kenya), 409
Tuna, 419
Tundra biome, 22
Turbidity, 230
  atmospheric (haze), 298, 299
Turbid water, 228–29
Turbogenerator, 447
Turnover rate of phytoplankton, 230

U

Ultraviolet light, 339–40
Ultraviolet (UV) radiation, 341
UNICEF, 148, 149
Union Carbide leak (Bhopal, India), 287, 297
United Nations Environmental Program (UNEP), 424
United Nations Food and Agriculture Organization, 110
United Nations Fund for Population Activities, 151
United Nations World Health Organization, 148, 151, 357, 367
United States:
  available surface water flow, 204
  declining groundwater levels, 205
  desertification by irrigation in, 178
  droughts in 1980s, 139
  Dust Bowl, 186–87
  energy sources and uses in, 444–51
    declining oil reserves and increasing importation of crude oil, 453, 456
    energy consumption in (1860–1988), 449
    oil production and consumption, 454
  fertility rates in, 131
  game animals in, 416–17
  government spending on energy research and development, 466, 516
  greenhouse effect on, 338
  land conservation measures in, 187
  per capita water consumption in, 209
  price support system and food surpluses in, 141
  solid waste output in, 427–28
  support for family planning, 151
United States National Academy of Sciences, 110
United States Peace Corps, 188
Units of measure, 563
University of California at Davis, 496
Uranium:
  common radioactive isotopes resulting from fission of, 476
  enrichment of, 469–70
Uranium-235, 468–69, 480
Uranium-238, 468, 469, 480
Urban homesteading, 539
Urban redevelopment, aiding, 538–39
Urban sprawl, 519–32
  environmental consequences of, 522–26
  origin of, 521–22
  slowing, 536–38
  social consequence of, 526–32
  vicious cycle of inner city decay and, 530–32
  (see also Cities)
Utterly dismal theorem, 141

V

Valdez oil spill (1989), 459–60
Value judgments, 10
Van den Bosch, Robert, 363
Vectors, 72
Vedalia (ladybird) beetles, 373
Vegetation in different biomes, 18–24
Vents, natural steam, 511
Verification, 6
Vested interests in status quo, 431–32

Veterans Administration, 521
Vietnam War, 129
Vigor, natural traits for, 397–98
Vincristine, 399
Virginia, phosphate ban in, 247
Virginia Institute of Marine Science, 227

W

Washington, D.C., advanced sewage
    treatment in, 255
Waste inventories or exchanges, 289
Wastes:
    leaching and runoff of animal, 240–41
    techniques for utilizing and recycling,
      144
    (see also Solid waste crisis; Toxic
      chemicals)
Wastewater, raw, 252
    (see also Sewage)
Water, 153, 154
    as balance or neutral point between
      acid and base, 323
    drinking water supplies, 284
    evaporation, condensation, and
      purification, 193
    over and through ground, 197–98
    interrelationships between air,
      minerals and, 47, 48
    leaching process, 157, 158
    management for pest control, 383
    molecules in, 45, 47
    needs, greenhouse effect and, 340, 341
    physical states of, 192–93, 194
    price of, 219
    resources, degradation by urban
      sprawl, 524–25
    separation of biomes and, 37
    temperature and evaporation of, 38
    water-holding capacity of soil, 158–59
Water balance, 160
Water cycle, 194–223
    human dependence and impacts on,
      200–219
      changing land use, effects caused by,
        210–15

      implementing solutions for, 219–23
      obtaining more vs. using less, 209
      overdraft of water resources,
        consequences of, 202–9
      potential for conservation and reuse
        of water, 209–10, 211
      sources and uses of fresh water, 200–
        202
      stormwater management, 215–19,
        220
    process of, 194–200
Water heating solar, 491, 495
Water-holding capacity of soil, 163, 164
Water hyacinth, 78, 79, 373, 376
Waterlogging, 160
Water pollution:
    groundwater (see Groundwater
      pollution)
    sewage and, 248–70
      handling and treatment, 251–67
      hazards of untreated sewage, 250–51
      monitoring for contamination, 269–
        70
    urban sprawl and, 524–25
Water power, 507–8, 509
Watershed, 210, 211, 217
Water table, 197–98, 199
    decreased infiltration and, 215
    falling, 204–7
Water vapor, 192, 194–95
    humidity, 192, 195, 196
Wave power, 512
Waybright, Richard, 505
Weathering, 157, 162
Weeds, 354, 373, 376
Wells, dry, 217
Western, David, 409
Westinghouse Corporation, 466
Wetlands:
    biota, 401
    loss of, 241–42
    preservation of, 246
    tidal, 241, 401, 404, 405
Whales:
    catches, 1942–1973, 421
    protection of, 419

    in St. Lawrence River, death of, 288–
      89
Wheat, Hessian fly and, 378, 379
Wheat rust, 383
Whooping crane, efforts to save, 418
Wild and Scenic Rivers Act of 1968, 202
Wildlife Conservation International, 409
Willamette River, 419
Wind, microclimate and, 39
Wind farms, 510
Wind power, 508–10
Windrows, 182, 260
Wind turbines, 509, 510
Women:
    education of, family planning and, 146,
      147
    status of, in industrial vs. agrarian
      societies, 132
    (see also Fertility)
Wood stoves, 504–5
Workability of soil, 163
World Bank, 143
World Health Organization, 148, 151,
    357, 367
World Population Institute, 151
World Resources Institute, 282
Worldwatch Institute, 153, 179, 187, 504
World Wildlife Federation, 407
World Wildlife Fund, 72, 409
Wortman, S., 141

Y

Yard wastes, 428
Yeasts, 57
Yellowstone National Park, 81, 511
Yucca Mountain, 477
Yukon River, 209

Z

Zones of stress, 35–36
Zoning laws, 536–38
Zoos, breeding of endangered animals in,
    419